KATAHDIN OSAR, WHALESBACK; AURORA. LOOKING SOUTHEAST.

The low pass by which the osar penetrates the hills is shown in the distance near the center. The ridge rises 100 feet above Union River, shown on the left.

# *Deglacial History and Relative Sea-Level Changes, Northern New England and Adjacent Canada*

Edited by

Thomas K. Weddle
Maine Geological Survey
22 State House Station
Augusta, Maine 04333
USA

and

Michael J. Retelle
Department of Geology
Bates College
Lewiston, Maine 04240
USA

**SPECIAL PAPER**

**351**

Geological Society of America
3300 Penrose Place
P.O. Box 9140
Boulder, Colorado 80301-9140
2001

Copyright © 2001, The Geological Society of America, Inc. (GSA). All rights reserved. GSA grants permission to individual scientists to make unlimited photocopies of one or more items from this volume for noncommercial purposes advancing science or education, including classroom use. Permission is granted to individuals to make photocopies of any item in this volume for other noncommercial, nonprofit purposes provided that the appropriate fee ($0.25 per page) is paid directly to the Copyright Clearance Center, 222 Rosewood Drive, Danvers, MA 01923, USA, phone (978) 750-8400, http://www.copyright.com (include title and ISBN when paying). Written permission is required from the Copyright Clearance Center for all other forms of capture or reproduction of any item in the volume including, but not limited to, all types of electronic or digital scanning or other digital or manual transformation of articles or any portion thereof, such as abstracts, into computer-readable and/or transmittable form for personal or corporate use, either noncommercial or commercial, for-profit or otherwise. Send permission requests to Copyright Clearance Center.

Copyright is not claimed on any material prepared wholly by government employees within the scope of their employment.

Published by The Geological Society of America, Inc.
3300 Penrose Place, P.O. Box 9140, Boulder, Colorado 80301
www.geosociety.org

Printed in U.S.A.

GSA Books Science Editor Abhijit Basu
GSA Books Editor Rebecca Herr
Cover design by Margo Good

**Library of Congress Cataloging-in-Publication Data**

Deglacial history and relative sea-level changes, northern New England and adjacent Canada / edited by Thomas K. Weddle and Michael J. Retelle.
    p. cm. — (Special paper / Geological Society of America ; 351)
  Based on papers from a symposium held at the 33rd Annual Meeting of the Geological Society of America, Northeastern Section in Portland, Me., Mar. 1998.
    Includes bibliographical references and index.
    ISBN 0-8137-2351-5
    1. Glacial epoch — New England. 2. Glacial epoch — Maritime Provinces. 3. Sea level — New England. 4. Sea level — Maritime Provinces. I. Weddle, Thomas K. II. Retelle, Michael J., 1953–. III. Geological Society of America. IV. Geological Society of America. Northeastern Section. Meeting (33rd : 1998 : Portland, Me.) V. Special papers (Geological Society of America) ; 351.

QE697.D44 2001
551.7'92'0974—dc21
                                                                                 00-061712

**Cover:** The Globe Delta in Washington, Maine. This deposit is typical of the many glaciomarine Gilbert-type deltas formed at the glacier margin during recession of the late Wisconsinan ice sheet from southern Maine. The exposure shows topset, foreset, and bottomset beds (photo and caption by Woodrow B. Thompson, Maine Geological Survey).

**Frontispiece and facing Dedication Page:** Line drawings from G.H. Stone, 1899, The glacial gravels of Maine and their associated deposits: U.S. Geological Survey Monograph 34, 499 p.

# Contents

*Preface* ............................................................................................................................................... vii

1. *Changing paradigms of surficial geology in Maine: From the biblical flood to glaciation* ..... 1
   Harold W. Borns, Jr.

2. *Deglaciation of the Gulf of Maine* ................................................................................................ 9
   Detmar Schnitker, Daniel F. Belknap, Tania S. Bacchus, Julie K. Friez, Barbara A. Lusardi, and Daniel M. Popek

3. *Glaciation and relative sea-level change in Maritime Canada* ............................................ 35
   Rudolph R. Stea, Gordon B.J. Fader, David B. Scott, and Patrick Wu

4. *Late Wisconsinan ice advances, ice extent, and glacial regimes interpreted from seismic data, sediment physical properties, and Foraminifera: Halibut Channel, Grand Banks of Newfoundland* ........................................................................................................................ 51
   Ann A.L. Miller, Gorder B.J. Fader, and Kathryn Moran

5. *Deglaciation of western Maine* ................................................................................................109
   Woodrow B. Thompson

6. *Late Quaternary morphogenesis of a marine-limit delta plain in southwest Maine* ...............125
   Anna K. Tary, Duncan M. Fitzgerald, and Ilya V. Buynevich

7. *Morainal banks and the deglaciation of coastal Maine* ......................................................151
   Lewis E. Hunter and Geoffrey W. Smith

8. *Atmospheric $^{14}C$ chronology for late Wisconsinan deglaciation and sea-level change in eastern New England using varve and paleomagnetic records* .......................................171
   John C. Ridge, Brandy A. Canwell, Meredith A. Kelly, and Sharon Z. Kelley

9. *Deglaciation and relative sea-level chronology, Casco Bay Lowland and lower Androscoggin River Valley, Maine* ..........................................................................................191
   Michael J. Retelle and Thomas K. Weddle

10. *Stratigraphy, paleoceanography, chronology, and environment during deglaciation of eastern Maine* ..........................................................................................................................215
    Christopher C. Dorion, Gregory A. Balco, Michael R. Kaplan, Karl J. Kruetz, James D. Wright, and Harold W. Borns, Jr.

11. *Late Wisconsinan glacial dynamics, deglaciation, and marine invasion in southern Québec* ..............................................................................................................................243
    Serge Occhietti, Michel Parent, William W. Shilts, Jean-Claude Dionne, Étienne Govare, and Dominique Harmand

12. *Relative sea-level changes in the St. Lawrence estuary from glaciation to present day* ..........271
    Jean-Claude Dionne

*Index* ..................................................................................................................................................285

BROAD OSAR PENETRATING A LOW PASS, WOODSTOCK.

# *Dedication*

The editors wish to dedicate this volume to some Friends of the Pleistocene who have been our mentors, both as teachers and professional colleagues: Joe Hartshorn, Dee Caldwell, Carl Koteff, Hal Borns, Bob Oldale, and the late Phil Schaefer. Throughout their careers with the U.S. Geological Survey or at their universities, they mapped and conducted detailed field studies of glacigenic deposits in New England, including the end moraines and outwash of Cape Cod, the record of Glacial Lake Hitchcock in the Connecticut River valley, the alpine zone on Mt. Katahdin and wildlands of northern Maine, and the glaciomarine deposits in coastal Maine. This work was greatly enhanced by their vast and ranging experiences in the modern glacigenic environments in Alaska, Greenland, and Antarctica, where they investigated subglacial zones, proglacial lakes, and calving margins. Their observations later served in classrooms and field discussions as vivid models of how things really worked, and provided students like us a more realistic context for our later studies. They have inspired us and a generation of Quaternarists to continue field work in New England, now armed with knowledge of modern processes and systems, and age determination techniques. We are fortunate to have had their formal and informal fellowship, the opportunities for discussions, exchange of ideas, and learning experiences. To them we fondly dedicate this volume.

# Preface

This volume is published shortly after the 100th anniversary of George H. Stone's monumental work, U.S. Geological Survey Monograph 34, *The glacial gravels of Maine and their associated deposits*. The monograph is an exquisitely illustrated text, primarily written about the eskers (or as Stone referred to them, osars) of Maine. However, Stone provided considerable discussion of the occurrence of raised glaciomarine deposits, some of which he recognized as ice-marginal deposits marking a retreating ice front in contact with the ocean. Today we are able to further this work aided by understanding of morphostratigraphy using process studies in modern environments, detailed surficial geological mapping and stratigraphic studies, and high-precision geochronology provided by accelerator mass spectrometry (AMS) radiocarbon age determinations. Stone's work is one of the first published in the late nineteenth century heralding the beginning of modern studies of Quaternary geology in Maine in the twentieth century. It is fitting that this Special Paper commemorates the earlier monograph published prior to the penultimate century change.

The manuscripts in this text are an outgrowth of a day-long symposium of the same title, "Deglacial history and relative sea-level changes, northern New England and adjacent Canada," held in Portland, Maine, in March 1998, at the 33rd Annual Meeting of the Northeastern Section of the Geological Society of America. At that session, 18 abstracts were presented, 12 of which are compiled as chapters in this volume. Along with the recently published issue of Geographie Physique et Quaternaire, "Late Quaternary History of the White Mountains, New Hampshire and adjacent southeastern Quebec" (edited by W.B. Thompson, B.K. Fowler, and P.T. Davis, this Special Paper provides an overview of recent work in the region between the north shore of the St. Lawrence River valley and the continental shelf from the Grand Banks of Newfoundland to Georges Bank off the coast of New England. The collection of papers in our volume addresses the climatic record based on terrestrial and marine geology from the southeastern sector of the area occupied by the Laurentide ice sheet.

In the first chapter, H.W. Borns, as the senior-ranking Quaternary geologist in Maine, gives a brief overview of Stone's contributions to Maine glacial geology. The next chapters cover the offshore record in the Gulf of Maine and eastern coastal Canada. D. Schnitker et al. present a model for a sequence of deglacial events recorded in deposits of the Gulf of Maine based on high-resolution seismic stratigraphy, sedimentology, micropaleontology, and AMS radiocarbon analyses. R.R. Stea et al. discuss glaciation, deglaciation, and relative sea-level fluctuation in Maritime Canada, linking the onshore and offshore records. The pattern of postglacial relative sea-level change represented in this chapter is controlled by rates of glacier retreat and varying ice loads over the region. The geologic record challenges the geophysical models of sea-level change that emphasize slow-response mantle deformation. A.A. Miller et al. provide a seismic-stratigraphic interpretation of the Halibut Channel region of the Grand Banks of Newfoundland. Geophysical, geotechnical, and micropaleontological data define a model for till-tongue deposition in the glaciomarine environment and a record of three glacial advances, one of which was centered offshore of Newfoundland.

The next chapters focus on the terrestrial record from northern New Hampshire and southern Maine, and provide new radiocarbon age analyses to the chronology for deglaciation. Several of these papers are outgrowths from surficial geologic mapping of 7.5 minute topographic quadrangles by the Maine Geological Survey, funded in part by the U.S. Geological Survey STATEMAP program. W.B. Thompson summarizes the surficial geology of the western Maine region north of the marine limit. He characterizes the style of deglaciation by deposits reflecting a pattern of progressive northward retreat of the ice margin, accompanied by a shift in ice-flow direction. Contrary to the deglacial model shown in the inset map *Inferred extent of ice cover during deglaciation* in the 1985 *Surficial geologic map of Maine*, the field evidence in western Maine does not support a west-facing ice margin as represented in that map.

A.K. Tary et al. detail the depositional history and internal structure of the Sanford sand plain, a coastal braid-plain delta, as determined largely by ground-penetrating radar. The braid plain represents deposition during the time of marine submergence as deglaciation of the region occurred. The Sanford plain was deposited at the maximum marine limit and is representative of similar plains found throughout the coastal region of Maine. These landforms are a record of deposition from near the ice margin to distal regressive-phase deltaic deposition.

L.E. Hunter and G.W. Smith compare modern processes in the coastal glacial environment in Alaska to the glaciomarine end moraines and associated deposits found along coastal Maine. Smith has had a long association with Maine glacial geology, in particular southwestern Maine. This chapter provides an overview of early twentieth century geologic work in Maine, and links that work and Smith's experience with Hunter's modern studies. One editorial note to this chapter regards the readvance locality at South Pond in Warren, Maine, ~7 km south of the well-known Waldoboro moraine, the largest moraine in southwestern Maine. Smith made mention of this readvance site in an unpublished field guide for field trips in conjunction with the meeting of the International Quaternary Association Commission on Genesis and Lithology of Quaternary Deposits, held at Orono, Maine, August 13–14, 1980. He noted that this locality was originally described by Bjorn Andersen (University of Oslo, Norway). By relations of crosscutting moraines and striation evidence, it records an apparent oscillation of the local ice margin over at least a 2 km distance (B. Andersen, unpublished surficial geology maps and report to the Maine Geological Survey, 1973). Moreover, a conventional radiocarbon analysis of a bulk-organic sediment sample from a core collected from Kalers Pond, south of the Waldoboro moraine and west of South Pond, gave an age for marine emergence of 13 240 ± 190 $^{14}$C yr B.P. (GX-23803, $\delta^{13}$C value −28.0‰, Voisin, 1998). As noted by J.C. Ridge, potential errors related to bulk sediment sample analysis could produce anomalously old ages. Although the $\delta^{13}$C value of the sample is within the acceptable range of terrestrial plants, this alone does not unequivocally prove that the sample is free of potential error. However, assuming several hundred years passed between moraine formation and emergence, these moraines most likely formed during the later part of Heinrich event 1, recognized from the deep-sea record to be between ca. 15 000 and 13 500 $^{14}$C yr B.P. Pending the validity of the age, it provides a minimum constraint on the timing of the South Pond readvance and the formation of the Waldoboro moraine.

Ridge has been responsible for rejuvenating the useful Antevs varve chronology in New England, linking it to atmospheric (terrestrial plant) radiocarbon ages in the Connecticut River valley and a paleomagnetic record established for northern New York. The significance of this chronology is of more than local interest, allowing comparisons of similar records in other parts of the world (e.g., Rittenour et al., year 2000). His work has established a well-documented deglacial chronology for southern New England, but one that is in conflict with ages of deglaciation in Maine based on marine shell dates uncorrected for the marine reservoir effect. Ridge et al. present a deglacial chronology comparing the varve record from the Connecticut River valley and the Merrimac River valley of New Hampshire to varve records at two sites in western Maine. The paleomagnetic record from one of the sites in Maine supports a correlation between it and the New Hampshire records, thus providing a link to the atmospheric radiocarbon chronology.

This apparent age disagreement between the atmospheric and marine chronologies is by as much as 1500 yr, and underscores the importance of determining a correction factor for the late Pleistocene Gulf of Maine to be able to compare the terrestrial record with the marine record, locally as well as globally.

M.J. Retelle and T.K. Weddle have established an internally consistent local chronology for deglaciation and relative sea-level emergence in a southwestern Maine embayment, based on uncorrected marine shell radiocarbon analyses. With application of a marine reservoir correction of −400 to −700 yr, as determined by others for the western coast of Norway, the deglacial and emergence chronology developed for the southwestern Maine embayment spans the time represented in deep-sea records by ice-rafting events in the North Atlantic (Heinrich event 1) and by a major eustatic sea-level rise (meltwater pulse 1-A). In addition, similar to the chapter by Thompson, field evidence in the form of ice-marginal deposits in the region can be reasonably correlated to show a progressively northward-retreating, south-facing ice margin, and not an east-facing ice margin, as is represented on the inset map on the 1985 *Surficial Geologic Map of Maine*.

C.C. Dorion et al. compiled recent work from eastern coastal Maine in a paper that presents an overview of timing of deglaciation of the region and paleoenvironmental analyses of the late glacial sea utilizing geomorphology, sedimentology, faunal assemblages, isotope geochemistry, and radiocarbon age analyses. Lake cores from areas above and below the marine limit provide support for a marine reservoir correction factor between −600 and −800 yr. During the GSA symposium in Portland, a paper was presented on the remains of a mammoth found in shallow-marine deposits in Scarborough, Maine. Since then, AMS radiocarbon age analyses on the mammoth have yielded reliable age estimates of ca. 12 200 $^{14}$C yr B.P. (Bruce Bourque, Maine State Museum, 1999, personal commun.). Age estimates reported by Retelle and Weddle on marine shells from similar elevations, but different locations from the mammoth site, yield ages of ca. 12 800 $^{14}$C yr B.P. Similar to Dorion et al., the age differences suggest at least −600 yr as a marine reservoir correction for the late Pleistocene Gulf of Maine.

The remaining two chapters focus on the glacial record in the Appalachian highland and the St. Lawrence lowland in southeastern Canada. S. Occhietti et al. provide a synthesis for glacial dynamics in the St. Lawrence lowland region of Quebec relating deglaciation to climatic and non-climatic factors. The lowland channelized a major ice stream, which caused thinning of adjacent ice masses, and generated the flow reversal in the northern margin of the Appalachian ice mass in northern Maine and southeastern Quebec. The sequence of flow is recorded by a complex striation record, and a chronology of deglaciation is presented using numerous radiocarbon analyses from the lowland. J.-C. Dionne, who has spent much of his career studying the record of the Goldthwait Sea, presents a detailed summary of the complicated record of relative sea-level changes in the St. Lawrence estuary. He provides a clear sequence from deglaciation to the present with emphasis on events during the middle Holocene, documenting a double transgression and regression, albeit complicated by the nonsynchronous nature of emergence between the southern and northern shores of the St. Lawrence lowland.

It has been more than 10 years since any similar compilation of current work in the region. We hope that this volume will be useful to other workers, both within and outside of the areas described. It, and the previously mentioned *GPQ* Volume 53 (Thompson et al., 1999), provide a closing compilation of the direction of geologic research over the last quarter of the twentieth century in this part of the world. Much of the work in the region has focused on topical studies and surficial geological mapping, which until recently has not been easily linked to the mid-continental glacial record, let alone the global climatic record. Yet, much of the glacial record in the mid-continent is clearly linked to the deglaciation of the St. Lawrence River valley and northern New England. With currently known details from oceanic and ice-core records, the glacial geology of the area covered in this volume can be understood in context with specific global climatic events of the late Pleistocene during the late Wisconsinan retreat of the Laurentide ice sheet, such as Heinrich event 1, meltwater pulse 1-A, and the Younger Dryas episode. We believe the papers in the volume provide a clearer image of the deglacial and relative sea-level history of northern New England and adjacent Canada, which in turn may provide more clues to a broad-scale understanding of the demise of the last ice sheets.

Thomas K. Weddle, Augusta, Maine
Michael J. Retelle, Lewiston, Maine

## REFERENCES CITED

Rittenhour, T.M., Brigham-Grette, J., and Mann, M.E., 2000, El Niño-like climate teleconnections in New England during the late Pleistocene: Science, v. 288, p. 1039–1042.

Stone, G.W., 1899, The glacial gravels of Maine and their associated deposits: U.S. Geological Survey Monograph 34, 499 p.

Thompson, W.B., and Borns, H.W., Jr., 1985, Surficial geologic map of Maine: Augusta, Maine Geological Survey, scale 1:500 000.

Thompson, W.B., Fowler, B.K., and Davis, P.T., 1999, Late Quaternary history of the White Mountains, New Hampshire and adjacent southeastern Quebec: Geographie Physique et Quaternaire, v. 53, 174 p.

Voisin, D.T. 1998, Late Quaternary post-glacial history of Kaler's Pond, Waldoboro, Maine [B.S. thesis]: Lewiston, Maine, Bates College, 85 p.

# Changing paradigms of surficial geology in Maine: From the biblical flood to glaciation

**Harold W. Borns, Jr.**
*Department of Geological Sciences and Institute for Quaternary and Climate Studies,
University of Maine, Orono, Maine 04469, USA*

## ABSTRACT

The first modern comprehensive publication devoted to the glaciation of Maine was *The Glacial Gravels of Maine and Their Associated Deposits*, by G.H. Stone, published in 1899. Stone's research of the glacial geology of Maine represents a major shift in paradigms from those of previous nineteenth-century works by C.T. Jackson and C.H. Hitchcock, who attributed the surficial geology of Maine to the biblical deluge. Stone's work stands as the most comprehensive and insightful document on the glacial history of Maine, and his observations still provide basic tools for researchers. Stone marshalled his observations into testable hypotheses and proceeded to test them. Unlike the earlier pivotal studies by Jackson and Hitchcock, Stone published the observations from which he drew his conclusions, in contrast to simply stating conclusions without presenting the observations. Among the firsts in Stone's publication are the statewide maps depicting the geography of the glacial-marine deposits and of ice-margin retreat. In addition, Stone was the first to present the idea of a late glacial ice cap over northern Maine, based on his observations in Maine and supported by those of R. Chalmers in the adjacent Eastern Townships of Quebec.

## INTRODUCTION

This is a short review of the major publications dealing with the surficial geology of Maine that document the major paradigm shift from one that attributed most of the surface geological features to the actions of the biblical flood to one that attributed the same features to glacial processes. In this change the publication *The Glacial Gravels of Maine* by George H. Stone (1899) (Fig. 1) is the first state-wide, comprehensive analysis of the results of glaciation with no mention of the biblical flood theory as a cause of the surficial geology of Maine.

## DISCUSSION

Louis Agassiz, a prominent Swiss naturalist, revealed his revolutionary new theory that advocated the existence of a glacier of continental proportions over northern Europe and the Alps in the recent past in his presidential address to the Swiss Society of Natural History at Neuchatel in 1837 (Carozzi, 1967). Agassiz subsequently provided his initial field evidence to support his theory in his monumental *Studies on the Glaciers* in 1840.

In the same year of Agassiz's historic presentation, the first State Geologist of Maine, Charles T. Jackson, published the *First Report on the Geology of the State of Maine* (1837) followed by the second and third reports in 1838 and 1839. In these reports, Jackson (1837, p. 110) synthesized his rather uncritical observations concerning the origin and distribution of various surficial sediments and erosional features and assigned their origin to the biblical flood in conjunction with the concept of iceberg drift in a cold sea.

Figure 1. Photo of George H. Stone, ca. 1875, compliments of Kents Hill School, Kents Hill, Maine (formally Maine Wesleyan Seminary, Readfield, Maine), where he taught natural sciences.

All the observations that have been made, tend to prove that a current of water has swept over the surface of the Globe, since the consolidation of all the rock formations, and the deposition of the tertiary marls and clay, and the current swept along with it, the loose masses of stone, gravel and sand, carrying them from the north or northwest toward the south or the southeast. . . . It is supposed that this rushing of water over the land, took place in the last great deluge, accounts of which have been handed down by tradition, and preserved in the archive of the people.

Seven years after Jackson first described the geology of Maine, Sir Charles Lyell, one of the great geologists of all time, made a significant journey to the United States, including Maine (1850). It was not his first visit to the continent, but it was certainly his most comprehensive. Lyell was much less sure of the geologic past than Jackson, probably because he had a more inquiring mind, and was nearer the actual debates over these changes. Lyell spent considerable time throughout his life testing the diluvian and deluge ideas, but he was prepared to let the land tell him its story. Lyell left Liverpool for New England on September 4, 1845. Lyell (1850, p. 5) demonstrated his deep interests in these questions when his vessel entered the area of the North Atlantic where icebergs appeared on a regular basis, and he saw his first iceberg.

To a geologist, accustomed to seek for the explanation of various phenomena in the British Isles and Northern Europe, especially the transportation of huge stones to great distances, and the polishing and grooving of the surfaces of solid rock, by referring to the agency of icebergs at remote reasons, when much of what is now land in the northern hemisphere was still submerged, it is no small gratification to see, for the first time, one of these icy masses floating so far to the southward.

On this second trip to North America, Lyell traveled to New Hampshire and Maine, where he was shown fossils of various kinds. In Gardiner, Maine, he met with local natural historians, where he was shown fossils of land mammals included in the marine blue clay, among which was a molar tooth which he judged to be from an extinct bison, exposed in the banks of the Kennebec River. This tooth was reexamined, and is from a modern cow (Borns and Ray, 1991). The location of the fossils, as well as the shape of several conical hills nearby, were all attributed by Lyell to the action of the flood and its recession.

The environmental affinities of the Gardiner fossils only confirmed his views that these animals had lived in a time that was much colder, but there was still no doubt in Lyell's mind that they were associated with the deposits of the deluge and recession of the floods of antiquity. The theory of the biblical flood continued to be invoked as the cause of much of the surficial geology of Maine and elsewhere into the mid-nineteenth century as seen, for example, in the reports of the geology of Maine in 1861 and 1862 by the state geologist, Charles H. Hitchcock. However, the glacial theory of surficial geology was gaining in adherents throughout the mid-nineteenth century, both abroad and in Maine. For example, J. DeLaski (1862, 1864) dismissed drifting icebergs as the cause of the striae and large-scale roche moutonnees in the Penobscot Bay region of Maine. He argued effectively, based upon critical field observations, for extensive glaciation as the cause and even suggesting that this glacier may have covered all of New England. In the same decade, A.S. Packard (1866) concluded that at least the entire coastal zone from Labrador southward into New England had been overrun by a glacier of continental proportions. In additions to these examples, other contemporaries of C.H. Hitchcock, including L. Agassiz (1867), adopted glaciation as being primarily responsible for many of the landscape features, not only in Maine, but throughout New England, the Atlantic provinces, and the middle part of the continent.

A continued major paradigm shift is recorded in George H. Stone's (1899) monumental publication, *Glacial Gravels of Maine*. In this book he totally rejected the flood theory by attributing the surficial geology to statewide glaciation. Most of the field observations for this pioneering publication were made in the 1870s and 1880s, only a decade or so after the publication of C.H. Hitchcock's (1862) *Geology of Maine*. In DeLaski's and Packard's work and certainly in Stone's work, there was a major paradigm shift from a cold biblical flood to unconditional recognition of widespread glaciation to account for most of the recent landscape of Maine.

Stone's approach to writing scientific papers, and perhaps to scientific investigation as well, was greatly advanced as compared to his predecessors who compiled earlier syntheses of Maine geology (Jackson, 1837, 1838, 1839; Hitchcock, 1861, 1862). Stone's (1899) publication on the glacial gravels first laid out the basis for the work in terms of a description of the bedrock geology, the geography, and the topography of Maine. He then proceeded to elucidate the processes of erosion, transportation, and deposition of sediments as they were generally known at that time. Having set the stage, Stone thoroughly described the glacial gravels and associated deposits based upon field observations from which he drew conclusions on glacial processes and history. Stone also noted the constraints put upon his study by limited field time, general lack of appropriate topographic maps, and the lack of adequate road access, especially to the forested northern portions of Maine.

Stone (1899, p. 1–2) recognized in his introduction to *The Glacial Gravels of Maine* (1899) the need to study the activities of modern glaciers in order to more adequately understand the results of past glaciers, thus implying that the present is the key to the past.

> The investigation made slow progress, not only because there were several thousand miles to be carefully explored, but especially because the nature of the subject renders such an investigation exceedingly difficult. The scout of the Western frontier who undertakes to guide a body of troops in pursuit of hostile Indians—to follow the trail, and, from the traces left behind, to give a history of the enemy's performances from day to day—has a difficult task before him; but in thus reconstructing history he has the advantage of knowing, from direct observation, the habits of the Indians. In his study of glacial deposits the glacialist labors under the disadvantage of not knowing, by observation, the exact nature of the geological work going on beneath and within an ice sheet. It is comparatively easy to theorize regarding the probable behavior of such a body of ice, and, if properly held in check, imagination is of the greatest use in such an investigation, but the chances for error are very great. The method here adopted has been to collect as large a body of facts as possible, and then carefully to test various hypotheses by the facts, rejecting or holding in abeyance all theories not supported by positive field evidence. Glacialists are exploring a comparatively untrodden field, and it behooves them to proceed cautiously and to avoid dogmatism and denunciation.

To my knowledge Stone never had the opportunity to study the actions of a modern glacier, yet he was remarkably accurate in deducing the glacial processes responsible for the features he observed in Maine based upon modern understanding of these processes published by others (e.g., Russell, 1893). In comparison with the earlier major publications on Maine (Jackson, 1837, 1838, 1839; Hitchcock, 1861, 1862), Stone achieved several scientific firsts that set the format for future studies leading to the present understanding of Maine's surficial geology as exemplified by the *Surficial Geologic Map of Maine* (Thompson and Borns, 1985).

Most important, Stone was a keen and critical observer, recorded his observations in his publications, and then drew his hypotheses and conclusions based upon these facts. This approach was not generally used by earlier writers. Many prominent earlier workers simply offered opinions and made hard statements on the geology without providing the reader the critical observations upon which they based their conclusions. In contrast, Stone's work can be reviewed by the modern geologist, not only in his publication, but by visiting many of his well-identified sites and repeating his observations and drawing conclusions based upon modern understanding of glacial action. In most cases where modern geologists have taken an opportunity to do this, they have realized that his descriptions were extremely accurate and that his conclusions were excellent for his times and differed little, if at all, from modern interpretations.

Stone's observations, for the first time, were compiled into summary statewide surficial geologic maps which today are still remarkably accurate for the scale he used. The maps include the distribution of the marine clays of the state (Fig. 2) and 15 individual county maps showing the distribution of the glacial gravels, primarily including eskers, glacial lacustrine and marine deltas, and outwash. In addition, he recognized mixed gravel and till deposits as forming along the ice margin and was the first to accurately identify and map end moraines, notably the Androscoggin and Waldoboro moraines (Stone, 1887). Stone (1880) also produced the first statewide map of kames and kame plains, and the first map depicting lines of ice-marginal recession across the entire state toward southeastern Quebec (Fig. 3). This synthesis was based upon his evaluation of synchronous glacial deposits that he documented so well (Fig. 4). He was remarkably accurate in his estimates in the southern half of the state, considering that his work predated the advent of radiocarbon dating.

Stone also, in this early time of acceptance of the glacial theory, observed features that suggested to him the possibility of at least two phases of glaciation in Maine. This was the first contribution to the continuing debate over whether New England had undergone two glaciations, and if so, the age of the first episode. Researchers refer to this as the "two till problem" (e.g., Flint, 1961).

In 1890 Stone also presented evidence from northern Maine of the presence of an ice cap, centered over the northern section of the St. John River valley, which flowed north, east, and south. This was in agreement with the conclusions of Chalmers (1890), who based his conclusions on evidence of ice flowing generally north and northwestward into the Eastern Townships of Quebec from adjacent Maine. However, Stone could not decide whether the northern Maine ice cap had existed throughout the glacial period or was present only during late glacial time, separated over Maine by the rising Champlain Sea as it melted up the valley of the St. Lawrence River.

It is not clear whether Chalmers or Stone first suggested

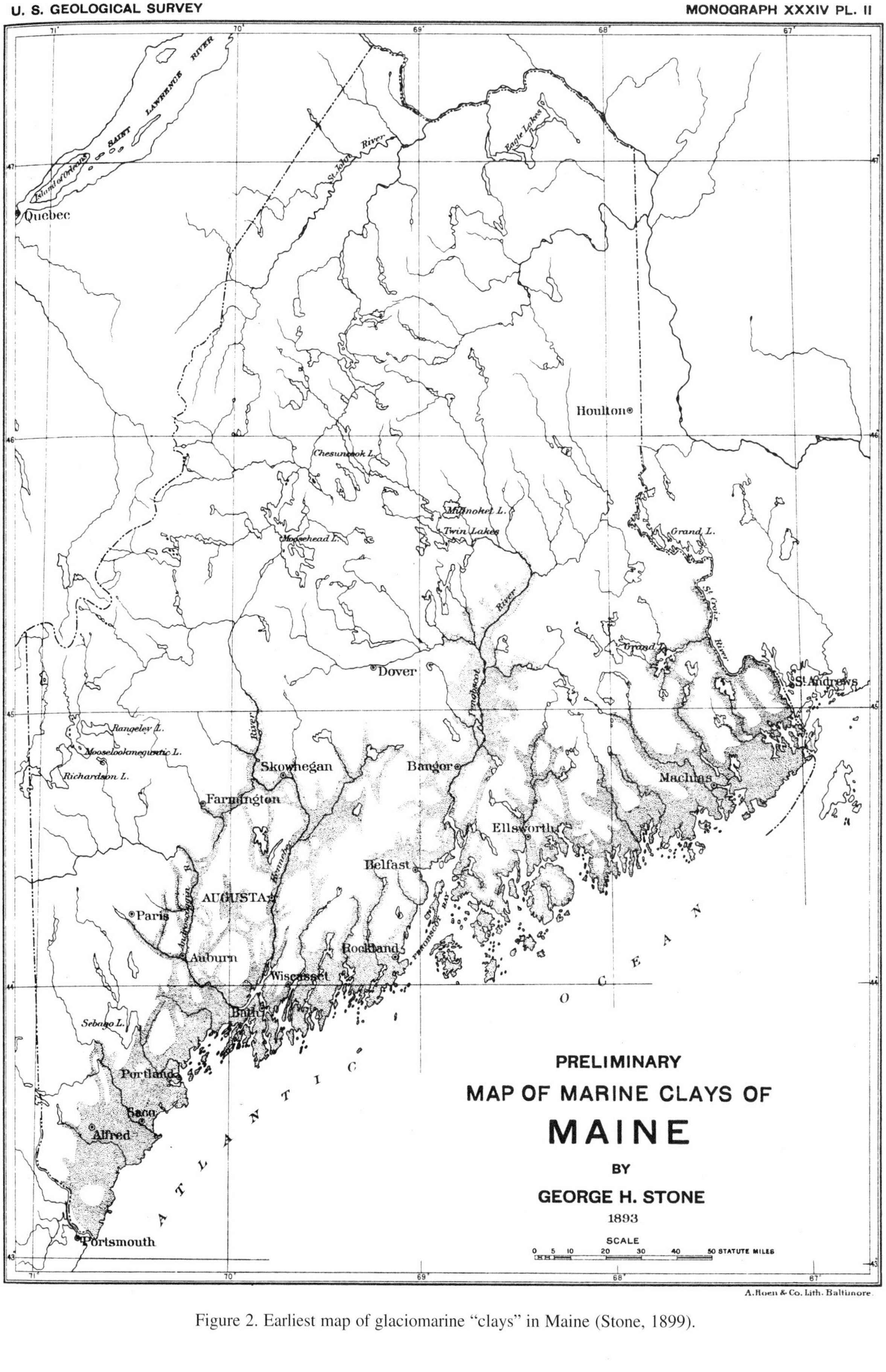

Figure 2. Earliest map of glaciomarine "clays" in Maine (Stone, 1899).

U. S. GEOLOGICAL SURVEY MONOGRAPH XXXIV PL. XXXI

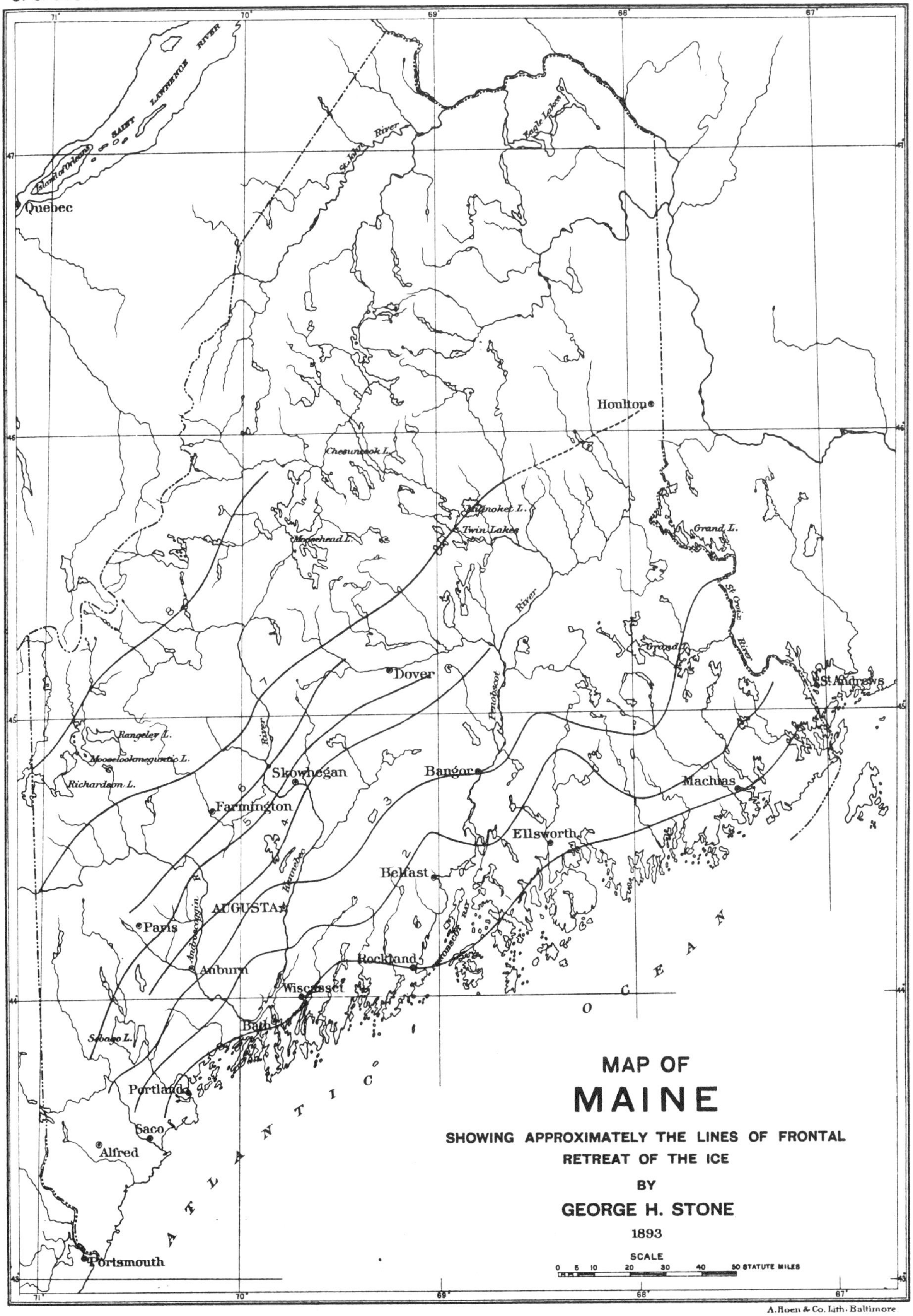


A. Hoen & Co. Lith. Baltimore

Figure 3. Earliest map of ice-margin retreat lines in Maine (Stone, 1899).

*List of approximately synchronous glacial deposits.*

| Localities. | Kind of deposit. |
|---|---|
| **FIRST SERIES.** | |
| Machias, near Little Kennebec Bay | Terminal moraine. |
| Jonesboro and Jonesport | Gravel massives, apparently passing into the sea. |
| Lamoine | Delta. |
| Waldoboro | Terminal moraine. |
| North part of Alna | Delta. |
| Stevens Plain, Deering | Do. |
| **SECOND SERIES.** | |
| Meddybemps-Dennysville | Delta. |
| Old Stream Plains, in part | Do. |
| Mont Eagle Plains | Do. |
| Deblois and vicinity | Do. |
| Otis and Aurora | Deltas. |
| Monroe | Delta. |
| Waldo and Brooks | Do. |
| North Searsmont, Liberty, and Appleton | Deltas. |
| Washington | Delta. |
| Windsor | Do. |
| Litchfield Plain | Do. |
| West Bowdoin (Pine Nursery) | Do. |
| West Cumberland | Do. |
| Gorham and North Scarboro | Deltas and gravel massives. |
| **THIRD SERIES.** | |
| Codyville | Marine glacial delta. |
| Old Stream Plains | Delta. |
| Race Ground near Machias River | Do. |
| Greenbush, Greenfield, and Milford | Do. |
| Dixmont, Unity, and Thorndike | Deltas. |
| Belgrade, Sidney, and Augusta | Delta. |
| Sabatisville | Terminal moraine. |
| New Gloucester and Gray | Delta. |
| Standish-Limington | Do. |
| **FOURTH SERIES.** | |
| Orneville-La Grange | Delta? |
| Harmony, valley of Half Moon Stream | Valley drift and marine beds. |
| Readfield | Terminal moraine. |

Figure 4. Example of tabulated and correlated ice-margin features that allowed G.H. Stone to establish first map of ice-margin recessional positions across Maine (Stone, 1899).

the idea of a northern Maine ice cap. They probably reached the same conclusion at about the same time, each basing their conclusion upon their respective work in Quebec and Maine. This idea was reborn in the paper by Borns (1963), and followed by considerable work in Maine by Kite (1983), Lowell (1987), and by others in Maine and adjacent Canada (cf. Borns et al., 1985; Kite et al., 1988). The distribution and chronology of the separated ice cap through time in northern Maine continues as a research goal.

## SUMMARY

The distribution and history of the surficial geologic features of Maine evolved from a model totally embracing the biblical flood or deluge invoked in C.T. Jackson's reports on the geology of Maine (1837, 1838, 1839), and continuing as the theme in the subsequent reports on the geology of Maine by C.H. Hitchcock (1861, 1862). However, Hitchcock introduced the possibility of some glaciation associated with the deluge. Contemporaneously with Hitchcock's studies, John DeLaski (1864) was the first naturalist who clearly saw and documented the evidence of glaciation, and dismissed the biblical flood as being responsible for the surficial geology of Maine. Thereafter the invocations of the flood being responsible for the surficial deposits rapidly diminished in the literature while there was increasing adherence to the glacial theory first proposed by Agassiz and to the concept that Maine had indeed been glaciated over its total land area.

The major paradigm shift from flood to glaciation in Maine began to appear in the literature between the extensive statewide publications of the State Geologist C.H. Hitchcock (1861, 1862), that hinted of some glaciation in Maine but adhered to the deluge concept, to the geographically extensive and critical work of G.H. Stone (1899), *The Glacial Gravels of Maine*, in which there is no mention of the deluge and all surficial deposits are attributed to glaciation.

In addition, Stone (1890) was the first to present field evidence from Maine in support of the hypothesis that a radically flowing ice cap, centered over the St. John River valley of northern Maine, existed during the last glaciation. However, *The Glacial Gravels of Maine* remains the major work that began and has underpinned much of the subsequent modern research on the glacial history of Maine in which the work of many researchers is summarized in the *Surficial Geologic Map of Maine* (Thompson and Borns, 1985). In many important ways this map resembles the first glacial geology maps of the state made by Stone (1899).

## REFERENCES CITED

Agassiz, L., 1840, Études sur les glaciers: Neuchatel, Switzerland, H. Nicolet, 346 p.
Agassiz, L., 1867, Glacial phenomena in Maine: Atlantic Monthly, v. 19, p. 211–220, 281–287.
Borns, H.W., Jr., 1963, Preliminary report on the age and distribution of the late Pleistocene ice in north central Maine: American Journal of Science, v. 261, p. 738–740.
Borns, H.W., Jr., LaSalle, P., and Thompson, W.B., eds., 1985, Late Pleistocene history of northeastern New England and adjacent Quebec: Geological Society of America Special Paper 197, 159 p.
Borns, H.W., Jr., and Ray, C., 1991, Historic bovid teeth from Gardiner, Maine: Current Research in the Pleistocene, v. 8, p. 85–86.
Carozzi, A.V., ed., 1967, Discourse of Neuchatel, July 24, 1837, *in* Agassiz, L. (Carozzi, A.V., translator), Studies of glaciers: New York, Hafner Publishing Co., 213 p.
Chalmers, R., 1890, The glaciation of the cordillera and the Laurentide: American Geologist, v. 6, p. 324–325.
DeLaksi, J., 1862, A letter to G.L. Goodale, *in* Seventh Annual Report of the Secretary of the Maine Board of Agriculture: Augusta, Maine, Stevens and Sayward, p. 382–388.
DeLaksi, J., 1864, Glacial action about Penobscot Bay: American Journal of Science, ser. 2, v. 37, p. 335–344.
Flint, R.F., 1961, Two tills in southern Connecticut: Geological Society of America Bulletin, v. 72, p. 1687–1692.
Hitchcock, C.H., 1861, General report upon the geology of Maine, *in* Sixth Annual Report of the Secretary of the Maine Board of Agriculture: Augusta, Maine, Stevens and Sayward, p. 146–328.
Hitchcock, C.H., 1862, Geology of Maine, *in* Seventh Annual report of the Secretary of the Maine Board of Agriculture: Augusta, Maine, Stevens and Sayward, p. 223–430.
Jackson, C.T., 1837, First report on the geology of the State of Maine: Augusta, Maine, Smith and Robinson, 127 p.
Jackson, C.T., 1838, Second annual report on the geology of the public lands belonging to the two states of Massachusetts and Maine: Boston, Massachusetts, Dutton and Wentworth, 93 p.
Jackson, C.T., 1839, Third annual report on the geology of the State of Maine: Augusta, Maine, Smith and Robinson, 276 p.
Kite, J.S., 1983, Late Quaternary glacial, lacustrine, and alluvial geology of the upper St. John River basin, northern Maine and adjacent Canada (Ph.D. thesis): University of Wisconsin, Madison, Wisconsin, 362 p.
Kite, J.S., and Lowell, T.V., 1988, Icestream dynamics and topography, major controls of the deglaciation in Maine and adjacent Canada: American Quaternary Association Program and Abstracts of the Tenth Biennial Meeting, p. 127.
Lyell, C.T., 1850, A second visit to the United States of America: London, Murray, v. 1, 368 p.
Lowell, T.V., 1987, Late Wisconsin stratigraphy, glacial ice-flow and deglaciation style of northwestern Maine [Ph.D. thesis]: Buffalo, State University of New York, 182 p.
Packard, A.S., Jr., 1866, Results of observations on the drift phenomenon of Labrador and the Atlantic coast southward: American Journal of Science, ser. 2, v. 40, p. 30–32.
Russell, I.C., 1893, Malaspina Glacier: Journal of Geology, v. 1, p. 219–245.
Stone, G.H., 1880, The kames of Maine: Boston Society of Natural History Proceedings, v. 20, p. 430–469.
Stone, G.H., 1890, Classification of the glacial sediments in Maine: American Journal of Science, 3rd Series, v. 40, p. 122–144.
Stone, G.H., 1887, Terminal moraines in Maine: American Journal of Science, 3rd series, v. 33, p. 378–385.
Stone, G.H., 1899, The Glacial Gravels of Maine: U.S. Geological Survey, Monograph 35, 489 p.
Thompson, W.B., and Borns, H.W., Jr., 1985, Surficial geologic map of Maine: Maine Geological Survey, Augusta, scale 1:500 000.

MANUSCRIPT ACCEPTED BY THE SOCIETY JUNE 9, 2000

# Deglaciation of the Gulf of Maine

**Detmar Schnitker**
*School of Marine Sciences, University of Maine, Orono, Maine 04469, USA*
**Daniel F. Belknap**
*Department of Geological Sciences, University of Maine, Orono, Maine 04469, USA*
**Tania S. Bacchus**
*Department of Environmental and Health Sciences, Johnson State College, Johnson, Vermont 05656, USA*
**Julie K. Friez**
*Maine Center for Osteoporosis Research and Education, St. Joseph Hospital, 360 Broadway, Bangor, Maine 04401, USA*
**Barbara A. Lusardi**
*Minnesota Geological Survey, 2642 University Avenue, St. Paul, Minnesota 55114-1057, USA*
**Daniel M. Popek**
*Tennessee Department of Transportation, 6601 Centennial Blvd., Nashville, Tennessee 37209, USA*

## ABSTRACT

High-resolution seismic stratigraphy, sedimentology, and micropaleontology of accelerator mass spectrometer (AMS) radiocarbon-dated piston cores showed that deglaciation started very early in the Gulf of Maine. Glaciomarine deposition began as early as 18 ka in the southeastern gulf (Georges Basin) and reached the present coastline about 14 ka. The retreat of the ice margin occurred in a sequence of distinct steps. (1) Marine water entered the basin beneath the ice, creating an ice shelf. (2) This ice shelf remained stable as long as its frontal edge was buttressed against sills or ledges. (3) Once thinning and calving freed the ice edge from its support, it calved back quickly until the next sill was reached. In this fashion, the ice retreated from basin to basin until shallow water and the coastline was reached. Following the collapse of an ice shelf over a basin, the initial marine environment there was one of thick, year-round sea ice and of icebergs, prohibiting surface productivity. This year-round sea-ice phase was followed by a phase of brief summer open water during which an arctic flora developed. Further ice retreat and increasing warming led to lengthier periods of summer ice-free conditions, and with upwelling of nutrient-rich water, highly productive surface conditions prevailed. This sequence of events can be traced in the sedimentary record laterally from the southeast to the northeast, and vertically within each basin.

## INTRODUCTION

The southeastern edge of the late Pleistocene Laurentide ice sheet met the ocean in the Gulf of Maine. Thus, glacial history and paleoceanography are intertwined in this region. Glacial geology holds that the late Wisconsin Laurentide ice overrode the Maritime Provinces, New England, and the Gulf of Maine, terminating on the Scotian Shelf, Browns Bank, and Georges Bank, an outlet glacier reaching the open ocean through the deep Northeast Channel (Mayewski et al., 1981; Hughes et al., 1985). The retreat of Laurentide ice from the coastal environments of the Gulf of St. Lawrence and the

Schnitker, D., Belknap, D.F., Bacchus, T.S., Friez, J.K., Lusardi, B.A., and Popek, D.M., 2001, Deglaciation of the Gulf of Maine, *in* Weddle, T. K., and Retelle, M.J., eds., Deglacial History and Relative Sea-Level Changes, Northern New England and Adjacent Canada: Boulder, Colorado, Geological Society of America Special Paper 351, p. 9–34.

Scotian Shelf has been well documented (Dyke and Prest, 1987; King and Fader, 1986; King, 1996), its retreat from the Gulf of Maine less so. It is the purpose of this study to examine the transition of the Gulf of Maine from Late Pleistocene glacial conditions to the interglacial environment and the ensuing evolution of the biologically highly productive body of water during the course of the Holocene. The information on this development can be retrieved from the geometry of the sediment deposits on the bottom of the Gulf of Maine, the physical and chemical nature of the sediments, and the microfossils that they contain. The seismic stratigraphy was developed in greater detail by Bacchus (1993) and Bacchus and Belknap (1997). This chapter concentrates on paleoenvironmental interpretation based on microfossil assemblages and dates from the cores.

## PHYSIOGRAPHY AND OCEANOGRAPHIC SETTING

The Gulf of Maine is a marginal shelf basin of about 90 000 km$^2$, partially enclosed by Nova Scotia and New Brunswick to the northeast, Maine to the northwest, and Massachusetts to the southwest. It is open to the North Atlantic Ocean to the southeast, but much of this marine connection is across the shallow (<50 m) Georges and Browns Banks. The only deep connection (232 m sill depth) is through the Northeast Channel. Within the Gulf of Maine there are many prominent bedrock ledges and rises that sill deep intervening basis (Fig. 1). The present floor of the Gulf of Maine has been shaped by repetitive Pleistocene glacial scouring and deepening of a late Tertiary fluvial system on Paleozoic through early Tertiary deposits (Oldale and Uchupi, 1970; Ballard and Uchupi, 1975). Early seismic exploration (Hoskins and Knott, 1961; Oldale et al., 1973) established the distribution and thickness of the overlying late Pleistocene and early Holocene sedimentary units, consisting of morainal deposits, glaciofluvial, glaciomarine, and postglacial marine units. Sediments accumulate primarily within the basins and troughs, and ridges and swells are sites of nondeposition or erosion. Sedimentological studies have characterized surface sediments (Schlee, 1973; Folger et al., 1975), while the glaciomarine nature of the coarse-grained lower subsurface sediments and the postglacial nature of fine-grained upper subsurface sediments was established by Tucholke and Hollister (1973). The Canadian sector of the Gulf of Maine has been investigated in detail with high-resolution seismic reflection profiling by King and Fader (1986), who established that the basic three-part division of sedimentary units of the Scotian Shelf is present in the Gulf of Maine as well. The sedimentary sequence envisioned by them is, from bottom to top: (1) a basal glacial deposit, the Scotian Shelf Drift, (2) a glaciomarine, poorly sorted clayey and sandy silt, the Emerald Silt (which is further subdivided into facies A, B, and C), and (3) a postglacial, soft, silty-clay unit, the LaHave Clay. Oldale and Bick (1987) and Oldale and Wommack (1987) found corresponding units in the nearshore basins of the western Gulf of Maine. Extensive high-resolution seismic profiling on the inner shelf along the coast of Maine has elucidated the late Quaternary seismic stratigraphy of the glaciomarine and overlying marine and littoral deposits (Belknap, 1987; Belknap et al., 1986, 1987, 1989; Belknap and Shipp, 1991; Kelley and Belknap, 1988; Kelley et al., 1986, 1987a, 1987b; Knebel and Scanlon, 1985; Knebel, 1986) and allowed correlation of marine units with the excellent record of Wisconsin age glacial and glaciomarine events exposed on the lowlands of Maine. Bloom (1963) elucidated sea-level and paleoenvironmental controls on the deposition of widespread glaciomarine sediments in southwestern Maine and named the unit Presumpscot Formation. Continuing investigations of this unit (e.g., Stuiver and Borns, 1975; Smith, 1985; Thompson, 1982; Thompson and Borns, 1985) have provided a concept to be used in the interpretation of the less accessible correlative offshore units.

At present, the Gulf of Maine waters are from three principal sources: (1) deep, warm (6–8 °C), and saline (>34‰ S) continental slope water (SW), which enters the Gulf of Maine through the Northeast Channel; (2) Scotian Shelf water (SSW) of lower salinity (<32‰ S), which enters the gulf as a surface current around the southern tip of Nova Scotia; and (3) freshwater runoff from the surrounding watersheds of the Maritime Provinces, Maine, New Hampshire, and Massachusetts. A counterclockwise baroclinic circulation results from the contrast between shallow, cool, fresh waters of Scotian and nearshore origin and deep, comparatively warm, and salty slope water that enters as an intermittent deep flow (Brooks, 1985). The vertical circulation is of the estuarine type, where saline water enters near the bottom and less saline mixed waters exit near the surface (Ramp et al., 1985; Mountain and Jessen, 1987).

## MATERIALS AND METHODS

More than 2900 km of high-resolution Huntec Deep Tow System and ORE Geopulse seismic reflection profiles and 42 piston cores (Table 1) were collected from the Gulf of Maine during cruises of the RVs *Eastward* (1979), *Cape Hatteras* (1990), and *Endeavor* (1991) (Fig. 2). Bacchus (1993) and Bacchus and Belknap (1997) analyzed portions of the seismic profiles from the 1990 and 1991 cruises and documented the thick sequences of late glacial and postglacial sediments that accumulated in the basins of the eastern Gulf of Maine. The remaining records demonstrate the existence of similar sediment sequences in the basins of the western Gulf of Maine.

The whole-core magnetic susceptibility of all cores taken in 1990 and 1991 was measured on board immediately upon recovery with a pull-through magnetometer. All cores were opened, photographed, described, and sampled in the laboratory shortly after the respective cruises. Samples for water content, foraminiferal, palynological, and diatom micropaleontology were taken at 10 cm intervals over the entire length of each

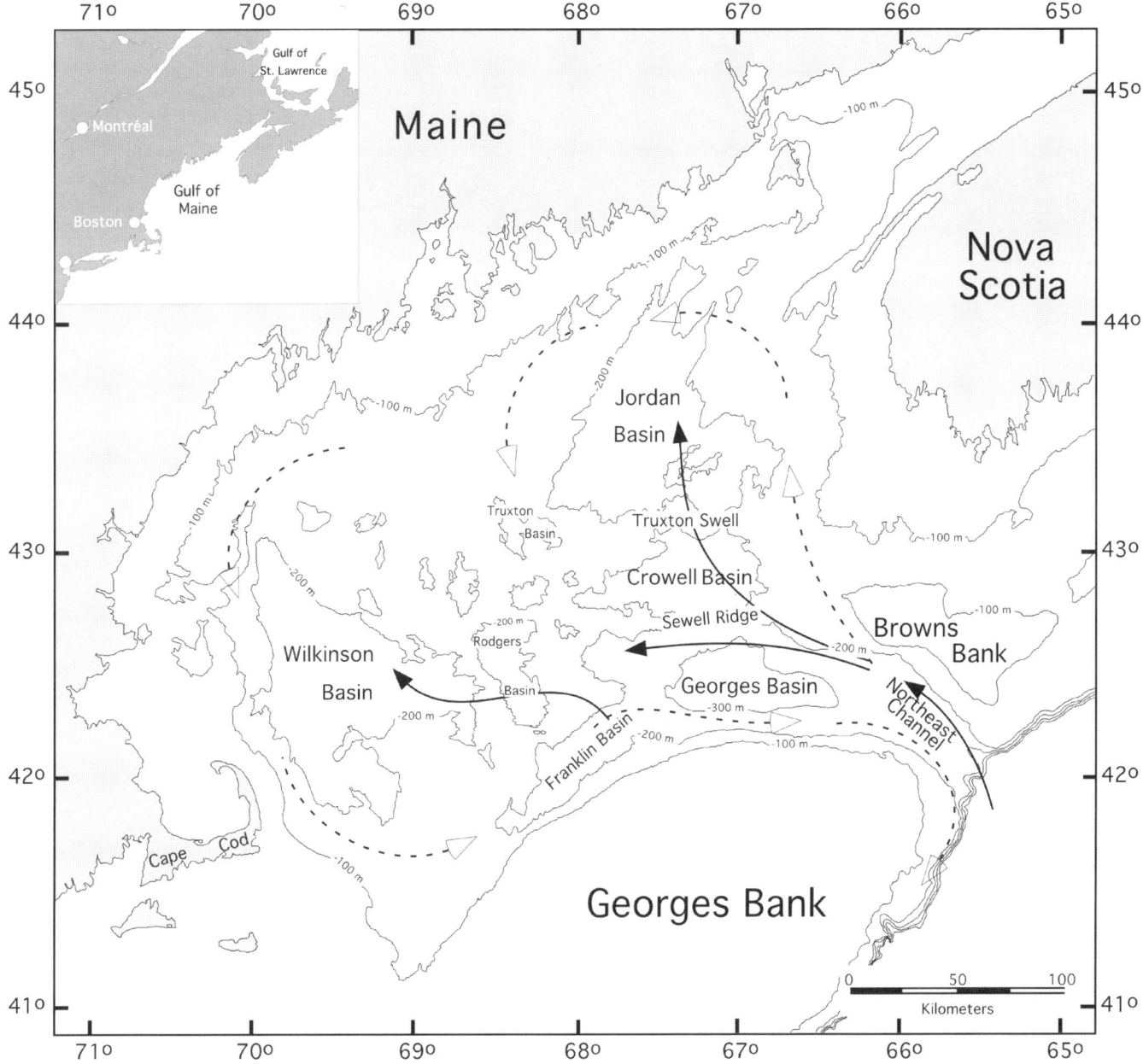

Figure 1. Generalized bathymetry and major named features of Gulf of Maine region. Solid arrows indicate general flow of deep water circulation, open and dashed arrows indicate surface circulation.

core, and samples for grain-size analyses were taken at 30 cm intervals. Water content was determined from weight loss of an ~10 cm³ subcore upon drying at 50 °C for at least 12 hr. No adjustment for salt content was made. The sediment was wet sieved at 62 µm, and the retained coarse fraction was dry sieved into gravel, sand, and silt fractions. The sand was further analyzed in an automated settling tube while the fine fraction was analyzed in a Sedigraph 5000T (Bacchus, 1993). Foraminifers from 19 cores were analyzed from the >125 µm coarse fraction residue from sample sizes that yielded at least 250 specimens. Diatom slides were prepared using a slightly modified cleaning method of Schrader (1974). Abundance estimates were recorded as "valves/traverse" and a minimum of 200 diatoms per slide was counted (except for barren samples). Accelerator mass spectrometry (AMS) radiocarbon dates were obtained on 33 foraminiferal samples from 7 cores. A set of 54 foraminiferal

TABLE 1. GULF OF MAINE PISTON CORE INVENTORY

| Core Number | Location | Latitude (N) | Longitude (W) | Depth (m) | Length (m) | Magnetism | H$_2$O | Grain | Forams | Diatoms | $^{14}$C | AAR | Isotopes |
|---|---|---|---|---|---|---|---|---|---|---|---|---|---|
| E2F-79-1 | Murray Basin | 42° 26.7' | 69° 51.3' | 229 | 12.28 | | | | | | | | |
| E2F-79-2 | Murray Basin | 42° 26.4' | 69° 51.8'2 | 200 | 16.84 | | | X | X | X | X | | X |
| E2F-79-3 | Murray Basin | 42° 27.6' | 69° 43.5' | 270 | 6.04 | | | | | | | | |
| E2F-79-4 | Murray Basin | 42° 31.0' | 69° 39.1' | 244 | 1.48 | | | | | | | | |
| E2F-79-5 | Georges Basin | 42° 27.0' | 67° 03.3' | 339 | 9.77 | | | | | | | | |
| E2F-79-6 | Georges Basin | 42° 27.0' | 67° 03.5' | 352 | 11.13 | | | | | | | | |
| E2F-79-7 | Georges Basin | 42° 27.2' | 67° 03.4' | 350 | 11.42 | | | | | | | | |
| E2F-79-8 | Georges Basin | 42° 27.5' | 67° 03.9' | 350 | 18.28 | | | | | | | | |
| E2F-79-10 | Crowell Basin | 42° 53.5' | 67° 23.1' | 236 | 2.00 | | | | | | | | |
| E2F-79-11 | Crowell Basin | 42° 53.5' | 67° 23.1' | 239 | 11.87 | | | X | X | X | X | | X |
| E2F-79-12 | Crowell Basin | 42° 54.4' | 67° 28.6' | 258 | 12.19 | | | X | X | X | X | | X |
| E2F-79-13 | Jordan Basin | 43° 26.1' | 67° 50.3' | 248 | 11.40 | | | | | | | | |
| E2F-79-14 | Jordan Basin | 43° 25.7' | 67° 43.7' | 247 | 1.62 | | | | | | | | |
| E2F-79-15 | Jordan Basin | 43° 25.7' | 67° 43.6' | 243 | 4.49 | | | | | | | | |
| E2F-79-16 | Platt's Basin | 43° 21.2' | 69° 22.5' | 175 | 9.85 | | | X | X | X | X | | X |
| E2F-79-17 | Platt's Basin | 43° 28.5' | 69° 23.5' | 155 | 8.70 | | | | | | | | |
| E2F-79-18 | Platt's Basin | 43° 31.4' | 69° 09.1' | 128 | 5.91 | | | | | | | | |
| CH10-90-1 | Jordan Basin | 43° 22.83' | 67° 45.63' | 274 | 8.23 | X | X | X | | | | | |
| CH10-90-1A | Jordan Basin | 43° 22.78' | 67° 45.71' | 274 | 12.02 | X | X | X | X | | | | X |
| CH10-90-2 | Jordan Basin | 43° 31.63' | 67° 44.05' | 238 | 4.94 | | | | | | | | |
| CH10-90-2A | Jordan Basin | 43° 31.65' | 67° 43.92' | 238 | 7.84 | X | X | X | X | | | | |
| CH10-90-3 | Jordan Basin | 43° 41.39' | 67° 41.44' | 240 | 16.10 | X | X | X | X | X | X | X | X |
| CH10-90-4 | Crowell Basin | 42° 56.26' | 67° 21.50' | 255 | 12.08 | X | X | X | X | X | X | X | |
| CH10-90-5 | Georges Basin | 42° 28.17' | 67° 06.62' | 345 | 18.19 | X | X | X | X | | | X | X |
| CH10-90-6 | Georges Basin | 42° 26.76' | 67° 06.52' | 368 | 11.42 | X | X | X | X | | | | X |
| CH10-90-7 | Wilkinson Basin | 42° 48.98' | 69° 57.57' | 213 | 16.78 | X | X | X | X | | | | |
| CH10-90-8 | Jeffreys Basin | 42° 00.91' | 69° 08.95' | 173 | 12.04 | X | X | X | X | | X | | X |
| CH10-90-9 | Jeffreys Basin | 43° 14.21' | 70° 07.02' | 139 | 11.82 | X | X | X | X | | X | | |
| CH10-90-10 | Wilkinson Basin | 42° 40.25' | 69° 39.73' | 285 | 16.18 | X | X | X | | | | | |
| CH10-90-12 | Rodgers Basin | 42° 27.94' | 68° 28.68' | 217 | 4.61 | X | X | X | X | | | | |
| CH10-90-12A | Rodgers Basin | 42° 27.95' | 68° 28.68' | 217 | 11.52 | X | X | X | X | | | | |
| CH10-90-13 | Jordan Basin | 43° 53.09' | 67° 37.79' | 210 | 4.77 | X | X | X | X | | | | |
| EN226-1 | Rodgers Basin | 42° 28.54' | 68° 27.68' | 220 | 15.23 | X | X | X | X | X | X | | X |
| EN226-2 | Franklin Basin | 42° 00.62' | 68° 11.14' | 228 | 11.24 | X | X | X | X | | | | X |
| EN226-3 | Lindenkohl Basin | 42° 33.87' | 67° 55.22' | 226 | 14.28 | X | X | X | X | | | | X |
| EN226-4 | Georges Basin | 42° 24.51' | 67° 06.56' | 365 | 13.57 | X | X | X | X | | | | X |
| EN226-5 | Georges Basin | 42° 26.34' | 67° 07.83' | 378 | 13.75 | X | X | X | X | X | X | | X |
| EN226-6 | Crowell Basin | 42° 56.91' | 67° 19.20' | 262 | 11.37 | X | X | X | X | | | | X |
| EN226-7 | Truxton Basin | 43° 07.73' | 68° 23.08' | 209 | 12.05 | X | X | X | X | X | X | | X |
| EN226-8 | Jordan Basin | 43° 31.93' | 68° 17.86' | 195 | 12.00 | X | X | X | X | | | | |
| EN226-9 | Platts Basin | 43° 21.40' | 69° 27.34' | 186 | 15.78 | X | X | X | X | X | X | | X |
| EN226-10 | Wilkinson Basin | 41° 55.32' | 69° 27.34' | 218 | 11.69 | X | X | X | X | | | | X |

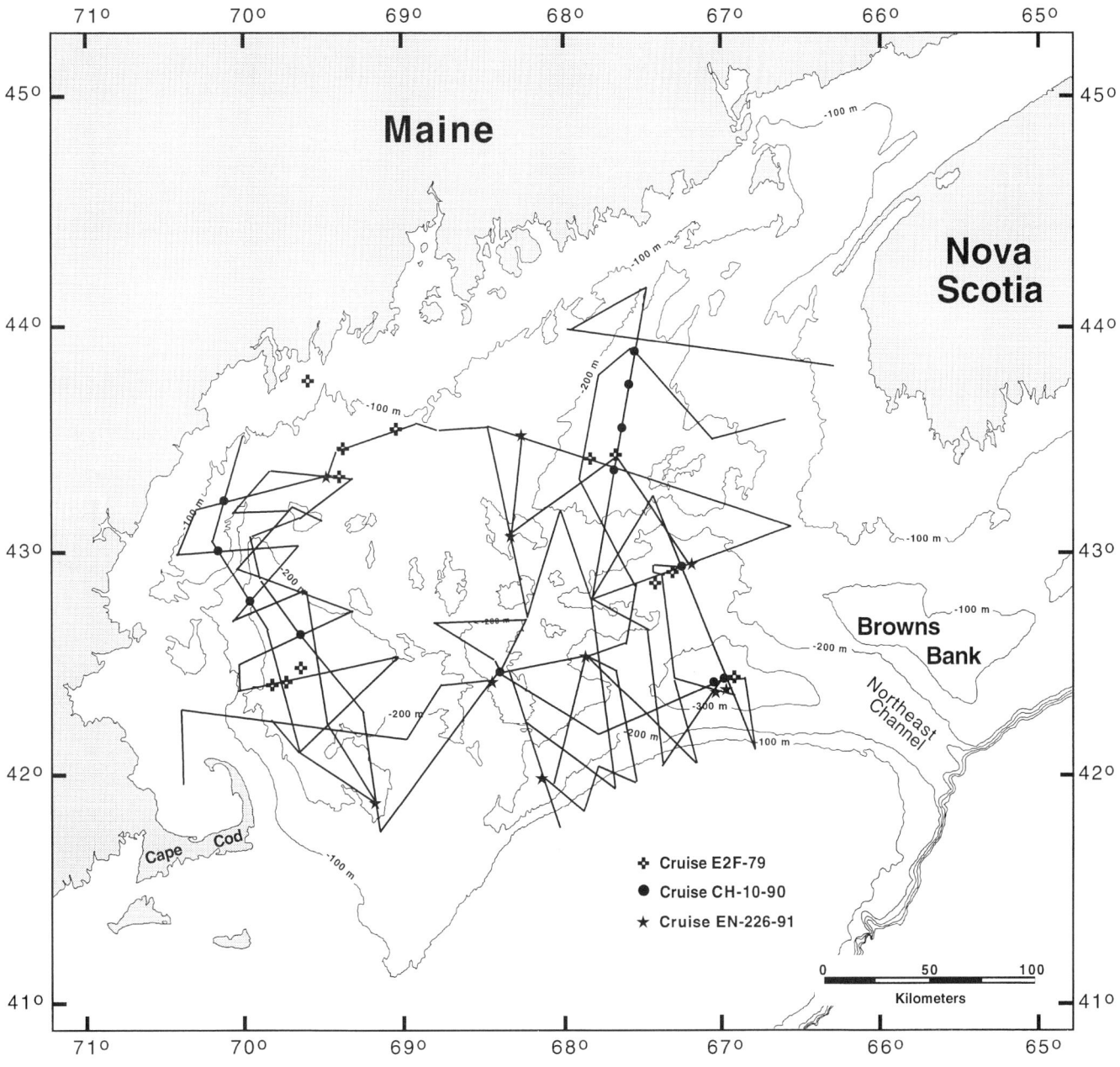

Figure 2. Location of high-resolution seismic lines and piston cores. One symbol may represent more than one core, where several cores were taken in close proximity to each other.

samples from 5 cores was subjected to amino acid racemization (AAR) chronology, using a slight modification of the technique described by Belknap (1979).

## RESULTS

### Seismic stratigraphy

The seismic facies and their relationships that were recognized in earlier investigations on the Scotian Shelf (e.g., King and Fader, 1986) are developed in very similar fashion in the basins of the Gulf of Maine as well. Whereas King and Fader (1986) assigned geographic names to the various seismic facies, Bacchus (1993) and Bacchus and Belknap (1997) suggested the use of designations similar to those developed by Belknap and Shipp (1991) for the nearshore. These designations attempt to indicate their likely genesis, as determined from their geometry and from their recovery in piston cores. These facies, from bottom to top, are proximal glaciomarine (PGM facies), transitional glaciomarine (TGM facies), distal glaciomarine (DGM

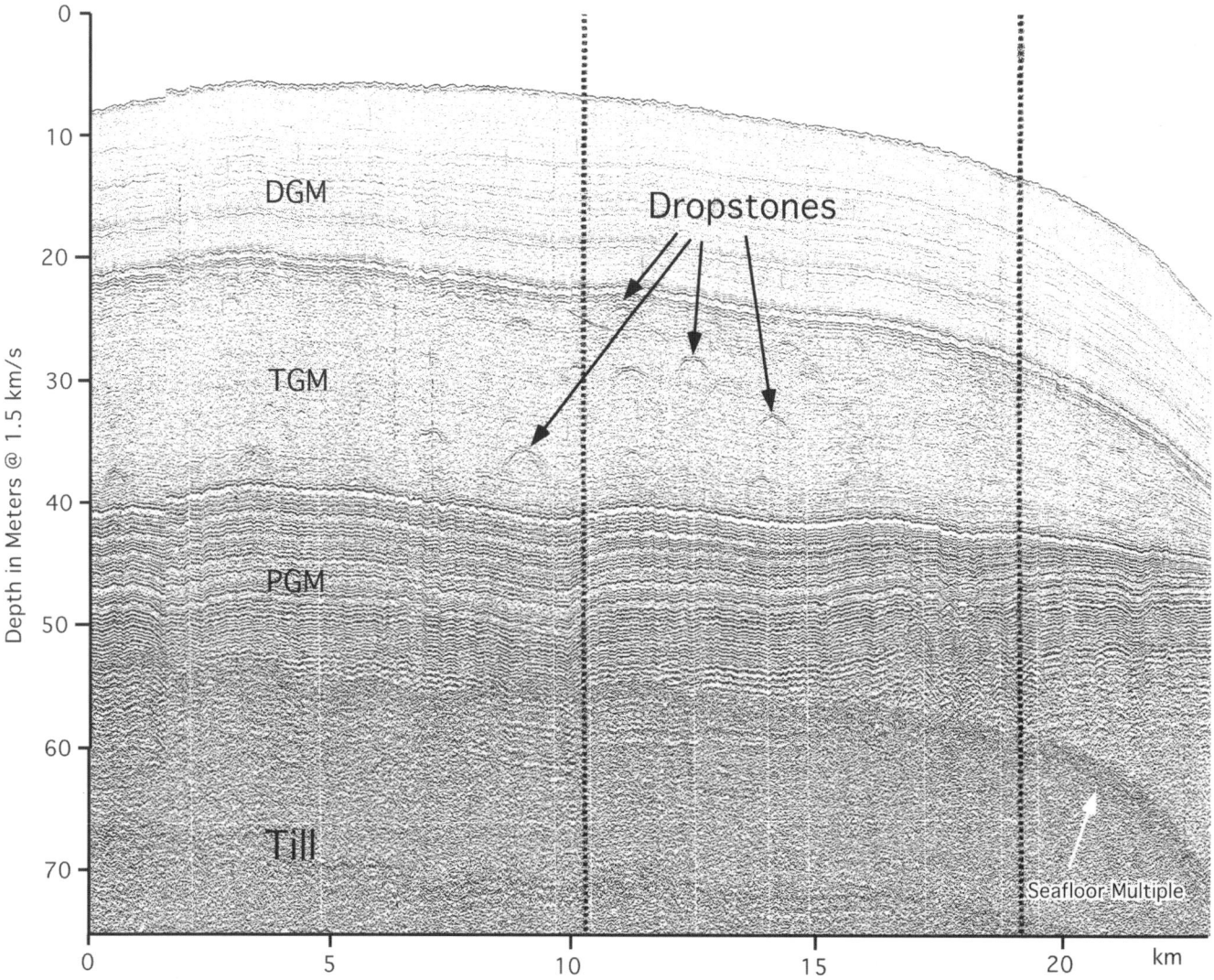

Figure 3. Typical seismic facies of Georges Basin. Note abundance of dropstones in transitional glaciomarine (TGM) facies. Rapid lateral thinning of TGM and distal glaciomarine (DGM) units attest to significant influence of currents. Postglacial mud (M) facies is not present. PGM—proximal glaciomarine. Water depth = 356 m, line CH-90-15 at 14:04h.

facies), and postglacial (M facies). This assemblage of marine units variously overlies an unstratified sedimentary unit with chaotic internal structure—most probably till, or on bedrock.

Figure 3 shows an example of the seismic facies appearance and relationship from the Georges Basin. Relatively thick units of facies PGM, TGM, and DGM overlie a till unit. The PGM unit maintains its thickness almost perfectly, while the later units, TGM and especially DGM, thin sharply over a relatively short distance. Within the basin, all three units are remarkably uniform and consistent over long distances. Postglacial sediments, seismic unit M, are very thin or absent over much of the floor of the Georges Basin. This stratigraphy extends into the Crowell Basin, just north of the Georges Basin, beyond the Sewell Ridge.

Figure 4 shows an example of seismic facies relationships in the Jordan Basin. The PGM facies is often very well stratified, and evenly draped over bedrock or till undulations. It is noteworthy that in this basin the TGM is absent, while the DGM is very thick and well stratified in many places. As in the Georges and Crowell Basins, while the PGM facies is most often rather uniform in thickness across topographic highs and lows, the DGM facies is thickest in depressions and thins toward topographic highs. The M facies are much more common in the Jordan Basin, and may attain considerable thicknesses in deep depressions.

Swells and rises shallower than ~200 m depth are characterized by till and iceberg-remolded glaciomarine sediments. The hummocky appearance of these swells is due to deglacial

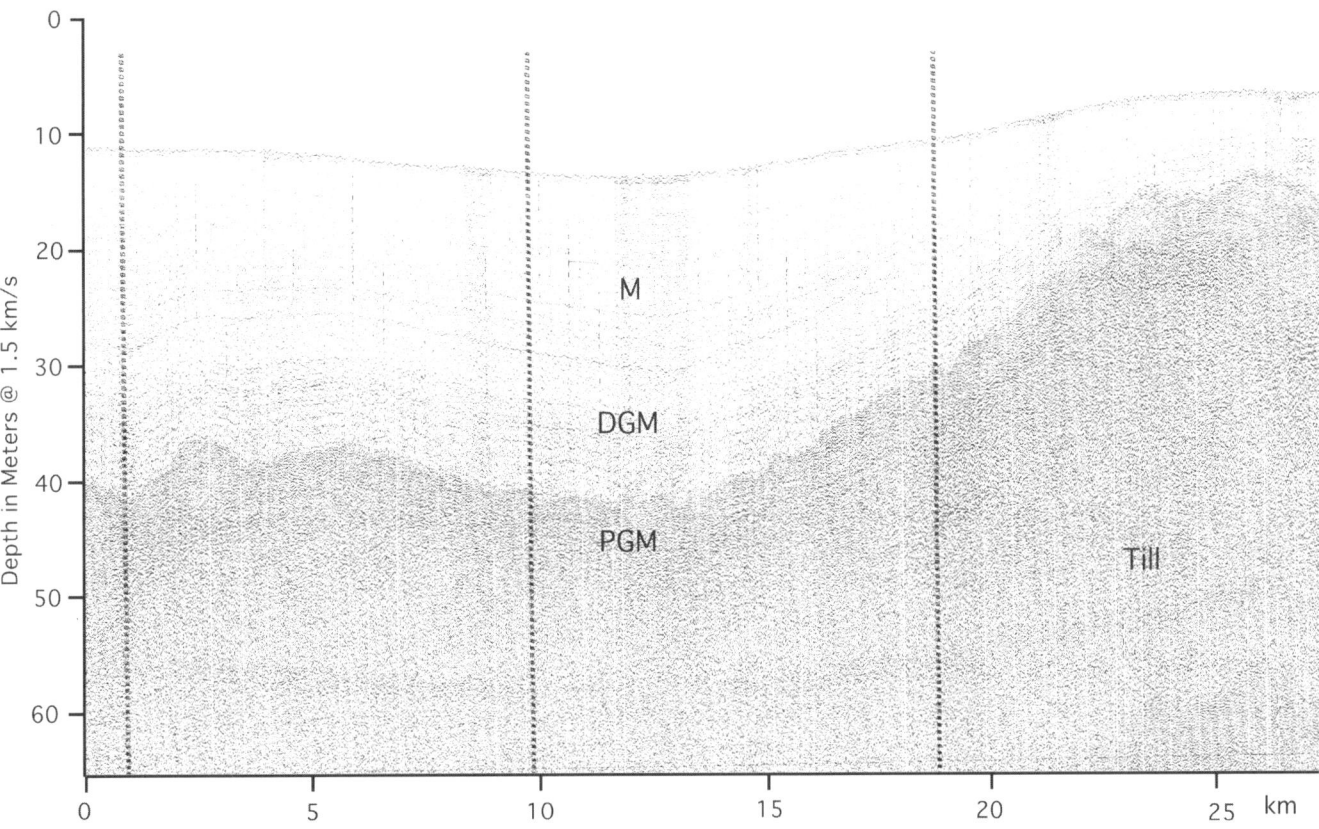

Figure 4. Typical seismic facies of Jordan Basin. Surface undulations of PGM facies (see Fig. 3 for abbreviations) indicate ice contact (iceberg plowmarks). Note upslope thinning (ponding) of DGM and M facies units. Water depth = 248 m, line CH-90-04 at 22:46h.

iceberg furrowing (King and Fader, 1986; Belknap and Schnitker, 1994, 1995).

PGM facies profiles are characterized by densely spaced, high-amplitude internal reflectors that are conformably draped over the bottom topography. Some individual reflectors can be traced across an entire basin. A feature seen in many PGM facies sections of all of the deeper basins of the Gulf of Maine are wedge-shaped occurrences of unstratified deposits within stratified (proximal) glaciomarine deposits (Fig. 5) were first recognized and named till tongues by King and Fader (1986). They were interpreted by them to be composed of till and to represent grounding-line features indicating ice advances and retreats. The undisturbed and continuous, almost conformable nature of the stratification below and above the till tongues indicates that the ice margin rested lightly on the previously deposited glaciomarine deposits, that the ice was in near-floating equilibrium, and that the release rate of sediment near the ice margin was no higher than it was farther into the basin. Sediment delivery was apparently more or less uniformly distributed over wide areas of the basin floor. Thick mounds of till, surrounded by and overlain by stratified PGM facies, also are present, primarily along basin margins, and were called lift-off moraines by King and Fader (1986). Their deposition is thought to have occurred along an ice front with little or no free-flowing subglacial water (Powell and Alley, 1997).

The internal reflectors of the TGM facies are more widely spaced and of lesser amplitude than those of the PGM facies, but may also extend for tens of kilometers. TGM reflectors are conformable at their lower bounding surface. Small hyperbolic reflectors, indicating the presence of iceberg dumps or of large dropstones, are abundant in TGM facies. They occur on occasion in the upper portions of the PGM facies but are rare or absent in DGM facies. The TGM facies thickens noticeably across depressions and thins over topographic highs.

The transition from TGM facies to DGM facies is most often abrupt and recognizable over long distances. The DGM facies is characterized by relatively faint internal reflectors that are conformable to the underlying topography. The DGM facies exhibits prominent ponding in areas of uneven topography. The upper bounding surface of the DGM facies is rarely well defined by seismic character. In the Georges and Crowell Basins, the transition from the DGM to M facies is somewhat apparent from the seismic profiles, whereas in the Jordan Basin it is recognized primarily on the basis of microfossil and sedimentologic data from the cores. The M facies is commonly strongly ponded or mounded into drift deposits.

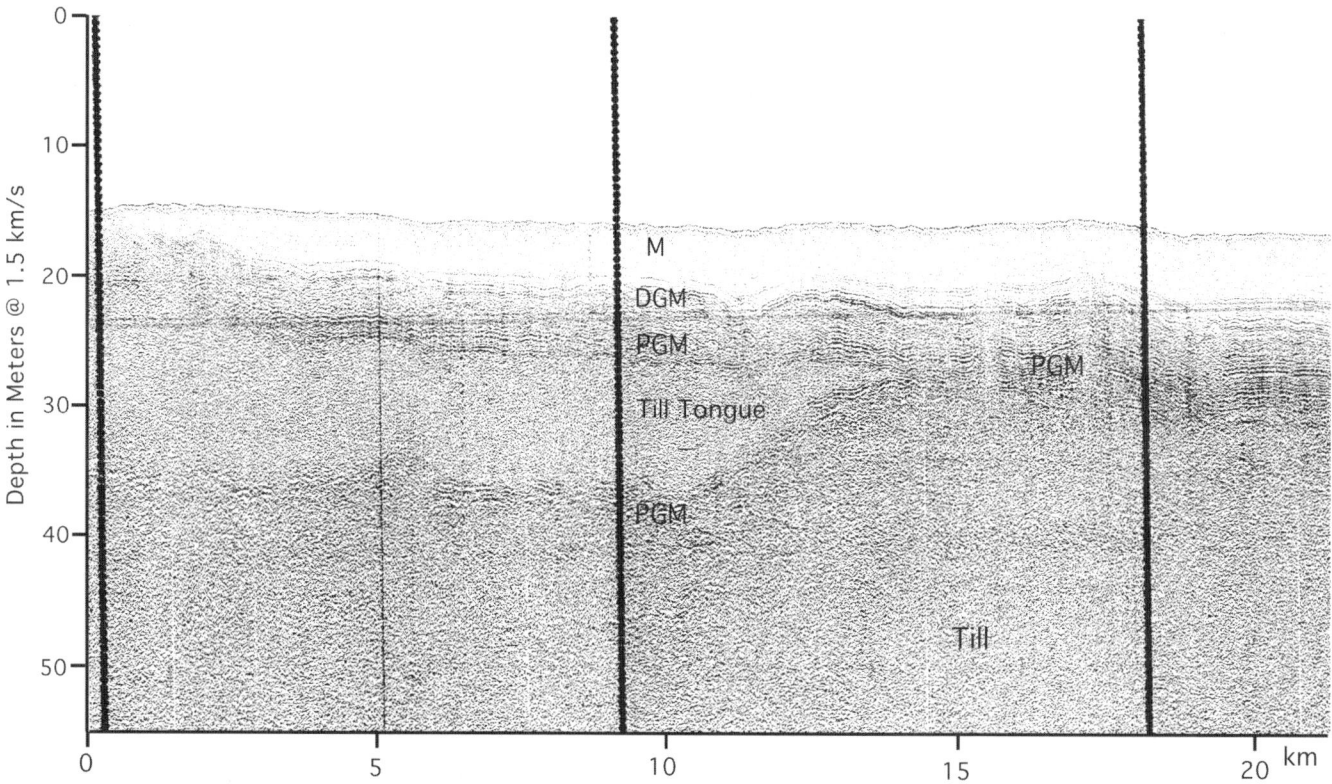

Figure 5. Example of till tongue from Wilkinson Basin, representing wedge of ice-contact sediment between well-stratified PGM sediments. See Figure 3 for abbreviations. Water depth = 255 m, line CH-90-25 at 18:53h.

These examples describe the general appearance of glaciomarine sediments in nearly all of the Gulf of Maine basins, and as a result some general conclusions can be drawn from them. The occurrence of lift-off moraines indicates that, on occasion, ice-margin sedimentation analogous to tidewater glacier sedimentation took place during the early deglaciation of the Gulf of Maine. However, the widespread distribution, rather uniform thickness, and even stratification of the lowermost glaciomarine unit, the PGM facies, largely independent of bottom topography (draping), indicates that this unit was emplaced in quiet water and by an agent that was competent to distribute very coarse as well as fine-grained sediment evenly and over long distances. The TGM facies provides clear evidence for discontinuous iceberg-drop deposits as well as continuous well-stratified beds. This suggests a cyclical calving, advancing, and retreating ice shelf. The deposition of the DGM sediments, while still dominated by glacial input, is obviously more influenced by water motion.

## Sediments

Glaciomarine sediments recovered in piston cores are generally clays with a significant component of silt, sand, and some gravel. In the proximal PGM facies, the coarse fraction (>125 µm) often reaches 50% or more of the total. Recovery of pebbles is not unusual, and occasional damage to the corers nose cone testifies to the encounter of large boulders. The TGM and DGM units are progressively finer grained, so that the DGM facies is composed essentially of green-gray or brown-gray clayey silt. Sand-sized, or larger, particles are extremely rare in the DGM facies. The M facies is composed of mostly bluish-gray or olive-gray silty clay.

The occurrence of glacial sediments within the cores is reflected precisely by high magnetic susceptibility values as well as low water-content values (Fig. 6). Susceptibility of sediments is an expression of water content, sediment grain size, sediment sorting, and mineralogy.

## Chronology

Radiocarbon dating (AMS) of 38 samples of benthic foraminifers (Table 2) from 10 selected cores provided a chronology for the seismic stratigraphy and the sedimentological, faunal, and floral events that were observed in the cores. A reservoir-age correction of 600 yr was applied to the reported AMS ages. Kashgarian and Southon (1993) established that due to coastal upwelling shallow-water reservoir ages are usually 100 yr older than typical values for open ocean waters. Gulf of

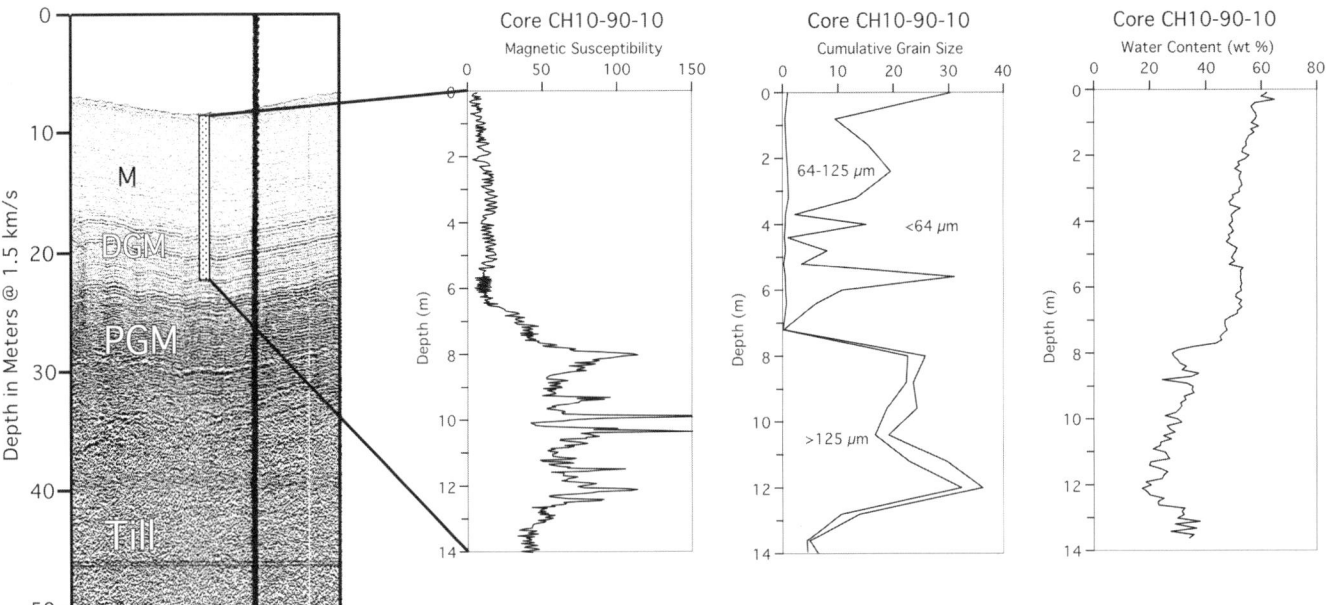

Figure 6. Correlation of seismic facies, downcore magnetic susceptibility, coarse fraction abundance, and water content of core CH10-90-10 from Wilkinson Basin. Abbreviations as in Figure 3.

Maine mollusks, collected between 1870 and 1879, indicate a reservoir-age correction of 600 yr (Kashgarian, 1999, personal commun.).

As is evident from the data in Table 2, the radiocarbon dating did not always result in a clear and unambiguous chronology. In four cores age inversions occurred when samples from the upper portion of the DGM facies produced a greater age than the samples below it from the lower portion of the DGM or even TGM facies (Fig. 7). D.B. Scott (Dalhousie University) pointed out (personal commun., 1997) that similarly anomalous dates were obtained from the Scotian Shelf and suggested that the benthic environment during the late DGM facies deposition was very dynamic and that partial reworking of older deposits most likely caused contamination. Jennings et al. (1998) also encountered age inversions in their glaciomarine core sections from Hudson Strait, which they ascribed to reworked faunas admixed into in situ faunas. A suggestion for such reworking is present in cores EN226-1, EN226-7, EN226-9, and E2F-79-2: *Cibicides lobatulus*, a species closely associated with the earliest deglacial environments encountered in Gulf of Maine cores (see following), reappears or increases slightly in abundance at the levels of the anomalously old radiocarbon dates. This evidence is ambiguous, however, because reworking implies strengthened bottom currents, an environment also favorable for the success of *C. lobatulus* (see following).

An unusual example of unexpected $^{14}$C ages was obtained from core CH-10-90-3 from the Jordan Basin. As shown in Figure 8, the ages of the glaciomarine section of this core increase rapidly and linearly with depth, reaching 40 740 yr B.P. ($\pm$1660 yr). We subjected 27 foraminiferal samples from this same core to AAR analysis (Table 3) (Lusardi, 1992). The age regression resulting from these analyses fails to reproduce the extreme ages of the AMS $^{14}$C dates, arriving at an approximate age of 17 000 yr B.P. for the bottom of the core instead of ca. 41 000 yr B.P. In this case, the results of the AAR analyses speak against the suggestion of reworking of older material, because both dating methods were performed on identical faunal elements from the same core material. Faunal contamination should have resulted in similar and not diverging time lines. Unfortunately, no single species was available for dating throughout the length of the core, thus interspecies calibrations make the AAR trend less certain (Lusardi, 1992).

*Paleontology*

All marine sediments recovered in cores contained a significant microfauna of foraminifers. The foraminiferal contents of glaciomarine and early postglacial sediments is relatively low, usually between 30 and 100 specimens per gram of dry sediment. Foraminiferal numbers increase by an order of magnitude well after cessation of glacial conditions in the Gulf of Maine (Fig. 9). This increase is in part due to enhanced biological productivity, and also due to a decrease in the rate of sedimentation. The compositions of the foraminiferal microfaunas contained in the glacial and early deglacial sediments follow very similar patterns in all cores. Core to core differences are primarily due to recovery of different segments of the deglacial sequence (Fig. 7). Only one core, from the southeast-

## TABLE 2. GULF OF MAINE AMS RADIOCARBON DATING

| Core Number | Date No. | Depth (m) | Weight (mg) | Fraction Modern | $^{14}$C Age (ka) | ± | Corrected Age (ka) |
|---|---|---|---|---|---|---|---|
| CH10-90-3 | AA-7044 | 3.20–3.22 | 7.79 | 0.5931 ± 0.0051 | 4195 | 70 | 3595 |
| CH10-90-3 | AA-7045 | 11.60–11.62 | 18.01 | 0.1880 ± 0.0042 | 13425 | 180 | 12825 |
| CH10-90-3 | AA-7079 | 12.80–12.85 | 8.906 | 0.0707 ± 0.0019 | 21285 | 210 | 20685 |
| CH10-90-3 | AA-7080 | 13.40–13.45 | 7.34 | 0.0434 ± 0.0016 | 25210 | 290 | 24610 |
| CH10-90-3 | AA-7081 | 15.60–15.70 | 9.022 | 0.0058 ± 0.0012 | 41340 | 1605 | 40740 |
| CH10-90-4 | OS-1844 | 2.8 | 75.6 | | 14200 | 50 | 13600 |
| CH10-90-4 | OS-1856 | 6 | 13.31 | | 14900 | 50 | 14300 |
| CH10-90-4 | OS-1859 | 10.8 | 14.13 | | 15050 | 45 | 14450 |
| CH10-90-5 | OS-1841 | 2.95 | 10.69 | | 11500 | 100 | 10900 |
| CH10-90-5 | OS-1849 | 10.8 | 30.96 | | 14150 | 55 | 13550 |
| CH10-90-5 | OS-1842 | 13.6 | 20.24 | | 14850 | 60 | 14250 |
| CH10-90-8 | AA-9105 | 0.4 | 9.64 | 0.7327 ± 0.0050 | 2500 | 55 | 1900 |
| CH10-90-8 | AA-9106 | 11.50–11.53 | 9.15 | 0.2593 ± 0.0028 | 10845 | 85 | 10245 |
| CH10-90-9 | AA-9107 | 9.83 | 20.62 | 0.1593 ± 0.0042 | 14760 | 215 | 14160 |
| EN226-1 | AA-9108 | 0.5 | 7.25 | 0.7013 ± 0.0063 | 2850 | 70 | 2250 |
| EN226-1 | OS-1855 | 5.6 | 9.84 | | 5120 | 45 | 4520 |
| EN226-1 | AA-9109 | 11.5 | 22.177 | 0.2018 ± 0.0024 | 12855 | 95 | 12255 |
| EN226-1 | AA-9110 | 12.50–12.53 | 11.33 | 0.1455 ± 0.0021 | 15480 | 115 | 14880 |
| EN226-1 | OS-1840 | 13.6 | – | | 14400 | 55 | 13800 |
| EN226-5 | OS-1857 | 0.5 | 12.43 | | 5510 | 40 | 4910 |
| EN226-5 | OS-1858 | 4.8 | 15.22 | | 8120 | 40 | 7520 |
| EN226-5 | OS-1846 | 9 | 15.41 | | 10700 | 60 | 10100 |
| EN226-5 | OS-1852 | 12.8 | 13.59 | | 12900 | 45 | 12300 |
| EN226-7 | OS-1854 | 1 | 11.77 | | 1460 | 25 | 860 |
| EN226-7 | OS-1847 | 8 | – | | 13700 | 55 | 13100 |
| EN226-7 | OS-1853 | 8.8 | 19.99 | | 16750 | 85 | 16150 |
| EN226-7 | OS-1848 | 11.2 | 23.86 | | 14750 | 50 | 14150 |
| EN226-9 | OS-1864 | 1.6 | 9.68 | | 5400 | 40 | 4800 |
| EN226-9 | OS-1845 | 8 | 34.03 | | 13400 | 45 | 12800 |
| EN226-9 | OS-1843 | 10 | 16.5 | | 19750 | 115 | 19150 |
| EN226-9 | OS-1863 | 13.2 | 10.66 | | 15200 | 50 | 14600 |
| E2F-79-2 | RIDDL-810 | 2 | ~10 | | 6020 | 160 | 5420 |
| E2F-79-2 | RIDDL-811 | 6.22 | ~10 | | 12600 | 180 | 12000 |
| E2F-79-2 | AA-9111 | 8.19–8.23 | 12.78 | 0.0837 ± 0.0020 | 19925 | 190 | 19325 |
| E2F-79-2 | AA-9112 | 9.09–9.15 | 10.581 | 0.2111 ± 0.0025 | 12494 | 95 | 11894 |
| E2F-79-2 | RIDDL-812 | 10.8 | ~10 | | 12200 | 150 | 11600 |
| E2F-79-2 | RIDDL-813 | 12.40–12.43 | ~10 | | 15420 | 120 | 14820 |
| E2F-79-2 | RIDDL-814 | 15.09–15.45 | ~10 | | 17660 | 130 | 17060 |

ernmost and deepest of the basins (Georges Basin), reached the proximal glaciomarine sequence (PGM facies) that can be observed on the seismic profiles. Other cores did not recover at their top the last portion of the deglacial sedimentary sequence.

As shown by Schnitker (1976, 1986), Lusardi (1992), and Friez (1993), an almost unvarying succession of only six common species characterizes the deglacial sedimentary sequences, species that are well known from relatively shallow arctic and cold-water environments. In almost all cores, these six species occur in three covarying pairs, each with a discrete stratigraphic range. *Cibicides lobatulus* and *Islandiella algida* reach their common abundance peaks in the oldest glaciomarine sediments (PGM). These species are followed in succession by *Cassidulina reniforme* and *Elphidium excavatum* (forma *clavatum*) to the top of the glaciomarine sediments (DGM), essentially dominating the entire transitional and distal glaciomarine sequence. While common, or even relatively abundant in the glaciomarine sediments, *Islandiella helenae* and *Nonionellina labradorica* become the dominant species immediately upon cessation of ice rafting, i.e., in the lowermost portion of the postglacial M unit (Fig. 10). Agglutinating species are entirely absent from Gulf of Maine glaciomarine sediments, perhaps surprisingly so in view of their abundance in modern periglacial-marine environments. Their absence is probably due to the poor preservation potential of the organic cement they use to construct their tests.

The occurrence of diatoms, investigated by Schnitker and Jorgensen (1990) and Popek (1993), also follows a consistent sequence through the deglacial sediment section of all cores examined. Most noteworthy is the absence, or extremely low abundance, of diatoms in all PGM sediments. The TGM sediments of the southeastern basins carry a very sparse flora of poorly preserved frustules of *Thalassiothrix longissima*. The abundance of diatom frustules and spores increases somewhat in the distal glaciomarine sediments, but they become plentiful only at the end of ice rafting, at the glacial-postglacial transition, often reaching their maximal numbers during early postglacial time. The first diatom to appear in significant numbers in the upper glaciomarine, and especially lowermost postglacial sediments, is *Thalassiosira gravida*, mostly in the form of

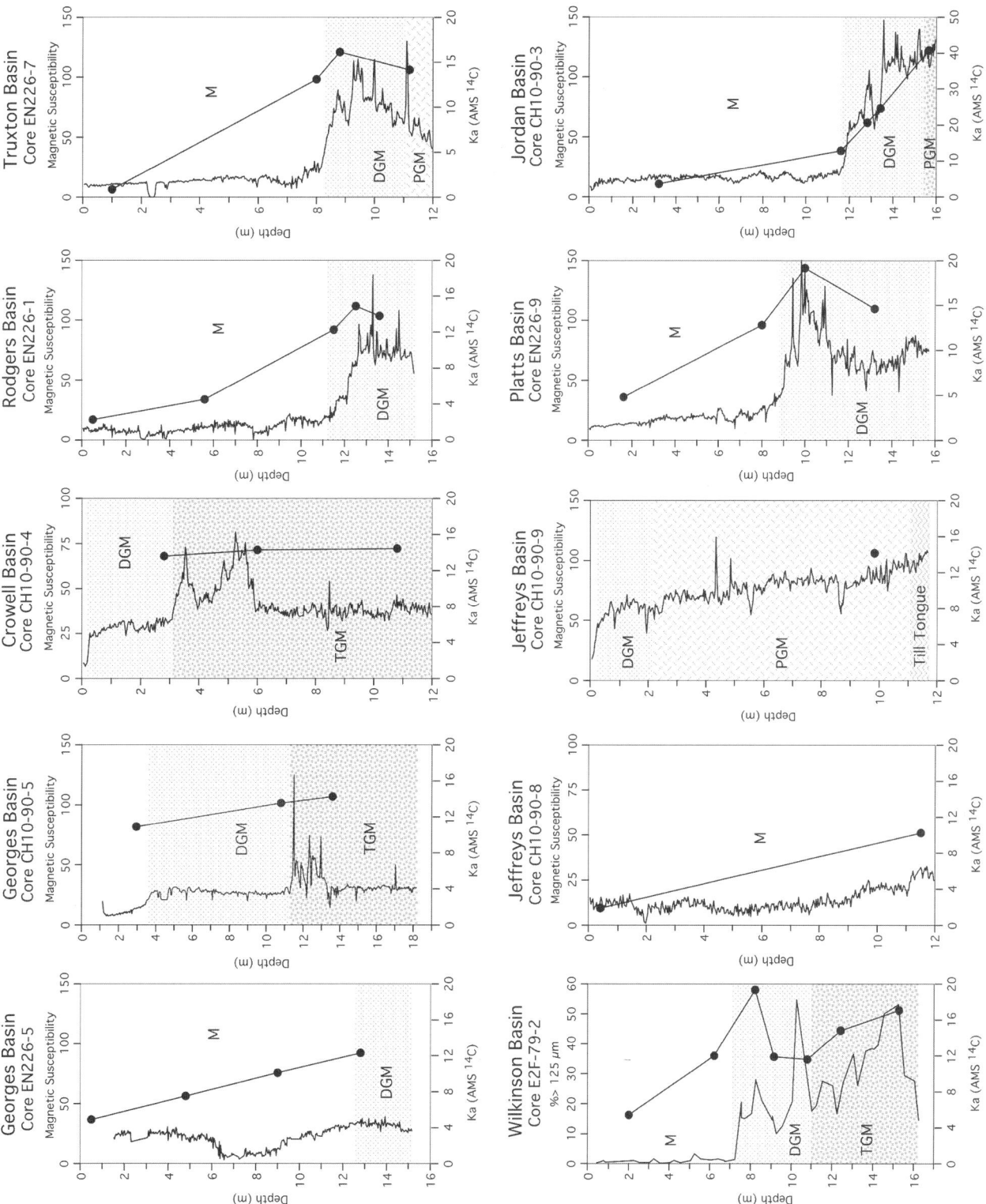

Figure 7. Comparisons of magnetic susceptibility, seismic facies, and (corrected) radiocarbon ages in 10 Gulf of Maine cores. Out of sequence old ages (age inversions) occur frequently near top of glaciomarine sequences, indicating increased sediment reworking at termination of glacial conditions. See Figure 3 for abbreviations.

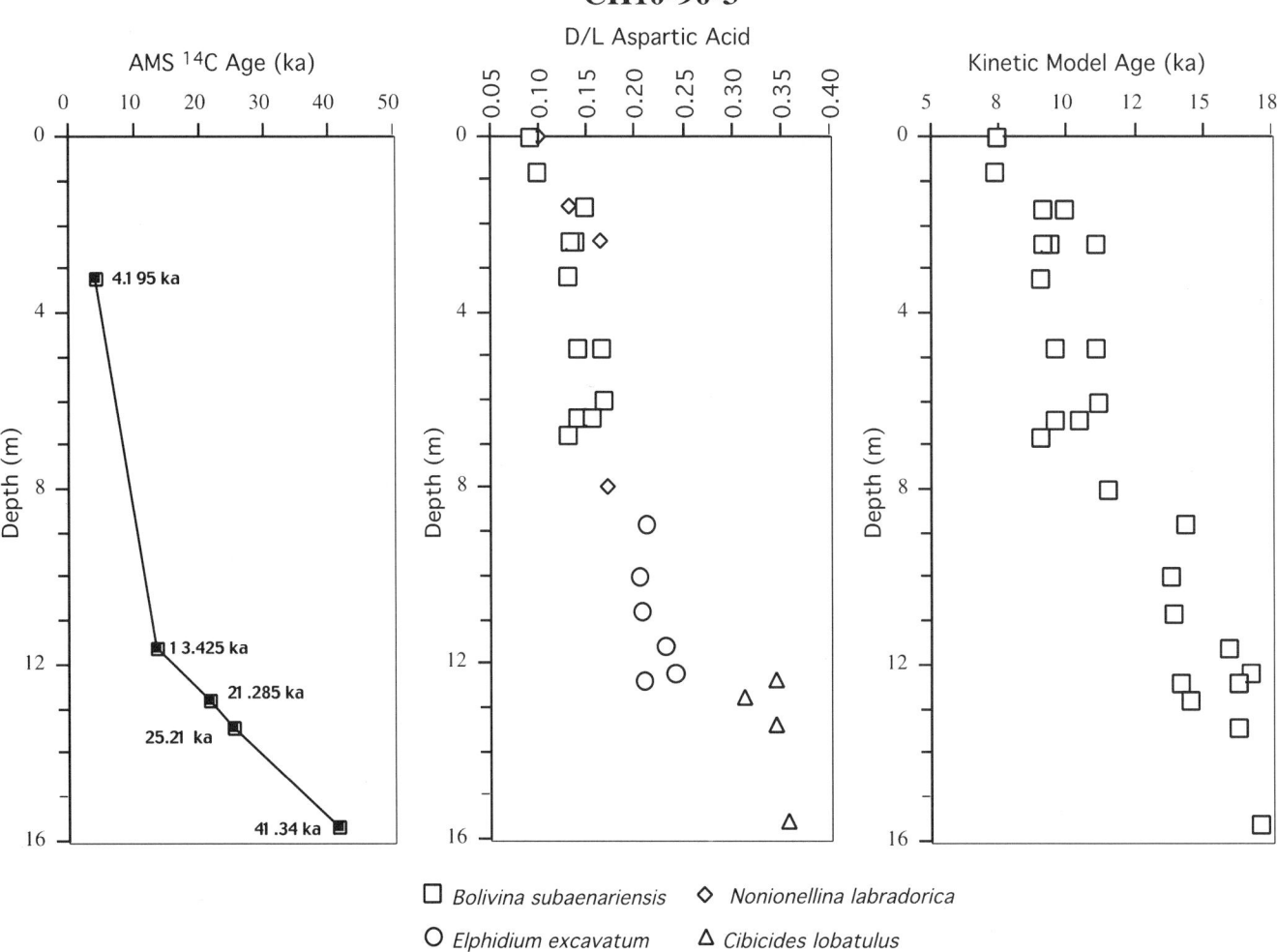

Figure 8. Comparison of accelerator mass spectrometer (AMS) radiocarbon and amino acid racemization (AAR) dating. AAR age model, although generated from same microfossil assemblages, fails to reproduce very rapid radiocarbon-age increases within glaciomarine section of core. ASP = D/L aspartic acid ratios.

spores. Various species of *Chaetoceros* become the dominant diatom spores in the lower postglacial sediments, immediately following the abundance peaks of *Thalassiosira gravida* (Fig. 11).

## DISCUSSION

The presence of widespread unstratified glacial sediment (till) attests to the presence of grounded ice throughout the Gulf of Maine region during the last glacial maximum. According to Schlee (1973), this ice extended out onto the Georges and Browns Banks, with an ice-stream exiting through Northeast Channel (Fig. 12). The appearance of stratified sediments indicates the incursion of marine waters into the Gulf of Maine. Although the contact of unstratified-stratified sediments is commonly very clear on seismic profiles, it has not been recovered in sediment cores. The timing of this transition has to be estimated through extrapolation of the sedimentation rates from the lowest $^{14}$C date to the depth of the lowest clear record of glaciomarine sediment on the seismic profile of the coring sites; thus an approximate sequence of deglaciation and successive marine stages can be developed. Deglaciation started in the Georges Basin probably by 22 000 yr B.P., but not later than 17 500 yr B.P.; between 19 000 and 18 000 yr B.P. in the Crowell and Rogers Basins and possibly southern Wilkinson Basin; and near 17 000 yr B.P. in the northern Wilkinson-Jeffreys and Platts Basins (Table 4). In the absence of better dating, the deglaciation in the Jordan Basin is assumed to have occurred after deglaciation of the Crowell Basin (ca. 18 000 yr B.P.) and before the appearance of marine deposits ca. 13 800 yr B.P. on the coast of southwestern Maine and ca. 13 200 yr B.P. in eastern Maine (Smith, 1985; Smith and Hunter, 1989; Kaplan, 1999).

TABLE 3. AMINO ACID RACEMIZATION DATES,
CORE PC-90-3, JORDAN BASIN

| Depth (m) | ASP Ratio | Species | Age | +ka | −ka |
|---|---|---|---|---|---|
| 0.01 | 0.09 ± 0.010 | B. subaenariensis | 7450 | 2 | 2 |
| 0.01 | 0.10 ± 0.030 | N. labradorica | 7450 | 2 | 2 |
| 0.8 | 0.098 ± 0.001 | B. subaenariensis | 7350 | 2 | 1 |
| 1.6 | 0.123 ± 0.006 | N. labradorica | 9100 | 2 | 2 |
| 1.6 | 0.146 ± 0.006 | B. subaenariensis | 9890 | 3 | 2 |
| 2.4 | 0.164 ± 0.020 | N. labradorica | 10990 | 3 | 2 |
| 2.4 | 0.136 ± 0.010 | B. subaenariensis | 9320 | 2 | 2 |
| 2.4 | 0.131 ± 0.020 | B. subaenariensis | 9040 | 2 | 2 |
| 3.2 | 0.13 ± 0.010 | B. subaenariensis | 8990 | 2 | 2 |
| 4.8 | 0.164 ± 0.020 | B. subaenariensis | 10990 | 3 | 2 |
| 4.8 | 0.139 ± 0.010 | B. subaenariensis | 9490 | 2 | 2 |
| 6.0 | 0.166 ± 0.008 | B. subaenariensis | 11110 | 3 | 2 |
| 6.4 | 0.14 ± 0.010 | B. subaenariensis | 9550 | 2 | 2 |
| 6.4 | 0.154 ± 0.008 | B. subaenariensis | 10370 | 3 | 2 |
| 6.8 | 0.13 ± 0.020 | B. subaenariensis | 8990 | 2 | 2 |
| 8.0 | 0.172 ± 0.045 | N. labradorica | 11500 | 3 | 2 |
| 8.8 | 0.212 ± 0.045 | E. clavatum | 14320 | 4 | 3 |
| 10.0 | 0.204 ± 0.023 | E. clavatum | 13740 | 3 | 3 |
| 10.8 | 0.206 ± 0.009 | E. clavatum | 13870 | 4 | 3 |
| 11.6 | 0.232 ± 0.003 | E. clavatum | 15890 | 4 | 3 |
| 12.2 | 0.241 ± 0.010 | E. clavatum | 16640 | 4 | 3 |
| 12.4 | 0.209 ± 0.001 | E. clavatum | 14090 | 4 | 3 |
| 12.4 | 0.344 ± 0.019 | C. lobatulus | 16220 | 4 | 3 |
| 12.8 | 0.312 ± 0.015 | C. lobatulus | 14470 | 4 | 3 |
| 13.4 | 0.344 ± 0.001 | C. lobatulus | 16220 | 4 | 3 |
| 15.6 | 0.358 ± 0.005 | C. lobatulus | 17060 | 4 | 3 |

Note: Kinetic model ages based on Wehmiller et al. (1988); EQT is 6 °C.

### Proximal glaciomarine environment

The oldest sediments (late PGM or earliest TGM facies) reached by our piston cores contain a sparse foraminiferal fauna essentially composed of two species: *Cibicides lobatulus* and *Islandiella algida*. *C. lobatulus* is a common epifaunal species, living attached to rocks, pebbles, and even macroorganisms (Svavarsson and Davíðsdóttir, 1994). This mode of life requires relatively coarse substrate and possibly some bottom currents to keep the attached foraminifers from being smothered by silt and clay. Schafer and Cole (1986) found *C. lobatulus* in ice-proximal sediments of Baffin Island, and Jennings and Helgadottir (1994) found it in coarse-grained sediments off Nansen Fjord, southeast Greenland. *Islandiella algida* (*Islandiella islandica* of authors; Miller et al., 1996) has been reported from the arctic and subarctic regions of the North Atlantic. On the Labrador Shelf, *Islandiella algida* is commonly associated with *C. lobatulus* in areas of relatively coarse substrates that are influenced by bottom currents (Vilks and Deonarine, 1988). Both species are found only in waters of normal salinities of >32‰–33‰.

The virtual absence of diatoms from these glaciomarine sediments indicates either that primary productivity did not occur in the overlying waters, or that silica dissolution was more intensive during PGM deposition than during subsequent deposition phases. Mayer et al. (1991) demonstrated that dissolution of diatom frustules from aluminosilicate sediments is enhanced when they are subjected to frequent redox transitions. The estuarine circulation ensures that bottom waters entering the Gulf of Maine are well oxygenated. Changes in redox conditions are therefore essentially a function of high organic matter input into the benthic environment, i.e., high surface productivity. Evidence for such is lacking in PGM sediments. The most probable inhibitor of primary productivity in such setting is the presence of a permanent ice cover that prevented the transmission of light into the water. The coarse sediment composition and the widespread and even appearance of seismic stratification strongly suggest that deposition of the PGM facies took place underneath an ice shelf.

The presence of benthic foraminifers that prefer coarse substrates, possibly some bottom currents, and that require fully marine salinities, suggests that normal marine water was drawn through the Northeast Channel as a bottom current. This estuarine circulation was driven by meltwater-induced subice outflow of mixed water of lowered salinity (Fig. 13). This influx of deep water brought in sufficient organic matter to sustain a sparse fauna of benthic foraminifers and invertebrates, the former presence of which is attested to by the very few observations of bioturbation and even fewer macrofossils in PGM sediments. This situation was similar to, but much less extreme than, the one described for the bacterial flora and metazoan fauna encountered underneath the 420-m-thick Ross Ice Shelf at Site J-9, 400 km inward of the ice edge (Azam et al., 1979; Lipps et al., 1979).

### Transitional and distal glaciomarine environment

The remaining glaciomarine sediments, the TGM facies in the Georges and Crowell Basins and the DGM facies in all other basins, contain an abundant fauna of benthic foraminifers that is dominated by *Cassidulina reniforme* and *Elphidium excavatum*, forma *clavatum*. These two species are ubiquitous in arctic periglacial marine environments (Miller, 1996). *Cassidulina reniforme* is indicative of ice-marginal environments, characterized by very cold and muddy waters with high (>33‰) salinities (Mudie et al., 1983; Schafer and Cole, 1986). Jennings and Helgadottir (1994) found *C. reniforme* to be the dominant calcareous species within the innermost reaches of Kangerdlugssuaq Fjord, an outlet of the Greenland ice sheet and of several tidewater glaciers. *Elphidium excavatum* forma *clavatum* occupies similar environments, possibly preferring slightly shallower depths and somewhat lower salinities (Hald et al., 1994; Miller, 1996). It may be an indicator of the presence of meltwater.

A persistent but sparse and poorly preserved flora of *Thalassiothrix longissima* frustules is the first indication of primary productivity in the late glacial Gulf of Maine; they occur in the TGM facies of the southeastern Gulf of Maine, or at the base of the DGM facies elsewhere. The appearance of diatom frustules and spores in the DGM facies, *Thalassiosira gravida*

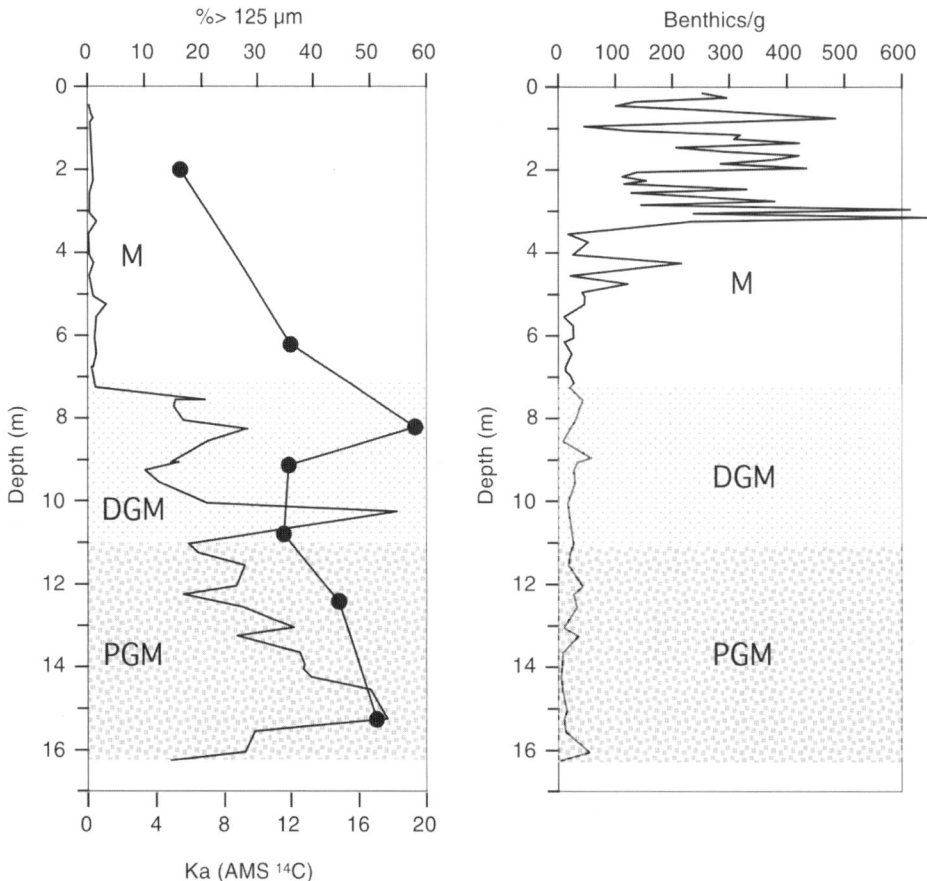

Figure 9. Abundance of benthic foraminifers in sediments of core E2F-79-2 from Wilkinson Basin. Note that foraminiferal numbers remain rather uniform throughout different glaciomarine phases, and increase sharply only well after cessation of glacial conditions. See Figure 3 for abbreviations.

(spores) being the dominant species, is sharply coincident with the appearance of the *C. reniforme* and *E. clavatum* foraminiferal faunas.

The appearance of diatoms in the sedimentary sequence marks the onset of primary productivity in surface waters of the Gulf of Maine. It marks the breakup of the ice shelf and the beginning of a summer season of ice-free waters or thin sea ice. *T. longissima* is known from arctic and boreal oceanic environments (Reid and Surey-Gent, 1980; Hasle and Semina, 1987); thus its presence in the deglaciating Gulf of Maine indicates that the surface waters were not strongly diluted by meltwater. In view of the very low abundance of the *T. longissima* flora within the sediments, surface productivity was probably still very minor. The postulated environment for this assemblage is a calving bay, where icebergs are floating in a sea still covered with heavy pack ice. Open water is available only briefly during summer and when wind stress opens leads in the sea ice.

The subsequent rapid increase in diatom abundance is almost exclusively due to a burgeoning of *Thalassiosira gravida*. This species is found in similar abundance in arctic environments along the edge of melting winter ice pack in western Baffin Bay (Williams, 1986, 1990), or within a very narrow zone along the edge of the pack ice in Fram Strait (MIZEX experiment; Smith et al., 1987). Upwelling along such ice edges creates zones of very cold water ($-1.5$ to $+1$ °C) of lowered salinity (27‰–31‰) and elevated productivity.

The environment to account for these observations is that of a calving bay, initially intensely cold, filled with icebergs and pack ice nearly year-round, so that only minor primary productivity could occur in the surface waters. The fining of the sediments testifies to a change from sediment meltout from the underside of the ice shelf to meltout from drifting icebergs and a widespread distribution of turbid meltwater in an estuarine-style circulation near the surface. Settling of fine-grained muds to the seafloor led to the dominance of the benthic *C. reniforme* fauna. Increased melting, combined with strong katabatic winds from the ice sheet, decreased the surface salinities and increased the area and duration of open water within the

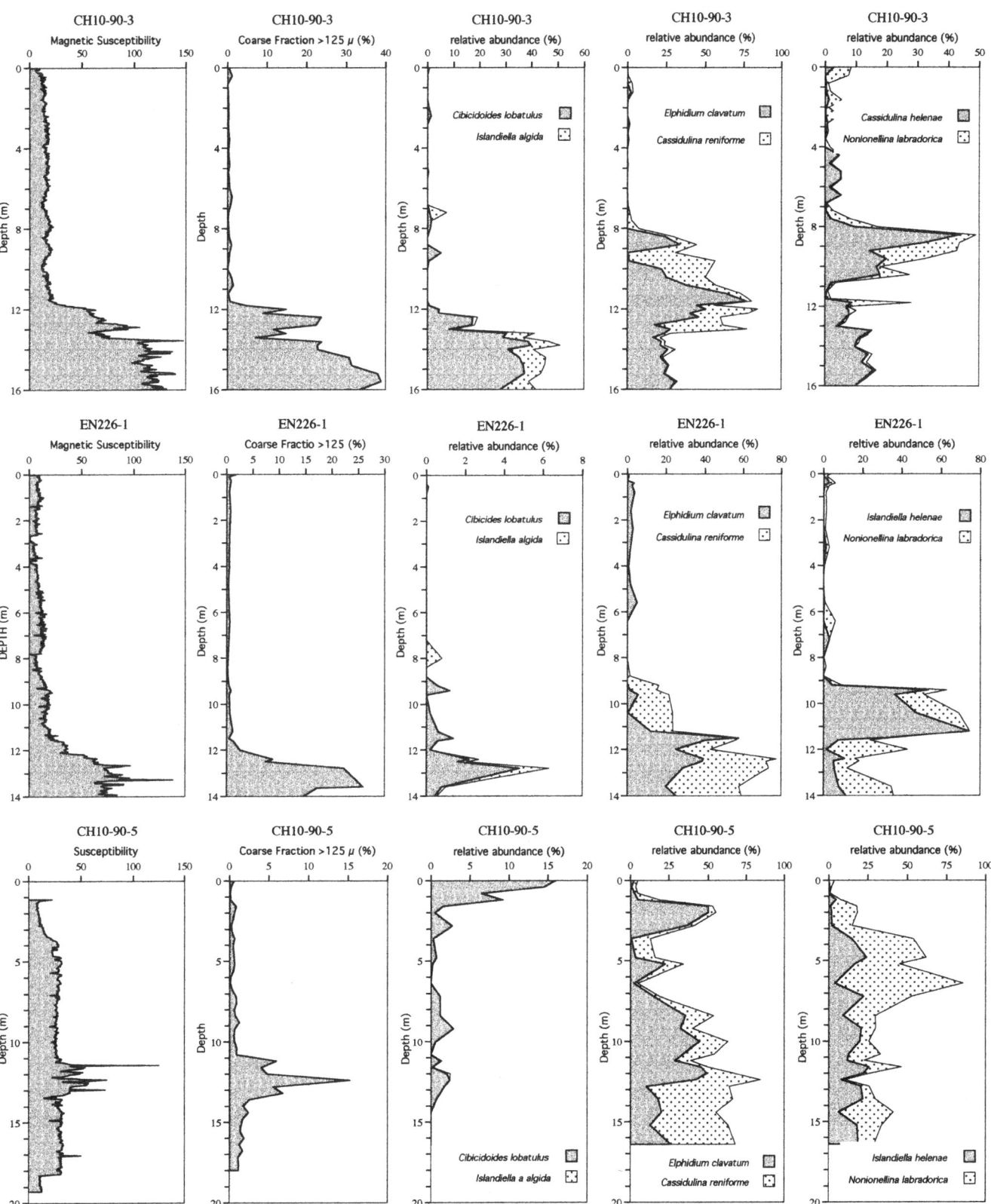

Figure 10. Examples of downcore distribution of *Cibicides lobatulus* and *Islandiella algida* in PGM facies (see Fig. 3 for abbreviations); *Elphidium excavatum* and *Cassidulina reniforme* in DGM facies; and of *Islandiella helenae* and *Nonionellina labradorica* in lowermost M facies. (Note that PGM facies was not reached by core CH10-90-5 and that *C. lobatulus* peak near top of core is secondary appearance of this species, when bottom currents increased at coring site during Holocene.)

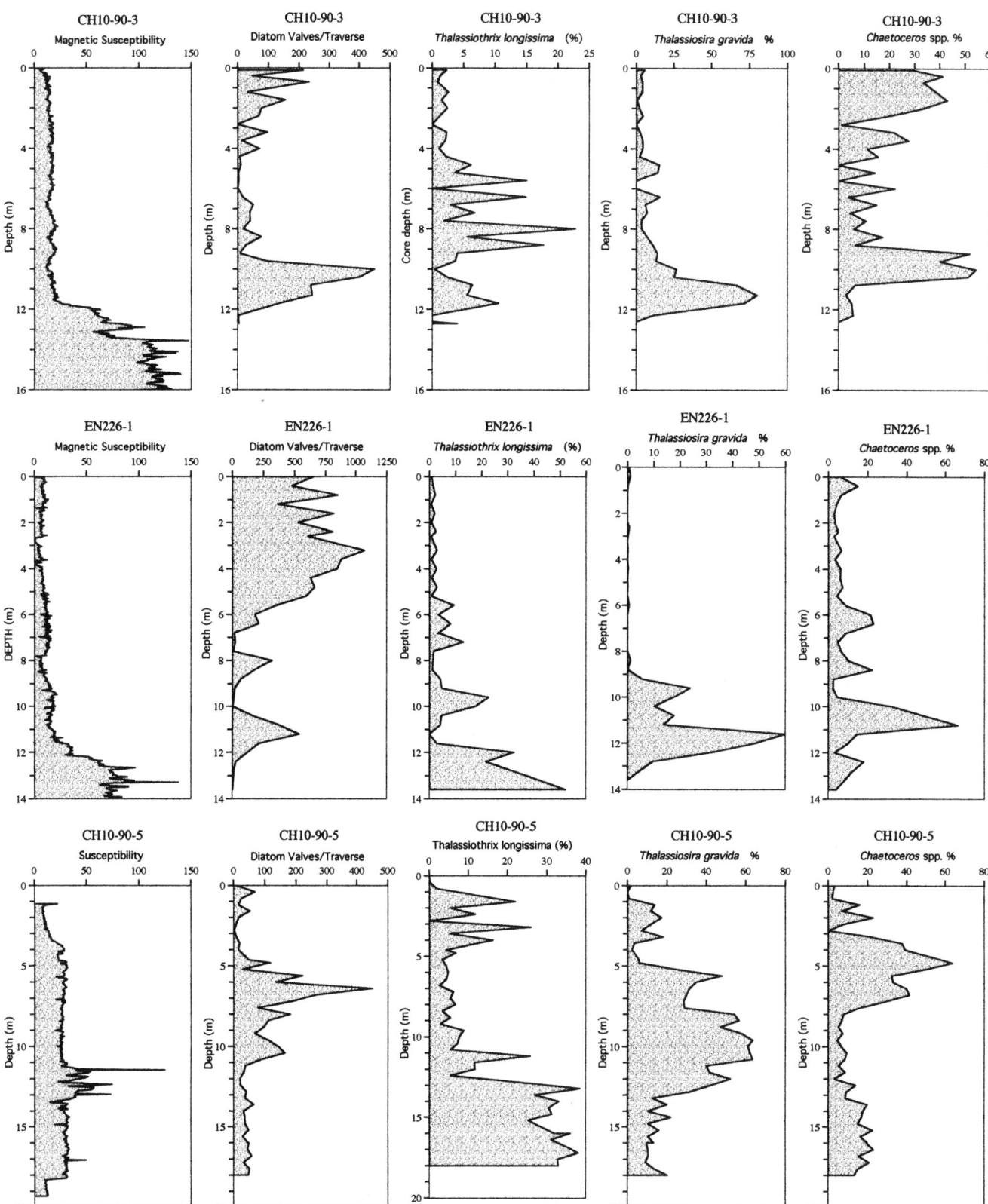

Figure 11. Examples of downcore distribution of diatom species *Thalassiothrix longissima*, *Thalassiosira gravida*, and *Chaetoceros* ssp.

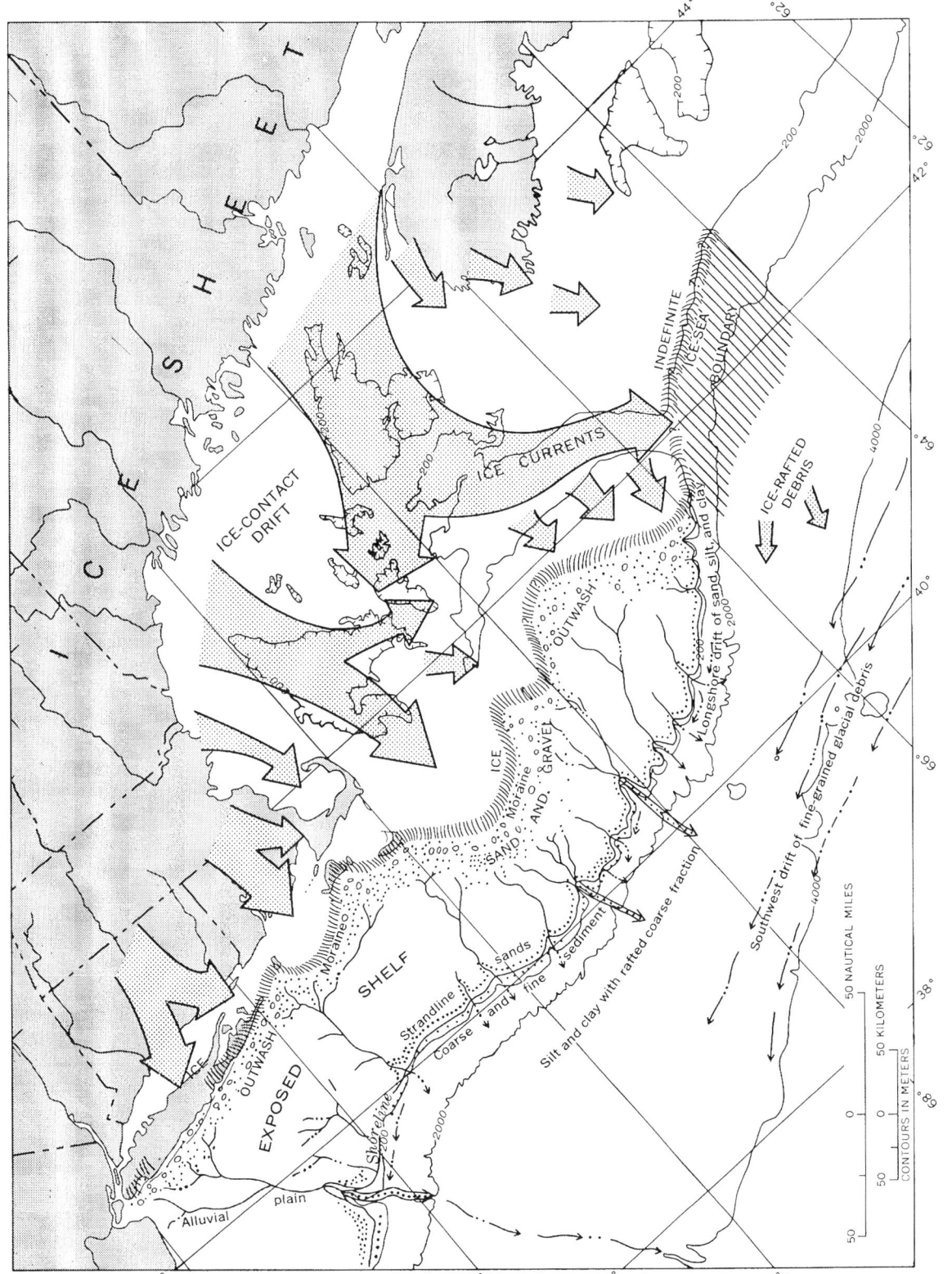

Figure 12. Margin of continental ice sheet across Gulf of Maine–New England region at time of last glacial maximum, as reconstructed by Schlee (1973).

TABLE 4. EXTRAPOLATED INITIATION AGES OF GLACIOMARINE SEDIMENTATION

| Basin | Core | | |
|---|---|---|---|
| Georges Basin | EN226-5 | Extrapolated bottom age: | 22.5 ka |
| Georges Basin | CH10-90-5 | Extrapolated bottom age: | 22 ka |
| Crowell Basin | CH10-90-4 | Extrapolated bottom age: | 18 ka |
| Rodgers Basin | EN226-1 | Extrapolated bottom age: | 19 ka |
| Truxton Basin | EN226-7 | Extrapolated bottom age: | 16.5 ka (to till tongue) |
| Wilkinson Basin | E2F-79-2 | Extrapolated bottom age: | 20 ka |
| Jeffreys Basin | CH10-90-8 | Extrapolated bottom age: | 19 ka |
| Platts Basin | EN226-9 | Extrapolated bottom age: | 17 ka |

calving bay, leading to the dominance of the *T. gravida* flora. Initially, the estuarine-style circulation maintained high salinities at the bottom, as indicated by *C. reniforme* dominance. At later stages of the deglacial interval the increase in the rate of melting led to somewhat reduced salinities even at the bottom of the basins, as indicated by intervals of *Elphidium excavatum* forma *clavatum* dominance.

### End of ice rafting

The cessation of ice rafting is indicated by the drastic change in sediment regime. In most cores this transition is the time of greatest primary productivity, as indicated by the abundance of diatom valves in the sediment. However, the environment did not change significantly at this transitional stage; the benthic fauna (*E. excavatum* forma *clavatum* and *C. reniforme*) and planktonic flora (*T. gravida*) persisted. The end of ice-margin disintegration (calving) did not signify the end of cold conditions. Sea ice with significant summer open water endured for a considerable time past the change in iceberg and sedimentary regime. The eventual change is indicated by a drastic change of both the benthic fauna and the planktonic flora and fauna.

The benthic foraminiferal fauna becomes dominated by *Islandiella helenae* and *Nonionellina labradorica*. Both are arctic cold-water species. *I. helenae* occurs abundantly in environments such as the Beaufort Sea (Canadian Arctic) (Vilks et al., 1979), the Labrador Shelf (Vilks and Deonarine, 1988; Vilks et al., 1989), or the eastern Baffin Shelf (Osterman and Nelson, 1989), where it prefers water temperatures of −2 to 4 °C and salinities of 32‰–33.5‰. *N. labradorica* also occurs commonly on the arctic and subarctic Canadian and Greenland shelves, but it appears to prefer to live at greater depths, such as at the bottoms of shelf basins or the upper continental slope, where it encounters warmer, more saline slope water (Mudie et al., 1983; Rodrigues and Hooper, 1982; Vilks, 1980; Williamson, 1982). These two species indicate more stable bottom water conditions; however, the frequent dominance of one over the other is a measure of the intensity of bottom water influx. A dominance of *N. labradorica* over *I. helenae* suggests increased indrafting of relatively warm and saline continental slope water through the Northeast Channel.

The diatom flora changes almost simultaneously with the benthic foraminiferal changeover to a flora that is dominated by spores of various species of *Chaetoceros*. Such an assemblage marks a general warming of surface waters, a lengthening of the ice-free season, and above all, a significant increase in surface productivity (Lapointe, 1998; Sancetta, 1981).

The first planktonic foraminifers appear during this stage in the evolution of the marine environment. *Neogloboquadrina pachyderma* (left coiling) and *Turborotalita quinqueloba*, sometimes alternatingly dominant, compose nearly 100% of the fauna. Monospecific occurrences of *N. pachyderma* (left coiling) are typical of high arctic environments where summer temperatures do not exceed 4 °C, such as is currently encountered in the East Greenland Current (Kohfeld et al., 1996). *T. quinqueloba* tolerates higher surface temperatures, but is less tolerant of salinity fluctuations, preferring environments with low vertical stratification (Hilbrecht, 1996).

Together, these sedimentary, faunal, and floral regimes suggest subarctic or boreal environmental conditions, cold enough

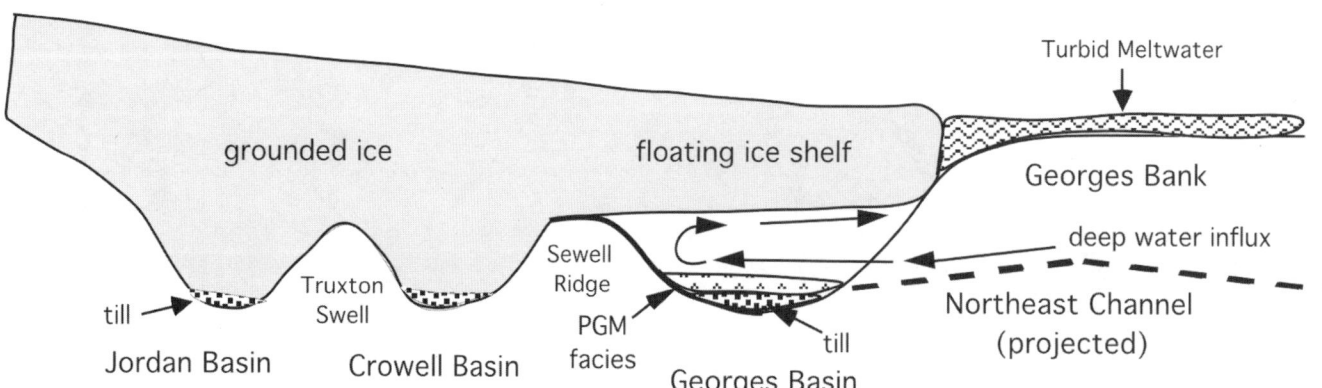

Figure 13. Stylized northwest-southeast cross section of Gulf of Maine, ca. 17 500 yr B.P., showing grounded ice in inner basins and ice shelf over Georges Basin. Meltwater sets up estuarine inflow of saline slope water through Northeast Channel and outflow of turbid water with reduced salinity near surface.

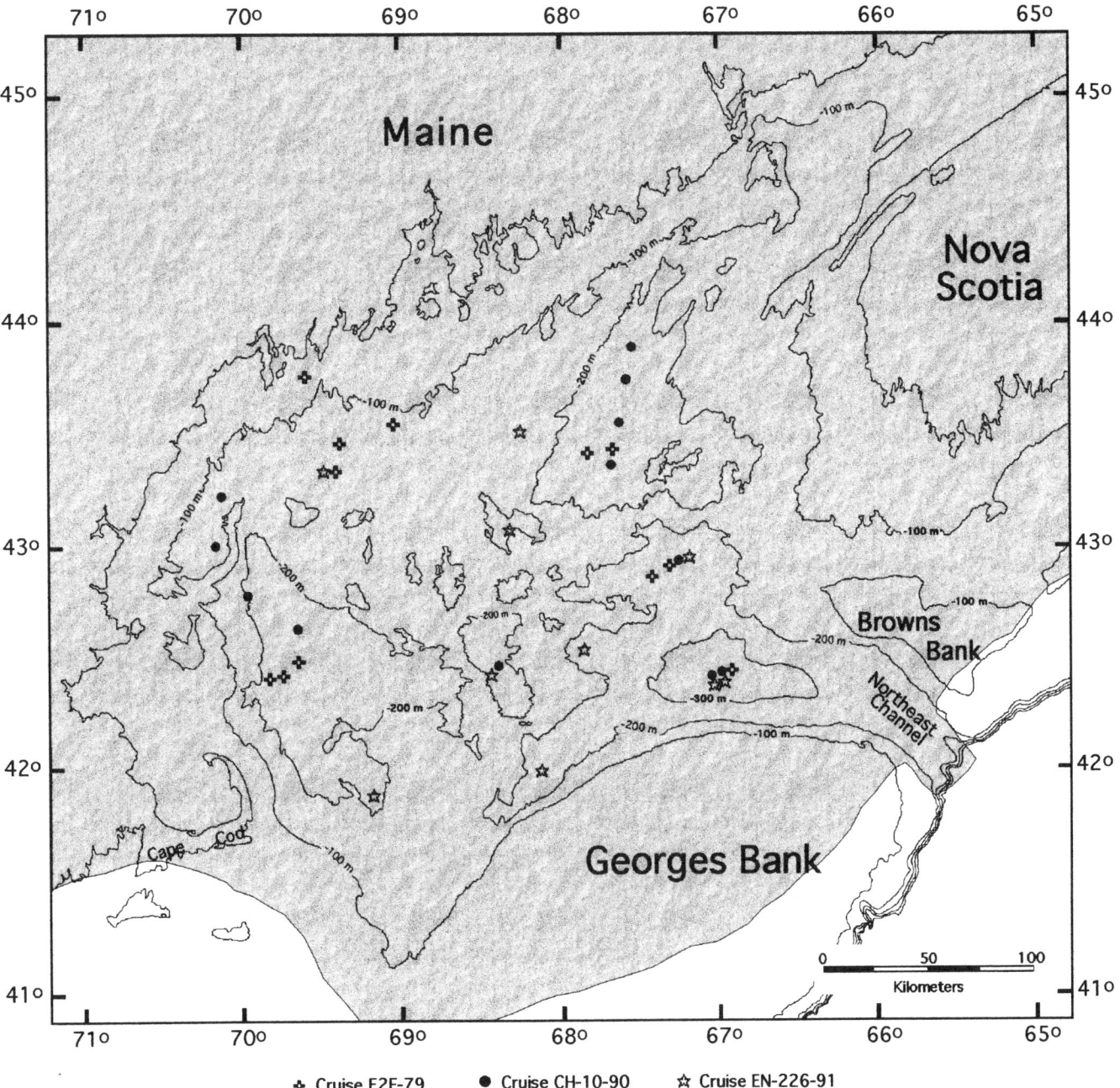

Figure 14. Map of Gulf of Maine region at about last glacial maximum. Gulf of Maine is filled with grounded ice.

for the development of a winter sea-ice cover, but warm enough for ice-free late spring, summer, and early fall conditions. Wind- and/or meltwater-induced upwelling provided the nutrients for abundant surface productivity.

### Sequential withdrawal

These successive deglacial environments occurred throughout the Gulf of Maine, but in a separate time frame in each basin. On the basis of the chronology presented herein, the following scenarios for the progress of deglaciation and the development of early postglacial marine conditions can be developed. The shoreline of the late glacial sea against the Georges and Browns Banks is very uncertain. Schlee (1973) suggested that the Georges Bank, and by inference, the Browns Bank, were emergent during the last glacial maximum. Freshwater peat dredged from the Georges Bank (Emery et al., 1965, 1967) indicate that those areas were exposed at least as late as 11 ka.

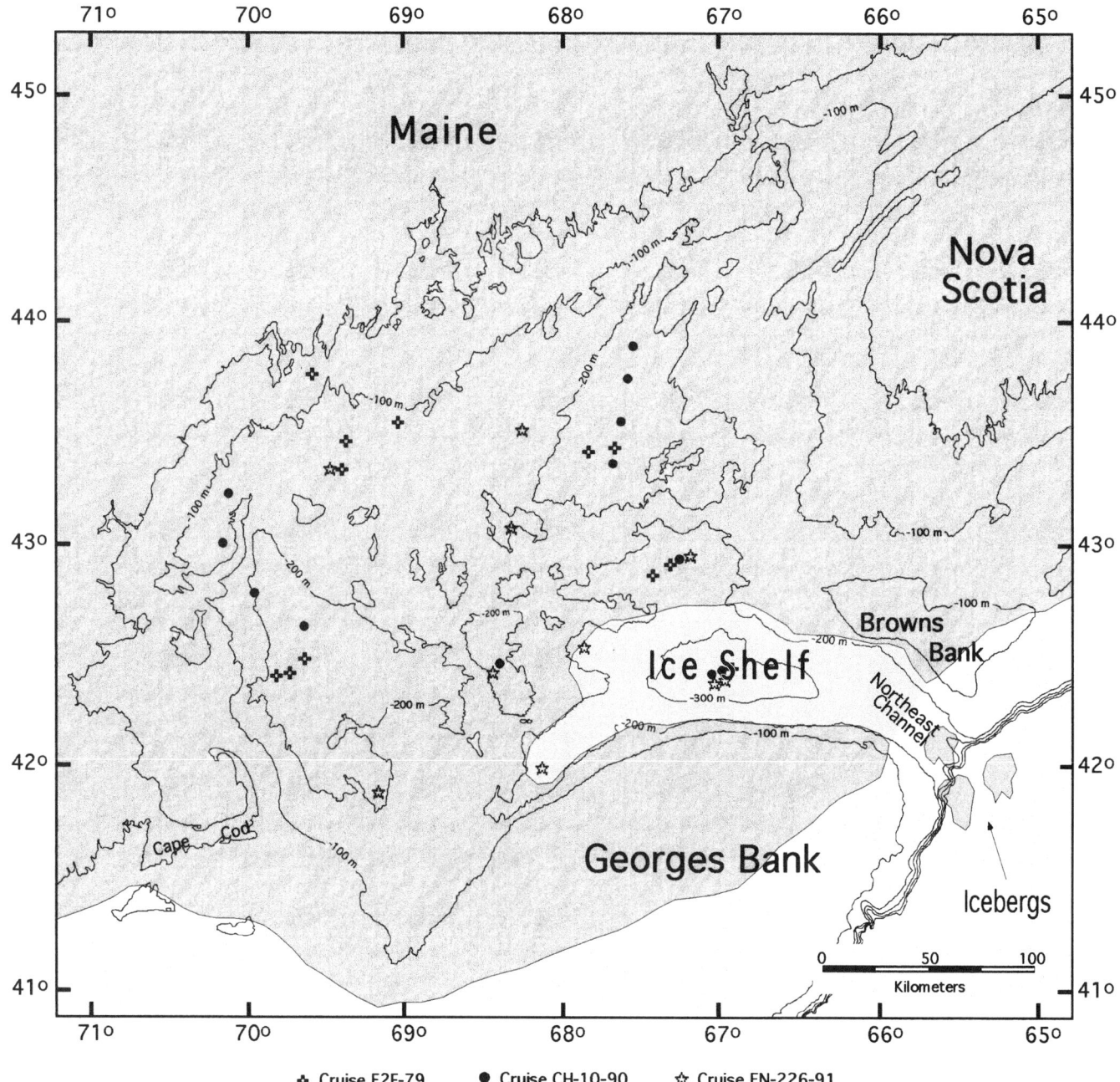

Figure 15. Map of Gulf of Maine region ca. 20 000 yr B.P. Ice shelf has formed above Georges Basin; calving occurs through Northeast Channel.

Exactly when isostatic subsidence and eustatic sea-level rise caused the flooding of these banks is not known.

***Ca. 22 000 yr B.P.*** Following the model of Schlee (1973) and of Hughes et al. (1985), we picture the Gulf of Maine filled with grounded ice. Ice terminated on the Georges and Browns Banks, and an ice stream exited through the Northeast Channel (Fig. 14).

***Before 20 000 yr B.P.*** Reduced ice supply or increased melting or ablation thinned the far edge of the ice sheet sufficiently to allow water to penetrate through the Northeast Channel and to lift it off bottom of the deep Georges Basin. This created an ice-shelf environment and initiated the deposition of the proximal glaciomarine sediments in the Georges Basin. A calving ice margin probably developed in the Northeast Channel (Fig. 15). Based on extrapolation of radiocarbon dates, this phase may have started as early as 22 000 yr B.P., and probably lasted until 17 000 yr B.P.

***Before 18 000 yr B.P.*** Calving removed the ice shelf from

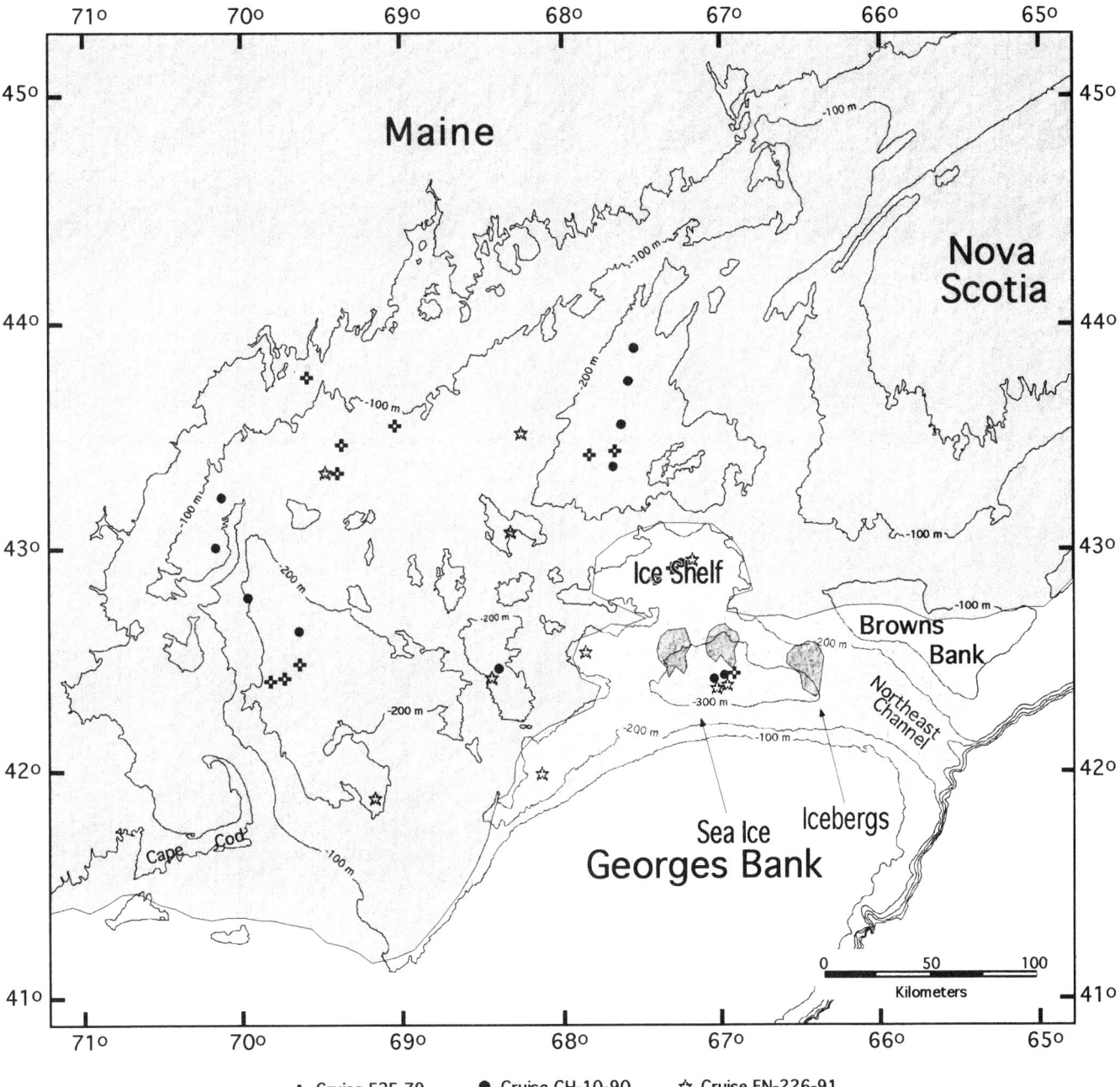

Figure 16. Map of Gulf of Maine region ca. 18 000 yr B.P. Georges Basin has developed into calving bay, filled with icebergs and heavy sea ice. Ice shelf is covering Crowell Basin.

the Georges Basin. The edge of the ice was now pinned on Sewell Ridge, inhibiting or slowing further retreat. The waters over the Georges Basin were filled with icebergs and covered by heavy sea ice, still inhibiting phytoplankton growth. Progressive thinning of the ice sheet allowed water to penetrate across Sewell Ridge and underneath the ice sheet over the Crowell Basin, thus creating a new ice shelf there (Fig. 16). The Crowell Basin now underwent the proximal glaciomarine environment, ~1–2 k.y. later than the Georges Basin. Near the end of this stage, the first growth of diatoms (*Thalassiothrix longissima*) occurred in the Georges Basin.

***Before 17 000 yr B.P.*** Calving caused the edge of the grounded ice to recede from the entire Georges-Franklin Basin. Calving also thinned the ice shelf over the Crowell Basin so that it no longer abutted against Sewell Ridge and as a consequence disintegrated and retreated rapidly. Progressive thinning

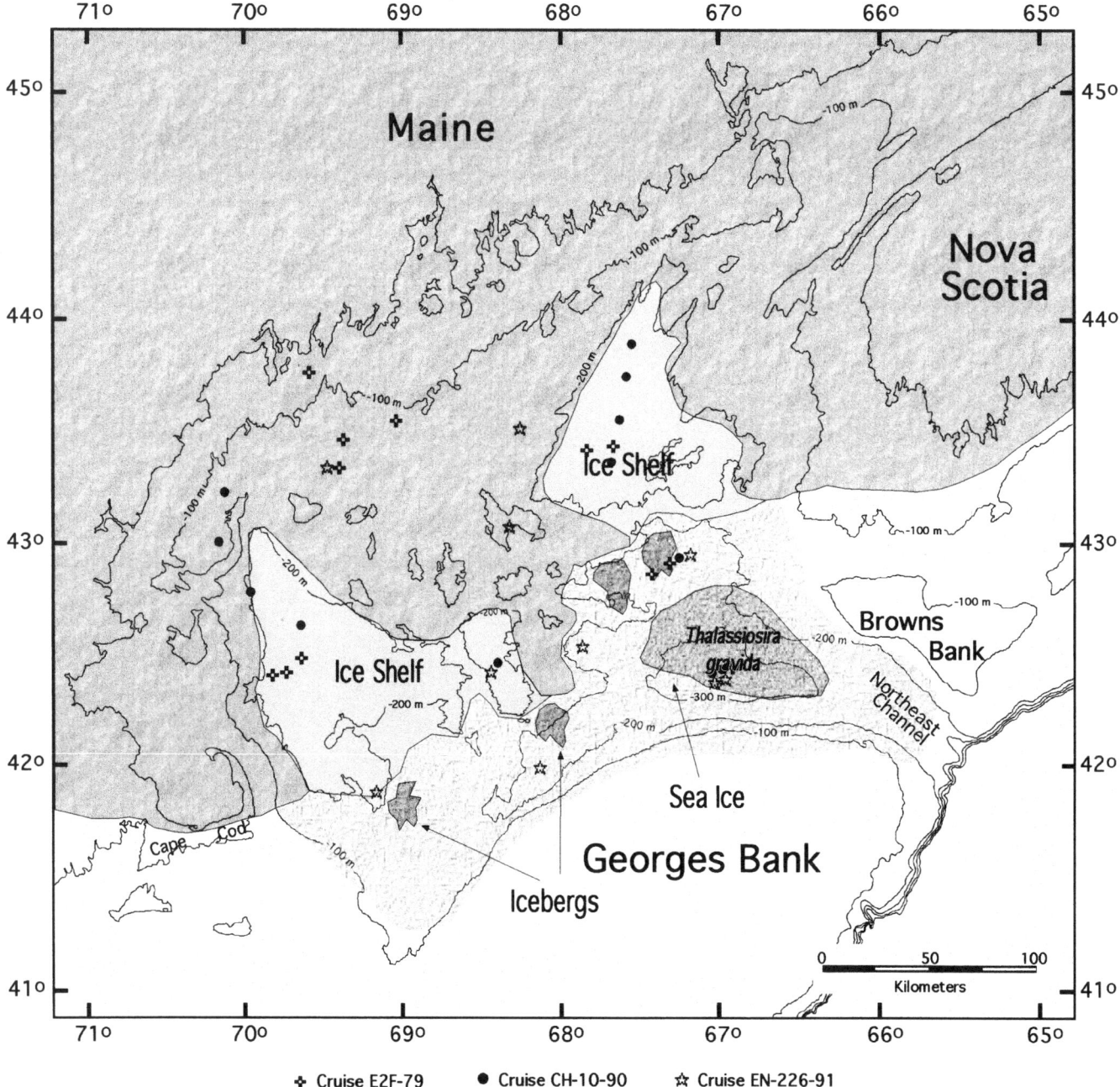

Figure 17. Map of Gulf of Maine region ca. 17 000 yr B.P. Georges and Crowell Basins are calving bays, filled with icebergs and sea ice. Over Georges Basin open water occurs during summer to allow phytoplankton growth of *Thalassiosira gravida*. Wilkinson and Jordan Basins are covered by ice shelf.

of the ice sheet allowed water to penetrate across the Truxton swell and underneath the ice sheet over the Jordan Basin, thus creating an ice shelf there. To the west, the sills between the Franklin and Rodgers, and Rodgers and Wilkinson Basins were also topped by water, floating the ice off bottom and creating an ice shelf there as well. During summer, meltwater-induced upwelling and/or off-ice katabatic winds created sufficient open water over the outer and central Georges Basin so that bursts of phytoplankton growth (*Thalassiosira gravida*) could occur (Fig. 17).

***Ca. 14 000 yr B.P.*** Once reduced ice supply, ablation, and calving caused the edge of the ice shelf to retreat from a buttressing sill, the free edge of the ice shelf succumbed to rapid calving (Hughes, 1987) and retreated quickly. Following the removal of the ice shelf from the Wilkinson and Jordan Basins, the ice margin retreated relatively quickly toward the New

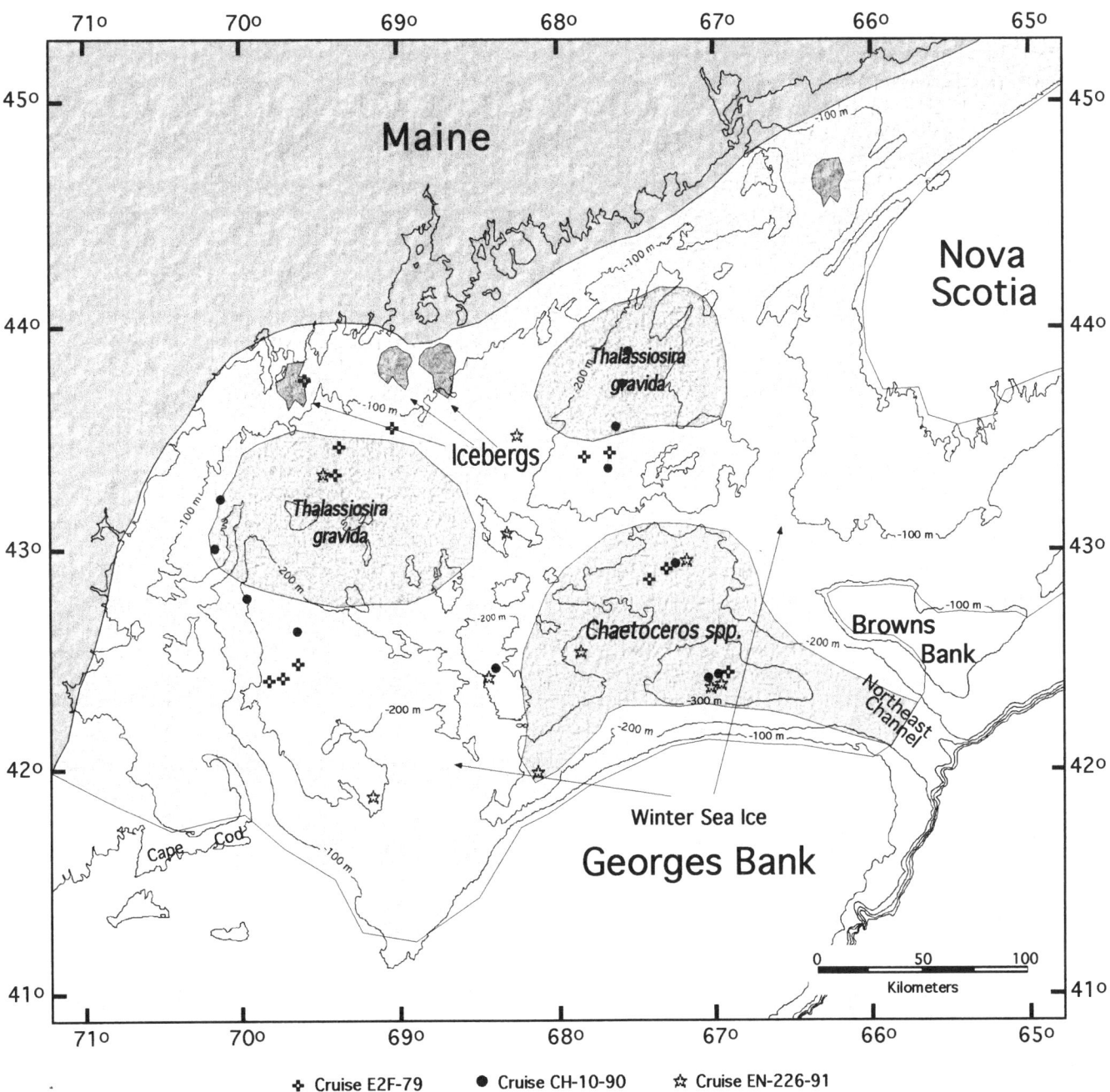

Figure 18. Map of Gulf of Maine region ca. 14 000 yr B.P. Grounded ice has disappeared completely from Gulf of Maine. Open water occurs briefly during summer in nearshore regions, supporting flora of *Thalassiosira gravida*, while in southeastern Gulf of Maine summer open-water season is longer, allowing growth of abundant flora of *Chaetoceros* spp.

England coastline and up the Bay of Fundy. By this time, the southeastern Gulf of Maine was covered by sea ice only during the cold season; largely open water occurred during the warm season. This allowed the establishment of a burgeoning primary productivity of various species of the diatom genus *Chaetoceros* (Fig. 18). This high productivity was supported by nutrients provided by wind and meltwater-induced upwelling. Over basins closer to the ice margin, cold winds off the ice sheet and abundant icebergs kept the water cold enough to allow only a short summer period of diatom growth, exploited by the species *Thalassiosira gravida*.

Further ice retreat onto the mainland (and its marine invasion) and continued warming brought extended ice-free conditions into the inner Gulf of Maine, so that sometime after 12 500 yr B.P. the high-productivity diatom flora and planktonic foraminiferal fauna occurred over the inner basins as well. The influx of sediments to the southeastern basins, especially the Georges Basin, became so reduced that bottom currents kept the floor of these basins essentially free of postglacial sediments.

## CONCLUSIONS

The history of deglaciation of the Gulf of Maine is recorded in great detail within its sediments. In each of the many basins contained within the Gulf of Maine the sedimentological, floral, and faunal records demonstrate the successive appearance and disappearance of distinct environments. Till was deposited when the basins were filled with grounded ice at the time of maximal extension of the Laurentide ice sheet. Coarse proximal glaciomarine sediments, containing a sparse foraminiferal fauna but no phytoplankton, were deposited when seawater incursion occurred underneath a thick ice shelf. These deposits are in turn overlain by either a transitional glaciomarine or a distal glaciomarine sedimentary facies, containing finer grained sediments, a distinct foraminiferal fauna, and a sparse diatom flora. This sedimentary facies originated in a calving bay environment, where the plume of meltwater and icebergs introduced glacial sediments, and a thick cover of nearly year-round sea ice limited phytoplankton productivity. The last sedimentary sequence included in this study no longer contains glaciomarine sediments, but harbors flora and fauna that are clearly still of arctic, periglacial, affinity, indicating the presence of substantial sea ice at least during the colder parts of the year.

While the environmental characterizations can be made in great detail and with considerable confidence, the narrative becomes more tentative concerning the chronology of events. Repeated occurrence of age inversions, improbable ages, and mutually incompatible results of AMS $^{14}$C dating and AAR age determinations clearly indicate that much work needs to be done to determine the appropriate material for dating. The data allow only the determination of the general chronology of an initial marine incursion into the Georges Basin as early as 22 000 yr B.P., and a stepwise, basin to basin retreat of the ice margin toward the mainland, which it reached nearly 14 000 yr B.P.

## ACKNOWLEDGMENTS

We thank the officers and crews of the research vessels *Eastward* (Duke University), *Cape Hatteras* (Duke University), and *Endeavor* (University of Rhode Island) for their aid and support during the research cruises. John Peck and Ellen Mecray performed the onboard magnetic measurements. This work benefited greatly from the careful reviews of the manuscript by Anne Jennings and Karl Kreutz. The research was supported by grants from the U.S. National Science Foundation (grants OCE-8911533 and OCE-9202148).

## REFERENCES CITED

Azam, F., Beers, J.R., Campbell, L., Carlucci, A.F., Holm-Hansen, O., Reid, F.M.H., and Karl, D.M., 1979, Occurrence and metabolic activity of organisms under the Ross Ice Shelf, Antarctica, at Station J-9: Science, v. 203, p. 451–453.

Bacchus, T.S., 1993, Late Quaternary stratigraphy and evolution of the eastern Gulf of Maine [Ph.D. thesis]: Orono, University of Maine, 347 p.

Bacchus, T.S., and Belknap, D.F., 1997, Glacigenic features and shelf basin stratigraphy of the eastern Gulf of Maine, in Davies, T.A., et al., eds., Glaciated continental margins, an atlas of acoustic images: New York, Chapman & Hall, p. 213–216.

Ballard, R.D., and Uchupi, E., 1975, Triassic structure in Gulf of Maine: American Association of Petroleum Geologists Bulletin, v. 59, p. 1041–1072.

Belknap, D.F., 1979, Application of amino acid geochronology to stratigraphy of late Cenozoic marine units of the Atlantic coastal plain [Ph.D. thesis]: Newark, University of Delaware, 550 p.

Belknap, D.F., 1987, Presumpscot Formation submerged on the Maine inner shelf, in Andrews, D.W., et al., eds., Geologic and geotechnical characteristics of the Presumpscot Formation: Symposium Volume: Augusta, Maine Geological Survey, p. 185–195.

Belknap, D.F., and Schnitker, D., 1994, Iceberg furrows and sand waves in the Gulf of Maine: Gulf of Maine News, Regional Association for Research on the Gulf of Maine, Summer 1994, p. 6–7.

Belknap, D.F., and Schnitker, D., 1995, In situ sampling of iceberg furrows, outcropping glacial and glaciomarine sediments and sedimentary bedforms in the Gulf of Maine: Avery Point, University of Connecticut, NOAA-National Undersea Research Center, Project UCAP 94-04 Report, 24 p.

Belknap, D.F., and Shipp, R.C., 1991, Seismic stratigraphy of the glacial-marine units, Maine inner shelf, in Anderson, J.B., and Ashley, G.M., eds., Glacial marine sedimentation; paleoclimatic significance: Geological Society of America Special Paper 261, p. 137–157.

Belknap, D.F., Shipp, R.C., and Kelley, J.T., 1986, Depositional setting and Quaternary stratigraphy of the Sheepscot Estuary, Maine: A preliminary report: Géographie Physique et Quaternaire, v. 40, p. 55–69.

Belknap, D.F., Kelley, J.T., and Shipp, R.C., 1987, Quaternary stratigraphy of representative Maine estuaries: Initial examination by high-resolution seismic reflection profiling, in FitzGerald, D.M., and Rosen, P.S., eds., Treatise on glaciated coasts: New York, Academic Press, p. 177–207.

Belknap, D.F., Shipp, R.C., Kelley, J.T., and Schnitker, D., 1989, Depositional sequence modeling of late Quaternary geologic history, west central Maine coast, in Tucker, R.D., and Marvinney, R.G., eds., Studies in Maine geology, Volume 5, Quaternary geology: Augusta, Maine Geological Survey, p. 29–46.

Bloom, A.L., 1963, Late Pleistocene fluctuations of sea level and postglacial crustal rebound in coastal Maine: American Journal of Science, v. 261, p. 862–879.

Brooks, D.A., 1985, Vernal circulation in the Gulf of Maine: Journal of Geophysical Research, v. 90, p. 4687–4705.

Dyke, A.S., and Prest, V.K., 1987, Late Wisconsinan and Holocene history of the Laurentide ice sheet: Géographie Physique et Quaternaire, v. 41, p. 237–263.

Emery, K.O., Wigley, R.L., and Rubin, M., 1965, A submerged peat deposit off the Atlantic coast of the United States: Limnology and Oceanography, v. 10, p. 97–102.

Emery, K.O., Wigley, R.L., Bartlett, A.S., Rubin, M., and Barghoorn, E.S., 1967, Freshwater peat on the continental shelf: Science, v. 158, p. 1301–1307.

Folger, D.W., O'Hara, C.J., and Robb, J.M., 1975, Maps showing bottom sediments on the continental shelf of the northeastern United States—Cape Ann, Massachusetts to Casco Bay, Maine: U.S. Geological Survey Miscellaneous Investigation Series Map I-839, scale 1:125 000.

Friez, J.K., 1993, Reconstruction of the postglacial paleoenvironmental evolution of the northwestern Gulf of Maine [M.S. thesis]: Orono, University of Maine, 121 p.

Hald, M., Steinsund, P.I., Dokken, T., Korsun, S., Polyak, L., and Aspeli, R., 1994, Recent and late Quaternary distribution of *Elphidium excavatum* f. *clavatum* in arctic seas, *in* Sejrup, H.P., and Knudsen, K.L., eds., Late Cenozoic benthic foraminifera: Taxonomy, ecology and stratigraphy: Cushman Foundation Special Publication 32, p. 141–154.

Hasle, G.R., and Semina, H.J., 1987, The marine planktonic diatoms *Thalassiothrix longissima* and *Thalassiothrix antarctica* with comments on *Thalassionema* spp. and *Synedra reinholdi*: Diatom Research, v. 2, p. 175–192.

Hilbrecht, H., 1996, Extant planktic foraminifera and the physical environment in the Atlantic and Indian Oceans. An atlas based on CLIMAP and Levitus (1982) data: Mitteilungen aus dem Geologischen Institut der Eidgenössischen Technischen Hochschule und der Universität Zürich, Neue Folge 300, p. 1–93.

Hoskins, H., and Knott, S.T., 1961, Geophysical investigations of Cape Cod Bay, Massachusetts, using the continuous seismic profiler: Journal of Geology, v. 69, p. 330–340.

Hughes, T., 1987, Ice dynamics and deglaciation models when ice sheets collapsed, *in* Ruddiman, W.F., and Wright, H.E., Jr., eds., North America and adjacent oceans during the last deglaciation: Boulder, Colorado, Geological Society of America, Geology of North America, v. K-3, p. 183–220.

Hughes, T., Borns, H.W., Fastook, J.L., Kite, J.S., Hyland, M.R., and Lowell, T.V., 1985, Models of glacial reconstruction and deglaciation applied to Maritime Canada and New England, *in* Borns, Jr., H.W., et al., eds., Late Pleistocene history of northeastern New England and adjacent Quebec: Geological Society of America Special Paper 197, p. 139–150.

Jennings, A.E., and Helgadottir, G., 1994, Foraminiferal assemblages from the fjords and shelf of eastern Greenland: Journal of Foraminiferal Research, v. 24, p. 123–144.

Jennings, A.E., Manley, W.F., MacLean, B., and Andrews, J.T., 1998, Marine evidence for the last glacial advance across eastern Hudson Strait, eastern Canadian Arctic: Journal of Quaternary Science, v. 13, p. 501–514.

Kaplan, M.R., 1999, Retreat of a tidewater margin of the Laurentide ice sheet in eastern coastal Maine between ca. 14000 and 13000 $^{14}$C yr B.P.: Geological Society of America Bulletin, v. 111, p. 620–632.

Kashgarian, M., and Southon, J.R., 1993, Radiocarbon reservoir ages in coastal surface waters [abs.]: Eos (Transactions, American Geophysical Union), supplement, p. 185.

Kelley, J.T., and Belknap, D.F., 1988, Geomorphology and sedimentary framework of the inner continental shelf of west central Maine: Maine Geological Survey Open-File Report 88-6, 51 p.

Kelley, J.T., Kelley, A.R., Belknap, D.F., and Shipp, R.C., 1986, Variability in the evolution of two adjacent bedrock-framed estuaries in Maine, *in* Wolfe, D.A., ed., Estuarine variability: New York, Academic Press, p. 21–42.

Kelley, J.T., Belknap, D.F., and Shipp, R.C., 1987a, Geomorphology and sedimentary framework of the inner continental shelf of south-central Maine: Maine Geological Survey Open-File Report 87-19, 76 p.

Kelley, J.T., Shipp, R.C., and Belknap, D.F., 1987b, Geomorphology and sedimentary framework of the inner continental shelf of southwestern Maine: Maine Geological Survey Open-File Report 87-5, 86 p.

King, L.H., 1996, Late Wisconsinan ice retreat from the Scotian Shelf: Geological Society of America Bulletin, v. 108, p. 1056–1067.

King, L.H., and Fader, G.B.J., 1986, Wisconsinan glaciation of the Atlantic continental shelf of southeast Canada: Geological Survey of Canada Bulletin 363, 72 p.

Knebel, H.J., 1986, Holocene depositional history of a large glaciated estuary, Penobscot Bay, Maine: Marine Geology, v. 73, p. 215–236.

Knebel, H.J., and Scanlon, K.M., 1985, Sedimentary framework of Penobscot Bay, Maine: Marine Geology, v. 65, p. 305–324.

Kohfeld, K.E., Fairbanks, R.G., Smith, S.L., and Walsh, I.D., 1996, *Neogloboquadrina pachyderma* (sinistral coiling) as paleoceanographic tracers in polar oceans: Evidence from Northeast Water Polynya plankton tows, sediment traps, and surface sediments: Paleoceanography, v. 11, p. 679–699.

Lapointe, M., 1998, Assemblages diatomologiques et paléoenvironnements au Quaternaire supérieur de l'estuaire maritime et du golfe du Saint-Laurent (Québec, Canada) [Ph.D. thesis]: Montréal, Université du Québec, 477 p.

Lipps, J.H., Ronan, T.E., and DeLaca, T.E., 1979, Life below the Ross Ice Shelf: Science, v. 203, p. 447–449.

Lusardi, B.A., 1992, Late glacial to postglacial paleo-environmental reconstruction in the eastern Gulf of Maine [Masters thesis]: Orono, University of Maine, 154 p.

Mayer, L.M., Jorgensen, J., and Schnitker, D., 1991, Enhancement of diatom frustule dissolution by iron oxides: Marine Geology, v. 99, p. 263–266.

Mayewski, P., Denton, G.H., and Hughes, T.J., 1981, Late Wisconsin ice sheets in North America, *in* Denton, G.H., and Hughes, T.J., eds., The last great ice sheets: New York, Wiley Interscience, p. 67–178.

Miller, A.A.L., 1996, Late Quaternary foraminiferal biostratigraphy of three shallow geotechnical boreholes in Halibut Channel, western Grand Banks of Newfoundland: Geological Survey of Canada Open-File Report 3369, 131 p.

Miller, A.A.L., Nomura, R., and Osterman, L.E., 1996, *Islandiella algida* (Cushman, 1944); senior subjective* synonym of *Islandiella islandica* (Nørvang): Journal of Foraminiferal Research, v. 26, p. 209–212.

Mountain, D.G., and Jessen, P.F., 1987, Bottom waters of the Gulf of Maine, 1978–1983: Journal of Marine Research, v. 45, p. 319–345.

Mudie, P.J., Keen, C.E., Hardy, I.A., and Vilks, G., 1983, Multivariate analysis and quantitative paleoecology of benthic foraminifera in surface and late Quaternary shelf sediments, northern Canada: Marine Micropaleontology, v. 8, p. 283–313.

Oldale, R.N., and Bick, J., 1987, Geology of the inner continental shelf, Massachusetts Bay, Massachusetts: U.S. Geological Survey Miscellaneous Field Studies Map MF-1923, scale 1:125 000.

Oldale, R.N., and Uchupi, E., 1970, The glaciated shelf off northeastern United States: U.S. Geological Survey Professional Paper 700-B, p. 167–173.

Oldale, R.N., and Wommack, L.E., 1987, Geology of the inner continental shelf, Cape Ann to New Hampshire: U.S. Geological Survey Miscellaneous Field Studies Map MF-1892, scale 1:125 000.

Oldale, R.N., Uchupi, E., and Prada, K.E., 1973, Sedimentary framework of the western Gulf of Maine and the southeastern Massachusetts offshore area: U.S. Geological Survey Professional Paper 757, 10 p.

Osterman, L.E., and Nelson, A.R., 1989, Latest Quaternary and Holocene paleoceanography of the eastern Baffin Island continental shelf, Canada: Benthic foraminiferal evidence: Canadian Journal of Earth Sciences, v. 26, p. 2236–2248.

Popek, D.M., 1993, Diatom paleoecology and paleoceanography of the lateglacial to Holocene Gulf of Maine [M.S. thesis]: Orono, University of Maine, 283 p.

Powell, R.D., and Alley, R.B., 1997, Grounding-line systems: Processes, glaciological inferences and the stratigraphic records, *in* Barker, P.F., and Cooper, A.K., eds., Geology and seismic stratigraphy of the Antarctic Margin, 2: American Geophysical Union Antarctic Research Series, v. 71, p. 169–187.

Ramp, S.R., Schlitz, R.J., and Wright, W.R., 1985, The deep flow through the Northeast Channel, Gulf of Maine: Journal of Physical Oceanography, v. 15, p. 1790–1808.

Reid, P.C., and Surey-Gent, S.C., 1980, *Thalassiothrix longissima*, an oceanic indicator species in the North Sea?: British Phycology Journal, v. 15, p. 199–206.

Rodrigues, C.G., and Hooper, K., 1982, Recent benthonic foraminiferal associations from offshore environments in the Gulf of St. Lawrence: Journal of Foraminiferal Research, v. 12, p. 327–352.

Sancetta, C., 1981, Diatoms as hydrographic tracers: Example from Bering Sea sediments: Science, v. 211, p. 279–281.

Schafer, C.T., and Cole, F.E., 1986, Reconnaissance survey of benthonic foraminifera from Baffin Island fjord environments: Arctic, v. 39, p. 232–239.

Schlee, J., 1973, Atlantic continental shelf and slope of the United States—Sediment texture of the northeastern part: U.S. Geological Survey Professional Paper 529L, 64 p.

Schnitker, D., 1976, Late glacial to recent paleoceanography of the Gulf of Maine. 1st International Symposium Continental Margin Benthic Foraminifera, Part B: Maritime Sediments Special Publication, v. 1, p. 385–392.

Schnitker, D., 1986, Ocean basin with a past, a cryptic history: Breaking the code, discerning the future: Explorations, v. 3, p. 29–37.

Schnitker, D., and Jorgensen, J., 1990, Late glacial and Holocene diatom successions in the Gulf of Maine: Response to climatologic and oceanographic change, in South, G.R., and Garbary, D., eds., Evolutionary biogeography of the marine algae of the North Atlantic. NATO ASI Series G: Ecological Sciences, Volume 22: Berlin, Heidelberg, Springer Verlag, p. 33–53.

Schrader, H.J., 1974, Proposal for a standardized method of cleaning diatom bearing deep-sea and land-exposed marine sediments: Nova Hedwigia, Beihefte, v. 6, p. 131–155.

Smith, G.W., 1985, Chronology of late Wisconsinan deglaciation of coastal Maine, in Borns, H.W., et al., eds., Late Pleistocene history of northeastern New England and adjacent Quebec: Geological Society of America Special Paper 197, p. 29–44.

Smith, G.W., and Hunter, L.E., 1989, Late Wisconsinan deglaciation of coastal Maine, in Tucker, R.D., and Marvinney, R.G., eds., Studies in Maine geology, Volume 6: Quaternary Geology, p. 13–32.

Smith, W.O., Baumann, M.E.M., Wilson, D.L., and Aletsee, L., 1987, Phytoplankton biomass and productivity in the marginal ice zone of Fram Strait during the summer 1984: Journal of Geophysical Research, v. 92, p. 6777–6786.

Stuiver, M., and Borns, H.W., Jr., 1975, Late Quaternary marine invasion in Maine: Its chronology and associated crustal movement: Geological Society America Bulletin, v. 86. p. 99–104.

Svavarsson, J., and Davíðsdóttir, B., 1994, Foraminiferan (Protozoa) epizoites on Arctic isopods (Crustacea) as indicators of isopod behaviour?: Marine Biology, v. 118, p. 239–246.

Thompson, W.B., 1982, Recession of the late Wisconsinan ice sheet in coastal Maine, in Larson, G.J., and Stone, B.D., eds., Late Wisconsinan glaciation of New England: Dubuque, Iowa, Kendall/Hunt Publishing Company, p. 211–228.

Thompson, W.B., and Borns, H.W., Jr., 1985, Surficial geologic map of Maine: Augusta, Maine Geological Survey, Scale 1:500 000.

Tucholke, B.E., and Hollister, C.D., 1973, Late Wisconsin glaciation of the southwestern Gulf of Maine: New evidence from the marine environment: Geological Society of America Bulletin, v. 84, p. 3279–3296.

Vilks, G., 1980, Postglacial basin sedimentation on Labrador shelf: Geological Survey of Canada Paper 78-28, 28 p.

Vilks, G., and Deonarine, B., 1988, Labrador shelf benthic foraminifera and stable oxygen isotopes of *Cibicides lobatulus* related to the Labrador Current: Canadian Journal of Earth Sciences, v. 25, p. 1240–1255.

Vilks, G., Wagner, F.J.F., and Pelletier, B.R., 1979, The Holocene marine environment of the Beaufort Shelf: Geological Survey of Canada Bulletin 303, 43 p.

Vilks, G., MacLean, B., Deonarine, B., Currie, C.G., and Moran, K., 1989, Late Quaternary paleoceanography and sedimentary environments in Hudson Strait: Géographie Physique et Quaternaire, v. 43, p. 161–178.

Wehmiller, J.F., Belknap, D.F., Boutin, B.S., Mirecki, J.E., Rahaim, S.D., and York, L.L., 1988, A review of the aminostratigraphy of Quaternary mollusks from United States Atlantic coastal plain sites, in Easterbrook, D.J., ed., Dating Quaternary sediments: Geological Society of America Special Paper 227, p. 69–110.

Williams, K.M., 1986, Recent arctic marine diatom assemblages from bottom sediments in Baffin Bay and Davis Strait: Marine Micropaleontology, v. 10, p. 327–341.

Williams, K.M., 1990, Late Quaternary paleoceanography of the western Baffin Bay region: Evidence from fossil diatoms: Canadian Journal of Earth Sciences, v. 27, p. 1487–1494.

Williamson, M.A., 1982, Distribution of recent foraminifera on the Nova Scotian shelf and slope, in Mamet, B., and Copeland, M.J., eds., Third North American Paleontological Convention, Montreal, August 5–7, Proceedings: Toronto, Business and Economic Service Ltd., p. 579–584.

MANUSCRIPT ACCEPTED BY THE SOCIETY JUNE 9, 2000

# Glaciation and relative sea-level change in Maritime Canada

**Rudolph R. Stea***
*Nova Scotia Department of Natural Resources, P.O. Box 698, Halifax, Nova Scotia B3J 2T9, Canada*
**Gordon B.J. Fader**
*Geological Survey of Canada (Atlantic), P.O. Box 1006, Dartmouth, Nova Scotia B2Y 4A2, Canada*
**David B. Scott**
*Dalhousie University, Halifax, Nova Scotia B3H 3J5, Canada*
**Patrick Wu**
*University of Calgary, Department of Geology and Geophysics, Calgary, Alberta T2N 1N4, Canada*

## ABSTRACT

Relative sea levels in Maritime Canada have fluctuated between 70 m above and 100 m below present since the last interglacial period. Patterns of relative sea-level change are controlled both by the rates of ice retreat and the distribution of ice loads. Late-melting ice caps produced two types of anomalous sea-level response: (1) unmanifested emergence with low or no highstands in areas of thick ice, and (2) diachronous lowstand shorelines separated by regions not recording transgression.

Deglaciation history must be considered when evaluating the Earth's response to former ice loads from the relative sea-level curves. In Maritime Canada, a high-amplitude, high-frequency relative sea-level response directly after deglaciation cannot be attributed to forebulge migration using current models of mantle viscosity due to the rapidity of relative sea-level fall. The apparent discrepancy between model and field data may be explained by initial direct isostatic recovery from thick Appalachian ice caps and late-melting ice caps. The migration of a peripheral forebulge appears to be recorded by regional uplift after 11 ka, resulting in slowing of relative sea-level rise and reversal points. The last interglacial marine deposits to 30 m above mean sea level suggest the possibility of long-term crustal subsidence and sea-level rise after glaciations.

## INTRODUCTION

Abandoned shorelines of Quaternary age are found as high as 70 m above present sea level to nearly 100 m below the present coastline of Maritime Canada. What caused ocean levels to fluctuate so dramatically during the past 100 k.y.? The location, elevation, and tilt of these shorelines are controlled by the interaction of two main factors: (1) uptake of ocean water for the formation of continental glaciers, and later expulsion as meltwater; and (2) depression of the Earth's lithosphere under the great weight of glaciers and rebound after melting of the ice. Today, sea level is rising in Maritime Canada at a rate of ~20–30 cm/100 yr, under the influence of long-term climatic change and crustal subsidence (Scott et al., 1995). A summary of the region's climate and glacial history over the past 100 k.y. is essential to understand relative sea-level changes in Maritime Canada.

Two opposing glaciation concepts for Maritime Canada have been debated since the glacial theory was first proposed (Fig. 1). The maximum model invokes continental ice masses

---
*E-mail: rrstea@gov.ns.ca.

Stea, R.R., Fader, G.B.J., Scott, D.B., and Wu, P., 2001, Glaciation and relative sea-level change in Maritime Canada, *in* Weddle, T.K., and Retelle, M.J., eds., Deglacial History and Relative Sea-Level Changes, Northern New England and Adjacent Canada: Boulder, Colorado, Geological Society of America Special Paper 351, p. 35–49.

Figure 1. Location of study area, sections, and major morainal systems on land and offshore. Previous models of extent and configurations of late Wisconsinan ice masses (after Grant, 1977, 1994; Denton and Hughes, 1981): 1, Joggins, Nova Scotia; 2, Squally Point; 3, Five Islands Delta. SSEMC is Scotian Shelf end moraine complex. 4, Halifax inner Scotian Shelf study area (Fig. 6). 5, George's Bay sections. 6, Bay St. Lawrence section.

extending out to the continental shelf edge stemming from the Laurentide ice center in Quebec (Goldthwait, 1924; King, 1969; Flint, 1971; Denton and Hughes, 1981). The minimum model evolved from the recognition of autonomous local glaciers in Maritime Canada (Chalmers, 1895; Prest, 1896; MacNeil and Purdy, 1951; Prest and Grant, 1969) into a concept of thin, late Wisconsinan ice restricted to lowland terrestrial areas (Grant, 1977, 1989; Dyke and Prest, 1987; Grant and King, 1984).

Mapping of glacial landforms in New Brunswick and Nova Scotia has produced a more complex model of Wisconsinan glaciation than previously envisioned in both the minimum or maximum models (Wightman, 1980; Stea, 1982, 1983, 1984; Rampton et al., 1984; Grant, 1994; Stea and Mott, 1998; Stea et al., 1998), with ice thicknesses more in line with the maximum model (e.g., Scott et al., 1987a). Glaciomarine deltas on land that were thought to form the margins of the minimum model in the Bay of Fundy were dated to 14 ka rather than 18–20 ka (Stea and Wightman, 1987). Radiocarbon dating of glaciomarine sediment at the Scotian Shelf–continental shelf edge and mid-shelf basins has confirmed the shelf-wide extent of late Wisconsinan ice (Mosher et al., 1989; Gipp, 1989, 1994; Gipp and Piper, 1989; King, 1996).

In this overview we describe some of these Quaternary shorelines, and present empirical and theoretical models for shoreline deformation after glaciation. This chapter can be considered an update of D.R. Grant's (1980) pivotal work on the Quaternary shorelines in Atlantic Canada.

## SANGAMONIAN SEA LEVELS (120–80 KA)

The earliest record of former sea levels in Maritime Canada is an abrasion platform, 4–6 m above sea level, overlain by peat and wood beyond the range of radiocarbon dating, and several Wisconsinan till sheets (Grant, 1980; Myers and Stea, 1986; Stea et al., 1992; Figs. 2 and 3). This feature has not been accurately surveyed, and elevation estimates are generally made from a high-tide berm datum (Grant, 1980). Sea stacks and potholes have been found on the surface of this platform and in some areas it is striated (Grant, 1980; Rampton et al., 1984). Well-sorted, massive sand beds of possible marine origin directly overlie the abrasion platform in northern Nova Scotia and attain elevations of 25 m (Stea et al., 1992). Grant (1980) interpreted the emerged rock bench as an erosional feature relating to higher sea levels during the last interglacial period (Sangamonian interglaciation) and postulated that the bench records a former equilibrium state of sea-level rise due to glacier melting and crustal subsidence after a major glaciation.

The Salmon River sand is a fossiliferous marine sand bed that is found between two tills in a coastal section in southwestern Nova Scotia at an elevation of ~6 m (Grant, 1989). It was inferred to be interglacial in age based on a warm-water (temperate) fauna and the extinct Sangamonian gastropod *Atractodon stonei* (Clarke et al., 1972). Wehmiller et al. (1988) dated the mollusc shells from the sand using amino acid racemization ratios calibrated by kinetic age modeling. They compared same species ratios from correlative pre-Wisconsinan beds throughout New England and Nova Scotia and calibrated the ratios based on the U/Th age of a solitary coral from the

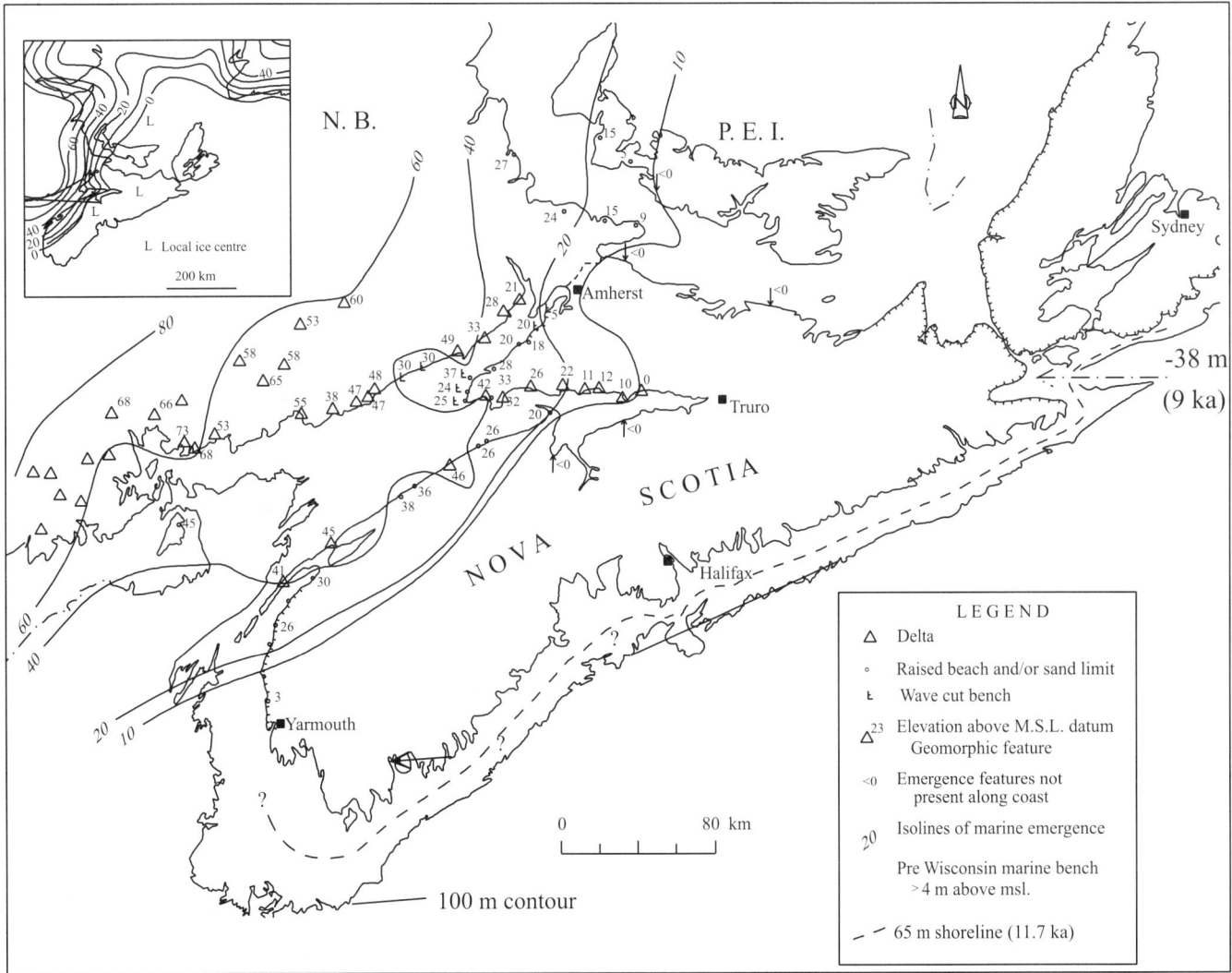

Figure 2. Isopleths (lines of equal emergence) of elevations of marine limit as represented by raised beaches and wave-cut terraces and deltas in Gulf of St. Lawrence region. Locations of interglacial shorelines and lowstand shorelines are shown (after Wightman and Cooke, 1978; Grant, 1980; Stea et al., 1987).

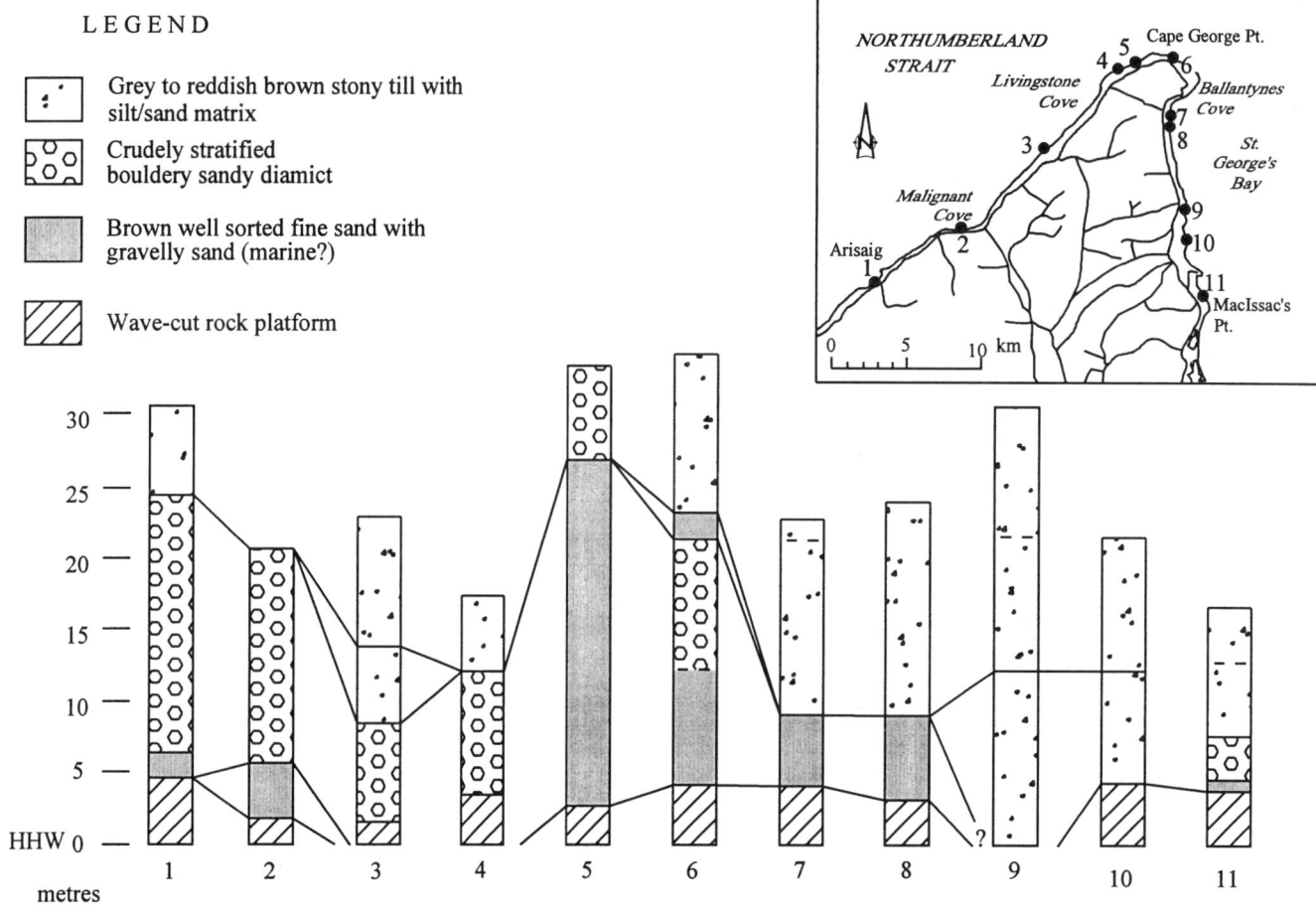

Figure 3. Stratigraphic sections along George's Bay (Fig. 1) showing Sangamon marine platform and overlying strata, including sand of possible marine origin.

Sankaty sand in Massachusetts (133 ka; Oldale et al., 1982). Stea et al. (1992) continued this work to the correlative beds in northern Nova Scotia, and obtained ages spanning marine isotope stage 5. From the molluscan fauna of the Salmon River sand, Wagner (1977) inferred water depths of 16–31 m. Grant and King (1984) suggested that the bed was thrust into an anomalous position, but there is little structural evidence to support this hypothesis. Similar sand beds in northern Nova Scotia lying just above the abrasion platform at 25 m elevation, are further evidence for an in situ origin of the Salmon River sand (Fig. 3).

The magnitude of relative sea-level rise during the last interglacial is of great importance, because it may be indicative of future sea levels. Most of the present sea-level rise in Maritime Canada is attributed to crustal subsidence due to a collapsing peripheral forebulge (Quinlan and Beaumont, 1981; Scott et al., 1987a). At the present rate of relative sea-level rise (0.2–0.3 m/100 yr) it would take another 2 k.y. to raise relative sea level by 6 m (the maximum average height of the abrasion platform) and an additional 10 k.y. to reach 30 m (the maximum interglacial relative sea-level estimate from the Salmon River sand). These are minimum time estimates only, because the rates of subsidence are expected to decrease exponentially (Pirazzoli, 1996). Estimates of the duration of the last interglacial (marine isotope stage 5e) can be made assuming similar previous glacioisostatic conditions. Using the maximum estimate of former interglacial sea levels from field data, and assuming today's rates of relative sea-level rise, the last interglacial would have lasted less than 20 k.y., nearly twice the duration of previous estimates using orbitally tuned marine oxygen isotope data (ca. 13 ka; Imbrie et al., 1989). This long chronology resembles the age duration estimates of the last interglacial based on ice-core data (Dansgaard et al., 1993), cave-calcite records (Winograd et al., 1992), and age-dated raised coral reefs in Hawaii (Szabo et al., 1994). Clearly, precise age dating of the interglacial marine deposits will be necessary to resolve these problems.

## LATE WISCONSINAN SEA LEVELS (20–10 KA)

During the Wisconsinan glaciation a succession of local ice domes and divides developed in Atlantic Canada, collectively termed the Appalachian ice complex (Prest and Grant, 1969). The earliest phases of ice flow from local ice centers (Caldonia and Escuminac ice flow phases, 70–20 ka; Stea et al., 1998) extended out to the continental shelf edge with calving margins in >300 m water depth (Fig. 4). During the initial deglaciation phase (ca. 20–17 ka) ice retreated rapidly out of the Gulf of Maine–Bay of Fundy, isolating an ice mass over Nova Scotia that later became an active center of outflow (Scotian ice divide–Scotian phase; Stea et al., 1998). At Sable Island, outwash deltas, which graded to a sea level of −85 m, formed at the margin of the Scotian Ice Divide (Scott et al., 1989). During a later stage of retreat (ca. 17–15 ka) over isostatically depressed terrain, submarine moraines formed at the grounding line of the Scotian ice divide (Fig. 4). As ice from the Scotian ice divide was drawn out of the Bay of Fundy, outwash deltas, wave-cut terraces, and raised marine beaches were formed (ca. 15–13 ka; Fig. 5, A and B).

If we were to assume rapid northwestward retreat of a large Laurentide ice sheet ("maximum" model), then the isobase pattern would be a series of parallel lines trending northeast-southwest across the region. This simple pattern is prevalent in Maine, where shoreline elevations generally increase toward the northwest (and toward the thicker Laurentide ice). In Maritime Canada, however, the pattern of shoreline deleveling is more complex. For example, there is an anomalous northeastward trend to lower marine limits in the Bay of Fundy, from +40 m at the mouth to <0 m near the head of the Bay (Fig. 2). These shorelines are tilted downward toward a late local ice center, which was directing flow out of the Bay of Fundy (Fig. 4). Along the south shore of the Bay of Fundy, the Gulf of St. Lawrence, and southeast New Brunswick, shoreline tilts also do not reflect differential ice loading (Fig. 2).

One of the main arguments for the minimum model was the lack of raised shorelines around the Gulf of St. Lawrence. Raised shorelines in a glaciated terrane result from rapid uplift after ice retreat and higher shorelines imply a greater ice load. Lower elevation shorelines or the lack thereof imply thin, peripheral ice. The 0 isobase was therefore considered as a proxy for the late Wisconsinan ice margin in the Gulf of St. Lawrence (Grant, 1977; Figs. 1 and 2). This hypothesis was based on the twin assumptions of synchroneity of marine limits and rapid ice retreat (Wightman and Cooke, 1978; Grant, 1980; Grant, 1989, p. 430; Tushingham and Peltier, 1993). Prest (1970) and Stea (1982, 1988) advanced the opposing notion that raised shorelines along the Bay of Fundy coast are diachronous, and marine limit elevation is not a function of ice thickness but of protracted ice retreat to local ice centers (e.g., Gray, 1995; Dyke, 1998). In support of this view, widely differing ages have been obtained on marine deposits on both sides of the Bay of Fundy and toward the head of the bay (Stea and Wightman, 1987; Scott et al., 1987b; Stea and Mott, 1998). The lack of raised shorelines in the Gulf of St. Lawrence can best be explained by slow glacial retreat and the persistence of a carapace of stagnant ice, under which much of the rebound had already occurred before shorelines could be established (Stea et al., 1998). This concept can be applied to many regions of Maritime Canada where local ice centers persisted well after 12.5 ka, including northern Nova Scotia and Prince Edward Island (Stea et al., 1994; Anderson, 1985; Stea and Mott, 1998). For example, the relatively low marine limits evident on the emergence contour map (Fig. 2) over Gaspésie, the Magdalen Shelf, and northern and eastern Nova Scotia probably indicate late-melting ice, rather than thin, rapidly retreating glaciers. New results of lake basin accelerator mass spectrometry radiocarbon dating, consistently younger than conventional methods (1–2 ka), have radically changed the existing notions of glacier retreat in eastern Canada (Richard et al., 1997; Stea and Mott, 1998). This subice rebound model predicts higher marine limits in the regions deglaciated earliest, such as the Bay of Fundy and Gulf of Maine, and possibly near the mouth of the Laurentian Channel. The Magdalen Islands are located in a crucial position to test this theory, near enough to the deep Laurentian Channel to have been deglaciated earlier than much of the Magdalen shelf area (Fig. 1). Undated, raised marine features (at +20 m) of possible late Wisconsinan age have been described in the Magdalen Islands (Dredge et al., 1992), suggesting that some emergence had taken place in the Gulf of St. Lawrence.

At 11.6 ka, relative sea level along the inner Scotian Shelf of Nova Scotia stood at −65 m (Stea et al., 1994; Fader et al., 1997). A shoreline developed, marked by the abrupt transition from stark, unmodified bedrock topography to a zone of truncation and littoral zone features. Figure 6 is a swath bathymetric image of the inner Scotian Shelf off Halifax (Fig. 1) showing features at the lowstand shoreline. The shoreline is marked by littoral zone erosional features such as abrasion notches and wave-cut platforms, as well as depositional features such as shoreface deltas and paleospits and lagoons (Stea et al., 1994; Fig. 6). Above this lowstand shoreline is a zone of muted topography, with estuarine peat found at −34 m (Forbes et al., 1991). In contrast to the tilted, emerged shorelines described earlier, this inner shelf lowstand shoreline can be traced at the same depth along most of the northeast Atlantic coast and perhaps into the Gulf of St. Lawrence (Loring and Nota, 1973; Fig. 2). Exceptions to this rule are found in embayments such as Chedabucto Bay (Fig. 2), where glaciomarine muds and apparently unmodified glacial landforms are found above −70 m and below −38 m (Shaw et al., 1995; Miller, 1996). In Chedabucto Bay, glaciers were active after 10.6 ka (Stea and Mott, 1989), suggesting the possibility that glacier ice held off an encroaching ocean until after 11 ka, at a higher relative sea level.

Recently collected seismic reflection data from Browns Bank (Fig. 1) show prominent bank-edge deltas with upper lapout contacts between 40 and 70 m. Preservation of iceberg scours in glaciomarine sediments on the bank indicates that the maximum fall of sea level on the southwestern Scotian Shelf did not exceed 85–90 m (Fig. 6). The relative sea-level history is more complex in the Bay of Fundy, the lowstand shoreline

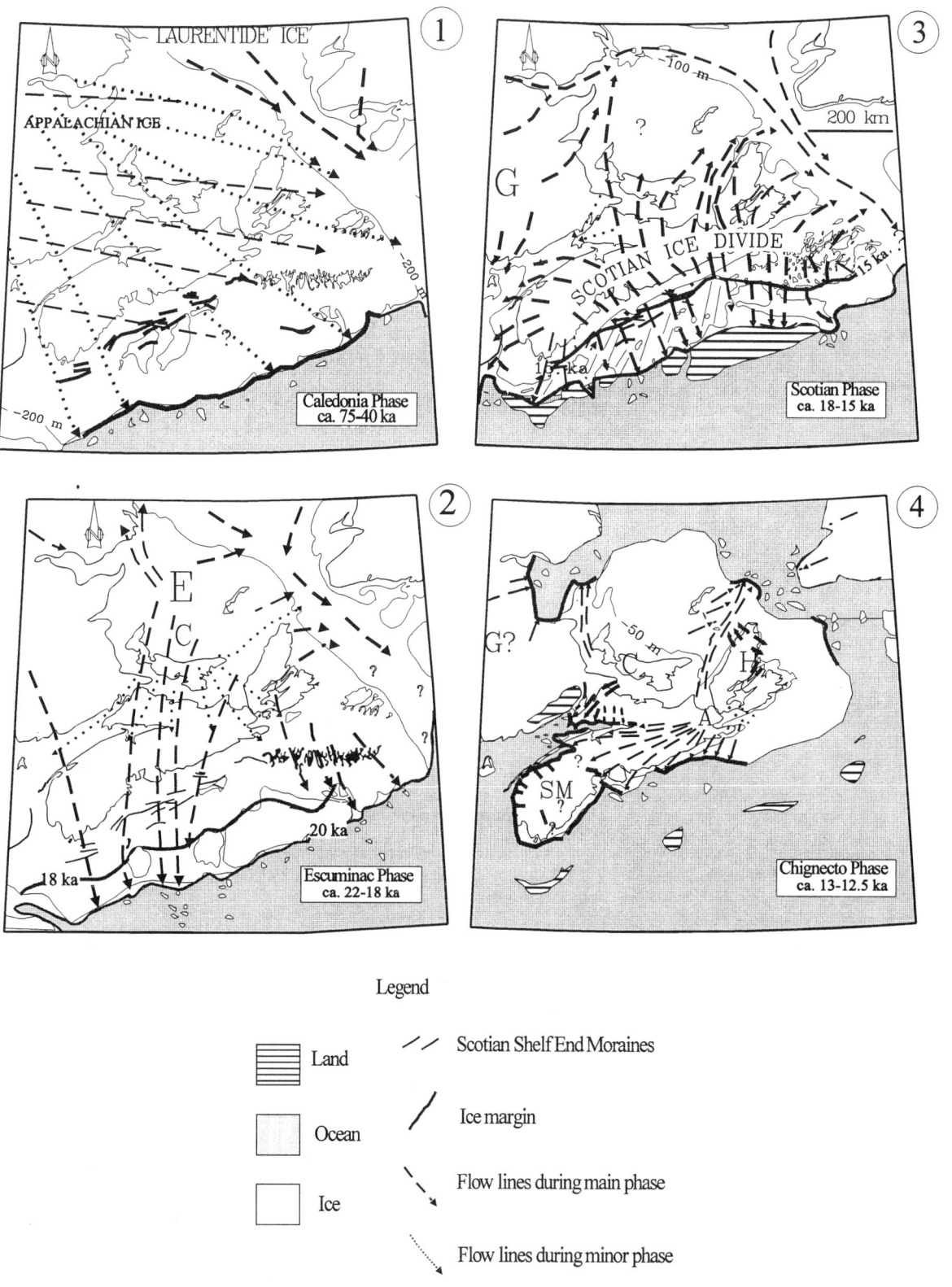

Figure 4. Evolution (advance and retreat) of ice divides and domes over Maritime Canada during Wisconsinan (75–10 ka). White is ice, dots are ocean, dashed arrows are flow lines, and dotted arrows are later flow lines. 1, Caledonia phase in Atlantic Canada and margins (early eastward flow designated phase 1a). 2, Escuminac phase from Escuminac ice center and divide (E) on Magdalen Plateau and Gaspereau ice center in New Brunswick (G). Chignecto glacier (C) was active for short time prior to phase 3. 3, Scotian phase (advance and retreats) (Scotian ice divide); lined area indicates emergent marine landscapes. 4, Chignecto phase from local centers over Antigonish Highlands and Chedabucto Bay (A); South Mountain (SM); Cape Breton Highlands (H), and Prince Edward Island (C).

Figure 5. A, Raised beach at Squally Point near head of Bay of Fundy (Fig. 1), +38 m. B, Outwash delta at Five Islands, Nova Scotia. Note tripartite Gilbert-type delta with topset-foreset contact at +22 m.

shallowing from southwest to northeast (Fader, 1989). The sea-level history of the Bay of Fundy is being addressed with the collection and interpretation of new multibeam bathymetry and seismic reflection data (G.B. Fader, personal commun., 1998).

## HOLOCENE SEA LEVELS (10–0 KA)

The first comprehensive work on changing Holocene relative sea level in Maritime Canada was that of Grant (1970), who documented many sites with $^{14}C$ dating of former sea levels during the Holocene. This work demonstrated that relative sea level during the Holocene was rising in all areas, largely due to crustal subsidence. Later work by Scott and Medioli (1980; 1982), Scott and Greenberg (1983), Scott et al. (1987a, 1987b, 1989, 1995), Smith et al. (1984), Shaw and Forbes (1990), and Forbes et al. (1991) provided rates of Holocene relative sea-level rise for more than 20 locations around Maritime Canada. A composite relative sea-level curve from the Atlantic coast of Nova Scotia reveals a more complex relative sea-level response than simple submergence throughout the Holocene (Fig. 7). Initial emergence after deglaciation (14–12 ka) culminated in the lowstand nadir (11.6 ka), when rates of isostatic uplift and eustatic sea-level rise were in equilibrium, forming a prominent shoreline (Fig. 6). The total amount of emergence is unknown and possibly underestimated in areas of Nova Scotia (including the Atlantic Coast) deglaciated well after 12 ka. Rebound occurred under thin ice, and may have been

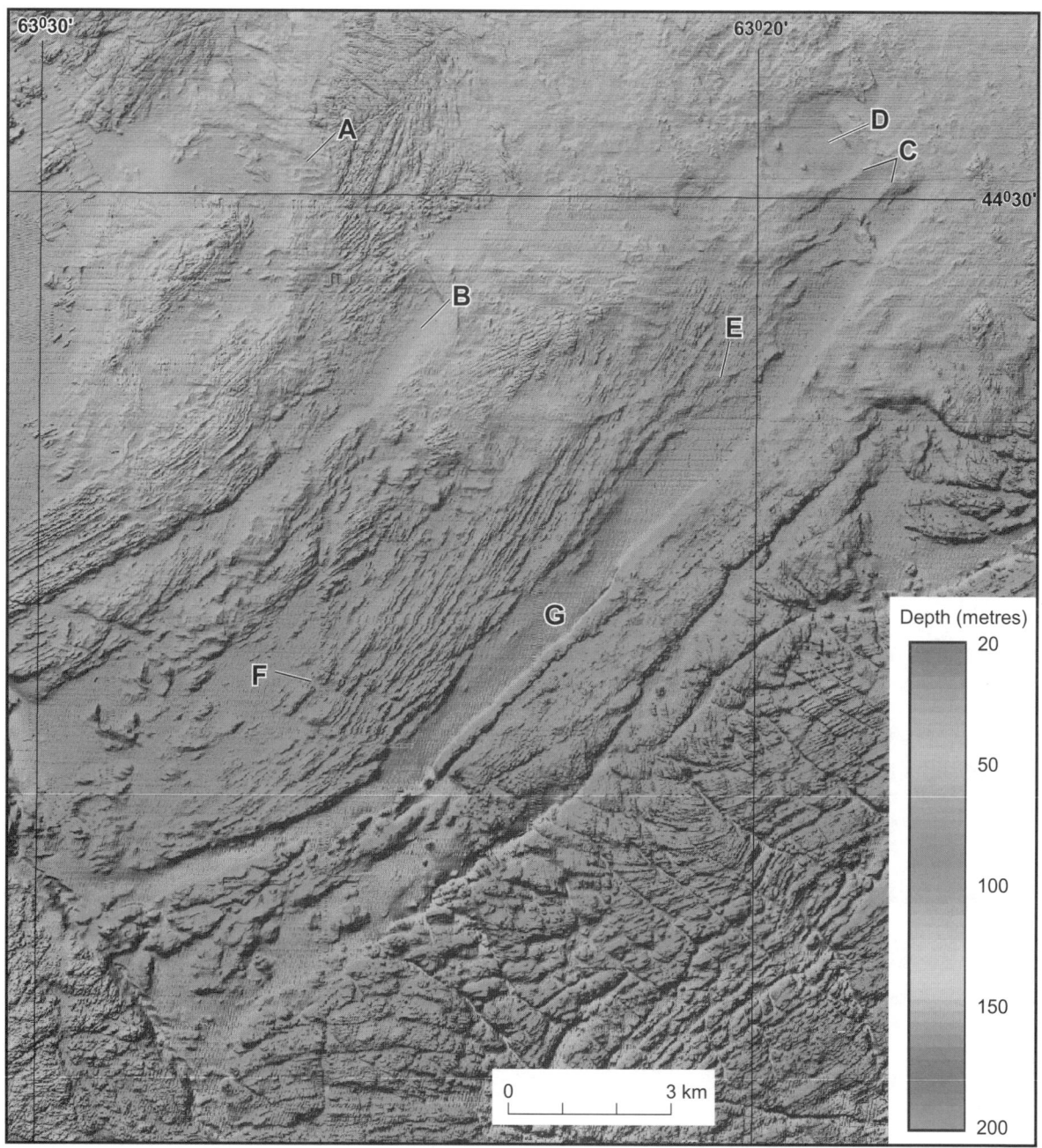

Figure 6. Multibeam bathymetric image of seafloor off Halifax, Nova Scotia (adapted from color image). Lowstand shoreline is at −65–70 m level marked by seafloor features A, B, and C, and general muting of seafloor topography (Stea et al., 1994). (For complete color image with depth scale, see http://agc.bio.ns.ca/mregion/ocean/MultibeamBathy/CRUISES/Halifax/outer_hbr/Full.html.) Note following seafloor features: A, progradational beach ridges; B, sambro delta where mussel shell hash in well-sorted sand at −65 m produced radiocarbon age of 11.6 ka; C, paleobarriers; D, paleolagoon; E, till morainal ridge; F, sharp bedrock strike ridges below lowstand, attenuated above lowstand; G, fault-valley filled with Holocene LaHave clay below 70 m, attenuated by erosional-depositional processes above lowstand.

interrupted by glacier readvances. The lowstand nadir was followed by rapidly rising relative sea level between 11.6 and 11.0 ka, then a marked slowing and falling relative sea level from 11 to 8 ka (Forbes et al., 1991). This was followed again by rising sea levels after 7 ka. An inflection point between the periods of rapid and slower relative sea-level rise occurs ca. 5 ka (Scott and Greenberg, 1983; Smith et al., 1984).

Three locations show a reversal from falling to rising sea level after the lowstand (Scott and Greenberg, 1983; Scott and Medioli, 1982; Edgecombe et al., 1999). The relative sea-level

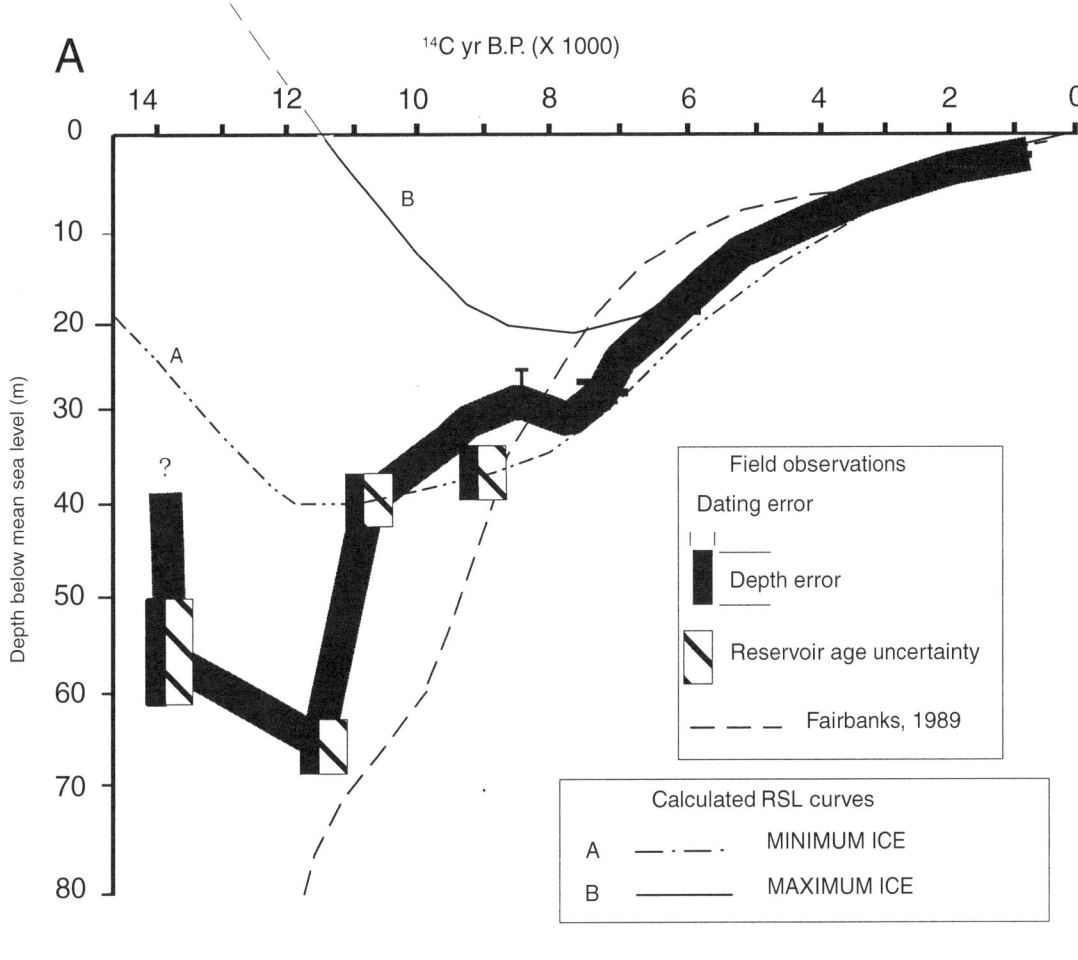
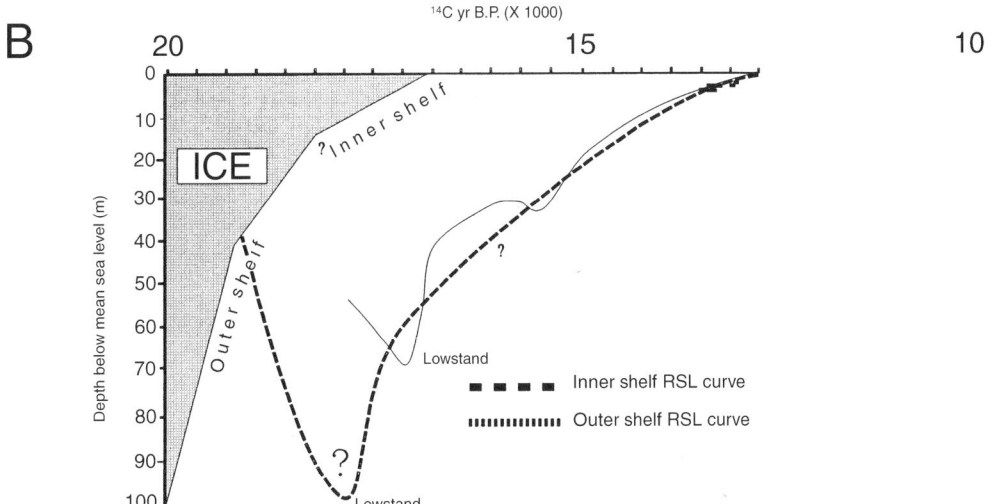

Figure 7. A, Relative sea-level curve for inner shelf and Atlantic coast compiled from Scott and Medioli (1982), Forbes et al. (1991), Piper and Fehr (1991), Stea et al. (1994), and Edgecombe et al. (1999). Model curve A represents calculated relative sea-level (RSL) curve based on Grant (1977) minimum ice model. Curve B is maximum ice model from Quinlan and Beaumont (1981). Fairbank's (1989) Barbados sea-level curve is also shown for comparison. B, Relative sea-level curves for outer banks and inner shelf (modified after Amos and Miller, 1990; Stea et al., 1994).

reversal points are based on a stratigraphic sequence in inner shelf cores that consists of an inorganic red mud overlain by a layer of intertidal salt marsh and then shoreface sand deposits. The red mud is interpreted as estuarine based on age-dated foraminiferal fauna and hence represents deeper water than the salt marsh deposits, but the lack of organic content is puzzling. Two of these reversal points are on the Atlantic coast and one is at the head of the Bay of Fundy. They occur at about the same time (7–8 ka) but at varying depths; the shallowest is at ~−25 m in Halifax Harbour (Edgecombe et al., 1999), −30 m in Lunenburg Bay (Scott and Medioli, 1982), to −37 m at the head of the Bay of Fundy (Scott and Greenberg, 1983). We believe that these reversal points and the preceding marked relative sea-level slowing represent the migration and collapse of a peripheral bulge as modeled by Quinlan and Beaumont (1982) and calibrated in Scott et al. (1987a). The Halifax Harbour site (Edgecombe et al., 1999) provides an idea of the actual magnitude of the sea-level fall or bulge amplitude, i.e., 10 m.

Tide-gauge data from Halifax Harbour for the years 1896–1905 and 1920–1989 indicate that mean sea level has increased by ~0.3 m over the past 95 years and rose at an average rate of 3.6 mm/yr (equivalent to 3.6 m/k.y.) between 1920 and 1988 (Shaw and Forbes, 1990). This is a more rapid rate of rise than the average for the past few thousand years but much slower than occurred at the time of most rapid transgression earlier in the Holocene. The rate of relative sea-level rise seems to be slowing in recent decades (Shaw and Forbes, 1990). The rates of relative sea-level rise tend to be higher along the Atlantic coast of Nova Scotia and inside the Bay of Fundy, and show a trend of decreasing rates to the northeast, i.e., the rate of relative sea-level rise decreases toward the former ice center. However, in most areas of Maritime Canada the rate of relative sea-level rise is anomalously high compared to areas away from the former ice margin (i.e., Florida, South Carolina, Bermuda) and comparable to those in New England (Redfield, 1967). This anomalous rate of rise is explained by the continuing collapse of the peripheral forebulge; the magnitude of this collapse is currently highest on the coast of Maritime Canada, and decreases to the south.

## GEOPHYSICAL MODEL COMPARISON (WHERE, HOW FAST, AND HOW BIG IS THE PREDICTED FOREBULGE?)

The anomalous relative sea-level rise during the Holocene and the patterns of deleveled former shorelines in the region are a result of the Earth's response to the imposition of ice loads. A model of sea-level change based on the MAXWELL viscoelastic Earth model responding to loading of the Earth during glaciation (Farrell and Clark, 1976; Peltier and Andrews, 1976) has been the best attempt at synthesizing this complex sea-level history. Sea-level changes around Nova Scotia and farther down the U.S. East Coast are mainly determined by (1) the eustatic rise in global sea levels due to the addition of meltwater into the ocean; (2) the gravitational attraction between the ice mass and the ocean water that resulted in accumulation of seawater and a geoid high near the edge of the ice sheet at glacial maximum and a rapid fall in sea level after deglaciation; and (3) the vertical motion of the ground due to the migrating and/or collapsing peripheral forebulge, which is formed of sublithospheric material extruded from accreting and isostatically subsiding ice sheets (Quinlan and Beaumont, 1981).

The behavior of the bulge is controlled by the rheology of the mantle (e.g., Cathles, 1980). For example, deep mantle flow under the lithosphere is characterized by an inward-migrating and collapsing forebulge (Peltier, 1974; Wu and Peltier, 1983). However, an outward-migrating bulge is produced when mantle flow is restricted to channels (e.g., a low-viscosity asthenosphere overlying a high-viscosity mantle or a low-viscosity upper mantle overlying a high-viscosity lower mantle, whereby the viscosity differences are a factor of 10 or more). If the rheology of the mantle is nonlinear, then the forebulge will just collapse in place (Wu, 1993, 1998).

In Figure 8 we have modeled the forebulge response using the ICE 3G model for an Earth with 150-km-thick lithosphere overlying a uniform $10^{21}$ Pa s linear mantle (Tushingham and Peltier, 1993). The curve marked RD in Figure 8C is the isostatic (uplift) portion of the model showing the effect of forebulge migration as it approached Halifax then started to collapse, after 8 ka. The curve marked RD + Eust takes into account the motion of the land and the rise in eustatic sea level. The resultant curve labeled RSL includes the effect of the gravitational attraction between the ice and water (Farrell and Clark, 1976). The minimum ice model was used in the calculation of these curves. If the maximum ice model was used, the forebulge would have arrived 2–3 k.y. later (e.g., Fig. 7). It is important to note that the sea-level reversal points on the inner Scotian Shelf that mark the crest of the bulge occur at 7–8 ka, between the modeled arrivals using the maximum and minimum ice configurations (Fig. 7). Figure 8B shows the evolution of lithospheric displacement, and Figure 8A shows the migration of the peripheral forebulge at 2 k.y. increments after the glacial maximum. At 18 ka, the zero displacement line was about 500–600 km from the Laurentide-Appalachian ice margin. Note again that the ice model used is a variant of the minimum model centered in Hudson's Bay, and hence does not represent our present concepts of ice thickness and extent in Atlantic Canada (e.g., Stea et al., 1998). For the purposes of this argument, however, the ice model can be considered as a minimum estimate of forebulge extent. We eventually hope to model the forebulge using the maximum extent of the Appalachian ice complex as now envisioned (Mosher et al., 1989; Stea et al., 1998). The results of the model show that the forebulge migrates inward toward the northwest from 18 to 8 ka, but collapses in place after ca. 8 ka. The peak amplitude of the forebulge was about 50 m at 12 ka. The rate of migration was time dependent, but reached 80 km/k.y. ca. 12 ka.

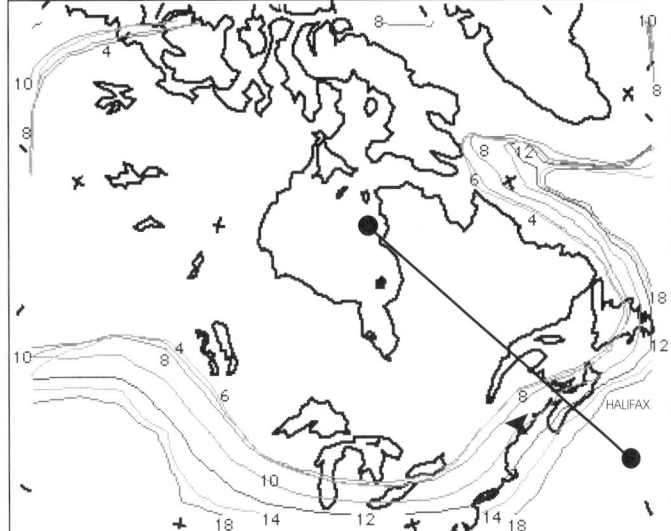
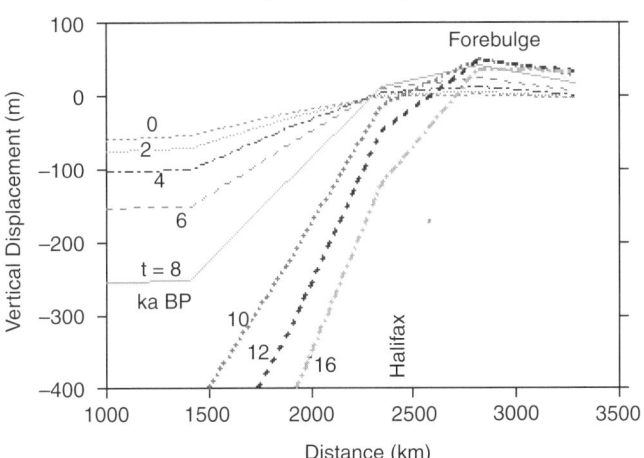
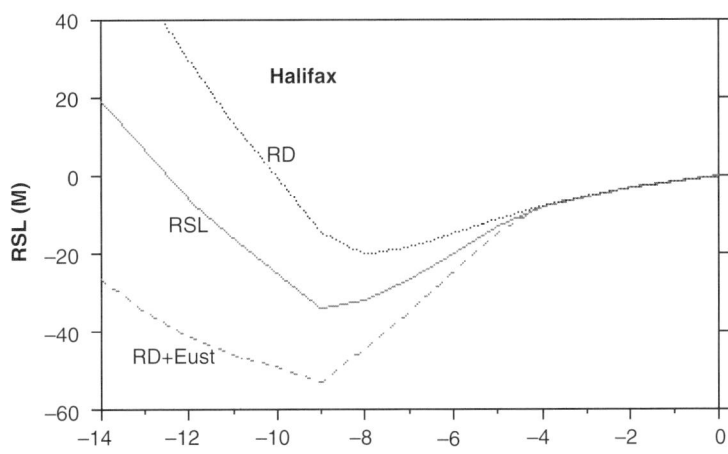

Figure 8. A, Map showing position of leading edge of forebulge at various time increments (k.y.) and location of vertical displacement profile shown below. B, Vertical displacement profile from Ottawa Island to Halifax through hypothetical minimum ice mass centered in Hudson Bay and terminating just off coast of Nova Scotia (see Tushingham and Peltier, 1993). C, Relative sea-level (RSL) curves generated from Halifax area based on ICE 3G model. RD is isostatic (uplift) record of forebulge passage. RD + eustatic (Eust) takes into account rise in eustatic sea level. RSL represents resultant curve with gravitational attraction included. Note difference in lowstand depths with this model and empirical data in Figure 7. Quinlan and Beaumont's (1982) minimum ice curve (Fig. 7) is very similar to RD + eust curve in this study. Note: Model ages in sidereal time (ka B.P.).

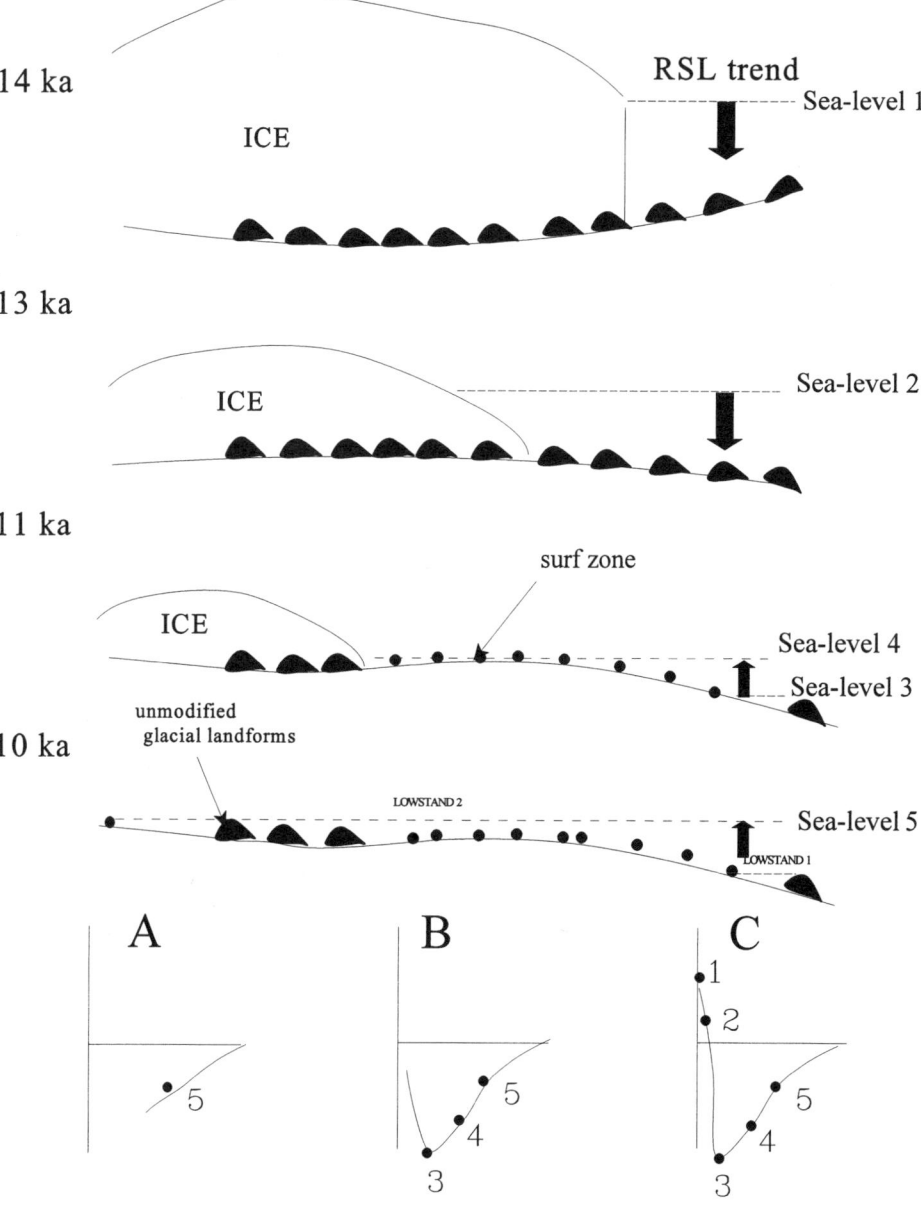

Figure 9. Schematic showing effects of late-melting ice caps on relative sea-level (RSL) records in shelf-margin areas. RSL records of areas first cleared of ice (e.g., Bay of Fundy) are compared with records of areas last deglaciated (e.g., Chedabucto Bay). After ca. 15 ka, ice sheet is removed from calving bays such as Bay of Fundy and first beaches form. These outer regions record complete RSL record from regression to transgression (C). Ice remains in land and some marine areas of Nova Scotia until 11–10 ka, and in these transitional areas uplift (regression) record is not recorded, having occurred under ice (A, B). These areas record just transgression, and in some narrow embayments with late-melting ice, transgression skips over unmodified landforms preserved under ice.

What is the field geological evidence for a migrating peripheral bulge in Maritime Canada? Is it represented by uplift and subsidence soon after deglaciation or uplift thousands of years later? In Newfoundland, Liverman (1994) described the northeastward trend toward progressively younger dates of marine submergence and interpreted this as bulge migration. Similarly, Barnhardt et al. (1995) interpreted the time-transgressive lowstands in eastern North America from Nova Scotia to Quebec as passage of the forebulge crest. The high amplitude and short wavelength of the relative sea-level curves in Maine and Nova Scotia imply a much faster migration than predicted in the uniform viscosity model. According to our model, at the maximum ice configuration the bulge crest is located ~500 km beyond the ice margin (located in the minimum ice model near the Nova Scotia coast). Migrating at 80 km/k.y. it takes ~7 k.y. for the leading edge of the bulge to reach the Nova Scotia coastline (150-km-thick lithosphere).

Using the minimum ice estimate, the high-amplitude response of the relative sea-level curves in Maine (Belknap et al., 1987) and Nova Scotia (Stea et al., 1994) between 14 and 11 ka is not explained by the linear rheology model alone, because most of the uplift occurs well before the modeled bulge even reaches the coastal zone. In a maximum ice model, the forebulge would take even longer to reach the region, resulting in a further delay in emergence (Fig. 7; Quinlan and Beaumont, 1981, 1982). As explained earlier, the modeled relative sea-level curve before 7 ka (Fig. 8A) is determined by both forebulge migration and the local sea-level change (including eustatic rise and ice-water interaction by gravity). As the motion of the land is viscously damped, rapid late-glacial changes of sea level cannot be attributed to forebulge migration. Elastic isostatic recovery from thick local ice loads may explain part of the early, rapid relative sea-level fall and lowstand formation in the coastal areas of Maine and Nova Scotia. Elastic recovery is instantaneous, however, and the regression record of both Maine and Nova Scotia suggests a rapid but finite rate of relative sea-level fall and some periods of relative sea-level slowing and shoreline formation (Stea et al., 1987). The magnitude of early relative sea-level fall and lowstand depth is also difficult to reconcile with the linear rheology model alone. Late-glacial advances such as the Chignecto phase (Stea et al., 1998) can greatly affect local sea levels, especially by superimposing smaller cycles of uplift and subsidence on the larger Laurentide-Appalachian isostatic recovery regime. Quinlan and Beaumont (1981) modeled the effect of a significant late-glacial readvance on relative sea level in the Maritimes and showed both periods of relative sea level slowing in the regression phase in areas adjacent to the ice cap, and greater rates of relative sea-level fall (and lower lowstands) away from the former ice cap. Sea-level regression in this region of large local ice caps marginal to the Laurentide ice sheet may be affected by a series of relative sea-level jumps followed by periods of stability due to glacier readvances. Higher resolution modeling based on our present knowledge of the extent and duration of local ice is clearly needed in order to resolve these problems. In summary, the relative sea-level curves in Maine and Nova Scotia between 11 and 7 ka record a time of relative sea-level slowing (plateau) and reversal points, which are in agreement with the model forebulge migration predictions.

There is contradictory evidence of both shallow and deep submerged shorelines off the Atlantic coast and in Gulf of St. Lawrence (Kranck, 1972; Loring and Nota, 1973; Syvistki, 1992; Shaw et al., 1995). Although this may be a result of misinterpretation, a similar paradox in emerged shorelines (high versus low or absent) suggests that persistence of local ice sheets in Maritime Canada may have also affected the history of shorelines in marine areas below present sea level. We suggest that two separate lowstands can be produced, separated by a region that doesn't appear to have been transgressed. In local embayments, late-melting ice caps can temporarily hold off ocean water during a period of rapid relative sea-level rise. Figure 9 demonstrates how delayed or late deglaciation can produce differing relative sea-level histories in regions with similar total ice loads during the glacial maximum. The essential point is that areas underneath late-melting ice caps will have an unmanifested emergence record. The strong geological evidence for a major late Wisconsinan ice center in the Gulf of St. Lawrence and other local ice caps (Rampton et al., 1984; Stea et al., 1998), therefore, can explain much of the complexity of relative sea-level records in eastern North America (Fig. 4). As stated earlier, glacier advances ca. 13 and 11 ka (Stea and Mott, 1989, 1998) may have induced rapid relative sea-level effects, including uplift well into the Holocene (Quinlan and Beaumont, 1981).

The emerged Sangamon interglacial shorelines in Maritime Canada represent the end result of a previous interglacial cycle of relative sea-level change. The present interglacial (Holocene) can help with predictions of future sea-level changes. In many stable areas away from former ice sheets, such as the east coast of South America or the west coast of Africa, sea level was higher than present during the Holocene, ca. 5 ka (Pirazzoli, 1996). At present, eastern Canada is still undergoing relative sea-level rise related to the continuing collapse of the peripheral forebulge. How much longer will the subsidence last? This depends on the Earth model. For the uniform viscosity $10\ E^{21}$ Pa s mantle model (Fig. 8), relative sea-level will rise for another 2 k.y., assuming no more glacier melting and thermal expansion of ocean water. With a higher viscosity upper part of the lower mantle, it can reach 10 ka or more, but this model is not preferred (Peltier, 1998). Long-term isostatic recovery from a much larger ice sheet in the Illinoian may, in part, explain sea levels of +30 m in Nova Scotia during the last interglacial.

## REFERENCES CITED

Amos, C.L., and Miller, A.A.L., 1990, The Quaternary stratigraphy of southwest Sable Island Bank: Geological Society of America Bulletin, v. 102, p. 915–934.

Anderson, T.W., 1985, Late-Quaternary pollen records from eastern Ontario, Quebec, and Atlantic Canada, in Bryant, M., and Holloway, R.G., eds., Pollen records of late Quaternary North American sediments: American Association of Stratigraphic Palynologists Foundation, p. 281–326.

Barnhardt, W.A., Gehrels, W.R., Belknap, D.F., and Kelley, J.T., 1995, Late Quaternary relative sea-level change in the western Gulf of Maine: Evidence for a migrating glacial forebulge: Geology, v. 23, p. 317–320.

Belknap, D.F., Andersen, B.G., Anderson, R.S., Anderson, W.A., Borns, H.W., Jr., Jacobson, G.L., Kelley, J.T., Shipp, R.C., Smith, D.C., Stuckenrath, R., Jr., Thompson, W.B., and Tyler, D.A., 1987, Late Quaternary sea-level changes in Maine, in Nummedal, D., et al., eds., Sea-level fluctuations and coastal evolution: Society of Economic Paleontologists and Mineralogists Special Publication 41, p. 71–85.

Cathles, L.M., 1980, Interpretation of postglacial isostatic adjustment phenomena in terms of mantle rheology, in Mörner, N.A., ed., Earth rheology, isostasy and eustasy: London, John Wiley and Sons, p. 11–43.

Chalmers, R., 1895, Report on the surface geology of eastern New Brunswick, northwestern Nova Scotia and a portion of Prince Edward Island: Geological Survey of Canada Annual Report 1894, v. 1, no. 7, 144 p.

Clarke, A.H., Grant, D.R., and MacPherson, E., 1972, The relationship of *Atractodon stonei* (Pilsbry) (Mollusca, Buccinidae) to Pleistocene stratigraphy and paleoecology of southwestern Nova Scotia: Canadian Journal of Earth Sciences, v. 9, p. 1030–1038.

Dansgaard, W., Johnsen, S.J., Clausen, H.B., Dahl-Jensen, D., Gundestrup, N.S., Hammer, C.U., Hvidberg, C.S., Steffensen, J.P., Sveinbjornsdottir, A.E., Jouzel, J., and Bond, G., 1993, Evidence for general instability of past climate from a 250-kyr ice-core record: Nature, v. 364, p. 218–220.

Denton, G.H., and Hughes, T.J., 1981, The last great ice sheets: Toronto, John Wiley and Sons, Inc., 484 p.

Dredge, L., Mott, R.J., and Grant, D.R., 1992, Quaternary stratigraphy, paleoecology, and glacial geology, Iles de la Madelaine, Quebec: Canadian Journal of Earth Sciences, v. 29, p. 1981–1996.

Dyke, A.S., 1998, Holocene delevelling of Devon Island, Arctic Canada: Implications for ice sheet geometry and crustal response: Canadian Journal of Earth Sciences, v. 35, p. 885–904.

Dyke, A.S., and Prest, V.K., 1987, Late Wisconsinan and Holocene history of the Laurentide ice sheet, in Fulton, R.J., and Andrews, J.T., eds., The Laurentide Ice Sheet: Géographie Physique et Quaternaire, v. 41, p. 237–264.

Edgecombe, R.S., Scott, D.B., and Fader, G.B.J., 1999, New data from Halifax Harbour: Paleoenvironment and a new sea-level curve for the inner Scotian Shelf: Canadian Journal of Earth Sciences, v. 36, p. 805–817.

Fader, G.B.J., 1989, A late Pleistocene low sea-level stand of the southeast Canadian offshore, in Scott, D.B., et al., eds., Late Quaternary sea level correlations and applications: Dordrecht, Netherlands, Kluwer Academic Publishers, p. 71–103.

Fader, G.B.J., Courtney, R., and Stea, R.R., 1997, A seabed drumlin field on the inner Scotian Shelf, Canada, in Coldseis Atlas, INQUA Project on the seismic facies analysis of glaciomarine sediments: London, Chapman and Hall, p. 50–51.

Fairbanks, R.G., 1989, A 17,000-year glacio-eustatic sea level record: Influence of glacial melting rates on the Younger Dryas even and deep ocean circulation: Nature, v. 342, p. 637–642.

Farrell, W.E., and Clark, J.A., 1976, On post-glacial sea level: Royal Astronomical Society Geophysical Journal, v. 46, p. 647–657.

Flint, R.F., 1971, Glacial and Quaternary geology: New York, John Wiley and Sons, 892 p.

Forbes, D.L., Boyd, R., and Shaw, J., 1991, Late Quaternary sedimentation and sea-level changes on the inner Scotian Shelf: Quaternary Shelf Research, v. 11, p. 1155–1179.

Gipp, M.R., 1989, Late Wisconsinan deglaciation of Emerald Basin [M.S. thesis]: St. Johns, Newfoundland, Memorial University, 219 p.

Gipp, M.R., 1994, Late Wisconsinan deglaciation of Emerald Basin, Scotian Shelf: Canadian Journal of Earth Sciences, v. 31, p. 554–566.

Gipp, M.R., and Piper, D.J.W., 1989, Chronology of late Wisconsinan glaciation, Emerald Basin, Scotian Shelf: Canadian Journal of Earth Sciences, v. 26, p. 333–335.

Goldthwait, J.W., 1924, Physiography of Nova Scotia: Geological Survey of Canada Memoir 140, p. 60–103.

Grant, D.R., 1970, Recent coastal submergence of the Maritime Provinces: Canadian Journal of Earth Sciences, v. 7, p. 676–689.

Grant, D.R., 1977, Glacial style and ice limits, the Quaternary stratigraphic record, and changes of land and ocean level in the Atlantic Provinces, Canada: Géographie Physique et Quaternaire, v. 31, p. 247–260.

Grant, D.R., 1980, Quaternary sea-level change in Atlantic Canada as an indication of crustal delevelling, in Mörner, N.A., ed., Earth rheology, isostasy and eustasy: London, John Wiley and Sons, p. 201–214.

Grant, D.R., 1989, Quaternary geology of the Atlantic Appalachian region of Canada, in Fulton, R.J., ed., Quaternary geology of Canada and adjacent Greenland: Geological Survey of Canada, Geology of Canada no. 1, p. 393–440.

Grant, D.R., 1994, Quaternary geology, Cape Breton Island: Geological Survey of Canada Bulletin 482, 159 p.

Grant, D.R., and King, L.H., 1984, A stratigraphic framework for the Quaternary history of the Atlantic Provinces, in Fulton, R.J., ed., Quaternary stratigraphy of Canada—A Canadian Contribution to IGCP Project 24: Geological Survey of Canada Paper 84-10, p. 173–191.

Gray, J.M., 1996, Glacio-isostasy, glacio-eustasy and relative sea level change, in Menzies, J., ed., Past glacial environments, sediments, forms and techniques: Oxford, Butterworth-Heinemann, p. 315–333.

Imbrie, J., McIntyre, A., and Mix, A., 1989, Oceanic response to orbital forcing in the late Quaternary; observational and experimental strategies, in Berger, A., et al., eds., Climate and geo-sciences: Boston, Kluwer, p. 121–164.

King, L.H., 1969, Submarine end moraines and associated deposits on the Scotian Shelf: Geological Society of America Bulletin, v. 80, p. 83–96.

King, L.H., 1996, Late Wisconsinan ice retreat from the Scotian Shelf: Geological Society of America Bulletin, v. 108, p. 1056–1067.

Kranck, K., 1972, Geomorphological development and post-Pleistocene sea level changes, Northumberland Strait, Maritime Provinces: Canadian Journal of Earth Sciences, v. 9, p. 835–844.

Liverman, D.G.E., 1994, Relative sea-level history and isostatic rebound in Newfoundland, Canada: Boreas, v. 23, p. 217–230.

Loring, D.H., and Nota, D.J.G., 1973, Morphology and sediments of the Gulf of St. Lawrence: Fisheries Research Board of Canada Bulletin, v. 182, 147 p.

MacNeill, R.H., and Purdy, C.A., 1951, A local glacier in the Annapolis-Cornwallis Valley [abs.]: Nova Scotian Institute of Science Proceedings, v. 23, p. 111.

Miller, A.A.L., 1996, Foraminiferal distribution on selected bank areas of the Scotian Shelf and Holocene history of inner Chedabucto Bay: Foraminiferal evidence, as a contribution to assessment of the marine aggregate potential of the Scotian Shelf: Geological Survey of Canada Open File Report 3258, 57 p.

Mosher, D.C., Piper, D.J.W., Vilks, G.V., Aksu, A.E., and Fader, G.B., 1989, Evidence for Wisconsinan glaciations in the Verrill Canyon area, Scotian Slope: Quaternary Research, v. 31, p. 27–40.

Myers, R.A., and Stea, R.R., 1986, Surficial mapping results, Pictou, Guysborough, and Antigonish Counties, Nova Scotia, in Mines and Minerals Branch Report of Activities: Bates, J.L., ed., Nova Scotia Department of Mines and Energy, p. 189–194.

Oldale, R.N., Valentine, P.C., Cronin, T.M., Spiker, E.C., Blackwelder, B.W., Belknap, D.F., Wehmiller, J.F., and Szabo, B.J., 1982, Stratigraphy, structure, absolute age and paleontology of the upper Pleistocene deposits at Sankaty Head, Nantucket Island, Massachusetts: Geology, v. 10, p. 246–252.

Peltier, W.R., 1974, The impulse response of a Maxwell earth: Reviews of Geophysics, v. 12, p. 649–669.

Peltier, W.R., 1998, Postglacial variations in the level of the sea: Implications for climate dynamics and solid-earth geophysics: Review of Geophysics, v. 36, p. 603–689.

Peltier, W.R., and Andrews, J.T., 1976, Glacial-isostatic adjustment—I, The forward problem: Royal Astronomical Society, Geophysical Journal v. 46, p. 605–646.

Piper, D.J.W., and Fehr, S.D., 1991, Radiocarbon chronology of late Quaternary sections on the inner and middle Scotian Shelf, south of Nova Scotia, in Current research, Part E: Geological Survey of Canada Paper 91-1E, p. 321–325.

Pirazzoli, P.A., 1996, Sea-level changes: The last 20,000 years: New York, John Wiley & Sons, 207 p.

Prest, V.K., 1970, Quaternary geology of Canada, in Geology and economic minerals of Canada: Energy Mines and Resources, Canada, Economic Geology Report No. 1, p. 676–764.

Prest, V.K., and Grant, D.R., 1969, Retreat of the last ice sheet from the Maritime Provinces–Gulf of St. Lawrence region: Geological Survey of Canada Paper 69-33, 15 p.

Prest, W.H., 1896, Glacial succession in central Lunenburg, Nova Scotia: Nova Scotian Institute of Science, Proceedings and Transactions, v. 9, p. 158–170.

Quinlan, G., and Beaumont, C., 1981, A comparison of observed and theoretical relative sea levels in Atlantic Canada: Canadian Journal of Earth Sciences, v. 18, p. 1146–1163.

Quinlan, G., and Beaumont, C., 1982, The deglaciation of Atlantic Canada as reconstructed from the postglacial relative sea-level record: Canadian Journal of Earth Sciences, v. 19, p. 2232–2246.

Rampton, V.N., Gauthier, R.C., Thibault, J., and Seaman, A.A., 1984, Quaternary geology of New Brunswick: Geological Survey of Canada Memoir 416, 77 p.

Redfield, A.C., 1967, Postglacial change in sea level in the western North Atlantic Ocean: Science, v. 157, p. 687–691.

Richard, P.J.H., Veillette, J.J., Larouche, A.C., Hétu, B., Gray, J.T., and Gangloff, P., 1997, Chronologie de la Déglaciation en Gaspésie: Novelles Données et implications: Géographie Physique et Quaternaire, v. 51, p. 163–184.

Scott, D.B., and Greenberg, D.A., 1983, Relative sea-level rise and tidal development in the Fundy tidal system: Canadian Journal of Earth Sciences, v. 20, p. 1554–1564.

Scott, D.B., and Medioli, F.S., 1980, Post-glacial emergence curves in the Maritimes determined from marine sediments in raised basins: Proceedings of Coastlines '80 (Canada): Ottawa, Canadian National Science and Engineering Research Council, p. 428–446.

Scott, D.B., and Medioli, F.S., 1982, Micropaleontological documentation for early Holocene relative sea-level fall on the Atlantic coast of Nova Scotia: Geology, v. 10, p. 278–281.

Scott, D.B., Boyd, R., and Medioli, F.S., 1987a, Relative sea-level changes in Atlantic Canada: Observed level and sedimentological changes vs. theoretical models, in Numendal, D., et al., eds., Sea-level rise and coastal evolution: Society of Economic Paleontologists and Mineralogists Special Publication 41, p. 87–96.

Scott, D.B., Medioli, F.S., and Miller, A.A.L., 1987b, Holocene sea levels, paleoceanography, and late glacial ice configurations near the Northumberland Strait, Maritime Provinces: Canadian Journal of Earth Sciences, v. 24, p. 668–675.

Scott, D.B., Boyd, R., Douma, M., Medioli, F.S., Yuill, Y., Leavitt, E., and Lewis, C.F.M., 1989, Holocene relative sea-level changes and Quaternary glacial events on a continental shelf edge, Sable Island Bank, in Scott, D.B., et al., eds., Late Quaternary sea-level correlations and applications: Dordrecht, Netherlands, Kluwer, p. 105–120.

Scott, D.B., Brown, K., Collins, E.S., and Medioli, F.S., 1995, A new sea-level curve from Nova Scotia, evidence for a rapid acceleration of sea-level rise in the late mid-Holocene: Canadian Journal of Earth Sciences, v. 32, p. 2071–2080.

Shaw, J., and Forbes, D.L., 1990, Short- and long-term relative sea-level trends in Atlantic Canada: Proceedings, Canadian Coastal Conference 1990 (Kingston, Ontario): Ottawa, National Research Council Canada, p. 291–305.

Shaw, J., Forbes, D.L., Ceman, J.A., Asprey, K.A., Beaver, D.E., Wile, B., Froebel, D., and Jodrey, F., 1995, Cruise report 94-138, marine geological surveys in Chedabucto and St. Georges Bays, Nova Scotia and Bay of Islands, Newfoundland: Geological Survey of Canada Open File Report 3230, 13 p.

Smith, D., Scott, D.B., and Medioli, F.S., 1984, Marsh foraminifera in the Bay of Fundy: Modern distribution and application to sea-level determinations: Maritime Sediments and Atlantic Geology, v. 20, p. 127–142.

Stea, R.R., 1982, The properties, correlation and interpretation of Pleistocene sediments in central Nova Scotia [M.S. thesis]: Halifax, Nova Scotia, Dalhousie University, 215 p.

Stea, R.R., 1983, Surficial geology of the western Part of Cumberland County, Nova Scotia, in Current research, Part A: Geological Survey of Canada Paper 83-1A, p. 197–202.

Stea, R.R., 1984, The sequence of glacier movements in northern mainland Nova Scotia determined through mapping and till provenance studies, in Mahaney, W.C., ed., Correlation of Quaternary chronologies: Norwich, U.K., Geo Books, p. 279–297.

Stea, R.R., 1988, Holocene sea levels, paleooceanography and late glacial ice configuration near the Northumberland Strait, Maritime Provinces: Discussion: Canadian Journal of Earth Sciences, v. 25, p. 348–350.

Stea, R.R., and Mott, R.J., 1989, Deglaciation environments and evidence for glaciers of Younger Dryas age in Nova Scotia, Canada: Boreas, v. 18, p. 169–187.

Stea, R.R., and Mott, R.J., 1998, Deglaciation of Nova Scotia: Stratigraphy and chronology of lake sediment cores and buried organic sections: Géographie Physique et Quaternaire, v. 41, p. 279–290.

Stea, R.R., and Wightman, D.M., 1987, Age of the Five Islands Formation, Nova Scotia, and the deglaciation of the Bay of Fundy: Quaternary Research, v. 27, p. 211–219.

Stea, R.R., Scott, D.B., Kelley, J.T., Kelley, A.R., Wightman, D.M., Finck, P.W., Seaman, A., Nicks, L., Bleakney, S., Boyd, R., and Douma, M., 1987, Quaternary glaciations, geomorphology and sea-level changes, Bay of Fundy region: NATO Advanced Course on Sea-level Correlations and Applications, Halifax, Nova Scotia: Halifax, Dalhousie University, Field Trip Guidebook, 79 p.

Stea, R.R., Mott, R.J., Belknap, D.F., and Radtke, U., 1992, The pre-late Wisconsinan chronology of Nova Scotia, Canada, in Clark, P.U., and Lea, P.D., eds., The last interglaciation/glaciation transition in North America: Geological Society of America Special Paper 270, p. 185–206.

Stea, R.R., Boyd, R., Fader, G.B.J., Courtney, R.C., Scott, D.B., and Pecore, S.S., 1994, Morphology and seismic stratigraphy of the inner continental shelf off Nova Scotia, Canada: Evidence for a −65 m lowstand between 11,650 and 11,250 $^{14}$C yr B.P.: Marine Geology, v. 117, p. 135–154.

Stea, R.R., Piper, D.J.W., Fader, G.B.J., and Boyd, R., 1998, Wisconsinan glacial and sea-level history of Maritime Canada, a correlation of land and sea events: Geological Society of America Bulletin, v. 110, p. 821–845.

Syvitski, J.P.M., 1992, Marine geology of the Baie Des Chaleurs: Géographie Physique et Quaternaire, v. 46, p. 331–348.

Szabo, B.J., Ludwig, K.R., Muhs, D.R., and Simmons, K.R., 1994, Thorium-230 ages of corals and duration of the last interglacial sea-level high stand on Oahu, Hawaii: Science, v. 266, p. 93–96.

Tushingham, A.M., and Peltier, W.R., 1993, Ice 3G, A new global model of late Pleistocene deglaciation based upon geophysical predictions of postglacial relative sea-level change: Journal of Geophysical Research, v. 96, p. 4497–4523.

Wagner, F.J.E., 1977, Paleoecology of marine Pleistocene Mollusca, Nova Scotia: Canadian Journal of Earth Sciences, v. 14, p. 1305–1323.

Wehmiller, J.F., Belknap, D.F., Boutin, B.S., Mirecki, J.E., Rahaim, S.D., and York, L.L., 1988, A review of the aminostratigraphy of Quaternary molluscs from United States Atlantic Coastal Plain sites, in Easterbrook, D.J., ed., Dating Quaternary sediments: Geological Society of America Special Paper 227, p. 69–110.

Wightman, D.M., 1980, Late Pleistocene glaciofluvial and glaciomarine sediments on the north side of the Minas Basin, Nova Scotia [Ph.D. thesis]: Halifax, Nova Scotia, Dalhousie University, 426 p.

Wightman, D.M., and Cooke, H.B.S., 1978, Post-glacial emergence in Atlantic Canada: Geoscience Canada, v. 5, p. 61–65.

Winograd, I.J., Coplen, T.B., Landwehr, J.M., Riggs, A.C., Ludwig, K.R., Szabo, B.J., Kolesar, P.T., and Revesz, K.M., 1992, Continuous 500,000-year climate record from vein calcite in Devils Hole, Nevada: Science, v. 258, p. 255–260.

Wu, P., 1993, Post-glacial rebound in a power-law medium with axial symmetry and the existence of the transition zone in relative sea-level data: Geophysical Journal International, v. 114, p. 417–432.

Wu, P., 1998, Postglacial rebound modelling with power-law rheology, in Wu, P., ed., Dynamics of the Ice Age Earth: A modern perspective: Trans Tech Publications, GeoResearch Forum, v. 3–4, p. 365–382.

Wu, P., and Peltier, W.R., 1983, Glacial isostatic adjustment and free air gravity anomaly as a constraint on deep mantle viscosity: Royal Astronomical Society Geophysical Journal, v. 74, p. 377–450.

MANUSCRIPT ACCEPTED BY THE SOCIETY JUNE 9, 2000

*Late Wisconsinan ice advances, ice extent, and glacial regimes interpreted from seismic data, sediment physical properties, and Foraminifera: Halibut Channel, Grand Banks of Newfoundland*

Ann A.L. Miller*
*Department of Geology, George Washington University, Washington, D.C., 20052, USA*
**Gordon B.J. Fader**
**Kathryn Moran**[†]
*Geological Survey of Canada-Atlantic, Bedford Institute of Oceanography, P.O. Box 1006, Dartmouth, Nova Scotia, B2Y 4A2, Canada*

## ABSTRACT

Outer Halibut Channel, central Grand Banks of Newfoundland, contains an isolated pocket of thick, glacigenic, Pleistocene sediment within a regional setting of thin, surficial sediments overlying Paleozoic or Tertiary bedrock, which has been studied using seismic reflection techniques, and borehole drilling and sampling. Nine lithostratigraphic units, containing six foraminiferal assemblages, attributed to three glacial advances, are recognized. Surficial sands and infilled channels contain a high-energy, shallow bank, *Islandiella algida* assemblage (assemblage 1). The channels cross cut two till tongues, which are interbedded with and overly glaciomarine mud. The till tongues represent the youngest advance, Late Wisconsinan (13–11 ka). The channel fill, upper till tongue, and uppermost glaciomarine sediments contain "*Glabratella*" and "*Cibicides*" species with dominant glaciomarine species *Cassidulina reniforme* or *Cribroelphidium excavatum* (assemblage 2), indicative of high-energy, shallow shelf conditions. Assemblages 3 (with increased percent occurrence of *Haynesina paucilocula*), 4 (*Stainforthia* spp.), and 5 (*Islandiella helenae*), all with dominant *C. excavatum* or *C. reniforme*; occur in sequence (deglacial assemblage sequence) in the glaciomarine sediments and the lower till tongue, indicative of the Late Wisconsinan meltwater influx. A *Stainforthia* spp. interval within the upper channel fill indicates another meltwater influx during the Younger Dryas chronozone. Underlying are a proximal glaciomarine unit, then a deformation till Middle to Late Wisconsinan (post-41 ka), both contain assemblage 6, a *C. excavatum–C. reniforme* fauna. The lowermost unit is interpreted acoustically as a lodgement or melt-out till, Middle Wisconsinan or older (pre-41 ka B.P.).

---

*Current address: marine g.e.o.s., 3824 Novalea Drive, Halifax, Nova Scotia, B3K 5K5, Canada

[†]Current address: Graduate School of Oceanography and Dept. of Ocean Engineering, University of Rhode Island, Narragansett, Rhode Island, 02882, USA

Miller, A.A.L., Fader, G.B.J., and Moran, K., 2001, Late Wisconsinan ice advances, ice extent, and glacial regimes interpreted from seismic data, sediment physical properties, and Foraminifera: Halibut Channel, Grand Banks of Newfoundland, *in* Weddle, T.K., and Retelle, M.J., eds., Deglacial History and Relative Sea-Level Changes, Northern New England and Adjacent Canada: Boulder, Colorado, Geological Society of America Special Paper 351, p. 51–107.

# INTRODUCTION

The timing, duration, and extent of Wisconsinan glaciation across Newfoundland and its continental margin (Fig. 1) is not clearly understood. A model of Laurentide ice sweeping across the entire land and marine region of eastern Canada in Late Wisconsin time, to near the edge of the continental shelf, became widespread in the 1920s (Prest, 1984) and continued until the 1980s (Denton and Hughes, 1981). Those working on reconstructions of global ice-sheet limits and volumes during the last glacial maximum, ca. 18 ka (CLIMAP Project Members, 1976; Denton and Hughes, 1981; Budd and Smith, 1987), envisioned a massive ice sheet, with ice streaming southeastward over the Maritime Provinces and through the Gulf of St. Lawrence, with a contiguous southeastern salient over Newfoundland (Brookes, 1982). However, there were those (land-based mappers) who supported the view of Newfoundland-centered ice, beginning with Bell (1884) and continuing until the work of Brookes (1982, and references therein), Grant and King (1984), Grant (1989), and Miller (1999, and references therein). Mapping and subsequent reconstructions have produced a wide variety of Late Wisconsinan proposed terminal ice positions; from north and west of the Avalon Peninsula (Jenness, 1960), to south (Grant, 1975) and east (Slatt, 1977) of Newfoundland, well across the continental shelf. There is also evidence that ice was slow to retreat throughout Atlantic Canada and there were local Late Wisconsinan resurgences (Liverman and Batterson, 1995; Stea and Mott, 1998; Stea et al., 1998). Brookes recognized the Robinson's Head readvance in southwest Newfoundland and dated it as 12.6 ka (Brookes, 1969, 1977), although the interpretation of the stratigraphy at this site is not without controversy, and another interpretation places the advance at younger than 12.6 ka (Liverman and Batterson, 1995). Grant (1969) recognized the Ten Mile Lake readvance near the northern tip of the island; the readvance has been firmly dated as 11 ka (Grant, 1992, 1994a), and is recognized throughout Newfoundland as a readvance or cold period from 12 to 11 ka (Liverman and Batterson, 1995, and references therein). Bonifay and Piper (1988) recognized a late (ca. 12–11 ka) ice surge, that they interpreted as originating from southeastern Newfoundland, extending south to the shelf edge through Halibut Channel, and to the slope in the vicinity of St. Pierre Bank (Fig. 1). In Nova Scotia there were resurgences during the Chignecto phase (13–12.5 ka) and the Collins Pond phase (10.8 ka) (Stea and Mott, 1989, 1998), both of which had ice domes centered over Prince Edward Island and the Northumberland Strait (Grant, 1994b; Stea and Mott, 1998; Stea et al., 1998) directly west of the Laurentian Channel and St. Pierre Bank. There are

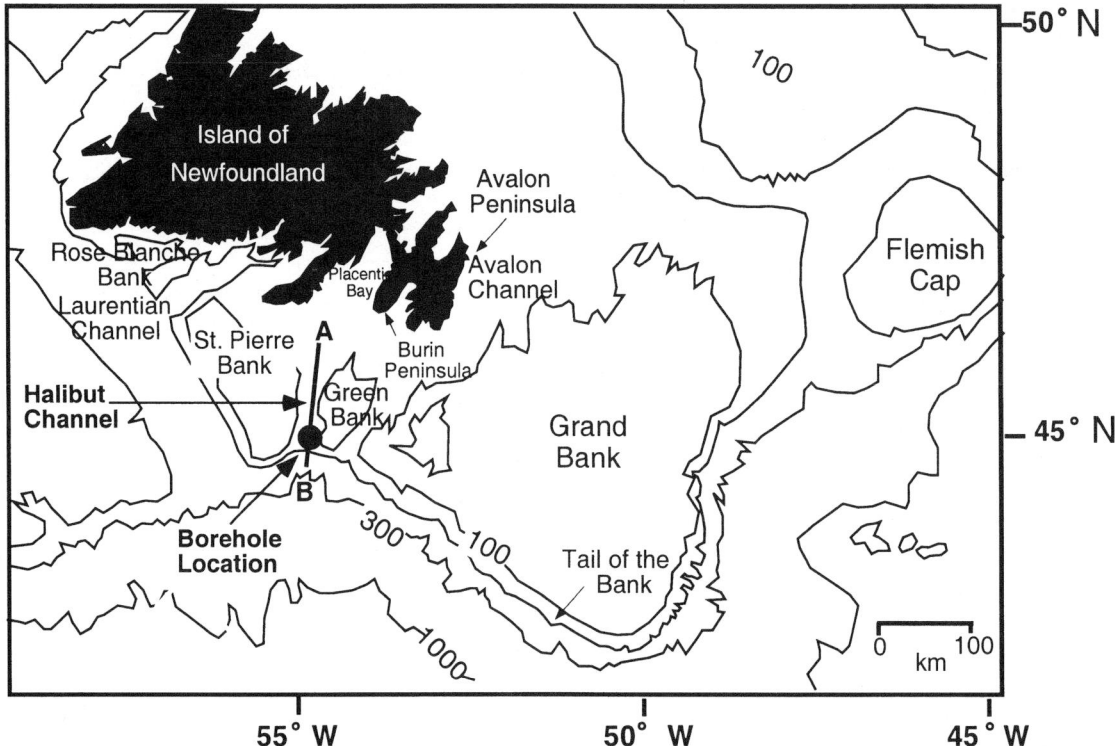

Figure 1. Index map of Halibut Channel and Grand Banks of Newfoundland, from Rose Blanche Bank in east to Grand Bank in west. Three boreholes are located within black circle. Track of interpreted composite section (Fig. 2) is shown as black line (A-B) (modified from Miller et al., 1994, 1995a; Moran and Fader, 1997).

also those who postulated numerous smaller ice domes centered on the Avalon Peninsula or offshore (Grant, 1975, 1989, 1994b; Tucker and McCann, 1980; Fader and Miller, 1986; Catto et al., 1995; Catto, 1998; Miller, 1999). Debate continues as to the relative roles of Labrador (Laurentide) versus Newfoundland ice, the areal and vertical extent of ice sheets, and the number and ages of separate advances (Grant, 1989).

The study of continental-shelf glacigenic sediments has, until recent years, been limited by the inadequacy and failure of techniques that can successfully sample coarse ice-proximal and subglacial sediments (King, 1993; Syvitski, 1993), often through thick deglacial and postglacial sequences (Syvitski et al., 1996). However, the study of glacigenic sediments has been facilitated by the development (Hutchins et al., 1976) and first application (King and Fader, 1976) of modern high-resolution acoustic equipment, allowing the collection of subbottom acoustic information. This has not been without controversy, because units exhibiting the same acoustic characteristics have been variously interpreted as lodgement till, subglacial meltout till, or ice-loaded glaciomarine sediment (Syvitski et al., 1996, 1997; Stravers and Powell, 1997; Miller, 1999). On the southeastern Canadian continental margin subglacial and ice-proximal sediments have been defined and interpreted largely on the basis of acoustic signature and stratigraphic position in seismic sections (King and Fader, 1986; King et al., 1991; Syvitski, 1991, 1993; Syvitski et al., 1996, 1997). Borehole drilling and sampling adds much-needed ground-truth (King, 1993; Syvitski, 1993), and numerous boreholes have been successfully drilled and sampled on the depressed Norwegian Shelf and in the central and northern North Sea, where active subsidence has allowed accumulation of thicker, more complete glacigenic sequences (King et al., 1987, 1991; Sejrup et al., 1987; Knudsen and Ásbjörnsdóttir, 1991) than are present on the Canadian margin, and funding has been available to finance these operations.

## Regional setting

Newfoundland's continental margin is one of the largest continental shelves in the world. It consists of the Northeast Newfoundland Shelf offshore of the northeastern coast, and the Grand Banks of Newfoundland along the southern and eastern margins, a 160 000 km² series of shallow banks extending from Rose Blanche Bank in the southwest to Grand Bank in the east. The banks are separated from one another and the Island of Newfoundland by deeper channels or enclosed basins (Fader and Miller, 1986; Fader and Piper, in Piper et al., 1990; Dalrymple et al., 1992). The inner shelf occurs as a narrow coast-parallel zone <10 km in width; the surficial sediment cover is thin to absent. The central-shelf area is to the east and northeast of the Avalon Peninsula and is a broad, fairly flat area 100–200 m in depth, developed over Paleozoic indurated lithologies (King et al., 1986). The bank areas are all part of the outer shelf; most of them are south and east of Newfoundland. These banks are referred to as the western banks (Fader et al., 1982); here the shelf break occurs at ~120 m and is located more than 200 km from land (Dalrymple et al., 1992). To the east of this archipelago of banks is the largest, Grand Bank. The banks are very flat and have typical water depths of 60–80 m, though St. Pierre Bank is only 31 m deep at its shallowest point (Fader et al., 1982) and Green Bank is 54 m deep (Dalrymple et al., 1992). Between St. Pierre and Green Banks is Halibut Channel, a shelf-crossing linear depression 10–30 km wide and 120–170 m deep that connects Placentia Bay in the north with the shelf edge to the south (Fig. 1).

During late Quaternary relative (eustatic and isostatic) sea-level changes (Quinlan and Beaumont, 1982; Fader, 1989; Liverman, 1994; Shaw and Forbes, 1995) much of the surficial sediment was reworked by shore-face and shallow-marine processes (Dalrymple et al., 1992). It was believed that sea level in the offshore regions had fallen to a $-90$ to $-110$ m lowstand ca. 15–13 ka because isostatic uplift preceded much of eustatic sea-level rise (Quinlan and Beaumont, 1982; Fader, 1989; Liverman, 1994; Shaw and Forbes, 1995) and only the deepest parts of the interbank channels and isolated depressions were below sea level. Modeling by J. Shaw (GSC-A, 2000, personal commun.) suggests that the lowstand for the outer shelf south of Newfoundland was at $\sim -45$ m at 10 ka; and, due to the interplay between eustatic and isostatic levels, and migration of the peripheral bulge, relative sea level was $-40$ to $-50$ m from 12 to 9 ka. Modern sediment input is almost negligible and is mostly biogenic material. Thick Holocene deposits occur only below the lowstand water depth, and consist of fine sands, silts, and clays that have been winnowed and transported from shallower areas. Above the lowstand depth the Holocene sediments consist of a thin, shelly, gravel lag on which are scattered mobile bedforms composed of medium to coarse sand (Dalrymple et al., 1992).

## Glacigenic features and surficial geology of outer Halibut Channel

A regional transect (Fig. 2) interpreted from seismic data sets along Halibut Channel suggests a multiple-till sequence in the southern end of the channel interbedded with glaciomarine sediments (Fader, 1986; Moran, 1987; Fader and Moran, 1988; Moran and Fader, 1997). Six tills (sensu lato, Dreimanis, 1982) occur in the complete section (Fig. 2), and represent more than 160 m of sediment thickness. The two uppermost units interpreted as till are confined only to this area; they terminate in till tongues thinning in both northerly and southerly directions, and are truncated by a regional erosional unconformity (Fig. 2). At the base of the section are two massive units that extend continuously from the nearshore (to the north) to the shelf edge (in the south). The co-occurrence of till tongues thinning in opposite directions, rooted in the same unit, suggests that in Late Wisconsinan time, offshore-centered ice was grounded on

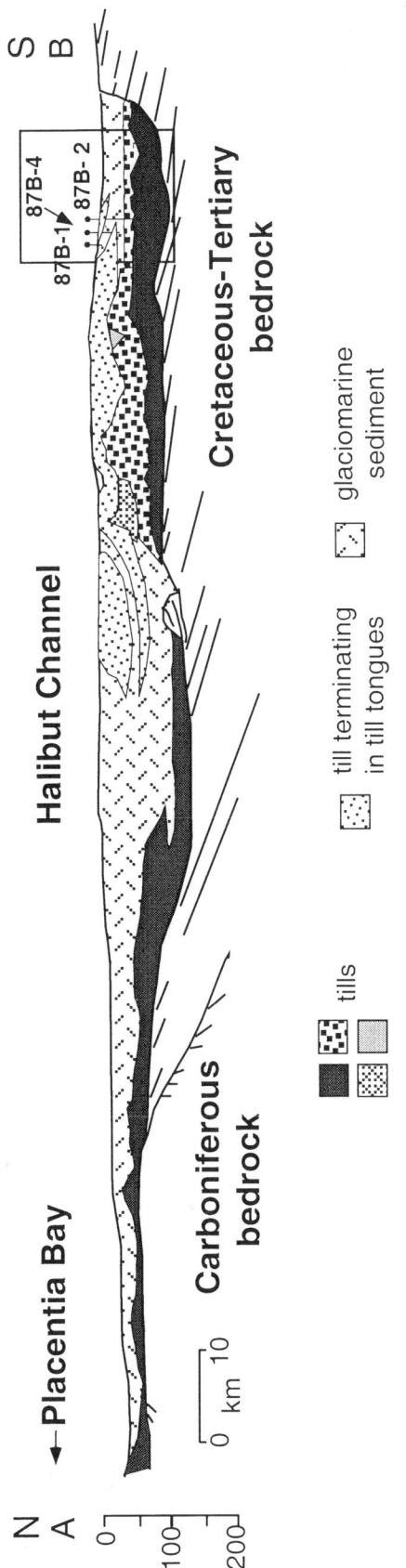

Figure 2. Regional interpretation of glacigenic sequence from Placentia Bay in north (N) to outer Halibut Channel in south (S) (line A-B in Fig. 1), based on both airgun and Huntec Deep Tow Boomer reflection profiles (Figs. 4 and 3, respectively). Till tongues were originally rooted in uppermost, massive, incoherent unit, and possibly similar unit above, both of which were truncated by regional seabed unconformity; only isolated remnants remain. Note that till tongues interdigitate to both north and south. Boxed area represents location of Huntec Deep Tow Boomer reflection profile illustrated in Figure 3 (modified from Fader and Moran, 1988; Miller et al., 1994, 1995a; Moran and Fader, 1997; Miller, 1999).

the outer banks and separated from Newfoundland-centered ice either by floating ice or open water (Fader and Moran, 1988; Miller et al., 1994, 1995a; Moran and Fader, 1997; Miller, 1999).

Till tongues are wedge-shaped deposits of sediment (acoustically, a uniform dense pattern of incoherent reflections), and are interpreted as discrete stratigraphic units deposited near the margins of marine ice sheets (King and Fader, 1986). Most till tongues are lateral extensions of massive till deposits and are rooted in large iceberg-furrowed constructional linear moraines hundreds of kilometers in length (King and Fader, 1986; King et al., 1991). They interdigitate with stratified ice-proximal glaciomarine deposits. The morphology and distribution of till tongues varies widely, and they appear as simple wedges or as complex, stacked successions (King et al., 1991). King and Fader (1986) and King et al. (1987) proposed that till tongues represent the migration paths of the grounding (buoyancy lines) of glaciers, where till is deposited from the meltout of debris immediately proximal to the lift-off point of an ice sheet (i.e., waterlain till). In the original model, King and Fader (1986) inferred that the ice was in contact, or at least closely coupled, with the seabed and that the ice terminated as a floating ice shelf. The contact between the till tongue and glaciomarine sediment was believed to be the result of successive local advances and retreats of the grounding line. Till-tongue roots were believed to be strictly depositional features; however, King et al. (1991) recognized that some of the truncations may be erosional in origin; ice-proximal and glaciomarine sediments deposited in front of the grounding line during retreat may be eroded during the next advance. Consequently, till tongues may not be strictly subglacial or entirely depositional. King et al. (1991) modified the interpretation of till tongues to include undermelt tills (Dreimanis and Lundqvist, 1984; Gravenor et al., 1984). Till tongues may also form through the successive accumulation of local and small-scale sediment gravity-flow deposits (flow tills) near the grounding line, stretching the till tongue downslope and forming a till apron. They appear to be very similar to the reworked grounding-line sediments of Orheim and Elderhøi (1981), the subaquatic flow till of Vorren et al. (1983), the till delta of Alley et al. (1989), the grounding-line fan of Vorren et al. (1989), either the frontal moraine or grounding-line fan of Syvitski (1991), and the diamict apron of Hambrey et al. (1992). The King and Fader (1986) interpretation and subsequent modifications (King et al., 1991) are not without controversy (Syvitski, 1991; Stravers and Powell, 1997). Syvitski (1991, p. 928) suggested that till tongues occur close to basin margins, and are associated with downslope (seaward) moraines. The till tongues found in Halibut Channel do not support Syvitski's (1991) model. Regardless of their differences, all models relate formation of these wedge-shaped constructional features to grounded ice-front positions.

Another major erosional event is recorded by channel downcutting of the till tongues and underlying sediments in Halibut Channel, observed on air-gun and high-resolution seismic reflection profiles (Figs. 3 and 4). These channels are infilled by sediments with continuous, conformable, coherent reflections (Fader and Miller, 1986) interpreted as glaciomarine and reworked glaciomarine (postglacial) deposits. Three types of channels (based on size and geometry) have been observed in the near surface on the Grand Banks (Miller, 1999); those seen on the seismic profiles in Halibut Channel (Figs. 3 and 4) are the smallest (type 3). They were referred to by Wingfield (1990) as minor incisions with open-ended geometry. King (1993, his Fig. 5) illustrates channels on this scale on the Scotian Shelf. Type 2 channels (broad with gently sloping sides) have also been observed on the Scotian Shelf (Boyd et al., 1988; Scott et al., 1989; King, 1993).

Utilizing acoustic data coupled with seabed surface and near-surface samples, a formal lithostratigraphy has been established for Halibut Channel. The following descriptions are compiled from King (1980), King and Fader (1986), Fader and Miller (1986), and Dalrymple et al. (1992). Three units, the Grand Banks Drift, the Downing Silt, and the Adolphus Sand are present.

The Grand Banks Drift is very dark grayish-brown, cohesive, poorly sorted diamict containing angular gravel to boulders, and may acoustically exhibit stratification. It conformably overlies the bedrock surface. It is interpreted as an ice-contact deposit (a primary basal till) and may also include stratified drift (subglacial to ice-marginal waterlain till), secondary till, and gravity-flow till. It is confined to the deeper basins, channels, and saddles between offshore banks. Acoustically it appears on the Huntec DTS (Deep Tow System) profiles as a uniform dense gray pattern of incoherent reflections, with scattered point-source reflections. Where the diamict is thick (i.e., >35 m) and penetration with the DTS is limited, small air gun seismic-reflection systems were used to define the character.

The Downing Silt overlies and is interbedded with the Grand Banks Drift. It is a dark grayish-brown, poorly sorted clayey and sandy silt, with some angular gravel, interpreted as a glaciomarine deposit. Acoustically it can show well-developed to weak stratification, and it exhibits high- to low-amplitude continuous coherent reflections. In some places it is highly conformable with the substrate; in others it shows some degree of ponding.

The Adolphus Sand is a sublittoral deposit of fine-grained sand and muddy sand (grading locally to a gravelly sand) that occurs in the peripheral areas of the banks (>100 m water depth), in the interbank channels, and on the upper continental slope, below the sea-level lowstand. Acoustically it is semitransparent to transparent, and in places appears faintly laminated. It is similar to the Downing Silt in that it can also be highly conformable with the substrate, or show some degree of ponding.

## Objectives

The objectives of this study were to continuously drill and sample the thick glacigenic sequence in Halibut Channel, representing the first time such an undertaking had been attempted on the eastern Canadian continental margin. Regional (low resolution) and high-resolution seismic-reflection data were collected (Fader, 1986; Moran, 1987), and three drilling sites were chosen on the basis of these records (Moran, 1987). The lithology, sediment physical properties, and foraminiferal content of the sediments were analyzed in order to understand more about the depositional mechanisms and environments of these units, particularly the tills and the till tongues. Accelerator mass spectrometer (AMS) $^{14}$C dates were obtained from mollusc shells and foraminifera to provide chronological control. Based on the interpreted environments of deposition, coupled with chronological control, a depositional history has been constructed, and insights into the timing, extent, and glacial regimes of Late Wisconsinan ice south of Newfoundland postulated.

## METHODS

### Geophysical and geotechnical methods

Three extensive seismic data sets were collected across the study area (Fader, 1986). Single channel 10 in$^3$ and 40 in$^3$ air gun records were collected utilizing a Bolt Associates model 600 marine-profiler airgun (Fader et al., 1982; Syvitski et al., 1996). High-resolution deep-towed seismic-reflection system (Huntec DTS) (Hutchins et al., 1976) profiles were also collected. This system (a boomer with a narrow pulse) allows resolution of acoustic events to a precision of 0.3 m and penetration of as much as 200 m in surficial material (Piper, in Piper et al., 1990; Syvitski et al., 1996; Josenhans, 1997). It utilizes two hydrophone systems, a single internal hydrophone mounted within the boomer fish (that gives a 0.5–1 m resolution, but with generally less penetration) and a six-element external hydrophone array towed aft of the fish that allows deeper penetration with lower resolution (3–5 ms; 2–4 m) (Milliman et al., 1990). Additional seismic profiles and borehole samples were collected on a cruise of the M/V *Balder Challenger* (Moran, 1987). Boreholes were drilled with a modified Failing 1500S drill rig with associated drill pipe and sampling tools; the sampling procedure was fully described in Moran (1987). General core-analysis procedures, from shipboard handling, labeling and storage, through splitting, describing, photographing, X-ray radiographing, subsampling, and preparation for long-term storage, were fully described in Mudie et al. (1984). Some pertinent detailed procedures can be found in Moran (1987). Physical properties measured include acoustic compressional wave (p-wave) velocity, undrained shear strength, and index properties (bulk density, water content, and porosity) (Moran, 1987). One-dimensional consolidation tests were also conducted on a few samples. These mechanical and physical properties are related to the type of sediment, the sedimentation rate, and the consolidation history of the sediment. The Dalhousie University–Geological Survey of Canada—Atlantic (GSC-A) velocimeter was used to measure acoustic compressional wave velocity on the split working half of the core. Measurements were

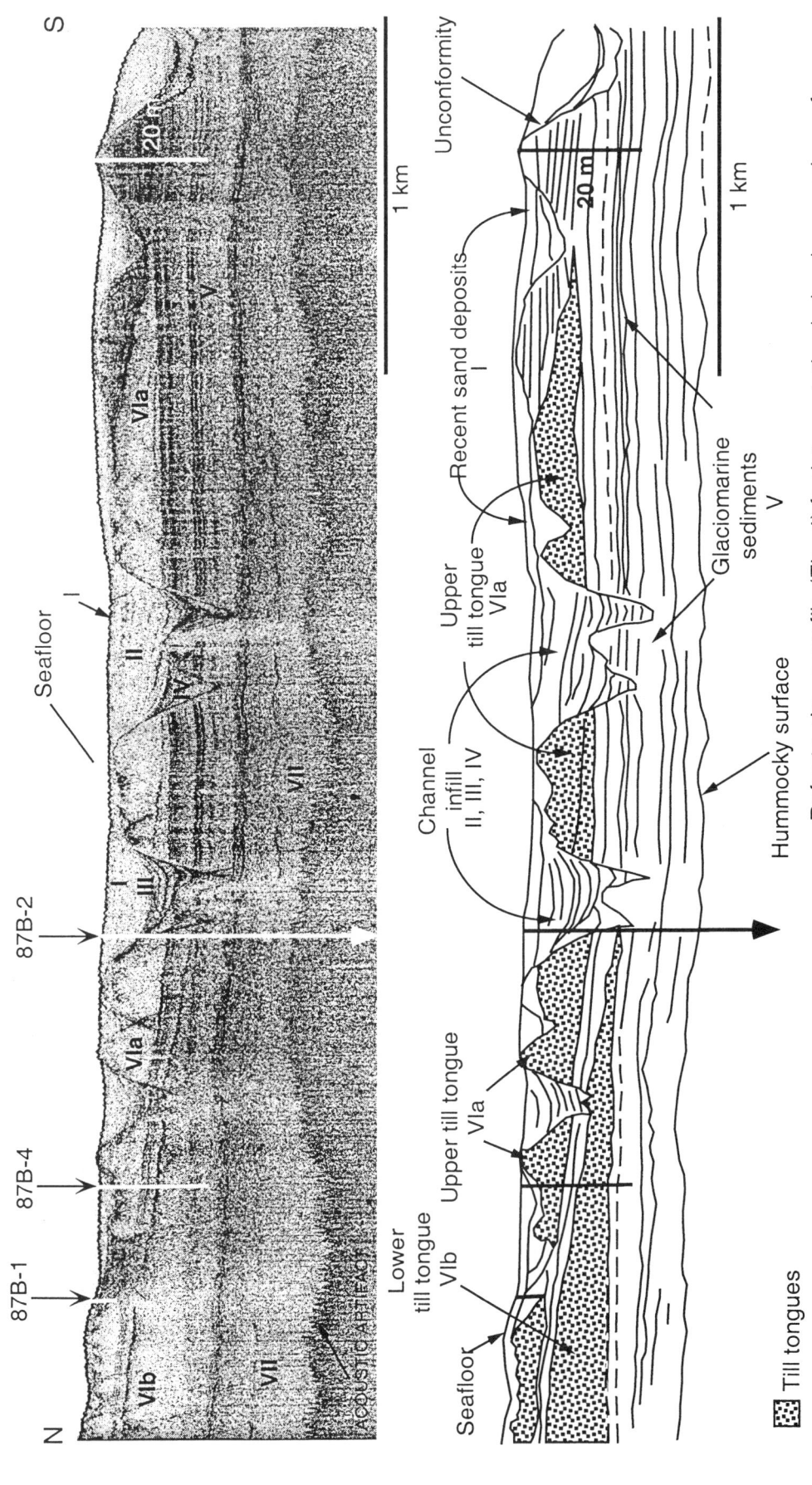

Figure 3. Huntec Deep Tow seismic reflection profile and line-drawing interpretation through borehole locations. Huntec profile resolves only upper 32 m, and 87B-2 penetrates to 76 m. Note infilled channels. Upper till tongue extends across section, pinching out to south, and is downcut in places by channels. Depths are based on water velocity (1500 m/s). (Seismic reflection profile modified from Miller et al., 1994, 1995a; Moran and Fader, 1997; Miller, 1999. Line drawing interpretation modified from Moran and Fader, 1997; Miller, 1999.)

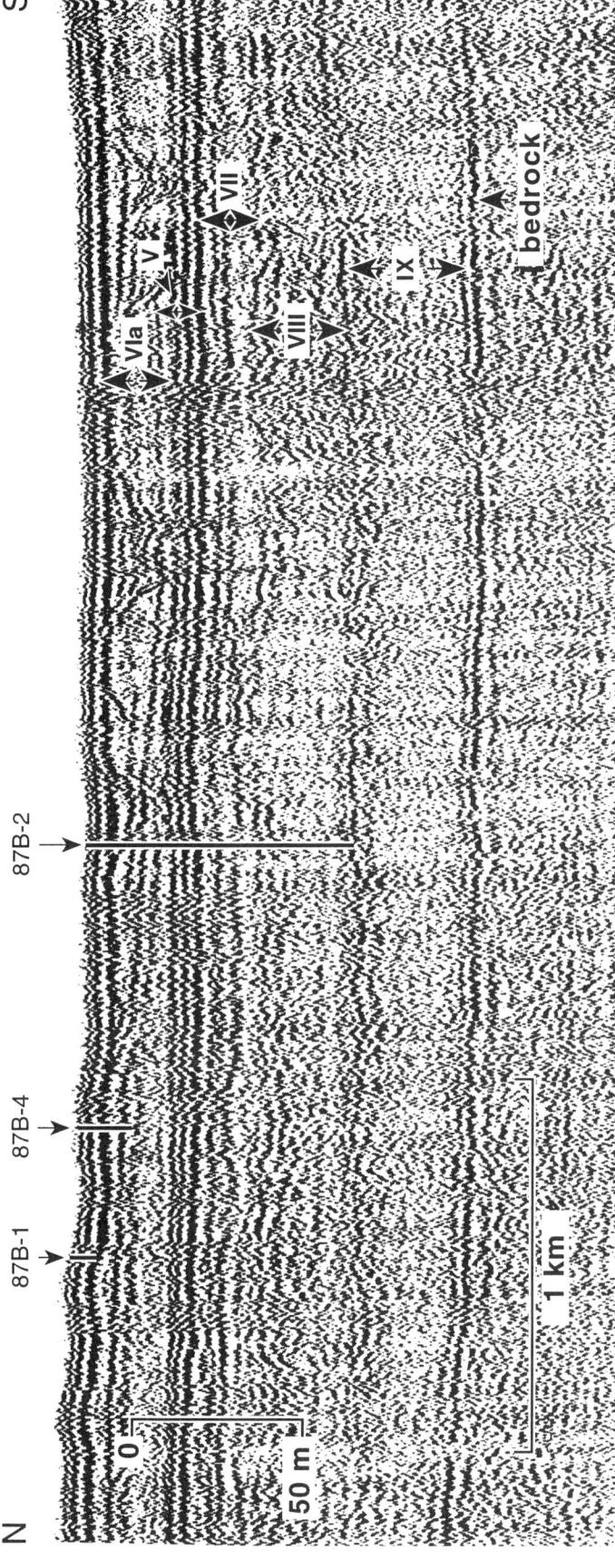

Figure 4. Low-resolution 40 in³ airgun profile, extending from north (left) to south (right), through outer Halibut Channel (modified from Miller, 1999).

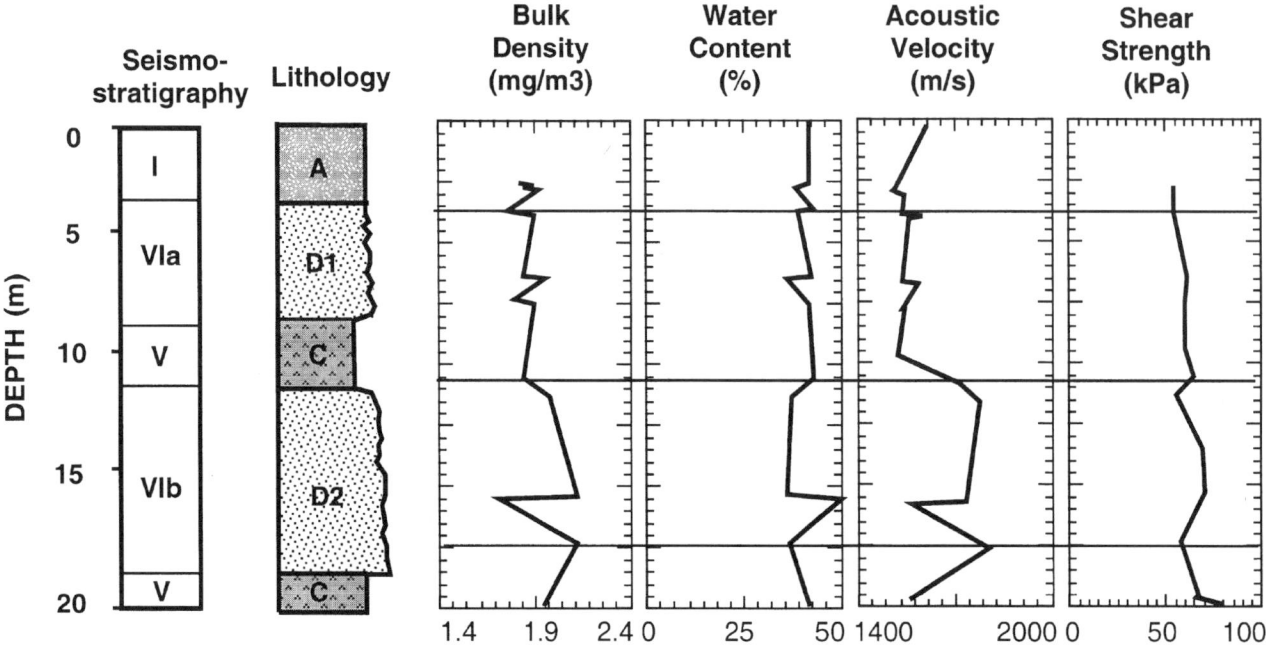

Figure 5. Lithostratigraphy (lithology and seismostratigraphic units) and physical properties of borehole 87B-4. Roman numerals in column on far left represent seimostratigraphic units, and letters in second column represent lithologic units (both described in text). Note relatively low and gradually increasing shear-strength values throughout borehole, suggesting normal consolidation in all units. Water content, bulk density, and velocity data are consistent with shear-strength values (adapted from Moran, 1987).

made at intervals of 10–20 cm for all samples with the exception of the samples that were predominantly sand (including all of 87B-1), where the samples would partially drain during testing. Due to the scattering of acoustic energy, measurements could not be made in some of the gravelly clay sequences. A longitudinal measurement was made at each depth interval. Undrained shear strength was measured at an interval of one test per core section using the GSC-A motorized miniature vane device. Subsamples for bulk density, water content, and porosity were taken from the split core at the same intervals as the velocity measurements.

### Foraminiferal Analyses

***Sample processing.*** On site samples of ~35 cm$^3$ (bulk sediment) were placed in plastic vials and covered with a solution of seawater and $CaCl_2$ buffer and stored until processing. Basic methods of sample processing were outlined in Murray (1973). All samples were washed with 500 and 63 μm stainless steel sieves, the 63–500 μm fraction retained, and dried overnight at 30 °C. A full discussion about the choice of lower sieve size (63 μm) can be found in Miller (1999). When dried, the foraminifera in sandy samples were then concentrated by adding the sample to a 10:4 solution of bromoform and acetone (Gibson and Walker, 1967), for flotation. The separation took place in about 1 min, after which the float was washed into filter paper, rinsed with acetone, and dried.

***Identification and analyses of foraminifera.*** A full discussion of the systematic paleontology was given in Miller (1999). The taxonomy of the indicator and diagnostic species in these boreholes is given in the Appendix. The generic classification is based on Loeblich and Tappan (1988); exceptions were given in Miller (1999). Note that quotation marks are used to delineate, under one genus name, closely related species previously known (Loeblich and Tappan, 1964) by only one or two genus names, and now assigned to many new genera (Loeblich and Tappan, 1988). Extensive literature surveys, and the study of types and illustrations, indicate that there are many species names for forms from various geographic locations that appear to be conspecific. Consequently, there has been extensive synonymization, and some specific names used here are not the most commonly cited names (e.g., Miller et al., 1996).

At least four components of a foraminiferal fauna may be present: Quaternary indigenous and introduced, Quatery in situ reworked, Quaternary transported reworked, and reworked (R) Tertiary-Cretaceous (K-T) forms (Miller, 1999). The extinct pre-Quaternary species present are well-known Cretaceous species and identified as such (i.e., species of *Heterohelix, Gavelinella, Praebulimina*, and *Pyramidina*). Problems arise when there is the potential for reworked species, still extant today, to be Tertiary in age. The Tertiary-Quaternary transported reworked, and Tertiary in situ and transported reworked are usually environmental erratics and are not consistent environmentally with the overall (interpreted as indigenous) Quaternary benthonic (QB) fauna. The composition of the Quaternary (Q) fauna, the size of the K-T component, and the state of preservation of the specimen were all considered when deciding which component of the fauna to assign a specimen (Q indigenous, Q in situ reworked, Q transported, or K-T fauna). The only genera potentially represented in both the Quaternary and reworked fauna were rare specimens of *Gyroidina-Gyroidinoides*. Small pristine specimens identified as *Gyroidinoides quinqueloba* or *Gyroidinoides nipponicus* are considered Quaternary, and all other *Gyroidinas-Gyroidinoides* are considered reworked.

The absolute number of specimens examined generally varies between 200 and 1000 per sample. However, most workers count ~300 specimens, based on studies of Phleger (1960) and Buzas (1990). The binomial standard error is appropriate for estimating confidence limits for a single sediment sample, and approaches cluster confidence limits when species proportions are small (i.e., p = 0.01). When p < 0.1, confidence limits are meaningless, and the presence of rare species should simply be recorded (Buzas, 1990).

Samples containing abundant foraminifera were dry split with a microsplitter. Between 300 and 800 QB (Q indigenous, Q in situ reworked, and Q transported) specimens (when present) and the accompanying Quaternary planktonic (QP) and R (K-T) foraminifera were subsequently identified and counted. If a counted split contained less than the appropriate numbers of Q forms, but abundant R foraminifera, additional splits were examined counting only the Q (B and P) specimens. Data for QB foraminifera were quantified as relative species abundance, or percent occurrence. Data for the QP and R foraminifera are presented as absolute abundances per unit volume.

We analyzed 67 samples (7 from 87B-1, 13 from 87B-4, and 47 samples from 87B-2). For each sample, the number, depth, size of the split, total number (TN) of foraminifera in each category present (QB, QP, and R), and abundance data are listed in data tables (Appendix, and Appendix C of Miller, 1999). A schematic core log, with core lithology, the absolute number of QB and R foraminifera, and the relative species abundances of the dominant and diagnostic QB species groups are plotted for each borehole. Samples are curated at GSC—Atlantic, Bedford Institute of Oceanography, Dartmouth. (Key slides are curated with A.A.L. Miller, marine g.e.o.s., Halifax, Nova Scotia.)

## RESULTS

### Seismostratigraphy

The upper 32 m of the regional seismic section (Fig. 2) are interpreted from high-resolution Huntec DTS data (Fig. 3). Here the section mainly consists of medium- to high-amplitude, continuous, coherent reflections and is interpreted and subdivided into six seismostratigraphic units. At the top of the section, unit I is characterized by weak reflections made by the surficial sands. The next three units are found within the channels above the erosional unconformity formed by channel downcutting. Unit II shows an interval of weak to moderate, discontinuous, incoherent reflections, and has acoustic characteristics similar to the overlying unit I. A narrow interval forms the third unit (unit III) and is represented by a series of stacked high-amplitude reflection events that form the boundary zone of the channel. Unit IV exhibits high-amplitude, coherent, acoustically stratified reflections and is at the channel base. Units III and IV show acoustic characteristics similar to those of the underlying units, and have a different acoustic signature than the upper channel-fill sediments (unit II).

Stratigraphically below the unconformity, unit V is acoustically stratified, and exhibits three high-amplitude reflection groupings interbedded with medium-amplitude reflections. The two till tongues (unit VI) are seen on the north side of the section. The shallowest till tongue is identified seismostratigraphically as unit VIa, and the deeper is identified as unit VIb, separated by unit V. The till tongues exhibit high-amplitude reflections and a distinct geometry. A regional undulating, high-amplitude reflection, observed as a hummocky surface (Fig. 3), marks the top boundary of an underlying unit (unit VII) that extends past the penetration of the Huntec DTS system.

Below 32 m, the seismostratigraphy is interpreted using lower resolution 40 in$^3$ airgun data (Moran, 1987), and is subdivided into three seismostratigraphic units (Fig. 4). All three units are characterized by incoherent reflections, separated by continuous, coherent, reflections. Unit VII is 16 m thick, with an undulatory upper surface and relief of ~2 m. Unit VIII is 27 m thick and its upper surface is highly undulatory with relief of 6–8 m. Unit IX is 34 m thick, and the upper surface of this unit is flat with a well-defined high-amplitude coherent reflection that decreases in intensity to the north, and then disappears. All three of these units were initially interpreted as till on the basis of their acoustic signatures and relationships to each other (Moran and Fader, 1997). At the base of the section, a bedrock unit is defined by a broad regional unconformity on seaward dipping, coherent, high-intensity reflections.

### Lithology

Boreholes were positioned to sample the major seismostratigraphic intervals in a water depth of 167 m, at the seaward end of the two till tongues (Figs. 2, 3, and 4). Borehole 87B-1 drilled through 11 m of postglacial sediment only and is the highest in the stratigraphic section. This hole became unstable during drilling and had to be abandoned. Sampling was limited to the upper 6.5 m (Moran, 1987). Borehole 87B-4 was drilled to 20 m and sampled to that depth. It penetrates the two upper till tongues. The deepest borehole (87B-2) was positioned to penetrate the lower part of the stratigraphic section, including one of the infilled channels. It was drilled to 76 m and sampled to 75 m.

The lithology of 87B-1 is as follows. From 0 to 4 m, there is laminated, dark olive-gray fine sand with occasional coarse sand and shell debris interlamination (unit A). Isolated clasts occur throughout. At 4 m, there is a gravel layer. From 4 to 11 m, the lithology is well-sorted shelly sand (unit B1). At 11 m, a gravel-shell hash was encountered, loosely packed, with an abundance of lithic fragments (unit B2). No sampling was done below 6.5 m. Physical properties were not measured on these coarse, sandy sediments.

Boring 87B-4 sampled four lithologies. From 0 to 3 m, the lithology is fine to medium, loose sand (unit A). From 3 to 8 m, the sediment is characterized as a massive, stiff, red clay (unit D1). From 8 to 11 m the sediment is silty clay and is slightly bioturbated (unit C2). From 11 to 18 m gravel clasts are present in the massive clay (unit D2). At 18–20 m, gravel is absent (unit C2). Undrained shear strengths ranging from 20 to 50 kPa (Fig. 5) suggest normal consolidation throughout (Moran, 1987). An increase in acoustic velocity and coarsening differentiates unit D from unit C.

Boring 87B-2 intersected and sampled most of the units identified seismostratigraphically. This borehole drilled through eight lithologic units and sampled seven (Fig. 6). At the surface (0–3 m), a layer of loose, well-sorted, dark olive-gray fine sand (unit A) was recovered. From 3 to 14 m the borehole sampled an upper, unsorted, silty sand deposit from 3 to 11 m (unit B1), and a similar lower deposit, with an increase in gravel, from 12 to 14 m (unit C1) separated by a shell-hash layer at 11 m (unit B2). Undrained shear strengths in these units are very low, 10–40 kPa (Fig. 6) (Moran, 1987). There is a peak in the percent water content in the upper part of unit B (Moran, 1987). From 14 to 32 m, the sediments are a highly bioturbated, clayey silt, with occasional mud clasts and lithic clasts (unit C2). Undrained shear strengths range from 10 to 90 kPa, indicating normal consolidation (Moran, 1987). From 32 to 38 m, the sediment is characterized as a massive, brick-red, silty clay, with increasing sand laminae downsection (unit E). The physical properties (shear strength, 20–90 kPa, and bulk density, 1.85–2 mg/m$^3$) suggest normal consolidation (Moran, 1987). The acoustic velocity is low, consistent with the other physical properties. Changes in the acoustic velocity at the upper and lower bounds of this unit result in the seismic reflections that define this interval. The sediment from 38 to 48 m is a gray silty clay, lacking any sedimentary structures, that has similar physical properties to the overlying unit, except for a slightly higher acoustic velocity (unit F). This unit is normally consolidated. The borehole bottoms out in a unit from 48 to 75 m (unit G), a dark gray clay with frequent clasts, interdigitated with well-sorted, very fine sand at the top, laminated in sections. The amount of gravel and/or dropstones increases with depth. This is an overconsolidated deposit, with high shear strength (100–200 kPa) and high acoustic velocity (1650–2100 m/s) (Fig. 6). Samples taken from this unit and tested result in consolidation ratios of 2, confirming the interpretation of overconsolidation (Moran, 1987; Moran and Fader, 1997). Changes in the acoustic velocity (to 1950 m/s) at 75 m result in a seismic reflector defining the boundary between this unit and the underlying (unsampled) one (unit H).

*Lithostratigraphy*

A lithostratigraphy has been developed for the borehole sequences by combining the lithology and physical properties with the seismostratigraphy (e.g., A/I), and placing these units within the context of the local stratigraphic framework (Fader and Miller, 1986). The lithostratigraphic sequence at each borehole is shown graphically and schematically in Figures 7 (87B-1), 5 (87B-4), and 6 (87B-2).

Boring 87B-1 (Fig. 7) contains the surficial sands (unit A/I), the upper channel fill (B1/II), and the shell hash (B2/III), all of which are included in the Adolphus Sand. It occurs below the depth of the low sea-level stand and was not subject to transgressive processes.

At the surface of 87B-4 (Figs. 5 and 8) is the Adolphus Sand (unit A/I), underlain by the uppermost till tongue (D1/V1a) of Grand Banks Drift (ice-contact to ice-proximal deposit); the two are separated by an erosional unconformity. The next unit is Downing Silt (C2/V), a glaciomarine sediment that separates the two till tongues, followed by the lower till tongue (D2/VIb) of Grand Banks Drift. At the base of the borehole is Downing Silt (C2/V).

Boring 87B-2 penetrates the following sequence (Figs. 6 and 9). The $^{14}$C ages have been obtained at seven levels (Table 1). The upper 12 m are Adolphus Sand, consists of surficial sand (unit A/I), and part of the channel complex, the upper channel fill (B1/II), and shell hash (B2/III). The lower channel fill (C1/IV) and the underlying sediments downcut by the channel are Downing Silt (C2/VI). Seismostratigraphic unit VII can be subdivided into two lithostratigraphic units: unit E, a red unit, and a gray unit, unit F. Both units are interpreted as Downing Silt. Unit E contains more coherent reflections than unit F. They are separated by an erosional unconformity with 2 m of hummocky relief. Underlying unit F is a muddy diamict, unit G/VIII, which acoustically appears to be a till (Grand Banks Drift). It is separated from unit F by an erosional unconformity with 6–8 m of relief. The borehole bottomed out at the upper surface of the underlying third unit (unit H/IX), also Grand Banks Drift, a subglacial deposit with high shear strength and very high acoustic velocity (Fig. 6) (Moran, 1987; Moran and Fader, 1997).

*Foraminiferal analyses*

***Delineated foraminiferal assemblages and their environmental interpretation.*** The occurrences of some living benthonic foraminiferal species are known and reported in the lit-

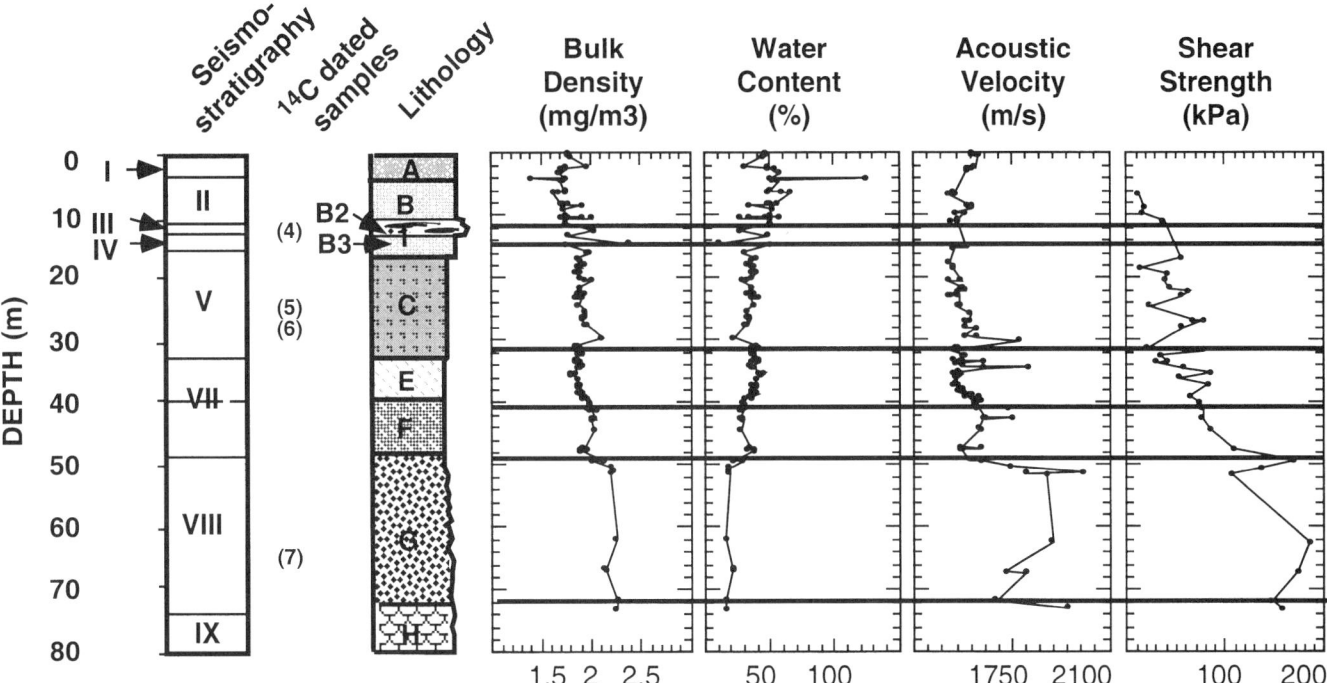

Figure 6. Lithostratigraphy (lithology and seismostratigraphic units, as in Fig. 5) and physical properties of borehole 87B-2. Stratigraphic positions of $^{14}$C dated samples are marked by numbers in brackets. $^{14}$C dates are listed in Table 1. Note large increase in acoustic velocity, shear strength, and bulk density at boundary between unloaded glaciomarine sediments (E/VII and F/VII) and ice-loaded material (G/VIII). Note change in acoustic velocity at base of borehole, indicating upper surface of basal till (H/IX) (adapted from Moran, 1987; Miller et al., 1994, 1995a; modified from Moran and Fader, 1997).

erature to be related to specific environmental conditions (Phleger, 1960; Murray, 1973, 1991; Haynes, 1981). Species known to be related to a specific, or a combination of specific parameters, are known as diagnostic or key species. Benthonic foraminiferal assemblages have been delineated in this study based on the co-occurrences of key species and specific ranges of percent occurrence for each key species (explained fully in Miller, 1999). Six benthonic foraminiferal assemblages have been recognized. Some of these assemblages are closely related to one another, in terms of species composition and inferred environment of deposition. With the hope of defining these assemblages and their environments of deposition clearly and succinctly, closely related assemblages will be discussed as a group. There are three assemblage groups: they are summarized in Table 2.

The aim here is to define each assemblage and discuss its environment of deposition independently, or in terms of the other assemblages within its specific assemblage group. With this information on the inferred environmental parameters of each assemblage, the complete downcore sequences of foraminiferal assemblages for each of the boreholes is presented and discussed. The nature of the substrate is a constituent of the environment of deposition; therefore, there is often a strong correlation between foraminiferal assemblages and their host sediments.

*Assemblage Group 1: Bank, calcareous assemblages: assemblages 1 and 2.* The assemblage group 1 faunas occur in the surface and near-surface sediments. The sediments are the Adolphus Sand, including the upper part of the channel fill. Assemblage 1 is characterized by the presence of *Islandiella algida* (Cushman) (Miller et al., 1996). In assemblage 1 *I. algida* occurs as 5%–30% of the assemblage and it is accompanied by species known to inhabit modern shelf environments (i.e., *Angulogerina angulosa*, *Cibicides* spp., *Cribroelphidium* spp., *Discorbis-Rosalina* spp., *Fissurina* spp., *Nonionella* spp.); it occurs in the uppermost (near surface) samples in 87B-1 and 87B-4; in 87B-4 at 3–8 m, and in 87B-2 at 0–2 m.

Sen Gupta and McMullen (1969) and Sen Gupta (1971) found *I. algida* to be the most common species on the sandy areas of the Tail of the Bank. Williamson et al. (1984) reported a monospecific *I. algida* assemblage prevalent on isolated outer banks and isolated depressions on the Scotian Shelf, where it replaces a *Cibicides* assemblage. Williamson et al. (1984) attributed this replacement to conditions of higher salinity and an increase in the gravel content. In Halibut Channel the *I. algida* assemblage is interpreted as modern. The lack of a glacial

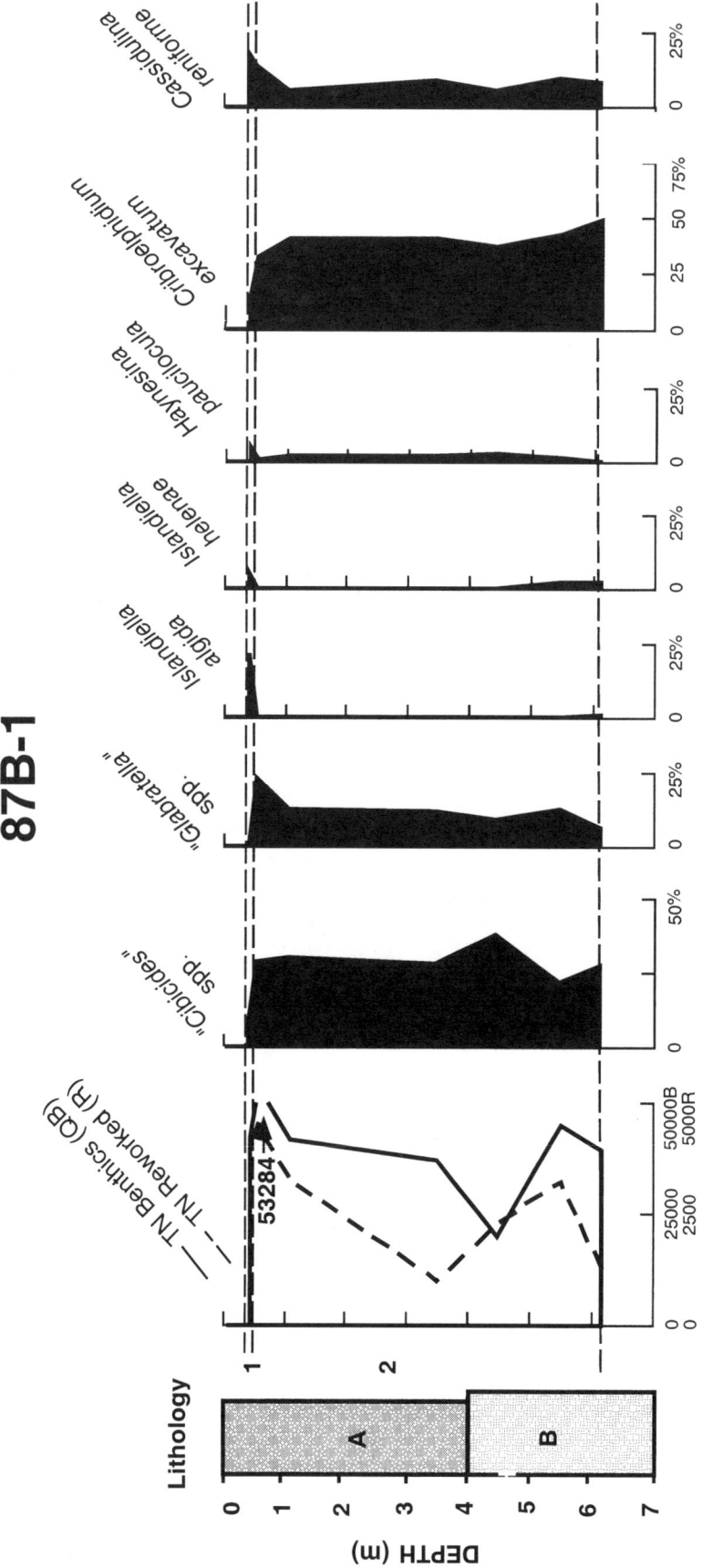

Figure 7. Lithostratigraphy and foraminiferal data and assemblage sequence for borehole 87B-1, based on seven foraminiferal samples. Absolute abundances (total numbers, TN) of Quaternary benthonic (QB), and reworked (R) Cretaceous-Tertiary (K-T) foraminifera. QB includes all Quaternary forms: forms interpreted as Quaternary indigenous, Quaternary in situ reworked, and Quaternary transported reworked. Percent occurrence data are given for key QB species—those species interpreted as diagnostic or indicative of specific environmental conditions, as used to define assemblages. Delineated foraminiferal assemblages (defined in text and summarized in Table 2) are numbers immediately to left of TN column, and assemblage boundaries are marked by solid black lines across plots (adapted from Miller, 1996; modified from Miller, 1999).

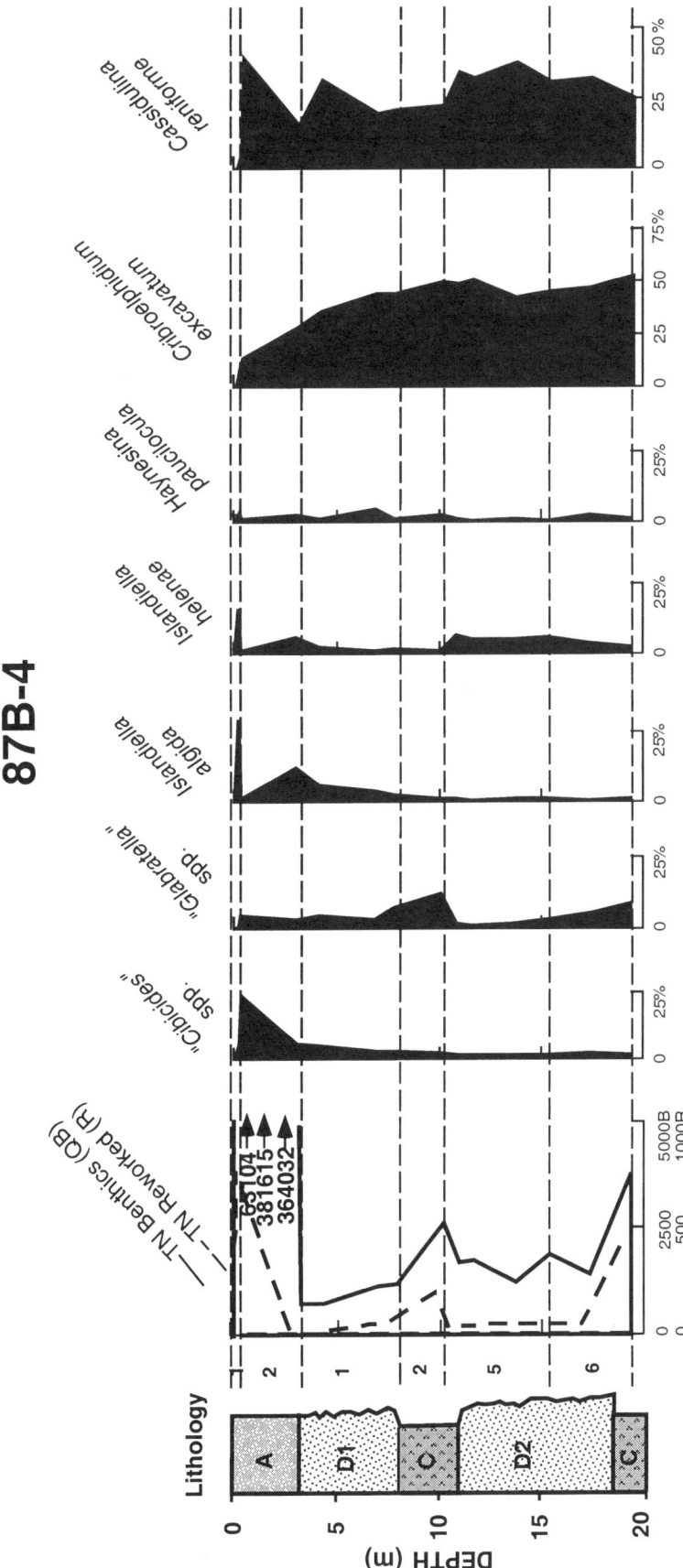

Figure 8. Lithostratigraphy (as in Fig. 5) and foraminiferal data and assemblage sequence (as for Fig. 7) for borehole 87B-4, based on 13 samples (adapted from Miller et al., 1994, 1995a; Miller, 1996; modified from Miller, 1999). Abbreviations as in Figure 7.

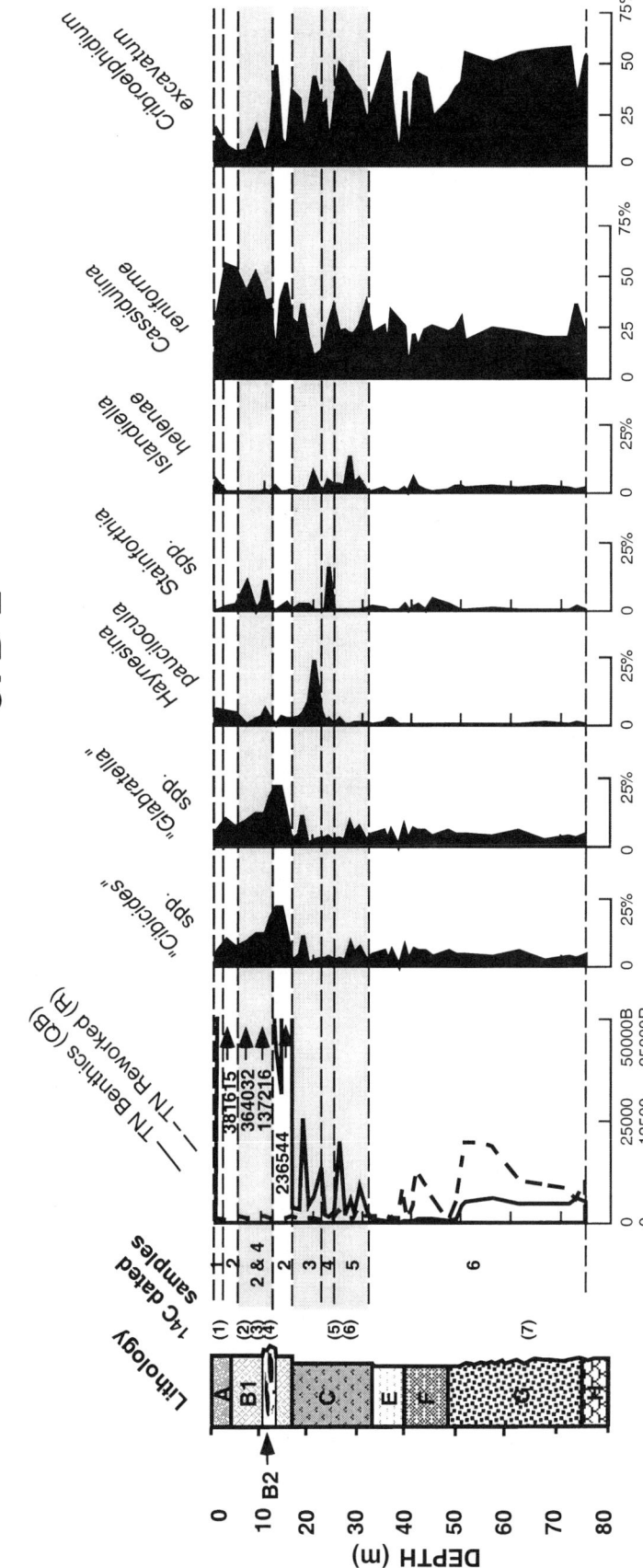

Figure 9. Lithostratigraphy (as in Fig. 6) and foraminiferal data and assemblage sequence (as for Fig. 7) for borehole 87B-2, based on 46 samples. Reduced-salinity foraminiferal assemblage in channel-fill sediments (assemblage 4 within assemblage 2) and deglacial foraminiferal assemblage sequence in underlying glaciomarine sediments (assemblages 3, 4, and 5) are highlighted in gray. Stratigraphic position of $^{14}$C dated samples are marked by numbers in brackets. $^{14}$C dates are listed in Table 1 (adapted from Miller et al., 1994, 1995a; Miller, 1996; modified from Miller, 1999). Abbreviations as in Figure 7.

TABLE 1. ¹⁴C DATED SAMPLES: BOREHOLE 87B-2

| Sample number* | Borehole depth (m) | ¹⁴C Age† (yr B.P.) | Lithostratigraphic unit | Material | Laboratory number | |
|---|---|---|---|---|---|---|
| | | | | | Beta/Isotrace | ETH number |
| 1 | 0.21 | 2735 ± 75 | Surficial sands (A) | Mixed foraminifera predominantly *Globobulimina auriculata* *Nonionellina labradorica* *Cribroelphidium bartletti* | 28274 | 4724 |
| 2 | 7.95 | 6150 ± 80 | Channel fill (B1) | Mixed foraminifera# and shell fragments§ | 28275 | 4725 |
| 3 | 10.21 | 8020 ± 90 | Channel fill (B1) | *Yoldia hyperborea* (complete) | 28276 | 4726 |
| 4 | 12.00 | 10780 ± 95 | Channel fill (B2) | *Macoma calcarea* (2 single valves) | 28277 | 4727 |
| 5 | 26.01 | 14870 ± 120 | Downing Silt (C) | Shell fragments§ | 28278 | 4728 |
| 6 | 29.56 | 19120 ± 140 | Downing Silt (C) | *Macoma* sp. fragments | 28279 | 4729 |
| 7 | 66.61 | 41080 ± 1020 | Grand Banks Drift (G) | Shell fragments§ | 28280 | 4730 |

*These numbers show sample depths on Figure 9.
†No corrections have been applied to these dates.
#Foraminifera are not differentiated.
§Shell fragments too small to be identified.

faunal component suggests that there has been some Holocene sediment input now covering the glaciomarine sediments. The other cold-water shelf species present may be indigenous or relict.

Assemblage 2 contains diagnostic species of the "*Cibicides*" species group and/or "*Glabratella*" species group. This assemblage occurs in 87B-1 at 1–7 m, in 87B-4 at 1–3 m and 8–10 m, and in 87B-2 at 2–18 m. It is delineated by 5%–25% "*Glabratella*" spp. and 2%–45% "*Cibicides*" spp. However, *Cassidulina reniforme–Cribroelphidium excavatum* are the two major components of the fauna (the environmental implications of these two species are discussed under assemblage 6). The occurrences of *C. excavatum* and *C. reniforme* fall within 2 different ranges, either *C. excavatum* is dominant (30%–50%) with 5.5%–20% *C. reniforme*, or *C. reniforme* is dominant (30%–55%) with *C. excavatum* a minor component of this fauna (<15%, with one exception).

"*Cibicides*" spp. includes primarily *Cibicides io*, *Cibicides reflugens*, and *Lobatula lobatula*; other *Cibicides*-related species encountered in all the material studied include rare occurrences of *Cibicicoides* sp. B, *Cibicidoides corpulentus*, *C. floridanus*, *C. pseudoungerianus*, *Heterolepa subhaidingerii*, and *Planulina vilksae*. Sen Gupta and McMullen (1969) found *L. lobatula* (sensu lato, which includes *C. io*, *C. reflugens*, and *L. lobatula*) at 90% of the stations on the Tail of the Bank with a mean percent occurrence of 12.5%. Williamson et al. (1984) found a *Cibicides lobatulus* (sensu lato) assemblage patchily distributed on the outer banks of the Scotian Shelf, including Banquereau. Substrate type (coarse sands, bedrock, and boulders), and high hydraulic energy (strong bottom currents) appear to be the controlling distribution factors (Miller, 1999, and glacial-marine references therein). In the modern glaciomarine environment *C. lobatulus* (= *C. io* and *L. lobatula*) has been found in ice-proximal settings in Baffin Island and Greenland fjords (Schafer and Cole, 1986; Jennings and Helgadóttir, 1994; Hald and Korsun, 1997) and at ice-distal locations in Spitsbergen (Nagy, 1965). The increase in the percent occurrence of "*Cibicides*" spp. is taken to indicate a period of shallower conditions with stronger currents.

"*Glabratellina*" spp. includes predominantly *G. lauriei*, *G. wrightii*, and *Rotaliella chasteri*, with rare *G. arcuata* and *Trichohyalus kingi*. Murray (1973) described *Glabratella* (sensu lato) as typical of the normal marine inner shelf. Schafer and Cole (1978) and F.E. Cole (1984, oral commun.) have found it dominant in assemblages from the north shore of the Bay of Chaleur (eastern Canada). F.E. Cole (1984, oral commun.) suggested that this is a shallow intertidal to subtidal (shallow shelf) species that prefers a very sandy-gravelly substrate and very turbulent water. This is a genus often attached to the substrate, and it seems to replace *L. lobatula* (sensu lato) in some environments, particularly when there is no plant life (e.g., seaweed and algae) (F.E. Cole, 1984, oral commun.). Murray (1991) attributed these species to a high-energy environment. Østby and Nagy (1982) described a *Cibicides-Rosalina* (includes "*Glabratella*") assemblage where these two genera are diagnostic, although not dominant. They illustrated specimens of *Rosalina* and *Glabratella*, which include the species referred to here as *Glabretellina lauriei*, *G. wrightii*, *Rosalina globularis*, and *Rotaliella chasteri*, and they described the environment as that of an open bank. Culver et al. (1996) reported *Glabratella* sp. as diagnostic of intermediate and deep tidal-channel environments. The occurrence of *Rotaliella chasteri* deserves further mention (see Appendix, and Miller, 1999). The only previous report of this, or related species, along the eastern Canadian continental margin was by Cole (1981). This genus has never been reported from middle to outer shelf environments prior to this study. This may be due to the fact that most specimens are in the 32–63 μm size range and are lost in the sample processing and sieving process (Pawlowski, 1991).

All of diagnostic species in assemblage group 1 (assemblages 1 and 2, *Islandiella algida*, *Cibicides io*, *C. reflugens*, *Glabratellina lauriei*, *G. wrightii*, and *R. chasteri*) are indicative of bank and/or channel environments, coarse substrates, and high hydraulic energy. Their occurrence is interpreted as

**TABLE 2. SUMMARY DIAGRAM OF THE DELINEATED FORAMINIFERAL ASSEMBLAGES AND THEIR ENVIRONMENTAL INTERPRETATION AND SIGNIFICANCE**

| Group and assemblage | | Dominant and diagnostic species | Interpretation and significance |
|---|---|---|---|
| 1 | 1 | *Islandiella algida* (5%–30%) | Calcareous bank faunas<br>Sandy/gravelly substrate; higher salinity counterpart to *Cibicides* spp. |
| | 2 | *Cibicides* spp. (2%–45%)<br>*Glabratellina* spp. (5%–25%)<br>*Cassidulina reniforme* (5%–20%)<br>*Cribroelphidium excavatum* (30%–50%)<br>OR<br>*Cassidulina reniforme* (30%–55%)<br>*Cribroelphidium excavatum* (<15%) | Coarse sands/gravel/boulder/bedrock substrate high hydraulic energy |
| 2 | 3 | *Haynesina pauciloculla* (6.5%–24%) | Meltwater deglacial sequence<br>Shallow water (<50 m) paleosalinity indicator (<32‰) channelized meltwater flow? |
| | 4 | *Stainforthia* spp. (1%–16%) | Indicator species, deep estuarine environments salinities <32‰<br>tolerates low dissolved oxygen levels, <2 ml/l |
| | 5 | *Islandiella helenae* (1%–13%) | Paleosalinity indicator (32.0‰–33.5‰) intermediate water depths (70–140 m) early meltwater indicator |
| 3 | 6 | *Cribroelphidium excavatum* (30%–60%)<br>*Cassidulina reniforme* (20%–30%)<br>reworked Q fauna<br>reworked K/T fauna | Glaciomarine fauna<br>Ice-marginal to ice-proximal environments<br>shallow water salinities <33‰ no mixing of oceanic and meltwaters high turbidity/Suspended particulate matter |

being present in postglacial to modern environments, although it is impossible to determine if many of the accompanying cold-shelf and/or glaciomarine species are relict, indigenous, or both. The absence, in the assemblage, of species known to occur in both glacial and nonglacial environments, indicates postglacial sediment accumulation; however, the presence of these species does not necessarily indicate a relict component.

*Assemblage Group 2: The meltwater–deglacial–early post-glacial sequence, calcareous assemblages: assemblages 3–5.* Assemblage 3 is delineated by the presence (4%–24%) of *Haynesina pauciloculla*. The dominant to codominant species are *C. excavatum* and "*Cibicides*" spp. This assemblage occurs in 87B-2 at 18–23 m. There is also a small increase in its percent occurrence, to 6.5%, in the 10–11 m interval.

*H. pauciloculla* was originally described by Cushman (1944, p. 24, *Nonion pauciloculum*) from shallow coastal waters around New England. *Haynesina pauciloculla* includes junior subjective synonyms *Elphidium subarcticum* (Cushman, 1944, p. 27) from the same location, and *Nonion pauciloculum* subspecies *albiumbilicatum* Weiss (1954) from the interglacial Gardiner's Clay in New York (Appendix, and Miller, 1999). Rottgardt (1952), Haake (1962), and Lutze (1965, 1974) reported the species as common in modern boreal shallow-water environments, and that it tolerates very low salinities and large fluctuations in temperature. Austin and Sejrup (1994) also reported it as common today on the western Norwegian Shelf, in water depths >39 m, temperatures ~3 °C and salinities ~32‰.

Other evidence suggests that *H. pauciloculla* represents significant meltwater influxes (Weiss, 1954; Mangerud et al., 1981; Knudsen, 1982; Bergsten and Dennegård, 1988; Guilbault, 1989, 1993), and it has been found in infilled meltwater channels (Knudsen, 1978, 1993; Jensen and Knudsen, 1988; Knudsen and Sejrup, 1993) where it is probably indicative of channelized flow. Knudsen (1978, 1993) found a *C. albiumbilicatum* zone in deposits of channel fill in buried tunnel valleys in northwest Germany. Mangerud et al. (1981) found a *C. albiumbilicatum* zone in gravel deposits (abundance ~45%), correlated it to the end of the Eemian interglacial, and estimated the paleowater depth as 0–10 m. Similar occurrences in late glacial sediments have been reported (Miller, 1999, and references therein).

Assemblage 4 occurs twice in 87B-2; the more pronounced interval is from 23 to 24 m, and is marked by an occurrence to 16% of *Stainforthia* spp., *Stainforthia fusiformis*, *S. rotundata*, *S. concava*, and *S. pauciloculata*. *C. reniforme* has an occurrence of 25%, and *C. excavatum* drops to 5.5%, although its occurrence is ~30% above and below this interval. *I. helenae* is also present (4.5%). There are also two smaller increases in the percent occurrence of *Stainforthia* spp. in the upper channel-fill sediments, at 6–7 m (10.5%) and 10–11 m (11.5%), occurring within the "*Cibicides*" spp. and "*Glabratella*" spp. assemblage 2 interval.

There has been much confusion and inconsistency in the literature over the identification of these species, in particular the recognition of *Stainforthia fusiformis* (sensu stricto) (see Appendix, and Miller, 1999). *S. fusiformis* (sensu lato) is well known in modern environments. Recent studies relate the common (modern) occurrence of *S. fusiformis* to anoxic conditions initiated by human-induced environmental alterations (urban and/or organic pollution) (Alve, 1994, 1995; Alve and Murray,

1995). It can tolerate salinities of ~30‰ (Alve, 1990), and low dissolved oxygen levels (<2 ml/l), even surviving brief periods of anoxia (Alve, 1994). It is also found in modern glacial-marine environments (Nagy, 1965; Elverhøi et al., 1980). It is considered an indicator species in deep estuarine environments of eastern Canada (Schafer and Cole, 1978; Scott et al., 1980; Miller et al., 1982a), where salinities are generally 30‰–33‰. *S. fusiformis* is common in surface assemblages on the Labrador Shelf (Vilks et al., 1984) and in surface sediments in the Canadian Arctic (MacLean et al., 1989), where it directly overlies the *C. excavatum* fauna. It is dominant (Murray, 1985, 1992) in surface assemblages in the North Sea. *S. fusiformis* was reported as very abundant in the Skagerrak (Alve and Murray, 1995), and common at water depths of 40–145 m in the southern Kattegat (Conradsen, 1993; Conradsen et al., 1994). It is associated with muddy sediments of high organic content and oxygen depletion (Sturrock and Murray, 1981; Conradsen, 1993; Conradsen et al., 1994). Hald et al. (1994) looked at the distribution of *S. fusiformis* in the Skagerrak and correlated it to salinities >30‰ and a very fine sand substrate.

*Fursenkoina (=Stainforthia) fusiformis* has also been found in muddy, near-surface sediments of late Quaternary age on the New Jersey Shelf (Lagoe et al., 1997). The high-resolution seismic stratigraphy and sedimentologic and foraminiferal analyses of selected short piston and Vibrocores from this area (Davies and Austin, 1997; Lagoe et al., 1997) suggest that this muddy unit is analogous to both the New Jersey "Mud Patch" (Knebel et al., 1979; Twichell et al., 1981; Milliman et al., 1990; Ashley et al., 1991) and the Tail of the Bank Mud (Miller, 1999). Lagoe et al. (1997) may have included other species of *Stainforthia* within their concept of *F. fusiformis*. Lagoe et al. (1997) believed that the occurrence of *F. fusiformis* is depth and substrate controlled, rather than related to any deglacial conditions; and that because *F. fusiformis* prefers a mud and clay substrate, this species must have been deposited adjacent to the slope, in middle shelf to outer shelf water depths.

*Stainforthia* spp. often is a minor to codominant component in the *Islandiella helenae* extreme distal glaciomarine and/or postglacial assemblage (assemblage 5; see Vilks and Rashid, 1976; Scott et al., 1984). Miller et al. (1995b) found its highest percent occurrence peak, in Emerald Basin, Scotian Shelf, within the *I. helenae* zone there, following the maximum occurrence of *I. helenae*. This same trend is observed in Halibut Channel and in the Tail of the Bank Mud (Miller, 1999). Coincident with the occurrence of *Stainforthia* spp. there is a marked decrease in the occurrence of *C. excavatum* in Halibut Channel, the Tail of the Bank Mud, and Emerald Basin, Scotian Shelf (Scott et al., 1984; Miller et al., 1995b; Miller, 1999). *Stainforthia* spp. has been found in late glacial sediments with variable percent occurrence frequencies and variable species associations; consequently, its presence has been interpreted as indicating ice shelf (Scott et al., 1984), proximal glaciomarine (Osterman, 1984; Osterman and Andrews, 1983; Osterman and Nelson, 1989), distal glaciomarine (Vilks and Rashid, 1976; Vilks et al., 1989), or extreme distal glaciomarine and/or post glacial (Miller et al., 1995b) environments.

Assemblage 5 is marked by the occurrence (1%–13%) of *Islandiella helenae* accompanied by *C. reniforme* and *C. excavatum*. It is present in 87B-4 at 10–16 m, and in 87B-2 at 24–31 m.

*I. helenae* is found living today in deep outer estuaries of eastern Canada (Schafer and Cole, 1978; Scott et al., 1980), the Gulf of St. Lawrence (Rodrigues and Hooper, 1982), and on the Labrador Shelf in salinities of 32.5‰–33.5‰ (Vilks et al., 1982, 1989; Vilks and Deonarine, 1988). In the Gulf of St. Lawrence it is found in the intermediate water depths of ~70–140 m (Rodrigues and Hooper, 1982). It is also a minor component in modern glaciomarine environments (Jennings and Helgadóttir, 1994; Korsun et al., 1995).

*I. helenae* is considered a marker species (Vilks, 1980, 1981; Vilks et al., 1982, 1984) and paleosalinity indicator (32‰–33.5‰) indicating extreme distal glaciomarine and/or early postglacial conditions and an influx of meltwater on eastern North American margins (Miller, 1999, and references therein). It has seldom been reported from similar environments on the eastern North Atlantic margin (for exceptions see Mangerud et al., 1981; Scourse et al., 1990).

After careful evaluation of the reported occurrences in the literature, coupled with the observations made here, it is proposed that *Stainforthia* spp. when in sequence with *Islandiella helenae* represents the acme of a meltwater plume, particularly in deeper water, finer grained sediments with high organic content, and in areas of nonchannelized flow (Miller, 1998a, 1998b, 1999). The counterpart of this *Stainforthia* spp. assemblage in coarser grained substrates lacking a high organic content, and where channelized flow is strongest and water depths shallower, is *H. paucilocula* (assemblage 3).

All three of these assemblages indicate that there was a marked decrease in salinity, attributed to the presence of substantial quantities of meltwater, during the time interval this assemblage sequence was present. Based on modern analogues, *I. helenae* (assemblage 5) indicates that the water depth was probably in the range of 70–140 m and salinities ~32‰–33.5‰, and meltwater and oceanic waters were well mixed. The *H. paucilocula* (assemblage 3) and the *Stainforthia* spp. interval (assemblage 4) are believed to indicate a decrease in salinity (to 32‰ or less). *Stainforthia* spp. may also indicate decrease in dissolved oxygen content. In deeper water, finer grained sediments, this may have been triggered by zones of upwelling (common in the distal glaciomarine environment on the outer continental shelf; Boulton, 1990), sustaining high rates of biological productivity and rich benthonic and planktonic faunas, which in turn deplete the dissolved oxygen supply. *H. paucilocula* indicates a period of shallower water and strongest channelized flow.

*Assemblage Group 3: The glacial calcareous Cassidulina reniforme and Cribroelphidium excavatum assemblage: assemblage 6.* Both *C. excavatum*– and *C. reniforme*–dominated faunas are present in modern ice-marginal to ice-proximal glacial regimes, and the occurrences of each need to be clearly distinguished. The crucial factor may be the mixing of meltwater and oceanic waters, and the amount of suspended particulate matter present. Assemblage 6 is characterized by 30%–60% *C. excavatum* and 20%–30% *C. reniforme. Cribroelphidium excavatum* has been reported as common to dominant in the inner fjord ice-marginal to ice-proximal glacial regime, in very shallow water, in salinities <33‰ and where there is no mixing of meltwater and oceanic waters (Nagy, 1965; Hansen and Knudsen, 1992; Hald et al., 1994; Jennings and Helgadóttir, 1994; Korsun et al., 1995; Hald and Korsen, 1997). It often occurs with *H. orbiculare* and *Cibicides* spp. Miller et al. (1982b) compiled reported modern occurrences of *E. excavatum* (= *C. excavatum*), including *C. excavatum* forma *clavata* (Cushman), and stated that it occurs today as a minor constituent in high diversity arctic-shelf faunas. Hald et al. (1994) have also looked at the modern, Holocene, and late Quaternary distribution of *C. excavatum* forma *clavatum* (sic). They concluded that three oceanographic parameters directly influence its occurrence: fluctuating salinity and slightly hyposaline conditions, turbidity/suspended particulate matter, and seasonal ice cover. They also believe that it is a meltwater indicator; its percent occurrence in assemblages increases at a point that stratigraphically correlates to the first meltwater pulse from the Barents and Fennoscandian ice sheets (Jones and Keigwin, 1988; Sarnthein et al., 1992). Long et al. (1986), Hald and Vorren (1987), Sejrup et al. (1987), and Haynes et al. (1995) suggested that a dominance of *C. excavatum* indicates meltwater events. Seidenkrantz and Knudsen (1993) suggested that the morphology of *C. excavatum* may be a salinity (and meltwater) indicator. Miller et al. (1995b) and Hall et al. (1998) suggested that *C. excavatum*–dominated assemblages can represent ice-distal (*C. excavatum* percent occurrence 25%–60%), ice-proximal (60%–88%), and possibly ice-marginal and/or floating glacier-tongue front (88%–94%) environments in isolated basins with high sedimentation rates (i.e., Emerald Basin), where the environment of deposition would be analogous to the modern environments where *C. excavatum* is currently found.

*C. reniforme* is reported in modern ice-marginal to ice-proximal glacial regimes, accompanied by an agglutinated fauna (Elverhøi et al., 1980; Schafer and Cole, 1986; Hald and Korsun, 1997) in environments where there is deeper water, high suspended particulate matter, salinities >33‰ and mixing of meltwater and oceanic waters. *C. reniforme* may be replaced ice distally by *L. lobatula*, particularly if the substrate becomes coarser, water shallower, current strength increases, and SPM decreases (Nagy, 1965; Schafer and Cole, 1986; Hald et al., 1994; Jennings and Helgadóttir, 1994; Haynes et al., 1995; Hald and Korsun, 1997).

Assemblage 6 occurs in 87B-4 at 16–20 m and in 87B-2 at 31–75 m, and contains large numbers of reworked (R) Cretaceous-Tertiary (K-T) specimens present with the Quaternary component, and also has scattered occurrences of Quaternary deeper water species, i.e., *Pseudoparella zhengae, Epistominella nipponica, Eponides pusillus, Ioanella tumidula, Oridosalis umbonatus, Tosaia hanzawaia*, and species of "*Uvigerina.*" These have all been recognized as outer shelf and slope species (Cole, 1981; Poag, 1981; Williamson et al., 1984; Scott, 1987) and are believed to be environmental erratics (Miller, 1999), extant species, but environmentally out of place due to sediment transport and reworking. The pre-Quaternary forms and the environmental erratics are interpreted as having been incorporated in the ice mass from the K-T bedrock (some of it is distinctively red colored) and/or older Quaternary sediments exposed in Placentia Bay, on the bank areas and in inner Halibut Channel (Fader et al., 1982; King et al., 1986), and transported east-west or south as the ice advanced.

Assemblages dominated by *C. excavatum* and *C. reniforme* are ubiquitous in near-surface and subsurface glacigenic (Late Wisconsinan) sediments throughout the eastern Canadian margin (Miller, 1999, and references therein). Foraminiferal faunas dominated by *C. excavatum* have also been found in Late Weichselian and older glacigenic sediments along the Scandinavian margin and in the North Sea, and in the sediments of the U.K. continental shelves (Miller, 1999, and references therein).

***Foraminiferal assemblage sequences and their significance.*** The uppermost part of the sequence is generally characterized by group 1 faunas, indicative of postglacial and modern bank environments with high hydraulic energy. The surficial sediments (unit A) in all three boreholes contain a fauna characterized by *Islandiella algida* (assemblage 1). This indicates modern bank conditions and some modern sediment accumulation in outer Halibut Channel. The presence of relict *C. excavatum* and *C. reniforme* in assemblage 1 indicates the substantial role of reworking in the formation of the modern surficial Adolphus Sand.

Assemblage 2 is present in the subsurface Adolphus Sand (unit A), throughout the channel-fill sediments (units B1, B2, and C1), in the uppermost part of the glaciomarine sequence (unit C2), and in the upper till tongue (unit D1); the assemblage is characterized by "*Cibicides*" spp. and "*Glabratella*" spp., indicating shallower water and stronger bank conditions than are present today. "*Cibicides*" spp. and "*Glabratella*" spp. are consistent with strong bottom currents and a coarse sand or gravel substrate. These species are generally well preserved, which indicates that they probably have not been subjected to severe reworking, although the occurrence of these species in the lower channel fill may be wholly or in part the result of reworking of the upper till tongue and underlying glaciomarine sediments. The high occurrences of *Cribroelphidium excavatum* and *Cassidulina reniforme* in assemblage 2 are interpreted as both indigenous and relict (penecontemporaneous, reworked in situ and transported) and are present in the surface sediments due to strong currents and low sediment accumulation rates. The dominant species in assemblage 2 alternates between *C. excavatum* and *C. reniforme*, and is believed to be a local reworking effect. Williamson et al. (1984) observed *C. excavatum* in some areas on the Scotian Shelf, and concluded, as a result of living total foraminiferal distribution studies, that the presence of *C. excavatum* on the Scotian Shelf was largely relict, and not in equilibrium with modern hydrographic conditions. Sen Gupta and McMullen (1969) found a very high diversity fauna, with *C. excavatum* present in the total population only, at 80% of the stations, to a percent occurrence of 59%, on the Tail of the Bank. The presence of this relict component confirms that the Adolphus Sand is a product of reworking, and that glaciomarine sediment served as the source. In addition to assemblage 2 within the upper channel-fill sediments (unit B1) are two short intervals of assemblage 4, with increases in the percent occurrence of *Stainforthia* spp. In the middle section of the Downing Silt (unit C2) there is a sequence of group 2 assemblages (3, 4, and 5), with characteristic *Haynesina paucilocula*, *Stainforthia* spp., and *Islandiella helenae*. In the lower till tongue, there is also an increase in the occurrence of *Islandiella helenae*. All three of these assemblages are indicators of reduced salinity conditions. Assemblage 3 and all occurrences of assemblage 4 appear as thin intervals, each interpreted as representing a brief time period and/or event within the units. The uppermost occurrences of *Stainforthia* spp. are believed to indicate a discrete period of meltwater influx, or increased freshwater runoff during channel infilling. The complete assemblage sequence, lower in the section, indicates a progressive decrease in salinity (upsection) during the interval represented by these sediments. Based on modern analogues, *I. helenae* (assemblage 5) indicates that the water depth was probably in the range of 70–140 m and salinities were ~32‰–33.5‰, and meltwater and oceanic waters were well mixed. The *Stainforthia* spp. interval (assemblage 4) is believed to indicate a decrease in salinity (<32‰), possibly coupled with a decrease in dissolved oxygen content. This may have been triggered by zones of upwelling (Boulton, 1990). The *H. paucilocula* interval (assemblage 3) is believed to indicate the time of strongest local meltwater influence: shallower water depth, salinity minimum, and minimum mixing of meltwater and oceanic waters due to the strongest channelized flow, which in turn may have scoured out the overlying channels. The presence of *I. helenae* in the lower till tongue is interpreted as being largely reworked from the underlying glaciomarine sediments.

The lower till tongue (unit D2) and the overlying glaciomarine sediment (unit C2), the remainder of glaciomarine units E and F, and the upper unit of Grand Banks Drift, unit G, contain assemblage 6. This assemblage is dominated by *C. excavatum* with *C. reniforme* subdominant, and is also distinguished by the the presence of reworked pre-Quaternary foraminifera. The pre-Quaternary forms and the foraminiferal environmental erratics are interpreted as having been glacially incorporated (Amos and Miller, 1990) in the ice mass from the Cretaceous and Tertiary bedrock units exposed in Placentia Bay, on the bank areas, and in inner Halibut Channel (Fader et al., 1982; King et al., 1986). The low numbers of Quaternary benthonic specimens and the high numbers of reworked specimens in assemblage 6 support an ice-proximal glaciomarine environment for these deposits. The dominance of *C. excavatum* in assemblage 6 indicates low salinities, high suspended particulate matter, and an absence of meltwater and oceanic water mixing.

The lowermost lithostratigraphic unit, unit G, was originally deposited as a glaciomarine sediment (Downing Silt), as indicated by the presence of assemblage 6. The geotechnical properties show that this unit is overconsolidated compared to the overlying units (Moran, 1987). The acoustic velocity and shear strength increase across the unit boundary and remain high (Fig. 6). Subsequent to deposition of this unit, and prior to deposition of the overlying units, this sediment was overconsolidated, possibly by ice loading by a subsequent ice advance. An estimate of the minimum ice thickness, based on shear strength, is 430 m (Moran, 1987). However, the glaciomarine sediments and foraminiferal faunas are intact, and there is no lithologic evidence in the sampled sediments, other than consolidation, that they have been glaciotectonically deformed.

## INTERPRETATION AND DISCUSSION

### Variations in glacial regimes and associated depositional mechanisms

The sequence in Halibut Channel is interpreted as containing sediments indicative of three glacial advances, each producing and preserving different facies attributable to variations in ice-sheet dynamics, glacial regime, and sea level.

Overconsolidation is the main criteria utilized here for distinguishing the action of grounded ice (Casagrande, 1936; Moran, 1987; Moran and Fader, 1997); the resultant deposit is either a basal till (if the overconsolidation occurred during primary deposition), or a deformation (secondary) till (INQUA commission, Dreimanis, 1989) if it was subsequently ice loaded and now exhibits lithologic characteristics and physical properties similar to those of a primary basal till (Boulton, 1996). High shear strengths, high acoustic velocity, and incoherent acoustic character recorded for the two basal units in 87B-2 (Fig. 6) are the main criteria employed here in making the interpretation that unit H/IX is a basal till, and unit G/VIII is a glaciomarine sediment that was subsequently ice loaded. The distinct paleontological, lithological, geotechnical, and acoustic differences of these units from the overlying deposits imply different glacial regimes and different source areas.

The basal unit (unit H/IX) acoustically identified as till directly overlies bedrock and is probably the product of bedrock erosion. If this is a lodgement till (Chamberlain, 1895), both net freezing and net melting have occurred, freezing incorporating debris into the ice mass (marginally) and melting releasing it (from the basal core). Alternatively, this could be a melt-out till (Boulton, 1970), formed during deglaciation.

By virtue of its physical properties and acoustic characteristics, the overlying unit (unit G/VIII) is interpreted as having been overridden by ice. The unit was deposited as a glaciomarine sediment, containing glaciomarine foraminiferal assemblage 6, and then remoulded into a deformation till (Elson, 1957; Boulton, 1987, 1990, 1996). In a glaciomarine section this thick (27 m) evidence of an ice-distal to ice-proximal (or vice versa) transition in the foraminiferal faunas is expected (i.e., a systematic variation upsection in the percent occurrence of *C. excavatum* and *C. reniforme*), but not observed here. In similar scenarios Hald et al. (1990, 1991) and Austin and McCarroll (1992) found that remoulding the sediments can remix the faunas to the extent that any evidence of a proximal to distal, or distal to proximal, environmental transition has been destroyed.

Discrete events from different source areas may be responsible for the distinct lithologies of, and distinct lithologic boundary between, units E/VII and F/VII. The hummocky 2 m of relief of the surface separating the two units (E/VII and F/VII) may indicate a period of iceberg groundings and/or turbation. The source area of the red sediments (E/VII) is different from the underlying material (F/VIII) and is probably the same source as that of the red sediments seen in the lower till tongue (unit D1/VIa). Brick-red sediments have been attributed by others to meltwater plumes carrying sediment eroded from Carboniferous bedrock exposed at the seafloor in the Gulf of St. Lawrence (Conolly et al., 1967; Loring and Nota, 1973; Bond and Lotti, 1995); Carboniferous bedrock exposed in Placentia Bay and south of the Burin Peninsula (Fader et al., 1982; King et al., 1986) may be the source of these red sediments in Halibut Channel.

Three possible depositional mechanisms are suggested for normally consolidated units F/VII and E/VII and both the till tongues (D1/VIa, D2/VIb). They could be the result of ice-proximal to ice-distal glaciomarine sedimentation by rainout and suspension settling from a grounded calving front or grounded ice sheet with a floating-tongue front. They may also be the result of ice-marginal local slumping or debris flows down the bank slopes and channel flanks, triggered by large volumes of sediment being released quickly by the retreating ice margin and resulting in high sedimentation rates. Alternatively, it is possible that subglacial deposition of water-saturated sediment with high pore pressures occurred. Fine-grained sediments, however, even under restricted drainage conditions, will consolidate with time. If drainage is limited and the sediments remain soft and normally to underconsolidated, rapid movement of the ice sheet (Boulton, 1979; Boulton and Jones, 1979; Boulton and Hindmarsh, 1987; Alley et al., 1989) is implied, and sedimentation processes similar to debris-flow deposition. Based on the facies relationships interpreted from the regional seismic data, either there was expansion of small residual ice centers remaining from the breakup of the larger caps during deglaciation, or new local caps formed. This scenario is analagous to the numerous caps forming along the south coast and on the Avalon Peninsula at that time (Catto et al., 1995; Catto, 1998). The foraminifera are consistent with all three possible interpretations. The fluctuations in foraminiferal percent occurrences in unit E/VII suggest discrete short-lived depositional events, rather than continuous sedimentation. The presence of *I. helenae* throughout the lower till tongue is interpreted, in part, as being the result of reworking of the underlying glaciomarine sediments. The upper till tongue contains foraminifera of assemblage 2; "*Cibicides*" spp. and "*Glabratella*" spp. suggest higher salinities, less SPM, and higher current velocities than would be expected in a predominantly deglacial environment. The environment was one of transition from glacial to nonglacial conditions, and even though retreating ice was nearby, regional conditions prevailed over local ones.

The presence of subglacial meltwater channels incised into till tongues is not often observed. It is believed that they formed subglacially, or proximal to the ice margin, when the sediments can no longer drain the volumes of meltwater present (Boulton and Dobbie, 1993; Boulton et al., 1995; Piotrowski, 1997). The presence of channels is believed to indicate temperate, wet-based ice (Boulton and Jones, 1979), a tidewater front (Powell, 1984; Pfirman and Solheim, 1989), and a water depth of <75 m (King et al., 1991; King, 1993). The channel geometry observed here (type 3 channels, Miller, 1999) is much smaller than that of type 1 (tunnel valley) or type 2 channels (broad with gently sloping sides, i.e., Boyd et al., 1988; Scott et al., 1989), but is similar to that of channels in the subsurface of Sable Island Bank (King, 1993, Fig. 5). In addition, these channels do not appear to have the massive basal chaotic infill reported for tunnel valleys and megachannels (Wingfield, 1990; Lagoe et al., 1994). These channels are believed to have been formed by deglacial processes at or near a rapidly retreating calving tidewater margin, by the ice-sheet margin decoupling from the bed in response to marine transgression after glacial isostatic loading (depression) and rebound (Boulton and Hindmarsh, 1987), similar to the deglacial response exhibited by the south coast of Newfoundland (Liverman, 1994; Liverman and Batterson, 1995; Liverman and Bell, 1996; Batterson and Janes, 1997; Catto, 1998). Powell (1984), Eyles and McCabe (1989), and Syvitski (1993) all postulated this mechanical trigger and feedback loop (rather than a climatically induced one) as the agent responsible for ice-sheet collapse.

The shell-hash boundary layer sampled within the meltwater channel and in the subsurface of the seafloor has been dated as 10.8 ka, coincident with the Younger Dryas chronozone. The acoustic characteristics of the upper and lower channel-fill sediments are quite different (Fig. 3), suggesting that the shell-hash layer represents a time (ca. 11–10 ka) when the environment of deposition underwent a marked change. The shell-hash layer may be a lag deposit; it is possible that there was local expansion of residual ice, or that small local caps formed on the submerging banks, eroding them and causing the lag deposit to form on the channel floor. A regional erosional event occurred at some point after deposition of the till tongues, as evidenced by the unconformity forming their upper surface (Fig. 2); this erosional event may also be responsible for deposition of the shell-hash layer.

The upper channel-fill sediments contain an interval with

increased occurrence of *Stainforthia* spp., indicating another decrease in ocean-water salinity. This salinity decrease is attributed to increased freshwater runoff from Newfoundland, including seasonal runoff and meltwater from the final deglaciation of Newfoundland. This salinity decrease continued throughout the early Holocene, as runoff from Newfoundland continued to be funnelled through the channels, between the emerged banks, to the shelf edge. This prolonged period of reduced salinity is not surprising; the Gulf of St. Lawrence has *I. helenae*–dominated faunas today, indicating salinities of 32‰–33.5‰ (Rodrigues and Hooper, 1982), and in cores collected from three locations in the gulf, Rodrigues et al. (1993) recognized *I. helenae* zones that they attributed to reduced salinity values throughout much of postglacial time. Between 10 and 8 ka the Green and St. Pierre Banks became fully submerged, and ocean waters would have flowed unimpeded over the bank tops. Salinity increased, and modern shelf and bank conditions began to prevail.

## Chronology of events

The lithostratigraphy (as interpreted from the lithology, physical properties, and seismostratigraphy) and environments of deposition (as interpreted from the foraminiferal assemblages), coupled with age control based on AMS $^{14}$C dating (Table 1), allow a chronology of events to be constructed for the preserved sedimentary sequence in the Halibut Channel area.

A broad unconformity on bedrock (Carboniferous and Tertiary) was developed by glacial erosion, before 41 ka. Prior to 41 ka a basal unit (unit H/IX) was deposited on the bedrock surface. This unit is acoustically identified as till, implying that ice had retreated from the area. Although the last cross-shelf ice advance may be older than Wisconsinan (Piper et al., 1994), Grant (1989) believed it may be early Middle Wisconsinan age (oxygen isotope stage 4) when the most extensive Wisconsinan glaciation of the Atlantic provinces took place (Caledonia phase; Rampton et al., 1984; Stea et al., 1998) and ice extended across the continental shelf (Fader et al., 1982; Fader and Miller, 1986; Piper, *in* Piper et al., 1990). The direction of ice movement is not known, although it probably advanced southward as an extension of Newfoundland-centered ice. This unit in outer Halibut Channel may correlate with the till that overlies bedrock in inner Halibut Channel and Placentia Bay to the north (Fig. 2) (Moran and Fader, 1997).

During oxygen isotope stages 4 and 3 (ca. 75–25 ka; Grant, 1989), glaciomarine sediment (unit G/VIII) with foraminiferal assemblage 6 was deposited over the basal unit, either during retreat of the ice that deposited unit H, or in front of the ice of another advance. These glaciomarine sediments were subsequently loaded, probably by a pre-Late Wisconsinan ice advance estimated to be more than 400 m thick (Moran, 1987; Moran and Fader, 1997). This unit (unit G/VIII) became a deformation (secondary) till (INQUA commission, Dreimanis, 1989), as indicated by its shear-strength record (Fig. 6). This loading occurred prior to the deposition of the overlying units F and E (which do not show ice-loading effects) well before 20 ka (Table 1). The regional seismic interpretation (Fig. 2) shows unit G facies equivalents to the north to be glaciomarine sediments, indicating that Placentia Bay was ice free at this time. This suggests that ice caps developed on exposed St. Pierre and Green Banks and advanced offshore over Halibut Channel (Miller et al., 1994, 1995a; Moran and Fader, 1997; Miller, 1999).

Prior to 19 ka, the loading was followed by the deposition of glaciomarine sediment, Downing Silt (units F/VII and E/VII), containing foraminiferal assemblage 6. Acoustically this unit is incoherent and difficult to interpret because it is at the limit of Huntec DTS penetration (Fig. 3). Lithologically it is a stiff, silty clay and consists of at least two distinct depositional events from possibly two different source areas.

This was followed by continued deposition of Downing Silt (unit C2/VI) prior to 19 ka until after 15 ka. The three Assemblage Group 2 foraminiferal assemblages within this unit (Table 2; Fig. 9) indicate that the depositional environments were undergoing rapid change. Ocean salinity began decreasing due to an increased influx of meltwater, as indicated by the *I. helenae* assemblage (assemblage 5). The fauna changed to a short-lived interval of *Stainforthia* spp. (assemblage 4) ca. 15 ka, indicating a period with a further reduction in both salinity and the dissolved oxygen content of the water mass. The short interval of increased occurrence of *H. paucilocula* (assemblage 3) immediately following is interpreted as representing a period of salinity minimum, perhaps because the meltwater influx reached its maximum, there was channelized flow, and the waters were cold and not mixing with the open ocean waters.

After ca. 13 ka but prior to 11 ka the third ice advance occurred, represented by the two till tongues. This advance is coincident with the Robinson's Head readvance in Newfoundland (Brookes, 1969, 1977) and the Chignecto phase in Nova Scotia (Stea and Mott, 1998; Stea et al., 1998). The lower till tongue (unit D2/VIb) contains the *Islandiella helenae* assemblage (assemblage 5) and the *C. excavatum–C. reniforme* assemblage (assemblage 6). There were local osscillations of the grounding line, retreat as evidenced by the deposition of Downing Silt (unit C2/V), followed by an advance forming the upper till tongue (unit D1/VIa). The sediments contain foraminifera of assemblage 2, with "*Cibicides*" spp. and "*Glabratella*" spp., suggesting higher salinities, less SPM, and higher current velocities than would be expected in a predominantly deglacial environment. During this third advance, eustatic sea level was rising (Quinlan and Beaumont, 1982; Fader, 1989; Liverman, 1994; Shaw and Forbes, 1995), but relative sea level was still dropping (J. Shaw, University of Alberta [GSC-A], 2000, personal commun.), the shallow banks were still partly emergent, and meltwater flow from both offshore and onshore directed through the channels to the shelf edge. Rates of melting and the meltwater influx from both onshore and offshore ice caps increased to the point where subglacial meltwater could not discharge as groundwater, and subglacial meltwater channels formed (Boulton and Dobbie, 1993; Boulton et al., 1995; Piotrowski, 1997). The lower channel-fill sediments also contain foraminiferal assemblage 2, with "*Cibicides*" spp. and "*Glabratella*" spp. The presence of assemblage 2 in both the upper till tongue and lower channel-fill sediments suggests that either the two units are contemporaneous with each other, or that the upper till-tongue sediments were eroded and redeposited in the lower channel.

A $^{14}$C age on shell material from the shell hash, the boundary layer within the channel-fill sediments, indicates an age of 10.8 ka, so the channel began to infill while ice was still present, as the ice was retreating, or soon thereafter. The shell-hash layer was deposited at the time of the Younger Dryas chronozone, and is coincident with the Ten Mile Lake readvance in northern Newfoundland (Grant, 1969, 1992, 1994a), the Collins Pond phase in Nova Scotia (Stea and Mott, 1989, 1998; Stea et al., 1998), and the Younger Dryas ice advance recognized elsewhere in offshore Atlantic Canada (Amos and Miller, 1990; Miller, 1993; King, 1994, 1996; Stea et al., 1998). Relative sea level reached its lowstand ($\sim -45$ m) ca. 10 ka (J. Shaw, University of Alberta [GSC-A], 2000, personal commun.).

Following deposition of the shell-hash layer, from ca. 10.8 ka until after 6 ka, the channel continued to infill and there was another increase in the occurrence of *Stainforthia* spp., which suggests another decrease in salinity through the early and middle Holocene. This salinity decrease is probably attributable to an increase in freshwater runoff, including meltwater influx from the final deglaciation of Newfoundland. Halibut Channel would have continued to act as a conduit until about 8 ka, when Green Bank (54 m) and St. Pierre Bank (31 m) would have become fully submerged (J. Shaw, University of Alberta [GSC-A], 2000, personal commun.).

The environment changed to present-day conditions ca. 6 ka, with the submergence of the banks. There has been very little sediment accumulation throughout the late Holocene, as evidenced by the thin layer of surficial sands throughout the study area (Fig. 3), the thin interval of modern foraminiferal faunas (assemblage 1) at the borehole locations, and the $^{14}$C date of 2.7 ka in the near-surface (0.21 m) sediments.

## SUMMARY AND CONCLUSIONS

The thick sediment sequence in Halibut Channel, with its glacigenic features, including interpreted tills, till tongues, and meltwater channels, has been continuously drilled and sampled to 75 m. The lithology, sediment physical properties, and foraminiferal content of the sediments have been analyzed in order to understand more about the depositional mechanisms and environments of these units, particularly the tills and the till tongues. AMS $^{14}$C dates were obtained from mollusc shells and foraminifera to provide chronological control.

The sequence shows evidence of three glacial advances; each advance produced very different facies interpreted as the result of different glacial regimes and sea-level stands.

The oldest advance is represented by an acoustically recognized basal till, interpreted as a lodgement or meltout till. It is Middle Wisconsinan or older (pre-41 ka) in age, and occurred during the Caledonian phase recognized throughout Atlantic Canada.

The second advance is represented by a deformation till with a *C. excavatum–C. reniforme* foraminiferal fauna; it is post-41 ka in age and is attributed to a Middle Wisconsinan advance.

Overlying these two tills is a thick glaciomarine section. The lower part also contains a glaciomarine *C. excavatum–C. reniforme* fauna, and the upper part contains a deglacial foraminiferal assemblage sequence; three foraminiferal assemblages, characterized by *Haynesina pauciloculla*, *Stainforthia* spp., and *Islandiella helenae*, respectively, are all indicative of increasing meltwater influence upsection and the development of channelized flow.

Evidence of the youngest advance suggests that it took the form of numerous small discrete ice caps on the outer shelf banks. It is dated as 13–11 ka, and is correlated to the Robinson's Head readvance in Newfoundland and the Chignetco phase in Nova Scotia. This advance deposited two normally consolidated till tongues that have been crosscut by meltwater channels, suggesting deposition by warm, wet-based ice. The meltwater channels and upper till tongue contain diagnostic foraminiferal species of the "*Cibicides*" spp. and "*Glabratella*" spp. groups, indicative of shallow bank and strong current conditions.

There is also evidence of events and major oceanographic changes during early postglacial time and the Holocene. There is a shell-hash layer, which may be a lag deposit coincident with the Younger Dryas chronozone; it may be the result of local residual ice caps expanding, or new ones forming, on the emergent banks. There is a major erosional event recorded; the event truncated the upper surface of the till tongues and the moraine in which they are rooted, and this may also have formed the lag deposit. There is a period of increased freshwater runoff (as indicated by the *Stainforthia* spp. assemblage in the channel-fill sediments) through the Early and Middle Holocene. Halibut Channel continued to act as a conduit until the banks became fully submerged; after submergence, shelf and bank conditions began to prevail.

## ACKNOWLEDGMENTS

We thank the officers, crew, and scientific personnel of CSS *Hudson* site-survey cruise 86-018 and M/V *Balder Challenger* cruise 87–400, personnel from Jacques, Whitford and Associates, Ltd., and Logan Drilling for the collection of geophysical data and borehole drilling; and scientific and technical staff from the Geological Survey of Canada—Atlantic (GSC-A, BIO, Dartmouth, Nova Scotia) for shipboard sample processing and analyses, laboratory analyses, and assistance with manuscript preparation.

The manuscript had numerous drafts; these drafts have greatly benefited from thorough, objective, and constructive reviews by P.J. Mudie, C.F.M. Lewis, and J. Shaw (GSC-A), F.R. Hall (University of New Orleans, Louisiana), D. Schnitker (University of Maine, Orono) and T. Weddle (Maine Geological Survey, Augusta, Maine). We also acknowledge volume reviews by T. Bacchus (Johnson State College, Johnson, Vermont) and J.L. Cullen (Salem State College, Salem, Massachusetts). We thank T. Weddle (Maine Geological Survey, Augusta, Maine) and M.E. Retelle (Bates College, Lewiston, Maine) for including this work in this volume, and for all their effort in organizing and overseeing this project throughout.

The foraminiferal analyses and the data compilation and interpretation are part of Miller's Ph.D. dissertation, completed as a cooperative project between The George Washington University, Washington, D.C., and the GSC-Atlantic. Miller sincerely thanks all committee members for their roles in the completion of this work; and D.J.W. Piper (GSC-A) and J. Shaw (University of Alberta, Edmonton) for knowledgable and helpful discussions. Financial support was provided by both graduate scholarships and teaching assistantships at The George Washington University from 1988 to 1990, and a National Sciences and Engineering Research Council of Canada Postgraduate Scholarship from 1987 to 1990. This is GSC contribution 2000056.

## REFERENCES CITED

Alley, R.B., Blankenship, D.D., Rooney, S.T., and Bentley, C.R., 1989, Sedimentation beneath ice shelves—The view from ice stream B, in Powell, R.D., and Elverhøi, A., eds., Modern glacimarine environments: Glacial and marine controls of modern lithofacies and biofacies: Marine Geology, v. 85, p. 101–120.

Alve, E., 1990, Variations in estuarine foraminiferal biofacies with diminishing oxygen conditions in Drammensfjord, SE Norway, in Hembleben, C., et al., eds., Paleoecology, biostratigraphy, paleoceanography and taxonomy of agglutinated foraminifera: NATO ASI Series C: Mathematical and Physical Sciences, v. 327, p. 661–694.

Alve, E., 1994, Opportunistic features of the foraminifer *Stainforthia fusiformis* (Williamson): Evidence from Frierfjord, Norway: Journal of Micropalaeontology, v. 13, p. 24.

Alve, E., 1995, Benthic foraminiferal distribution and recolonization of formerly anoxic environments in Drammensfjord, southern Norway: Marine Micropaleontology, v. 25, p. 169–186.

Alve, E., and Murray, J.W., 1995, Benthic foraminiferal distribution and abundance changes in Skagerrak surface sediments: 1937 (Höglund) and 1992/1993 data compared: Marine Micropaleontology, v. 25, p. 269–288.

Amos, C.L., and Miller, A.A.L., 1990, The Quaternary stratigraphy of southwest Sable Island Bank: Geological Society of America Bulletin, v. 102, p. 915–934.

Ashley, G.M., Wellner, R.W., Esker, D., and Sheridan, R.E., 1991, Clastic sequences developed during late Quaternary glacio-eustatic sea-level fluctuations on a passive margin: Example from the inner continental shelf near Barnegat Inlet, New Jersey: Geological Society of America Bulletin, v. 103, p. 1607–1621.

Austin, W.E.N., and McCarroll, D., 1992, Foraminifera from the Irish Sea glacigenic deposits at Aberdaron, western Lleyn, North Wales: Paleoenvironmental implications: Journal of Quaternary Science, v. 7, p. 311–317.

Austin, W.E.N., and Sejrup, H.P., 1994, Recent shallow water benthonic foraminifera from western Norway: Ecology and paleoecological significance, in Sejrup, H.P., and Knudsen, K.L., eds., Late Cenozoic benthonic foraminifera: Taxonomy, ecology and stratigraphy: Cushman Foundation for Foraminiferal Research Special Publication 32, p. 103–125.

Batterson, M.J., and Janes, J., 1997, Stratigraphy of Late Quaternary sediments exposed in coastal cliffs, west of Stephenville, in Current Research, 1997: Newfoundland Department of Mines and Energy, Geological Survey Branch, Report 97-1, p. 151–165.

Bell, R.A., 1884, Observations on the geology, mineralogy, zoology and botany of the Labrador coast, Hudson's Strait and Bay: Geological Survey of Canada Annual Reports 1882–1884, v. DD, p. 5–62.

Bergsten, H., and Dennegård, B., 1988, Late Weichselian–Holocene foraminiferal stratigraphy and paleohydrodynamic changes in Gothenburg area, southwestern Sweden: Boreas, v. 17, p. 229–242.

Bond, C.G., and Lotti, R., 1995, Iceberg discharges into the North Atlantic on millennial time scales during the last glaciation: Science, v. 267, p. 1005–1010.

Bonifay, D., and Piper, D.J.W., 1988, Probable Late Wisconsinan ice margin on the upper continental slope off St. Pierre Bank, eastern Canada: Canadian Journal of Earth Sciences, v. 25, p. 853–865.

Boulton, G.S., 1970, On the deposition of subglacial and melt-out tills at the margin of certain Svalbard glaciers: Journal of Glaciology, v. 9, p. 231–245.

Boulton, G.S., 1979, Processes of glacier erosion on different substrata: Journal of Glaciology, v. 23, p. 15–38.

Boulton, G.S., 1987, A theory of drumlin formation by subglacial sediment deformation, in Menzies, J., and Rose, J., eds., Drumlin symposium: Rotterdam, A.A. Balkema, p. 25–80.

Boulton, G.S., 1990, Sedimentary and sea level changes during glacial cycles and their control on glacimarine facies architecture, in Dowdeswell, J.A., and Scourse, J.D., eds., Glacimarine environments, processes and sediments: Geological Society [London] Special Publication 53, p. 15–52.

Boulton, G.S., 1996, Theory of glacial erosion, transport and deposition as a consequence of subglacial sediment deformation: Journal of Glaciology, v. 42, p. 43–62.

Boulton, G.S., and Dobbie, K.E., 1993, Consolidation of sediments by glaciers: Relations between sediment geotechnics, soft-bed glacier dynamics and subglacial ground-water flow: Journal of Glaciology, v. 39, p. 26–44.

Boulton, G.S., and Hindemarsh, R.C.A., 1987, Sediment deformation beneath glaciers: Rheology and geological consequences: Journal of Geophysical Research, v. 92, p. 9059–9082.

Boulton, G.S., and Jones, A.S., 1979, Stability of temperate ice caps and ice sheets resting on beds of deformable sediment: Journal of Glaciology, v. 24, p. 29–43.

Boulton, G.S., Caban, P.E., and Van Gijssel, J., 1995, Groundwater flow beneath ice sheets: Part I—Large scale patterns: Quaternary Science Reviews, v. 14, p. 545–562.

Boyd, R., Scott, D.B., and Douma, M., 1988, Glacial tunnel valleys and Quaternary history of the outer Scotian Shelf: Nature [London], v. 333, p. 61–64.

Brookes, I.A., 1969, Late-glacial marine overlap in western Newfoundland: Canadian Journal of Earth Sciences, v. 6, p. 1397–1404.

Brookes, I.A., 1977, Radiocarbon age of the Robinson's Head moraine, west Newfoundland, and its significance for post-glacial sea level changes: Canadian Journal of Earth Sciences, v. 14, p. 2121–2126.

Brookes, I.A., 1982, Ice marks in Newfoundland: A history of ideas: Géographie Physique et Quaternaire, v. 36, p. 139–163.

Budd, W.F., and Smith, I.N., 1987, Conditions for growth and retreat of the Laurentide ice sheet: Géographie et Physique Quaternaire, v. 41, p. 279–280.

Buzas, M.A., 1990, Another look at species confidence limits for species proportions: Journal of Paleontology, v. 64, p. 842–843.

Casagrande, A., 1936, The determination of pre-consolidation load and its practical significance, in Proceedings, Annual Engineering Geology and Soils Engineering Symposium, International Conference on Soil Mechanics and Foundation Engineering, Volume 1: Cambridge, Massachusetts, Harvard University Graduate School of Engineering, p. 60–64.

Catto, N., 1998, The patterns of glaciation on the Avalon Peninsula of Newfoundland: Géographie Physique et Quarternaire, v. 52, p. 23–45.

Catto, N., Hamlyn, C., and Catto, G., 1995, Eastern Avalon Field Trip Guide: Canadian Quaternary Association–Canadian Geomorphological Research Group, Programme, Abstracts, Field guides, St. John's, Newfoundland, Geography Department, Memorial University of Newfoundland, p. EC1–EC9.

Chamberlain, T.C., 1895, Recent glacial studies in Greenland: Geological Society of America Bulletin, v. 6, p. 199–220.

CLIMAP Project Members, 1976, The surface of the Ice Age Earth: Science, v. 191, p. 1131–1137.

Cole, F.E., 1981, Taxonomic notes on bathyal zone foraminiferal species off north-east Newfoundland: Bedford Institute of Oceanography Report Series BI-R-81-7, 122 p.

Conolly, J.R., Needham, H.D., and Heezen, B.C., 1967, Late Pleistocene and Holocene sedimentation in the Laurentian Channel: Journal of Geology, v. 75, p. 131–147.

Conradsen, K., 1993, Recent benthic foraminifera in the southern Kattegat, Scandanavia: Distributional pattern and controlling parameters: Boreas, v. 22, p. 367–382.

Conradsen, K., Bergsten, H., Knudsen, K.L., Nordberg, K., and Seidenkrantz, M.-S., 1994, Recent benthic foraminiferal distribution in the Kattegat and the Skagerrak, Scandinavia, in Sejrup, H.P., and Knudsen, K.L., eds., Late Cenozoic benthonic foraminifera: Taxonomy, ecology and stratigraphy: Cushman Foundation for Foraminiferal Research Special Publication 32, p. 53–67.

Culver, S.J., Woo, H.J., Oertel, G.F., and Buzas, M.A., 1996, Foraminifera of coastal depositional environments, Virginia, U.S.A.: Distribution, and taphonomy: Palaios, v. 11, p. 459–486.

Cushman, J.A., 1944, Foraminifera from the shallow water of the New England Coast: Cushman Laboratory for Foraminiferal Research Special Publication 12, 37 p.

Dalrymple, R.W., LeGresley, E.M., Fader, G.B.J., and Petrie, B.D., 1992, The western Grand Banks of Newfoundland: Transgressive Holocene sedimentation under the combined influence of waves and currents: Marine Geology, v. 105, p. 95–118.

Davies, T.A., and Austin, J.A., Jr., 1997, High-resolution seismic reflection and coring techniques applied to late Quaternary deposits on the New Jersey Shelf, in Syvitski, J.P.M., et al., eds., COLDSEIS (seismic facies of glacigenic deposits): Marine Geology, v. 143, p. 137–149.

Denton, G.H., and Hughes, T.J., 1981, The Arctic Ice Sheet: An outrageous hypothesis, in Denton, G.H., and Hughes, T.J., eds., The last great ice sheets: New York, John Wiley and Sons, p. 437–467.

Dreimanis, A., 1982, Work group (1)—Genetic classification of tills and criteria for their differentiation: Progress report on activities 1977–1982, and definitions of glaciogenic terms, in Schlüchter, C., ed., INQUA Commission on genesis and lithology of Quaternary deposits: Report on activities 1977–1982: Zurich, Switzerland, International Union for Quaternary Research, p. 12–31.

Dreimanis, A., 1989, Tills: Their genetic terminology and classification, in Goldthwait, R.P., and Matsch, C.L., eds., Genetic classification of glacigenic deposits: Final report of the Commission on Genesis and Lithology of Glacial Quaternary Deposits of the International Union for Quaternary Research (INQUA): Rotterdam, Balkema Publishers, p. 17–84.

Dreimanis, A., and Lundqvist, J., 1984, What should be called till?, in Königsson, L.K., ed., Ten years of Nordic till research: Striae, v. 20, p. 5–10.

Elson, J.A., 1957, Striated boulder pavements of southern Manitoba: Geological Society of America Bulletin, v. 68, p. 1722.

Elverhøi, A., Liestøl, O., and Nagy, J., 1980, Glacial erosion, sediments and microfauna in the inner part of Kongsfjorden, Spitsbergen: Norsk Polarinstitutt Skrifter 172, p. 33–61.

Eyles, N., and McCabe, A.M., 1989, Glaciomarine facies within subglacial tunnel valleys: The sedimentary record of glaciostatic downwarping in the Irish Sea Basin: Sedimentology, v. 36, p. 431–448.

Fader, G.B.J., 1986, Cruise Report 86-017, CSS Hudson: Geological Survey of Canada, Atlantic Geoscience Centre, Bedford Institute of Oceanography, 32 p.

Fader, G.B.J., 1989, A Late Pleistocene low sea level stand of the south-east Canadian offshore, in Scott, D.B., et al., eds., Late Quaternary sea level correlation and applications: NATO ASI Series C: Mathematical and Physical Sciences, v. 256, p. 71–104.

Fader, G.B.J., and Miller, R.O., 1986, A reconnaissance of the surficial and shallow bedrock geology of the southeastern Grand Banks of Newfoundland, in Current research, Part B: Geological Survey of Canada Paper 86-1B, p. 591–604.

Fader, G.B.J., and Moran, K., 1988, A preliminary interpretation of the results of boreholes through a till tongue succession on the Grand Banks of Newfoundland: Geological Society of America Abstracts with Programs, v. 20, no. 1, p. 17.

Fader, G.B.J., King, L.H., and Josenhans, H.J., 1982, Surficial geology of the Laurentian Channel and western Grand Banks of Newfoundland: Marine Sciences Paper 21: Geological Survey of Canada Paper 81-22, 37 p.

Gibson, T.G., and Walker, D.W., 1967, Flotation methods for obtaining foraminifera from sediment samples: Journal of Paleontology, v. 41, p. 1294–1297.

Grant, D.R., 1969, Late Pleistocene re-advance of piedmont glaciers in western Newfoundland: Maritime Sediments, v. 5, p. 123–125.

Grant, D.R., 1975, Glacial features of the Hermitage–Burin Peninsula area, Newfoundland, in Report of activities, Part C: Geological Survey of Canada Paper 75-1C, p. 333–334.

Grant, D.R., 1989, Quaternary geology of the Atlantic Appalachian region of Canada, in Fulton, R.J., ed., Quaternary geology of Canada and Greenland: Geological Survey of Canada, Geology of Canada, no. 1, p. 393–440.

Grant, D.R., 1992, Quaternary geology of St. Anthony–Blanc Sablon area, Newfoundland: Geological Survey of Canada Map 1664A, scale 1:250 000.

Grant, D.R., 1994a, Quaternary geology of Port Saunders map area, Newfoundland: Geological Survey of Canada Paper 91-20, 59 p.

Grant, D.R., 1994b, Quaternary geology, Cape Breton Island, Nova Scotia: Geological Survey of Canada Bulletin 482, 159 p.

Grant, D.R., and King, L.H., 1984, A stratigraphic framework of the Quaternary history of the Atlantic Provinces, in Fulton, R.J., ed., Quaternary stratigraphy of Canada—A Canadian contribution to IGCP Project 24: Geological Survey of Canada Paper 84-10, p. 173–191.

Gravenor, C.P., Von Brunn, V., and Dreimanis, A., 1984, Nature and classification of waterlain glaciogenic sediments, exemplified by Pleistocene, late Paleozoic and late Precambrian deposits: Earth Science Reviews, v. 20, p. 105–166.

Guilbault, J.P., 1989, Foraminiferal distribution in the central and western parts of the Late Pleistocene Champlain Sea basin, eastern Canada: Géographie Physique et Quaternaire, v. 43, p. 3–26.

Guilbault, J.P., 1993, Quaternary foraminiferal stratigraphy in sediments of the eastern Champlain Sea basin, Québec: Géographie Physique et Quaternaire, v. 47, p. 43–68.

Haake, F.-W., 1962, Untersuchungen an der Foraminiferen-Fauna im Wattgebiet zwischen Lageoog und dem Festland: Meyniana, Veroffentlichugen aus dem Geologischen Institut der Universität Kiel, v. 12, p. 25–64.

Hald, M., and Korsen, S., 1997, Distribution of modern benthic foraminifera from fjords of Svalbard, European Arctic: Journal of Foraminiferal Research, v. 27, p. 101–122.

Hald, M., and Vorren, T.O., 1987, Foraminiferal stratigraphy and environments of Late Weichselian deposits on the continental shelf off Troms, northern Norway: Marine Micropaleontology, v. 12, p. 129–160.

Hald, M., Sættem, J., and Nesse, E., 1990, Middle and Late Weichselian stratigraphy in shallow drillings from southwestern Barents Sea: Norsk Geologisk Tidsskrift, v. 70, p. 241–270.

Hald, M., Vorren, T.O., Danielson, T.K., Nesse, E., Lorentzen, S., Poole, D., and Steinsund, P.I., 1991, Quaternary benthic foraminiferal distributions in the western Barents Sea: Striae, v. 34, p. 153–157.

Hald, M., Steinsund, P.I., Dokken, T., Korsun, S., Polyak, L., and Aspeli, R., 1994, Recent and late Quaternary distribution of *Elphidium excavatum* f. *clavatum* in arctic seas, *in* Sejrup, H.P., and Knudsen, K.L., eds., Late Cenozoic benthonic foraminifera: Taxonomy, ecology and stratigraphy: Cushman Foundation for Foraminiferal Research Special Publication 32, p. 141–154.

Hall, F.R., Miller, A.A.L., and Moran, K., 1998, Deglaciation of the central Scotian Shelf: Foraminifera, physical properties and acoustic records from Emerald Basin: Geological Society of America Abstracts with Programs, v. 30, no. 1, p. 23.

Hambrey, M.J., Barrett, P.J., Ehrmann, W.U., and Larsen, B., 1992, Cenozoic sedimentary processes on the Antarctic continental margin and record from deep drilling: Zeitschrift für Geomorphologie, supplimentband, v. 86, p. 77–103.

Hansen, A., and Knudsen, K.L., 1992, Recent foraminifera in Freemansundet, eastern Svalbard, *in* Møller, P., Hjort, C., and Ingolfsson, Q., eds., Weichselian and Holocene glacial and marine history of east Svalbard: Preliminary report on the PONNAM fieldwork in 1991: LUNDQUA, Report 35: Lund, Norway, Lund University, Department of Quaternary Geology, p. 177–189.

Haynes, J.R., 1981, Foraminifera: London, Macmillian Publishers Ltd., 433 p.

Haynes, J.R., McCabe, A.M., and Eyles, N., 1995, Microfaunas from Late Devensian glaciomarine deposits in the Irish Sea basin: Irish Journal of Earth Sciences, v. 14, p. 81–103.

Hutchins, R.W., McKeown, D.L., and King, L.H., 1976, A deep towed high resolution seismic system for continental shelf mapping: Geoscience Canada, v. 3, p. 95–100.

Jenness, S.E., 1960, Late Pleistocene glaciations of eastern Newfoundland: Geological Society of America Bulletin, v. 71, p. 161–180.

Jennings, A.E., and Helgadóttir, G., 1994, Foraminiferal assemblages from the Greenland fjords and shelf of Eastern Greenland: Journal of Foraminiferal Research, v. 24, p. 123–144.

Jensen, K.A., and Knudsen, K.L., 1988, Quaternary foraminiferal stratigraphy in boring 81/29 from the central North Sea: Boreas, v. 17, p. 273–287.

Jones, G.A., and Keigwin, L.D., 1988, Evidence from Fram Strait (78° N) for early deglaciation: Nature [London], v. 336, p. 56–59.

Josenhans, H., 1997, Simultaneous use of multiple seismic reflection systems for high resolution and deep penetration, *in* Davies, T.A., et al., eds., Glaciated continental margins: An atlas of acoustic images: New York, Chapman and Hall, p. 31–32.

King, L.H., 1980, Aspects of regional surficial geology related to site investigation requirements—Eastern Canadian Shelf, *in* Ardus, D.A., ed., Offshore site investigation: London, Graham and Trottman, p. 37–59.

King, L.H., 1993, Till in the marine environment: Journal of Quaternary Science, v. 8, p. 347–358.

King, L.H., 1994, Proposed Younger Dryas of the eastern Scotian Shelf: Canadian Journal of Earth Sciences, v. 31, p. 401–417.

King, L.H., 1996, Ice retreat from the Scotian Shelf: Geological Society of America Bulletin, v. 108, p. 1056–1067.

King, L.H., and Fader, G.B., 1976, Application of the Huntec deep tow high-resolution seismic system to surficial and bedrock studies—Grand Banks of Newfoundland, *in* Report of activities, Part C: Geological Survey of Canada Paper 76-1C, p. 5–7.

King, L.H., and Fader, G.B.J., 1986, Wisconsinan glaciation of the continental shelf, southeast Atlantic Canada: Geological Survey of Canada Bulletin 363, 72 p.

King, L.H., Fader, G.B.J., Jenkins, W.A.M., and King, E.L., 1986, Occurrence and regional geological setting of Paleozoic rocks on the Grand Banks of Newfoundland: Canadian Journal of Earth Sciences, v. 23, p. 504–526.

King, L.H., Rokoengen, K., and Gunleiksrud, T., 1987, Quaternary seismostratgraphy of the Mid-Norwegian Shelf, 65°–67°30′ N—A till tongue stratigraphy: Trondheim, Norway, Continental Shelf and Petroleum Technology Research Institute Publication 114, 58 p.

King, L.H., Rokoengen, K., Fader, G.B.J., and Gunleiksrud, T., 1991, Till-tongue stratigraphy: Geological Society of America Bulletin, v. 103, p. 637–659.

Knebel, H.J., Wood, S.A., and Spiker, E.C., 1979, Hudson River: Evidence for extensive migration on the exposed continental shelf during Pleistocene time: Geology, v. 7, p. 254–258.

Knudsen, K.L., 1978, Middle and Late Weichselian marine deposits at Nørre Lyngby, northern Jutland, Denmark, and their foraminiferal faunas: Danmarks Geologiske Undersøgelse II, Raekke 112, 44 p.

Knudsen, K.L., 1982, Foraminifers, *in* Olausson, E., ed., The Pleistocene/Holocene boundary in southwestern Sweden: Sveriges Geologiska Undersökning, ser. C, Ayhandlingar och Undersökning 794, p. 148–177.

Knudsen, K.L., 1993, Late Elsterian–Holsteinian foraminiferal stratigraphy in boreholes in the lower Elbe area, NW Germany: Geologisches Jahrbuch, ser. A, v. 138, p. 97–119.

Knudsen, K.L., and Asbjörnsdóttir, L., 1991, Plio-Pleistocene foraminiferal stratigraphy and correlation in the central North Sea: Marine Geology, v. 101, p. 113–124.

Knudsen, K.L., and Sejrup, H.-P., 1993, Pleistocene stratigraphy in the Devils Hole area, central North Sea: Foraminiferal and amino-acid evidence: Journal of Quaternary Science, v. 8, p. 1–14.

Korsen, S.A., Pogodina, I.A., Forman, S.L., and Lubinski, D.J., 1995, Recent foraminifera in glaciomarine sediments from three arctic fjords of Novaja Zemlja and Svalbard: Polar Research, new series, v. 14, p. 15–31.

Lagoe, M.B., Eyles, C.H., and Eyles, N., 1994, Foraminiferal biofacies and paleoenvironments in a Pliocene megachannel of the glaciomarine Yakataga Formation, Gulf of Alaska, *in* Sejrup, H.P., and Knudsen, K.L., eds., Late Cenozoic benthonic foraminifera: Taxonomy, ecology and stratigraphy: Cushman Foundation for Foraminiferal Research Special Publication 32, p. 127–140.

Lagoe, M.B., Davies, T.A., Austin, J.A., Jr., and Olson, H.C., 1997, Foraminiferal constraints on very high-resolution seismic stratigraphy and Late Quaternary glacial history, New Jersey continental shelf: Palaios, v. 12, p. 249–266.

Liverman, D.G.E., 1994, Relative sea-level history and isostatic rebound in Newfoundland: Boreas, v. 23, p. 217–230.

Liverman, D., and Batterson, M., 1995, West Coast Newfoundland Field Trip Guide: Canadian Quaternary Association–Canadian Geomorphological Research Group, Programme, Abstracts, Field guides, St. John's, Newfoundland, Geography Department, Memorial University of Newfoundland, p. WC1–WC78.

Liverman, D., and Bell, T., 1996, Late Quaternary glacial and glaciomarine sediments in southern St. George's Bay, *in* Current Research, 1996: Newfoundland Department of Natural Resources, Geological Survey Branch Report 96-1, p. 29–40.

Loeblich, A.R., Jr., and Tappan, H., 1964, Sarcodina, chiefly "Thecamoebians" and foraminifera, *in* Moore, R.C., ed., Treatiste on invertebrate paleontology, Part C, Protista 2: Lawrence, Kansas University Press, 899 p.

Loeblich, A.R., Jr., and Tappan, H., 1988, Foraminifera genera and their classification: New York, Van Nostrand Reinhold Company, 970 p.

Long, D., Bent, A., Harland, R., Gregory, D.M., Graham, D.K., and Morten, A.C., 1986, Late Quaternary paleontology, sedimentology and geochemistry of a vibrocore from the Witch Ground Basin, central North Sea: Marine Geology, v. 73, p. 109–123.

Loring, D.H., and Nota, D.J.G., 1973, Morphology and sediments of the Gulf of St. Lawrence: Fisheries Research Board of Canada Bulletin 182, 147 p.

Lutze, G., 1965, Zur Foraminiferen-Fauna der Ostsee: Meyniana, Veroffentlichugen aus dem Geologischen Institut der Universität Kiel, v. 15, p. 75–147.

Lutze, G., 1974, Foraminiferin der Kieker Bucht (Westiche Ostsee): 1. "Haugartengebeit" des Sonderforschungbereiches 95 der Universität Kiel: Meyniana, Veroffentlichugen aus dem Geologischen Institut der Universität Kiel, v. 26, p. 9–22.

MacLean, B., Sonnichsen, G., Vilks, G., Powell, C., Moran, K., Jennings, A., Hodgson, D., and Deonarine, B., 1989, Marine geological and geotechnical investigations in Wellington, Byam Martin, Austin, and adjacent channels, Canadian Arctic Archipeligo: Geological Survey of Canada Paper 89-11, 69 p.

Mangerud, J., Sønstegaard, E., Sejrup, H.-P., and Haldorsen, S., 1981, A continuous Eemian–Early Weichselian sequence containing pollen and marine fossils at Fjøsanger, western Norway: Boreas, v. 10, p. 137–208.

Miller, A.A.L., 1993, Quaternary foraminiferal biostratigraphy of three shallow geotechnical boreholes, Balmoral, Cohasset and Panuke wellsites, western Sable Island Bank: marine g.e.o.s. Technical Report 6: Geological Survey of Canada Open-File Report 2710, 87 p.

Miller, A.A.L., 1996, Late Quaternary foraminiferal biostratigraphy of three shallow geotechnical boreholes in Halibut Channel, western Grand Banks of Newfoundland: marine g.e.o.s. Technical Report 8: Geological Survey of Canada Open-File Report 3369, 131 p.

Miller, A.A.L., 1998a, All deglaciations are not alike: Insights into deglacial marine environments from indicator foraminiferal species, Atlantic Canada: Geological Society of America Abstracts with Programs, v. 30, no. 1, p. 62.

Miller, A.A.L., 1998b, All deglaciations are not alike: Insights into deglacial marine environments from indicator foraminiferal species, Atlantic Canada: Geological Society of America Abstracts with Programs, v. 30, no. 7, Late breaking abstract no. 60012.

Miller, A.A.L., 1999, The Quaternary sediments and seismostratigraphy of the Grand Banks of Newfoundland and the Northeast Newfoundland Shelf: Foraminiferal refinements and constraints [Ph.D. dissertation]: Washington, D.C., Columbian School of Arts and Sciences, The George Washington University, 972 p.

Miller, A.A.L., Mudie, P.J., and Scott, D.B., 1982a, Holocene history of Bedford Basin, Nova Scotia: Foraminifera, dinoflagellate and pollen records: Canadian Journal of Earth Sciences, v. 19, p. 2342–2367.

Miller, A.A.L., Scott, D.B., and Medioli, F.S., 1982b, *Elphidium excavatum* (Terquem): Ecophenotypic versus subspecific variation: Journal of Foraminiferal Research, v. 12, p. 116–144.

Miller, A.A.L., Moran, K., and Fader, G.B.J., 1994, Quaternary history of the central Grand Banks of Newfoundland: Foraminiferal content, seismostratigraphy and physical properties of sediments in Halibut Channel: Geological Association of Canada Program with Abstracts, v. 19, p. A76.

Miller, A.A.L., Moran, K., and Fader, G.B.J., 1995a, Quaternary history of the central Grand Banks of Newfoundland: Foraminiferal content, seismostratigraphy and physical properties of sediments in Halibut Channel: Halifax, Nova Scotia, Fifth International Conference on Paleoceanography, Program and Abstracts, p. 108.

Miller, A.A.L., Moran, K., and Hall, F.R., 1995b, Deglaciation of the central Scotian Shelf: Foraminifera, physical properties and acoustic records from Emerald Basin: Halifax, Nova Scotia, Fifth International Conference on Paleoceanography, Program and Abstracts, p. 107.

Miller, A.A.L., Nomura, R., and Osterman, L.E., 1996, *Islandiella algida* (Cushman, 1944), senior subjective synonym of *Islandiella islandica* (Nørvang, 1945): Journal of Foraminiferal Research, v. 26, p. 209–212.

Milliman, J.D., Jiezao, Z., and Ewing, J.I., 1990, Late Quaternary sedimentation on the outer and middle New Jersey continental shelf: Results of two local deglaciations?: Journal of Geology, v. 98, p. 966–976.

Moran, K., 1987, Cruise Report 87400, M/B Balder Challenger: Dartmouth, Canada, Geological Survey of Canada, Atlantic Geoscience Centre, Bedford Institute of Oceanography, 26 p.

Moran, K., and Fader, G.B.J., 1997, Glacial and glaciomarine sedimentation: Halibut Channel, Grand Banks of Newfoundland, *in* Davies, T.A., et al., eds., Glaciated continental margins: An atlas of acoustic images: New York, Chapman and Hall, p. 217–220.

Mudie, P.J., Piper, D.J.W., Rideout, K., Robertson, K.R., Schafer, C.T., Vilks, G., and Hardy, I.A., 1984, Standard methods for collecting, describing and sampling Quaternary sediments at the Atlantic Geoscience Centre: Geological Survey of Canada Open-File Report 1044, 47 p.

Murray, J.W., 1973, Distribution of living benthic Foraminifera: London, Heinemann Educational Books, 274 p.

Murray, J.W., 1985, Recent foraminifera from the North Sea (Forties and Ekofisk areas) and the continental shelf west of Scotland: Journal of Micropalaeontology, v. 4, p. 117–125.

Murray, J.W., 1991, Ecology and palaeoecology of benthic foraminifera: Harlow, United Kingdom, Longman Scientific and Technical, 397 p.

Murray, J.W., 1992, Distribution and population dynamics of benthic foraminifera from the southern North Sea: Journal of Foraminiferal Research, v. 22, p. 114–128.

Nagy, J., 1965, Foraminifera in some bottom samples from the shallow waters in Vestspitsbergen: Norsk Polarinstitutt Årbok (1963), p. 109–125.

Orheim, O., and Elverhøi, A., 1981, Model for submarine glacial deposition: Annals of Glaciology, v. 2, p. 123–128.

Østby, K., and Nagy, J., 1982, Foraminiferal distribution in the western Barents Sea, recent and Quaternary: Polar Research, no. 1, p. 53–96.

Osterman, L.E., 1984, Benthic foraminiferal zonation of a glacial/interglacial transition from Frobisher Bay, Baffin Island, North West Territories, Canada, *in* Oertli, H.J., ed., Benthos '83, 2nd International Symposium on Benthic Foraminifera, Pau, 1983: Pau and Bordeaux, France, Elf Aquitaine, Esso REP, and Total CFP, p. 471–476.

Osterman, L.E., and Andrews, J.T., 1983, Changes in glaciomarine sedimentation in core HU 77-159, Frobisher Bay, Baffin Island, N.W.T.: A record of proximal, distal and ice-rafting glaciomarine environments, *in* Molnia, B.J., ed., Glaciomarine sedimentation: New York, Plenum Press, p. 451–494.

Osterman, L.E., and Nelson, A.R., 1989, Latest Quaternary and Holocene paleoceanography of eastern Baffin Island continental shelf, Canada: Benthic foraminiferal evidence: Canadian Journal of Earth Sciences, v. 26, p. 2236–2248.

Pawlowski, J., 1991, Distribution and taxonomy of some benthic tiny foraminifers from the Bermuda Rise: Micropaleontology, v. 37, p. 163–172.

Pfirman, S.L., and Solheim, A., 1989, Subglacial meltwater discharge in the open marine tidewater glacier environment: Observations from Nordaustlandet, Svalbard Archipelago: Marine Geology, v. 86, p. 265–281.

Phleger, F.B, 1960, Ecology and distribution of recent Foraminifera: Baltimore, Maryland, John Hopkins Press, 297 p.

Piotrowski, J.A., 1997, Subglacial hydrology in north-western Germany during the last glaciation: Groundwater flow, tunnel valleys and hydrological cycles: Quaternary Science Reviews, v. 16, p. 169–185.

Piper, D.J.W., Mudie, P.J., Fader, G.B.J., Josenhans, H.W., MacLean, B., and Vilks, G., 1990, Quaternary geology, *in* Keen, M.J., and Williams, G.L., eds., Geology of the continental margin of eastern Canada: Geological Survey of Canada, Geology of Canada, no. 2, p. 475–607.

Piper, D.J.W., Mudie, P.J., Aksu, A.E., and Skene, K.I., 1994, A 1 Ma record of sediment flux south of the Grand Banks used to infer the development of glaciation in southeastern Canada: Quaternary Science Reviews, v. 13, p. 23–37.

Poag, C.W., 1981, Ecologic atlas of benthic foraminifera of the Gulf of Mexico: Woods Hole, Massachusetts, Marine Sciences International, 174 p.

Powell, R.D., 1984, Glacimarine processes and inductive lithofacies modelling of ice shelf and tidewater glacier sediments based on Quaternary examples, *in* Bornhold, B.D., and Guilcher, A., eds., Sedimentation on high-latitude continental shelves: Marine Geology, v. 57, p. 1–52.

Prest, V., 1984, The Late Wisconsinan glacier complex, *in* Fulton, R.J., ed.,

Quinlan, G., and Beaumont, C., 1982, The deglaciation of Atlantic Canada as reconstructed from the postglacial relative sea-level record: Canadian Journal of Earth Sciences, v. 19, p. 2232–2246.

Rampton, V.N., Gauthier, R.C., Thibault, J., and Seaman, A.A., 1984, Quaternary geology of New Brunswick: Geological Survey of Canada Memoir 416, 77 p.

Rodrigues, C., and Hooper, K., 1982, Recent benthonic foraminiferal associations from offshore environments in the Gulf of St. Lawrence: Journal of Foraminiferal Research, v. 12, p. 327–352.

Rodrigues, C.G., Ceman, J.A., and Vilks, G., 1993, Late Quaternary paleoceanography of deep and intermediate water masses off Gaspé Peninsula, Gulf of St. Lawrence: Foraminiferal evidence: Canadian Journal of Earth Sciences, v. 30, p. 1390–1403.

Rottgardt, D., 1952, Mickropaläontologische wichtige Bestandteile recenter brackische Sedimente an den Kusten Schleswig-Holsteins: Meyniana, Veroffentlichungen aus dem Geologischen Institut der Universität Kiel, v. 1, p. 169–228.

Sarnthein, M., Jansen, E., Arnold, M., Duplessy, J.C., Erlenkeuser, H., Flatøy, A., Veum, T., Vogelsang, E., and Weinelt, M.S., 1992, $\delta^{18}O$ time-slice reconstruction of melt water anomalies at termination 1 in the North Atlantic between 50° and 80° N, in Bard, E., and Broecker, W.S., eds., The last deglaciation: Absolute and radiocarbon chronologies: NATO ASI Series I: Global Environmental Change, v. 2, p. 184–200.

Schafer, C.T., and Cole, F.E., 1978, Distribution of Foraminifera in Chaleur Bay, Gulf of St. Lawrence: Geological Survey of Canada Paper 77-30, 55 p.

Schafer, C.T., and Cole, F.E., 1986, Reconnaissance survey of benthonic Foraminifera from Baffin Island fiord environments: Journal of the Arctic Institute of North America, v. 39, p. 232–239.

Scott, D.B., 1987, Quaternary benthic foraminifers from Deep Sea Drilling Project Sites 612 and 613, Leg 95, New Jersey Transect, in Poag, C.W., et al., Initial reports of the Deep Sea Drilling Project; Leg 95: Washington, D.C., U.S. Government Printing Office, p. 313–337.

Scott, D.B., Schafer, C.T., and Medioli, F.S., 1980, Eastern Canadian estuarine foraminifera: A framework for comparison: Journal of Foraminiferal Research, v. 10, p. 205–234.

Scott, D.B., Mudie, P.J., Vilks, G., and Younger, D.C., 1984, Latest Pleistocene–Holocene paleoceanographic trends on the continental margin of eastern Canada: Foraminiferal, dinoflagellate and pollen evidence: Marine Micropaleontology, v. 9, p. 181–218.

Scott, D.B., Boyd, R., Douma, M., Medioli, F.S., Yuill, S., Leavitt, E., and Lewis, C.F.M., 1989, Holocene relative sea level changes and Quaternary glacial events on a continental shelf edge: Sable Island Bank, in Scott, D.B., et al., eds., Late Quaternary sea level correlation and applications: NATO ASI Series C: Mathematical and Physical Sciences, v. 256, p. 105–120.

Scourse, J.D., Austin, W.E.D., Bateman, R.M., Catt, J.A., Evans, C.D.R., Robinson, J.E., and Young, J.R., 1990, Sedimentology and micropaleontology of glacimarine sediments from the central and southwestern Celtic Sea, in Dowdeswell, J.A., and Scourse, J.D., eds., Glacimarine environments, processes and sediments: Geological Society [London] Special Publication 53, p. 329–348.

Seidenkrantz, M-S., and Knudsen, K.L., 1993, Middle Weichselian to Holocene palaeoecology in the eastern Kattegat, Scandinavia: Foraminifera, ostracods and $^{14}C$ measurements: Boreas, v. 22, p. 299–310.

Sejrup, H.-P., Aarseth, I., Ellingsen, K.L., Reither, E., Jansen, E., Løvlie, R., Bent, A., Brigham-Grette, J., Larsen, E., and Stoker, M., 1987, Quaternary stratigraphy of the Fladen area, central North sea: A multidisciplinary study: Journal of Quaternary Science, v. 2, p. 35–58.

Sen Gupta, B.K., 1971, The benthonic foraminifera of the Tail of the Grand Banks: Micropaleontology, v. 17, p. 69–98.

Sen Gupta, B.K., and McMullen, R. M., 1969, Foraminiferal distribution and sedimentary facies on the Grand Banks of Newfoundland: Canadian Journal of Earth Sciences, v. 6, p. 475–487.

Shaw, J., and Forbes, D.L., 1995, The postglacial relative sea-level lowstand in Newfoundland: Canadian Journal of Earth Sciences, v. 32, p. 1308–1330.

Slatt, R.M., 1977, Late Quaternary terrigenous and carbonate sedimentation on the Grand Banks of Newfoundland: Geological Society of America Bulletin, v. 88, p. 1357–1367.

Stea, R., and Mott, R.J., 1989, Deglaciation environments and evidence for glaciers of Younger Dryas age in Nova Scotia, Canada: Boreas, v. 18, p. 169–187.

Stea, R., and Mott, R.J., 1998, Deglaciation of Nova Scotia: Stratigraphy and chronology of lake sediment cores and buried organic sections: Géographie Physique et Quaternaire, v. 52, p. 3–21.

Stea, R.R., Piper, D.J.W., Fader, G.B.J., and Boyd, R., 1998, Wisconsinan glacial and sea-level history of Maritime Canada and the adjacent continental shelf: A correlation of land and sea events: Geological Society of America Bulletin, v. 110, p. 821–845.

Stravers, J.A., and Powell, R.D., 1997, Glacial debris flow deposits on the Baffin Island Shelf: Seismic facies architecture of till-tongue-like deposits, in Syvitski, J.P.M., et al., eds., COLDSEIS (seismic facies of glacigenic deposits): Marine Geology, v. 143, p. 151–168.

Sturrock, S., and Murray, J.W., 1981, Comparison of low energy and high energy marine shelf foraminiferal faunas, Celtic Sea and western English Channel, in Neale, J.W., and Braisier, M.D., eds., Microfossils from recent and fossil shelf seas: Chichester, United Kingdom, Ellis Horwood Ltd., p. 250–260.

Syvitski, J.P.M., 1991, Towards an understanding of sediment deposition on glaciated continental shelves, in Amos, C.L., and Collins, M., eds., Proceedings of the Canadian continental shelf seabed symposium ($C^2S^3$): Continental Shelf Research, v. 11, p. 897–937.

Syvitski, J.P.M., 1993, Glaciomarine environments in Canada: An overview: Canadian Journal of Earth Sciences, v. 30, p. 354–371.

Syvitski, J.P.M., Lewis, C.F.M., and Piper, D.J.W., 1996, Paleoceanographic information derived from acoustic surveys of glaciated continental margins: Examples from eastern Canada, in Andrews, J.T., et al., eds., Late Quaternary palaeoceanography of the North Atlantic margins: Geological Society [London] Special Publication 111, p. 51–76.

Syvitski, J.P.M., Stoker, M.S., and Cooper, A.K., 1997, Seismic facies of glacial deposits from marine and lacustrine environments, in Syvitski, J.P.M., et al., eds., COLDSEIS (seismic facies of glacigenic deposits): Marine Geology, v. 143, p. 1–4.

Tucker, C.M., and McCann, S.B., 1980, Quaternary events on the Burin Peninsula, Newfoundland, and the islands of St. Pierre and Miquelon, France: Canadian Journal of Earth Sciences, v. 17, p. 1462–1479.

Twichell, D.C., McClennen, C.E., and Butman, B., 1981, Morphology and processes associated with the accumulation of the fine-grained sediment deposit on the southern New England Shelf: Journal of Sedimentary Petrology v. 51, p. 269–280.

Vilks, G., 1980, Post glacial sedimentation on Labrador Shelf: Geological Survey of Canada Paper 78-28, 28 p.

Vilks, G., 1981, Late glacial–postglacial foraminiferal boundary in sediments of eastern Canada, Denmark and Norway: Geoscience Canada, v. 8, p. 48–55.

Vilks, G., and Deonarine, B., 1988, Labrador Shelf benthic foraminifera and stable oxygen isotopes of *Cibicides lobatulus* related to the Labrador Current: Canadian Journal of Earth Sciences, v. 25, p. 1240–1255.

Vilks, G., and Rashid, M.A., 1976, Post-glacial paleo-oceanography of Emerald Basin, Scotian Shelf: Canadian Journal of Earth Sciences, v. 13, p. 1256–1267.

Vilks, G., Deonarine, B., Wagner, F.J., and Winters, G., 1982, Foraminifera and mollusca in surface sediments of southeastern Labrador Shelf: Geological Society of America Bulletin, v. 93, p. 225–238.

Vilks, G., Hardy, I., and Josenhans, H.W., 1984, Late Quaternary stratigraphy of the inner Labrador shelf, *in* Current research, Part A: Geological Survey of Canada Paper 84-1A, p. 57–65.

Vilks, G., MacLean, B., Deonarine, B., Currie, C.G., and Moran, K., 1989, Late Quaternary paleoceanography and sedimentary environments in Hudson Strait: Géographie Physique et Quaternaire, v. 43, p. 161–178.

Vorren, T.O., Hald, M., Edvardsen, M., and Lind-Hansen, O.-W., 1983, Glacigenic sediments and sedimentary environments on continental shelves: General principles with a case study from the Norwegian Shelf, *in* Ehlers, J., ed., Glacial deposits in north-west Europe: Rotterdam, Balkema Publishers, p. 61–73.

Vorren, T.O., Lebesbye, E., Andreassen, K., and Larsen, K.B., 1989, Glacigenic sediments on a passive continental margin as exemplified by the Barents Sea, *in* Powell, R.D., and Elverhøi, A., eds., Modern glacimarine environments: Glacial and marine controls of modern lithofacies and biofacies: Marine Geology, v. 85, p. 251–272.

Weiss, L., 1954, Foraminifera and origin of the Gardiners Clay (Pleistocene), eastern Long Island, New York: U.S. Geological Survey Professional Paper 254-G, p. 143–163.

Williamson, M.A., Keen, C.E., and Mudie, P.J., 1984, Foraminiferal distribution on the continental margin off Nova Scotia: Marine Micropaleontology, v. 9, p. 219–239.

Wingfield, R., 1990, The origin of major incisions within the Pleistocene deposits of the North Sea: Marine Geology, v. 91, p. 31–52.

MANUSCRIPT ACCEPTED BY THE SOCIETY JUNE 9, 2000

# Appendix: Systematic paleontology

The following are abbreviated synonymies for the dominant, and diagnostic or indicator Quaternary benthonic foraminiferal species found in the Halibut Channel boreholes. Complete synonymies and illustrations can be found in Miller (1999).

Suprageneric classification is in accordance with Lee (1990), except for the Family Rotaliellidae, which has been placed in the Superfamily Glabratellacea by Pawlowski et al. (1993). Generic classification is in accordance with Loeblich and Tappan (1988).

Family: **CASSIDULINIDAE** d'Orbigny, 1839
Genus: CASSIDULINA d'Orbigny, 1826
*Cassidulina reniforme* Nørvang

*Cassidulina crassa* d'Orbigny, CUSHMAN, 1948 (non *C. crassa* d'Orbigny, 1839), p. 74, pl. 8: 9. SCOTT ET AL., 1980, p. 226, pl. 4: 1–2.
*Cassidulina crassa* d'Orbigny var. *reniforme*, NØRVANG, 1945, p. 41, text-figs. 6c–h.
*Cassidulina islandica* Nørvang var. *minuta*, NØRVANG, 1945, p. 41, text-figs. 8a–c.
*Cassidulina islandica* Nørvang var. *nørvangi*, THALMANN in PHLEGER, 1952, p. 83, pl. 14: 30 (*nomen novum*).
*Cassidulina islandica* Nørvang, LOEBLICH and TAPPAN, 1953, p. 118, pl. 24: 1.
*Cassidulina reniforme* Nørvang, SEJRUP and GUILBAULT, 1980, p. 79–80, fig. 2: F–K. RODRIGUES ET AL., 1980, p. 58–59, pl. 2: 2, 4, 6; pl. 3: 3, 6, 9, 11–12; pl. 5: 10–12.
MILLER ET AL., 1982a, p. 2362, pl. 2: 8. FEYLING-HANSSEN, 1990a, p. 22, pl. 4, figs. 4–9.
*Cassidulina nørvangi* Thalmann, NOMURA, 1983a, pl. 23: 10–12; pl. 24: 1–3; figs. 45–46. NOMURA, 1983b, p. 53–55, pl. 4: 12–13.

*Remarks*: Sejrup and Guilbault (1980) clarified the taxonomy of *Cassidulina crassa, Cassidulina reniforme* and *Cassidulina obtusa*, and the following account is taken from their paper.

Nørvang (1945) described two species from Recent faunas in shallow waters off Iceland: *Cassidulina crassa* var. *reniforme* and *Cassidulina islandica*. After Nørvang designated his variety (*reniforme*), it, and *Cassidulina obtusa*, both appear to have been referred to, often, as just *C. crassa* (i.e., for *C. reniforme*: Cushman, 1948; Nørvang, 1958; Feyling-Hanssen, 1964; and for *C. obtusa*: Nørvang, 1945, 1958; Parker, 1948; Phleger and Parker, 1951). *C. crassa* var. *reniforme* was elevated to specific rank by Haynes (1973). Sejrup and Guilbault (1980) drew attention to the fact that most authors seemed to pay little attention to d'Orbigny's (1839) description and illustration of *C. crassa* (reproduced by Sejrup and Guilbault, 1980, Fig. 1). Heron-Allen and Earland (1932) published good illustrations of topotypic material of *C. crassa*, which clearly showed the test size and aperture to be quite different from Nørvang's variety, *reniforme*. Most references, then, of *C. crassa* by subsequent authors prior to 1980, are erroneous; and are actually references to *C. reniforme* or *C. obtusa* (Sejrup and Guilbault, 1980).

Genus: **ISLANDIELLA** Nørvang, 1958,
emended Nomura, 1983a
*Islandiella algida* (Cushman)

*Cassidulina algida*, CUSHMAN, 1944, p. 35, pl. 4: 24a–c. PARKER, 1952, p. 421, pl. 6: figs. 21a–b.
*Cassidulina islandica*, NØRVANG, 1945, p. 41, text-figs. 7, 8d–f. CUSHMAN, 1948, p. 75, pl. 8: 13a–c. LOEBLICH and TAPPAN, 1953, p. 118–120, pl. 24: 1a–b.
*Islandiella islandica* (Nørvang), NØRVANG, 1958, p. 27, pl. 6: 1–5. SEN GUPTA, 1971, p. 88, pl. 2: 11–12. NOMURA, 1983a, p. 47, pl. 1: 2a–c; pl. 11: 4–9, text-figs. 21–22. WILLIAMSON ET AL., 1984, p. 224, pl. 1: 20. FEYLING-HANSSEN, 1990b, p. 101, pl. 1: 15–17.
*Globocassidulina subglobosa* (Brady), AMOS and MILLER, 1990 (Appendix A).
*Islandiella algida* (Cushman), MILLER ET AL., 1996, p. 303–305, pl. 1: 1a–5.

*Remarks*: Cushman (1944) described a species from coastal Maine that he designated *Cassidulina algida*. A year later, Nørvang (1945) described *Cassidulina islandica* from off the coast of Iceland. In 1958 Nørvang erected the genus *Islandiella* for those Cassidulinid species having a radial wall texture and internal tooth, as distinct from other Cassidulinid species having a granular wall texture and no internal tooth (Nomura, 1983a). By these criteria *Cassidulina islandica* was transferred to the new genus and designated the type species. Sen Gupta (1972) realized that *C. algida* is the senior subjective synonym of *I. islandica* (Nørvang) and used the name *C. algida* (B. Sen Gupta, 1995, personal commun.). Miller et al. (1996) have shown that *C. algida* and *I. islandica* are conspecific, making *I. islandica* a junior subjective synonym of *C. algida*, though *I. islandica* remains the type species of *Islandiella*.

*Islandiella helenae* Feyling-Hanssen and Buzas

*Cassidulina laevigata* d'Orbigny, PARKER and JONES, 1865 (non *Cassidulina laevigata* d'Orbigny, 1826), p. 377, pl. 15: 2–4.
*Cassidulina teretis* Tappan, LOEBLICH and TAPPAN, 1953 (part), (non *Cassidulina teretis* Tappan, 1951), p. 121, pl. 24: 3–4.
*Islandiella helenae*, FEYLING-HANSSEN and BUZAS, 1976, p. 155, text-figs. 1–4. RODRIGUES ET AL., 1980, p. 49, pl. 1: 1, 3, 5; pl. 4: 3, 6, 9; pl. 6: 1–2. VILKS ET AL., 1982, p. 226, pl. 1: 14. WILLIAMSON ET AL., 1984, p. 224, pl. 1: 18. VILKS, 1989, p. 538, pl. 21-IV: 3–4.
*Islandiella teretis* (Tappan), SCOTT, 1987 (part), p. 328, pl. 2: 13 (only).

*Remarks*: Tappan (1951) described *Cassidulina teretis* from off Alaska; and Feyling-Hanssen and Buzas (1976) showed that *Cassidulina teretis* Tappan is actually two species, *Cassidulina teretis* (*sensu stricto*) and *Islandiella helenae*. Scott (1987) states that he combines *I. teretis, I. helenae* and *I. norcrossi*. Miller (1999) believes that *C. teretis* is a junior subjective synonym of *Cassidulina latacamerata* (Voloshinova, 1939).

Family: **STAINFORTHIIDAE** Reiss, 1963
Genus: STAINFORTHIA Hofker, 1956
*Stainforthia rotundata* (Parr)

"*Bulimina*" *fusiformis*, Williamson, HÖGLUND, 1947 (part, non *B. pupoides* var. *fusiformis*, Williamson, 1858), p. 232–235, text-figs. 219–224. SCOTT ET AL., 1980, p. 226, pl. 3: 9–10. VILKS, 1989, p. 543–544, pl. 21-V: 16–17.
*Virgulina rotundata*, PARR, 1950, p. 337, pl. 12: 14.
*Fursenkoina rotundata* (Parr), LOEBLICH and TAPPAN, 1994, p. 131, pl. 256: 6–13.
*Stainforthia feylingi*, KNUDSEN and SEIDENKRANTZ, 1994 (part), p. 5–7, pl. 1: 4, 5, 10, 12, 13, 21, 22, 27, 29, 31; pl. 2: 1–6, 8.
*Stainforthia complexa* (Sidebottom), MILLER, 1999, p. 501–502, pl. 9: 26–27.

*Remarks*: Sidebottom (1907, 16) illustrated and described a variety of forms, some of which (1907, pl. IV, fig. 1; text-fig. 3–5) appear to have the morphology of *Stainforthia rotundata* (Parr), but Sidebottom refers to the aperture in all forms as cribrate. Revets (1989) recounts the history of Sidebottom's species, and completed detailed structural and morphological studes; placing it in *Delosina* (Wiesner). See remarks below under *S. pauciloculata* re: *S. feylingi*. *S. feylingi* is considered here, in part, to be a junior subjective synonym of *S. rotundata*.

*Stainforthia concava* (Höglund)

*Virgulina concava*, HÖGLUND, 1947, p. 257, pl. 23: 3–4; pl. 32: 4–7, text-figs. 273–275. FEYLING-HANSSEN, 1964, p. 306–307, pl. 14: 9–11.
*Virgulina loeblichi*, FEYLING-HANSSEN, 1964, p. 308–309, pl. 14: 12–14.
*Stainforthia concava* (Höglund), VILKS, 1989, p. 538–539, pl. 21-VI: 7–8. KNUDSEN and SEIDENKRANTZ, 1994, p. 5–13, pl. 3: 12–13.

*Stainforthia pauciloculata* (Brady)

*Virgulina pauciloculata*, BRADY, 1884, p. 414, pl. 52: 4–5.
*Virgulina pontoni*, CUSHMAN, 1932, p. 17, pl. 3: 7.
*Virgulina schreibersania* (Cžjžek), FEYLING-HANSSEN, 1964 (non *V. schreibersania* Cžjžek, 1848), p. 309–310, pl. 14: 19–21.
*Fursenkoina pontoni* (Cushman), POAG, 1981, p. 66, pl. 33: 5; pl. 34: 5a–b.
*Stainforthia feylingi*, KNUDSEN and SEIDENKRANTZ, 1994 (part), p. 5–13; pl. 1: 1–3, 6–9, 11, 14–20, 23–26, 28, 30, 32.
*Fursenkoina pauciloculata* (Brady), LOEBLICH and TAPPAN, 1994, p. 131, pl. 256: 1–5.

*Remarks*: The taxonomic concept and position of *Fursenkoina schreibersania* (Cžjžek) has been discussed by Knudsen and Seidenkrantz (1994); who pointed out, that, according to Papp and Schmid (1985), *F. schreibersania* is a junior synonym of *F. acuta* (d'Orbigny, 1846). Knudsen and Seidenkrantz (1994) erected the species *Stainforthia feylingi* for the Quaternary specimens erroneously attributed to *F.* (or *S.*) *schreibersania*, with no mention of *Virgulina pauciloculata* Brady (1884), or *V. pontoni* Cushman (1932).

*S. feylingi* has a variable aperture and is here considered, in part, a junior subjective synonym of *S. rotundata*.

*Stainforthia fusiformis* (Williamson)

*Bulimina pupoides* var. *fusiformis*, WILLIAMSON, 1858, p. 63, pl. 5: 129–130.
"*Bulimina*" *fusiformis* Williamson, HÖGLUND (part), 1947, p. 232–235, pl. 20: 3; text-figs. 225–231.
*Virgulina fusiformis* (Williamson), FEYLING-HANSSEN, 1964, p. 307–308, pl. 14: 15–18.
*Fursenkoina fusiformis* (Williamson), WILLIAMSON ET AL., 1984, p. 224, pl. 1: 19.
*Stainforthia fusiformis* (Williamson), KNUDSEN and SEIDENKRANTZ, 1994, p. 5–13, pl.3: 1–7, 16–17.
*Stainforthia (Fursenkoina) fusiformis* (Williamson), AUSTIN and SEJRUP, 1994, p. 121, pl. 1: 9.

*Remarks*: See notes pertaining to *Stainforthia*. Miller (1999) followed Williamson's (1858) concept of the species and recognized as distinct those forms with a terminal aperture and placed them in *Fursenkoina*, i.e., as *F. fusiformis*, but recently has received communication and ESEM illustrations from Elizabeth Alve (1999, personal commun.) clearly showing forms with a terminal aperture that have been carefully broken open to reveal earlier formed chambers with a *Stainforthia*-type aperture. As a result, these forms with a terminal aperture are recognized as distinct and currently placed in *Stainforthia* as *S. fusiformis*.

Family: **ROTALIELLIDAE** Loeblich and Tappan, 1964
Genus: ROTALIELLA Grell, 1954
*Rotaliella chasteri* (Heron-Allen and Earland)

*Discorbina minutissima*, CHASTER, 1892 (part) (non *D. minutissima* Seguenza, 1880), p. 65, pl. 1: 15. WRIGHT, 1903, p. 174–175. GOUGH, 1906, p. 57, pl. 1: 8.
*Discorbina chasteri*, HERON-ALLEN and EARLAND, 1913, p. 128, pl. 13: 1–3.
*Glabratella arctica*, SCOTT and VILKS, 1991, p. 30, pl. 2: 10–12.
*Rotaliella chasteri* (Heron-Allen and Earland), PAWLOWSKI ET AL., 1992, p. 129, pl. 1: 1a–2. PAWLOWSKI and ZANINETTI, 1993, p. 232, pl. 1: 5–7. PAWLOWSKI ET AL., 1993, p. 278, pl. 2: 1.
*Rotaliella heronallenia*, PAWLOWSKI ET AL., 1992, p. 129–130, pl. 1: 3a–4b.

*Remarks*: Pawlowski et al. (1992) have recounted the history of this species. Heron-Allen and Earland erected the species *Discorbina chasteri* for specimens described by Chaster (1892) under the name *Discorbina minutissima* Chaster, preoccupied by *D. minutissima* Seguenza (1880). However, Pawloski et al. (1992) are of the opinion that Heron-Allen and Earland's specimens of *Discorbina chasteri* are actually two species, one is *Discorbina minutissima* Chaster (preoccupied by *D. minutissima* Seguenza, 1880) and it retains the name assigned to it by Heron-Allen and Earland (*D. chasteri*). The other species matches the illustrations of Heron-Allen and Earland (1913, pl. 13: 1–3) and is given the name *Rotaliella heronalleni* (Pawlowski et al., 1992, p. 129–130, pl. 1: 3a–4b). Pawlowski and Zaninetti (1993, p. 232) subsequently place both the name *R. chasteri* and Heron-Allen and Earland's (1913) illustrations (i.e., *R. heronallenia*) in synonymy together; inferring that they now (1993) consider the two species to be conspecific, though they don't list *R. heronallenia* as a junior synonym of *R. chasteri*. Pawlowski et al. (1993, p. 278) treat the two species as distinct. Specimens resembling both species are here placed in *R. chasteri*.

Cole (F. E. Cole, 1994, oral communication) has previously recognized this species; first as *Discorbis transluscens* Earland (1934) (Cole, 1981) then as *Discorbis chasteri* variety *spinosa* (Heron-Allen and Earland) (Scott et al., 1989), then as *Rotaliella chasteri*. Cole (1994, oral communication) has confirmed that the specimens identified here as *R. chasteri* are conspecific with those that she has referred to as *R. chasteri* Scott and Vilks (1991) illustrate this same species (confirmed by Vilks, 1991, oral communication) but they identify it as *Glabratella arctica (species novum)*.

Family: **GLABRATELLIDAE** Loeblich and Tappan, 1964
Genus: GLABRATELLINA Seiglie and Bermúdez, 1965

There are three species of (?) *Glabratellina* recognized in the Halibut Channel boreholes, *G. arcuata* (Seiglie and Bermúdez), *G. lauriei* (Heron-Allen and Earland) and *G. wrightii* (Brady). Previously *G. lauriei* and *G. wrightii* were placed in *Glabratella* by authors, but here this generic placement is questioned. The specimens are not small, coarsely perforate, or with inflated globular chambers, as per the generic description in Loeblich and Tappan (1988), nor do they appear to resemble specimens identified as *Glabratella* illustrated in Loeblich and Tappan (1988, 1994). They more closely resemble specimens illustrated as *Glabratellina* in Loeblich and Tappan (1988).

*Glabratellina arcuata* Seiglie and Bermúdez

*Glabratellina arcuata*, SEIGLIE and BERMÚDEZ, 1965, p. 40, pl. 8: 6–11. LOEBLICH and TAPPAN, 1988, p. 567, pl. 619: 7–9.

*Glabratellina lauriei* (Heron-Allen and Earland)

*Discorbina lauriei*, HERON-ALLEN and EARLAND, 1924, p. 633, pl. 36: 50–52; pl. 37: 53–55.
*Discorbis ornatissima*, CUSHMAN, 1925, p. 42, pl. 6: 11–12.
*Rosalina ornatissima* (Cushman), TODD and LOW, 1967, p. A30–A31, pl. 4: 23.
*Glabratella lauriei* (Heron-Allen and Earland), SCHNITKER, 1971, p. 200, pl. 6: 7a–c.
*Glabratella obtusa*, ROUVILLOIS, 1974, p. 15, pl. 2: 9; pl. 3: 1–15.

*Glabratellina wrightii* (Brady)

*Discorbina wrightii*, BRADY, 1881, p. 413, pl. 21: 6a–c.
*Rosalina wrightii* (Brady), FEYLING-HANSSEN, 1980a, p. 276, pl. 2: 11–12. FEYLING-HANSSEN, 1980b, p. 173, pl. 4: 16–18.
*Glabratella wrightii* (Brady), FEYLING-HANSSEN, 1990b, p. 104, pl. 2: 6–8.

Family: **PARRELLOIDIDAE** Hofker, 1956
Genus: CIBICIDOIDES Thalmann, 1939

*Remarks:* The group of species belonging to the genera *Cicicidoides*, *Cibicicoides*, *Cibicides*, and *Lobatula* have been very difficult to work with, for a number of reasons. The morphology can be extremely variable, in part due to the tendency for this species to attach itself to the substrate. Many of the North American east coast species were described by Cushman; and examination of his types at the United States National Museum of Natural History (Smithsonian Institution) reveals that he had both very broad and evolving concepts of the species falling into this general group. He also described species from the (Tertiary) Coastal Plain sediments and applied these names to modern forms; a practice some other workers have taken exception to (i.e. Poag, 1981, p. 53). Phleger and Parker (1951) also described species of the "*Cibicides*" group, and commented on Cushman's broad concepts. One of the earliest described and very widely distributed species in this group, *Lobatula lobatula* (Walker and Jacob) has traditionally been used as a "catch-all" name by those working on the eastern Canadian continental margin; and includes three of the forms recognized in this study. Some of the specimens recognized here agree with some author's concepts of recognized species, but do not match others; so no specific name has been applied in these cases.

*Cibicidoides corpulentus* (Phleger and Parker)

*Cibicides robustus*, PHLEGER and PARKER, 1951 (non *C. robustus* Le Calvez, 1949), p. 31, pl. 17: 1a–4b.
*Cibicides corpulentus* PHLEGER and PARKER, 1952, p. 14 (*nomen novum*).
*Cibicidoides corpulentus*, (Phleger and Parker), POAG, 1981, p. 52–53, pl. 31: 1; pl. 32: 1a–b.

*Cibicidoides floridanus* (Cushman)

*Truncatulina floridana*, CUSHMAN, 1919, p. 62, pl. 19: 2a–c.
*Cibicides floridana* (Cushman), CUSHMAN, 1931 (part), p. 122–123, pl. 23: 4a–c (only).
*Cibicides* aff. *floridanus* (Cushman), PHLEGER and PARKER, 1951 (part), p. 30, pl. 16: 1a–b.
*Cibicidoides "floridanus"* forma *bathyalis*, POAG, 1981, p. 53, pl. 29: 1; pl. 30: 1a–b.

*Remarks:* The early concept of this species was quite broad; Cushman's (1931) and Phleger and Parker's (1951) illustrated specimens probably belong to more than one species; but the species found in Halibut Channel agrees very well with Cushman (1931), pl. 23: 4a–c; and Phleger and Parker (1951), pl. 16: 1a–b. Poag (1981) comments that he believes the modern forms are related to, but not conspecific with, Cushman's (1919) Miocene species. Loeblich and Tappan (1994, p. 149) cite and illustrate a species they refer to as Cushman's species, by the name *Planulina floridana* (Cushman); which is not considered conspecific with the species illustrated here.

*Cibicidoides pseudoungerianus* (Cushman)

*Truncatulina pseudoungeriana*, CUSHMAN, 1922, p. 97, pl. 20: 9.
*Cibicides pseudoungeriana* (Cushman), CUSHMAN, 1931 (part), p. 123–124, pl. 23: 3–6 (only). MURRAY, 1971, p. 177, pl. 74: 1–6.
*Cibicides* cf. *pseudoungerianus* (Cushman), PFLUM and FRERICHS, 1976, p. 114, pl. 3: 1–2.

Family: **PLANULINIDAE** Bermúdez, 1952
Genus: PLANULINA d'Orbigny, 1826
*Planulina vilksae nomen novum*

*Discorbis pulchra*, CUSHMAN, 1947, p. 91–92, pl. 20: 9 (non *Reussella pulchra*, Cushman, 1945).

*Discorbinella minuta*, BUZAS ET AL., 1977 (part), p. 83–84, pl. 3: 23–25 (only) (non *Stetsonia minuta* Parker, 1954).
*Planulina plana*, BELFORD, 1966, p. 122, pl. 10: 14–19 (non *Discorbis plana* Heron-Allen and Earland, 1932). VAN MARLE, 1991, p. 205–206, pl. 22: 7–8.

*Remarks:* This species was difficult to identify; in part due to confusion surrounding this genus, which Belford (1966) has recounted and clarified. This species has an evolute, convex, spiral side; and partially evolute, plano-convex to slightly concave, umbilical side. Confusion has arisen as to which side is dorsal and which is ventral, and the relationship of the apertural position to these two sides. Loeblich and Tappan (1988, p. 580), refer to the umbilical and spiral sides, but do not distinguish between ventral and dorsal. Belford (1966, p. 13–15) recounts various critera employed by different authors to determine the dorsal and ventral sides. Belford (1966) has adopted Glaessner's (1945) criteria of regarding the apical surface as dorsal, and Belford (1966) himself regards the surface nearest the proloculus as the dorsal side, and therefore the greater part of the test lies on the ventral side of the plane passing through the periphery and the center of the proloculus. Another useful observation is that the dorsal side of a trochospiral species is usually the more perforate, or more coarsely perforate. Consequently, the dorsal side of this species is the perforate, plano-convex side, and the evolute spiral side becomes the ventral side. With these criteria established, this species fits reasonably well into the genus *Planulina*. A thick, imperforate, marginal, keel is not as well developed in this species, as it is in others of the genera.

The species itself agrees very well with both Cushman's (1947) description of *Discorbis pulchra* (though Cushman had the dorsal and ventral sides reversed), and Belford's (1966) description of *Planulina plana*. In particular, both authors note the coarsely perforate, partly evolute, plano-convex apertural side, and a very finely, sparsely, irregularly perforate other side. Both authors note the 8–9 chambers in the final whorl, and Belford (1966) makes note of the depressed spiral suture on the ventral side.

Family: **CIBICIDIDAE** Cushman, 1927

*Remarks:* See notes above under "*Cibicidoides*."

Genus: **CIBICIDES** de Montfort, 1808
*Cibicides grossa* Ten Dam and Reinhold

*Cibicides lobatulus* var. *grossa*, TEN DAM and REINHOLD, 1941, p. 62, pl. 5: 5; pl. 6: 1. FEYLING-HANSSEN, 1980b, p. 164, pl. 5: 1–6.
*Cibicides grossa* Ten Dam and Reinhold, FEYLING-HANSSEN, 1990a, p. 24, pl. 4: 24–26.

*Cibicides io* Cushman

*Cibicides pseudoungerianus* (Cushman) var. *io*, CUSHMAN, 1931 (part), p. 125, pl. 23: 1 (only).
*Cibicides io* Cushman, PHLEGER and PARKER, 1951, p. 30, pl. 16: 5a–16b. PARKER, 1954, p. 542, pl. 12: 6–7.
*Cibicidoides io* (Cushman), POAG, 1981, p. 54, pl. 31: 2, pl. 32: 2a–2b.
*Cibicides lobatulus* (Walker and Jacob), SCHAFER and COLE, 1978 (non *N. lobatulus* Walker and Jacob, *in* Kanmaker, 1798), p. 27, pl. 9: 1–2. SCOTT ET AL., 1980 (non *N. lobatulus* Walker and Jacob, *in* Kanmaker, 1798), p. 226, pl. 4: 8–9.
*Cibicidoides io* (Cushman) POAG, 1981, p. 54, pl. 31: 2, pl. 32: 2a–2b.

*Remarks:* This species is the one illustrated by Phleger and Parker (1951), Parker (1954), and Poag (1981) as *Cibicides io* (Cushman). Phleger and Parker (1951) discuss the confusion surrounding this species. Phleger and Parker (1951) point out that the two specimens illustrated by Cushman belong to different species, and in all probability, different genera. Phleger and Parker (1951) point out that Cushman's figure 2 is erroneously referred to as the holotype in Cushman's (1931) publication (plate 2 caption). Figure 2 appears to be a specimen related to *Anomalina*, and Phleger and Parker (1951) refer it to *Cibicides deprimus*, new species. Phleger and Parker (1951) state that the holotype is represented by Cushman's (1931) Figure 1. Pflum and Frerichs (1976, p. 28) refer to Cushman's species as *Anomalina io* (Cushman), with no mention that the original figured specimens may represent more than one species. Those working along the eastern Canadian continental margin have included this form under the name *Cibicides lobatulus* (Walker and Jacob) (see synonymies above). Counts of *Cibicides io* in the Halibut Channel boreholes include *Lobatula lobatula*.

*Cibicides reflugens* de Montfort

*Truncatulina reflugens*, DE MONTFORT, 1808.
*Cibicides lobatulus* (Walker and Jacob), COLE, 1981 (non *Nautilus lobatulus* Walker and Jacob, *in* Kanmaker, 1798), p. 104, pl. 12: 4. VILKS ET AL., 1982, p. 226, pl. 1: 2a–c. VILKS, 1989, p. 543, pl. 21-V: 14–15.
*Cibicides reflugens* de Montfort, LOEBLICH and TAPPAN, 1988, p. 582, pl. 634: 1–3. LOEBLICH and TAPPAN, 1994, p. 149, pl. 318: 7–9.
*Cibicides* sp. B, MILLER, 1999, p. 541–543, pl. 14: 4–6.

*Remarks: Cibicides reflugens* is the most commonly occurring species in the *Cibicides* group, along with *Cibicides io*. Many working along the eastern Canadian continental margin have a broad concept of *Cibicides* (now *Lobatula*) *lobatula* and have included this form in with this species (i.e. Schafer and Cole, 1978; Scott et al., 1980; Cole, 1981; Vilks et al., 1982; Vilks, 1989; see above) but this practice has not been followed here. It most resembles *Cibicides refluegens* de Montfort, cited and illustrated by Loeblich and Tappan (1994, p. 149, pl. 318: 7–9), but their concept of this species in this publication appear to be different from that illustrated by other authors (i.e. *Truncatulina reflugens* (de Montfort), Brady, 1884, p. 659, pl. 92: 7–9; *Cibicides reflugens* (de Montfort): Barker, 1960, p. 190, pl. 92: 7–9; Boltovskoy et al., 1980, p. 24–25, pl.9: 9–11; Ujiié, 1995, p. 68, pl. 11: 1a–c.).

Genus: **LOBATULA** Fleming, 1828
*Lobatula lobatula* (Walker and Jacob)

*Nautilus lobatulus*, WALKER and JACOB in KANMAKER, 1798, p. 642, pl. 14: 36.
*Lobatula vulgaris*, FLEMING, 1828, p. 232.
*Cibicides lobatulus* (Walker and Jacob), MURRAY, 1971, p. 175, pl. 73: 1–7.
*Lobatula lobatulus* (Walker and Jacob), LOEBLICH and TAPPAN, 1988, p. 583, pl. 637: 10–13. LOEBLICH and TAPPAN, 1994, p. 150, pl. 316: 8–11; pl. 319: 1–7.

*Remarks:* Counts of *Cibicides io* in the the Halibut Channel boreholes includes some specimens of *Lobatula lobatula*.

**FAMILY: NONIONIDAE** Schultze, 1854
GENUS: *Haynesina* Banner and Culver, 1978
*Haynesina paucilocula* (Cushman)

*Nonion paucilocula*, CUSHMAN, 1944, p. 24, pl. 3: 26.
*Elphidium subarcticum*, CUSHMAN, 1944, p. 27, pl. 3: 34–35.
*Nonion pauciloculum* Cushman subsp. *albiumbilicatum*, WEISS, 1954, p. 157; pl. 32: 1–2.
*Elphidium pauciloculum* (Cushman), BUZAS, 1965, p. 60, pl. 3: 3.
*Elphidium albiumbilicatum* (Weiss), FEYLING-HANSSEN, 1990a, p. 28, pl. 5: 14–15.
*Haynesina albiumbilicatula* (Weiss), (= *Haynesina germanica* (Ehrenberg) *sensu lato*), BANNER and CULVER, 1978, p. 188–189, p. 195–196.
*Elphidium frigidum* Cushman, SCHAFER and COLE, 1978 (non *E. frigidum* Cushman 1933), p. 27, pl. 10: 1.
*Haynesina paucilocula* (Cushman), MILLER ET AL., 1982a, p. 2362, pl. 2: 6.
*Elphidium subarcticum* Cushman, VILKS, 1989, p. 542, pl. 21-V: 10.
*Haynesina nivea* (Lafrenz), HANSEN and KNUDSEN, 1995, p. 222, Fig. 18: 5–7.

*Remarks:* *H. paucilocula* was originally described by Cushman (*Nonion pauciloculum*, 1944, p. 24) from shallow coastal waters around New England, but the name appears seldom in the literature (see Culver and Buzas, 1980). *Haynesina paucilocula* includes junior subjective synonyms *Elphidium subarcticum* (Cushman, 1944, p. 27) from the same location, and *Nonion pauciloculum* subspecies *albiumbilicatum* Weiss (1954) from the interglacial Gardiner's Clay in New York. Weiss (1954), makes no mention of Cushman's species, or how his variety differs from Cushman's. Cushman's holotypes of *N. pauciloculum* and *E. subarcticum* at the National Museum of Natural History–Smithsonian Institution, appear conspecific with Weiss's holotype (also at the Smithsonian). European workers (i.e. Feyling-Hanssen, 1990a) have identified *E. albiumbilicatum*; Weiss's publication probably came to their attention because they, too, were working in Pleistocene glacial sediments. They may not have been aware of Cushman's (1944) two species, because they were reported as modern ones. Culver and Buzas (1980) transferred *N. pauciloculum* to *Haynesina*, and have kept *H. paucilocula* distinct from *H. germanica*. Culver and Buzas (1980) also cite only one reference (modern occurrence) to *N. pauciloculum albiumbilicatum*, which they synonymize with *H. paucilocula*. Previous to this, Banner and Culver (1978) transferred *N. albiumbilicatum* to *Haynesina*, and referred to the species *H. albiumbilicata* under the heading of *H. germanica sensu lato*.


Family: **HETEROLEPIDAE** Gonzáles-Donoso, 1969
Genus: HETEROLEPA Franzenau, 1884
*Heterolepa subhaidingerii* (Parr)

*Cibicides subhaidingerii*, PARR, 1950, p. 364, pl. 15: 7. BARKER, 1960, p. 196, pl. 95: 7.
*Cibicides umbonatus*, PHLEGER and PARKER, 1951, p. 31, pl. 17: 7a–9b.
*Heterolepa subhaidingerii* (Parr), LOEBLICH and TAPPAN, 1994, p. 163, pl. 359: 1–13.


Family: **ELPHIDIIDAE** Galloway, 1933
Genus: CRIBROELPHIDIUM Cushman
and Brönnimann, 1948
*Cribroelphidium excavatum* (Terquem)

*Polystomella exacavata*, TERQUEM, 1876, p. 429, pl. 2: 2a–d.
*Polystomella striatopunctata* (Fichtel and Moll) variety, HERON-ALLEN and EARLAND, 1909, p. 695, pl. 21: 2a–c.
*Polystomella striatopunctata* var. *selseyensis*, HERON-ALLEN and EARLAND, 1911, p. 448.
*Elphidium incertum* (Williamson) var. *clavatum*, CUSHMAN, 1930, p. 20, pl. 7: 10a–b.
*Elphidium excavatum* (Terquem) forma *clavata* Cushman, MILLER ET AL., 1982b, p. 124–128, pl. 1: 5–8; pl. 2: 3–8; pl. 4: 1–6; pl. 5: 4–8, 14; pl. 6: 1–5.
*Elphidium excavatum* (Terquem) forma *excavata* (Terquem), MILLER ET AL., 1982b, p. 128, pl. 1: 9–12; pl. 2: 1–2; pl. 3: 1–2; pl. 4: 13–16; pl. 5: 15–16; pl. 6: 6–8, 14.
*Elphidium excavatum* (Terquem) forma *selseyensis* (Heron-Allen and Earland), MILLER ET AL., 1982b, p. 132–134, pl. 1: 13–16; pl. 5: 10–13; pl. 6: 9–13.

*Remarks:* Feyling-Hanssen (1972) was the first to recognize and document the polytypic nature of *Elphidium excavatum* (Terquem) (here transferred to the genus *Cribroelphidium* based on Loeblich and Tappan, 1988) and suggest the morphological (and possibly genetic) relationships among *E. excavatum*, *E. selsyensis* (Heron-Allen and Earland) and *E. clavatum* Cushman. Miller (1979) was the first to show intergradation between *E. excavatum* and *E. clavatum*; Miller et al. (1982b) expanded the study of Miller (1979) and showed intergradation between five morphotypes from a number of Recent, Holocene, and Pleistocene localities on each side of the North Atlantic. This was further expanded by Miller (1983) to include ten morphotypes from arctic and boreal environments; and included a comprehensive discriminant (multivariate) analyses of 15 parameters on 721 specimens from 20 arctic to boreal, Late Pleistocene to Recent localities.

Three of the formae recognized by Miller et al. (1982b) have been encountered on the Grand Banks and Northeast Newfoundland Shelf. *C. excavatum* formae *excavata*, *clavata* and *selseyensis*, though *C. excavatum* forma *clavata* is the dominant form. Comprehensive synonymic lists for each formae can be found in Miller et al. (1982b) and Miller (1983).

Miller et al. (1982b) also compiled reported modern occurrences of the various formae of *C. excavatum* ( = *E. excavatum*) including *C. excavatum* forma *clavata* (Cushman), the small, orange-brown disc-shaped translucent ecophenotype, with a smooth peripheral outline and lacking papillae in the sutures and umbilical region (Miller et al., 1982b); and stated that it occurs today as a minor constituent in high-diversity arctic shelf faunas. Hald et al. (1994) have also recently looked at the recent and Late Quaternary distribution of *C. excavatum* forma *clavatum (sic)*.


## REFERENCES CITED

Amos, C.L., and Miller, A.A.L., 1990, The Quaternary stratigraphy of southwest Sable Island Bank: Geological Society of America Bulletin, v. 102, p. 915–934.

Austin, W.E.N., and Sejrup, H.P., 1994, Recent shallow water benthonic foraminifera from western Norway: Ecology and paleoecological significance, *in* Sejrup, H.P., and Knudsen, K.L., eds., Late Cenozoic benthonic foraminifera: Taxonomy, ecology and stratigraphy: Cambridge, Massachusetts, Cushman Foundation for Foraminiferal Research Special Publication No. 32, p. 103–125, plates 1–2.

Banner, F.T., and Culver, S.J., 1978, Quaternary *Haynesina* n. gen. and Paleogene *Protelphidium* Haynes: their morphology, affinities, and distribution: Journal of Foraminiferal Research, v. 8, p. 177–207, plates 1–10.

Barker, R.W., 1960, Taxonomic notes on the species figured by H.B. Brady in his report of the foraminifera dredged by *H.M.S. Challenger* during the years 1873–1876: Tulsa, Oklahoma, Society of Economic Paleontologists and Mineralogists Special Publication No. 9, 239 p., 115 plates.

Belford, D.J., 1966, Miocene and Pliocene smaller foraminifera from Papua–New Guinea: Australia Bureau of Mineral Resources, Geology and Geophysics Bulletin No. 79, 306 p., 38 plates.

Bermúdez, P.J., 1952, Estudio sistemático de los Foraminíferos rotaliformes: Boletín de Geología Venezuela, v. 2, no. 4, p. 1–230, plates 1–35.

Boltovskoy, E. Giussani, G., Watanabe, S., and Wright, R., 1980, Atlas of benthic shelf foraminifera of the southwest Atlantic: The Hague, Netherlands, Dr. W. Junk Publishers, 147 p., 36 plates.

Brady, H.B., 1881, On some Arctic foraminifera from soundings obtained on the Austro-Hungarian North Polar Expedition of 1872–1874: Annals and Magazine of Natural History, series 5, v. 8, p. 393–418, plate 21.

Brady, H.B., 1884, Report on the foraminifera dredged by *H.M.S. Challenger*, during the years 1873–1876, *in* Report on the Scientific Results of the Voyage of the *H.M.S. Challenger* during the years 1873–1876: Zoology, v. 9, 836 p., 115 plates.

Buzas, M.A., 1965, Foraminifera from Late Pleistocene clay near Waterville, Maine: Smithsonian Miscellaneous Collections, v. 145, no. 8, p. 1–35, plates 1–5.

Buzas, M.A., Smith, R.K., and Beem, K.A., 1977, Ecology and systematics of foraminifera in two *Thalassia* habitats, Jamaica, West Indies: Smithsonian Contributions to Paleobiology No. 31, 139 p., 8 plates.

Chaster, G.W., 1892, Report upon the foraminifera of the Southport Society of Natural Science District: First Report Southport Society of Natural Science (1890–1891), appendix, p. 54–72, plate 1.

Cole, F.E., 1981, Taxonomic notes on bathyal zone foraminiferal species off north-east Newfoundland: Dartmouth, Canada, Bedford Institute of Oceanography, Report Series BI-R-81-7/June 1981, 122 p., 20 plates.

Culver, S.J., and Buzas, M.A., 1980, Distribution of recent benthic foraminifera off the North American Atlantic Coast: Smithsonian Contributions to the Marine Sciences, Number 6, 512 p.

Cushman, J.A., 1919, Some Pliocene and Miocene foraminifera from the Coastal Plain of the United States: United States Geological Survey Bulletin No. 676 (1918), p. 1–100, plates 1–31.

Cushman, J.A., 1922, The Byram Calcareous Marl of Mississippi and its foraminifera: United States Geological Survey Professional Paper 129E, p. 79–122, 15 plates.

Cushman, J.A., 1925, Recent foraminifera from British Columbia: Cushman Laboratory for Foraminiferal Research Contributions, v. 1, p. 38–47, plate 6, 7 (part).

Cushman, J.A., 1927, An outline of the re-classification of the foraminifera: Cushman Laboratory for Foraminiferal Research Contributions, v. 3, p. 1–105, plates 1–21.

Cushman, J.A., 1930, The foraminifera of the Atlantic Ocean. Part 7. Nonionidae, Camerinidae, Peneroplidae and Alveolinellidae: Bulletin of the United States National Museum, v. 104, no. 7, 86 p., plates 1–18.

Cushman, J.A., 1931, The foraminifera of the Atlantic Ocean. Part 8. Rotaliidae, Amphisteginidae, Calcarinidae, Cymbaloporettidae, Globorotaliidae, Anomalinidae, Planorbulinidae, Rupertiidae and Homotremidae: Bulletin of the United States National Museum, v. 104, no. 8, 190 p., plates 1–39.

Cushman, J.A., 1932, Notes of the genus *Virgulina*: Cushman Laboratory for Foraminiferal Research Contributions, v. 8, p. 7–23, plates 2–3.

Cushman, J.A., 1933, New Arctic foraminifera collected by Capt. R.A. Bartlett from Fox Basin and off the northeast coast of Greenland: Smithsonian Miscellaneous Collections, v. 89, no. 9, p. 1–8, plates 8–10.

Cushman, J.A., 1944, Foraminifera from the shallow water of the New England Coast: Sharon, Massachusetts, Cushman Laboratory for Foraminiferal Research Special Publication No. 12, 37 p., 4 plates.

Cushman, J.A., 1945, The species of the subfamily Reussellinae of the foraminiferal family Buliminidae: Cushman Laboratory for Foraminiferal Research Contributions, v. 21, p. 23–54, plates 5–8.

Cushman, J.A., 1947, New species and varieties of foraminifera from off the southeastern coast of the United States: Cushman Laboratory for Foraminiferal Research Contributions, v. 23, p. 86–93, plates 19–20.

Cushman, J.A., 1948, Arctic foraminifera: Sharon, Massachusetts, Cushman Laboratory for Foraminiferal Research Special Publication No. 23, 79 p., 8 plates.

Cushman, J.A., and Brönnimann, P., 1948, Some new genera and species of foraminifera from brackish waters off Trinidad. Cushman Laboratory for Foraminiferal Research Contributions, v. 24, p. 15–21, plates 7, 8 (part).

Cžjžek, J., 1848, Beitrag zur Kenntniss der fossilen Foraminiferen des Weinetr Beckens: Haidinger's Naturwissenschaftliche Abdhandlungen Wien, v. 2, no. 1, p. 137–150.

Dam, A. ten, and Reinhold, Th., 1941, Die Stratigraphische Gliederung des niederländischen Plio-Plestozäns nach Foraminiferen: Netherlands, Medeleelingen van de Geologische Stichting, series C, v. 5, no. 1, p. 5–66, 6 plates.

Earland, A. 1934, Foraminifera. Part 3. The Falklands sector of the Antarctic (excluding South Georgia): Discovery Reports, v. 10, p. 1–208, plates 1–10.

Feyling-Hanssen R.W., 1964, Foraminifera in Late Quaternary deposits from the Oslofjord area: Nørges Geologiske Undersoekelse No. 225, 383 p., 25 plates.

Feyling-Hanssen, R.W., 1972, The foraminifer *Elphidium excavatum* (Terquem) and its variant forms: Micropaleontology, v. 18, p. 337–354, plates 1–6.

Feyling-Hanssen, R.W., 1980a, An assemblage of Pleistocene foraminifera from Pigojoat, Baffin Island: Journal of Foraminiferal Research, v. 10, p. 266–285, plates 1–3.

Feyling-Hanssen, R.W., 1980b, Microbiostratigraphy of young Cenozoic marine deposits of the Qivituq Peninsula, Baffin Island: Marine Micropaleontology, v. 5, p. 153–184, plates 1–6.

Feyling-Hanssen, R.W., 1990a, Foraminiferal stratigraphy in the Plio-Pleistocene Kap København Formation, North Greenland: Meddelelser om Grønland Geoscience 24, 32 p., 7 plates.

Feyling-Hanssen, R.W., 1990b, A remarkable foraminiferal assemblage from the Quaternary of northeast Greenland: Bulletin of the Geological Survey of Denmark, v. 38, p. 101–107, plates 1–3.

Feyling-Hanssen, R.W., and Buzas, M.A., 1976, Emendation of *Cassidulina* and *Islandiella helenae* new species: Journal of Foraminiferal Research, v. 6, p. 154–158, text-figs. 1–4.

Fleming, J., 1828, A history of British animals, exhibiting the descriptive characters and systematic arrangement of the genera and species of quadrepeds, birds, fishes, mollusca and radiata of the United Kingdom: Edinburgh, Scotland, Bell and Bradfute, 226 p.

Frazzenau, A., 1884, *Heterolepa* egy uj genus a Foraminiferák rendjében: Természetrajzi Fürzetek Budapest, v. 8, p. 181–184, 214–217, plate 5.

Galloway, J.J., 1933, A manual of foraminifera: Bloomington, Indiana, Principia Press, 496 p., 42 plates.

Glaessner, M.F., 1945, Principles of micropaleontology: Melbourne University Press, Melbourne, Australia, 312 p., 14 plates.

Gonzáles-Donoso, J.M., 1969, Données nouvelles sur la texture et la structure du test de quelques foraminifères du Bassin de Grenade (Espagne): Revue de Micropaléontologie, v. 12, p. 3–8, 2 plates.

Gough, G.C., 1906, The foraminifera of Larne Lough and District: Scientific Investigations, Fisheries Branch, Department of Agriculture and Fisheries, Ireland (Eire), Dublin 3 (1905), p. 1–10, plate 1.

Grell, K.G., 1954, Der Generationswechsel der polythalamen Foraminifére *Rotaliella heterocaryotica*: Archiv für Protistenkunde, v. 100, p. 268–286.

Hald, M., Steinsund, P.I., Dokken, T., Korsun, S., Polyak, L., and Aspeli, R., 1994, Recent and Late Quaternary distribution of *Elphidium excavatum* f. *clavatum* in arctic seas, *in* Sejrup, H.P., and Knudsen, K.L., eds., Late Cenozoic benthonic foraminifera: Taxonomy, ecology and stratigraphy: Cambridge, Massachusetts, Cushman Foundation for Foraminiferal Research Special Publication No. 32, p. 141–154.

Hansen, A., and Knudsen, K.L., 1995, Recent foraminiferal distribution in Freemansundet and Early Holocene stratigraphy on Edgeøya, Svalbard: Polar Research, new series, v. 14, p. 215–238, figs. 17–18.

Haynes, J.R., 1973, Cardigan Bay Recent foraminifera (cruises of the R.V. Antur, 1962–1964). Bulletin of the British Museum (Natural History) (Zoology) Suppliment 4, p. 1–245, plates 1–33.

Heron-Allen, E., and Earland, A., 1909, On the recent and fossil foraminifera of the Shore Sands at Selsey Bill, Sussex. Part II: Journal of the Royal Microscopial Society, London 1909, p. 306–336, plates 15–16.

Heron-Allen, E., and Earland, A., 1911, On the recent and fossil foraminifera of the Shore Sands at Selsey Bill, Sussex, Part VII: Journal of the Royal Microscopial Society, London 1911, p. 298–343.

Heron-Allen, E., and Earland, A., 1913, Clare Island Survey, Part 64, Foraminifera. Proceedings of the Royal Irish Academy, v. 3, p. 1–188, plates 1–13.

Heron-Allen, E., and Earland, A., 1924, The foraminifera of Lord Howe Island, South Pacific: Journal of the Linnean Society, Zoology, v. 35, p. 599–647, plates 35–37.

Heron-Allen, E., and Earland, A., 1932, Foraminifera. Part 1. The ice free area of the Falkland Islands and adjacent seas: Discovery Reports, v. 4, p. 291–460, plates 6–17.

Hofker, J., 1956, Tertiary foraminifera of coastal Ecuador. Part 2. Additional notes on the Eocene species: Journal of Paleontology, v. 30, p. 891–958, 101 text-figs.

Höglund, H., 1947, Foraminifera in the Gullmar Fjord and Skagerrak: Zoologiska Bidrag Från Uppsala, v. 26, p. 1–328, plates 1–32, 308 text-figs.

Kanmacher, F., 1798, Adam's essays on the microscope: The Second Edition, with considerable additions and improvements: London, Dillon and Keating, 712 p.

Knudsen, K.L., and Seidenkrantz, M.-S., 1994, *Stainforthia feylingi* new species from arctic to subarctic environments, previously recorded as *Stainforthia schreibersania* (Cžjžek), in Sejrup, H.P., and Knudsen, K.L., eds., Late Cenozoic benthonic foraminifera: Taxonomy, ecology and stratigraphy: Cambridge, Massachusetts, Cushman Foundation for Foraminiferal Research Special Publication No. 32, p. 5–13, plates 1–3.

Le Calvez, Y., 1949, Révision des foraminiféres Lutétiens du Bassin de Paris. 2. Rotaliidae et familles affines: Mémoires du Service de la Carte Géologique Détaillée de la France, 54 p., 6 plates.

Lee, J.J., 1990, Chapter 29: Phylum Granuloreticulosa (Foraminifera), in Margulis, L., Corliss, J.O., Melkonian, M., and Chapman, D.J., eds., Handbook of Protoctista: Boston, Massachusetts, Jones and Bartlett Publishers, p. 524–548, 31 text-figs.

Loeblich, A.R., Jr., and Tappan, H., 1953, Studies in Arctic foraminifera: Smithsonian Institution Miscellaneous Collections, v. 121, no. 7, p. 1–150, plates 1–24.

Loeblich, A.R., Jr., and Tappan, H., 1964, Sarcodina, chiefly "Thecamoebians" and foraminifera, in Moore, R.C., ed., Treatise on invertebrate paleontology, Part C, Protista 2: Lawrence, Kansas University Press, 899 p., 653 text-figs.

Loeblich, A.R., Jr., and Tappan, H., 1988, Foraminifera genera and their classification, New York, Van Nostrand Reinhold Company, 970 p., 847 plates.

Loeblich, A.R., Jr., and Tappan, H., 1994, Foraminifera of the Sahul Shelf and Timor Sea: Cambridge, Massachusetts, Cushman Foundation for Foraminiferal Research Special Publication No. 31, 661 p., 393 plates.

Miller, A.A.L., 1979, Taxonomy, morphology and microprobe analyses of the Recent foraminifer *Elphidium excavatum* (Terquem) from a Labrador Shelf sediment core [B.Sc. thesis]: Kingston, Canada, Queen's University, 116 p., plates 1–7.

Miller, A.A.L., 1983, *Elphidium excavatum* (Terquem): paleobiological and statistical investigation of infraspecific variation [M.Sc. thesis]: Halifax, Canada, Dalhousie University, 383 p., plates 1–29.

Miller, A.A.L., 1999, The Quaternary sediments and seismostratigraphy of the Grand Banks of Newfoundland and the Northeast Newfoundland Shelf: Foraminiferal refinements and constraints [Ph.D. dissert.]: Washington, D.C., Columbian School of Arts and Sciences, The George Washington University, 972 p., inc. 17 plates.

Miller, A.A.L., Mudie, P.J., and Scott, D.B., 1982a, Holocene history of Bedford Basin, Nova Scotia: Foraminifera, dinoflagellate and pollen records: Canadian Journal of Earth Sciences, v. 19, p. 2342–2367, plates 1–3.

Miller, A.A.L., Scott, D.B., and Medioli, F.S., 1982b, *Elphidium excavatum* (Terquem): Ecophenotypic versus subspecific variation: Journal of Foraminiferal Research, v. 12, p. 116–144, plates 1–6.

Miller, A.A.L., Nomura, R., and Osterman, L.E., 1996, *Islandiella algida* (Cushman, 1944), senior subjective synonym of *Islandiella islandica* (Nørvang, 1945): Journal of Foraminiferal Research, v. 26, p. 209–212, plate 1.

Montfort, P.D. de, 1808, Conchyliologie Systématique et Classification Méthodique des Coquilles, v. 1: Paris, F. Schoell, 446 p.

Murray, J.W., 1971, An atlas of British Recent Foraminiferids: New York, American Elsevier Publishing Company, Inc., 244 p., 96 plates.

Nomura, R., 1983a, Cassidulinidae (Foraminiferida) from the uppermost Cenozoic of Japan. Part 1: Science Reports of the Tohoku University, Sendai, Second Series (Geology), v. 53, no. 1, pt. 1, 101 p., plates 1–25.

Nomura, R., 1983b, Cassidulinidae (Foraminiferida) from the uppermost Cenozoic of Japan. Part 2: Science Reports of the Tohoku University, Sendai, Second Series (Geology), v. 53, no. 1, pt. 2, 93 p., plates 1–6.

Nørvang, A., 1945, The zoology of Iceland, Foraminifera: v. 2, pt. 2: Copenhagen (Denmark) and Reykjavik (Iceland), Ejnar Munksgaard, 79 p., 14 text-figs.

Nørvang, A., 1958, *Islandiella*, n. g., and *Cassidulina* d'Orbigny: Dansk Naturhistorisk Horening Videnskabelige Meddelelser, v. 125, p. 25–41, plates 6–9.

d'Orbigny, A., 1826, Tableau méthodique de la classe des Céphalopodes: Paris, France, Crochard, Annales des Sciences Naturelles, Paris, series 1, v. 7, p. 245–314, atlas, plates 10–17.

d'Orbigny, A., 1839, Foraminifères, in de la Sagra, R., Histoire physique, politique et naturelle de l'île de Cuba: Paris, France, Arthus Bertran, 272 p., atlas, 12 plates.

d'Orbigny, A., 1846, Foraminifères fossiles du basin Tertiaire de Vienne (Austriche): Paris, France, Gide et Comp$^e$, 312 p., 21 plates.

Papp, A., and Schmid, M.E., 1985, Die fossilen Foraminiferen des Tertiaren Beckens von Wien. Revision der Monographie von Alcide d'Orbigny (1846): Abhandlungen der Geologischen Bundesanstalt, v. 37, p. 1–311, 16 text-figs., 102 plates.

Parker, F.L., 1948, Foraminifera of the continental shelf from the Gulf of Maine to Maryland: Harvard College, Bulletin of the Museum of Comparative Zoology, v. 100, no. 2, p. 213–253, plates 1–7.

Parker, F.L., 1952, Foraminiferal species off Portsmouth, New Hampshire: Harvard College, Bulletin of the Museum of Comparative Zoology, v. 106, no. 9, p. 391–423, plates 1–6.

Parker, F.L., 1954, Distribution of the foraminifera in the northeastern Gulf of Mexico: Harvard College, Bulletin of the Museum of Comparative Zoology, v. 110, no. 10, p. 453–588, plates 1–13.

Parker, W.K., and Jones, T.R., 1865, On some foraminifera from the North Atlantic and Arctic Oceans, including Davis Straits and Baffin's Bay: Philosophical Transactions of the Royal Society, v. 155, p. 325–441, plates 12–19.

Parr, W.J., 1950, Foraminifera: Reports of the British, Australian and New Zealand Antarctic Research Expedition 1929–1931, Series B (Zoology, Botany), v. 5, part 6, p. 232–392, plates 3–15.

Pawlowski, J., and Zaninetti, L., 1993, *Rossyatella*, N. Gen. (Foraminiferida) and other tiny Glabratellacea from the Mediterranean Sea: Journal of Foraminiferal Research, v. 23, p. 231–237, plates 1–3.

Pawlowski, J., Zaninetti, L., Whittaker, J., and Lee, J.J., 1992, The taxonomic status of the minute foraminifera *Discorbina minutissima* Chaster (1862), *D. chasteri* Heron-Allen and Earland (1913) and related species: Journal of Micropalaeontology, v. 11, p. 127–134, plates 1–2.

Pawlowski, J., Lee, J.J., and Gooday, A., 1993, Microforaminifera: Perspective on a neglected group of foraminifera: Archiv für Protistenkunde, v. 143, p. 271–284, plates 1–3.

Pflum, C.E., and Frerichs, W.E., 1976, Gulf of Mexico deep-water foraminifera: Washington, D.C., Cushman Foundation for Foraminiferal Research Special Publication No. 14, 125 p., 8 plates.

Phleger, F., 1952, Foraminiferal distribution in some sediment samples from the Canadian and Greenland Arctic: Cushman Foundation for Foraminiferal Research Contributions, v. 3, p. 79–89.

Phleger, F.B., and Parker, F.L., 1951, Ecology of foraminifera, Northwest Gulf of Mexico. Part 2. Foraminiferal species: Geological Society of America Memoir 46, 64 p., plates 1–20.

Phleger, F.B., and Parker, F.L., 1952, New names for northwestern Gulf of Mexico foraminifera: Cushman Foundation for Foraminiferal Research Contributions, v. 3, p. 14.

Poag, C.W., 1981, Ecologic atlas of benthic foraminifera of the Gulf of Mexico: Woods Hole, Massachusetts, Marine Sciences International, 174 p., 64 plates.

Reiss, Z., 1963, Reclassification of the perforate foraminifera: Bulletin of the Geological Survey of Israel, v. 35, p. 1–111, plates 1–8.

Revets, S.A., 1989, Structure and taxonomy of the genus *Delosina* Wiesner, 1931 (Protozoa: Foraminiferida): Bulletin of the British Museum of Natural History (Zoology), v. 55, p. 1–9.

Rodrigues, C.G., and Richard, S.H., 1986, An ecostratigraphic study of Late Pleistocene sediments of the western Champlain Sea Basin, Ontario and Quebec: Geological Survey of Canada Paper 85-22, 33 p., 6 plates.

Rodrigues, C.G., Hooper, K., and Jones, P.C., 1980, The apertural structures of *Islandiella* and *Cassidulina*: Journal of Foraminiferal Research, v. 10, p. 48–60, plates 1–6.

Rouvillois, A., 1974, Sur quelques espèces rares de Foraminifères dans l'Estuaire de la Rance: Laboratoire de Micropalèontologie de l'Ecole Pratique des Hautes Études, Cahiers de Micropaléontologie 1974, no. 3, p. 11–27, plates 1–4.

Saidova, Kh. M., 1975, Bentosnye Foraminifery Tikhogo Okeana [Benthonic foraminifera of the Pacific Ocean]: Institut Okeanologii *P.P. Shirshova*, Moscow, Akademiya Nauk SSR, 3 v., 875 p., 116 plates.

Schafer, C.T., and Cole, F.E., 1978, Distribution of foraminifera in Chaleur Bay, Gulf of St. Lawrence: Geological Survey of Canada Paper 77-30, 55 p., 13 plates.

Schnitker, D., 1971, Distribution of foraminifera on the North Carolina continental shelf: Tulane Studies in Geology and Paleontology, v. 8, p. 169–215, plates 1–10.

Schultze, M.S., 1854, Über den Organismus der Polythalamien (foraminiferen), nebst Bermerkungen überdie Rhizopoden im Allgemeinen: Leipzig, Germany, Wilhelm Engelmann, 68 p., 7 plates.

Scott, D.B., 1987, Quaternary benthic foraminifers from Deep Sea Drilling Project Sites 612 and 613, Leg 95, New Jersey Transect, *in* Poag, C.W., Watts, A.B., et al., eds., Initial Reports, Deep Sea Drilling Project:, Leg 95: Washington, D.C., U.S. Government Printing Office, p. 313–337, plates 1–2.

Scott, D.B., and Vilks, G., 1991, Benthonic foraminifera in the surface sediments of the deep-sea Arctic Ocean: Journal of Foraminiferal Research, v. 21, p. 20–38, plates 1–2.

Scott, D.B., Schafer, C.T., and Medioli, F.S., 1980, Eastern Canadian estuarine foraminifera: A framework for comparison: Journal of Foraminiferal Research, v. 10, p. 205–234, plates 1–4.

Scott, D.B., Mudie, P.J., Baki, V., MacKinnon, K.D., and Cole, F.E., 1989, Biostratigraphy and late Cenozoic paleoceanography of the Arctic Ocean: foraminiferal, lithostratigraphic and isotopic evidence: Geological Society of America Bulletin, v. 101, p. 260–277.

Seguenza, G., 1880, La formazioni Terziarie nella provincia di reggio (Calabria): Atti Reale Accademia dei Lincei, Roma, Memorie: Classe di Scienze Fisiche, Matematiche e Naturali, series 3, v. 6, p. 1–446, plates 1–17.

Seiglie, G.A., and Bermúdez, P.J., 1965, Monografia de la familia de foraminíferos Glabratellidae: Geos, Caracas, v. 12, p. 15–65, plates 1–14.

Sejrup, H.-P., and Guilbault, J.-P., 1980, *Cassidulina reniforme* and *C. obtusa* (Foraminifera), taxonomy, distribution, and ecology: Sarsia, v. 65, p. 79–85, 2 text-figs.

Sen Gupta, B.K., 1971, The benthonic foraminifera of the Tail of the Grand Banks: Micropaleontology, v. 17, p. 69–98, 2 plates.

Sen Gupta, B.K., 1972, Distribution of Holocene benthonic foraminifera on the Atlantic Continental Shelf of North America: Montréal, Canada, 24th International Geological Congress, Proceedings of Section 8, p. 125–134.

Sidebottom, H., 1907, Report on the Recent foraminifera from the coast of the Island of Delos (Grecian Archipeligo). Part IV: Memoirs and Proceedings of the Manchester Literary and Philosophical Society, v. 51, no. 9, p. 1–29, plates 1–4.

Tappan, H., 1951, Northern Alaska index foraminifera: Cushman Foundation for Foraminiferal Research Contributions, v. 2, p. 1–8, plate 1.

Terquem, O., 1876, Essai sur le classement des animaux qui vivent sur la plage et dans les environs de Dunquerque, partie 1: Mémoires de la Société Dunquerquoise pour l'Encouragement des Sciences et des Lettres et des Arts (1874–1875), v. 19, p. 405–457, plates 1–6.

Thalmann, H.E., 1939, Bibliography and index to new genera, species, and varieties of foraminifera for the year 1936: Journal of Paleontology, v. 13, p. 425–465.

Todd, R., and Low, D., 1967, Recent foraminifera from the Gulf of Alaska and southeastern Alaska: U.S. Geological Survey Professional Paper 573-A, p. i–iii, + A1-A46, 5 plates.

Ujiié, H., 1995, Benthic foraminifera common in the bathyal surface sediments of the Ryukyu Island Arc region, Northwest Pacific: University of the Ryukyus, Bulletin of the College of Science, No. 60, 111 p., plates 1–14.

Van Marle, L.J., 1991, Eastern Indonesian Late Cenozoic smaller benthic foraminifera: Verhandelingen der Koninklijke Nedlerlandse Akademie der Wetenschnappen, Afedling Natuurkunde, Eerste Reeks, Deel 34, Amsterdam, North Holland, 328 p., plates 1–25.

Vilks, G., 1989, Chapter 21: Ecology of recent foraminifera on the Canadian Continental Shelf of the Arctic Ocean, *in* Hermann, Y., ed., The Arctic seas: Climatology, oceanography, geology and biology: New York, Van Nostrand Reinhold, p. 497–569, plates 21: 1–9.

Vilks, G., Deonarine, B., Wagner, F.J., and Winters, G.V., 1982, Foraminifera and mollusca in surface sediments of southeastern Labrador Shelf: Geological Society of America Bulletin, v. 93, p. 225–238, plate 1.

Voloshinova, N.A., 1939, On foraminifera from the Tertiary deposits of Sakhalim and Kamchatka, *in* Fursenko, A.V., ed., Articles on Microfauna: Leningrad, Neftianyi geologo-razvedochnyi Institut, Trudy [Transactions of the Oil Geological Institute], series A, no. 116, p. 70–89, 3 plates.

Weiss, L., 1954, Foraminifera and origin of the Gardiners Clay (Pleistocene), eastern Long Island, New York: U.S. Geological Survey Professional Paper 254-G, p. 143–163, plates 32–33.

Williamson, M.A., Keen, C.E., and Mudie, P.J., 1984, Foraminiferal distribution on the continental margin off Nova Scotia: Marine Micropaleontology, v. 9, p. 219–239, plate 1.

Williamson, W.C., 1858, On the recent foraminifera of Great Britain: London, Ray Society, 127 p., 7 plates.

Wright, J., 1903, Foraminiferal high level boulder clay in the County of Dublin, and in Dumfrieshire and Ayrshire; with observations on the origins of the boulder clays: Dublin, Irish Naturalist, v. 12, p. 173–180.

# DATA

Tables A1–A3: Quantitative benthic foraminiferal data, reported as percent occurrence, and actual counts of Quaternary planktonic and reworked foraminifera, reported to the nearest 0.5 percent. X < 1 percent. Sample size 35 cc (bulk sample).

*indicates that the species was not encountered in the split counted, but was observed in the whole sample, usually when the sample was examined to choose specimens for ESEM photographing.

***indicates that counts for *Cibicides io* includes *Lobatula lobatulus*.

~~indicates that counts for *Nonionella turgida* includes *Nonionoides grateloupi*.

~~~indicates that counts for *Pulleniella osloensis* includes counts for *Haynesina germanica* (see Systematic Paleontology in Miller [1999] for discussion of these combinations).

MANUSCRIPT ACCEPTED BY THE SOCIETY JUNE 9, 2000

## TABLE A1. BOREHOLE 87B-1

| Sample number | 1 | 2 | 3 | 4 | 5 | 6 | 7 |
|---|---|---|---|---|---|---|---|
| Depth (m) | 0.4 | 0.53 | 1.06 | 3.46 | 4.42 | 5.46 | 6.1 |
| Total number Quaternary benthics | 42069 | 53248 | 41344 | 37056 | 19712 | 44928 | 39040 |
| Split counted | (3/256) | (1/128) | (1/128) | (1/64) | (1/64) | (1/128) | (1/128) |
| **AGGLUTINATED FORAMINIFERA** | | | | | | | |
| *"Adercotryma" group* | | | | | | | |
| Adercotryma glomerata | X | | | | | | |
| *"Eggerella" group* | | | | | | | |
| Rhumblerella humbolti | X | | | | | | |
| *"Recurvoides" group* | | | | | | | |
| Recurvoides scitulum | X | | | | | | |
| *"Saccammina" group* | | | | | | | |
| Lagenammina atlantica | X | | | | | | |
| *"Spiroplectammina" group* | | | | | | | |
| Spiroplectammina biformis | X | | | | | | |
| Spiroplectammina earlandi | X | | | | | | |
| *"Trochammina" group* | | | | | | | |
| Paratrochammina haynesi | X | | | | | | |
| **CALCAREOUS FORAMINIFERA** | | | | | | | |
| *"Bolivina" group* | | | | | | | |
| Bolivinella pseudopunctata | X | | | | | | |
| *"Buccella" group* | | | | | | | |
| Buccella frigida | | | | X | | | |
| Buccella hannai | 9.5 | 2.0 | 1.0 | 2.0 | | 4.0 | 2.0 |
| Buccella sp. A | | 1.0 | | X | X | | |
| *"Bulimina" group* | | | | | | | |
| Bulimina marginata / aculeata | | | | | X | | |
| *"Cassidulina—Islandiella" group* | | | | | | | |
| Cassidulina reniforme | 18.0 | 14.0 | 5.5 | 9.5 | 6.0 | 10.0 | 8.5 |
| Islandiella algida | 21.5 | X | | X | | X | 1.0 |
| Islandiella helenae | 5.0 | | X | X | | 2.0 | 2.0 |
| *"Cibicides" group* | | | | | | | |
| ***Cibicides io / Lobatula lobatula | 1.0 | 26.5 | 29.5 | 23.5 | 23.0 | 10.5 | 9.5 |
| Cibicides reflugens | 7.5 | 2.0 | 1.0 | 4.5 | 14.5 | 11.5 | 18.0 |
| Cibicidoides corpulentus | | X | | | | | |
| *"Discorbis—Rosalina" group* | | | | | | | |
| Eoeponidella pulchella | X | X | | | X | X | |
| Gavelinopsis praegeri | | | | | | X | 1.0 |
| Rosalina columbiensis | | X | | | | | |
| Rosalina globularis | | X | | | X | X | |
| Rotorbis auberi | | X | | | | | |
| *"Elphidium" group* | | | | | | | |
| Cribroelphidium asklundi | | 10.0 | 1.0 | X | | | |
| Cribroelphidium bartletti | 1.0 | | | | | | |
| Cribroelphidium excavatum | 13.5 | 32.5 | 40.0 | 40.0 | 36.5 | 42.0 | 49.0 |
| Cribroelphidium hallandense | | X | | | | | |
| *"Epistominella" group* | | | | | | | |
| Pseudoparella takayanagii | 2.0 | X | | | X | X | X |
| *"Fissurina" group* | | | | | | | |
| Fissurina aequilabialis | X | | | | | | |
| Fissurina marginata | X | X | | | | | |
| Fissurina stewartii | X | | X | X | | | |

## TABLE A1. BOREHOLE 87B-1 (continued)

| Sample number | 1 | 2 | 3 | 4 | 5 | 6 | 7 |
|---|---|---|---|---|---|---|---|
| Depth (m) | 0.4 | 0.53 | 1.06 | 3.46 | 4.42 | 5.46 | 6.1 |
| Total number Quaternary benthics | 42069 | 53248 | 41344 | 37056 | 19712 | 44928 | 39040 |
| Split counted | (3/256) | (1/128) | (1/128) | (1/64) | (1/64) | (1/128) | (1/128) |
| **"Glabratella" group** | | | | | | | |
| Glabratellina arcuata | X | 21.0 | 12.5 | 11.5 | 8.5 | 11.0 | 5.0 |
| Glabratellina wrightii | | 1.0 | X | X | X | 1.0 | 1.5 |
| Rotaliella chasteri | | 1.0 | | | | X | |
| **"Miliolinellid" group** | | | | | | | |
| Miliolinella circularis | | | X | X | | | |
| Triloculinella tegminus | | | X | | 1.0 | | |
| **"Nonion" group** | | | | | | | |
| Laminononion stellatum | 1.0 | 2.0 | 1.0 | 1.5 | X | 2.5 | 1.0 |
| Haynesina orbiculare | X | 1.5 | 4.5 | 2.0 | 1.5 | 2.0 | 2.0 |
| Haynesina paucilocula | 5.0 | 1.0 | 2.0 | 2.0 | 3.0 | 1.5 | |
| Nonionellina labradorica | 2.0 | X | | | | | |
| **"Oolina" group** | | | | | | | |
| Favulina melo | X | | | | | | |
| Homalohedra apiopleura | X | | X | X | | | |
| **"Quinqueloculina" group** | | | | | | | |
| Quinqueloculina lata | | | | X | | | |
| Quinqueloculina seminula | X | | | | X | X | |
| Siphonaperta aspera | X | | | | | | |
| Siphonaperta parvula | | | | | 1.0 | | |
| **"Stainforthia" group** | | | | | | | |
| Stainforthia fusiformis | 2.0 | | | | | X | |
| Stainforrthia rotundata | X | X | | | | | |
| **"Spiralina" group** | | | | | | | |
| Patellina corrugata | | X | | | | | |
| **"Uvigerina" group** | | | | | | | |
| Angulogerina angulosa/fluens | 5.0 | | | X | | X | |
| Uvigerina peregrina | | | | X | | | |
| **QUATERNARY PLANKTONICS** | | | | | | | |
| **Total number of specimens** | 768 | 896 | 640 | 192 | 448 | 640 | 256 |
| **Split counted** | (3/256) | (1/128) | (1/128) | (1/32) | (1/64) | (1/128) | (1/128) |
| Globigerina bulloides | | | | | | 256 | |
| Globogerinita uvula | 768 | | | | | | 128 |
| Neoglobquadrina pachyderma—sinistral | | 640 | 128 | 160 | | 256 | 128 |
| Neoglobquadrina pachyderma—dextral | | 128 | 512 | 32 | | | |
| Turborotalita quinqueloba—sinistral | | | | | | 128 | |
| Turborotalita quinqueloba—dextral | | 128 | | | | | |
| **Reworked** | | | | | | | |
| **Total number of specimens** | / | 768 | 128 | 96 | 192 | 512 | 384 |
| **Split counted** | | (1/128) | (1/128) | (1/32) | (1/64) | (1/128) | (1/128) |

## TABLE A2. BOREHOLE 87B-4

| Sample number | 63 | 62 | 64 | 65 | 66 | 67 | 69 | 70 | 71 | 72 | 73 | 74 | 76 |
|---|---|---|---|---|---|---|---|---|---|---|---|---|---|
| Depth (m) | 0.23 | 0.25 | 3.01 | 4.05 | 6.73 | 7.72 | 9.66 | 10.77 | 11.49 | 13.6 | 15.32 | 17.29 | 19.40 |
| Total number of Quaternary benthics | 42752 | 98560 | 736 | 757 | 1140 | 1212 | 2640 | 1700 | 1755 | 1256 | 1920 | 1428 | 3776 |
| Split counted | (1/128) | (1/256) | (1/2) | (9/16) | (1/4) | (1/4) | (1/8) | (1/4) | (3/16) | (1/4) | (3/16) | (1/4) | (1/8) |

### AGGLUTINATED FORAMINIFERA

**"Cribrostomoides" group**

| Species | 63 | 62 | 64 | 65 | 66 | 67 | 69 | 70 | 71 | 72 | 73 | 74 | 76 |
|---|---|---|---|---|---|---|---|---|---|---|---|---|---|
| *Labrospira jeffreysii* | | | | X | | | | | | | | | |

**"Eggerella" group**

| Species | 63 | 62 | 64 | 65 | 66 | 67 | 69 | 70 | 71 | 72 | 73 | 74 | 76 |
|---|---|---|---|---|---|---|---|---|---|---|---|---|---|
| *Rhumblerella humbolti* | | | | | | X | | 3.0 | | | X | | 1.0 |

**"Reophax" group**

| Species | 63 | 62 | 64 | 65 | 66 | 67 | 69 | 70 | 71 | 72 | 73 | 74 | 76 |
|---|---|---|---|---|---|---|---|---|---|---|---|---|---|
| *Reophax fusiformis* | | X | | | | | | | | | | | |
| *Subreophax guttifer* | X | | | | | | | | | | | | |

**"Saccammina" group**

| Species | 63 | 62 | 64 | 65 | 66 | 67 | 69 | 70 | 71 | 72 | 73 | 74 | 76 |
|---|---|---|---|---|---|---|---|---|---|---|---|---|---|
| *Hemispherammina bradyi* | | | | 1.0 | X | | | | | | | | |
| *Psammosphaera fusca* | | | | | X | | | | | | | | |

**"Spiroplectammina" group**

| Species | 63 | 62 | 64 | 65 | 66 | 67 | 69 | 70 | 71 | 72 | 73 | 74 | 76 |
|---|---|---|---|---|---|---|---|---|---|---|---|---|---|
| *Spiroplectammina biformis* | X | | | | | | | | | | X | | |

**"Trochammina" group**

| Species | 63 | 62 | 64 | 65 | 66 | 67 | 69 | 70 | 71 | 72 | 73 | 74 | 76 |
|---|---|---|---|---|---|---|---|---|---|---|---|---|---|
| *Lepidodeuterammina ochracea* | | | | | X | | | | | | | | |
| *Lepidodeuterammina ochracea sinuosa* | | X | | | | | | | | | | | |

### CALCAREOUS FORAMINIFERA

**"Bolivina" group**

| Species | 63 | 62 | 64 | 65 | 66 | 67 | 69 | 70 | 71 | 72 | 73 | 74 | 76 |
|---|---|---|---|---|---|---|---|---|---|---|---|---|---|
| *Aphelophragmina britannica* | | | | | | | | | | | | X | |
| *Aphelophragmina spathulata* | | | | | | | | X | | | X | | |
| *Bolivina aenariensis* | | | X | | | | | X | | | | | |
| *Bolivinella pseudopunctata* | | X | X | 1.0 | X | X | | | | | | | |

**"Buccella" group**

| Species | 63 | 62 | 64 | 65 | 66 | 67 | 69 | 70 | 71 | 72 | 73 | 74 | 76 |
|---|---|---|---|---|---|---|---|---|---|---|---|---|---|
| *Aubignyana perlucida* | | | | | | | | X | | | | X | X |
| *Buccella depressa* | | 1.0 | | X | X | | X | | | | | 2.0 | X |
| *Buccella frigida* | | | 9.5 | | | | | | | | | X | |
| *Buccella hannai* | | | X | | | | X | | | | | | |
| *Buccella karsteni* | 10.5 | 7.5 | | 9.0 | 8.5 | 5.0 | 4.0 | 1.0 | 1.5 | 2.0 | 1.5 | 1.0 | 2.0 |
| *Buccella sp. A* | | | 2.0 | 1.0 | 1.5 | 2.0 | 2.0 | 1.0 | X | | 1.0 | | 1.0 |
| *Trichohyalus kingi* | | | | | | | | | | | | X | |

**"Bulimina" group**

| Species | 63 | 62 | 64 | 65 | 66 | 67 | 69 | 70 | 71 | 72 | 73 | 74 | 76 |
|---|---|---|---|---|---|---|---|---|---|---|---|---|---|
| *Bulimina marginata/aculeata* | | | | X | | | | | | | | | |
| *Buliminella tenuis* | X | | X | | | | | | | | X | | |
| *Globobulimina auriculata* | | X | | | | | | | | X | | | |

**"Cassidulina—Islandiella" group**

| Species | 63 | 62 | 64 | 65 | 66 | 67 | 69 | 70 | 71 | 72 | 73 | 74 | 76 |
|---|---|---|---|---|---|---|---|---|---|---|---|---|---|
| *Cassidulina carinata* | | | X | | | | | | | | | | |
| *Cassidulina reniforme* | 5.0 | 38.5 | 14.5 | 30.5 | 19.0 | 20.0 | 21.5 | 33.5 | 31.5 | 36.5 | 30.0 | 31.0 | 24.5 |
| *Paracassidulina neocarinata* | | | X | | | | | | | X | | | |
| *Islandiella algida* | 29.0 | X | 11.5 | 5.5 | 3.5 | 2.0 | X | X | | X | X | | X |
| *Islandiella helenae* | 16.0 | 1.0 | 5.5 | 2.0 | X | 1.5 | 1.0 | 6.0 | 5.0 | 5.0 | 5.5 | 3.5 | 2.0 |
| *Islandiella norcrossi* | 1.0 | X | X | X | X | X | | X | X | | | | |
| *Lernella braziliensis* | | | | | | | | | | | X | | |

**"Cibicides" group**

| Species | 63 | 62 | 64 | 65 | 66 | 67 | 69 | 70 | 71 | 72 | 73 | 74 | 76 |
|---|---|---|---|---|---|---|---|---|---|---|---|---|---|
| ***Cibicides io/Lobatula lobatula* | 1.0 | 2.0 | 1.0 | X | | X | | X | | | X | X | |
| *Cibicides reflugens* | 4.0 | 20.5 | 2.0 | 3.0 | 2.0 | 1.0 | 1.0 | X | X | X | X | X | X |
| *Cibicidoides corpulentus* | | | 2.0 | | | X | X | | | | | | |
| *Planulina vilksae* | | X | | X | | | | | | | | X | |

**"Cornuspira" group**

| Species | 63 | 62 | 64 | 65 | 66 | 67 | 69 | 70 | 71 | 72 | 73 | 74 | 76 |
|---|---|---|---|---|---|---|---|---|---|---|---|---|---|
| *Coluspira sablensis* | | | | X | | X | | X | | | | | |
| *Cornuspira planorbis* | | | | | | | | | | | | | X |

## TABLE A2. BOREHOLE 87B-4 (continued)

| Sample number | 63 | 62 | 64 | 65 | 66 | 67 | 69 | 70 | 71 | 72 | 73 | 74 | 76 |
|---|---|---|---|---|---|---|---|---|---|---|---|---|---|
| Depth (m) | 0.23 | 0.25 | 3.01 | 4.05 | 6.73 | 7.72 | 9.66 | 10.77 | 11.49 | 13.6 | 15.32 | 17.29 | 19.40 |
| Total number of Quaternary benthics | 42752 | 98560 | 736 | 757 | 1140 | 1212 | 2640 | 1700 | 1755 | 1256 | 1920 | 1428 | 3776 |
| Split counted | (1/128) | (1/256) | (1/2) | (9/16) | (1/4) | (1/4) | (1/8) | (1/4) | (3/16) | (1/4) | (3/16) | (1/4) | (1/8) |
| **"Dentalina" group** | | | | | | | | | | | | | |
| Laevidentalina ittai | | | | | | X | | | | | | | |
| **"Discorbis—Rosalina" group** | | | | | | | | | | | | | |
| Altastellerella riveroae | | | X | | | | | | | | | | X |
| Eoeponidella pulchella | X | | X | | X | | 1.5 | 1.0 | | 1.5 | X | X | |
| Gavelinopsis praegeri | | | | | | X | | | | | | | X |
| Gavelinopsis sp. C | | X | | | | | | | | | | | |
| Lamarckina haliotidea | | | | | X | | | | X | | | | X |
| Neodiscorbinella plana | | | X | X | | X | X | | | | X | | X |
| Rosalina globularis | | 1.5 | X | 1.0 | X | X | X | | X | | X | | X |
| Rotorbis auberi | | | | | | | | | | | | X | |
| **"Elphidium" group** | | | | | | | | | | | | | |
| Cribroelphidium asklundi | | X | X | X | X | X | X | 1.0 | X | X | | | |
| Cribroelphidium bartletti | 1.5 | | X | X | X | | X | X | 1.0 | | X | X | |
| Cribroelphidium excavatum | 9.5 | 12.5 | 26.5 | 34.0 | 43.0 | 43.0 | 48.0 | 47.5 | 49.0 | 41.5 | 43.5 | 45.0 | 50.5 |
| Elphidiella hannai | | | | | | | | | X | | | | |
| Elphidiella rolfi | | | | | | | | | | | | | X |
| **"Epistominella" group** | | | | | | | | | | | | | |
| Pseudoparella naraensis | | | 1.0 | 3.0 | 2.0 | 3.0 | 3.0 | | | | X | X | X | 2.0 |
| Pseudoparella takayanagii | X | | X | | | X | X | | X | | | X | |
| **"Fissurina" group** | | | | | | | | | | | | | |
| Fissurina aequilabialis | | | 1.0 | | | | | X | X | | X | X | |
| Fissurina elliptica | X | | | | | | | | | | | | |
| Fissurina goreaui | | | | | | | | X | | | | | |
| Fissurina labeona | | | | | | | | X | | | | X | |
| Fissurina marginata | | | | | | | | X | X | X | | | |
| Fissurina pseudoglobosa | | | | | | | | | | | | X | |
| Fissurina quadracostulata | | | | | X | | | | | | | | |
| Fissurina stewartii | | | X | | | | | | | | X | | |
| Fissurina tricarinata | | | | | | | | | | | | | X |
| Irenita humila | | | | | | | | | | | | X | |
| Lagenosolenia inflataperforata | | | | | | | X | | | | | | |
| Lagenosolenia lagenoides | | X | | | | | | | | | | | |
| Parafissurina curvans | | | | | | X | | | | | | | |
| Parafissurina subcarinata | | X | | | | | | | | | | X | |
| Parafissurina himatiostoma | | | X | | | | | | | | | | |
| Parafissurina labiata | | | | X | | | X | | | | | | |
| Pseudoolina fissurinea | | | | | | | | | | | | X | |
| Wiesnerina minutiformis | | | | X | | | | | | | | | |
| **"Glabratella" group** | | | | | | | | | | | | | |
| Glabratellina arcuata | | 1.0 | 1.5 | 1.0 | 1.0 | 3.0 | 7.5 | 1.0 | X | | X | 2.5 | 6.0 |
| Glabratellina lauriei | | | X | | | X | X | | | X | X | | |
| Glabratellina wrightii | | 1.0 | | X | 1.0 | | X | | | X | X | 1.0 | X |
| Heronallenina parva | | X | | | | | | | | | | | |
| Rotaliella chasteri | | 2.0 | 1.0 | 3.0 | X | 3.5 | 3.0 | X | | | 1.0 | 1.0 | 2.0 |
| **"Gyroidina" group** | | | | | | | | | | | | | |
| Gyroidinoides nipponicus | | | X | | | | | | | | | | |
| **"Lagena" group** | | | | | | | | | | | | | |
| Lagena semilineata | | | X | | | | | | | | | | |
| Procerolagena meridionalis | | | | | | | X | | | | | | |
| **"Miliolinellid" group** | | | | | | | | | | | | | |
| Miliolinella circularis | | | | X | | | | X | | | | | |
| Miliolinella subrotunda | | | | | X | X | | | | | | | X |
| Triloculinella pyriformis | | | | | | | | | | | | | X |
| Triloculinella tegminus | | | | | | | X | | | | | | |

## TABLE A2. BOREHOLE 87B-4 (continued)

| Sample number | 63 | 62 | 64 | 65 | 66 | 67 | 69 | 70 | 71 | 72 | 73 | 74 | 76 |
|---|---|---|---|---|---|---|---|---|---|---|---|---|---|
| Depth (m) | 0.23 | 0.25 | 3.01 | 4.05 | 6.73 | 7.72 | 9.66 | 10.77 | 11.49 | 13.6 | 15.32 | 17.29 | 19.40 |
| Total number of Quaternary benthics | 42752 | 98560 | 736 | 757 | 1140 | 1212 | 2640 | 1700 | 1755 | 1256 | 1920 | 1428 | 3776 |
| Split counted | (1/128) | (1/256) | (1/2) | (9/16) | (1/4) | (1/4) | (1/8) | (1/4) | (3/16) | (1/4) | (3/16) | (1/4) | (1/8) |
| **"Nonion" group** | | | | | | | | | | | | | |
| Laminononion stellatum | | 2.0 | X | X | X | 1.0 | X | X | X | | X | | X |
| Haynesina orbiculare | 2.0 | 1.0 | 3.5 | 2.0 | 3.0 | 1.5 | 4.5 | 3.0 | 3.5 | 3.0 | 4.5 | 4.5 | 4.0 |
| Haynesina paucilocula | 3.0 | 1.0 | 2.0 | 1.5 | 4.0 | 1.0 | 2.0 | 1.0 | | 1.0 | | 2.0 | 1.0 |
| Nonionella lobsannensis | | X | | | | | | | | | | | |
| Nonionella stella | | | | | | | X | | | | | | |
| Nonionellina labradorica | 1.0 | 4.0 | 3.0 | 2.0 | 1.0 | 1.0 | X | 2.0 | 3.5 | 1.0 | 5.5 | 2.0 | X |
| ~~Nonionoides grateloupi/ Nonionella turgida | | X | | | | | | | | | | | |
| ~~Pulleniella osloensis/ Haynesina germanica | | | X | 1.0 | | X | | | | | X | | |
| **"Oolina" group** | | | | | | | | | | | | | |
| Favulina melo | | | X | | | | | | | | | X | X |
| Homalohedra apiopleura | | | X | | | | | | | | | | |
| **"Quinqueloculina" group** | | | | | | | | | | | | | |
| Quinqueloculina akneriana | | | | | | | | | X | | | | X |
| Quinqueloculina longa | | | | | | X | | | | | | | |
| Quinqueloculina seminula | | | | | X | | | | | | | | X |
| Siphonaperta parvula | | | | | X | | | | | | X | X | |
| **"Rotalia" group** | | | | | X | | | | | | | | |
| Valvulineria minuta | | | | X | | | | | | | | | |
| **"Stainforthia" group** | | | | | | | | | | | | | |
| Stainforthia fusiformis | | X | 1.0 | X | X | | | | | | X | | |
| Stainforthia pauciloculata | | | | | | X | X | | X | | | | |
| Stainforthia rotundata | X | | X | X | X | | | | | | | | |
| **"Spiralina" group** | | | | | | | | | | | | | |
| Patellina corrugata | | | 1.0 | X | 1.0 | X | | | X | | | | |
| **"Triloculina" group** | | | | | | | | | | | | | |
| Triloculina angulosa | | | | | | | | X | | | X | | |
| Triloculina tricarinata | X | | | | | | | | | | X | X | |
| **"Uvigerina" group** | | | | | | | | | | | | | |
| Angulogerina angulosa/ fluens | 14.0 | X | 4.5 | 1.0 | | | | | | X | 1.0 | X | |
| Euuvigerina brunnensis | | | | | X | | | | | | | | |
| **QUATERNARY PLANKTONICS** | | | | | | | | | | | | | |
| Total number of specimens | 384 | 8192 | 18 | 7 | 24 | 48 | 112 | 36 | 43 | 60 | 21 | 56 | 392 |
| Split counted | (1/128) | (1/256) | (1/2) | (9/16) | (1/4) | (1/4) | (1/8) | (1/4) | (3/16) | (1/4) | (3/16) | (1/4) | (1/8) |
| Globigerina bulloides | | 512 | | | 4 | 4 | | 8 | | | | | 16 |
| Globogerinita uvula | | 512 | 4 | | 4 | | | 4 | | | 8 | 5 | 8 |
| Globorotalia inflata | | 256 | | | | | | | | | | | |
| Neogloboquadrina dutertrei | | 256 | 2 | | 8 | | 8 | | | | | | |
| Neogloboquadrina pachyderma—sinistral | 256 | 4096 | 4 | 3 | 8 | 28 | 80 | 16 | 43 | 40 | 11 | 36 | 336 |
| Neogloboquadrina pachyderma—dextral | 128 | 1536 | 6 | 2 | | 4 | 8 | 4 | | 4 | | 4 | 32 |
| Turborotalita quinqueloba—sinistral | | 256 | 2 | | | 8 | 8 | 4 | | 8 | | | |
| Turborotalita quinqueloba—dextral | | 768 | | 2 | | 4 | 8 | | | | 5 | 8 | 8 |
| **Reworked** | | | | | | | | | | | | | |
| Total number of specimens | / | 768 | 16 | 11 | 52 | 64 | 208 | 28 | 48 | 60 | 53 | 56 | 464 |
| Split counted | | (1/256) | (1/2) | (9/16) | (1/4) | (1/4) | (1/8) | (1/4) | (3/16) | (1/4) | (3/16) | (1/4) | (1/8) |

## TABLE A3. BOREHOLE 87B-2

| Sample number | 8 | 9 | 10 | 11 | 12 | 13 | 14 | 15 | 16 | 17 | 18 | 19 | 20 | 21 | 22 | 23 | 24 |
|---|---|---|---|---|---|---|---|---|---|---|---|---|---|---|---|---|---|
| Depth (m) | 0.45 | 2.43 | 4.26 | 6.29 | 8.3 | 9.17 | 10.08 | 11.21 | 12.32 | 13.25 | 14.56 | 15.8 | 17.02 | 17.92 | 18.94 | 20.17 | 21.4 |
| Total number of Quaternary benthics | 63,104 | 381,615 | 364,032 | 73,536 | 137,216 | 86,016 | 65,664 | 62,080 | 41,984 | 31,317 | 236,544 | 3,584 | 3,157 | 25,088 | 4,171 | 6,480 | 13,080 |
| Split counted | (1/128) | (1/512) | (1/512) | (1/64) | (1/256) | (1/128) | (1/128) | (1/64) | (1/128) | (3/256) | (1/512) | (3/32) | (3/32) | (1/64) | (3/32) | (1/16) | (1/8) |
| **AGGLUTINATED FORAMINIFERA** | | | | | | | | | | | | | | | | | |
| **"Adercotryma" group** | | | | | | | | | | | | | | | | | |
| Adercotryma glomerata | | | | | | | | | | | | | | | 1.0 | | |
| **"Cribrostomoides" group** | | | | | | | | | | | | | | | | | |
| Labrospira crassimargo | | | | | | | | | | | | | | | | | |
| **"Eggerella" group** | | | | | | | | | | | | | | | | | |
| Rhumblerella humbolti | X | | | | | | X | | | | | | | | 2.0 | | 1.5 |
| **"Recurvoides" group** | | | | | | | | | | | | | | | | | |
| Recurvoides scitulum | | | | | | | | | | | | | | | X | | |
| **"Reophax" arctica** | | | | | | | | | | | | | | | | | |
| Cuneata arctica | X | | | | | | | | | | | | | | | | |
| **"Saccammina" group** | | | | | | | | | | | | | | | | | |
| Hemispherammina bradyi | | | | | | | | | | | | X | 5.0 | | | | X |
| Lagenammina atlantica | X | | | | | | | | | | | | | | | | |
| Psammosphaera fusca | | | | | | | | | | | | 2.5 | 2.0 | 2.5 | | | 5.0 |
| Saccammina sphaerica | | | | | | | | X | | | | | | | X | | |
| **"Spiroplectammina" group** | | | | | | | | | | | | | | | | | |
| Spiroplectammina biformis | | | | | | | | | | | | X | | | X | | |
| Spiroplectammina foliacea | X | | | | | | X | | | | | | | | | | |
| **"Trochammina" group** | | | | | | | | | | | | | | | | | |
| Lepidodeuterammina ochracea | 1.0 | | | | | X | 1.0 | | | 2.5 | 1.0 | | | 4.0 | 1.0 | | |
| Lepidodeuterammina ochracea sinuosa | | | | | | | | | X | | | | | | | | |
| Lepidoparatrochammina haynesi | X | X | 1.0 | X | X | X | X | | | 2.0 | X | | x | | X | | X |
| Lepidoparatrochammina lepida | | | | | | | | | | | | | | | | | |
| **CALCAREOUS FORAMINIFERA** | | | | | | | | | | | | | | | | | |
| **"Bolivina" group** | | | | | | | | | | | | | | | | | |
| Aphelophragmina britannica | | | | X | | | | | | | | | | | | | |
| Aphelophragmina spathulata | | X | X | X | X | X | X | | | | X | X | X | | | | |
| Aphelophragmina vadescens | | | | X | | | | | | | | | X | | | | |
| Bolivina aenariensis | | | X | X | X | | X | | | | | | | | | | |
| Bolivina decussata | | | | | | | | | | | | X | | | | | |
| Bolivinella pseudopunctata | X | X | X | X | X | X | X | X | | | X | | | | X | | |
| **"Buccella" group** | | | | | | | | | | | | | | | | | |
| Aubignyana perlucida | X | | | | | | | | | | | | | | | | |
| Buccella depressa | | | | X | | | X | X | | | | | | | | | |
| Buccella frigida | | | 1.0 | | | 2.0 | 5.5 | | | X | X | | X | | | | X |
| Buccella hannai | 2.5 | | | | X | X | X | X | | X | X | | X | | | | |
| Buccella karsteni | 2.5 | 4.5 | 2.5 | 5.0 | 7.5 | X | X | 4.5 | X | 3.5 | 5.0 | 1.0 | | 4.5 | 1.5 | X | 1.5 |
| Buccella oregonensis | 1.0 | | | | | | | | | | | | | | X | | |
| Buccella sp. A | 3.5 | 2.0 | 1.0 | 2.5 | 1.0 | 3.5 | X | 2.5 | 1.0 | X | 3.5 | 4.5 | 5.0 | 4.5 | 2.5 | 1.0 | 1.5 |
| Trichohyalus kingi | | | | | | | | | | | | | | | | 1.5 | 1.0 |

TABLE A3. BOREHOLE 87B-2 (continued)

| Sample number | 8 | 9 | 10 | 11 | 12 | 13 | 14 | 15 | 16 | 17 | 18 | 19 | 20 | 21 | 22 | 23 | 24 |
|---|---|---|---|---|---|---|---|---|---|---|---|---|---|---|---|---|---|
| Depth (m) | 0.45 | 2.43 | 4.26 | 6.29 | 8.3 | 9.17 | 10.08 | 11.21 | 12.32 | 13.25 | 14.56 | 15.8 | 17.02 | 17.92 | 18.94 | 20.17 | 21.4 |
| Total number of Quaternary benthics | 63,104 | 381,615 | 364,032 | 73,536 | 137,216 | 86,016 | 65,664 | 62,080 | 41,984 | 31,317 | 236,544 | 3,584 | 3,157 | 25,088 | 4,171 | 6,480 | 13,080 |
| Split counted | (1/128) | (1/512) | (1/512) | (1/64) | (1/256) | (1/128) | (1/128) | (1/64) | (1/128) | (3/256) | (1/512) | (3/32) | (3/32) | (1/64) | (3/32) | (1/16) | (1/8) |
| **"Bulimina" group** | | | | | | | | | | | | | | | | | |
| Bulimina marginata/aculeata | | | | 1.0 | | | | | | | | | | | | | |
| Buliminella elegantissima | | X | | | | | | | | | | | | | X | X | |
| Floresina milletti | | | | | | | | | | | | | | | | | |
| Globobulimina auriculata | 1.5 | | | | | | | | | | 2.0 | | | | | | |
| **"Cassidulina—Islandiella" group** | | | | | | | | | | | | | | | | | |
| Cassidulina carinata | | | | | | | | | | | | | | | | | |
| Cassidulina laevigata | | | | X | | X | X | X | | X | | | | | | | |
| Cassidulina latacamerata | | | | | | | X | | | | | | | | | X | |
| Cassidulina reniforme | 29.5 | 55.0 | 53.0 | 41.5 | 50.5 | 44.0 | 36.5 | 38.0 | 6.0 | 37.5 | 45.5 | 28.0 | 24.5 | 36.0 | 19.5 | 9.5 | 13.0 |
| Islandiella algida | 4.5 | 1.0 | 1.0 | X | 1.0 | X | | X | 1.0 | | 1.5 | 1.5 | X | | 1.5 | 1.0 | 1.0 |
| Islandiella helenae | 4.5 | X | | X | | | 1.0 | X | 2.0 | X | | 1.0 | | | 1.0 | 7.0 | 1.0 |
| Islandiella norcrossi | X | X | X | X | | | | | | | | | | | | | |
| Paracassidulina neocarinata | | X | | | X | X | X | | | | | | | | | | |
| **"Cibicides" group** | | | | | | | | | | | | | | | | | |
| Cibicicoides sp. B | | | | | | | | | | | | | | | | | |
| Cibicides grossa | 2.0 | X | | | | | X | | | | | | | | | | |
| ***Cibicides io/Lobatula | 3.5 | 9.0 | 6.5 | 8.0 | 11.0 | X | X | 13.5 | 18.0 | 3.5 | X | 2.0 | 4.0 | 10.0 | 1.0 | 2.0 | 3.0 |
| Cibicides reflugens | * | | | | X | 11.0 | 9.5 | 5.0 | 3.5 | 17.0 | 12.0 | | | | | | |
| Cibicidoides floridanus | | | | | | | | | | | | | | | | | |
| Heterolepa subhaidingeri | | | | | | | | | 2.0 | | | | | | | | |
| Planulina vilksae | | X | X | 1.0 | X | X | 2.0 | | X | 1.0 | X | | X | | | | |
| **"Cornuspira" group** | | | | | | | | | | | | | | | | | |
| Cornuspira involvens | | | | | | | | | | | | | | | | | |
| Cornuspira planorbis | | | | | | | | | | | | | | | | | |
| **"Discorbis—Rosalina" group** | | | | | | | | | | | | | | | | | |
| Altastellerella riveroae | | | | | | | | | | | | | | | | | |
| Eoeponidella pulchella | | X | 1.0 | X | X | X | X | | | | X | X | | | X | | X |
| Discorbinella bertheloti | X | | 1.0 | X | | | | X | X | | | | X | | X | | X |
| Gavelinopsis praegeri | | | | X | | | | | | X | | | | | X | | |
| Gavelinopsis sp. C | | | | | | | X | X | | | | X | | | X | | |
| Lamarckina haliotidia | | | | | | | | | | | | | | | X | | |
| Neodiscorbinella plana | X | X | X | | | | | | | | X | | | | | | |
| Neodiscorbinella transluscens | X | | 1.0 | 1.0 | | | | | | | | | X | | | | |
| Orbitina williamsoni | | X | X | | 2.0 | | | | | | | | | | | | |
| Rosalina columbiensis | | X | X | | | | | | | | | | | | | | |
| Rosalina globularis | | | 1.0 | 1.0 | | 1.0 | X | X | | X | 1.5 | X | X | | 1.5 | X | |
| **"Elphidium" group** | | | | | | | | | | | | | | | | | |
| Cribroelphidium asklundi | | | | X | | | | X | | | | | | 1.0 | | | 2.0 |
| Cribroelphidium bartletti | 5.5 | | | | | | | | X | | | | | | | 3.0 | |
| Cribroelphidium excavatum | 16.5 | 8.5 | 6.5 | 7.0 | 18.5 | 8.5 | 4.0 | 17.0 | 48.5 | 13.0 | 8.0 | 35.5 | 32.0 | 12.5 | 25.5 | 43.5 | 28.0 |
| Cribroelphidium frigidum | | | | X | | | | | | | | | | | | X | |
| Cribroelphidium hallandense | | | | | | | | | | | | | | | | X | |
| Elphidiella groenlandicum | | | | | | | | X | | | | | | | | | |
| **"Epistominella" group** | | | | | | | | | | | | | | | | | |
| Pseudoparella naraensis | 1.0 | | 1.0 | | 1.0 | | | | | X | X | X | X | | | | |
| Pseudoparella takayanagii | | X | X | X | X | X | | | | X | | 1.5 | 1.5 | | 1.0 | | 1.0 |
| Pseudoparella vitrea | | | X | | X | X | | | | | | | | | | | |
| Pseudoparella zhengae | | | | X | | | | | | | | | | | | | |
| Stetsonia arctica | | | | | | | X | | | | | | | | | | |

TABLE A3. BOREHOLE 87B-2 (continued)

| Sample number | 8 | 9 | 10 | 11 | 12 | 13 | 14 | 15 | 16 | 17 | 18 | 19 | 20 | 21 | 22 | 23 | 24 |
|---|---|---|---|---|---|---|---|---|---|---|---|---|---|---|---|---|---|
| Depth (m) | 0.45 | 2.43 | 4.26 | 6.29 | 8.3 | 9.17 | 10.08 | 11.21 | 12.32 | 13.25 | 14.56 | 15.8 | 17.02 | 17.92 | 18.94 | 20.17 | 21.4 |
| Total number of Quaternary benthics | 63,104 | 381,615 | 364,032 | 73,536 | 137,216 | 86,016 | 65,664 | 62,080 | 41,984 | 31,317 | 236,544 | 3,584 | 3,157 | 25,088 | 4,171 | 6,480 | 13,080 |
| Split counted | (1/128) | (1/512) | (1/512) | (1/64) | (1/256) | (1/128) | (1/128) | (1/64) | (1/128) | (3/256) | (1/512) | (3/32) | (3/32) | (1/64) | (3/32) | (1/16) | (1/8) |
| **"Fissurina" group** | | | | | | | | | | | | | | | | | |
| Fissurina aequilabialis | * | | | | | | | | | | | | | | | | |
| Fissurina alata | | X | | X | | | X | X | | | | | | | | | |
| Fissurina circula | | X | X | X | | X | | | | X | X | | | | | | |
| Fissurina circulara | | | | | | | | | | | | | | | X | | |
| Fissurina communa | | | | X | | | | | | | | | | | | | |
| Fissurina compressa | | | | | | | | | | | | | | | | | |
| Fissurina dancia | * | | | | | | | | | | | | | | | | |
| Fissurina elliptica | * | | | | | | | | | | | | | | | | |
| Fissurina goreaui | * | X | X | | | X | | X | | X | | | | | | | |
| Fissurina incipiens | * | | | | | | | | | | | | | | | | |
| Fissurina latiostoma | | | | | | | | | | | | | | | | | |
| Fissurina marginata | * | | X | X | X | | | X X | | | | | | | | X | X |
| Fissurina pseudoglobosa | * | | | X | | | X | X | | X | | X | | | | | |
| Fissurina quadracostulata | * | | X | X | | | X | | | | | | | | | | |
| Fissurina stewartii | * | | | | | | | X | | | | | | | | | X |
| Fissurina subchasteri | | | | | | | | | | | | | | | | | |
| Galwayella biangulata | | | | X | | | | | | | | | | | | | |
| Lagenosolenia inflataperforata | | | | | | | | | | | X | | | | | | |
| Lagenosolenia lagenoides | | | X | X | | | | X | | | | | | | | | |
| Parafissurina clavigera | | | | | | | | | | | | | | | | | |
| Parafissurina curvans | * | | | | | | | | | | | | | | | | |
| Parafissurina exiguiformis | | | | | | | | | | | | | | | | | |
| Parafissurina hamigera | | | | | | | * | | | | | | | | | | |
| Parafissurina himatiostoma | | | | | | | | | | | X | | | | | | |
| Parafissurina labiata | | | | | | | | | | | | | | | | | |
| Parafissurina subcarinata | * | | | | | | X | | | | | | | | | | |
| Parafissurina tectulostoma | | | | | X | | | | | | | | | | | | |
| Pseudoolina fissurinea | | | | | | | X | | | | | | | | | | |
| Wiesnerina minutiforma | X | | X | | | | | | | | | | | | | | |
| Wiesnerina remova | | | | X | | | | | | | | | | | | | |
| **"Glabratella" group** | | | | | | | | | | | | | | | | | |
| Conorbella pulvinata | | | | | | | | | | | | | X | | | | X |
| Glabratellina arcuata | | | | | | | | X | | 1.5 | 1.5 | | | 1.0 | X | 1.5 | 1.0 |
| Glabratellina lauriei | | | | | | | X | X | | X | X | | | | | 1.5 | 5.0 |
| Glabratellina wrightii | 1.0 | | X | X | | | | X | | | | | 5.0 | | | | 14.0 |
| Heronallenina parva | | | | X | | | | | | | | | | | | | |
| Rotaliella chasteri | 1.0 | 2.0 | 3.5 | 3.5 | 2.5 | 10.0 | 5.5 | 4.5 | 16.0 | 3.5 | 4.0 | 11.0 | 8.0 | 4.0 | 16.0 | X | 1.0 |
| Sphaeroidina bulloides | | | | | | | | | | | | | | | | | X |
| **"Gyroidina" group** | | | | | | | | | | | | | | | | | |
| Gyroidinoides nipponicus | | | | X | | | | | | | | | X | | X | | |
| Gyroidinoides quinqueloba | | | | | | | | | | | | | X | | | | |
| **"Lagena" group** | | | | | | | | | | | | | | | | | |
| Lagena semilineata | * | X | | | | | | | | | | | | | | | |
| Lagena substriata | * | | | | | | | | | | | | | | | | |
| Lagena cf. L. sulcata | * | | | | | | | | | | | | | | | | |
| Procerlagena distoma | * | | | | | X | | | | | | | | | | | |
| Procerolagena meridionalis | * | | | | | | | | | | | | | | | | |
| Pygmaeoseistron vulgaris | * | | | | | | | | | | | | | | | | |

## TABLE A3. BOREHOLE 87B-2 (continued)

| Sample number | 8 | 9 | 10 | 11 | 12 | 13 | 14 | 15 | 16 | 17 | 18 | 19 | 20 | 21 | 22 | 23 | 24 |
|---|---|---|---|---|---|---|---|---|---|---|---|---|---|---|---|---|---|
| Depth (m) | 0.45 | 2.43 | 4.26 | 6.29 | 8.3 | 9.17 | 10.08 | 11.21 | 12.32 | 13.25 | 14.56 | 15.8 | 17.02 | 17.92 | 18.94 | 20.17 | 21.4 |
| Total number of Quaternary benthics | 63,104 | 381,615 | 364,032 | 73,536 | 137,216 | 86,016 | 65,664 | 62,080 | 41,984 | 31,317 | 236,544 | 3,584 | 3,157 | 25,088 | 4,171 | 6,480 | 13,080 |
| Split counted | (1/128) | (1/512) | (1/512) | (1/64) | (1/256) | (1/128) | (1/128) | (1/64) | (1/128) | (3/256) | (1/512) | (3/32) | (3/32) | (1/64) | (3/32) | (1/16) | (1/8) |
| **"Miliolinellid" group** | | | | | | | | | | | | | | | | | |
| Milionella circularis | | | | | | | | | X | X | | X | X | | | | |
| Miliolinella subrotunda | | | | | | | | | | | | | X | | | X | |
| Pyrgo oblonga | | | | X | | | | | | | | | | | | | |
| Triloculinella differens | | | | | | | X | | | | | | | | | | |
| Triloculinella matthewense | | | | | | | * | | | | | | | | | | X |
| Triloculinella parisa | | | | | | | | | | | | | | | | | |
| Triloculinella pyriformis | | | | | | | | | | | | | | | | | |
| Triloculinella tegminus | | | | | | | X | | | | | | | | | | X |
| **"Nonion" group** | | | | | | | | | | | | | | | | | |
| Allomorphina fragilis | | | | | | | | | | | | | | | | | |
| Chillostomella oolina | | | | | | | | | | | | | | | | | |
| Astrononion stelligerum | 2.0 | 2.0 | | | 1.5 | X | X 2.5 | X | | X | 1.5 | X | | 1.0 | X | X | |
| Laminononion stellatum | | | | | | | | | | | | | | | | | |
| Haynesina depressula | | | | | | | | | | | | | | | X | | |
| Haynesina orbiculare | 11.0 | 5.0 | X 4.0 | 2.0 2.0 | 2.5 | 3.0 | 6.5 1.0 | X 3.0 | 1.0 | 3.5 | 4.0 | 1.5 3.0 | 1.0 3.0 | 4.5 | 1.0 8.5 | 7.0 23.5 | 3.0 5.0 |
| Haynesina paucilocula | X | | X | X | | | | | | | X | | | | X | | |
| Melonis barleeanus | | | | X | | | X | X | | | | | | | | | |
| Nonionella lobsannensis | | | | X | | | | | | | | | | | | | |
| Nonionella stella | 3.5 | 1.0 | 1.5 | 1.5 | 1.0 | 5.5 | X 4.5 | 2.0 | | 4.0 | X | X | | 3.5 | | X | 1.0 |
| Nonionellina labradorica | | | X | X | | | X | | | | X | | | | | | |
| Nonionoides grateloupi/Nonionella turgida | | 1.0 | 1.0 | 1.5 | X | X | 2.0 | | | | X | X | | 1.0 | | | X |
| Pulleniella osloensis/Haynesina germanica | | | | | | | | | | | | | | | | | |
| **"Oolina" group** | | | | | | | | | | | | | | | | | |
| Cushmanina striatopunctata | X | X | | | | | | | | | | | | | | | |
| Favulina hexagona | * | | | X | | | | | | | | | | | | | |
| Favulina melo | * | | | | | | | | | | | | | | | | |
| Favulina scalariformis | * | X | | | | | | | | | | | | | | | |
| Favulina squamosa | * | | | | | | | | | | | | | | | | |
| Homalohedra acuticostata | * | | | | | | | | | | | | | | | | |
| Homalohedra apiopleura | * | | | | | | | | | | | | | | | | |
| Homalohedra williamsoni | * | | | | | | | | | | | | | | | | |
| Oolina ampullineata | * | | | | | | | | | | | | | | | | |
| Oolina lineata | * | | | | | | | | | | | | | | | | |
| **"Polymorphina" group** | | | | | | | | | | | | | | | | | |
| Entopolymorphina oppositiforma | * | | | | | | | | | | | | | | | | |
| Laryngosigma hyalascidia | | | | | | | | | | | | | | | X | | |
| Laryngosigma lactea | * | | | | | | | | | X | | | | | | | |
| Laryngosigma williamsoni | * | | | | | | | | | | | | | | | | |
| Metapolymorphina ligua | | | | | | X | | | X | | | | | | | | |
| Pseudopolymorphina suboblonga | | | | X | | | | | | | | | | | | | |
| **"Quinqueloculina" group** | | | | | | | | | | | | | | | | | |
| Quinqueloculina akerniana | | | | | | | | | | | | | | | | | |
| Quinqueloculina crassicarinata | | | | | | | | | | | | | | | | | |
| Quinqueloculina obliquecamerata | | | | | | | | | | | | X | | | | | |
| Quinqueloculina seminula | | | | | | | | | | | | | X | | | | X |
| Quinqueloculina viennensis | | | | | | | | | | | | X | X | | X | X | |
| Siphonaperta parvula | | | | X | | | | X | | | | | | | | | |

## TABLE A3. BOREHOLE 87B-2 (continued)

| Sample number | 8 | 9 | 10 | 11 | 12 | 13 | 14 | 15 | 16 | 17 | 18 | 19 | 20 | 21 | 22 | 23 | 24 |
|---|---|---|---|---|---|---|---|---|---|---|---|---|---|---|---|---|---|
| Depth (m) | 0.45 | 2.43 | 4.26 | 6.29 | 8.3 | 9.17 | 10.08 | 11.21 | 12.32 | 13.25 | 14.56 | 15.8 | 17.02 | 17.92 | 18.94 | 20.17 | 21.4 |
| Total number of Quaternary benthics | 63,104 | 381,615 | 364,032 | 73,536 | 137,216 | 86,016 | 65,664 | 62,080 | 41,984 | 31,317 | 236,544 | 3,584 | 3,157 | 25,088 | 4,171 | 6,480 | 13,080 |
| Split counted | (1/128) | (1/512) | (1/512) | (1/64) | (1/256) | (1/128) | (1/128) | (1/64) | (1/128) | (3/256) | (1/512) | (3/32) | (3/32) | (1/64) | (3/32) | (1/16) | (1/8) |
| **"Rotalia" group** | | | | | | | | | | | | | | | | | |
| *Eponides pusillus* | | | | | | | | | | | | | | | | | |
| *Ioanella tumidula* | | | | | | | | | | | | | | | | | |
| *Nuttalides bradyi* | | | | | | | | | | | | | | | | | |
| *Oridosalis umbonatus* | | | | | | | | | | | | | | | | | |
| *Valvulineria minuta* | | X | 1.0 | X | | | | X | | X | | | | X | X | | X |
| **"Stainforthia" group** | | | | | | | | | | | | | | | | | |
| *Rutherfordoides mexicana* | | | | | | | | | | | | | | | | | |
| *Stainforthia concava* | | X | | | | | | | | | | | | | | | |
| *Stainforthia fusiformis* | | X | 1.5 | 7.5 | | 2.5 | 10.5 | X | | X | 2.5 | X | 1.0 | 1.0 | X | | X |
| *Stainforthia pauciloculata* | X | | | | | | | | | | | | X | | 1.0 | | |
| *Stainforthia rotundata* | | X | 1.0 | 3.0 | X | 1.0 | 1.0 | X | | 1.0 | X | | X | 1.0 | 1.0 | X | |
| *Tosaia hanzawaia* | | | | | | | | | | | | | | | | | |
| **"Spiralina" group** | | | | | | | | | | | | | | | | | |
| *Patellina corrugata* | 2.0 | X | 1.5 | X | 1.0 | 1.0 | 1.0 | X | | | X | | | | X | | |
| **"Triloculina" group** | | | | | | | | | | | | | | | | | |
| *Triloculina angulosa* | | | | X | X | X | X | X | | X | | X | | X | X | | |
| *Triloculina tricarinata* | | | | X | | | * | | | X | | | | | | | |
| *Triloculina gibba* | | | | | | | | | | | | | | | | | |
| **"Uvigerina" group** | | | | | | | | | | | | | | | | | |
| *Angulogerina angulosa/fluens* | 3.5 | X | X | X | X | X | X | X | | | X | 1.0 | X | X | 1.0 | X | X |
| *Neouvigerina canariensis* | | | | | | | | | | | | | | | | | |
| *Uvigerina peregrina* | | | | | | | | | | | | | | | | | |
| **QUATERNARY PLANKTONICS** | | | | | | | | | | | | | | | | | |
| Total number of specimens | 768 | 17,408 | 21,504 | 6,592 | 2,560 | 7,296 | 12,800 | 5,952 | 512 | 3,584 | 17,920 | 96 | 117 | 1,856 | 277 | 176 | 184 |
| Split counted | (1/128) | (1/512) | (1/512) | (1/64) | (1/256) | (1/128) | (1/128) | (1/64) | (1/128) | (3/256) | (1/512) | (3/32) | (3/32) | (1/64) | (3/32) | (1/16) | (1/8) |
| *Globigerina bulloides* | 512 | 512 | 1,536 | 192 | 512 | | 128 | 384 | 128 | 170 | 1,024 | | | 64 | | | |
| *Globogerinita uvula* | 256 | 3,072 | 1,536 | 320 | | 512 | 2,176 | 256 | | 86 | 1,536 | | | | 11 | | |
| *Globorotalia inflata* | | | | | | | | | | | | | | | | | |
| *Neogloboquadrina deutertri* | | 512 | | | | | | | | | | 32 | | | | | |
| *Neogloboquadrina pachyderma—sinistral* | | 9,216 | 12,288 | 2,560 | 1,028 | 2,816 | 6,016 | 2,816 | 384 | 1,280 | 8,192 | 53 | 64 | 1,344 | 180 | 144 | 96 |
| *Neogloboquadrina pachyderma—dextral* | | 3,072 | 4,096 | 1,600 | 256 | 1,438 | 3,456 | 832 | | 1,280 | 3,072 | 11 | 53 | 256 | 53 | 32 | 80 |
| *Turborotalita quinqueloba—sinistral* | | | 512 | 640 | 256 | 640 | 768 | 896 | | 256 | 1,024 | | | 192 | | | 8 |
| *Turborotalita quinqueloba—dextral* | | 512 | 1,536 | 1,280 | 256 | 1,920 | 256 | 768 | | 512 | 3,072 | | | | 11 | | |
| **Reworked** | | | | | | | | | | | | | | | | | |
| Total number of specimens | 128 | / | 1,024 | 384 | / | 512 | 640 | 448 | 384 | 170 | 512 | 587 | 267 | 192 | 299 | 176 | 24 |
| Split counted | (1/128) | | (1/512) | (1/64) | | (1/128) | (1/128) | (1/64) | (1/128) | (3/256) | (1/512) | (3/32) | (3/32) | (1/64) | (3/32) | (1/16) | (1/8) |

## TABLE A3. BOREHOLE 87B-2 (continued)

| Sample number | 25 | 26 | 27 | 28 | 29 | 30 | 31 | 32 | 33 | 34 | 36 | 37 | 38 | 39 | 40 | 41 | 42 |
|---|---|---|---|---|---|---|---|---|---|---|---|---|---|---|---|---|---|
| Depth (m) | 22.29 | 23.01 | 24.21 | 25.22 | 26.43 | 27.55 | 28.41 | 29.46 | 30.95 | 32.04 | 34.4 | 35.32 | 36.17 | 37.31 | 38.44 | 39.34 | 40.29 |
| Total number of Quaternary benthics | 1,723 | 1,092 | 2,656 | 19,712 | 1,717 | 5,504 | 2,064 | 8,640 | 2,816 | 606 | 262 | 137 | 989 | 11 | 233 | 63 | 280 |
| Split counted | (3/16) | (1/4) | (1/8) | (1/64) | (3/16) | (1/16) | (3/16) | (1/32) | (1/8) | (1/2) | / | / | (5/8) | / | / | / | / |
| **AGGLUTINATED FORAMINIFERA** | | | | | | | | | | | | | | | | | |
| **"Adercotryma" group** | | | | | | | | | | | | | | | | | |
| *Adercotryma glomerata* | | | | | | | | | | | | | | | | | |
| **"Cribrostomoides" group** | | | | | | | | | | | | | | | | | |
| *Labrospira crassimargo* | | | | | | | | | X | | | | | | | | |
| **"Eggerella" group** | | | | | | | | | | | | | | | | | |
| *Rhumblerella humboltii* | X | | X | | X | | X | | | | | | | | | | |
| **"Recurvoides" group** | | | | | | | | | | | | | | | | | |
| *Recurvoides scitulum* | X | | | X | | | | | | | | | | | | | |
| **"Reophax" arctica** | | | | | | | | | | | | | | | | | |
| *Cuneata arctica* | | | | | | | | | | | X | | | | | | |
| **"Saccammina" group** | | | | | | | | | | | | | | | | | |
| *Hemispherammina bradyi* | 11.5 | | | 2.0 | | | | | 1.0 | 14.0 | | | | | | | |
| *Lagenammina atlantica* | | | | | | | | | | | | | | | | | |
| *Psammosphaera fusca* | 13.0 | 6.0 | X | | | | 5.0 | | | 3.5 | | | | | | | |
| *Saccammina sphaerica* | | | | | | | | | | | | | | | | | |
| **"Spiroplectammina" group** | | | | | | | | | | | | | | | | | |
| *Spiroplectammina biformis* | | | X | | | | | | | | | | | | | | |
| *Spiroplectammina foliacea* | | | | | | | | | | | | | | | | | |
| **"Trochammina" group** | | | | | | | | | | | | | | | | | |
| *Lepidodeuterammina ochracea* | | 1.0 | 1.0 | | X | X | | | | | | | | | 1.0 | | |
| *Lepidodeuterammina ochracea sinuosa* | | | | | X | | | | | | | | | | | | |
| *Lepidoparatrochammina haynesi* | | | | X | | | | | | | | | X | | | | X |
| *Lepidoparatrochammina lepida* | | | | | | | X | | | X | | | | | | | |
| **CALCAREOUS FORAMINIFERA** | | | | | | | | | | | | | | | | | |
| **"Bolivina" group** | | | | | | | | | | | | | | | | | |
| *Aphelophragmina britannica* | | | | | | | | | X | | | | | | | | |
| *Aphelophragmina spathulata* | | X | X | | | | | | | | | | X | | | | |
| *Aphelophragmina vadescens* | | | | | | | | | | | | | | | | | |
| *Bolivina aenariensis* | | | | | X | | | | X | | | | | | | | |
| *Bolivina decussata* | | | | | | | | | | | | | | | | | |
| *Bolivinella pseudopunctata* | | X | X | | | | | | | | | | X | | X | | 1.0 |
| **"Buccella" group** | | | | | | | | | | | | | | | | | |
| *Aubignyana perlucida* | | | | | X | | | | | | | | | | | | |
| *Buccella depressa* | | | | | 1.0 | | | | X | | | 1.5 | | | 1.0 | | X |
| *Buccella frigida* | | | | | | | | | | | X | | | | | | |
| *Buccella hannai* | 1.0 | 1.0 | 2.0 | 2.0 | 1.0 | X | 2.0 | 2.0 | 3.0 | 1.0 | 3.5 | 1.5 | 2.0 | | 1.0 | P | 1.5 |
| *Buccella karsteni* | | | | | | | | | | | | | | | | | |
| *Buccella oregonensis* | 3.0 | 2.0 | 2.0 | 3.0 | 1.0 | X | 2.0 | 2.0 | 2.0 | 1.5 | 2.0 | 1.0 | X | | 2.0 | P | X |
| *Buccella sp. A* | | | | | | | | | | | | | | | | | |
| *Trichohyalus kingi* | | | | | | | | | | | | | | | | | |
| **"Bulimina" group** | | | | | | | | | | | | | | | | | |
| *Bulimina marginata/aculeata* | | X | | | | | | | | X | X | 1.5 | | | | | X |
| *Buliminella elegantissima* | | X | | | | | | | X | | | | | | | | |
| *Floresina milletti* | | | | | | | | | | | | | | | | | |
| *Globobulimina auriculata* | | | | | | | | | | | | 1.0 | | | | | |

TABLE A3. BOREHOLE 87B-2 (continued)

| Sample number | 25 | 26 | 27 | 28 | 29 | 30 | 31 | 32 | 33 | 34 | 36 | 37 | 38 | 39 | 40 | 41 | 42 |
|---|---|---|---|---|---|---|---|---|---|---|---|---|---|---|---|---|---|
| Depth (m) | 22.29 | 23.01 | 24.21 | 25.22 | 26.43 | 27.55 | 28.41 | 29.46 | 30.95 | 32.04 | 34.4 | 35.32 | 36.17 | 37.31 | 38.44 | 39.34 | 40.29 |
| Total number of Quaternary benthics | 1,723 | 1,092 | 2,656 | 19,712 | 1,717 | 5,504 | 2,064 | 8,640 | 2,816 | 606 | 262 | 137 | 989 | 11 | 233 | 63 | 280 |
| Split counted | (3/16) | (1/4) | (1/8) | (1/64) | (3/16) | (1/16) | (3/16) | (1/32) | (1/8) | (1/2) | / | / | (5/8) | / | / | / | / |
| **"Cassidulina—Islandiella" group** | | | | | | | | | | | | | | | | | |
| Cassidulina carinata | | | | | | | | | | | | | | | | | |
| Cassidulina laevigata | | | | | | | | X | | X | | | | | | | |
| Cassidulina latacamerata | 15.5 | 25.0 | 34.0 | 22.0 | 23.0 | 20.5 | 20.5 | 24.5 | 35.5 | 21.0 | 24.5 | 13.0 | 32.0 | | 25.0 | D | 20.5 |
| Cassidulina reniforme | | 1.0 | 1.0 | X | X | 1.0 | X | 1.0 | 2.0 | | X | 2.5 | 2.0 | | 1.5 | P | 2.0 |
| Islandiella algida | 1.0 | 4.5 | 3.0 | 3.0 | 2.0 | 13.5 | 3.0 | 5.0 | 1.0 | X | 1.5 | 1.0 | | | 1.5 | | 5.0 |
| Islandiella helenae | | | | | | | | 1.5 | | X | 1.0 | | X | | 1.0 | | 1.0 |
| Paracassidulina neocarinata | | | | | | | | | | | | | X | | | | |
| **"Cibicides" group** | | | | | | | | | | | | | | | | | |
| Cibicicoides sp. B | | | | | | X | | | | | | | | | | | |
| Cibicides grossa | | | | | | | | | | | | | | | | | |
| ***Cibicides io/Lobatula lobatula | 3.0 | X | | | | 4.0 | 2.0 | X | | | | 1.0 | | | | | X |
| Cibicides reflugens | | 3.0 | 2.0 | 3.0 | 2.0 | 4.0 | 3.0 | 6.5 | 2.0 | 4.5 | 2.0 | 1.0 | 6.0 | | 7.0 | P | 5.5 |
| Cibicidoides floridanus | | | | | | | | | | | | | | | | | |
| Heterolepa subhaidingeri | | | | | | | | | | | 3.5 | | | | | P | X |
| Planulina vilksae | | | | | | | | | | | X | | | | | | |
| **"Cornuspira" group** | | | | | | | | | | | | | | | | | |
| Cornuspira involvens | | | | | | | | | | | | | | | | | |
| Cornuspira planorbis | | | | | X | X | | | | | | | | | | | |
| **"Discorbis—Rosalina" group** | | | | | | | | | | | | | | | | | |
| Altastellerella riveroae | | | | | X | | 1.0 | | X | X | | | | | | | |
| Eoeponidella pulchella | X | X | 1.0 | | 2.0 | | | | X | X | | | | | | | X |
| Discorbinella bertheloti | | 1.0 | | | X | | | | | | | | | | | | |
| Gavelinopsis praegeri | | | | X | X | | | | | | | | | | | | |
| Gavelinopsis sp. C | | | | | | | | | | | | 1.5 | | | | | |
| Lamarckina haliotidia | | | X | | | | | | | | | | | | | | |
| Neodiscorbinella plana | | X | | X | | | | | | | X | | | | | | |
| Neodiscorbinella transluscens | | | | | | | | | | | | 1.0 | | | | | |
| Orbitina williamsoni | | | | | | | | | | | | | | | | | |
| Rosalina columbiensis | | | | | | | | | | | | | | | | | |
| Rosalina globularis | | 4.5 | 4.0 | | X | 1.5 | | | 1.0 | 1.0 | X | 2.5 | | | X | P | X |
| | | | | | | | | | | | | | | | 1.5 | | 2.0 |
| **"Elphidium" group** | | | | | | | | | | | | | | | | | |
| Cribroelphidium asklundi | | X | X | | | X | 2.0 | 2.0 | | 2.0 | | | X | | | | X |
| Cribroelphidium bartletti | | | | | | | | | | | | | | | | | |
| Cribroelphidium excavatum | 30.0 | 5.5 | 29.5 | 48.5 | 45.5 | 41.5 | 38.5 | 36.0 | 20.0 | 30.5 | 50.0 | 55.0 | 30.0 | D | 36.0 | C | 40.0 |
| Cribroelphidium hallandense | | | | | | | 1.0 | | | | | | 1.0 | | | | |
| Elphidiella groenlandicum | | | | | | | | | | | | | | | | | |
| **"Epistominella" group** | | | | | | | | | | | | | | | | | |
| Pseudoparella naraensis | X | | X | 1.0 | X | | X | 1.5 | X | 4.5 | X | 1.0 | 1.0 | | 2.0 | P | X |
| Pseudoparella takayanagii | | 2.0 | | 1.0 | 2.5 | X | | | 10.0 | | | | X | | 2.0 | | 2.0 |
| Pseudoparella vitrea | | X | | | | | | | | | | | | | | | |
| Pseudoparella zhengae | | 1.0 | | | | | | | | X | | | | | X | | X |
| Stetsonia arctica | | | | | | | | | | | | | | | | | |
| **"Fissurina" group** | | | | | | | | | | | | | | | | | |
| Fissurina aequilabialis | | | | | | | | | | | | | X | | | | |
| Fissurina alata | | | | | | | | | | | | | | | | | |
| Fissurina circula | | | | | | | | | | | | | X | | | | |
| Fissurina circulara | X | | | | | | | | | | | | | | | | |
| Fissurina communa | | | | | | | | | | | | | | | | | |
| Fissurina compressa | | | | | | | | | | | | | | | | | |
| Fissurina dancia | | | | | | | | | X | | | | | | | | |

## TABLE A3. BOREHOLE 87B-2 (continued)

| Sample number | 25 | 26 | 27 | 28 | 29 | 30 | 31 | 32 | 33 | 34 | 36 | 37 | 38 | 39 | 40 | 41 | 42 |
|---|---|---|---|---|---|---|---|---|---|---|---|---|---|---|---|---|---|
| Depth (m) | 22.29 | 23.01 | 24.21 | 25.22 | 26.43 | 27.55 | 28.41 | 29.46 | 30.95 | 32.04 | 34.4 | 35.32 | 36.17 | 37.31 | 38.44 | 39.34 | 40.29 |
| Total number of Quaternary benthics | 1,723 | 1,092 | 2,656 | 19,712 | 1,717 | 5,504 | 2,064 | 8,640 | 2,816 | 606 | 262 | 137 | 989 | 11 | 233 | 63 | 280 |
| Split counted | (3/16) | (1/4) | (1/8) | (1/64) | (3/16) | (1/16) | (3/16) | (1/32) | (1/8) | (1/2) | / | / | (5/8) | / | / | / | / |
| **"Fissurina" group (continued)** | | | | | | | | | | | | | | | | | |
| *Fissurina elliptica* | X | | | | | | | | | | | | | | | | |
| *Fissurina goreaui* | | | | | | | | | | | | | | | | | |
| *Fissurina incipiens* | | | X | | | | | | | | | | | | | | |
| *Fissurina latiostoma* | | | | | | X | | | X | | | | X | | | | |
| *Fissurina marginata* | | | | | | | | | | | X | | X | | X | P | |
| *Fissurina pseudoglobosa* | | | | | | | | | X | | | | | | | | X |
| *Fissurina quadracostulata* | | | | | | | | | | | | | | | | | |
| *Fissurina stewartii* | X | | | | | | | | | | | | | | | | |
| *Fissurina subchasteri* | X | | | | | | | | | | | | | | | | X |
| *Galwayella biangulata* | | | | | | | | | | | | | | | | | |
| *Lagenosolenia inflataperforata* | | | | | | | | | | | | | | | | | |
| *Lagenosolenia lagenoides* | | | | | | | | | | | | | | | | | |
| *Parafissurina clavigera* | | | | | | | | | | | | | | | | | |
| *Parafissurina curvans* | | | | | | | | | | | | | | | | | |
| *Parafissurina exiguiformis* | | | | | | | | | | | | 1.0 | | | | | |
| *Parafissurina hamigera* | | | | | | | | | | | | | | | | | |
| *Parafissurina himatiostoma* | | | | | | | | | | | | | | | | | X |
| *Parafissurina labiata* | | | | | | | | | | | | | X | | | | |
| *Parafissurina subcarinata* | | | | | | | | | | | | | | | | | |
| *Parafissurina tectulostoma* | | | | | | | | | | | | | | | | | X |
| *Pseudoolina fissurinea* | | | | | X | | | | | | | | | | | | |
| *Wiesnerina minutiforma* | | | | | | | | | | | | | | | | | |
| *Wiesnerina remova* | | | | | | | | | | | | | | | | | |
| **"Glabratella" group** | | | | | | | | | | | | | | | | | |
| *Conorbella pulvinata* | 1.0 | | | | | | | | | | | | | | | | |
| *Glabratellina arcuata* | X | X | | 2.5 | 2.5 | | 1.5 | | 1.0 | X | | | 1.0 | | | | X |
| *Glabratellina lauriei* | 9.0 | X | | | | X | 5.0 | 8.0 | 2.0 | | | | X | | | | |
| *Glabratellina wrightii* | | | | | | 1.0 | 1.0 | | | | X | | 6.0 | | | | |
| *Heronallenina parva* | | | | | | | | | | | | | | | | | |
| *Rotaliella chasteri* | 3.5 | | 11.0 | 2.0 | 6.0 | 4.5 | 5.0 | | 5.0 | 3.5 | | 1.5 | 4.0 | | 2.0 | | 2.0 |
| *Sphaeroidina bulloides* | | | | | | | | | | | | | | | | | |
| **"Gyroidina" group** | | | | | | | | | | | | | | | | | |
| *Gyroidinoides nipponicus* | | | | | | | | | | | | | | | | | |
| *Gyroidinoides quinqueloba* | | X | | | | | | | X | | | 1.0 | X | P | 1.5 | | X |
| **"Lagena" group** | | | | | | | | | | | | | | | | | |
| *Lagena semilineata* | | | | | | | | | | | | | | | | | |
| *Lagena substriata* | | | | | | | | | | | | | X | | | | |
| *Lagena* cf. *L. sulcata* | | | | | | | | | | | | | | | | | |
| *Procerolagena distoma* | | X | | | | | | | | | | | | | | | |
| *Pygmaeoseistron vulgaris* | | | | | | | | | | | | | | | | | |
| **"Miliolinellid" group** | | | | | | | | | | | | | | | | | |
| *Miliolinella circularis* | X | X | | X | X | X | X | | | | | | | | | | |
| *Miliolinella subrotunda* | | | | | | | | | | | | | | | | | |
| *Pyrgo oblonga* | | | | | | X | | | X | | | | | | | | |
| *Triloculinella differens* | | | | | | | | X | | | | | | | | | |
| *Triluculinella matthewense* | | | | | | | | | | X | | | | | | | |
| *Triloculinella parisa* | | | | | | | | | | | | | | | | | |
| *Triloculinella pyriformis* | | | | | | | | | | | | | | | | | |
| *Triloculinella tegminus* | | | | | | | | | | | | | | | | | |

## TABLE A3. BOREHOLE 87B-2 (continued)

| Sample number | 25 | 26 | 27 | 28 | 29 | 30 | 31 | 32 | 33 | 34 | 36 | 37 | 38 | 39 | 40 | 41 | 42 |
|---|---|---|---|---|---|---|---|---|---|---|---|---|---|---|---|---|---|
| Depth (m) | 22.29 | 23.01 | 24.21 | 25.22 | 26.43 | 27.55 | 28.41 | 29.46 | 30.95 | 32.04 | 34.4 | 35.32 | 36.17 | 37.31 | 38.44 | 39.34 | 40.29 |
| Total number of Quaternary benthics | 1,723 | 1,092 | 2,656 | 19,712 | 1,717 | 5,504 | 2,064 | 8,640 | 2,816 | 606 | 262 | 137 | 989 | 11 | 233 | 63 | 280 |
| Split counted | (3/16) | (1/4) | (1/8) | (1/64) | (3/16) | (1/16) | (3/16) | (1/32) | (1/8) | (1/2) | / | / | (5/8) | / | / | / | / |
| **"Nonion" group** | | | | | | | | | | | | | | | | | |
| Allomorphina fragilis | | | | | | | | | | | | | | | | | |
| Chillostomella oolina | | X | | | | | | | | | | | | | | | |
| Astrononion stelligerum | | 1.0 | | | | | X | 2.0 | X | | | 1.0 | X | | | | X |
| Laminononion stellatum | | | | | | | | 2.0 | X | X | | | X | | | | |
| Haynesina depressula | | | | | X | | | | | | | | | | | | |
| Haynesina orbiculare | X | 1.0 | 2.0 | 2.5 | | 4.5 | 3.0 | 2.0 | X | X | 4.0 | 1.5 | 1.0 | | 5.0 | P | 3.0 |
| Haynesina pauciloculata | 2.5 | 2.0 | X | 2.5 | 3.0 | X | 2.0 | 1.0 | 4.5 | | 1.0 | 2.0 | 2.0 | | X | | |
| Melonis barleeanus | | X | X | | | | | | | X | | | | | | | |
| Nonionella lobsannensis | | X | | | | | | | | | | | | | | | |
| Nonionella stella | | | | | | | | | | | | | | | | | |
| Nonionellina labradorica | 1.0 | 3.0 | X | 1.0 | X | 1.0 | X | | 1.0 | 5.0 | X | 3.5 | 3.0 | | 2.5 | P | 2.0 |
| ~~Nonionoides grateloupi/ Nonionella turgida | | | | | | | | | | | | | | | | | |
| ~~Pulleniella osloensis/ Haynesina germanica | | 1.0 | 1.0 | | | | | | X | | X | | | | 1.5 | | 1.5 |
| **"Oolina" group** | | | | | | | | | | | | | | | | | |
| Cushmanina striatopunctata | | | | | | | | | | | | | | | | | |
| Favulina hexagona | | | | | | | | | | | | | | | | | |
| Favulina melo | | | | | | | | | | | | | | | | | |
| Favulina scalariformis | | | | | | | | | | | | | | | | | |
| Favulina squamosa | | | | | | | | | | | | | | | | | |
| Homalohedra acuticostata | | | | | | | | | | | | | | | | | |
| Homalohedra apiopleura | | | | | | | | | | | | | | | | | |
| Homalohedra williamsoni | | | | | | | | | | | | | | | | | |
| Oolina ampulliineata | | | | | | | | | | | | 1.0 | | | | | |
| Oolina lineata | | | | | | | | | | | | | | | | | |
| **"Polymorphina" group** | | | | | | | | | | | | | | | | | |
| Entopolymorphina oppositiforma | | | | | | | | | | | | | | | | | |
| Laryngosigma hyalascidia | | | | | | | | | | | | | | | | | |
| Laryngosigma lactea | | | | | | | | | | | | | | | | | |
| Laryngosigma williamsoni | | | | | | | | | | | | | | | | | |
| Metapolymorphina ligua | | | | | | | | | | | | | | | | | X |
| Pseudopolymorphina suboblonga | | | | | | | | | | | | | | | | | |
| **"Quinqueloculina" group** | | | | | | | | | | | | | | | | | |
| Quinqueloculina akerniana | | | | | | | | | | | | | | | | | |
| Quinqueloculina crassicarinata | X | | X | | | | | | | | | | | | | | |
| Quinqueloculina obliquecamerata | | | | | | | | X | | | | | | | | | |
| Quinqueloculina seminula | | | | | | | | X | | | | | | | | | |
| Quinqueloculina viennensis | | | | | | | | | | | | | | | | | |
| Siphonaperta parvula | X | | | | | | | | | | | 1.0 | | | | | |

## TABLE A3. BOREHOLE 87B-2 (continued)

| Sample number | 25 | 26 | 27 | 28 | 29 | 30 | 31 | 32 | 33 | 34 | 36 | 37 | 38 | 39 | 40 | 41 | 42 |
|---|---|---|---|---|---|---|---|---|---|---|---|---|---|---|---|---|---|
| Depth (m) | 22.29 | 23.01 | 24.21 | 25.22 | 26.43 | 27.55 | 28.41 | 29.46 | 30.95 | 32.04 | 34.4 | 35.32 | 36.17 | 37.31 | 38.44 | 39.34 | 40.29 |
| Total number of Quaternary benthics | 1,723 | 1,092 | 2,656 | 19,712 | 1,717 | 5,504 | 2,064 | 8,640 | 2,816 | 606 | 262 | 137 | 989 | 11 | 233 | 63 | 280 |
| Split counted | (3/16) | (1/4) | (1/8) | (1/64) | (3/16) | (1/16) | (3/16) | (1/32) | (1/8) | (1/2) | / | / | (5/8) | / | / | / | / |
| **"Rotalia" group** | | | | | | | | | | | | | | | | | |
| Eponides pusillus | | | | | | | | | | | | | | | | | |
| Ioanella tumidula | | | | | | | | | | | | | | | 1.0 | | X |
| Nuttalides bradyi | | | | | | | | | | | | | | | X | | |
| Oridosalis umbonatus | | X | | | | | | | | X | X | | | X | | | |
| Valvulineria minuta | | | | | X | | | | | | | | | | | | |
| **"Stainforthia" group** | | | | | | | | | | | | | | | | | |
| Rutherfordoides mexicana | | | | | | | | | X | | | | | | | | |
| Stainforthia concava | | 7.5 | | | | | | | | | X | | | | | | |
| Stainforthia fusiformis | | 5.5 | X | | | | | | | X | X | | | | 2.0 | | X |
| Stainforthia pauciloculata | | 3.0 | | X | | | | | | X | | | | | | | |
| Stainforthia rotundata | | | | | | | | | X | X | | | X | | | | 1.0 |
| Tosaia hanzawaia | | | | | | | | | | | | 1.0 | | | | | |
| **"Spiralina" group** | | | | | | | | | | | | | | | | | |
| Patellina corrugata | | 2.0 | 1.0 | | | | | | X | X | | 1.0 | X | | 1.5 | | |
| **"Triloculina" group** | | | | | | | | | | | | | | | | | |
| Triloculina angulosa | X | X | | X | | | | | | | | | | | | | |
| Triloculina tricarinata | | | | | | | | | | | | | | | | | |
| Triloculina gibba | | | | | | | | | | | | | | | | | |
| **"Uvigerina" group** | | | | | | | | | | | | | | | | | |
| Angulogerina angulosa/fluens | | | | | | | | 1.0 | | | | | | | | | |
| Neouvigerina canariensis | | | | | | | | | | X | X | | X | | | | |
| Uvigerina peregrina | | | | | | | | | | | | | | | | | X |
| **QUATERNARY PLANKTONICS** | | | | | | | | | | | | | | | | | |
| Total number of specimens | 91 | 80 | 280 | 1,536 | 208 | 368 | 112 | 448 | 208 | 40 | 4 | 9 | 16 | 1 | 16 | 3 | 22 |
| Split counted | (3/16) | (1/4) | (1/8) | (1/64) | (3/16) | (1/16) | (3/16) | (1/32) | (1/8) | (1/2) | / | / | (5/8) | / | / | / | / |
| Globigerina bulloides | | 36 | 40 | 64 | 9 | | | 64 | 8 | 4 | | | 5 | | | | |
| Globogerinita uvula | | | | | | | 5 | 32 | 32 | | | | | | 1 | | 1 |
| Globorotalia inflata | | | | | | | | | | | | | | | | | |
| Neogloboquadrina deutertri | 5 | | | | | | 5 | 32 | 24 | 2 | | 1 | 1 | | 1 | | 7 |
| Neogloboquadrina pachyderma—sinistral | 81 | 16 | 216 | 1,408 | 148 | 288 | 70 | 224 | 88 | 10 | 2 | 4 | 8 | | 8 | 1 | 6 |
| Neogloboquadrina pachyderma—dextral | 5 | 24 | 16 | 64 | 46 | 80 | 16 | 96 | 48 | 12 | 1 | 4 | 1 | | 4 | 1 | 8 |
| Turborotalita quinqueloba—sinistral | | 4 | 8 | | 5 | | 11 | | 8 | | 1 | | | 1 | 1 | | |
| Turborotalita quinqueloba—dextral | | | | | | | 5 | | | 4 | | | 1 | | 1 | 1 | |
| **Reworked** | | | | | | | | | | | | | | | | | |
| Total number of specimens | 91 | 260 | 112 | 2,176 | 251 | 352 | 181 | 3,136 | 912 | 830 | 851 | 860 | 131 | 80 | 7,200 | 109 | 3,344 |
| Split counted | (3/16) | (1/4) | (1/8) | (1/64) | (3/16) | (1/16) | (3/16) | (1/32) | (1/8) | (1/2) | / | (1/2) | (5/8) | / | (1/16) | / | (1/16) |

TABLE A3. BOREHOLE 87B-2 (continued)

| Sample number | 43 | 44 | 45 | 46 | 47 | 48 | 49 | 51 | 52 | 53 | 54 | 55 | 56 |
|---|---|---|---|---|---|---|---|---|---|---|---|---|---|
| Depth (m) | 41.02 | 42.69 | 44.23 | 47.5 | 49.12 | 50.24 | 50.91 | 56.31 | 61.53 | 66.7 | 71.5 | 73.0 | 74.87 |
| Total number of Quaternary benthics | 827 | 676 | 870 | 129 | 782 | 4,192 | 4,928 | 5,728 | 4,688 | 4,464 | 4,528 | 5,773 | 4,915 |
| Split counted | (3/8) | (1/2) | (1/2) | / | (1/2) | (3/32) | (1/16) | (1/16) | (1/16) | (1/16) | (1/16) | (5/64) | (5/64) |
| **AGGLUTINATED FORAMINIFERA** | | | | | | | | | | | | | |
| **"Adercotryma" group** | | | | | | | | | | | | | |
| Adercotryma glomerata | | | | | | | | | | | | | |
| **"Cribrostomoides" group** | | | | | | | | | | | | | |
| Labrospira crassimargo | | | | | | | | | | | | | |
| **"Eggerella" group** | | | | | | | | | | | | | |
| Rhumblerella humbolti | | | | | | | | | | | | | |
| **"Recurvoides" group** | | | | | | | | | | | | | |
| Recurvoides scitulum | | | | | | | | | | | | | |
| **"Reophax" arctica** | | | | | | | | | | | | | |
| Cuneata arctica | | | | | | | | | | | | | |
| **"Saccammina" group** | | | | | | | | | | | | | |
| Hemispherammina bradyi | | | | | | | | | | | | | |
| Lagenammina atlantica | | | | | | | | | | | | | |
| Psammosphaera fusca | | | | | | | | | | | | | |
| Saccammina sphaerica | 2.0 | | | | | | | | | | | | |
| **"Spiroplectammina" group** | | | | | | | | | | | | | |
| Spiroplectammina biformis | | | X | | | | | | | | | | |
| Spiroplectammina foliacea | | | X | | | | | | | | | | |
| **"Trochammina" group** | | | | | | | | | | | | | |
| Lepidodeuterammina ochracea | | 1.0 | | 1.0 | 1.0 | X | | | | | | | X |
| Lepidodeuterammina ochracea sinuosa | | | | | | | | | | | | | |
| Lepidoparatrochammina haynesi | | | | | X | | | | | | | | |
| Lepidoparatrochammina lepida | | | | | | | | | | X | | | |
| **CALCAREOUS FORAMINIFERA** | | | | | | | | | | | | | |
| **"Bolivina" group** | | | | | | | | | | | | | |
| Aphelophragmina britannica | | | | | | | | | | | | | |
| Aphelophragmina spathulata | | X | X | | | | | X | X | | | X | X |
| Aphelophragmina vadescens | | | | | | | | | | | | | |
| Bolivina aenariensis | | | | | | | X | | | | | | |
| Bolivina decussata | | | | | | | | | | | | | |
| Bolivinella pseudopunctata | X | | 1.0 | 1.0 | | | | | | 1.0 | X | X | X |
| **"Buccella" group** | | | | | | | | | | | | | |
| Aubignyana perlucida | | | | | | | | | | | | | |
| Buccella depressa | | | | | | X | | X | | | | X | |
| Buccella frigida | | | | | | | | | | | | X | |
| Buccella hannai | | | | | | | | | | X | | 1.0 | X |
| Buccella karsteni | 1.5 | 3.0 | 6.0 | 5.0 | 3.5 | 2.0 | 1.5 | 1.5 | 2.0 | 2.0 | 1.5 | 2.5 | 2.0 |
| Buccella oregonensis | 2.0 | | 2.0 | | 1.0 | 2.5 | 1.0 | 1.0 | 1.0 | | X | | 1.0 |
| Buccella sp. A | | | | | | | | | | 3.0 | X | 3.0 | 1.0 |
| Trichohyalus kingi | | | | | | | | | | | | | |
| **"Bulimina" group** | | | | | | | | | | | | | |
| Bulimina marginata/aculeata | 1.0 | | 1.0 | | | 1.0 | X | | X | X | X | X | X |
| Buliminella elegantissima | X | | | | X | | | | | | | X | |
| Floresina milletti | | | | | | | | | | | | | |
| Globobulimina auriculata | | | | | | | | | | | | | |

TABLE A3. BOREHOLE 87B-2 (continued)

| Sample number | 43 | 44 | 45 | 46 | 47 | 48 | 49 | 51 | 52 | 53 | 54 | 55 | 56 |
|---|---|---|---|---|---|---|---|---|---|---|---|---|---|
| Depth (m) | 41.02 | 42.69 | 44.23 | 47.5 | 49.12 | 50.24 | 50.91 | 56.31 | 61.53 | 66.7 | 71.5 | 73.0 | 74.87 |
| Total number of Quaternary benthics | 827 | 676 | 870 | 129 | 782 | 4,192 | 4,928 | 5,728 | 4,688 | 4,464 | 4,528 | 5,773 | 4,915 |
| Split counted | (3/8) | (1/2) | (1/2) | / | (1/2) | (3/32) | (1/16) | (1/16) | (1/16) | (1/16) | (1/16) | (5/64) | (5/64) |
| **"Cassidulina—Islandiella" group** | | | | | | | | | | | | | |
| Cassidulina carinata | X | | | | | | | | | X | | | |
| Cassidulina laevigata | | | | | | | | | | X | | | |
| Cassidulina latacamerata | 14.0 | 23.0 | 25.0 | 22.0 | 24.0 | 28.0 | 17.0 | 24.0 | 22.0 | 19.0 | 19.0 | 35.5 | 20.0 |
| Cassidulina reniforme | X | 1.0 | 1.0 | | X | 2.0 | X | X | 1.0 | 2.0 | X | 2.0 | X |
| Islandiella algida | 2.5 | 1.0 | X | 1.0 | 2.5 | 2.0 | 1.5 | 2.0 | 1.5 | 2.5 | 1.5 | 1.0 | 1.5 |
| Islandiella helenae | X | 1.0 | X | | | X | | X | X | | 2.0 | X | |
| Islandiella norcrossi | | | | | | | | | | | | | |
| Paracassidulina neocarinata | | | | | | | | | | | | | |
| **"Cibicides" group** | | | | | | | | | | | | | |
| Cibicicoides sp. B | X | | | | | | | | | | | | |
| Cibicides grossa | | | | | | | | | | | | | |
| ***Cibicides io/Lobatula lobatula | 5.5 | 3.5 | 3.0 | 6.0 | 3.5 | 4.0 | 4.5 | 3.0 | 5.0 | 1.0 | 3.5 | 3.0 | 4.5 |
| Cibicides reflugens | | | | | | X | | X | 1.0 | 1.0 | | | |
| Cibicidoides floridanus | | | | | | 4.0 | | 3.0 | | | | | |
| Heterolepa subhaidingeri | | 2.5 | X | | X | | | | | 1.0 | | | |
| Planulina vilksae | | | | | | | | | | | | | |
| **"Cornuspira" group** | | | | | | | | | | | | | |
| Cornuspira involvens | | | | | | | | | | | | | |
| Cornuspira planorbis | | | | | | | | | | | | | |
| **"Discorbis—Rosalina" group** | | | | | | | | | | | | | |
| Altastellerella riveroae | | | | | | X | | | | | | | |
| Eoeponidella pulchella | | 1.0 | | 3.0 | X | X | X | 1.0 | X | | X | 1.0 | |
| Discorbinella bertheloti | X | | | | X | | | | | | | X | |
| Gavelinopsis praegeri | | | 1.0 | | | | | | | | | | |
| Gavelinopsis sp. C | | | | | | | | | | | | | |
| Lamarckina haliotidia | | | | | | X | | | | | | | |
| Neodiscorbinella plana | | | X | | | | X | X | | | | X | X |
| Neodiscorbinella transluscens | | | | | | | | | | | | | |
| Orbitina williamsoni | 1.0 | 1.0 | | | X | | | | | | | | |
| Rosalina columbiensis | 1.5 | 1.0 | 1.5 | 3.0 | X | 2.5 | X | 1.0 | X | 1.0 | 1.0 | 2.0 | |
| Rosalina globularis | | | | | | | | | | | | | |
| **"Elphidium" group** | | | | | | | | | | | | | |
| Cribroelphidium asklundi | X | | | 1.0 | | X | | 1.5 | 1.0 | | X | X | 1.5 |
| Cribroelphidium bartletti | | | | | | | | | | | X | | |
| Cribroelphidium excavatum | 44.5 | 42.0 | 23.0 | 31.5 | 37.5 | 40.5 | 54.0 | 50.0 | 54.5 | 56.0 | 57.5 | 31.0 | 53.5 |
| Cribroelphidium frigidum | | | | | | | | | | | | | |
| Cribroelphidium hallandense | | | | | | X | X | X | | | | | |
| Elphidiella groenlandicum | | | | | | | | | | | | | |
| **"Epistominella" group** | | | | | | | | | | | | | |
| Pseudoparella naraensis | 1.0 | | 2.5 | | | | | | | X | | X | |
| Pseudoparella takayanagii | X | 9.0 | 6.0 | 3.5 | 12.0 | 2.5 | X | | X | 2.0 | 1.5 | 1.0 | 1.5 |
| Pseudoparella vitrea | | | | | | | 2.0 | | 1.5 | X | | | |
| Pseudoparella zhengae | | | | | | | | | | | | | |
| Stetsonia arctica | | | X | | | | | | | | | | |

## TABLE A3. BOREHOLE 87B-2 (continued)

| Sample number | 43 | 44 | 45 | 46 | 47 | 48 | 49 | 51 | 52 | 53 | 54 | 55 | 56 |
|---|---|---|---|---|---|---|---|---|---|---|---|---|---|
| Depth (m) | 41.02 | 42.69 | 44.23 | 47.5 | 49.12 | 50.24 | 50.91 | 56.31 | 61.53 | 66.7 | 71.5 | 73.0 | 74.87 |
| Total number of Quaternary benthics | 827 | 676 | 870 | 129 | 782 | 4,192 | 4,928 | 5,728 | 4,688 | 4,464 | 4,528 | 5,773 | 4,915 |
| Split counted | (3/8) | (1/2) | (1/2) | / | (1/2) | (3/32) | (1/16) | (1/16) | (1/16) | (1/16) | (1/16) | (5/64) | (5/64) |
| **"Fissurina" group** | | | | | | | | | | | | | |
| Fissurina aequilabialis | | | | | | | | | | | | | X |
| Fissurinaa alata | | | | | | | X | X | X | | | | |
| Fissurina circula | | | | | | X | | X | | | | | |
| Fissurina circulara | | | | | | | | | | | | | |
| Fissurina communa | | | | | | | | | | X | | | |
| Fissurina compressa | | | | | | | | | | | | | |
| Fissurina dancia | | | | | | | | | | 1.0 | | | |
| Fissurina elliptica | | | | | | | | | | | | | |
| Fissurina goreaui | | | | | | | | | | | | | |
| Fissurina incipiens | | | | | | | | | | | | | |
| Fissurina latiostoma | | | | | | | | | | | | | |
| Fissurina marginata | | | | | | | | | | | | | |
| Fissurina pseudoglobosa | | | X | | | | | | | | | | |
| Fissurina quadracostulata | | | | | | | | | | | | X | |
| Fissurina stewartii | X | | | | 1.0 | | | | | | | | |
| Fissurina subchasteri | | | | | | | | | | | | | |
| Galwayella biangulata | | | | | | | | | | | | | |
| Lagenosolenia inflataperforata | | | | | | | | | | | | | |
| Lagenosolenia lagenoides | | | | | | | | | | | | | |
| Parafissurina clavigera | | | | | | | | | | | | | |
| Parafissurina curvans | | | | | | | | | | | | | |
| Parafissurina exiguiformis | | | | | | | | | | | | | |
| Parafissurina hamigera | X | | | | | | | | | | | | |
| Parafissurina himatiostoma | | | | | | | | | | | | X | |
| Parafissurina labiata | | | | | | | | | | | | X | |
| Parafissurina subcarinata | | | | | | | | X | | | | | |
| Parafissurina tectulostoma | | | | | | | | | | | | | |
| Pseudoolina fissurinea | | | | | | | | | | | | | |
| Wiesnerina minutiforma | | | | 1.0 | | | | | | X | X | | |
| Wiesnerina remova | | | | | | | | | | | | | |
| **"Glabratella" group** | | | | | | | | | | | | | |
| Conorbella pulvinata | | | | | | | | | | | | | |
| Glabratellina arcuata | | | 1.5 | 6.0 | | 1.0 | 1.0 | | 1.0 | 3.0 | | 1.5 | |
| Glabratellina lauriei | | | | | | | | | 1.0 | | X | | |
| Glabratellina wrightii | | | | | | | | | | | 1.0 | | |
| Heronallenina parva | | | | | | | | | | | | | |
| Rotaliella chasteri | 3.5 | X | 3.0 | 6.0 | 2.0 | 6.5 | 2.5 | 4.0 | 1.0 | | 2.0 | 5.0 | 3.0 |
| Sphaeroidina bulloides | | | | | | | | | | | | | |
| **"Gyroidina" group** | | | | | | | | | | | | | |
| Gyroidinoides nipponicus | X | 1.5 | 1.0 | 1.0 | 1.0 | | | | | | | | |
| Gyroidinoides quinqueloba | | 2.0 | 1.0 | | 1.5 | X | | X | 1.0 | | | | |
| **"Lagena" group** | | | | | | | | | | | | | |
| Lagena semilineata | | | | | | | | | | | | | |
| Lagena substriata | | | | | | | | | | | | | |
| Lagena cf. L. sulcata | | | | | | | | | | | | | |
| Procerlagena distoma | | | | | | | | | | | | | |
| Procerolagena meridionalis | | | | | | | | | | | | | |
| Pygmaeoseistron vulgaris | | | | | | | | | | | | | |

## TABLE A3. BOREHOLE 87B-2 (continued)

| Sample number | 43 | 44 | 45 | 46 | 47 | 48 | 49 | 51 | 52 | 53 | 54 | 55 | 56 |
|---|---|---|---|---|---|---|---|---|---|---|---|---|---|
| Depth (m) | 41.02 | 42.69 | 44.23 | 47.5 | 49.12 | 50.24 | 50.91 | 56.31 | 61.53 | 66.7 | 71.5 | 73.0 | 74.87 |
| Total number of Quaternary benthics | 827 | 676 | 870 | 129 | 782 | 4,192 | 4,928 | 5,728 | 4,688 | 4,464 | 4,528 | 5,773 | 4,915 |
| Split counted | (3/8) | (1/2) | (1/2) | / | (1/2) | (3/32) | (1/16) | (1/16) | (1/16) | (1/16) | (1/16) | (5/64) | (5/64) |
| **"Miliolellid" group** | | | | | | | | | | | | | |
| *Miliolinella circularis* | | X | X | | | | | | | | | | |
| *Miliolinella subrotunda* | | | | | | | | | | | | | |
| *Pyrgo oblonga* | | | | | | | | | | | | | |
| *Triloculinella differens* | | | | | | | | | | | | | |
| *Triloculinella matthewense* | | | | | | | | | | | X | | |
| *Triloculinella parisa* | | | | | | | | | | | | | |
| *Triloculinella pyriformis* | | | | | | | | | | | | | |
| *Triloculinella tegminus* | X | | | | | | | | | | | | |
| **"Nonion" group** | | | | | | | | | | | | | |
| *Allomorphina fragilis* | X | | | | | | | | | | | | |
| *Chillostomella oolina* | | | | | | | | | | | | | |
| *Astrononion stelligerum* | 1.0 | X | 1.0 | | | | 1.0 | | X | 1.0 | X | 2.0 | 1.0 |
| *Laminononion stellatum* | | | | | | | | | | | | | |
| *Haynesina depressula* | | | | | 2.5 | 1.0 | | | | | | | |
| *Haynesina orbiculare* | 1.0 | 1.0 | 1.0 | 1.0 | | | 4.0 | 3.5 | 1.5 | 1.0 | 1.0 | X | 2.0 |
| *Haynesina paucilocula* | X | 1.5 | X | | X | | X | X | X | 1.0 | X | 1.5 | X |
| *Melonis barleeanus* | X | | | 1.0 | | X | | X | | | | | |
| *Nonionella lobsannensis* | | | | | | | | | | | | | |
| *Nonionella stella* | | | | | | | | | | | | | |
| *Nonionellina labradorica* | 2.0 | 1.0 | 3.5 | 4.0 | X | X | 2.0 | 1.5 | 1.0 | 2.0 | 1.0 | 2.0 | 1.5 |
| ~~ *Nonionoides grateloupi/Nonionella turgida* | | | | | | | | | | | | | |
| ~~ *Pulleniella osloensis/Haynesina germanica* | 1.0 | 1.0 | 2.0 | | 1.0 | X | | | 1.0 | 1.0 | X | | |
| **"Oolina" group** | | | | | | | | | | | | | |
| *Cushmanina striatopunctata* | | | | | | | | | | | | | |
| *Favulina hexagona* | | | | | | | | | | | | | |
| *Favulina melo* | | | | | | | | | | | | | |
| *Favulina scalariformis* | | | | | | | | | | | | | |
| *Favulina squamosa* | | | | | | | | | | | | | |
| *Homalohedra acuticostata* | | | | | | | | | | | | | |
| *Homalohedra apiopleura* | | | | | | | | | | | | | |
| *Homalohedra williamsoni* | | | | | | | | | | | | | |
| *Oolina ampullilineata* | | | | | | | | | | | | | |
| *Oolina lineata* | | | | | | | | | | | | | |
| **"Polymorphina" group** | | | | | | | | | | | | | |
| *Entopolymorphina oppositiforma* | | | | | | | | | | | | | |
| *Laryngosigma hyalascidia* | | | X | | | | | | | | | | |
| *Laryngosigma lactea* | | | X | | | | | | | | | | |
| *Laryngosigma williamsoni* | | X | | | | | | | | | | | |
| *Metapolymorphina ligua* | | | | | | | | | | | | | |
| *Pseudopolymorphina suboblonga* | | | | | | X | | | | | | | |
| **"Quinqueloculina" group** | | | | | | | | | | | | | |
| *Quinqueloculina akerniana* | | | | | | | | | | | | | |
| *Quinqueloculina crassicarinata* | | | | | | | | | | | | | |
| *Quinqueoculina obliquecamerata* | | | | | | | | | | | | | |
| *Quinqueloculina seminula* | | | | | | | | | | | | | X |
| *Quinqueloculina viennensis* | | | | | | | | | | X | | | |
| *Siphonaperta parvula* | | | | | | | | | | | | | |

## TABLE A3. BOREHOLE 87B-2 (continued)

| Sample number | 43 | 44 | 45 | 46 | 47 | 48 | 49 | 51 | 52 | 53 | 54 | 55 | 56 |
|---|---|---|---|---|---|---|---|---|---|---|---|---|---|
| Depth (m) | 41.02 | 42.69 | 44.23 | 47.5 | 49.12 | 50.24 | 50.91 | 56.31 | 61.53 | 66.7 | 71.5 | 73.0 | 74.87 |
| Total number of Quaternary benthics | 827 | 676 | 870 | 129 | 782 | 4,192 | 4,928 | 5,728 | 4,688 | 4,464 | 4,528 | 5,773 | 4,915 |
| Split counted | (3/8) | (1/2) | (1/2) | / | (1/2) | (3/32) | (1/16) | (1/16) | (1/16) | (1/16) | (1/16) | (5/64) | (5/64) |
| **"Rotalia" group** | | | | | | | | | | | | | |
| Eponides pusillus | | | X | | | | | | | | | | |
| Ioanella tumidula | X | 1.5 | 2.0 | | 1.0 | | | | | | | | X |
| Nuttalides bradyi | | | | | | | | | | | | | X |
| Oridosalis umbonatus | | | X | | X | | | | | | | | |
| Valvulineria minuta | | | | | | | | X | | | X | | |
| **"Stainforthia" group** | | | | | | | | | | | | | |
| Rutherfordoides mexicana | X | | | | | | | | | | | | |
| Stainforthia concava | | | | | X | | | X | | | | | |
| Stainforthia fusiformis | X | | 3.0 | 1.0 | X | X | | | X | | | 1.0 | |
| Stainforthia pauciloculata | | | | | | | | | | | | | |
| Stainforthia rotundata | 1.0 | | 1.0 | 1.0 | X | | X | X | | | | X | |
| Tosaia hanzawaia | | 1.0 | X | | | | | | | | | | |
| **"Spiralina" group** | | | | | | | | | | | | | |
| Patellina corrugata | X | X | X | | | | | 1.0 | | 1.0 | X | | 1.0 |
| **"Triloculina" group** | | | | | | | | | | | | | |
| Triloculina angulosa | | | X | | | | | | | | | | |
| Triloculina gibba | | | | | | | | | | | | | |
| Triloculina trihedra | | | | | | | | | | | | | |
| **"Uvigerina" group** | | | | | | | | | | | | | |
| Angulogerina angulosa/fluens | | | | | | X | X | | | | | | |
| Neouvigerina canariensis | X | X | | | | | | | | | | | |
| Uvigerina peregrina | | | | | | | | | | | | | |
| **QUATERNARY PLANKTONICS** | | | | | | | | | | | | | |
| Total number of specimens | 83 | 28 | 64 | 11 | 46 | 352 | 1,280 | 96 | 144 | 32 | 240 | 180 | 64 |
| Split counted | (3/8) | (1/2) | (1/2) | / | (1/2) | (1/16) | (1/64) | (1/16) | (1/16) | (1/16) | (1/16) | (5/64) | (5/64) |
| Globigerina bulloides | 4 | | 6 | 1 | 2 | | 64 | | | | | | |
| Globogerinita uvula | 6 | | 22 | 1 | 10 | | 128 | 16 | | | 16 | 26 | 14 |
| Globorotalia inflata | | | | | | | | | | | 16 | | |
| Neogloboquadrina deutertri | 20 | | 8 | | | 64 | 192 | | 48 | 16 | 64 | | |
| Neogloboquadrina pachyderma—sinistral | 20 | 14 | 22 | 7 | 24 | 192 | 576 | 64 | 32 | | 64 | 90 | 26 |
| Neogloboquadrina pachyderma—dextral | 24 | 6 | 18 | 2 | 6 | 80 | 320 | 16 | 48 | 16 | 48 | 51 | 26 |
| Turborotalita quinqueloba—sinistral | 4 | 2 | 4 | | | 16 | | | 16 | | 16 | 13 | |
| Turborotalita quinqueloba—dextral | 3 | 6 | 2 | | 4 | | | | | | 16 | | |
| **Reworked** | | | | | | | | | | | | | |
| Total number of specimens | 12,144 | 8,624 | 5,376 | 71 | 4,520 | 13,024 | 19,008 | 18,048 | 9,632 | 8,592 | 7,616 | 4,774 | 9,312 |
| Split counted | (1/16) | (1/16) | (1/16) | / | (1/8) | (1/16) | (1/64) | (1/32) | (1/16) | (1/16) | (1/32) | (5/64) | (1/32) |

# Deglaciation of western Maine

**Woodrow B. Thompson**
*Maine Geological Survey, 22 State House Station, Augusta, Maine 04333-0022, USA*

## ABSTRACT

Surficial geologic mapping of western Maine has provided evidence concerning the recession of the Laurentide ice sheet in this part of the state. End moraines, heads of outwash, and meltwater channels indicate the orientation of the ice margin and the direction of its retreat. Although these features are unevenly distributed, they consistently show that the ice sheet receded generally northward from the inland marine limit to the Canadian border. In the southern and central regions of the study area, striation patterns reveal a shift in late glacial ice-flow direction from southeast to south or south-southwestward. Meltwater channels and outwash heads in these regions record late-glacial northward ice retreat. North of the Mahoosuc Range, late Wisconsinan ice flow was dominantly southeast, with topographically controlled ice-margin recession to the north or northwest. The degree to which ice flow was maintained during deglaciation is difficult to ascertain, but shifting ice-flow directions, moraines and glaciotectonic structures, and morphosequences of ice-contact sand and gravel imply that active ice flow persisted across much of the study area until late stages of deglaciation. The few limiting radiocarbon ages from western Maine indicate that ice retreat occurred between ca. 13 500 and 11 500 $^{14}$C yr B.P.

## INTRODUCTION

This chapter summarizes what is currently known about the deglaciation of western Maine; during the past 15 years there has been a rapid accumulation of new data concerning late Wisconsinan ice recession. The Maine Geological Survey's programs of surficial geologic quadrangle mapping and delineation of sand and gravel aquifers have provided much of this information. Additional findings have resulted from research projects on sea-level change and deglaciation chronology. The study area is bounded to the south by a belt of glaciomarine deltas along the inland limit of late glaciomarine submergence, and extends north to the Canadian border. For the purposes of discussion, western Maine is divided into southern, central, and northern regions, as shown in Figure 1. The field localities mentioned here are located according to the 7.5-minute quadrangles in which they occur (Fig. 2).

Surficial deposits in Maine's coastal lowland show that late Wisconsinan deglaciation was marked by the retreat of a grounded ice margin in shallow-marine waters. The ice sheet remained active as it withdrew from this part of the state, and there were frequent minor oscillations of the ice margin accompanied by deposition of end moraines, ice-contact glaciomarine deltas, and submarine fans (Smith and Hunter, 1989; Thompson et al., 1989). Inland from the marine limit, the mode of ice retreat is not so clear. During the mid 1900s it was widely believed that deglaciation of the mountainous interior of New England occurred primarily by stagnation and downwasting of the last ice sheet (e.g., Leavitt and Perkins, 1935; Goldthwait, 1938). However, little evidence was published supporting either stagnation or active-ice deglaciation models in western Maine (Borns, 1989). Investigations in coastal Maine since the early 1970s, and the evolution of the morphosequence concept in central and southern New England (Koteff and Pessl, 1981), suggest that evidence of active ice retreat might be found in the uplands between the marine limit and the borders with New Hampshire and Quebec.

Several methods have been used to reconstruct the pattern

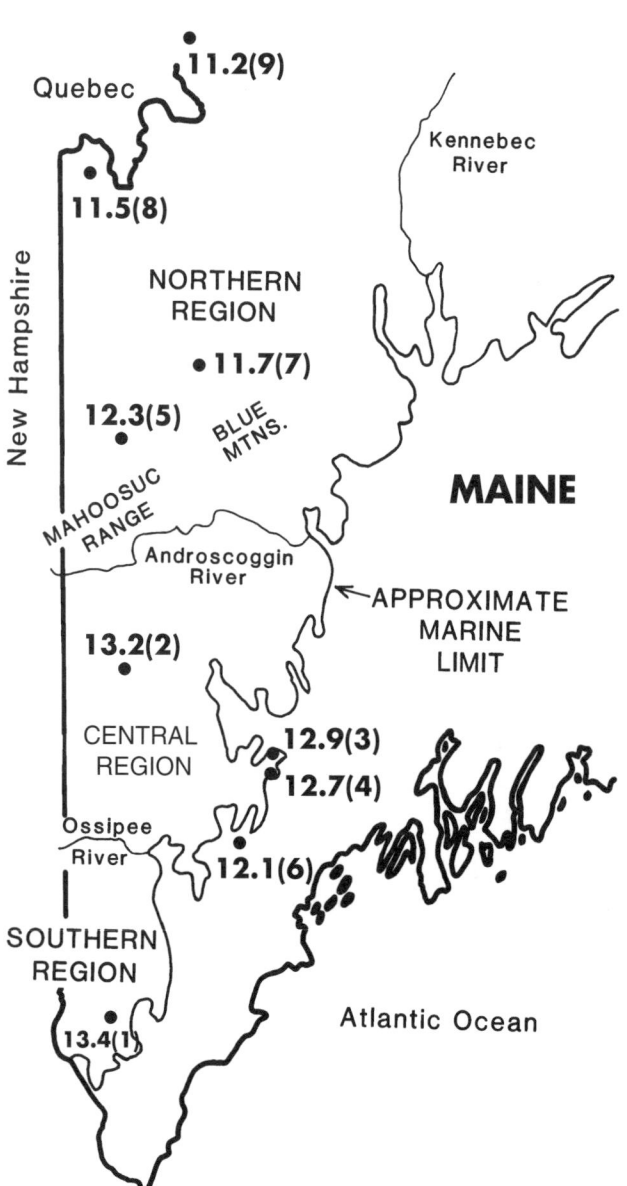

Figure 1. Map showing location of study area, regions of western Maine discussed in text, and radiocarbon-dated sites listed in Table 1.

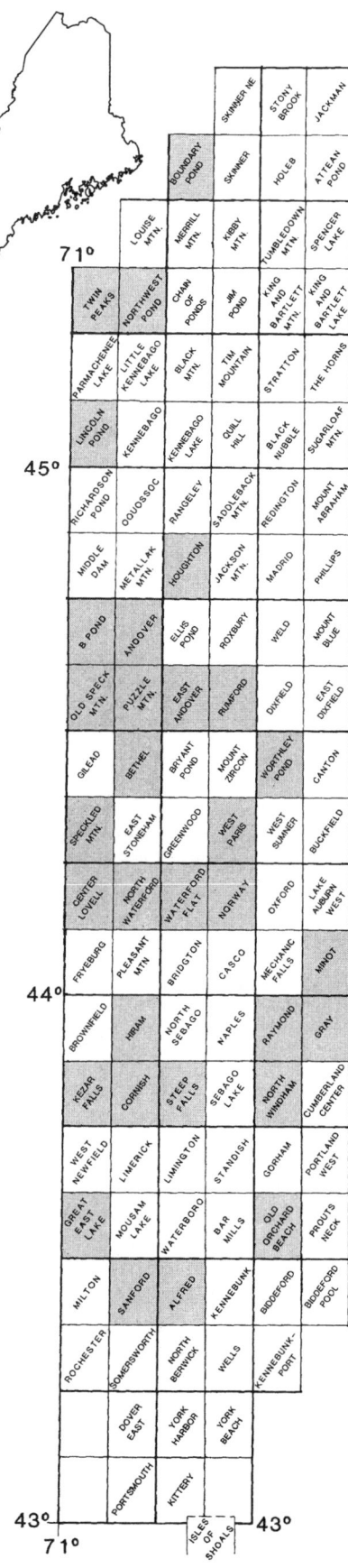

Figure 2. Index to 7.5-minute quadrangles in study area. Quadrangles mentioned in text are shaded.

of late Wisconsinan deglaciation in western Maine. Striations on bedrock outcrops indicate shifting ice-flow directions in late glacial time. Successive positions of the retreating ice margin have been inferred from end moraines and (more commonly) from the ice-contact heads of stratified-drift morphosequences. Meltwater channels that formed along the fronts and sides of valley ice tongues have also been used to delineate ice margins.

The locations of spillway channels associated with successive levels of ice-dammed glacial lakes are helpful in tracing ice retreat. In some places these spillways can be correlated with glaciolacustrine deltas to provide an exceptionally clear record of deglaciation. This technique has been applied in the southern part of the study area, where ephemeral glacial lakes were ponded against the ice margin in north-sloping valleys (e.g., Davis and Holland, 1997a). Each lake drained through the lowest available col in the surrounding hills, and ice recession opened new and lower outlets. Gilbert-type deltas composed of sand and gravel were graded to these successive lake stages, enabling the elevations of delta tops to be matched with corresponding spillways.

Most of the evidence of ice flow in the marginal zone of the receding ice sheet is indirect. Late glacial ice shove structures are present in the study area but are rarely exposed. However, end moraines occur locally in valley bottoms in the southern part of the area. These moraine ridges probably resulted from ice flow conveying debris to the edge of the glacier. Meltwater streams issuing from the ice sheet deposited large volumes of sand and gravel in this part of Maine. In some places these deposits can be grouped into contemporaneous sets called morphosequences, which are useful for tracing successive ice-margin positions (Koteff and Pessl, 1981). It is likely that active ice was needed to supply the great quantity of sediments in these deposits.

Geologic mapping in the study area has shown that the preceding features are not distributed evenly across the region. Glaciolacustrine deltas are very abundant in the Saco, Ossipee, and Little Ossipee River basins in the southern region, but are smaller and less common to the north. Possible reasons for this are discussed in the following. Stratified drift deposits throughout the region are concentrated in the valleys, where sedimentation was focused by glacial meltwater action. For this reason it is often possible to determine ice-margin positions from heads of outwash in valleys, but correlations across the intervening uplands are apt to be speculative at best. Exceptions occur locally where meltwater channels enable the ice margin to be traced across the high terrains.

## PREVIOUS WORK

The first in-depth study of Maine's Quaternary deposits was conducted by G.H. Stone (1899; see also Borns, this volume). Stone's investigation of the glacial gravels of Maine was largely concerned with esker systems and related ice-contact features. He discussed the origin of Maine eskers (osars) at great length, and described the esker ridges of western Maine. Stone (1899) also produced a map of successive ice-margin positions (lines of frontal retreat) during deglaciation of the state. This compilation does not extend completely across southwestern Maine, but it does indicate northwestward recession over much of the area discussed here.

Stone believed that the last ice sheet continued to flow actively during at least the early stages of thinning and recession (Stone, 1899, p. 271). He often compared his models with observations of modern glaciers. Stone described the process of end-moraine deposition in the coastal lowland as resulting from stillstands of an active glacier margin, to which debris was conveyed from a thin basal zone of debris-rich ice (Stone, 1899, p. 275–276). He also contributed to understanding the deglaciation of the upper Androscoggin Valley through his discovery and description of large end moraines on the Maine–New Hampshire border (Stone, 1880, 1899).

The next study of western Maine glacial deposits was published by Leavitt and Perkins (1934, 1935). Their three-volume series combined a statewide survey of road materials with descriptions of glacial sediments. These authors also compiled an accompanying map showing the surficial geology of Maine. Leavitt and Perkins (1935, p. 193–195) cited "kame terrace" sequences as evidence that "In the hilly western portion of the State, the last ice remained as narrow tongues along the valley floors, and wasted away in place." This statement reflects the changing views of New England glacial geologists resulting from the controversy, which peaked in the 1930s, regarding the mode of glacial retreat. Previous assumptions that an active ice margin had receded northward from New England were replaced for several decades by the model of widespread simultaneous stagnation and downwasting of the ice sheet (e.g., Flint, 1929; Goldthwait, 1938).

Although they favored the stagnation model for deglaciation of western Maine, Leavitt and Perkins (1935) noted the preferred distribution of ice-contact glacial lake deposits in north-draining tributaries of the Saco and Ossipee River valleys. Similar evidence of ice damming in north-draining tributaries of the Androscoggin and other valleys in western Maine led to the conclusion (Thompson and Borns, 1985b) that the late Wisconsinan ice margin receded generally northward, although clear evidence of sustained ice flow during deglaciation is rarely found. Thompson and Fowler (1989) confirmed the existence of a late glacial active ice tongue that deposited end moraines in the upper Androscoggin valley. They mapped the Androscoggin moraine complex in detail and discovered numerous moraine ridges in addition to those originally described by Stone (1880, 1899).

Since the late 1900s, surficial geologic mapping within the study area has been carried out by the Maine Geological Survey. Reconnaissance-level mapping of 15-minute quadrangles in the southern region (Fig. 1) was conducted by G.W. Smith in the late 1970s, and incorporated into the *Surficial Geologic Map of Maine* (compiled by Thompson and Borns, 1985a). The latter map also includes work by D.W. Caldwell, who mapped the surficial geology of numerous 15-minute quadrangles in the northern region of western Maine (Caldwell, 1974). Since the late 1980s, detailed mapping of 7.5-minute quadrangles in southwestern Maine has been done by numerous individuals under U.S. Geological Survey–Maine Geological Survey cooperative programs, the most recent being the STATEMAP program. Investigations of sand and gravel aquifers have yielded much new information on the boundaries and stratigraphy of glacial sedimentary units in the study area. Aquifer mapping in part of the study area was done by the U.S. Geological Survey (Prescott, 1979, 1980), followed by detailed work by the Maine

Geological Survey, U.S. Geological Survey, and other cooperating state agencies (Williams et al., 1987; Nichols et al., 1995).

Information on the deglaciation of the study area resulting from these investigations has been published in professional journals (e.g., Borns and Calkin, 1977); abstracts for meetings such as those of the Geological Society of America (Boothroyd, 1995; Davis et al., 1998; Thompson, 1991, 1998; Thompson and Koteff, 1995); and field guides for the Northeastern Friends of the Pleistocene (Thompson et al., 1995), Geological Society of Maine (Thompson, 1985), and the New England Intercollegiate Geological Conference (Holland, 1986; Thompson, 1986, 1989).

Radiocarbon ages limiting the time of ice retreat from Maine have been obtained from organic debris extracted from the basal portions of lake sediment cores. Davis and Jacobson (1985) used these ages to infer successive positions of the retreating late Wisconsinan ice margin in northern New England. Other investigators have recently obtained ages from ponds in western Maine (Fig. 1).

## TRANSITION FROM MARINE TO TERRESTRIAL DEGLACIATION

Large glaciomarine deltas were deposited along the inland limit of late glaciomarine submergence in southwestern Maine. Thompson et al. (1989) investigated deltas throughout the zone of marine transgression, and found that these deltas are the Gilbert type, consisting of sand and gravel that accumulated on the delta plains (topset beds) and prograding delta fronts (foreset beds) (Fig. 3). Many of them formed in an ice-contact environment, often in association with esker systems. The eskers record locations of subglacial tunnels through which sediment was carried to the delta heads. Ashley et al. (1991) noted that deltas in eastern Maine developed as submarine fans built upward to the ocean surface, and this was likely the origin of some of the western Maine deltas. Others (leeside deltas) were deposited where meltwater streams carried sediments through gaps in hills and then deposited them in the sea (Thompson et al., 1989).

The concentration of large deltas along the marine limit may reflect two factors associated with ice retreat: (1) slowing or stillstand of retreat when the ice margin withdrew from the sea and no longer underwent iceberg calving, and/or (2) sustained delivery of sediment to the sea from the inland glacial drainage network. Boothroyd (1995, 1997) proposed that the ice-tunnel system marked by the Pine River esker, which crosses the Maine–New Hampshire border and trends southeast through the Great East Lake quadrangle, simultaneously conveyed sediment along its full length and thus was able to supply the great quantity of sand and gravel where the esker terminates in a large glaciomarine delta complex. Ashley et al. (1991) applied the same argument to the large esker-fed marine deltas in eastern Maine.

Thompson et al. (1989) precisely measured the elevations of contacts between topset and foreset units in the glaciomarine deltas, and found that deltas in southwestern Maine record postglacial uplift and crustal tilt of 0.53 m/km (higher to the northwest). The actual tilt probably was somewhat greater, but the gradient of delta elevations was reduced by eustatic sea-level rise as the sequence of deltas was deposited.

Recent work by Thompson and Koteff (unpublished data) has refined the survey data for deltas in southwestern Maine, indicating a tilt of 0.51 m/km in the region between Sanford and Augusta. In the southernmost part of the state, Koteff et al. (1993) found that glaciomarine delta elevations record a steeper

Figure 3. Gravelly topset beds overlying foreset beds in New Dam Road delta, Sanford (Alfred quadrangle; see Fig. 2). Pit face is ~8 m high.

crustal tilt of 0.89 m/km, in close agreement with western New England. These authors proposed that in the latter areas, crustal uplift was delayed until after deglaciation. They also reported evidence of a very rapid (50–100 yr) 7–10 m rise in late glacial eustatic sea level, based on an abrupt northward increase in delta elevations near Wells.

## RECESSION OF THE LATE WISCONSINAN ICE SHEET FROM THE MARINE LIMIT TO THE CANADIAN BORDER

Several interrelated aspects of deglaciation are considered here, including directions of late glacial ice flow, the record of ice recession, and whether the marginal part of the ice sheet remained active during deglaciation. The regional variation of these parameters is the framework for discussion. The northern region (north of the Mahoosuc Range; Fig. 1) differs from the others with respect to ice-flow history, while the region south of lat 44°N is distinguished by a much greater abundance of glaciofluvial and glaciolacustrine sand and gravel deposits.

### Ice-flow directions during last glacial maximum and deglaciation

Streamlined hills and stoss-and-lee erosional features throughout the study area suggest that ice-flow directions during the late Wisconsinan glacial maximum ranged from southeast to south-southeast. However, striations on glacially abraded bedrock indicate different late glacial flow directions north and south of the Mahoosuc mountain range (Fig. 4; Thompson and Koteff, 1995). At sites where relative ages of multiple striation directions could be determined, only the youngest trend was plotted in Figure 4. Otherwise, all directions were plotted for each site.

Striations in the northern region show a strong orientation toward the southeast, in agreement with streamlined landforms. In the central and southern regions, they record a more southerly to south-southwestward flow (Fig. 4). Where age relations are known for multiple striation sets, the younger striations in the latter regions usually trend more southerly than the older sets. South-trending striations of probable late glacial age were found both in lowland areas such as the Saco Valley in Fryeburg (Thompson, 1999b) and on the highest mountains. For example, on the south spur of Speckled Mountain (at the site marked with a 2623 ft elevation on the Speckled Mountain quadrangle) striations trending 180° crosscut an older set trending 162°.

Vose (1868) published the first striation map including part of western Maine. It is interesting to note that his map likewise shows south-trending striations on Speckled Mountain, as well as eastward channeling of ice flow along the Androscoggin Valley. Thompson and Fowler (1989) discussed this topographic deflection of the ice tongue that deposited the Androscoggin Moraine system. Elsewhere in the study area, there are rare east-trending striations that predate southeast sets and thus may in-

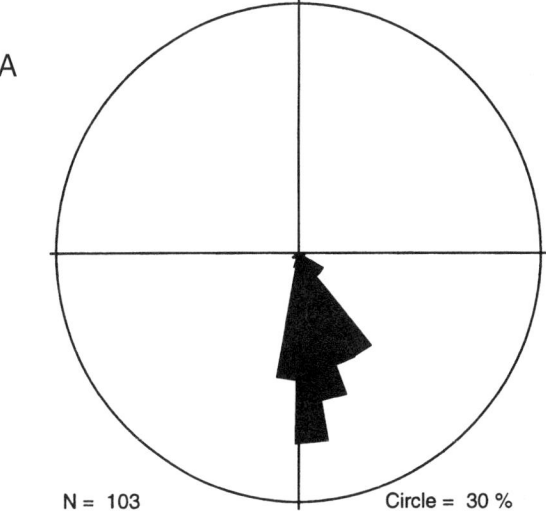

## Striation data south of Mahoosuc Range

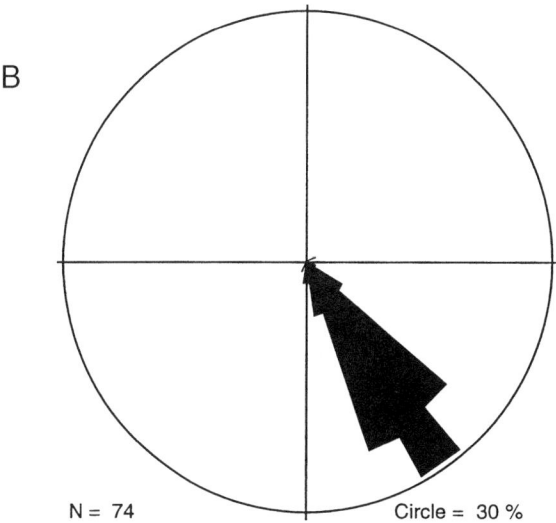

## Striation data north of Mahoosuc Range

Figure 4. Rose diagrams showing youngest ice-flow directions inferred from striated bedrock outcrops: A, south of Mahoosuc Range; B, north of Mahoosuc Range.

dicate a phase of ice flow prior to the late Wisconsinan glacial maximum.

### Ice-marginal deposits

**End moraines.** Inland from the marine limit, a variety of tills and waterlaid sediments were deposited during recession of the late Wisconsinan ice sheet. Sandy to gravelly ablation till containing abundant lenses of washed sediment forms an irregular blanket over much of the area, but exposures of this material rarely indicate where it was deposited relative to the ice margin. In a few places there are bouldery ridges which, judging from shallow surface exposures, consist of ablation till and are interpreted to be end moraines (Fig. 5). These ridges are commonly 3–20 m high and several hundred meters long. Most occur in valley bottoms and suggest north to north-northwestward ice retreat.

Moraines are rare in the central and northern regions. Two notable exceptions are the Androscoggin moraine and the Frontier moraine. The Androscoggin moraine (Fig. 6) is an arcuate series of large bouldery moraine ridges that crosses the Androscoggin River valley on the Maine–New Hampshire border. These moraines were deposited by a tongue of the Laurentide ice sheet issuing from the northern White Mountains (Thompson and Fowler, 1989). A few short moraines have been found elsewhere in the central region, south of Worthley Pond in Peru (Worthley Pond quadrangle; Eusden, 1980) and in the Crooked River valley (Norway and Waterford Flat quadrangles; Thompson, 2000a, 2000b). Field reconnaissance suggests that other moraines exist in the area, but they have not been examined closely. These bouldery ridges usually are located in forested areas and are only revealed by detailed mapping.

In the northern region, the Frontier moraine system generally follows the southwest part of the Maine-Quebec border, principally on the Canadian side. The Frontier moraine includes ridges composed of till and stratified drift, as well as ice-contact deltas (Shilts, 1981). Several segments of the moraine system extend into Maine, including those mapped by Shilts (1981) and a moraine ridge just north of Lower Black Pond in Oxbow (Twin Peaks quadrangle). Deposition of this moraine system was favored by active ice impinging against the north slopes of the Boundary Mountains while remnant ice masses stagnated in the lee of the mountains.

Caldwell (1974, 1975) reported a large end moraine in the Andover quadrangle, which he named the C Pond Moraine. He said that the moraine is composed of bouldery till and "apparently records the last position of active glacial ice in northwestern Maine" (Caldwell, 1974, p. 6). This deposit is located in a narrow valley on the divide between the headwaters of Dead Cambridge River to the west and Sawyer Brook to the east. Caldwell proposed that the moraine was deposited by an ice tongue flowing eastward. His map of the moraine (Caldwell, 1975) shows an area of hummocky topography that rises abruptly from the east side of C Pond to a height of 85 m and extends 2.25 km eastward across the divide. The C Pond moraine evidently consists of a diffuse area of irregular morainic topography, rather than a single distinct moraine ridge. I have found bedrock outcrops in the southwestern part of the moraine complex shown in Caldwell's map, and it is possible that some of the other knobs in this area are likewise bedrock.

***Hummocky and ribbed moraine.*** Surficial quadrangle mapping in southwestern Maine has delineated areas of hummocky moraine and ribbed moraine. These deposits consist of bouldery diamictons locally interstratified with variable amounts of sand and gravel. They are concentrated in the southern region of the study area. Holland (1986) described ribbed moraine as ridges of coarse glacial diamicton. These ridges selectively occur in the bottoms of valleys oriented parallel to

Figure 5. End moraine in Crooked River valley (Norway quadrangle; see Fig. 2). Moraine ridge is ~3 m high.

Figure 6. View southwest across Androscoggin Valley on Maine–New Hampshire border, showing part of Androscoggin moraine complex (ridge crests marked by dashed lines). Ridge projecting from Hark Hill is 750 m long and up to 30 m high.

glacial ice flow, and are generally normal to the valley axes. Davis (in Thompson et al., 1995; Davis and Holland, 1997b) compared ribbed moraine in the Kezar Falls quadrangle with similar deposits elsewhere, such as the ribbed moraine south of Mount Katahdin in central Maine and the Rogen moraine in Sweden. Davis found that the till fabric in one moraine shows a strong orientation normal to the ridge axis, and concluded that the deposit probably formed by subglacial lodgement or melt-out. If this is true, the moraine ridges are not reliable indicators of recessional ice-margin positions. Davis and Holland (1997b) also mapped and described end moraines and hummocky moraine in the Kezar Falls quadrangle.

Holland (1986) contrasted ribbed moraine with irregular knobby deposits of ice-disintegration moraine, composed of mixtures of diamicton and stratified sediment, which he interpreted as having formed in a dead-ice environment. However, the degree of ice activity associated with deposition of these sediments is usually uncertain, so they have been given the morphologic name hummocky moraine in recent geologic maps of southwestern Maine. Boothroyd's (1997, p. 3) description of hummocky moraine in the Great East Lake quadrangle applies throughout the study area: "irregularly-shaped mounds and small hills with up to 40 feet (12 m) of internal relief and 120 feet (37 m) of total relief that are comprised of interstratified debris-flow till and ice-marginal fluvial and lacustrine deposits. . . . Internal structure shows abundant evidence of collapse and remobilization of sediment after initial deposition, probably by melting of buried ice."

Hummocky moraine usually occurs at relatively low elevations on the floors or sides of valleys. Judging from its topography and composition, it probably formed by supraglacial sedimentation on debris-rich valley ice tongues along the receding margin of the last ice sheet. Melting of the ice resulted in topographic inversion of small debris-filled basins, leaving them as hummocks. Distinct ridges among some of these deposits, for example in the Hiram quadrangle (Thompson and Holland, 1999), may have formed at the glacier margin. Areas of hummocky moraine associated with ice-contact stratified drift morphosequences have been used to help define ice-margin trends in the Salmon Falls River valley and elsewhere in the Great East Lake quadrangle (Boothroyd, 1997).

A remarkable belt of hummocky moraine occurs in the northeast part of the Raymond quadrangle, where it was mapped by Retelle (1997) as a succession of two fluvial ice-contact morphosequences (Notched Pond deposits) because of the inclusion of abundant waterlaid sediments. These deposits occupy a low area bounded to the west by an abrupt transition to glacially streamlined till hills. Retelle suggested that the Notched Pond sequences resulted from stagnation of a valley ice tongue, and he correlated them with deposits that are clearly ice marginal in a valley to the west.

***Morphosequences deposited by glacial meltwater.*** The southern region of the study area shows extensive sand and gravel deposited in proglacial meltwater streams and lakes. Meltwater was temporarily ponded in north-sloping valleys blocked by glacial ice, and in south-draining valleys obstructed by plugs of glacial sediment or remnant ice masses. Glacial streams entering these lakes deposited Gilbert-type deltas. Many glacial lakes totally filled with deltaic sand and gravel, with the result that fine-grained lake-bottom sediments are buried beneath prograding deltas. Detection of these fine materials in well and test boring logs helps confirm the former lakes where the surface delta plains may only suggest fluvial sequences. Lake-bottom sediments also may be overlain by postglacial alluvial and stream-terrace deposits (Fig. 7).

Koteff and Pessl (1981) coined the term "morphosequence" (sequence in abbreviated form) for a group of essentially contemporaneous waterlaid glacial deposits graded to a common base level, such as a particular lake level. Following the classification scheme of these authors, many sand and gravel deposits in the southern region are morphosequences deposited entirely or partially in glacial lakes. They include: (1) simple ice-contact deltas (lacustrine ice-contact sequences); (2) ice-contact deltas that filled lakes and were buried beneath aggrading outwash fans (lacustrine-fluvial ice-contact sequences); and (3) deltas fed by subaerial outwash systems terminating in glacial lakes (fluvial-lacustrine sequences).

Lacustrine-fluvial sequences are common in the southern region of the study area. Boothroyd (1997) and Meglioi and Thompson (1997) described examples deposited in Glacial Lake Mousam, in the present Mousam Lake and Little Ossipee River basins. Some of them were aggraded to the extent that Boothroyd (1997) mapped them as fan deposits, although the deeper portions are deltaic. The most useful deposits for mapping ice margins in southwestern Maine are series of ice-contact lacustrine and lacustrine-fluvial sequences formed as glacial retreat opened progressively lower spillways for associated lakes.

Figure 7. Varved lake-bottom deposits of Glacial Lake Cornish (center) (Cornish quadrangle; Stop 2 in Holland, 1986).

Ideally, the preserved head of outwash in each ice-contact sequence shows topographic and sedimentologic evidence of where the glacier margin formerly stood. These features may include a kettled (collapsed) zone, esker feeders, bouldery gravels, and an ice-contact slope.

Surficial geologic mapping in the southwest part of the Steep Falls quadrangle (Gosse and Thompson, 1999) provides an example of reconstructing deglaciation from sequence analysis (Fig. 8). In this area, Glacial Lake Town Farm was dammed by the ice margin in a north-sloping tributary valley of the Saco River. Ice-contact deltas and associated meltwater deposits (map units $1t_{1-3}$) formed successively as northward ice recession uncovered lower outlets for the lake. Figure 8 shows the approximate location of the ice-margin during each stage of Lake Town Farm. Continued ice retreat opened the Saco River valley to the north, where a delta (unit 1d) formed in a younger lake. This lake was dammed by ice-contact and marine deltaic deposits that choked a narrow part of the Saco valley.

### *Direction and mode of ice-margin recession*

Ice-marginal deposits show that glacial recession in the southern region occurred in a generally northward direction, with local northwest or northeast retreat controlled by the orientation of valleys. Direct evidence of active ice retreat is limited to a few end moraines. However, the great volume of ice-contact sand and gravel together with thick deposits of ribbed and hummocky moraine suggest that ice flow continued to transport debris to the ice margin during deglaciation. Much sediment was conveyed in esker tunnels to lacustrine or marine depositional environments.

The relative abundance of glacial sediment in the southern region also may be a function of the erodibility of local bedrock. Most of this region is underlain by the Sebago pluton and other granitic rocks, and Silurian-Devonian metasediments of the Rindgemere Formation (Osberg et al., 1985). Hanson and Caldwell (1989) documented the relationship between topography and lithologic and structural properties of rock formations in Maine. They noted that easily eroded plutonic rocks commonly underlie lowlands, including the area of the Sebago pluton. Preglacial weathering and glacial excavation of the coarse granitic rock in the latter pluton probably were important factors in generating the surficial deposits of the southern region.

In the central region, the trends of the youngest striations suggest ice-margin retreat to the north or north-northeast. A section at the former Bryant Hill Pit in the Center Lovell quadrangle showed clear evidence of sustained ice activity in late glacial time (Thompson, 1994, 1999a). At this locality, late Wisconsin till and proglacial fan gravel were offset by thrust faults when the ice sheet pushed against the hillside. The faults indicate ice shoving from the northeast, compatible with nearby striations trending 184° and 195°.

Glaciofluvial deposits in the central region include outwash in the Crooked River and Little Androscoggin River valleys and several south-trending esker systems (Thompson and Borns, 1985a). There are few morphosequences among these deposits to indicate distinct heads of outwash, with the exception of scattered lacustrine fans and ice-contact deltas confined to valleys (e.g., Hildreth, 1997a, 1997b). The relatively small volume of meltwater deposits in this region may reflect reduced late glacial ice activity in the lee of the high mountains composing the Mahoosuc Range.

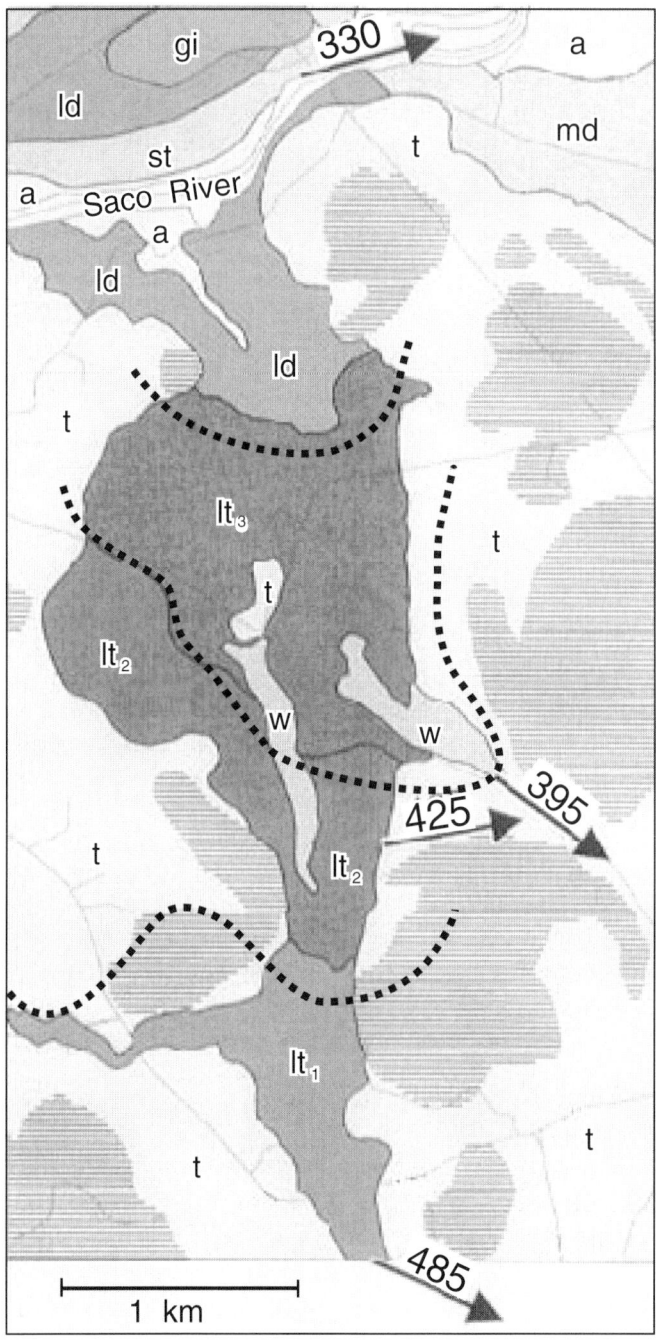

Figure 8. Surficial geologic map of southwest part of Steep Falls quadrangle (see Fig. 2), showing lacustrine sequences deposited in Glacial Lake Town Farm ($lt_{1-3}$) and Saco River valley (ld). Arrows indicate spillway locations and elevations (in feet, to conform with topographic map) associated with these sequences. Dashed lines show successive ice-margin positions when Lake Town Farm deposits formed. Other map abbreviations: a, alluvium; gi, ice-contact sand and gravel; md, marine delta; st, Saco River terrace; t, till; w, wetland. Ruled pattern shows upland areas with numerous bedrock outcrops. (Modified from Gosse and Thompson, 1999.)

The positions of glaciolacustrine deposits and proglacial meltwater channels are further evidence of northerly ice retreat in the central region. Descending flights of channels locally show both the thinning and recession of ice tongues among the hills (Fig. 9). They are useful indicators of ice-margin positions in upland areas where stratified drift deposits are sparse or lacking. Glaciolacustrine deposits occur in some of the north-sloping valleys that were blocked by the ice margin. Several valleys that contained ice-dammed lakes are tributary to the Androscoggin River between West Bethel and Peru (Fig. 10). These valleys contain sandy lake-bottom sediments, whereas the tributaries on the north side of the Androscoggin Valley were free-draining and contain glaciofluvial deposits.

Late glacial reorganization of ice flow occurred as far north as the Mahoosuc Range. Outcrops just south of Plumbago Mountain (East Andover quadrangle) show striations trending 180°–207°, which in places appear to be younger than a set averaging 170°. At higher elevations on Plumbago Mountain, only striations in the 160°–170° range have been seen. It is inferred that a late south-southwest–flowing ice tongue occupied the lower terrain south of the mountain. Meltwater flowed southwest from this ice tongue and cut deep channels across a till ridge in the adjacent part of the Puzzle Mountain quadrangle, and may have contributed sediment to the Stony Brook fan at the confluence of the Bear River with the Androscoggin (Bethel quadrangle). Exposures in the Chadbourne gravel pit in Newry have shown that the Stony Brook fan unconformably overlies collapsed ice-contact deposits (Thompson, 1989). The fan gravel is not deformed, so remnant ice in the lower Bear River valley must have melted prior to deposition of the fan. However, it is not certain whether the Stony Brook fan is glacial outwash or a younger (perhaps paraglacial) deposit formed by brooks draining the steep south face of Plumbago Mountain.

In the northern region, where limited reconnaissance has been conducted, outwash and glaciolacustrine deposits indicate ice-margin recession in directions between north and west. Eskers and other meltwater deposits are largely confined to a few major valleys (Caldwell, 1974; Thompson and Borns, 1985a). As in the central region, meltwater channels are locally important in marking ice retreat. One of the best examples is a set of channels on the east side of Aziscohos Lake in the Magalloway River valley (Lincoln Pond quadrangle; Fig. 11). The channels are deeply incised in till and record both thinning and retreat of a valley ice tongue.

Meltwater deposits in the upper Magalloway, Cupsuptic, and Kennebago River valleys show topographically controlled northward ice recession; however, unlike areas farther south, there is no evidence of a widespread shift in late glacial ice-flow direction. Near the Quebec border, meltwater channels and an ice-contact delta in the vicinity of Massachusetts Bog (Northwest Pond quadrangle) indicate northward retreat just prior to deposition of the Frontier moraine described by Shilts (1981).

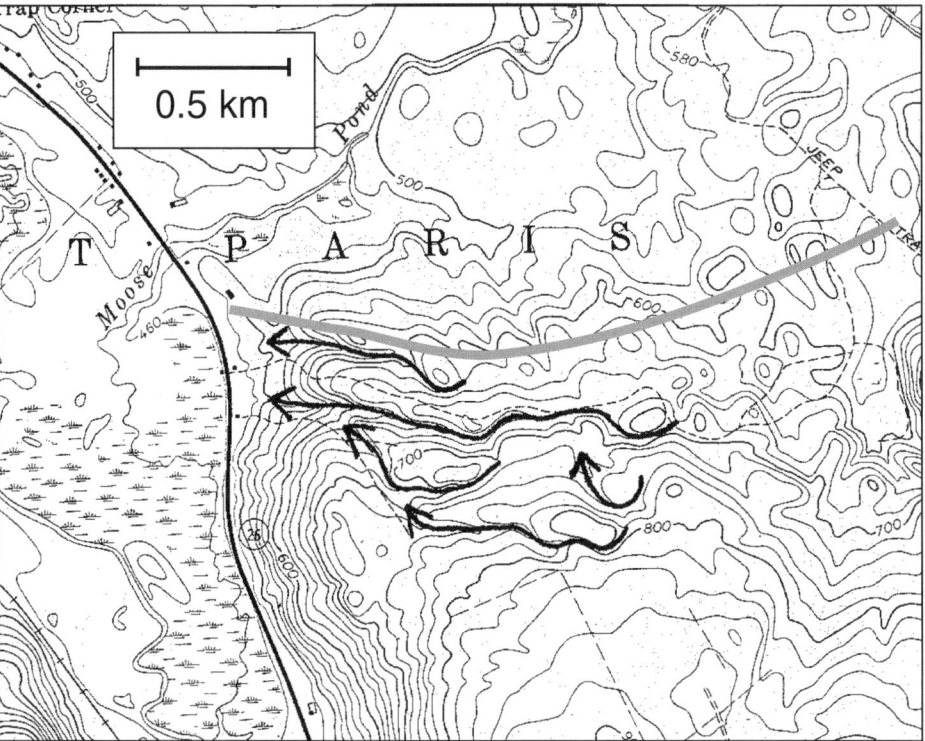

Figure 9. Proglacial meltwater channels and hummocky moraine deposits formed during northward ice-margin recession on hillside east of Route 26 in West Paris (West Paris quadrangle; see Fig. 2). Gray line shows location of ice margin when youngest channel formed.

Glacial lakes existed in a few places in the northern region, where there were temporary ice or drift dams. The deposits in some of these lakes indicate the deglaciation sequence. One of the clearer examples is north of Grafton Notch (B Pond and Old Speck Mountain quadrangles). A high level of ice-dammed Glacial Lake Cambridge spilled south through the notch until northward ice retreat opened lower outlets. The second lake stage drained east into the Ellis River basin, followed by probable westward drainage into the upper Androscoggin Valley in New Hampshire (Leavitt and Perkins, 1935; Thompson and Fowler, 1989).

## CHRONOLOGY OF DEGLACIATION

There are few radiocarbon dates from the study area to limit the time of deglaciation (see Fig. 1 and Table 1; all ages discussed here are in uncalibrated radiocarbon years). Two accelerator mass spectrometer (AMS) radiocarbon ages were obtained from terrestrial organics in the southern region. Basal organic sediment from Deering Pond, located above the marine limit in the Sanford quadrangle, yielded an age of 13 350 ± 75 yr B.P. (OS-12839; Dorion, 1997). A poplar twig from marine(?) mud in the Tandberg sand and gravel pit, slightly seaward of the marine limit in the North Windham quadrangle, was dated as 12 100 ± 110 yr B.P. (OS-4416; Thompson et al., 1995). Compared to the ages previously obtained at nearby Sinkhole Pond and Poland Spring Pond (Table 1; Davis and Jacobson, 1985), the latter age seems too young to indicate the time of deglaciation.

Three radiocarbon ages for the central and northern regions are considered minimum limits on deglaciation (Thompson et al., 1996). From south to north, these are: (1) 13 150 ± 50 yr B.P. (OS-7122) at Cushman Pond, North Waterford quadrangle; (2) 12 250 ± 55 yr B.P. (OS-7119) at Surplus Pond, Andover quadrangle; and (3) 11 500 ± 50 yr B.P. (OS-7123) at Lower Black Pond, Twin Peaks quadrangle. The latter ages are consistent with those obtained at Spencer Pond and Boundary Pond in the northern region (Table 1). These five sites show a south to north progression of pond-bottom ages from about 13.2 ka to 11.2 ka. Lower Black Pond is on the south side of a previously unrecognized moraine ridge belonging to the Frontier moraine system (Shilts, 1981), and thus the age from this site may closely approximate when the ice margin stood at the Maine-Quebec border.

The ages obtained by Thompson et al. (1996) are generally consistent with other workers' results from ponds in central and northern New Hampshire and adjacent Quebec. They indicate ice recession to the Canadian border by ca. 11.5 ka. Thompson (1998) inferred from these data that the late Wisconsinan ice margin withdrew from western Maine at an average rate of ~70 m/yr. However, the Deering Pond age is only 200 years older than the sample from Cushman Pond, located 84 km to the north. More ages are needed before the deglaciation rate inland from the marine limit can be inferred with confidence.

South and southeast of the study area, uncorrected dates from glaciomarine shells in coastal Maine indicate ice recession to the marine limit by 13.0 ka (Davis and Jacobson, 1985), and

Figure 10. Glacial lake basin in north-draining Pleasant River valley, West Bethel (Bethel quadrangle). Androscoggin River valley is seen in upper right.

radiocarbon ages as old as 14.8 ka have been obtained from shells in glaciomarine clay at the Scarborough mammoth site near Portland, Old Orchard Beach quadrangle (H.W. Borns, Jr., 1999, personal commun.). The marine dates have led some workers to propose an earlier deglaciation chronology than would be expected from the terrestrial ages discussed herein (see discussion by Retelle and Weddle, this volume). Uncertainty in comparing the two sets of data has resulted from not knowing the reservoir correction factor that should be applied to radiocarbon ages from marine fossils.

Ridge et al. (this volume) develop a deglaciation chronology for southernmost Maine based on the relationship of lacustrine varve sequences to radiocarbon ages and paleomagnetic declinations in New England. These authors advocate a younger chronology of ice retreat, with northward recession from the Sanford area beginning ca. 12.6 ka. If correct, this model requires corrections of as much as 1 k.y. or more for the marine radiocarbon ages. It also implies that deglaciation of western Maine occurred more rapidly than Thompson's (1998) estimate, assuming the ice margin reached the Canadian border by 11.5 ka.

There is no indication that the late glacial residual ice cap in northern Maine and adjacent Quebec extended southwest to the present study area. All of the known field evidence shows that the margin of the Laurentide ice sheet receded generally northward to the Quebec border in this part of the state. Thompson et al. (1999) reached the same conclusion for ice retreat in northern New Hampshire. Field studies in southern Quebec, summarized by Parent and Occhietti (1988, 1999), demonstrated continued northward recession of this ice sheet to the Sherbrooke area prior to marine transgression in the St. Lawrence Lowland.

## CONCLUSIONS

Field investigations have established the deglaciation pattern of the late Wisconsinan ice sheet in western Maine. Meltwater channels and ice-marginal deposits record progressive northward recession of the ice margin, accompanied by a shift in ice-flow direction from southeast to south or south-southwest in the region south of the Mahoosuc Range. This shift probably occurred as the Mahoosucs obstructed the flow of the thinning ice sheet. The ice remaining to the south of these mountains developed a surface gradient with a more southerly slope than during the glacial maximum.

Davis and Jacobson's (1985) reconstruction shows the Mahoosuc Range protruding from the thinning ice sheet at 14.0 ka. They inferred that by 13.0 ka there was a semidetached ice lobe extending southward across southwestern Maine (on the lee side of the high mountains). However, there is no field evidence of spreading flow to the east or west, as might be expected along the lateral margins of such an ice lobe. In both terrestrial and glaciomarine environments the receding ice margin in southwestern Maine had a generally east-west trend with only small-scale embayments (Thompson and Borns, 1985a; Retelle and Weddle, this volume). Moreover, Boothroyd (1995, 1997) noted that the southeast trend of eskers in this part of the state is incompatible with the ice lobe proposed by Davis and Jacobson (1985). My geologic mapping suggests that if a west-facing ice margin existed, it was passive and limited to a small part of the central region. Remnants of stagnant ice may have blocked the Androscoggin Valley near Rumford, impounding Glacial Lake Bethel to the west.

The internal dynamics of the receding late Wisconsinan ice sheet in western Maine have been inferred from meager field

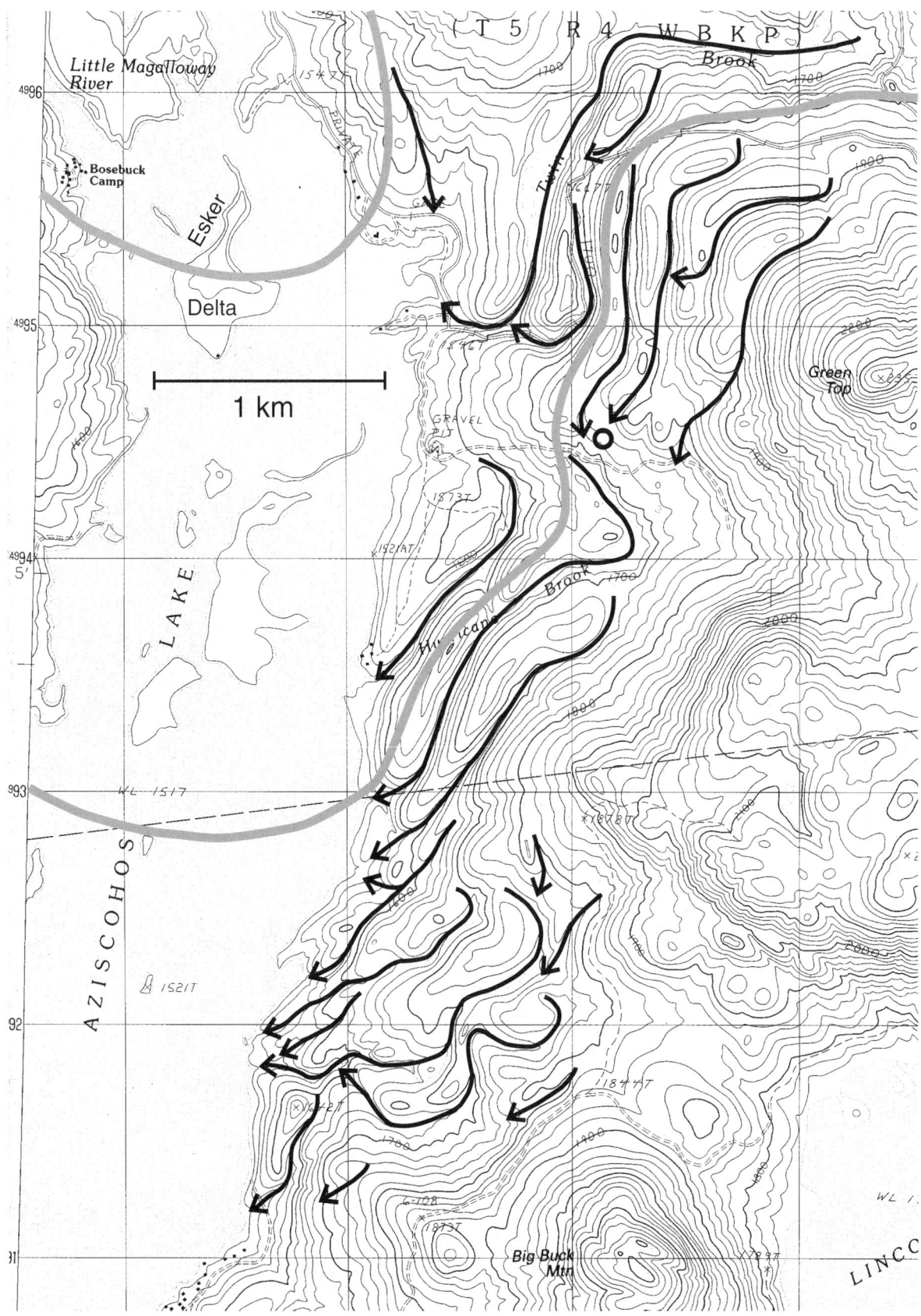

Figure 11. Flights of meltwater channels cut in till on east side of Aziscohos Lake valley (Lincoln Pond quadrangle; see Fig. 2). Gray lines show representative ice-margin positions during channel formation. Circle indicates plunge pool eroded in hillside at lower end of one of the channels.

TABLE 1. TERRESTRIAL RADIOCARBON AGES CONSTRAINING TIME OF DEGLACIATION IN WESTERN MAINE

| Site number/name | Quadrangle | Lab Number | Age ($^{14}$C yr B.P.) | Reference |
|---|---|---|---|---|
| 1 Deering Pond | Sanford | OS-12839 | 13 350 ± 75 | Dorion (1997) |
| 2 Cushman Pond | North Waterford | OS-7122 | 13 150 ± 50 | Thompson et al. (1996) |
| 3 Poland Spring Pond | Minot | SI-4656 | 12 860 ± 325 | Davis and Jacobson (1985) |
| 4 Sinkhole Pond | Gray | SI-4657 | 12 710 ± 125 | Davis and Jacobson (1985) |
| 5 Surplus Pond | Andover | OS-7119 | 12 250 ± 55 | Thompson et al. (1996) |
| 6 Tandberg Pit | North Windham | OS-4416 | 12 100 ± 110 | Thompson et al. (1995) |
| 7 Spencer Pond | Houghton | AA-9506 | 11 665 ± 85 | C. Dorion, pers. comm., 1999 |
| 8 Lower Black Pond | Twin Peaks | OS-7123 | 11 500 ± 50 | Thompson et al. (1996) |
| 9 Boundary Pond | Boundary Pond | GSC-1248 | 11 200 ± 200 | Shilts (1981) |

*Note:* Site numbers refer to locations shown on Figure 1.

evidence (compared to the coastal moraine belt), and thus are not well understood. Moraines and other features described here suggest that ice flow persisted, at least locally, during deglaciation. This model contrasts with the proposal of Leavitt and Perkins (1935) that deglaciation occurred chiefly by stagnation and downwasting of the ice sheet. Most indicators of late ice flow have been found in the southern and central regions of the study area, and likely formed when ice still covered most or all of the mountains to the north. A greater degree of stagnation may have resulted as thinning ice masses were detached in valleys among the Mahoosuc and Boundary Mountains. The sparse amount of sand and gravel in the latter region implies that less sediment was delivered to meltwater drainage systems by active ice.

A firm chronology of deglaciation awaits the determination of additional radiocarbon ages, which in most cases will have to be obtained from basal lake and pond sediments. Currently available terrestrial ages from pond bottoms indicate deglaciation of western Maine over a span of time between about 13.5 and 11.5 ka (Fig. 1). The oldest age reported from the southern region (ca. 13.4 ka; Table 1) would be ~800 years too early, according to the chronology proposed by Ridge et al. (this volume). Recent AMS ages from north and west of the Mahoosuc Range vary between 12.5 and 11.5 ka (Thompson et al., 1996), and these are consistent with recent work on the varve chronology in the upper Connecticut River valley by Ridge et al. (1999).

## ACKNOWLEDGMENTS

This summary of western Maine deglaciation has benefited greatly from the Maine Geological Survey's surficial geology and aquifer mapping programs, conducted under the auspices of State Geologist Robert Marvinney and emeritus State Geologist Walter Anderson. Funding for much of this work was provided by the U.S. Geological Survey through the COGEO-MAP and STATEMAP cooperatives. Quadrangle mapping projects by numerous individuals have contributed to our present state of knowledge. I particularly wish to acknowledge the following people who have done recent work in western Maine and participated in many discussions in the field: Jon Boothroyd, Dee Caldwell, Tom Davis, Chris Dorion, Carol Hildreth, Bill Holland, Carl Koteff, Carolyn Lepage, Cheryl Marvinney, Andres Meglioli, Craig Neil, Bob Newton, Mike Retelle, Tom Weddle, and Thom Wilch. Dee Caldwell and Carl Koteff provided helpful review comments for this paper.

The National Science Foundation (NSF), through a grant to the University of Maine under the EPSCoR Program, has supported a joint University of Maine–Maine Geological Survey project to acquire and compile data on the timing of deglaciation throughout the state. This project funded the Maine radiocarbon age determinations discussed here, which were obtained from the National Ocean Sciences AMS Facility in Woods Hole, Massachusetts with support from NSF Cooperative Agreement OCE-801015.

## REFERENCES CITED

Ashley, G.M., Boothroyd, J.C., and Borns, H.W., Jr., 1991, Sedimentology of late Pleistocene (Laurentide) deglacial-phase deposits, eastern Maine; an example of a temperate marine grounded ice-sheet margin, *in* Anderson, J.B., and Ashley, G.M., eds., Glacial marine sedimentation; paleoclimatic significance: Geological Society of America Special Paper 261, p. 107–125.

Boothroyd, J.C., 1995, The Pine River and other esker systems of southwestern Maine and south-central New Hampshire: Implications for regional deglaciation: Geological Society of America Abstracts with Programs, v. 27, no. 1, p. 31.

Boothroyd, J.C., 1997, Surficial geology of the Great East Lake 7.5-minute quadrangle, York County, Maine: Maine Geological Survey Open-File 97-61, 9 p.

Borns, H.W., Jr., 1989, Changing perspectives of the Quaternary surficial geology of Maine, *in* Tucker, R.D., and Marvinney, R.G., eds., Studies in Maine geology—Volume 6: Quaternary geology: Augusta, Maine Geological Survey, p. 1–11.

Borns, H.W., Jr., and Calkin, P.E., 1977, Quaternary glaciation, west-central Maine: Geological Society of America Bulletin, v. 88, p. 1773–1784.

Caldwell, D.W., 1974, Surficial materials of the wildlands of Northwestern Maine: Maine Geological Survey Open-File 74-13, 32 p.

Caldwell, D.W., 1975, Reconnaissance surficial geology of the Old Speck Mountain quadrangle, Maine: Maine Geological Survey Open-File 75-15, scale 1:62 500.

Davis, P.T., and Holland, W.R., 1997a, Surficial geology of the Brownfield quadrangle: Maine Geological Survey Open-File 97-48, scale 1:24 000.

Davis, P.T., and Holland, W.R., 1997b, Surficial geology of the Kezar Falls 7.5-minute quadrangle, Oxford and York Counties, Maine: Maine Geological Survey Open-File 97-67, 14 p.

Davis, P.T., Holland, W.R., and Thompson, W.B., 1998, Glacial Lakes and deglacial morphosequences in the Kezar Falls and Brownfield quadrangles, southwestern Maine [abs.]: Geological Society of America Abstracts with Programs, v. 30, no. 1, p. 13.

Davis, R.B., and Jacobson, G.L., Jr., 1985, Late glacial and early Holocene landscapes in northern New England and adjacent areas of Canada: Quaternary Research, v. 23, p. 341–368.

Dorion, C.C., 1997, Report on core analyses from two ponds in southwest Maine: Deering Pond and Littlefield Pond: Augusta, Maine Geological Survey Report, 30 p.

Eusden, J.D., Jr., 1980, Surficial geology and late Wisconsinan history of the Worthley Pond 7.5-minute quadrangle, Maine [senior thesis]: Lewiston, Maine, Bates College, 53 p.

Flint, R.F., 1929, The stagnation and dissipation of the last ice sheet: Geographical Review, v. 19, p. 256–289.

Goldthwait, J.W., 1938, The uncovering of New Hampshire by the last ice sheet: American Journal of Science, ser. 5, v. 36, p. 345–372.

Gosse, J.C., and Thompson, W.B., 1999, Surficial geology of the Steep Falls quadrangle: Maine Geological Survey Open-File 99-102, scale 1:24 000.

Hanson, L.S., and Caldwell, D.W., 1989, The lithologic and structural controls on the geomorphology of the mountainous areas in north-central Maine, in Tucker, R.D., and Marvinney, R.G., eds., Studies in Maine geology—Volume 5: Quaternary geology: Augusta, Maine Geological Survey, p. 147–167.

Hildreth, C.T., 1997a, Surficial geology of the Naples 7.5-minute quadrangle, Cumberland County, Maine: Maine Geological Survey Open-File 97-65, 9 p.

Hildreth, C.T., 1997b, Surficial geology of the Naples quadrangle: Maine Geological Survey Open-File 97-50, scale 1:24 000.

Holland, W.R., 1986, Features associated with the deglaciation of the upper Saco and Ossipee River basins, northern York and southern Oxford Counties, Maine, in Newberg, D.W., ed., Guidebook for field trips in southwestern Maine: New England Intercollegiate Geological Conference, guidebook for 78th annual meeting, Trip B1: Lewiston, Maine, Bates College, p. 98–123.

Koteff, C., and Pessl, F., Jr., 1981, Systematic ice retreat in New England: U.S. Geological Survey Professional Paper 1179, 20 p.

Koteff, C., Robinson, G.R., Goldsmith, R., and Thompson, W.B., 1993, Delayed postglacial uplift and synglacial sea levels in coastal central New England: Quaternary Research, v. 40, p. 46–54.

Leavitt, H.W., and Perkins, E.H., 1934, A survey of road materials and glacial geology of Maine—Volume 1, Part I: A survey of road materials of Maine: Their occurrence and quality: Orono, Maine Technology Experiment Station Bulletin 30, 487 p.

Leavitt, H.W., and Perkins, E.H., 1935, A survey of road materials and glacial geology of Maine—Volume 2: Glacial geology of Maine: Orono, Maine Technology Experiment Station Bulletin 30, 232 p.

Meglioli, A., and Thompson, W.B., 1997, Surficial geology of the Mousam Lake quadrangle, York County, Maine: Maine Geological Survey Open-File 97-74, 6 p.

Newton, R.M., 1997a, Surficial geology of the West Newfield 7.5-minute quadrangle, York County, Maine: Maine Geological Survey Open-File 97-47, 7 p.

Nichols, W.J., Jr., Neil, C.D., and Weddle, T.K., 1995, Hydrology and water quality of significant sand and gravel aquifers in parts of Franklin, Oxford, and Somerset Counties, Maine: Maine Geological Survey Open-File 95-3, 89 p.

Osberg, P.H., Hussey, A.M., II, and Boone, G.M., eds., 1985, Bedrock geologic map of Maine: Augusta, Maine Geological Survey, scale 1:500 000.

Parent, M., and Occhietti, S., 1988, Late Wisconsinan deglaciation and Champlain Sea invasion in the St. Lawrence Valley, Quebec: Géographie Physique et Quaternaire, v. 42, p. 215–246.

Parent, M., and Occhietti, S., 1999, Late Wisconsinan deglaciation and glacial lake development in the Appalachians of southeastern Quebec, in Thompson, W.B., et al., eds., Late Quaternary history of the White Mountains, New Hampshire, and adjacent southeastern Quebec: Géographie Physique et Quaternaire, v. 53, p. 117–135.

Prescott, G.C., Jr., 1979, Royal, upper Presumpscot, and upper Saco River basins (Maine) area: U.S. Geological Survey, Maine Hydrologic-Data Report 10, Ground-Water Series, 53 p.

Prescott, G.C., Jr., 1980, Ground-water availability and surficial geology of the Royal, upper Presumpscot, and upper Saco River basins, Maine: U.S. Geological Survey Water-Resources Investigations 79-1287, 3 maps.

Retelle, M.J., 1997, Surficial geology of the Raymond quadrangle: Maine Geological Survey Open-File 97-57, scale 1:24 000.

Ridge, J.C., Besonen, M.R., Brochu, M., Brown, S., Callahan, J.W., Cook, G.J., Nicholson, R. S., and Toll, N.J., 1999, Varve, paleomagnetic, and $^{14}$C chronologies for late Pleistocene events in New Hampshire and Vermont, U.S.A., in Thompson, W.B., et al., eds., Late Quaternary history of the White Mountains, New Hampshire, and adjacent southeastern Quebec: Géographie Physique et Quaternaire, v. 53, p. 79–106.

Shilts, W.W., 1981, Surficial geology of the Lac Megantic area, Quebec: Geological Survey of Canada Memoir 397, 102 p.

Smith, G.W., and Hunter, L.E., 1989, Late Wisconsinan deglaciation of coastal Maine, in Tucker, R.D., and Marvinney, R.G., eds., Studies in Maine geology—Volume 6: Quaternary geology: Augusta, Maine Geological Survey, p. 13–32.

Stone, G.H., 1880, Note on the Androscoggin glacier: American Naturalist, v. 14, p. 299–302.

Stone, G.H., 1899, The glacial gravels of Maine and their associated deposits: U.S. Geological Survey Monograph 34, 499 p.

Thompson, W.B., 1985, Glacial geology of the Fryeburg-Bethel region, southwestern Maine: Augusta, Maine Geological Survey, Guidebook for Geological Society of Maine field trip, July 27, 14 p.

Thompson, W.B., 1986, Glacial geology of the White Mountain foothills, southwestern Maine, in Newberg, D.W., ed., Guidebook for field trips in southwestern Maine: New England Intercollegiate Geological Conference, guidebook for 78th annual meeting, Trip C-1: Lewiston, Maine, Bates College, p. 275–288.

Thompson, W.B., 1989, Glacial geology of the Androscoggin River valley in Oxford County, western Maine, in Berry, A.W., Jr., ed., Guidebook for field trips in southern and west-central Maine: New England Intercollegiate Geological Conference, guidebook for 81st annual meeting, Trip B-2/C-2: Farmington, University of Maine, p. 165–182.

Thompson, W.B., 1991, Recession of the Laurentide Ice Sheet from southwestern Maine to the Quebec border [abs.]: Geological Society of America Abstracts with Programs, v. 23, no. 1, p. 138.

Thompson, W.B., 1994, Glacial fan deposits overridden by the late Wisconsinan ice sheet, southwestern Maine [abs.]: Geological Society of America Abstracts with Programs, v. 26, no. 3, p. 76.

Thompson, W.B., 1998, Deglaciation of western Maine and the northern White Mountains [abs.]: Geological Society of America Abstracts with Programs, v. 30, no. 1, p. 79.

Thompson, W.B., 1999a, Surficial geology of the Center Lovell 7.5-minute quadrangle, Oxford County, Maine: Maine Geological Survey Open-File 99-2, 12 p.

Thompson, W.B., 1999b, Surficial geology of the Center Lovell quadrangle: Maine Geological Survey Open-File 99-1, scale 1:24 000.

Thompson, W.B., 2000a, Surficial geology of the Norway quadrangle: Augusta, Maine Geological Survey, Open-File 00-135, scale 1:24 000.

Thompson, W.B., 2000b, Surficial geology of the Waterford Flat quadrangle: Augusta, Maine Geological Survey, Open-File 00-133, scale 1:24 000.

Thompson, W.B., and Borns, H.W., Jr., 1985a, Surficial geologic map of Maine: Augusta, Maine Geological Survey, scale 1:500 000.

Thompson, W.B., and Borns, H.W., Jr., 1985b, Till stratigraphy and late Wisconsinan deglaciation of southern Maine: A review: Géographie Physique et Quaternaire, v. 39, p. 199–214.

Thompson, W.B., and Fowler, B.K., 1989, Deglaciation of the upper Androscoggin River valley and northeastern White Mountains, Maine and New

Hampshire, *in* Tucker, R.D., and Marvinney, R.G., eds., Studies in Maine geology—Volume 6: Quaternary geology: Augusta, Maine Geological Survey, p. 71–88.

Thompson, W.B., and Holland, W.R., 1999, Surficial geology of the Hiram quadrangle: Maine Geological Survey Open-File 99-85, scale 1:24 000.

Thompson, W.B., and Koteff, C., 1995, Deglaciation sequence in southwestern Maine: Stratigraphic, geomorphic, and radiocarbon evidence [abs.]: Geological Society of America Abstracts with Programs, v. 27, no. 1, p. 87.

Thompson, W.B., Crossen, K.J., Borns, H.W., Jr., and Andersen, B.G., 1989, Glaciomarine deltas of Maine and their relation to late Pleistocene–Holocene crustal movements, *in* Anderson, W.A., and Borns, H.W., Jr., eds., Neotectonics of Maine: Maine Geological Survey Bulletin 40, p. 43–67.

Thompson, W.B., Davis, P.T., Gosse, J.C., Johnston, R.A., and Newton, R., 1995, Late Wisconsinan glacial deposits in the Portland–Sebago Lake–Ossipee Valley region, southwestern Maine: Guidebook for the 58th field conference of the northeastern Friends of the Pleistocene: Augusta, Maine Geological Survey, 71 p.

Thompson, W.B., Fowler, B.K., Flanagan, S.M., and Dorion, C.C., 1996, Recession of the late Wisconsinan ice sheet from the northwestern White Mountains, New Hampshire, *in* Van Baalen, M.R., ed., Guidebook to field trips in northern New Hampshire and adjacent regions of Maine and Vermont: Guidebook for 88th annual meeting of New England Intercollegiate Geological Conference, Trip B-4: Cambridge, Massachusetts, Harvard University, p. 203–234.

Thompson, W.B., Fowler, B.K., and Dorion, C.C., 1999, Deglaciation of the northwestern White Mountains, New Hampshire, *in* Thompson, W.B., et al., eds., Late Quaternary history of the White Mountains, New Hampshire, and adjacent southeastern Quebec: Géographie Physique et Quaternaire, v. 53, p. 59–77.

Vose, G.L., 1868, Traces of ancient glaciers in the White Mountains of New Hampshire: American Naturalist, v. 2, p. 281–291.

Williams, J.S., Tepper, D.H., Tolman, A.L., and Thompson, W.B., 1987, Hydrogeology and water quality of significant sand and gravel aquifers in parts of Androscoggin, Cumberland, Oxford, and York Counties, Maine: Maine Geological Survey, Open-File Report 87-1a, 121 p.

MANUSCRIPT ACCEPTED BY THE SOCIETY JUNE 9, 2000

*Late Quaternary morphogenesis of a marine-limit delta plain in southwest Maine*

Anna K. Tary*
Duncan M. FitzGerald
Ilya V. Buynevich
*Department of Earth Sciences, Boston University, Boston, Massachusetts 02215, USA*

## ABSTRACT

Detailed study of the Sanford-Kennebunk sand plain using sediment analysis of 125 samples from 20 locations and 37 km of ground-penetrating radar (GPR) indicates that much of the feature can trace its origins to regressive-phase deltaic deposition during the late Pleistocene.

The Sanford-Kennebunk sand plain is an ~125 km² elongate, west-east–trending landform overlying a large, deep bedrock trough. With the exception of the eastern coastal margin, the entire sand plain is bordered by bedrock uplands.

The plain is composed of coarsening-upward sand and minor gravel. Sediment sorting ranges from poor to moderate in the western plain, to moderate to good in the eastern sector. Grain size generally decreases from west to east.

Extensive ground-penetrating radar coverage shows that many areas display packages of clinoforms, the thicknesses of which range from shallow, <1 m thick, channel fill–type deposits to the more common 5–14-m-thick, rhythmic, inclined (4°–15° average dip) beds. Dip directions generally radiate outward from the west-central plain, and farther eastward, from the south-central plain. These GPR records in conjunction with the sediment analyses strongly suggest a deltaic origin for the orderly dipping beds of the sand plain.

A model for development and evolution of the sand plain encompasses deglaciation, associated marine incursion, and subsequent regression. As ice retreated from the area and isostatic recovery led to marine regression, sediments from the glacier's margin were transported in energetic meltwater channels to be deposited in a bedrock-enclosed, marine bathymetric low, forming the Sanford-Kennebunk sand plain.

## INTRODUCTION

The State of Maine has long provided opportunities for research of Quaternary glacial and deglacial processes and landforms; recent technological advances in field-data collection allow for reassessment, and often reclassification, of previously studied regions. This study refines earlier understanding of the deglacial morphogenesis of a broad sand plain found within

---

*Present address: Department of Natural Sciences, Bentley College, Waltham, Massachusetts 02452, USA.

Tary, A.K., FitzGerald, D.M., and Buynevich, I.V., 2001, Late Quaternary morphogenesis of a marine-limit delta plain in southwest Maine, *in* Weddle, T.K., and Retelle, M.J., eds., Deglacial History and Relative Sea-Level Changes, Northern New England and Adjacent Canada: Boulder, Colorado, Geological Society of America Special Paper 351, p. 125–149.

the area of Pleistocene marine incursion in southwest Maine.

The study area for this project is in southwestern Maine (Fig. 1, inset) within the towns of Alfred, Kennebunk, Lyman, Sanford, and Wells. The focus of this research is the Sanford-Kennebunk sand plain (Fig. 1), a broad west-east–trending body of silty sand, sand, and gravel exhibiting variable but generally flat to rolling topography. The sand plain is in the area of late glaciomarine submergence as mapped by Thompson and Borns (1985).

The Sanford-Kennebunk sand plain is one of many coarse-grained landforms mapped along the inland marine limit in the State of Maine (Thompson and Borns, 1985). Whereas much glacial and geomorphologic research has been performed in the study area, relatively few studies have specifically addressed the morphogenesis of the sand plain. The importance of such a study is in the fact that results may help determine, refine, or refute ideas of deglaciation and the sea-level history for a specific region. In addition, our understanding of general glacigenic processes may be increased. In this chapter we use ground-penetrating radar data, which show numerous areas of east- to northeast-dipping clinoform reflector patterns, in conjunction with sedimentologic data showing the coarsening upward and westward nature of the sand-plain sediments, to suggest that landforms such as the Sanford-Kennebunk sand plain may have formed more recently than previously believed, as a regressive deltaic feature deposited into a shallowing sea. This idea, in turn, sheds new light on late to postglacial hydrologic and sedimentologic conditions in coastal Maine.

## BACKGROUND

### Physical setting

The Sanford-Kennebunk sand plain occupies an area of ~125 km$^2$ in the coastal lowlands of southwestern Maine (Smith, 1985), entirely within the zone of late glaciomarine incursion. The area is underlain by metamorphosed Precambrian to Mesozoic stratified and plutonic bedrock with a dominantly northeast-southwest structural grain (Hussey, 1989).

The sand plain is bordered by the marshes and estuaries adjacent to the Atlantic Ocean to the east and the bedrock-controlled uplands of Sanford and Alfred to the west and northwest. The plain is incised by two larger rivers, the Mousam River and Branch Brook, and several smaller tributaries, in-

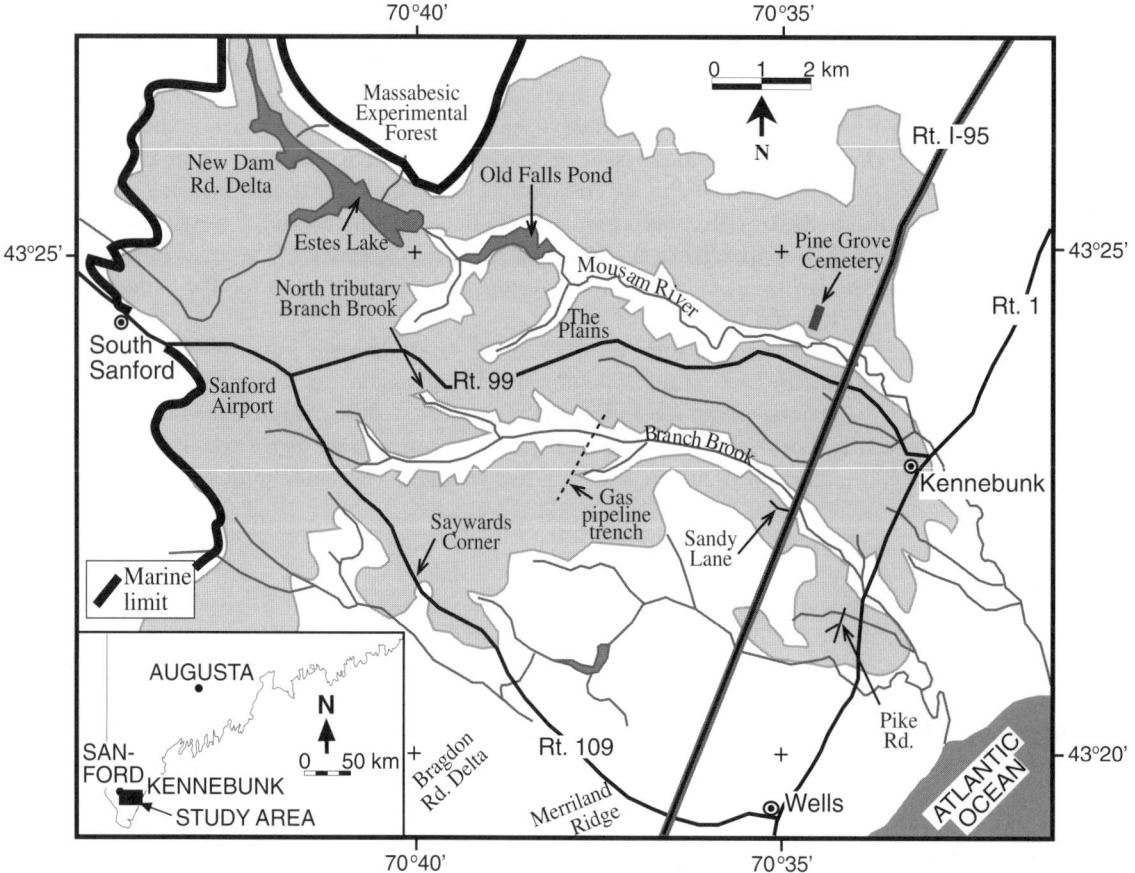

Figure 1. Sanford-Kennebunk sand plain, showing extent of plain (gray shading), area towns, roads, drainages, locations discussed in text, and maximum marine limit (modified from Smith, 1977, 1980; Thompson and Borns, 1985). Regional location is shown in inset.

cluding Day Brook, Cold Water Brook, and Perkins Marsh Brook. The two larger rivers flow roughly parallel to one another in arcuate paths from west to east to southeast, where they drain into the Gulf of Maine.

The Sanford-Kennebunk sand plain can be traced westward to bedrock-controlled valleys in the Sanford and Alfred areas (Bloom, 1960). At an upper elevation of 70 m, the plain fans out from these valleys in a sweeping east- to southeast arc; its eastern margin is a north-south–trending terminus at a minimum elevation of 6 m. Two small lobes extend from the eastern body of the plain, one to the north and the other to the south (Smith, 1977). The elevation of the western plain margin is coincident with the maximum marine limit elevation in the area (Bloom, 1960); from its inland limit the plain slopes gradually (4.2 m/km gradient) toward the sea.

## Glacial history

The Sanford-Kennebunk sand plain and associated sediments were deposited as a result of the recent Wisconsinan stage glaciation, during which time the Laurentide ice sheet expanded over New England, reaching its maximum extent on the continental shelf east and south of Massachusetts ca. 18.0 ka (Hughes et al., 1985; Oldale, 1989; Hunter et al., 1996). Retreat of this ice sheet began ca. 17–15 ka (Smith, 1982); the calving ice margin reached the present coastline of southwest Maine ca. 13.8 ka (Smith, 1985; Belknap et al., 1987; Crossen, 1991). Ice retreat was contemporaneous with marine incursion across the isostatically depressed crust (Bloom, 1960; Caldwell et al., 1985; Thompson et al., 1989).

Deglaciation and calving rates slowed temporarily as the ice front retreated from the deeper waters of the Gulf of Maine and encountered the shallow water and increased topography of present-day southwestern Maine (Smith, 1984; Crossen, 1991). Subsequent retreat by the calving tidewater glacier occurred in a northwest, parallel to the coast fashion (Smith, 1981) until 13.0–12.5 ka, at which time ongoing glacial unloading and isostatic rebound led to decoupling of the ice and sea (Smith, 1985; Belknap et al., 1987; Barnhardt et al., 1995). Prior to this time, the sea reached its maximum inland extent at ~70 m elevation (Koteff et al., 1993), and a series of glaciomarine deltas formed at the marine limit (Belknap et al., 1987; Thompson et al., 1989; Barnhardt et al., 1997). As the ice margin became terrestrial based and glacial retreat shifted to a more northerly direction (Thompson et al., 1995), the sea regressed rapidly due to isostatic uplift (Fig. 2), leading to emergence of coastal Maine by ca. 11.5 ka (Bloom, 1960; Smith, 1984; Thompson et al., 1989; Barnhardt et al., 1995). During the regressive phase, shoaling waters and meltwater streams traversed earlier ice-proximal deposits and reworked the surfaces of these features. New fluvial landforms were also being formed by channels debouching into the sea. Modification of both older ice-contact forms and newly deposited features resulted in stranded shorelines, spits, and terraces. Rebound outpaced eu-

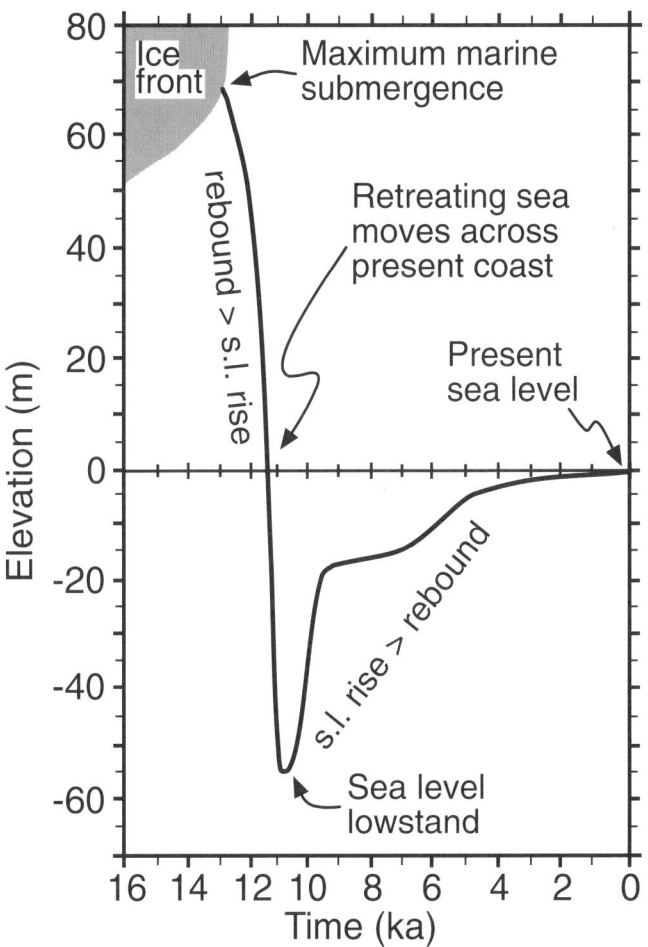

Figure 2. Relative sea-level curve for coastal Maine during late Quaternary; elevations are based on present sea level (after Barnhardt et al., 1995).

static rise until ca. 11–10.5 ka, when a sea-level lowstand of −55 m was reached (Barnhardt et al., 1995, 1997). From this time to the present, a decreasing rate of transgression has been evident throughout the Holocene (Gehrels et al., 1996).

## Previous studies

Most studies in the Sanford-Kennebunk area have pertained to regional deglaciation and sea-level history. Specific features used to develop sea-level or deglacial chronologies have included the many ice-contact features in the region, such as Merriland Ridge, the Bragdon Road delta, and the Jailhouse delta (Thompson et al., 1989; Crossen, 1991; Koteff et al., 1993), and numerous end moraines (Smith, 1982, 1984; Smith and Hunter, 1989). Relatively few studies provide absolute dates for landform morphogenesis due to the scarcity of radiocarbon-datable materials. Based on the few available dates (Stuiver and Borns, 1975; Smith, 1985; C. Neil, 1997, personal commun.), and on glaciomarine delta topset-foreset contacts, some reasonable deglacial chronologies have been developed

(Smith, 1985; Belknap et al., 1987; Smith and Hunter, 1989; Thompson et al., 1989; Koteff et al., 1993).

Interpretations of the sand plain's origin are vague. Bloom (1960) provided the first detailed assessment of the region, describing the plain as an outwash delta that likely formed during maximum marine submergence. Smith (1977) produced a surficial map of the Kennebunk 15-minute quadrangle, labeling the sand plain as an outwash unit of sand and gravel.

Few other studies have attempted to explain the origin of the Sanford-Kennebunk sand plain. Workers have continued to identify the plain as various features, allowing their labels to suggest an origin. D'Amore (1983), in his study of the hydrogeology and geomorphology of the area, referred to the "Great Sanford Outwash Plain," and Montello (1992), in a study of the Wells barrier system, referred to the feature as the "Sanford Submarine Fan." Thompson and Borns (1985) identified the sand plain simply as an undifferentiated proglacial coarse-grained glaciomarine unit. The most recent work on the Sanford-Kennebunk sand plain (Smith, 1990a, 1990b, 1990c, 1990d) has produced new surficial maps of the four 7.5′ quadrangles in the study area (Alfred, Kennebunk, North Berwick, Wells). The sand plain was mapped by Smith as two different units; the largest portion, including the western three-quarters of the plain, was mapped as a distal marine delta, but was interpreted in his discussion of the deposit as a mostly subaerial outwash plain. The eastern quarter of the plain was mapped as distal outwash, graded to sea level.

## METHODS

The database for the study includes sediment samples for stratigraphic analysis taken from a number of cores throughout the study area, and ground-penetrating radar transects taken in and around the study area ranging in length from <0.1 km to >2 km (Fig. 3).

The stratigraphy was determined using sediment samples taken from open pits, roadcuts, trenches (including a fall 1998 excavation for a gas pipeline; see Fig. 1 for location), and cores (Edelman auger and pulse auger cores). The database included 125 separate samples from 20 general locations (Fig. 3). A textural analysis was performed on the sediment samples. In addition to samples collected for this study, test hole and test boring logs from earlier investigations (Prescott and Drake, 1962; D'Amore, 1983; Tolman et al., 1982) and compiled by Smith (1990a, 1990b, 1990c, 1990d) were used.

Shallow subsurface studies using ground-penetrating radar (GPR) were performed in areas both within the boundaries of the Sanford-Kennebunk sand plain and in areas to the north, south, and west outside the plain margin (Fig. 3). The 37 km of records were obtained using a Geophysical Survey Systems Inc. chart recorder–120 MHz transceiver system. Profiles were run at a two-way traveltime of 300 ns. GPR record interpretation was based on correlation with cores, trenches, and open pits at or near profile lines. The interpretation of profile records was aided by criteria detailed in other GPR studies (Jol and Smith, 1991; Smith and Jol, 1992; Dominic et al., 1995; van Heteren et al., 1998). Terminology used to describe reflection configurations is that of van Heteren et al. (1998).

## RESULTS

### Sand-plain topography

***Description.*** For discussion purposes, the study area has been divided into three zones based on geomorphology, elevation, and local relief (Fig. 3). The three zones, from east to west, are: the coastal zone, a relatively narrow (<2.5 km wide) north-south strip along the eastern margin of the sand plain; the intermediate zone, trending north-south and ~6 km wide; and the 6.2-km-wide inland zone, extending from the western margin of the intermediate zone to the western border of the sand plain.

The coastal zone is marked by elevations of 0–30 m, and relief of <15 m. This zone generally contains broad, flat estuarine areas, and flat to rolling topography, often obscured by dense vegetation. The two larger streams in this zone, Branch Brook and the Mousam River, are contained in meandering, low-relief channels with moderate to strong flood-plain development. Short tributaries and steep-walled gullies, some fed by groundwater seeps, incise the valley walls of Branch Brook; some small-scale terracing along the same stream is evident.

The intermediate zone is characterized by elevations ranging from 30 to 60 m, and by the steepest relief found on the sand plain (15–27 m). This central zone contains several bedrock or till hills, and variable topography, ranging from hummocky surfaces to the broad, flat "Kennebunk Plains." Stream channels in the intermediate zone are deeply incised with numerous short, stubby tributary valleys. Some terracing is present in the upper channel sides of Branch Brook near the gas pipeline trench (Fig. 1). The base of a long, north-south–trending, indistinct scarp forms the eastern boundary of the intermediate zone.

The inland zone is distinguished by elevations >61 m, and relief ranging from 14 to 25 m. This zone contains broad, flat expanses leading to the western uplands, and exhibits several bedrock and till hills and numerous glacial features, including many ice-contact deltas, and some kettles, swamps, and drumlins. The inland zone contains a number of small, shallow channel heads, and two large, deeply incised streams, Branch Brook and the Mousam River.

***Interpretation.*** The low-altitude, subdued topography of the coastal zone suggests that this region underwent final emergence well away from the margin of the Laurentide ice sheet. The presence of meandering streams in wide shallow channels and estuarine features suggests a strong recent marine influence. The lack of ice-contact features indicates that the Laurentide ice margin did not exert a strong control over present geomorphology seen in the zone.

The intermediate zone contains the greatest relief and to-

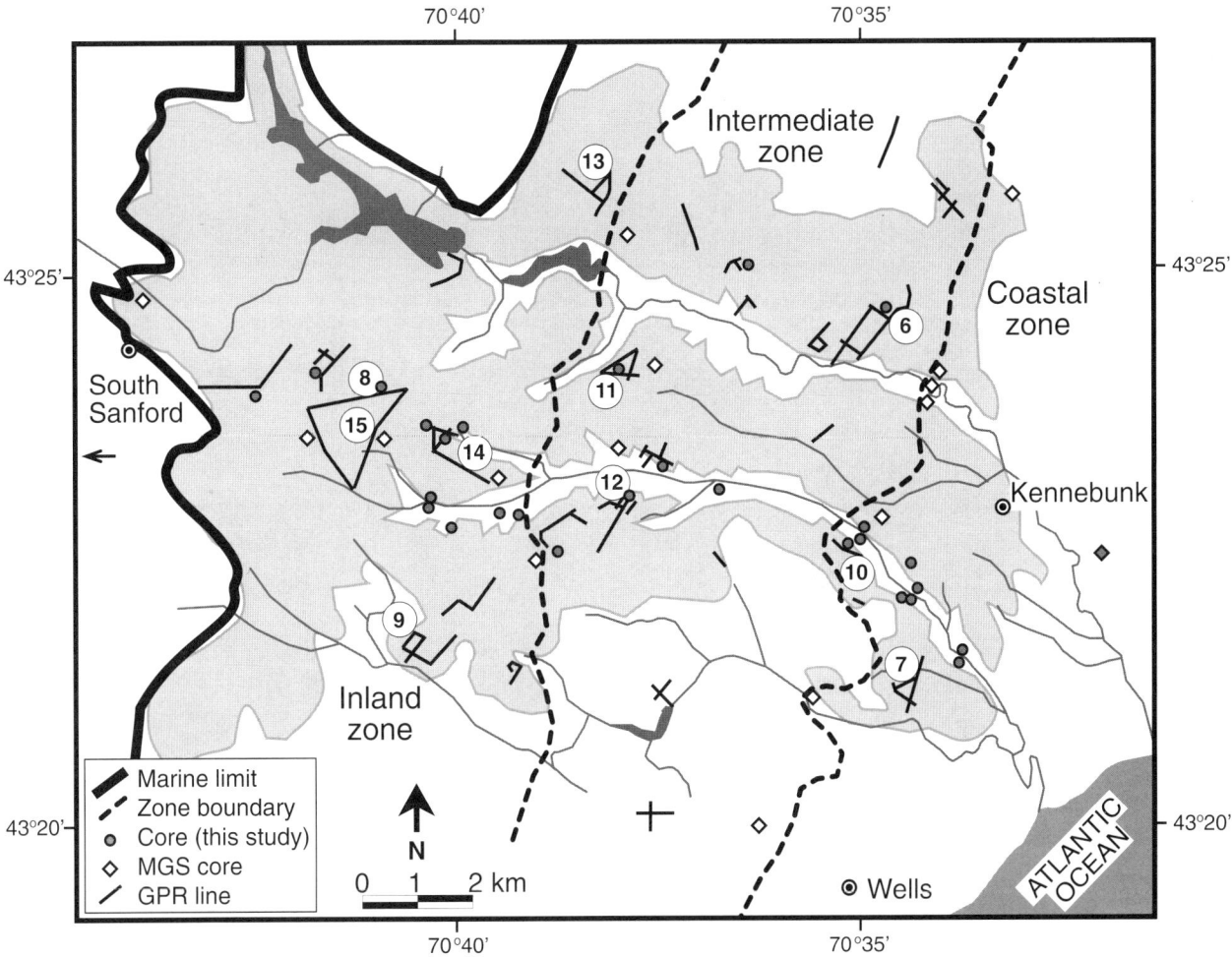

Figure 3. Locations of cores and other sampling sites and of ground-penetrating radar (GPR) transects. Cores from previous Maine Geological Survey (MGS) work are indicated by small diamonds (MGS core data are from Smith, 1990a, 1990b, 1990c, 1990d, after Prescott and Drake, 1962; Tolman et al., 1982). Dashed lines indicate geomorphic zone division boundaries discussed in text. Circled numbers indicate GPR transect locations discussed in text; numerals indicating text figure number are adjacent to corresponding profile line. West-pointing arrow below South Sanford indicates location of additional GPR profiles beyond sand-plain margin.

pographic variability found in the Sanford-Kennebunk sand plain. The central plain is also quite flat here, its surface disrupted only by occasional bedrock outcrops and by the steep valleys of the Mousam River and Branch Brook. Rare ice-contact features exist beyond the southern plain margin. It is inferred that, while ice was well to the northwest during final emergence, the melting ice sheet was close enough to provide large quantities of meltwater and sediment to the regressing sea, producing a large sand plain. Holocene modification of this deposit by grading streams formed stranded terraces in the upper plain surface, and ultimately, the steep valleys of Branch Brook and the Mousam River.

The inland zone contains the only major occurrence of ice-contact features found in the study area. While the flat expanse of the Sanford-Kennebunk sand plain forms much of the topography of this zone, the surface plane of this landform is interrupted by several bedrock and/or till highlands, various glaciomarine ice-contact deltas (Thompson et al., 1989), and occasional drumlins (at the plain's western margin). The geomorphology of this area is greatly affected by the underlying bedrock and by local ice-contact features. The inland zone is interpreted as ice proximal, on the basis of the existing morphology. Drumlins formed at the base of the ice sheet, and glaciomarine deltas and kettles formed in contact with the ice sheet or just proximal to the receding ice margin and its calved icebergs. Subsequent low-relief deposition of the sand plain over and around these features was accomplished by sediment-laden meltwater streams as ice retreated from the area.

## Depth to bedrock

***Description.*** The Sanford-Kennebunk sand plain is characterized by a large central depositional basin. An isopach map for the sand plain region (Fig. 4) was constructed from field

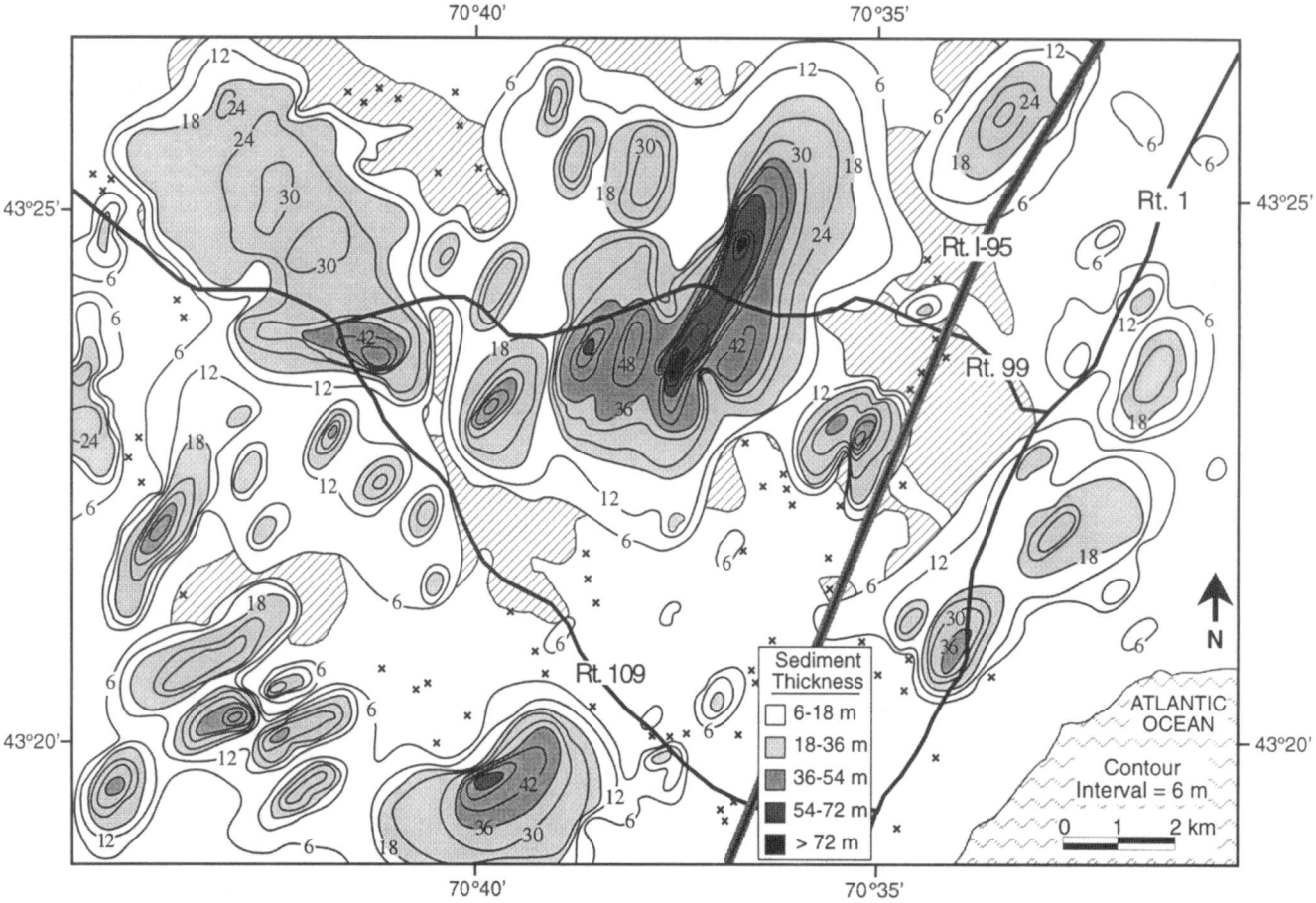

Figure 4. Isopach map of Sanford-Kennebunk sand plain and surrounding area. Map depicts total late Quaternary sediment thickness overlying bedrock. Outcrops are indicated by x symbol. Data are compiled from field observations, 7.5-minute U.S. Geological Survey topographic maps, and other sources (Prescott and Drake, 1962; D'Amore, 1983; Tolman et al., 1982; Smith, 1990a, 1990b, 1990c, 1990d; C.D. Neil, Maine Geological Survey, 1998, personal commun.). Sanford-Kennebunk sand plain (diagonal gray shading) and roads are shown for location.

reconnaissance and a preexisting database of well logs, seismic lines, and mapped outcrops (Tolman et al., 1982; Lanctot and Tolman, 1985; Tolman and Lanctot, 1985; C. Neil, Maine Geological Survey, 1998, personal commun.). Bedrock uplands generally control the topography along the southwestern, northwestern, and western edges of these basins.

The deepest structural depression underlies the central portion of the sand plain and has an areal extent of >40 km$^2$ and maximum sediment thickness exceeding 73 m. The basin directly underlies "The Plains" region and the steep-walled segments of both Branch Brook and the Mousam River. Several other smaller depressions underlie parts of the coastal and inland zones (Fig. 4); the depth of these smaller basins ranges from <12 m to >48 m. Many of the structural lows, including the central basin, are elongate in plan view, with long axes trending northeast-southwest.

*Interpretation.* The elongate orientation of the basins (Fig. 4) is an expression of the bedrock's overall northeast-southwest structural lineation (Osberg et al., 1985). Depositional centers, similar to those suggested by Crossen (1991), may have developed between strike ridges. These depocenters would have initially received ice-contact sediments, as adjacent bedrock ridges acted as pinning points to slow the glacier's retreat. Partial infilling of the depressions by till and other ice-contact deposits, and by later marine clays, would have occurred. As the ice sheet retreated to the bedrock uplands along the western and northwestern margin of the study area, proglacial sediments accumulated in the remaining depressions, further subduing the topography. Continued ice retreat and contemporaneous isostatic uplift led to decoupling of the ice and sea. The remaining topographic lows would have filled with sediment that was delivered to the area by meltwater and meteoric streams.

The Sanford-Kennebunk sand plain sediment volume was calculated using the isopach map in conjunction with stratigraphic information. Stratigraphy allowed determination of the possible depth of sand-plain sediments, thereby isolating volumetric calculations to include only the sandy facies of the plain and not the marine clay and ice-contact deposits. Calculations of this fraction indicate that the sediment volume of the sand plain is ~$1.5 \times 10^9$ m$^3$. This figure is similar to the $1.3 \times 10^9$ m$^3$ volume of the Merrimack River lowstand paleodelta (Oldale et al., 1983), and may also be similar to the deltaic facies volume of the $2.1 \times 10^9$ m$^3$ regressive and lowstand Kennebec River paleodelta (Barnhardt et al., 1997).

*Stratigraphy*

Stratigraphic facies of the Sanford-Kennebunk area are summarized in Table 1. Interpretations are based on correlation of geomorphologic and sedimentologic data with ground-penetrating radar profiles. A series of stratigraphic profiles (Fig. 5) from the three zones shows that similar stratigraphic trends are followed throughout the Sanford-Kennebunk sand plain, with some variation in grain size and unit thickness from one zone to another. In addition, some isolated features of various types are present in each zone (Table 1).

*Facies descriptions.* Unit BR is found at the base of the sequence; in some cases, the unit extends upward to the ground surface (profile D, Fig. 5). Such outcrops occur in some stream channels and at topographic highs. This unit is identifiable on GPR profiles by a hyperbolic signal (Fig. 6). Visual correlation with radar records is possible in areas where outcrops are found adjacent to the GPR transect. Unit BR consists of metamorphosed sedimentary and plutonic rocks with a highly irregular upper surface, and is interpreted as Paleozoic and Mesozoic bedrock (Hussey, 1989).

Overlying unit BR is unit T, a poorly sorted diamicton. This facies is characterized by a wide range of grain sizes, from boulder gravel to clay. In the Sanford-Kennebunk sand plain, unit T is generally found as a discontinuous drape overlying bedrock structures or as a fill in areas of structural lows. At several sites, this material covers or forms topographic highs, creating local features with as much as 25 m relief. On GPR records, the unit is identified by wavy or chaotic reflectors, which attenuate further signal penetration (Figs. 6 and 7). Unit T is interpreted as till deposited below or adjacent to the ice sheet, with thickness ranging from <1 m in areas of bedrock knobs to >10 m in structural valleys.

Unit Tm is composed of materials similar to those found in unit T, but forms distinct, isolated features. These landforms are variable in surface expression and form large, rounded hills (<1.2 km wide, <25 m high), smaller oblong hills (<800 m long, <25 m high), and narrow ridges (<1 to 5 km long, 3–12 m high). Ridges may contain some well-sorted fine sand and mud near the surface, as at a small ridge in the inland zone (Fig. 8). Merriland Ridge (Fig. 1), a stratified linear feature to the south of the sand plain (Bloom, 1960; Smith, 1990c), contains unit Tm deposits as its core (Miller, 1997). Unit Tm is inter-

**TABLE 1. STRATIGRAPHIC UNITS OF THE SANFORD-KENNEBUNK SAND PLAIN**

| Unit | Geomorphic setting | Sedimentology | Internal and external structures | Interpretation | GPR sections |
|---|---|---|---|---|---|
| (H)* | Adjacent to modern streams | Very well-sorted fine-grained sand and silt | Flat upper surfaces | Holocene floodplain deposits | No figures shown |
| (Brw) | Isolated small-scale landforms | Well-sorted, fine-grained stratified sand | Low-relief ridges | Beach ridges composed of reworked deltaic sediments | No figures shown |
| SDfc | Discontinuous layers, mostly in western plain | Moderately to poorly sorted, medium- to coarse-grained stratified sand, some gravel | Upper surface flat; dipping beds present and continuous with those of unit SDff | Coarse-grained foresets and topsets of regressive deltaic sequence | Figures 6, 13(?), 14, 15 |
| SDff | Extensive thin-to-thick blanket | Moderately to well-sorted, fine- to medium-grained stratified sands | Well-defined dipping beds; upper surface relatively flat | Fine- to medium-grained sandy delta foresets and topsets of regressive delta | Figures 6, 7, 8, 10, 11, 12, 14, 15 |
| GM | Ubiquitous; found in stream channels and at depth | Blue-gray clay, sometimes found as thin clay-silt-sand interbeds; no fossils | Massive; forms drape over some features; thick valley fills | Distal glaciomarine mud deposited by hypopycnal flow | Figures 6, 7(?), 8, 10 |
| (ICd) | Forms isolated topographic highs | Stratified, variably sorted, boulder gravel to sand, some clay | Beds generally dip away from ice retreat direction; some topsets present | Ice-contact deposits formed near bedrock pinning points | Figures 8, 9 |
| (Tm) | Forms isolated geomorphologic features | Poorly sorted, massive, gravel- to clay-sized diamicton | Several feature types, including narrow ridges, large rounded or small oblong hills | Till, forming moraines, drumlins and hills; occasionally overlain by Unit ICd | Figure 8 |
| T | Overlying bedrock and infilling structural basins | Poorly sorted, gravel- to clay-sized diamicton | Massive; some lenses of unit T found within other units | Till deposited in contact with basal ice | Figures 6, 7 |
| BR | Found at base of sequence, and at exposed outcrops | Metamorphosed Mesozoic and older stratified and plutonic rocks | High-relief valley-and-ridge structures; striations on some outcrops | Bedrock | Figure 6(?) |

*Note:* GPR is ground-penetrating radar.
*Parentheses indicate isolated or rare units.

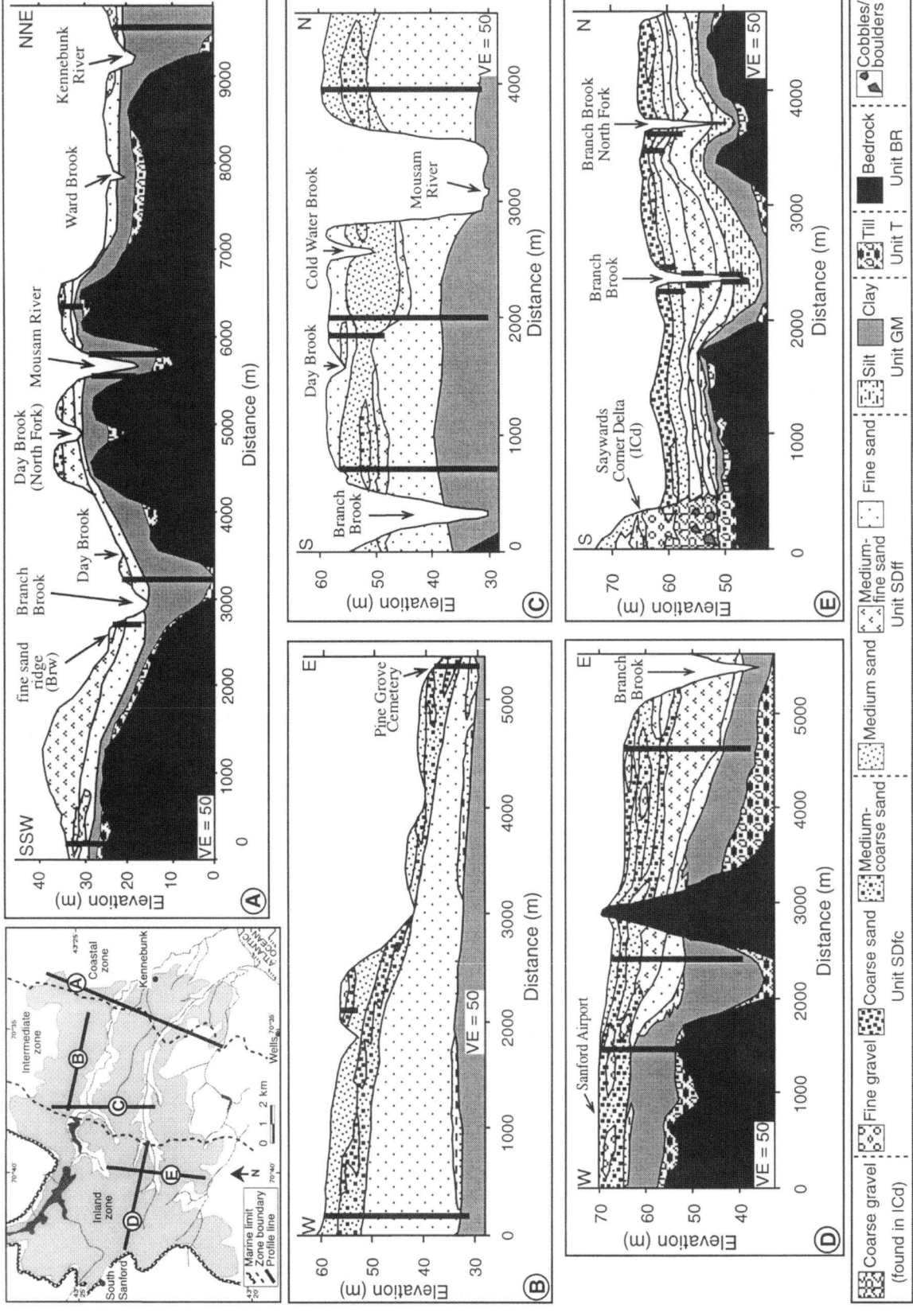

Figure 5. Stratigraphic profiles of Sanford-Kennebunk sand plain discussed in text. Inset shows profile locations (profile A, coastal zone; profiles B and C, intermediate zone; profiles D and E, inland zone). Thick vertical lines in profiles indicate core locations. Stream channels and some locations discussed in text are indicated in individual profile lines.

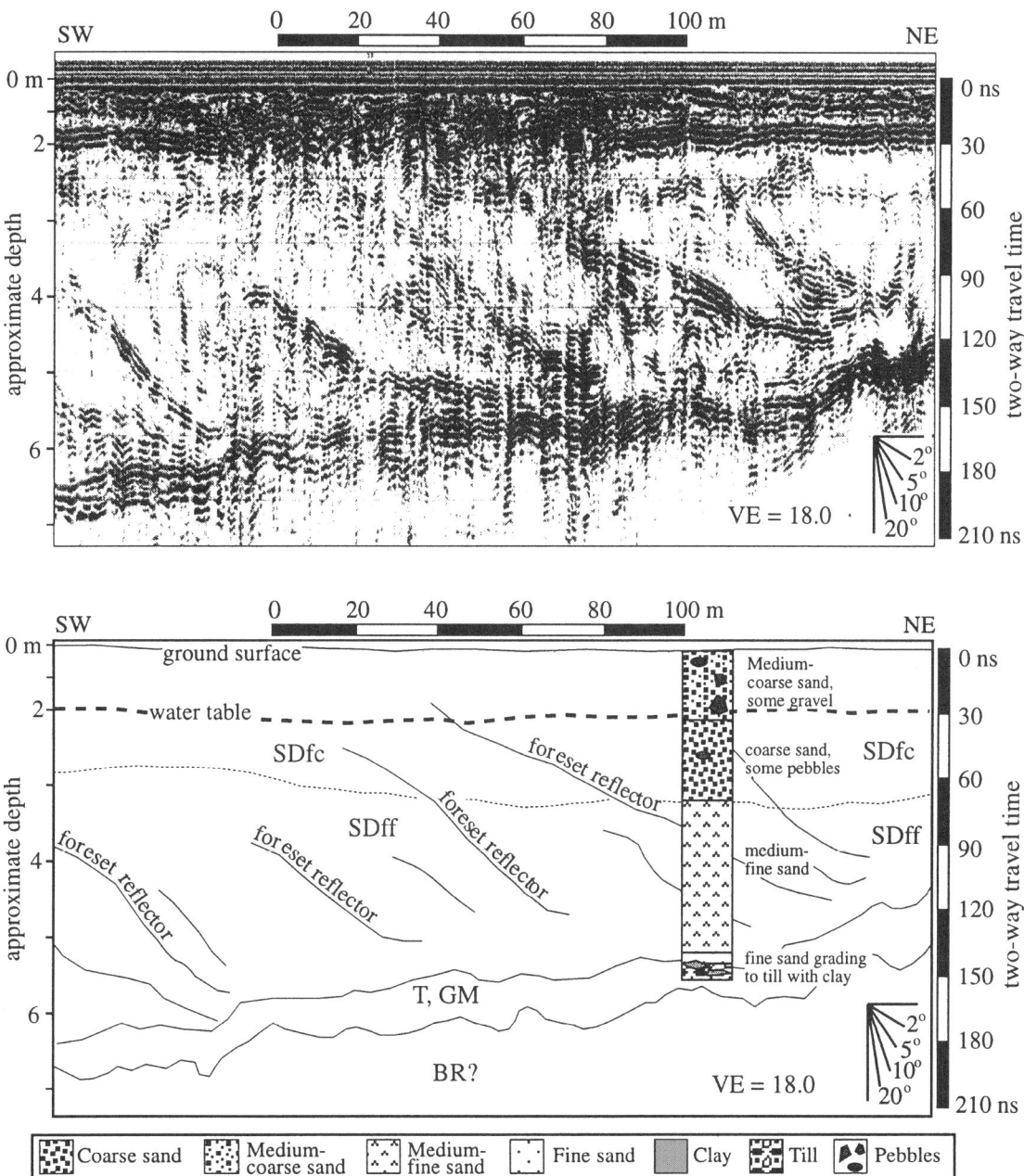

Figure 6. Ground-penetrating radar (GPR) record and interpretive sketch from Pine Grove Cemetery in northeast intermediate zone (location 6, Fig. 3). In this and other GPR profiles, where water table is indicated (dashed line), vertical exaggeration (VE) and dip angle schematic shown correspond to saturated portion of record. Approximate depth axis is adjusted for differential signal penetration above and below water table, if present. All GPR profiles were run at 300 ns. This profile shows a series of gently northeast-dipping reflectors (2°–7°) with basal terminations draped over lower bounding surface. Interpretation of radar profile is aided by correlation with core BB-49 on GPR path (shown in lower panel). Reflectors are interpreted as sandy foresets (unit SDfc and unit SDff) overlying a lower surface of till (T) and clay (GM), which in turn may drape bedrock (BR).

preted as till deposited as topographic highs (large hills) or drumlins (oval hills), or deposited as moraines (ridges) at the ice front. Any of these features may be bedrock cored. Wave reworking of till hills and ridges during marine regression may account for the stratification seen in some ridges. Merriland Ridge has been interpreted as a wave-reworked moraine (Smith, 1990c), as a moraine overlain by subaqueous outwash (Miller, 1997), or as a series of reworked till hills connected by littoral deposits (Bloom, 1960).

Unit ICd is found as isolated topographic highs in and be-

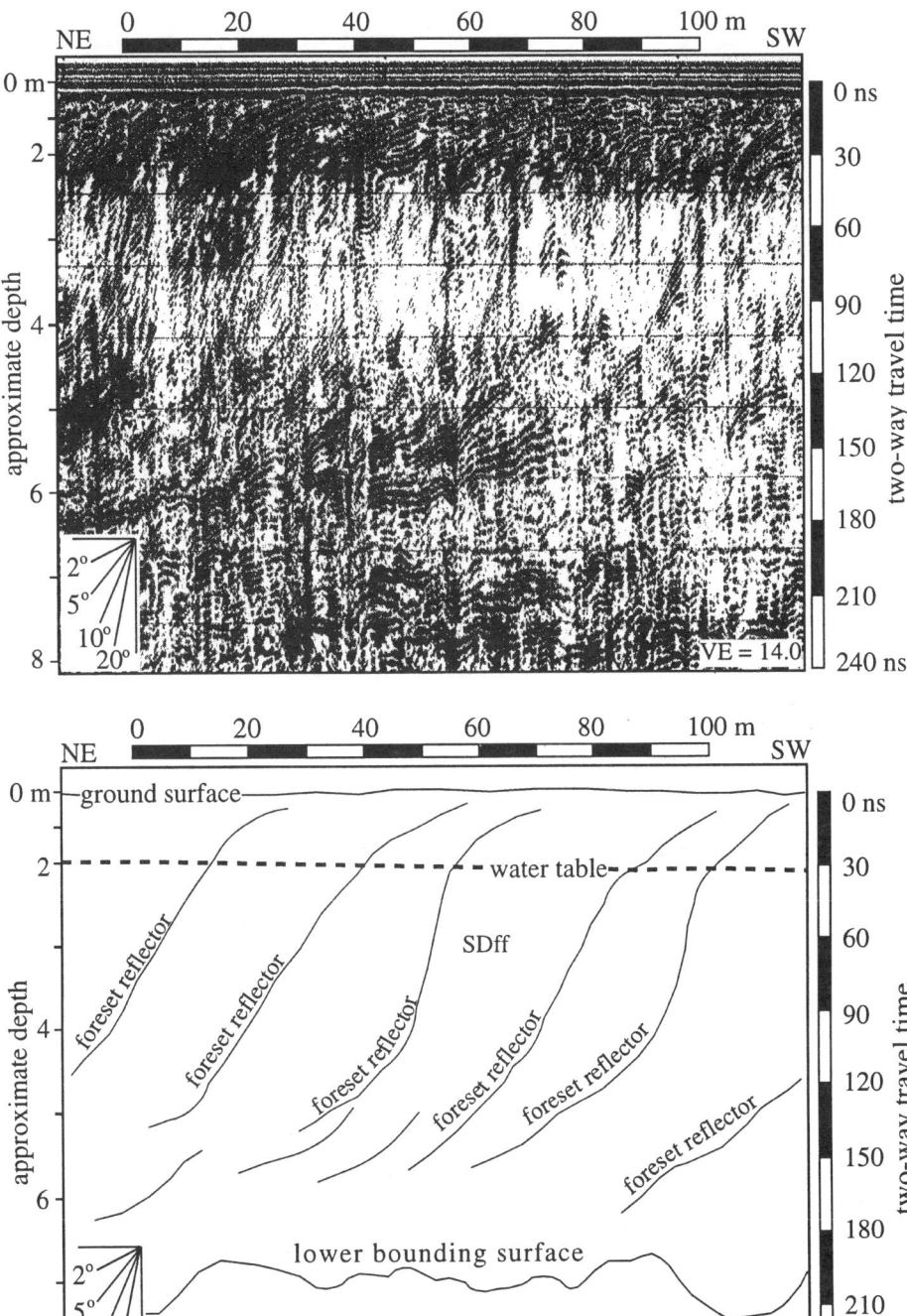

Figure 7. Ground-penetrating radar profile from Pike Road (location 7, Fig. 3). Rhythmic sigmoidal reflectors dipping to northeast at ~7°, interpreted as progradational foreset beds, are clearly visible on record, extending from ground surface to a depth of more than 6 m. Base of foreset bed unit (SDff?) is in contact with lower bounding surface of wavy parallel reflectors interpreted as till (T), possibly with some clay (GM). VE, vertical exaggeration.

yond the margins of the western sand plain (Table 1). One such landform, at the southeastern margin of the inland zone, is a small lobate feature extending southwest from Sayward's Corner (Fig. 1). This flat-topped rise (elevation 73 m) is between two other elevated features. A till hill of 85 m elevation is 1 km to the north, and a bedrock knob, also at 85 m elevation, is 1.6 km to the south. In contrast to other stratigraphy found throughout the Sanford-Kennebunk sand plain, at the Saywards Corner hill, >12 m of pebble, cobble, and minor boulder gravel are overlain by as much as 7 m of stratified medium to fine sands containing some silt and dipping to the south (profile E, Fig. 5). A series of inclined GPR reflectors were detected in the upper sands, dipping to the southwest at angles to 15° (Fig. 9). Unit ICd is interpreted as ice-contact stratified deposits of subaqueous origin; the Sayward's Corner feature has previously been mapped as a kettled ice-frontal marine delta (Smith,

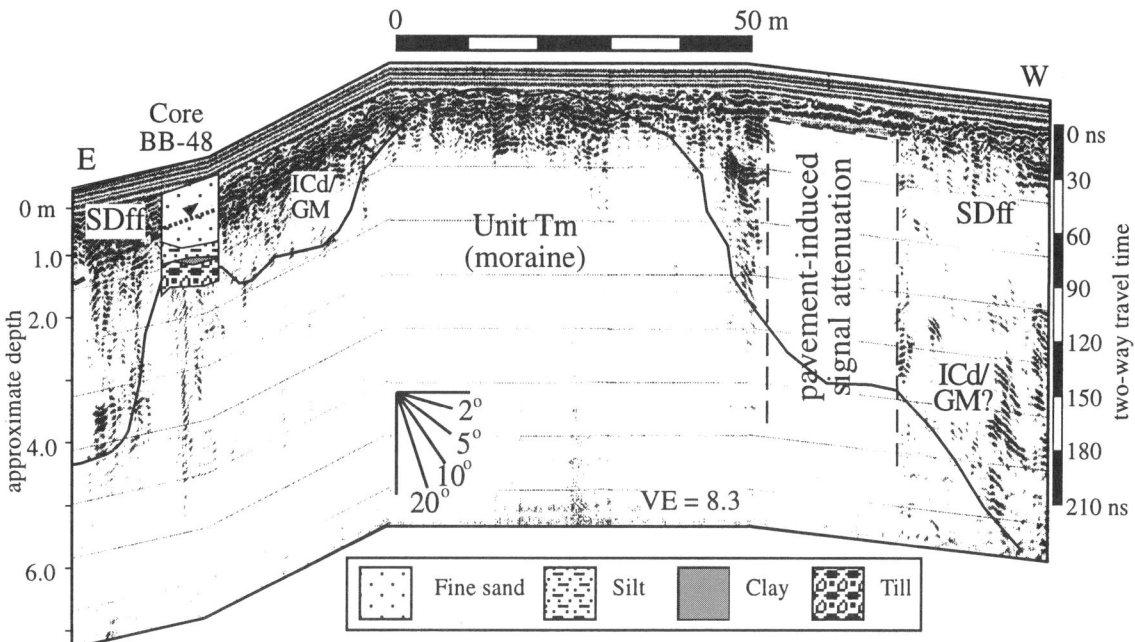

Figure 8. Ground-penetrating radar profile and core BB-48 from central inland zone (location 8, Fig. 3). Profile has been adjusted for topography. Dashed line in core indicates position of water table. Blank area (west side of profile) is due to signal attenuation while crossing road. Moraine (unit Tm) has been draped on its upper flanks by stratified ice-contact deposits (ICd), marine mud (GM) and some fine sand (SDff). VE, vertical exaggeration.

1990c). Other such landforms (Bragdon Road delta, New Dam Road delta; Fig. 1) to the south, west, and northwest of the sand plain were identified by Thompson et al. (1989). A GPR study of Merriland Ridge (Fig. 1) suggests that a moraine core (unit Tm) was overlain by stratified subaqueous deposits of unit ICd (Miller, 1997).

Unit GM forms an extensive, topography-subduing blanket found throughout the region. This facies overlies unit T and may be present as only a thin drape (<1 m) in locations where units T and BR are near surface. Sediments of unit GM may overstep features of unit Tm. Maximum thickness of unit GM recovered from cores in the study area has exceeded 20 m. The facies is characterized by massive blue-gray clay. The transition from marine clay to overlying silt and sand generally occurs in a zone of thin interbeds of fine sand, silt, and clay. The clay is often found at the base of modern stream channels in the region; due to its cohesiveness and impermeability, the facies is relatively resistant to fluvial incision. This unit is identified in GPR records as a lower bounding surface marked by wavy-parallel reflectors below which signals are absent (Fig. 10). Unit GM is interpreted as glaciomarine mud that was distributed into the marine environment by hypopycnal flows (Bates, 1953).

Unit SDff is the extensive sandy facies that forms the majority of the Sanford-Kennebunk sand plain. It overlies unit GM and oversteps preexisting features. The facies is composed of moderately to well-sorted, fine- to medium-grained stratified sands. Generally, the sands are mature, with subrounded to well-rounded grains. This facies is found throughout the sand plain in variable thicknesses ranging from <2 m to 19 m. Unit SDff is laterally continuous and conformably underlies unit SDfc; both units are incised by active and relict stream channels. GPR profiles display two reflection patterns in unit SDff. One such GPR configuration is characterized by a series of stacked, flat-lying parallel reflectors; the other is distinguished by rhythmic, inclined, parallel reflectors that may be relatively straight (Figs. 10–12) or sigmoidal (Figs. 6 and 7) in nature (tangential oblique and sigmoidal oblique, respectively, of van Heteren et al., 1998). Dip angles of inclined reflectors range from 4° to 23°; most angles are <15°. Signal penetration is generally deep in these areas; the reflector pattern often extends uninterrupted to depths beyond chart display range. This GPR record configuration has been correlated with unit SDff by coring (Figs. 6 and 11), as well as by visual inspection (Fig. 12). Unit SDff is interpreted as a prograding deltaic facies, deposited by sediment-laden meltwater streams debouching into a marine environment. The inclined reflectors seen in GPR records are interpreted as delta foresets and are indicative of progradational direction.

Unit SDfc is found on parts of the upper surface of the Sanford-Kennebunk sand plain, and is most common in the inland zone. This facies is composed of medium- to coarse-grained sand, and some fine gravel. Sediment grains are subangular to rounded, and form moderately to poorly sorted beds with total thickness of <2–9 m. Generally, unit SDff grades upward into this facies. Some laterally discontinuous gravel lenses are found within SDfc sand beds. Ground-penetrating

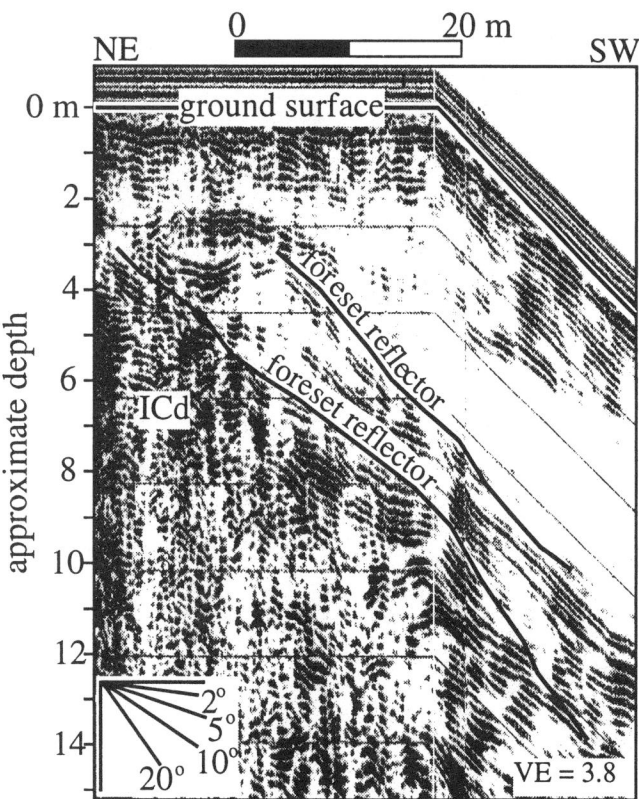

Figure 9. Ground-penetrating radar record from Saywards Corner (location 9, Fig. 3), previously identified as ice-contact delta (Smith, 1990c). Transect was taken down southwest slope of feature; record is corrected for topography. Medium- to fine-grained sandy foresets (unit ICd) dip to southwest at angles to 15°. VE, vertical exaggeration.

radar records (Figs. 13 and 14) suggest that unit SDfc is a continuation of the underlying SDff facies. Several profiles display near-surface dipping-reflector patterns that continue into the SDff unit (Fig. 14). Certain reflector configurations are prevalent only in the upper SDfc portion of GPR records. These include small-scale (<2 m vertical thickness), steeply inclined parallel reflectors that can be traced laterally for <20 m, and isolated patterns of sharply contrasting, concave-up reflectors (<30 m wide, <3 m deep) filled by parallel horizontal (Fig. 15), dipping, or chaotic reflectors. Unit SDfc is interpreted as the coarse-grained upper foresets and delta plain (topsets) of a regressive depositional sequence. The few sets of small-scale steeply inclined GPR reflectors are possible buried beach features or channel fills. The distinct, concave-up, reflector configurations (Fig. 15) that appear filled with various types of reflectors are interpreted as channel cut-and-fill structures.

Unit Brw, where present, consists of small-scale ridges on the surface of the plain. One example of such a feature is found in the coastal zone and extends across Sandy Lane (Fig. 1; profile A, Fig. 5), and above the inclined beds of unit SDff (Fig. 10). Here, Brw forms a 2.5-m-high, arcuate north-northeast–south-southwest–trending ridge ~150 m long. This narrow ridge is composed of thinly bedded, westward-dipping, medium- to fine-grained sand layers. The landform is interpreted as a beach ridge, formed during a minor stillstand of the regressing sea during late Pleistocene time. Unit Brw encompasses all well-sorted, medium- to fine-sand units on the plain surface that may be interpreted as distinctive, small-scale features formed by littoral reworking of Sanford-Kennebunk sand plain sands, including beach ridges, sand bars, and spits.

Unit H is included here to complete the overall stratigraphic classification of sediments found in the study area. This facies consists of flat-lying, well-sorted silt and fine-sand deposits alongside modern stream channels. Some sand shadows and ripples have been seen in recent deposits of this facies. Unit H is interpreted as aggradational Holocene floodplain deposits.

***Stratigraphic variation.*** Distinctions may be made between the three geomorphic zones of the Sanford-Kennebunk sand plain on the basis of morphostratigraphy (Table 2). Generally, stratigraphy becomes more complex westward. Fine-grained sand-plain sediments (unit SDff) are thickest in the intermediate zone and thinnest in the coastal zone. The most extensive deposits of unit SDfc are found in the inland zone and consist of complexly interbedded medium- to coarse-grained sand beds. Whereas the sand-plain architecture is generally similar throughout the region, some variation in progradational direction is present within the three zones (Fig. 16; Table 2), as well as some disparity in the overall thickness of foreset units.

***Coastal zone.*** Unit GM overlies unit T and forms a thick blanket (>20 m in some areas) at or near the surface over much of the coastal zone. The overlying unit SDff is thin compared to that found in the two more western zones, ranging from <1 m near present stream channels to >10 m at topographic highs. Unit SDff sands thin both seaward, where clay and modern estuarine sediments become prevalent, and toward the north and south margins of the Sanford-Kennebunk sand plain, where clay (unit GM) again dominates. GPR records from the coastal zone (Figs. 7 and 10) indicate that clinoforms in unit SDff dip to the east and northeast at angles of 4°–7°. Unit SDff is generally the uppermost regional facies found in the coastal zone; only minor amounts of unit SDfc have been found. Units Tm and ICd have not been found in the coastal zone; features of unit Brw are rare.

***Intermediate zone.*** The stratigraphy of the intermediate zone is somewhat more complex than that found in the coastal zone (Fig. 5). Sediment units are thicker (as shown by the isopach map, Fig. 4); clay (unit GM) is >20 m thick, unit SDff sands are up to 18 m thick, and unit SDfc deposits are up to 10 m thick. Sediment grain size in the intermediate zone is also more variable and coarsens upward more than that in the coastal zone (Fig. 5). GPR records from the intermediate zone (Figs. 6, 11, and 12) show that clinoforms in units SDff and SDfc dip at angles ranging from 4° to 15° and are generally oriented to the northeast (Fig. 16). Several hills of unit Tm are found ad-

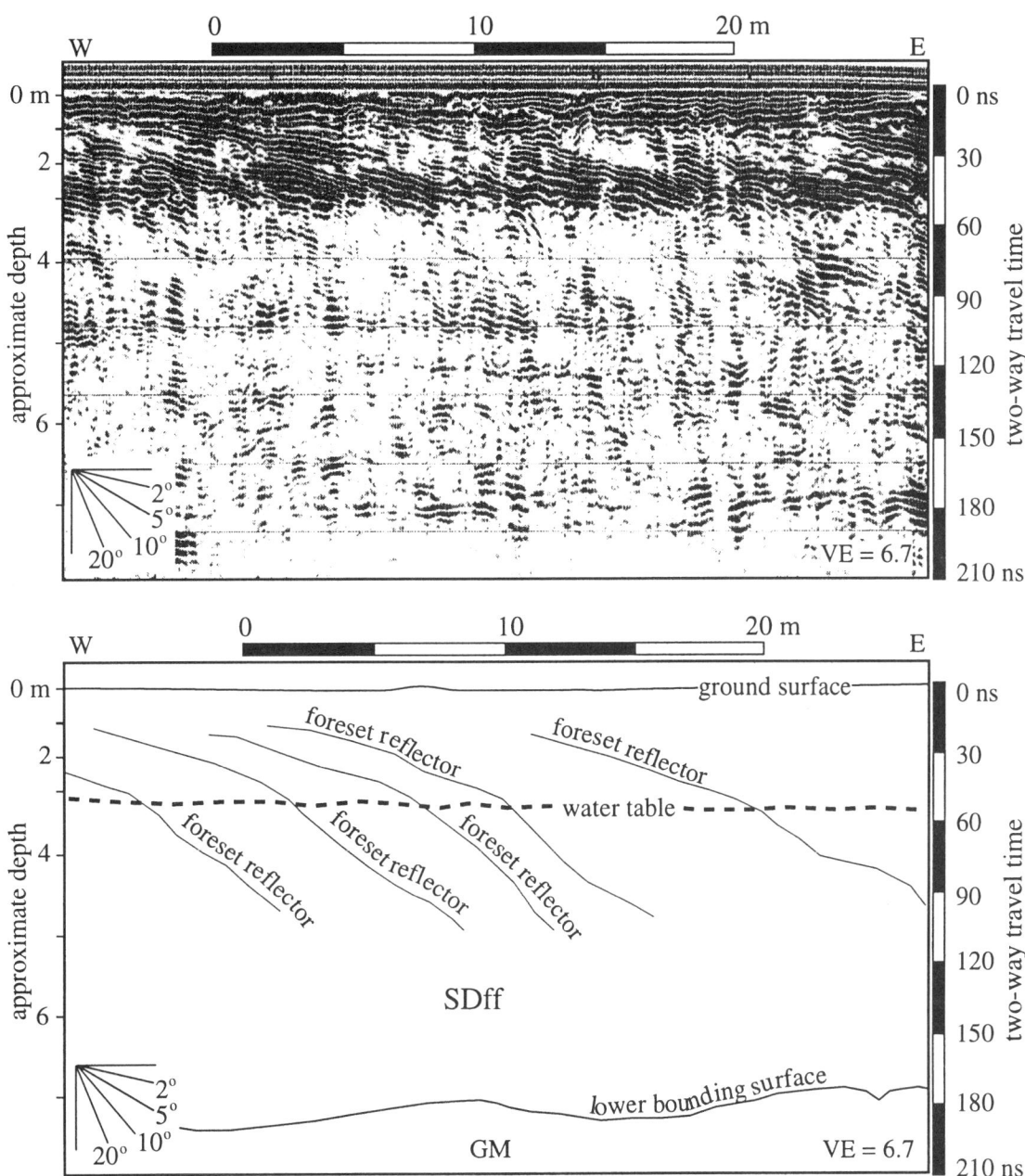

Figure 10. Sandy Lane ground-penetrating radar record and line-drawing interpretation (location 10, Fig. 3). East-dipping reflectors (dip angles 5°–7°) in this record are rhythmic and regular in facies SDff, with lower wavy surface of clay (unit GM) at 7 m depth. VE, vertical exaggeration.

jacent to the southern margin of the sand plain. Few relict beach deposits (unit Brw) have been found. No ICd landforms have been identified within the margins of this zone of the sand plain (Table 2); several ICd features exist to the south of the plain (upper portion of Merriland Ridge, Bragdon Road delta, Bauneg Beg delta).

***Inland zone.*** The regional stratigraphy of the inland zone follows the same general pattern seen in the coastal and intermediate zones (Fig. 5), but is coarser-grained and more complicated than the stratigraphy to the east. Unit GM thickness ranges from at least 7 to 13 m and may extend to a total thickness of 35 m or more. Unit SDff is thinner (generally <10 m) and slightly coarser-grained than the extensive fine-grained SDff deposits found in the intermediate zone. The upper coarse-grained sand and gravel beds of unit SDfc are thicker (2–10+ m total thickness), more laterally extensive, and more com-

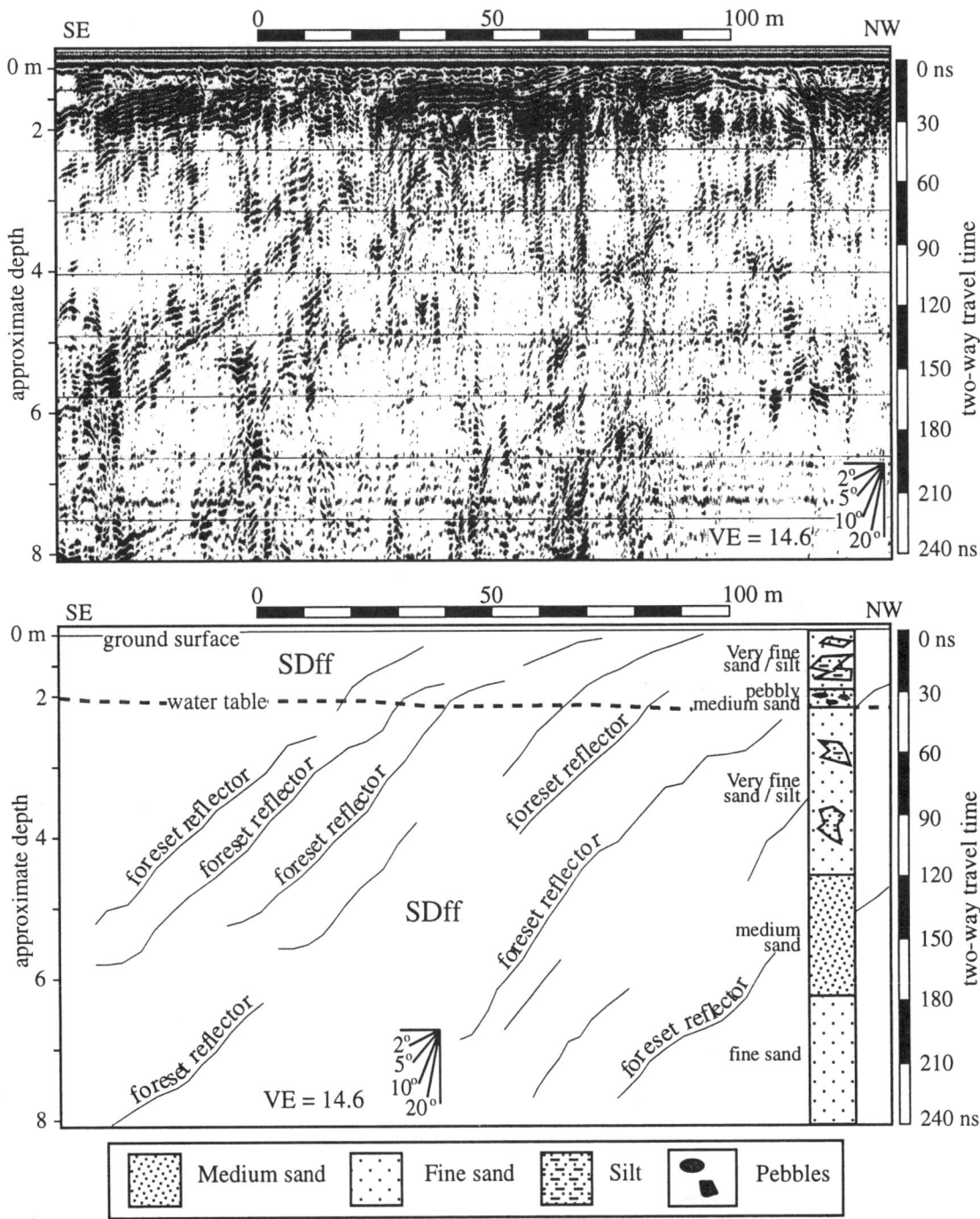

Figure 11. Ground-penetrating radar record from "The Plains" area in west-central intermediate zone (location 11, Fig. 3). Reflectors dip to southeast at ~4°; true dip in this area has been calculated as 6° with a bearing of 58°. Tangential reflectors extend from ground surface to depths beyond radar signal penetration (>8 m). Reflectors, correlated with core BB-50, are interpreted as large-scale sandy foresets of unit SDff. VE, vertical exaggeration.

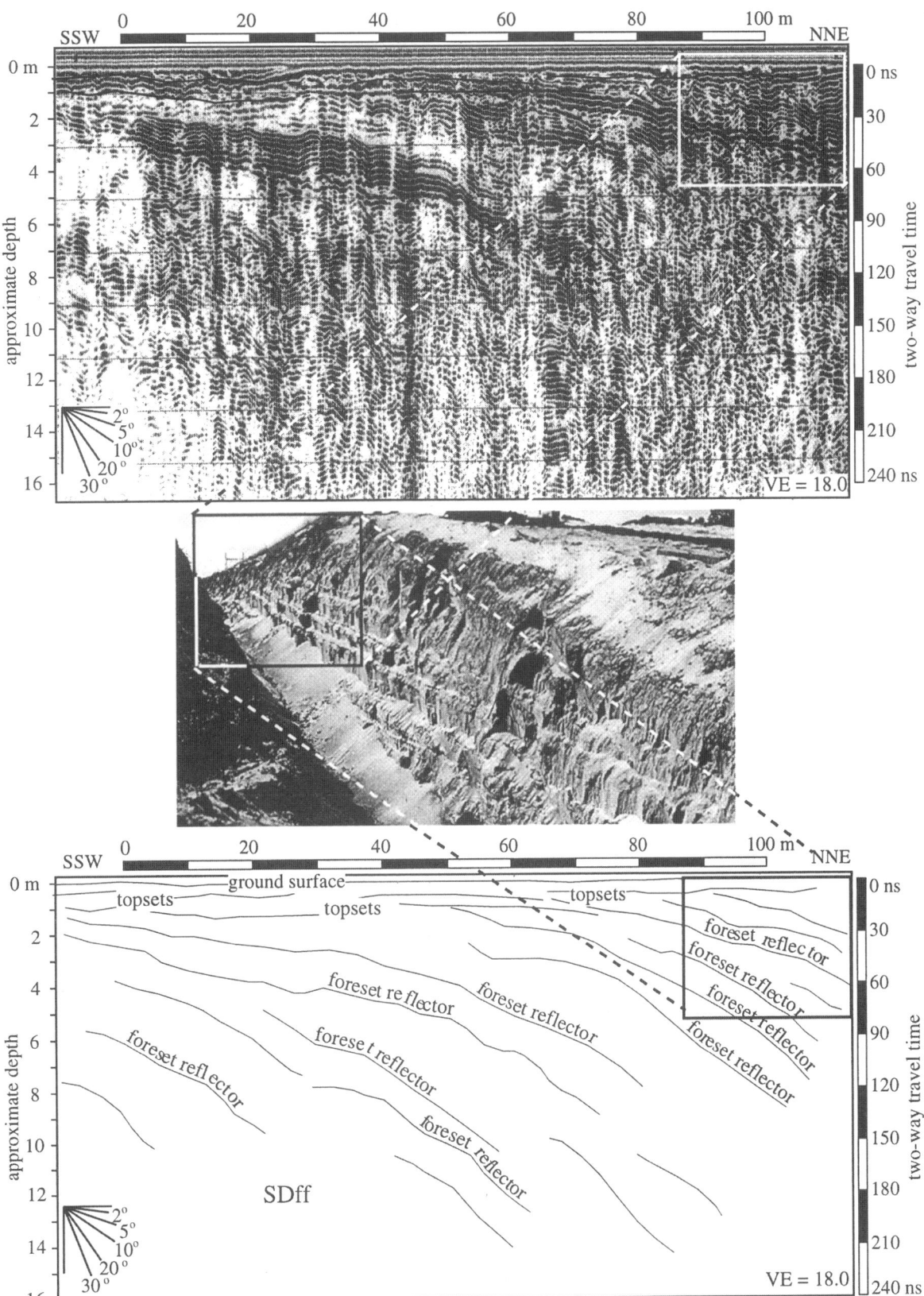

Figure 12. Ground-penetrating radar (GPR) record, photograph, and interpretive sketch from recent gas pipeline excavation immediately south of Branch Brook (location 12, Fig. 3). Boxes in upper right of GPR record and line drawing, and in upper left of photograph correlate photo view with radar profile and sketch. Photo view is to southwest from slightly below plain surface and was taken at an oblique angle. Horizontal land surface of plain can be seen at upper left corner of photograph. Beds in unit SDff dip at angles of ~12°–18°; GPR records indicate that true dip in region is ~15° and has a bearing of 74°. GPR image clearly shows topsets (uppermost reflectors) that roll over into shallowly dipping foresets, which become more steeply dipping with increased depth. VE, vertical exaggeration.

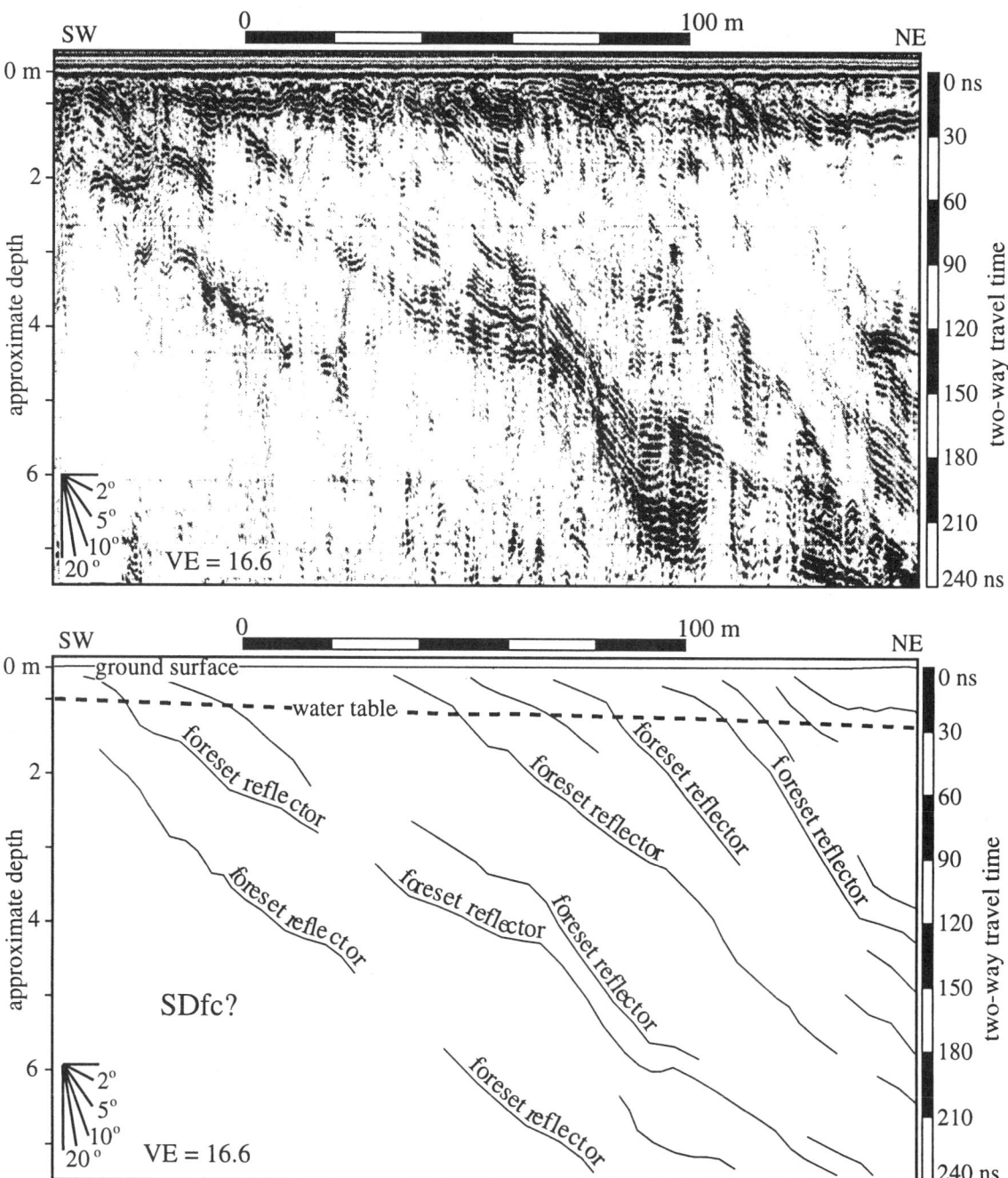

Figure 13. Ground-penetrating radar profile from northern inland zone (location 13, Fig. 3). Dipping reflectors (unit SDfc?) extend from near-surface water table to beyond radar signal recovery depth and are regular and rhythmic. Dip angles range from 3° to 5°; true dip is 6.5° and has a bearing of 81°. VE, vertical exaggeration.

plexly interbedded than elsewhere in the sand plain. This facies also contains lenses of gravel as well as rare interbedded finer-grained sands. GPR records in the inland zone (Figs. 13–15) indicate that deltaic progradation of plain sands generally proceeded to the east-southeast in the central inland zone (Fig. 16), with average dip angles of 7°–9°. This area, both within and beyond the Sanford-Kennebunk sand plain margin, contains the widest assortment of features consisting of unit Tm, including till hills, drumlins, and moraine segments. The inland zone is the only area containing ice-contact deltas (unit ICd) inside the plain boundaries. The abundance of these features contrasts with the scarcity of beach (Brw) deposits (Table 2).

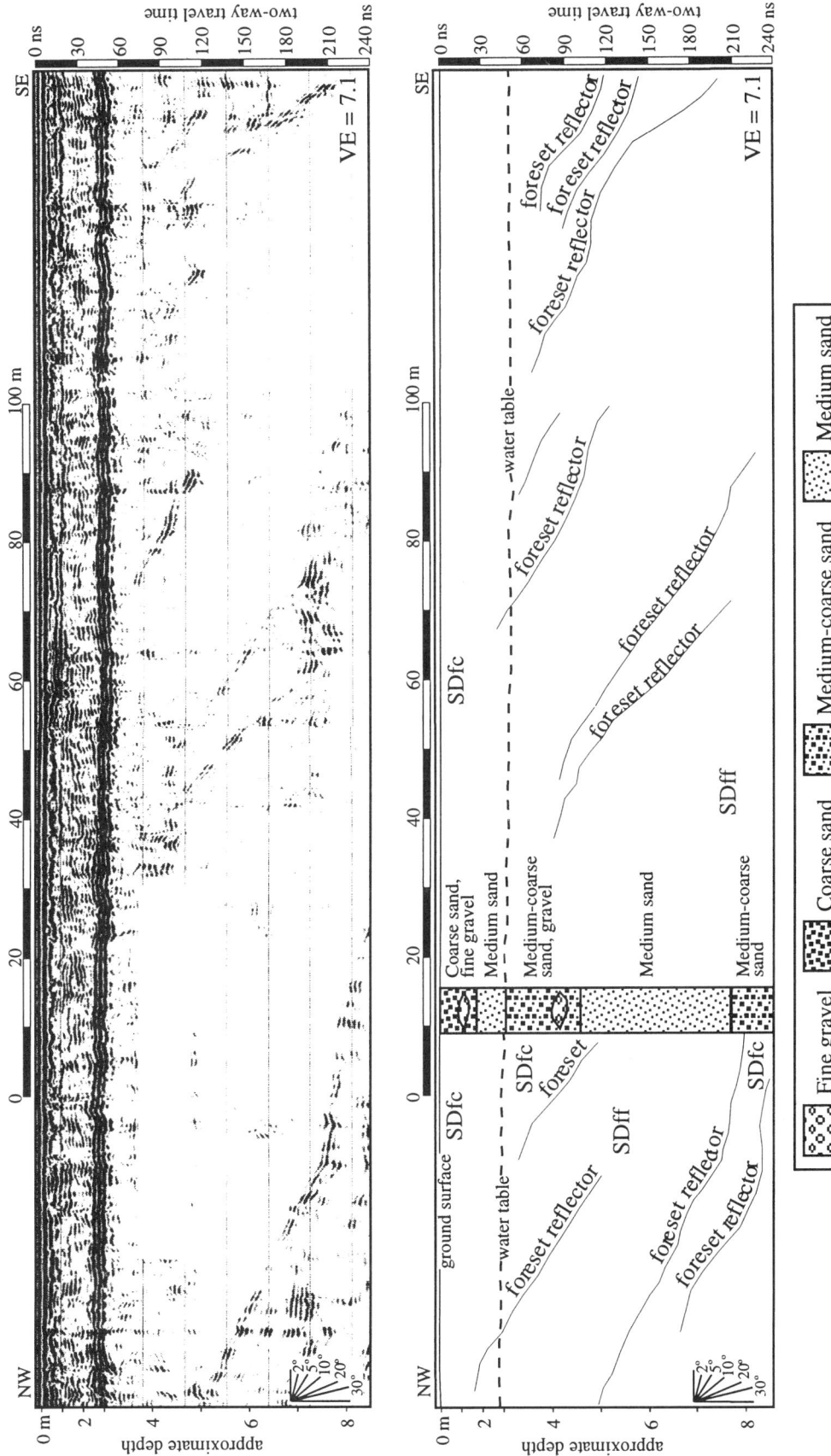

Figure 14. Ground-penetrating radar record and interpretive sketch from central inland zone (location 14, Fig. 3). Foreset reflectors appear widely spaced in this profile, but are regular, rhythmic stacked sets of clinoforms dipping at angles of 6°–10°. Correlation with core T-25 (from Smith, 1990a) suggests that foresets consist of cosets of medium-coarse sand (unit SDfc) alternating with medium sand (unit SDff). True dip is ~8.5° with a bearing of 119°. VE, vertical exaggeration.

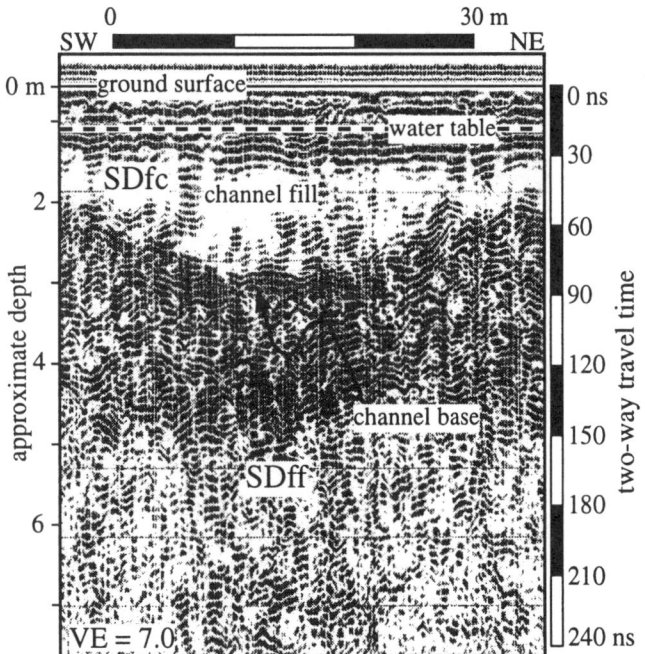

Figure 15. Ground-penetrating radar (GPR) record from central inland zone (location 15, Fig. 3). This profile was run to west of GPR shown in Figure 14, and perpendicular to that line. This record therefore shows strike section of Sanford-Kennebunk sand plain, with no visible clinoform reflectors. Instead, one small channel cut-and-fill structure is visible between 2 and 3 m depth, and is likely filled by unit SDfc (flat-lying beds infilling channel). VE, vertical exaggeration.

## DISCUSSION

### Facies depositional interpretation

The bedrock, till, and clay sequence at the base of the stratigraphic profile (Fig. 5) record preexisting topography and ice retreat under marine conditions. These units correspond to lower bounding surfaces seen in several GPR profiles (Figs. 6, 7, and 10). Unit T is interpreted as subglacial till, deposited by the retreating ice sheet and infilling depressions in the bedrock topography (unit BR). Features of unit Tm were also deposited at this time, under and adjacent to the ice margin. Similarly, primary features of unit ICd were deposited by emerging meltwater tunnels at ice-frontal positions in locations where bedrock acted as pinning points to slow the glacier's retreat (Crossen, 1991).

Unit GM is interpreted as part of the locally extensive Presumpscot Formation (Bloom, 1960), representing rock flour transported in suspension by hypopycnal flow into the sea (Bates, 1953; Ashley, 1995; Powell and Domack, 1995). The fine particles flocculated upon mixing with marine waters and settled some distance from the ice margin, blanketing the area with the massive blue-gray clay (Oldale, 1989). The deposits of unit GM are thickest in the intermediate zone, based on cores and isopach information. The large basin in the center of the Sanford-Kennebunk sand plain (Fig. 4) likely received some deposition of till, but would have remained a wide, deep depression as ice retreated from the area, allowing for the development of extensive proglacial clay deposits during continued ice-sheet recession. During this time, the lower elevations of landforms composed of units Tm and ICd would have begun to be buried by the marine mud, generating the overstepped stratigraphy seen around these features.

The medium- to fine-grained sands of unit SDff overlying the sequence of bedrock, till, and clay record a coarsening-upward trend interpreted as fluvial deposition of proglacial deltaic sediments. This interpretation is based on three factors. First, unit SDff is stratigraphically above the Presumpscot clay, and there is no interbedding of units GM and SDff. This indicates that deposition of the sands began after final deposition of the clay. Some intercalation of sandy layers and silty mud has been recognized, but such muds are not characteristic of the blue-gray clay of the Presumpscot Formation. Unit SDff also circumvents preexisting structures, continuing the overstep begun by the clay deposition (Fig. 8). Second, GPR records (Figs. 6–14) show packages of laterally extensive, rhythmic, dipping reflectors ranging in vertical thickness from 4 m to >14 m. Such patterns are similar to deltaic GPR patterns found by other workers, including van Heteren et al. (1998) in a study of coastal paraglacial features in New England, Smith and Jol (1992) in a GPR study of a Lake Bonneville delta, and Jol and Smith (1991) in a study of Canadian lacustrine deltas. Third, the dip angles and orientation of inclined beds based on GPR records (Figs. 6–14) and visual evidence (Fig. 12) are consistent with meltwater-transported deposition into a regressing sea. Generally, beds dip at relatively low angles (<15°), suggesting deltaic deposition from a buoyant jet, as opposed to the steeper angles of Gilbert-type deltas deposited by homopycnal flows (Bates, 1953). Such buoyant deposition would occur with proglacial streams emptying into a marine environment. The progradational direction suggested by GPR-derived foreset orientations (Fig. 16), is generally consistent with the late Pleistocene position of the ice front to the northwest and an ocean margin regressing to the east-southeast.

The medium- to coarse-grained sands and fine gravel of unit SDfc represent a continuation of the coarsening-upward trend seen in the underlying facies. This facies is rare in the coastal zone, and most extensive in the inland zone, suggesting that the source for these sediments became more removed from the area as the unit advanced eastward. Facies SDfc is interpreted as coarse-grained foresets (and topsets) of the Sanford-Kennebunk sand plain, based on the stratigraphy, which suggests relatively high energy deposition of coarse-grained sand and fine gravel; the facies thins eastward. As ice retreated from the region into eastern New Hampshire, high-discharge meltwater conditions (Boothroyd, 1995) meant that fluvial energy would have remained high, but coarse-grained sediments carried into the region would have become less plentiful with distance as they prograded across the Sanford-Kennebunk area,

## TABLE 2. ZONAL VARIATION IN THE SANFORD-KENNEBUNK SAND PLAIN

| Unit | Description | Coastal zone | Intermediate zone | Inland zone |
|---|---|---|---|---|
| (H)* | Holocene floodplain deposits | Found in wide floodplains of Branch Brook and Mousam River | Extensive at large floodplain of Mousam River; found at narrow floodplain of Branch Brook | Generally found at small floodplains of Branch Brook and Mousam River |
| (Brw) | Beach deposits | Rare beach ridges; shallow beach scarps | Sand bars; beach scarps; littoral reworking of Merriland Ridge | Small spit extending from glaciomarine delta |
| SDfc | Coarse-grained sandy foresets and topsets of regressive deltaic sequence | Rare coarse-grained sand caps at highest elevations | Laterally continuous, interbedded medium-grained to coarse-grained sands; fine sands at surface in "The Plains" region | Most extensive unit; complexly interbedded medium-coarse to coarse sands; some gravel lenses; more poorly sorted than other SDfc deposits |
| SDff | Fine-grained sandy foresets and topsets of regressive delta | Thinnest in this zone; continuous but incised by many stream channels; foresets dip to east and northeast | Laterally extensive; mostly fine sand; thickest deposits of facies found; thins to east; foresets generally dip to northeast | Laterally continuous but of varying thickness; mostly medium-fine sand; foreset dip directions range from east-northeast to east-southeast |
| GM | Glaciomarine mud | Laterally continuous; thins toward zone margins; present at or near surface and in most stream channels | Found at depth; laterally continuous and thickest in this zone; present in some stream channels | Found at depth and present in deeply incised stream channels; variable thickness |
| (ICd) | Ice-contact deposits | None found in zone | None found in zone; upper portion of Merriland Ridge to south of plain | Three ice-contact deltas within plain; numerous to south, west and northwest of plain boundaries† |
| (Tm) | Isolated landforms composed of till | None found in zone | Several till rises found adjacent to southern margin; also Merriland Ridge core to south of plain | Till hills adjacent to southern margin; small moraine in central zone; drumlins adjacent to western margin |
| T | Ground moraine till deposits | Unknown thickness | Unknown thickness | Unknown thickness |
| BR | Bedrock | Near surface; some outcrops present | Found at depth; some outcrops in stream channels (eastern zone) | Highly variable; present at depth; some outcrops; some outcrops striated |

*Parentheses indicate isolated or rare units.
†Thompson et al. (1989) provide detailed information of glaciomarine ice-contact deltas in region.

explaining the decrease in unit SDfc thickness from west to east. These sediments were deposited prior to reaching the delta front or at the top of the delta slope; the finer-grained sand of facies SDff would be transported farther into the nearshore environment on buoyant meltwater jets. A distant ice margin and sediment source is further indicated by the absence of kettled topography in the two sand-plain facies, which suggests that ice was well away from the area during the time of delta deposition.

The deltaic interpretation of unit SDfc is aided by evidence from GPR profiles, which often indicate concordant deposition of facies SDff and facies SDfc. GPR records from some areas show uninterrupted foreset reflectors passing through the interface between units SDfc and SDff (Fig. 14). This suggests continuous deposition; there are gradational contacts between the two units, which in turn may indicate that deposition of the upper facies occurred primarily in the subaqueous environment. Topsets of the deltaic package are indicated where coarse-grained deposits are shown to truncate upper foresets.

Unit Brw forms isolated small-scale landforms interpreted as beach deposits. After deposition of the deltaic sands into the regressing sea, some of the sands were reworked into wave-built features such as spits and beach ridges. These features formed in the nearshore environment during minor stillstands of sea level. Scarps cut into the sand plain are interpreted as wave-cut features, similar to those noted by Crossen (1991) in glaciomarine deltas in southern Maine.

Holocene floodplain deposits (unit H) form flat-lying, fine-grained aggradational surfaces along the banks of the two deeply incised stream channels within the margins of the Sanford-Kennebunk sand plain, Branch Brook, and the Mousam River. A more detailed discussion of Holocene stratigraphy in the Wells coastal area was provided by Gehrels et al. (1996).

### Sand-plain development

Based on stratigraphy and GPR records, the Sanford-Kennebunk sand plain is interpreted as a large delta formed after the Laurentide ice sheet had left the immediate region. With removal of ice from the area and concurrent isostatic rebound, decoupling of the ice sheet and highstand sea occurred, causing the shallow marine waters to regress over subglacial and marine highstand glacial deposits. During this time, sediment-laden glacial meltwater would have entered the eastward-moving nearshore environment of the receding Atlantic Ocean, depositing much of its sandy load over preexisting deposits. Rapid regression of sea level to lowstand would not have permitted development of an extensive braidplain such as those

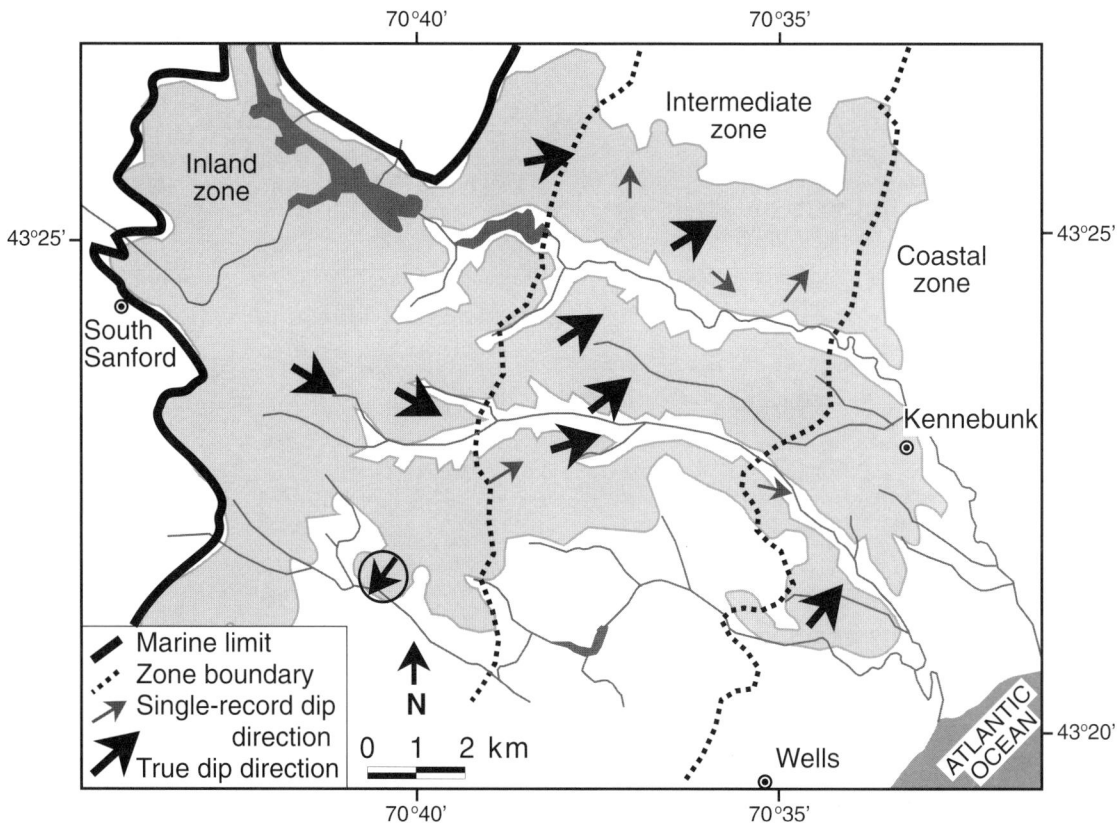

Figure 16. Generalized dip directions (arrows) for foresets found at various sites in Sanford-Kennebunk sand plain; dips are based on calculations from ground-penetrating radar (GPR) data and visual observation. True dip directions (calculated using two or more intersecting GPR lines) are shown where possible. Dip orientation of foresets indicates direction of delta progradation. Deposition of plain sands generally proceeded from northwest to east with some spreading of units northward from a central axis roughly parallel to modern Branch Brook channel. One anomalous dip orientation to southwest in inland zone (circled arrow) is due to inclusion of Saywards Corner ice-contact delta. Another aberrant dip orientation near southeast margin of coastal zone at Pike Road is indicative of a separate deltaic lobe formed by smaller tributary stream believed to have been deflected southward by two bedrock and/or till knobs, and then diverted to northeast by another bedrock and/or till knob to south of channel.

present in Alaskan outwash plains and Icelandic sandar today. Instead, relatively few large meltwater streams channeled by preexisting bedrock and recent subglacial deposits would have deposited large quantities of sorted sand to the plain area.

This interpretation is supported by the stratigraphy seen throughout the Sanford-Kennebunk sand plain, in which sediments deposited directly by the glacier (subglacial till) underlie clays of the Presumpscot Formation, which in turn are below the coarsening-upward plain sands. Had the sand plain been deposited during the highstand or earlier, clays would have been interbedded with the sandy foresets. Because marine clays have generally been found below the sand, or are interspersed with only distal bottomsets (C. Neil, 1997, personal commun.), the sand plain must have been deposited in the area after the Presumpscot, while the ice was retreating from the region, with continued clay deposition occurring farther seaward in deeper waters. This would suggest that the distant ice no longer exerted direct control over deposition of the sands, but that fluviodeltaic regimes would have had the greatest influence on deposition of glacially derived sediments.

A deltaic interpretation for formation of the Sanford-Kennebunk sand plain is further supported by the regularity of foreset reflectors seen in GPR records. Had active ice still been present in the area, the inclined reflectors would have undergone folding and shearing as the glacier margin fluctuated. There is no evidence of contorted reflectors in GPR profiles. A nearby melting ice margin would also indicate that kettles formed by drifting ice blocks should be present in at least the western portion of the sand plain. While kettles are abundant in several ice-contact features around the margin of the plain, few kettles were found on the surface of the plain.

The calculated dip angles of foreset beds are generally in the range expected for a delta prograding into a marine environment by hypopycnal flow. Dip angles generally range from

4° to 13°, with minimum and maximum angles of 2° and 17° obtained by GPR records. These angles are not as high as those found in ice-contact deltas (10°–30°, Thompson et al., 1989; 23°–27°, Ashley et al., 1991), nor are they as low as would be expected in a subaerial braid-delta environment, where surfaces dip at very shallow angles, often <1° (Nummedal and Boothroyd, 1976; McPherson et al., 1987). The dip angles found in GPR records in this area are also too low to represent a Gilbert-style lake delta with foresets which dip at angles of 25°–30°. This indicates that the plain could not have been deposited until ice had retreated from the area, nor was it deposited subaerially or by a homopycnal flow into a freshwater basin (Bates, 1953).

### Alternative explanations

One of the earliest interpretations of the Sanford-Kennebunk sand plain was by Bloom (1960), who identified the plain as an outwash delta. Our work suggests that Bloom's concept of a deltaic origin is correct; however, several problems exist with his interpretation of the plain as an outwash-derived highstand feature. Marine clays are not interbedded with the foreset beds, as would likely be found in a highstand delta; clays are instead found below the dipping sands and are stratigraphically distinct from the plain sediments. Similarly, little evidence for an outwash origin has been found. Foresets are only minimally truncated by topset beds; generally, <1–2 m of plain surface fluvial modification has been found. Bloom (1960) noted the lack of topsets in his "outwash delta"; in addition, GPR work in this study has uncovered relatively few reflectors indicative of cut-and-fill channel structures as would be found in a typical braided outwash environment.

Smith's (1977) map of the Kennebunk 15′ quadrangle shows the sand plain as an outwash unit of sand and gravel. As noted, little evidence of extensive topsets has been found. Smith's interpretation also does not explain the origin of large-scale dipping beds described by Bloom (1960) and verified for this study by GPR.

One alternative explanation for the dipping reflectors seen in the field and in GPR records might be to describe the Sanford-Kennebunk sand plain as a submarine fan (Montello, 1992), based on the overall shape of the feature and clinoform inclinations, which are only slightly higher than might be expected in a submarine fan environment. However, such an origin is not likely due to the regularity and continuity of the beds, as well as the previously mentioned lack of interbedding of marine clay and sand. Other interpretations for inclined progradational units, such as spits and channel-fill deposits, are not indicated here due to the thickness and lateral continuity of the dipping units. The sediments forming the plain are far too extensive both vertically and laterally to be interpreted as such small-scale features.

## EVOLUTIONARY MODEL FOR THE SANFORD-KENNEBUNK SAND PLAIN

A morphostratigraphic model for the development of the Sanford-Kennebunk sand plain and associated sediments is proposed on the basis of geomorphologic, stratigraphic, and geophysical analysis. This model incorporates nine stratigraphic facies, ranging from subglacially deposited till to recent surficial fluvial deposits. The sequence of deposition encompasses the late Pleistocene withdrawal of the Laurentide ice sheet from the study area, concomitant marine encroachment, and subsequent sea-level fall during isostatic recovery. Prior to this sequence, ice had begun to retreat from its terminal position in the Gulf of Maine. After emergence of the Sanford-Kennebunk region, continued sea-level fall to lowstand and later Holocene transgression had little effect on the regional stratigraphy in the area, other than to result in deep fluvial incision of the sand plain and to initiate Holocene stream meandering and the deposition of floodplain deposits (unit H).

Central to this model is the identification of two stratigraphic facies of deltaic origin, units SDff and SDfc. These units represent a gradational sequence of meltwater-transported proglacial sediments deposited into the regressing nearshore marine environment. These units do not signify simple subaerial outwash deposition, although it may be tempting to place the two facies into that category. Instead, deposition of the sand-plain sediments is believed to have occurred almost exclusively in the marine realm. This may lead to reevaluation of current sea-level chronologies and depositional histories in areas of similar large sand bodies near the maximum marine limit.

### Phase 1: Ice retreat (ca. 13.8 ka)

The Laurentide ice sheet had reached its maximum in the Gulf of Maine ca. 18 ka. As ice retreat progressed northwestward to the position of the present coastline, ocean waters transgressed in contact with the marine-based ice sheet. At this position, the glacier slowed in its retreat as it came in contact with the underlying bedrock topography (Smith and Hunter, 1989; Crossen, 1991). Coarse sediments were deposited subglacially and along the deteriorating ice margin (Fig. 17A). Massive, unsorted diamict (unit T) formed the base of these deposits. At the newly grounded ice margin, moraines (unit Tm) developed and subglacial meltwater tunnels began to disgorge crudely sorted grounding line fans (unit ICd), which spread small aprons of coarse sediment onto the basal till.

As ice-sheet retreat progressed into the central study area, deposition of these sediments continued, with the addition of more stratified ice-marginal features (unit ICd), which began to build into the abutting marine waters (Fig. 17B). Features such as submarine fans (unit ICd) formed in areas where bedrock pinning points slowed ice retreat long enough to allow the

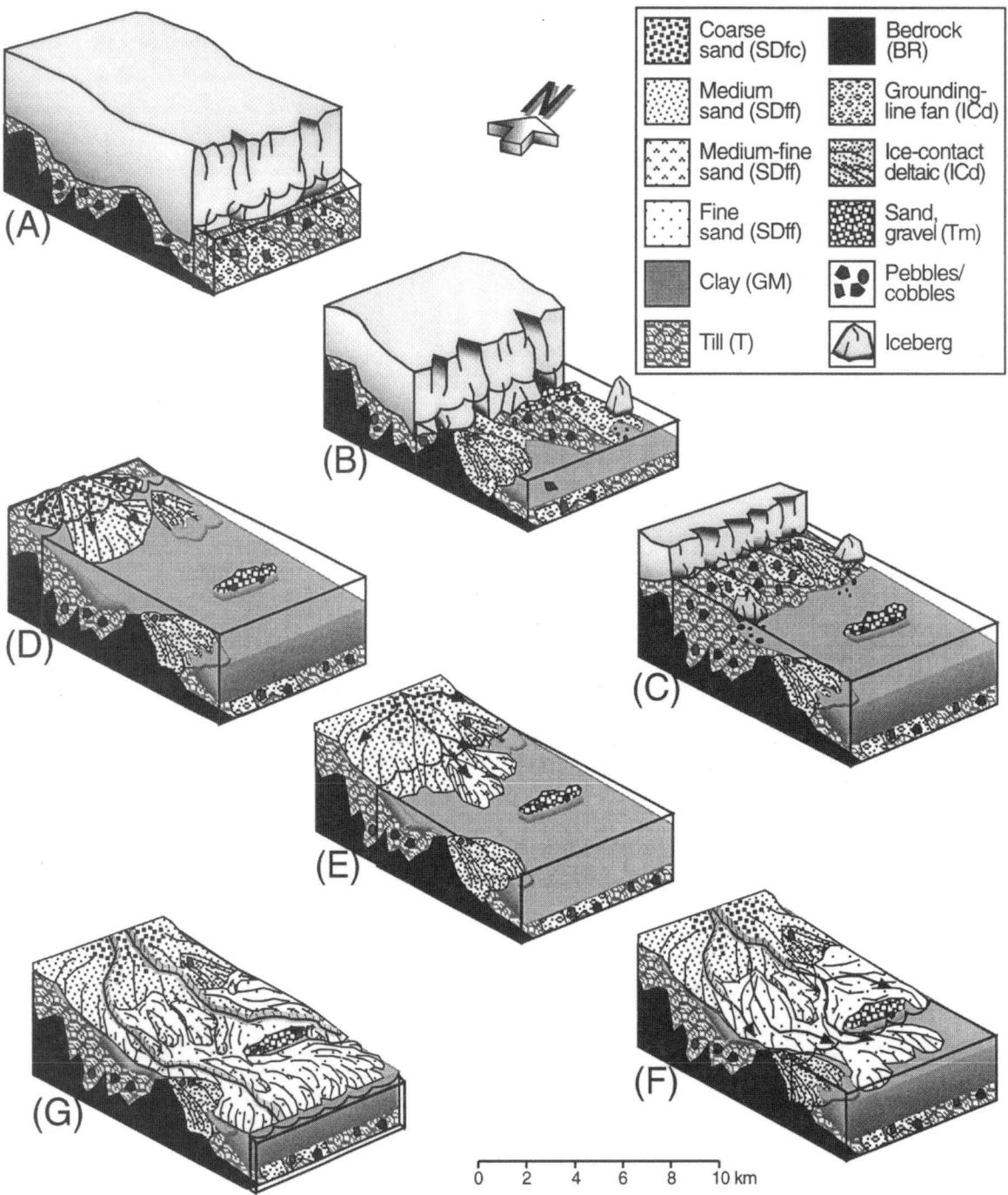

Figure 17. Seven-part model for evolution of Sanford-Kennebunk sand plain and its underlying stratigraphy. This model encompasses three phases: ice retreat (A, B), sea-level highstand (C, D), and early marine regression (E, F, G). See text for explanation.

buildup of stratified sediments (Belknap and Shipp, 1991; Crossen, 1991). In some cases, these forms became large enough that their upper limits reached the water surface, forming ice-contact deltas (ICd). A number of these deltas developed topsets; stranded ice blocks from the calving glacier margin left kettles in the delta tops. At this time, in the deeper waters to the east, distal glaciomarine mud began to blanket the till and fans; drifting icebergs deposited dropstones into the sediment. As deeper quiet waters advanced landward in conjunction with ice retreat, the lower reaches of the newly formed moraines, fans, and deltas in the area began to be draped with glaciomarine mud (unit GM).

*Phase 2: Highstand (ca. 12.6 ka)*

As the glacier front retreated to the northwest, abundant sediments were supplied to the ice front by erosion of the underlying bedrock. Deposition of till and ice-contact features continued, while earlier such deposits to the east and east-southeast became further buried by accumulating glaciomarine mud (Fig. 17C). At some point, glacial unloading triggered initial isostatic recovery of the crust in the Sanford-Kennebunk area. As the rate of rebound approached the rate of eustatic rise, local relative sea level remained relatively stationary for a period of time long enough to allow the growth of a series of highstand ice-contact deltas along the ice-sea boundary (Smith and Hunter, 1989; Thompson et al., 1989; Crossen, 1991; Koteff et al., 1993). As isostatic rebound overcame eustatic sea-level rise, the retreating ice became detached from the shallow-marine water and sea level began to regress rapidly across the area (Fig. 2). Proglacial meltwater streams carried large quantities of sorted sand and gravel to the east-southeast, depositing coarser grained sediments (unit SDfc) at or near the marine margin, and transporting finer grained sands (unit SDff) and silt in a plume into the retreating nearshore environment (Ashley, 1995), where the sand was deposited as delta foresets (Fig. 17D). These fluvial sediments were deposited on top of the recently deposited glaciomarine mud. As the deltaic package prograded into the regressing sea, sediment and meltwater discharge fluctuations led to the deposition of foresets composed of silty and sandy cosets.

*Phase 3: Regression (ca. 12.4 ka)*

As sea level continued to retreat from its highstand position, fluviodeltaic sedimentation progressed across the Sanford-Kennebunk region (Fig. 17E). During early regression, the glacier was still near enough to supply copious quantities of sediment-laden meltwater to the area. The earliest sand and gravel beds were deposited in the inland zone and around high-relief features in the area, such as bedrock knobs, till hills, moraines, and ice-contact deltas, resulting in the overstepped stratigraphic signature instituted by mud deposition. These early deposits were traversed by a small number of meltwater streams as deposition prograded eastward. As streams adjusted to the rapidly changing base level produced by continued isostatic uplift and attendant sea-level drop, little planation and channel shifting occurred across the area (cf. Hart and Long, 1996). As these streams entered the intermediate zone, they preferentially retained positions to complete infilling of the large central basin in that zone (Fig. 4).

Once deposition into the intermediate zone depocenter was complete, the streams once again traversed the new SDff and SDfc deposits to reach the receding shoreline (Fig. 17F). This fluvial process ultimately resulted in a small number of relatively large deltaic lobes prograding into the marine environment. As the ice sheet retreated further from the area into New Hampshire, sediment supply and fluvial energy decreased; the sands forming the delta foresets became finer grained and better sorted. As the deltaic lobes prograded into the coastal zone, the decrease in proglacial sediment supply led to some reworking of the sand plain and an overall decrease in vertical thickness of the delta package. Reworking of the deltaic sands took two forms. As the shoreline underwent minor stillstands, small beach ridges, spits, and beach scarps (unit Brw) developed in response to wave action. Meltwater channels also began to cannibalize the sand-plain deposits, deeply incising the plain and carrying the well-sorted sand to the delta front for redeposition (cf. Hart and Long, 1996). When the ice was so far removed that it could no longer supply plentiful sediment or meltwater to the area, and the sea had regressed past the present coast into the Gulf of Maine, delta growth ceased, and fluvial incision became the predominant geomorphic agent (Fig. 17G).

## CONCLUSIONS

The Sanford-Kennebunk sand plain in southwestern Maine is a large, elongate, west-east–trending sand body of ~15 km length, and 5–8 km in width. The plain, with an overall area of ~125 km$^2$, is entirely within the zone of Pleistocene marine submergence. The maximum depth of the sand plain is ~30 m, and the total sediment thickness exceeds 73 m in a large structural basin under the central plain.

Calculations of sediment volume for the Sanford-Kennebunk sand plain reveal a total volume of ~1.5 × 10$^9$ m$^3$ for the sand and gravel facies of the plain. This value is comparable to the 1.3 × 10$^9$ m$^3$ volume of the Merrimack River lowstand paleodelta (Oldale et al., 1983), and may also be similar to the deltaic facies volume of the 2.1 × 10$^9$ m$^3$ regressive-lowstand Kennebec River paleodelta (Barnhardt et al., 1997).

The stratigraphy of the Sanford-Kennebunk area may be subdivided into two broad categories: ice-retreat sediments overlain by regressive marine deltaic sediments. The regional stratigraphy displays a basal fining-upward sediment pattern of till and ice-contact features underlying glaciomarine clay, deposited as the Laurentide ice sheet retreated from the area. Overlying this sequence, a conformable, coarsening-upward succession of fine- to medium- to coarse-grained sand and fine gravel records the growth of a large delta complex composed of proglacial sands. Some variability in the thickness, lateral continuity, grain size, and sorting of deltaic sands is evident throughout the sand plain, as is some deviation in the types of landforms found in each part of the plain.

The subsurface architecture of the Sanford-Kennebunk sand plain as indicated by GPR records is made up of large-scale accretionary surfaces interpreted to be progradational deltaic lobes formed during regression of the Atlantic Ocean in the late Pleistocene. These lobes form a large delta complex built toward the east and northeast by glacial meltwater-transported sediment. Average dip angles of foresets within the complex range from 4° to 15°, which are comparable to dip angles found in shallow-marine deltas. Such angles are too high to represent submarine fan or terrestrial outwash deposition, and too low to represent classic, Gilbert-delta development.

A model developed for the formation and evolution of the Sanford-Kennebunk sand plain indicates that the body was deposited into the regressing shallow-marine environment. This differs from previous models for the area, which use ice-contact features to outline chronology, ignoring the broad sand plain, or which suggest that the sand plain was deposited subaerially, or as a submarine fan. Further study may be required to answer questions pertaining to local deglacial timing and the location of the ice margin in terms of sediment supply, and to the origin and location of the meltwater channels that delivered sediment to the area.

## ACKNOWLEDGMENTS

We thank the Maine Geological Survey, particularly Tom Weddle and Craig Neil, for resources, field assistance, and stimulating debate. We also thank D.W. Caldwell and David Marchant for helpful discussions, and the people of the Laudholm Trust in Wells, Maine, for allowing the use of their facilities during field seasons. This paper has benefited from reviews by Craig Neil, Tom Weddle, and Greg Miller.

## REFERENCES CITED

Ashley, G.M., 1995, Glaciolacustrine environments, in Menzies, J., ed., Modern glacial environments; processes, dynamics and sediments: Oxford, United Kingdom, Butterworth-Heinemann, p. 417–444.

Ashley, G.M., Boothroyd, J.C., and Borns, H.W., Jr., 1991, Sedimentology of late Pleistocene (Laurentide) deglacial-phase deposits, eastern Maine; an example of a temperate marine grounded ice-sheet margin, in Anderson, J.B., and Ashley, G.M., eds., Glacial marine sedimentation; paleoclimatic significance: Geological Society of America Special Paper 261, p. 107–125.

Barnhardt, W.A., Gehrels, W.R., Belknap, D.F., and Kelley, J.T., 1995, Late Quaternary relative sea-level change in the western Gulf of Maine; evidence for a migrating glacial forebulge: Geology, v. 23, p. 317–320.

Barnhardt, W.A., Belknap, D.F., and Kelley, J.T., 1997, Stratigraphic evolution of the inner continental shelf in response to late Quaternary relative sea-level change, northwestern Gulf of Maine: Geological Society of America Bulletin, v. 109, p. 612–630.

Bates, C.C., 1953, Rational theory of delta formation: American Association of Petroleum Geologists Bulletin, v. 37, p. 2119–2162.

Belknap, D.F., and Shipp, R.C., 1991, Seismic stratigraphy of glaciomarine units, Maine inner shelf, in Anderson, J.B., and Ashley, G.M., eds., Glacial marine sedimentation; paleoclimatic significance: Geological Society of America Special Paper 261, p. 137–157.

Belknap, D.F., Andersen, B.G., Anderson, R.S., Anderson, W.A., Borns, H.W., Jr., Jacobson, G.L., Kelley, J.T., Shipp, R.C., Smith, D.C., Stuckenrath, R., Jr., Thompson, W.B., and Tyler, D.A., 1987, Late Quaternary sea-level changes in Maine, in Nummedal, D., et al., eds., Sea-level fluctuation and coastal evolution: Society of Economic Paleontologists and Mineralogists Special Publication 41, p. 71–85.

Bloom, A.L., 1960, Late Pleistocene changes of sea level in southwestern Maine: Augusta, Maine Geological Survey, 143 p.

Boothroyd, J.C., 1995, The Pine River and other esker systems of southwestern Maine and south-central New Hampshire: implications for regional deglaciation: Geological Society of America Abstracts with Programs, v. 27, no. 1, p. 31.

Caldwell, D.W., Hanson, L.S., and Thompson, W.B., 1985, Styles of deglaciation in central Maine, in Borns, H.W., Jr., et al., eds., Late Pleistocene history of northeastern New England and adjacent Quebec: Geological Society of America Special Paper 197, p. 45–58.

Crossen, K.J., 1991, Structural control of deposition by Pleistocene tidewater glaciers, Gulf of Maine, in Anderson, J.B., and Ashley, G.M., eds., Glacial marine sedimentation; paleoclimatic significance: Geological Society of America Special Paper 261, p. 127–135.

D'Amore, D., 1983, Hydrogeology and geomorphology of the Great Sanford outwash plain, York County, Maine: Boston, Massachusetts, Boston University Department of Geology, Environmental Hydrogeology Center Technical Report 3, 146 p.

Dominic, D.F., Egan, K., Carney, C., Wolfe, P.J., and Boardman, M.R., 1995, Delineation of shallow stratigraphy using ground-penetrating radar: Journal of Applied Geophysics, v. 33, p. 167–175.

Gehrels, W.R., Belknap, D.F., and Kelley, J.T., 1996, Integrated high-precision analyses of Holocene relative sea-level changes: Lessons from the coast of Maine: Geological Society of America Bulletin, v. 108, p. 1073–1088.

Hart, B.S., and Long, B.F., 1996, Forced regressions and lowstand deltas; Holocene Canadian examples: Journal of Sedimentary Research, v. 66, p. 820–829.

Hughes, T., Borns, H.W., Jr., Fastook, J.L., Hyland, M.R., Kite, J.S., and Lowell, T.V., 1985, Models of glacial reconstruction and deglaciation applied to Maritime Canada and New England, in Borns, H.W., Jr., LaSalle, P., and Thompson, W.B., eds., Late Pleistocene history of northeastern New England and adjacent Quebec: Geological Society of America Special Paper 197, p. 139–150.

Hunter, L.E., Powell, R.D., and Smith, G.W., 1996, Facies architecture and grounding-line fan processes of morainal banks during the deglaciation of coastal Maine: Geological Society of America Bulletin, v. 108, p. 1022–1038.

Hussey, A.M., II, 1989, Geology of southwestern coastal Maine, in Anderson, W.A., and Borns, H.W., Jr., eds., Neotectonics of Maine; studies in seismicity, crustal warping, and sea-level change: Maine Geological Survey Bulletin 40, p. 149–155.

Jol, H.M., and Smith, D.G., 1991, Ground-penetrating radar of northern lacustrine deltas: Canadian Journal of Earth Sciences, v. 28, p. 1939–1947.

Koteff, C., Robinson, G.R., Goldsmith, R., and Thompson, W.B., 1993, Delayed postglacial uplift and synglacial sea levels in coastal central New England: Quaternary Research, v. 40, p. 46–54.

Lanctot, E.M., and Tolman, A.L., 1985, Hydrogeologic data for significant sand and gravel aquifers in part of York County, Maine, map 2: Maine Geological Survey Open-File 85-91, scale 1:50 000.

McPherson, J.G., Shanmugam, G., and Moiola, R.J., 1987, Fan-deltas and braid-deltas; varieties of coarse-grained deltas: Geological Society of America Bulletin, v. 99, p. 331–340.

Miller, G.T., 1997, Deglaciation of Wells Embayment, Maine; interpretation from seismic and side-scan sonar data [M.S. thesis]: Orono, University of Maine, 231 p.

Montello, T.M., 1992, The stratigraphy and evolution of the Wells barrier system, Wells, Maine [M.A. thesis]: Boston, Massachusetts, Boston University, 220 p.

Nummedal, D., and Boothroyd, J.C., 1976, Morphologic and hydrodynamic characteristics of terrestrial fan environments: Kingston, University of Rhode Island Department of Geology, Coastal Research Division Technical Report 10-CRD, 61 p.

Oldale, R.N., 1989, Timing and mechanisms for the deposition of the glaciomarine mud in and around the Gulf of Maine: A discussion of alternative models, in Tucker, R.D., and Marvinney, R.G., eds., Studies in Maine geology; Volume 5, Quaternary geology: Augusta, Maine Geological Survey, p. 1–10.

Oldale, R.N., Wommack, L.E., and Whitney, A.B., 1983, Evidence for a postglacial low relative sea-level stand in the drowned delta of the Merrimack River, western Gulf of Maine: Quaternary Research, v. 19, p. 325–336.

Osberg, P.H., Hussey, A.M., II, and Boone, G.M., 1985, Bedrock geologic map of Maine: Augusta, Maine Geological Survey, scale 1:500 000.

Powell, R., and Domack, E., 1995, Modern glaciomarine environments, in Menzies, J., ed., Modern glacial environments; processes, dynamics and sediments: Oxford, United Kingdom, Butterworth-Heinemann, p. 445–486.

Prescott, G.C., and Drake, J.A., 1962, Records of selected wells, test holes, and springs in southwestern Maine: Groundwater series, Southwestern Area: U.S. Geological Survey, Maine Basic Data Report 1, 35 p.

Smith, D.G., and Jol, H.M., 1992, Ground-penetrating radar investigation of a Lake Bonneville delta, Provo level, Brigham City, Utah: Geology, v. 20, p. 1083–1086.

Smith, G.W., 1977, Surficial geology of the Kennebunk quadrangle, Maine: Maine Geological Survey Open-File Map 77-13, scale 1:62 500.

Smith, G.W., 1980, End moraines and glaciofluvial deposits, Cumberland and York Counties, Maine: Augusta, Maine Geological Survey map, scale 1:250 000.

Smith, G.W., 1981, Kennebunk glacial readvance: A reappraisal: Geology, v. 9, p. 250–253.

Smith, G.W., 1982, End moraines and the pattern of last ice retreat from central and south coastal Maine, in Larson, G.J., and Stone, B.D., eds., Late Wisconsinan glaciation of New England: Dubuque, Iowa, Kendall/Hunt Publishing, p. 195–209.

Smith, G.W., 1984, Glaciomarine sediments and facies associations, southern York County, Maine, in Hanson, L.S., ed., Guidebook for field trips in the coastal lowlands, Boston, Massachusetts to Kennebunk, Maine; New England Intercollegiate Geological Conference No. 76: Salem, Massachusetts, Salem State College, p. 352–369.

Smith, G.W., 1985, Chronology of late Wisconsinan deglaciation of coastal Maine, in Borns, H.W., Jr., et al., eds., Late Pleistocene history of northeastern New England and adjacent Quebec: Geological Society of America Special Paper 197, p. 29–44.

Smith, G.W., 1990a, Surficial geology of the Alfred quadrangle, York County, Maine: Maine Geological Survey Open-File 90-38, 9 p.

Smith, G.W., 1990b, Surficial geology of the Kennebunk quadrangle, York County, Maine: Maine Geological Survey Open-File 90-39, 9 p.

Smith, G.W., 1990c, Surficial geology of the North Berwick quadrangle, York County, Maine: Maine Geological Survey Open-File 90-40, 8 p.

Smith, G.W., 1990d, Surficial geology of the Wells quadrangle, York County, Maine: Maine Geological Survey Open-File 90-41, 8 p.

Smith, G.W., and Hunter, L.E., 1989, Late Wisconsinan deglaciation of coastal Maine, in Tucker, R.D., and Marvinney, R.G., eds., Studies in Maine geology; Volume 6, Quaternary geology: Augusta, Maine Geological Survey, p. 13–32.

Stuiver, M., and Borns, H.W., Jr., 1975, Late Quaternary marine invasion in Maine: Its chronology and associated crustal movement: Geological Society of America Bulletin, v. 86, p. 99–104.

Thompson, W.B., and Borns, H.W., Jr., eds., 1985, Surficial geologic map of Maine: Augusta, Maine Geological Survey, scale 1:500 000.

Thompson, W.B., Crossen, K.J., Borns, H.W., Jr., and Andersen, B., 1989, Glaciomarine deltas of Maine and their relation to late Pleistocene–Holocene crustal movements, in Anderson, W.A., and Borns, H.W., Jr., eds., Neotectonics of Maine; studies in seismicity, crustal warping, and sea-level change: Maine Geological Survey Bulletin 40, p. 43–63.

Thompson, W.B., Hildreth, C.T., Johnston, R.A., and Retelle, M.J., 1995, Glacial geology of the Portland–Sebago Lake area, in Hussey, A.M., II, and Johnston, R.A., eds., Guidebook for field trips in southern Maine and adjacent New Hampshire; New England Intercollegiate Geological Conference No. 87: Brunswick, Maine, Bowdoin College, p. 51–70.

Tolman, A.L., and Lanctot, E.M., 1985, Hydrogeologic data for significant sand and gravel aquifers in parts of York and Cumberland Counties, Maine, map 4: Maine Geological Survey Open-File 85-93, scale 1:50 000.

Tolman, A.L., Tepper, D.H., Prescott, G.C., and Gammon, S.O., 1982, Hydrogeology of significant sand and gravel aquifers, northern York and southern Cumberland Counties, Maine: Maine Geological Survey Open-File 83-1, 4 plates, map, scale 1:125 000.

van Heteren, S., FitzGerald, D.M., McKinlay, P.A., and Buynevich, I. V., 1998, Radar facies of paraglacial barrier systems, coastal New England, USA: Sedimentology, v. 45, p. 181–200.

MANUSCRIPT ACCEPTED BY THE SOCIETY JUNE 9, 2000

# Morainal banks and the deglaciation of coastal Maine

**Lewis E. Hunter**
*U.S. Army Cold Regions Research and Engineering Laboratory, Hanover, New Hampshire 03755, USA*
**Geoffrey W. Smith**
*Department of Geological Sciences, Ohio University, Athens, Ohio 45701, USA*

## ABSTRACT

Morainal banks (end moraines and associated ice-frontal features) are found in abundance below the limit of late glacial marine submergence throughout the coastal zone of Maine. These features formed during retreat of the late Wisconsinan Laurentide ice and record successive grounding-line positions. These thick sequences of glaciofluvial sediment document temperate ice conditions, and internal deformation records regular annual advance-retreat cycles of the ice margin. The morainal banks range from sharp-crested ridges to flat-topped ice-contact deltas. Composition varies from diamicton to glaciofluvial sediment, although most morainal banks are composed of admixtures of both. Regardless of size, shape, and composition, these features share a common origin with morainal banks forming at the grounding line of modern tidewater glaciers. Detailed study of these banks has led to the recognition of several lithofacies and lithofacies assemblages, and the development of a model for morainal bank formation. This model incorporates deposition related to meltwater jet trajectories at the glacier grounding line, remobilization of sediment by gravity-flow mechanisms, and glacial deformation associated with seasonal fluctuations of the grounding line. The distribution of morainal banks defines the pattern of the last ice retreat from the coastal zone, and indicates that retreat was regular and roughly parallel to the present coastline. Rates of ice retreat, calculated on the basis of morainal bank spacing, range from 0.05 to 0.13 km/yr, the highest rates occurring close to the present coast and lower rates occurring near the marine limit.

## INTRODUCTION

### Late Wisconsinan deglaciation of coastal Maine

Late Wisconsinan Laurentide ice advanced from the northwest across the Maine coastal zone to a terminal position on the continental shelf by ca. 18.0 ka (Borns, 1973; Hughes, 1987). Withdrawal of late Wisconsinan ice from its terminal position was underway between 17.0 and 15.0 ka, and the ice margin had retreated across the Gulf of Maine to a position roughly parallel, but some distance offshore of, the present coastline by 14.0 ka (Smith, 1985; Smith and Hunter, 1989). At 13.8 ka, the glacier margin had reached a position ~20 km south of Saco Bay (Kelley et al., 1986), and it continued to retreat at a reduced rate into the vicinity of Great Hill and Kennebunk (Fig. 1; Smith, 1985).

Continued ice retreat across the Maine coastal zone was accompanied by marine submergence. Retreat rates were reduced as a function of glacier stillstands or minor readvances as ice became grounded in shallow water. Ice withdrawal across the coastal zone was accompanied by rapid sedimentation and the production of morainal features, and esker-fed and beheaded ice-contact deltas (Gluckert, 1975; Smith, 1985; Crossen, 1991). These features outline in detail the pattern of ice withdrawal from the coastal zone (Smith, 1982).

The retreating ice front had reached a position above the

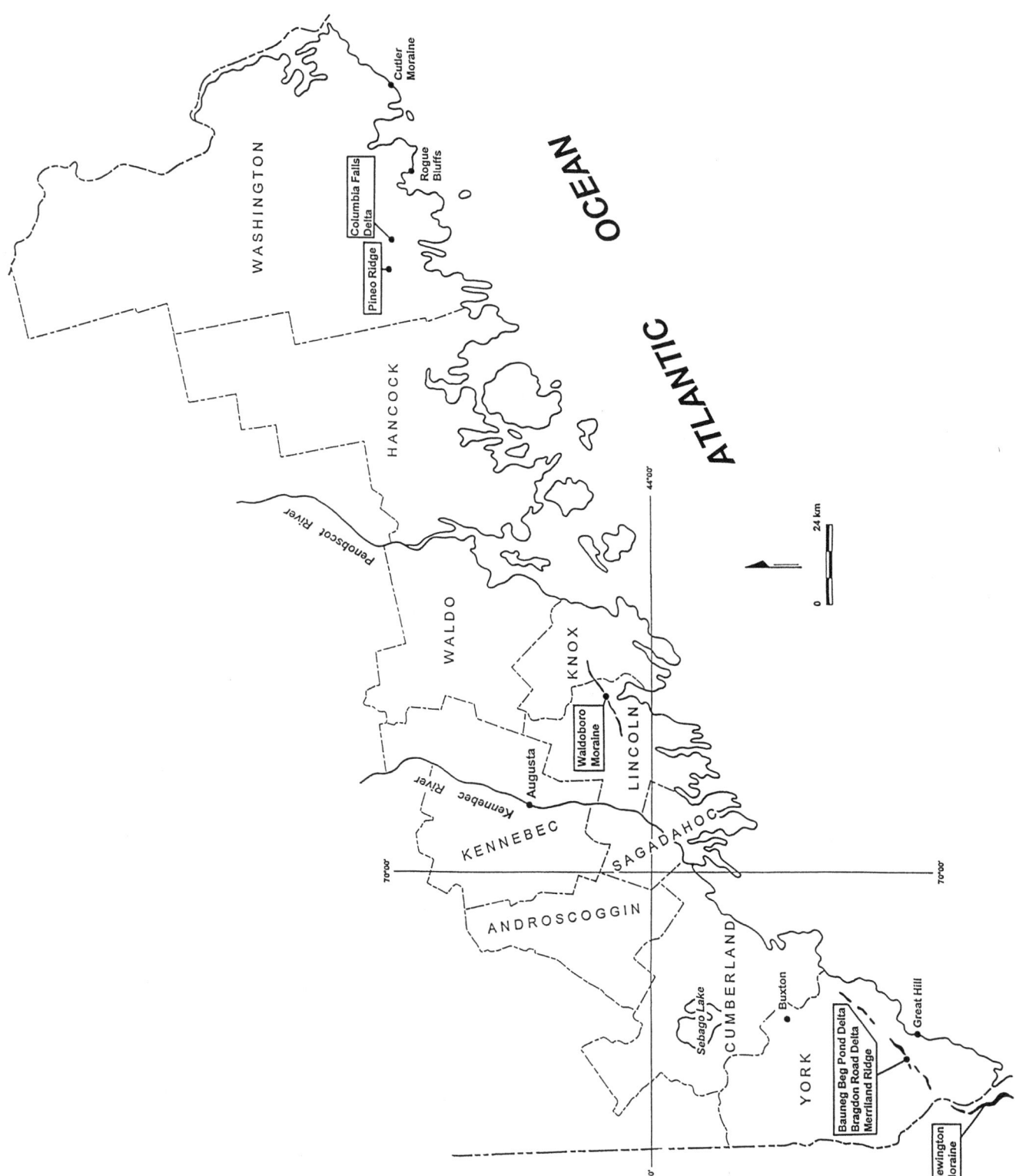

Figure 1. General location map of coastal Maine. Highlighted (boxed) features are discussed in detail and are illustrated elsewhere in text.

marine limit along its entire extent between 12.6 and 12.4 ka. Late glaciomarine submergence attained its maximum extent during this interval of time. With continued ice retreat and isostatic recovery, the coastal zone gradually became emergent. This phase of deglaciation was complete ca. 12.0 ka in eastern Maine and 11.5 ka in southern Maine.

This chapter provides a summary of current information regarding the formation of ice-marginal features during the late Wisconsinan deglaciation of coastal Maine. It also proposes the use of these features in calculating rates of last ice retreat from the coastal zone.

## Definition of morainal banks

A variety of features formed along the grounding line of the Laurentide ice sheet as it retreated across coastal Maine. These features exhibit complex sedimentary frameworks (i.e., composition and facies relationships) that record ice-proximal sediment and ice dynamics. Many of these features have been mapped and described as end moraines that range in size and shape from simple straight linear ridges to broad flat-topped deltas. Composition varies from till dominated (Bingham, 1981) to stratified sand and gravel (Smith, 1980, 1982; Thompson and Smith, 1988; Smith and Hunter, 1989; Retelle and Bither, 1989). To accommodate these differences, a wide variety of terms has been used to describe these features. The smaller moraines have been described as washboard moraines, De Geer moraines, and minor moraines, and the larger moraines have been called stratified moraines, delta moraines, ice-contact (or ice-frontal) deltas, and partially developed deltas.

We suggest that virtually all of these features share a common origin, having been constructed along the grounding line of a tidewater ice cliff during the last glacial retreat from the Maine coastal zone (Hunter et al., 1996c). Variations in size and shape simply reflect differences in the duration of grounding-line stability, rates of ice-cliff retreat, and rates of sediment delivery through the glaciofluvial system. Textural and structural variations in the sediments reflect proximity to primary sediment sources and the associated sediment dynamics with diverse debris entrainment and delivery mechanisms, especially near the grounding line.

We further suggest that, because these morainal features share a common origin, they should be referred to by a common term, i.e., morainal bank. There is clear precedent for using this term in the writings of Goldthwait (1925, 1949) and Leavitt and Perkins (1935), who described the features as "moraine banks." Moreover, the term morainal bank has been applied by a number of workers (e.g., Powell, 1981, 1984; Hunter et al., 1996a, 1996b; Seramur et al., 1997) to describe similar features produced at the grounding line of modern glaciers with tidewater termini. In accord with the usage of Powell (1981, 1984), we consider a morainal bank to be any sediment pile that accumulates at the grounding line, through a wide variety of depositional and deformational processes.

In this chapter we offer a summary of the state of current understanding of the morainal banks of coastal Maine, including their occurrence, composition, and genesis. We emphasize the idea that all of the features actually represent a continuum of grounding-line features with a common origin.

## Background

End moraines and associated ice-frontal features have been recognized in coastal Maine for nearly a century. Early observations include those of Stone (1899), who described the occurrence of the Waldoboro moraine (system) in central coastal Maine (Fig. 2), and suggested that it was possibly constructed along a marine-based ice front. Stone recognized a variety of glacial geomorphic features that he considered to record frontal positions of a retreating ice margin, many having been constructed into the late glacial sea. Among the features he described are terminal moraines, gravel massives, deltas, marine glacial deltas, valley drift, marine beds, overwash gravels, kames, and osars (eskers). On the basis of the occurrence and distribution of these features, Stone (1899, pl. XXXI) constructed a map (Fig. 3) depicting approximate lines of ice-frontal retreat (lines of synchronal retreat, p. 392) for much of the state of Maine. Although Stone's model of (last) ice retreat might evoke some debate among modern glacial geologists, the map he produced is remarkably close to one that is described herein and by others (Ridge et al., this volume; Smith, 1985). Implicit in Stone's map is the idea that deglaciation from coastal Maine was relatively simple, the ice margin being roughly parallel to the present coastline. Stone's map also depicts the development of calving embayments in major river valleys.

In a reconnaissance study, Katz and Keith (1917) described several discrete segments of morainal material (including till and washed sediments) between Saco-Biddeford, Maine, and Newbury, Massachusetts. They considered separate segments as part of a recessional moraine that they called Newington moraine (Fig. 4). Katz and Keith recognized that moraine segments were ice marginal, constructed in front of a retreating ice front in contact with the sea. They further considered the moraine to be late Wisconsinan age. Included among the segments of the Newington moraine are Merriland Ridge and the ice-marginal features to the north and west (Bragdon Road, Bauneg Beg Pond; Fig. 5). While there has been argument regarding the glaciomarine origin of this part of the Newington moraine, there is strong evidence that Merriland Ridge and the ridge segments at Bauneg Beg Pond and Bragdon Road are ice-contact marine (partial) deltas, and that they record grounding-line positions of late Wisconsinan ice (Smith, 1982). Each of the Newington moraine segments in Maine has been subsequently mapped (1970s–1980s) as ice-contact or morainal deposits (e.g., Hildreth, 1990; Smith, 1977a, 1977b, 1990a, 1990b, 1994b). There is little question that the separate segments mapped by Katz and Keith within the Newington moraine are in fact ice-marginal glaciomarine features that define

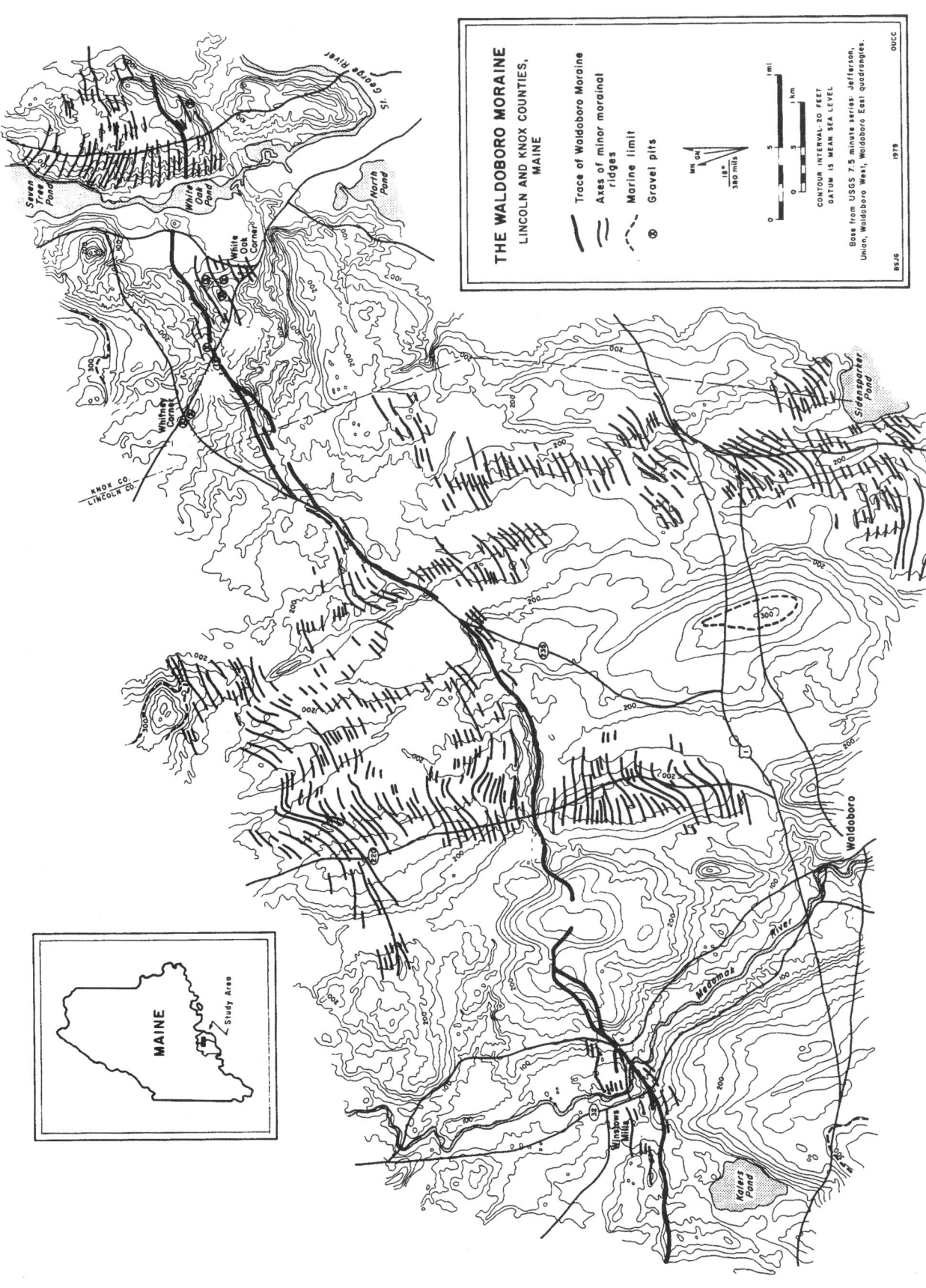

Figure 2. Waldoboro moraine, Lincoln and Knox Counties, Maine (from Smith, 1982).

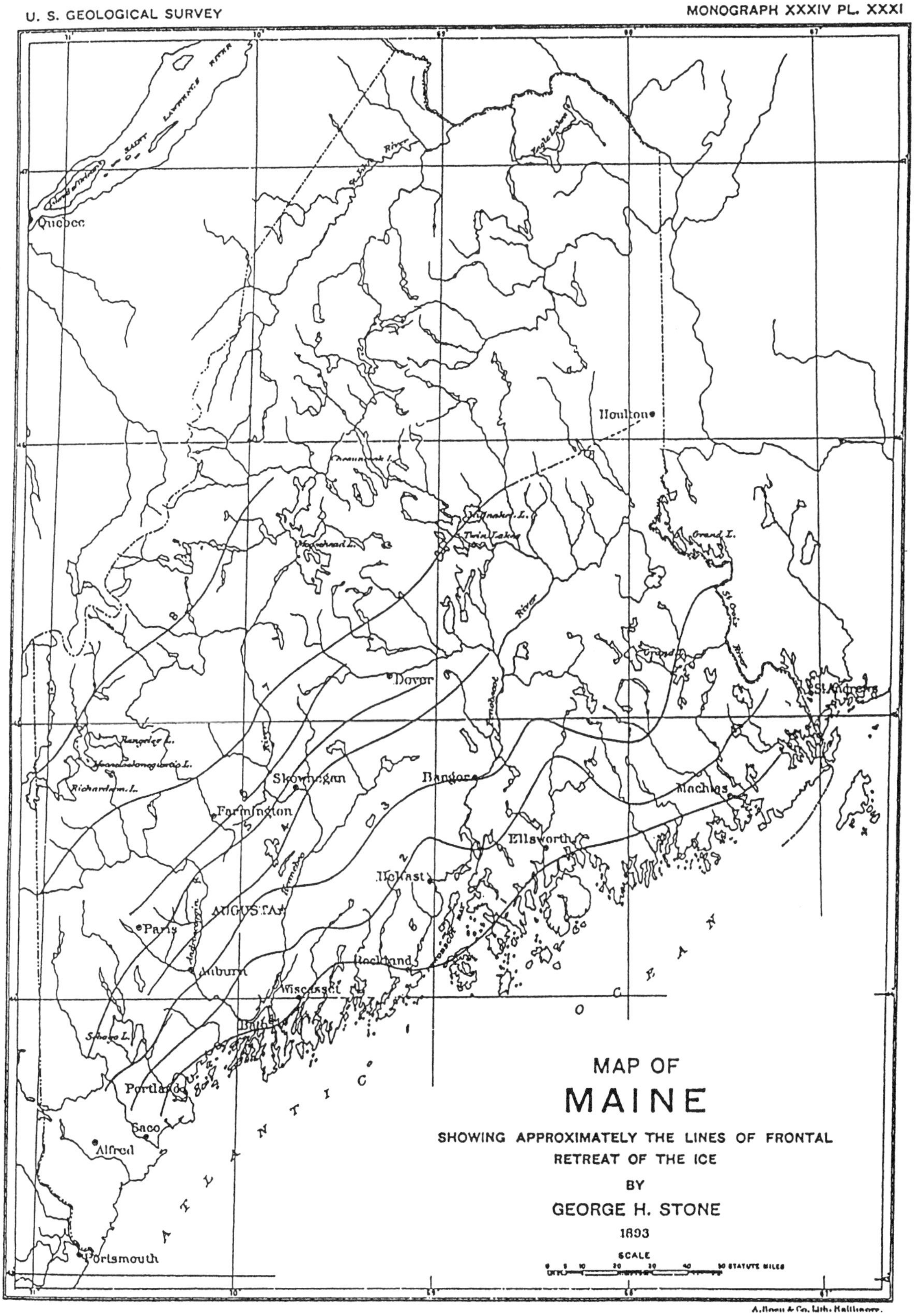

Figure 3. Ice-marginal positions of Laurentide ice during last retreat from coastal Maine (from Stone, 1899).

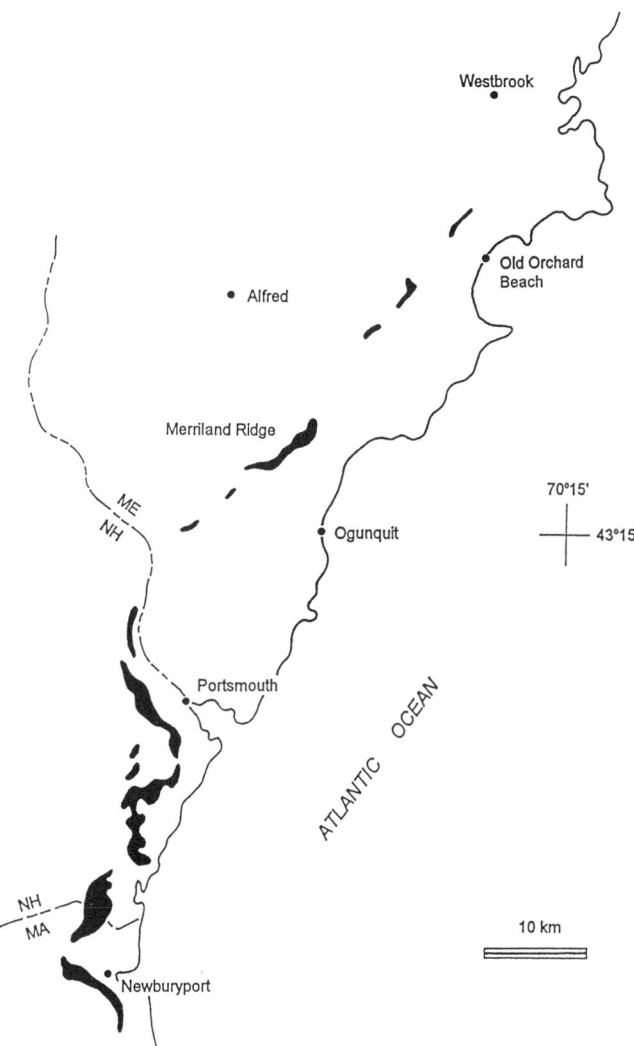

Figure 4. Occurrence of Newington moraine (in black) in coastal Maine, New Hampshire, and Massachusetts (after Katz and Keith, 1917).

several temporally separate recessional grounding-line positions during retreat of late Wisconsinan ice.

In a comprehensive study of the glacial materials in Maine, Leavitt and Perkins (1935) noted the occurrence of several features that they termed "terminal moraines," as well as a variety of features that they called "moraine banks." They described the Waldoboro moraine as being more extensive than previously noted, and they discussed the occurrence of the morainal features of eastern Maine (e.g., Pineo Ridge, the Columbia Falls delta, the Cutler moraine, the moraine at Roque Bluffs). Leavitt and Perkins also reevaluated the Newington moraine, and described several segments (Merriland Ridge, Bauneg Beg Pond moraine, Sebago Lake moraine) as moraine banks constructed along the grounding line of a tidewater ice front. Leavitt and Perkins were the first to recognize that most of the moraines in coastal Maine were composed of glaciofluvial sediments, not of till, as in the mid-continent.

Several of the more prominent features mapped and described by these early workers have been reexamined as part of the inventory mapping for the Maine Geological Survey (e.g., Pineo Ridge: Borns, 1973, 1980; Borns and Hughes, 1977; Miller, 1986; Merriland Ridge: Bloom, 1960; Smith, 1982; Waldoboro moraine: Thompson and Smith, 1988; Smith et al., 1982; Stemen, 1979; Jong, 1980). These features have been described and discussed in terms of ice-marginal deposits (moraines), and in some cases have been attributed to significant glacial readvances (Borns and Hughes, 1977). We now recognize that all of these features record deposition along the grounding line of a retreating tidewater ice margin. Compositional and morphological differences represent glacial and sedimentologic controls (Hunter et al., 1996c) on moraine construction that produce the spectrum of morainal deposits seen throughout coastal Maine. These moraines record the progressive regular retreat of a well-defined calving tidewater ice front without evidence of significant ice stagnation, and retreat of the margin without interruption by climatic changes that would have produced significant glacial readvance (Bloom, 1960; Smith, 1981; Borns, 1973; Miller, 1986).

Most of the features referred to as moraine banks (Goldthwait, 1925, 1949; Leavitt and Perkins, 1935) occur in eastern and southwestern Maine, and virtually all of them occur very close to the marine limit. Such features are absent from the central coastal zone, where DeGeer moraines are prevalent (Smith, 1982, 1985). Clusters of small moraines, previously also called washboard moraines, minor moraines, and DeGeer moraines, were first described in Maine by Bloom (1960, washboard moraines) in the vicinity of Buxton in southwestern Maine, where the moraines occur in clusters. Similar features were subsequently described by Borns (1980) in eastern coastal Maine, and were delineated in early open-file maps of the Maine Geological Survey (Fig. 6).

The abundance of small moraines was documented by Smith and Andersen (1975) in the Waldoboro area of central Maine. Subsequent mapping by Smith (1977b; Smith and Thompson, 1977; Thompson and Smith, 1978) showed them to be ubiquitous throughout the coastal zone. Although there remains some disagreement as to how widespread the occurrence of small moraines is in coastal Maine, it is clear that they are common and that they formed below the limit of late glaciomarine submergence (Smith, 1980, 1982). Stratigraphic relationships between morainal and marine sediments (Fig. 7) further demonstrate that the moraines were formed in the ice-proximal marine environment. Compositional variation within the DeGeer moraines, especially the significant amounts of stratified fluvial sediment, indicates that the retreating ice was warm based and discharged large volumes of sediment and water that fed bank-building processes (Smith and Hunter, 1989; Powell, 1981, 1991; Retelle and Bither, 1989; Hunter et al., 1996a, 1996c).

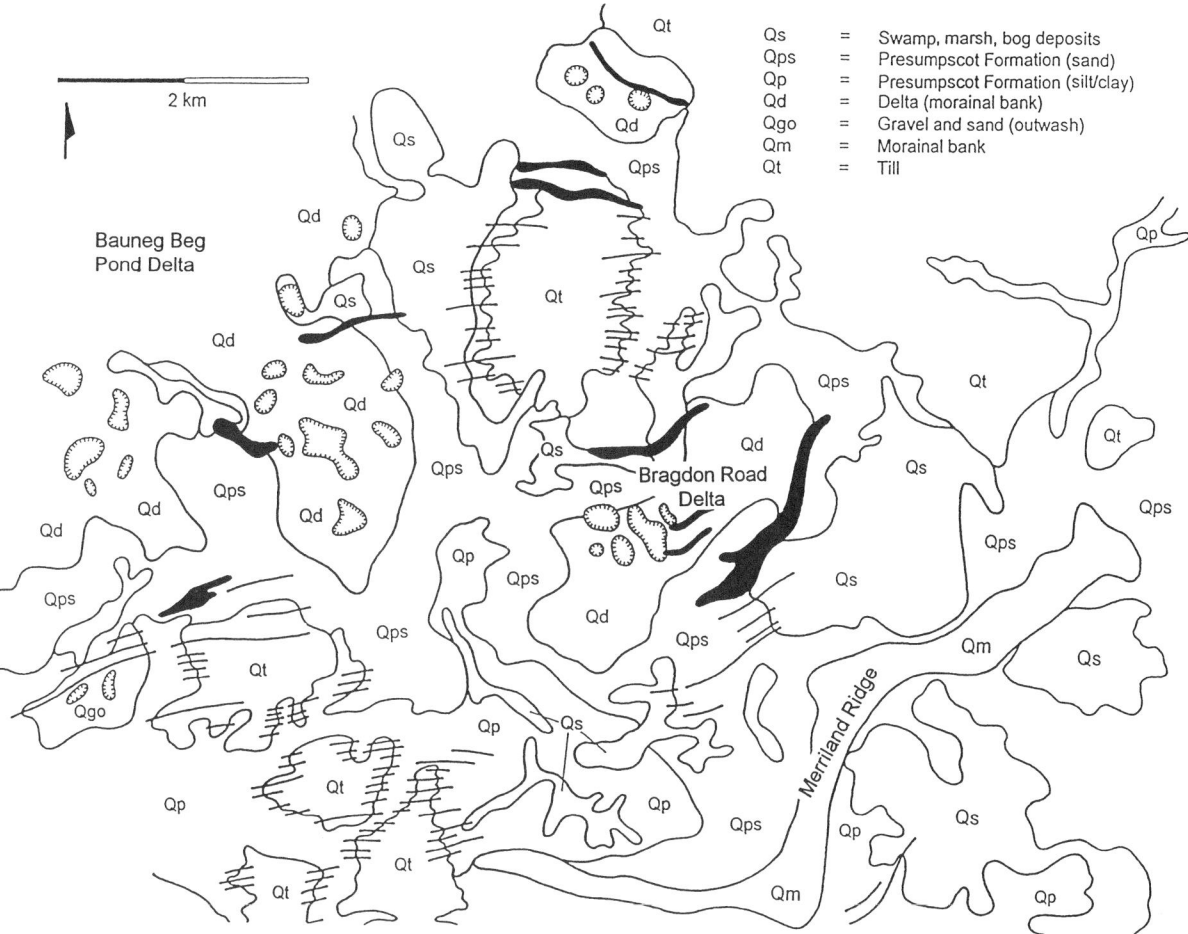

Figure 5. Surficial geology of portion of North Berwick 7.5-minute quadrangle, Maine (Smith, 1990a), showing occurrence of Merriland Ridge, Bragdon Road delta (morainal bank), and Bauneg Beg Pond delta (morainal bank).

## Coastal geomorphology

The Maine coastal zone is considered in this chapter to be roughly coincident with the limit of late glaciomarine submergence (Thompson and Borns, 1985). The zone is narrow in southwestern (20–25 km) and eastern (35–40 km) Maine, but widens to 175 km in central coastal Maine in the Penobscot River valley. Although much of this area is not coastal today, it is the area flooded during the marine transgression that accompanied retreat of late Wisconsinan ice. The topography of this area as it emerged from under the ice determined grounding-line water depth and the position of pinning points and calving embayments, and it likely influenced the pattern of subglacial drainage.

Original preglacial topography was buried by a blanket of glacial, proglacial, and late glaciomarine sediments as ice retreated from the area. However, that the position of the maximum extent of late glacial submergence coincides with the present Kennebec River and Penobscot River valleys shows that these were major preglacial valleys.

# MORAINAL BANKS OF COASTAL MAINE

## Distribution of morainal banks

The distribution of morainal banks throughout coastal Maine was described in previous publications (e.g., Smith, 1981) and is depicted on the map of the *Glacial Geology of Maine* (Thompson and Borns, 1985). Notable aspects of the distribution of these features are as follows.

1. The morainal banks, regardless of size, occur exclusively below the limit of late glaciomarine submergence. Stratified sand and gravel deposits that compose morainal banks are typically interbedded with, and overlain and/or underlain by marine deposits (e.g., the Presumpscot Formation; Bloom, 1960).

2. Linear sharp-crested morainal banks are most abundant, and most conspicuously developed in the central coastal zone (Lincoln, Knox, Sagadahoc Counties). Here, large and small morainal banks are interspersed with one another (Fig. 8). Morainal bank lithologies (e.g., till, sand and gravel) are

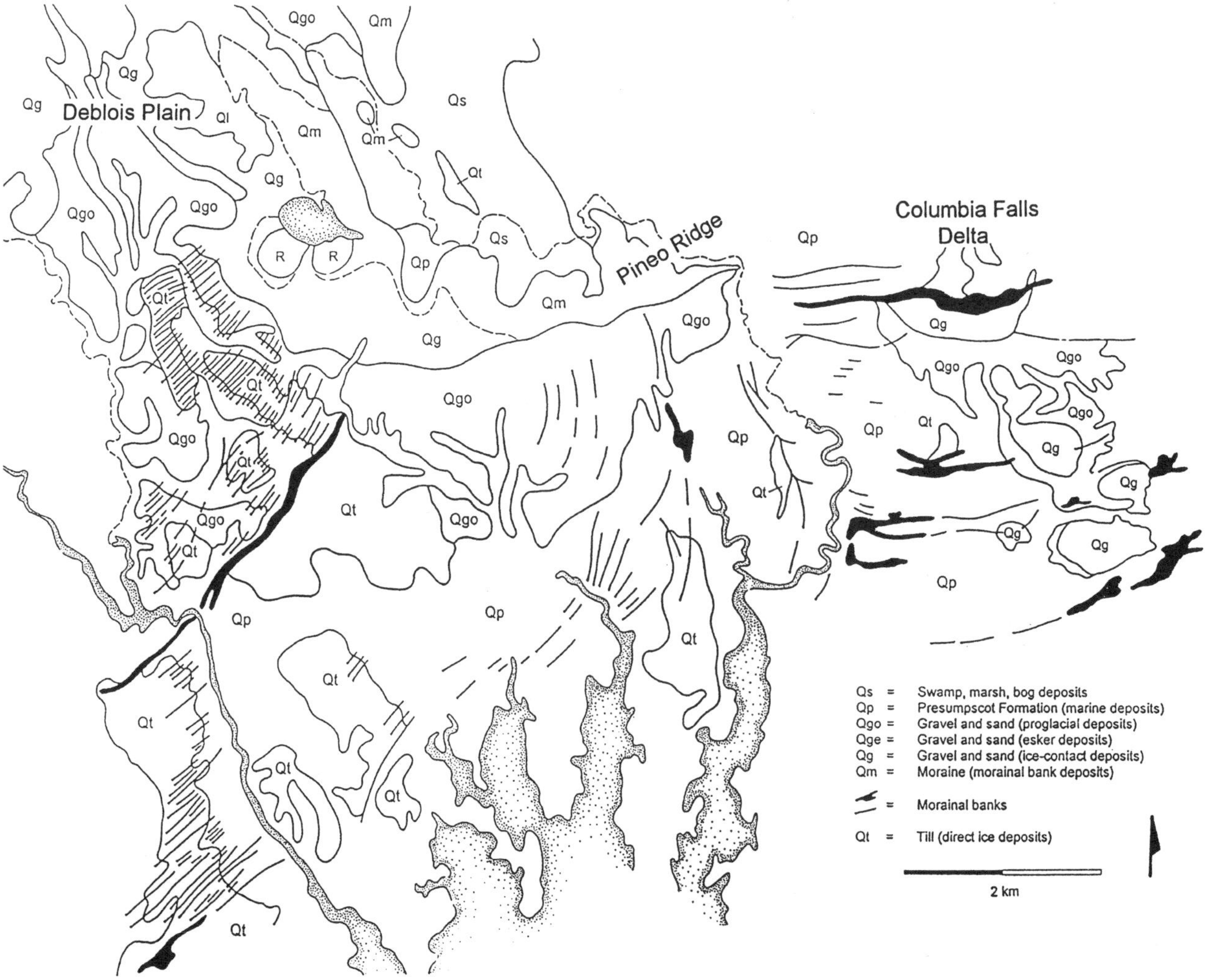

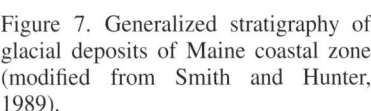

Figure 6. Surficial geology of portion of Cherryfield and Columbia Falls 7.5-minute quadrangles, Maine (Borns, 1975; Borns and Andersen, 1982), showing location of Pineo Ridge and Columbia Falls delta (morainal bank).

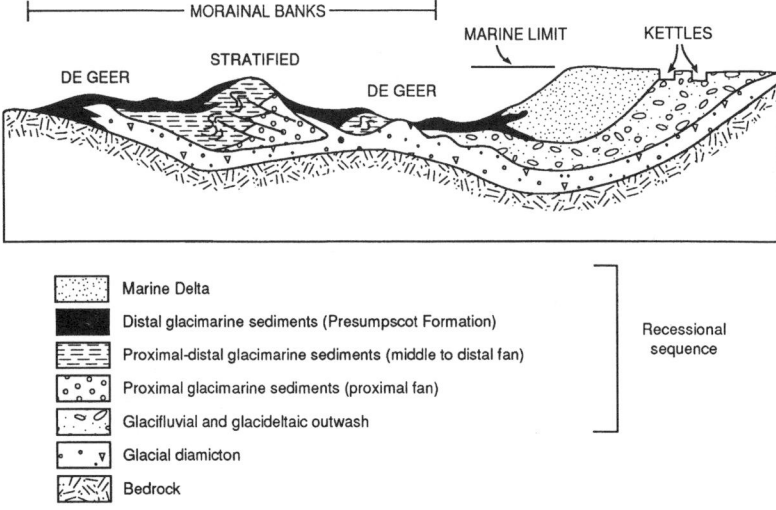

Figure 7. Generalized stratigraphy of glacial deposits of Maine coastal zone (modified from Smith and Hunter, 1989).

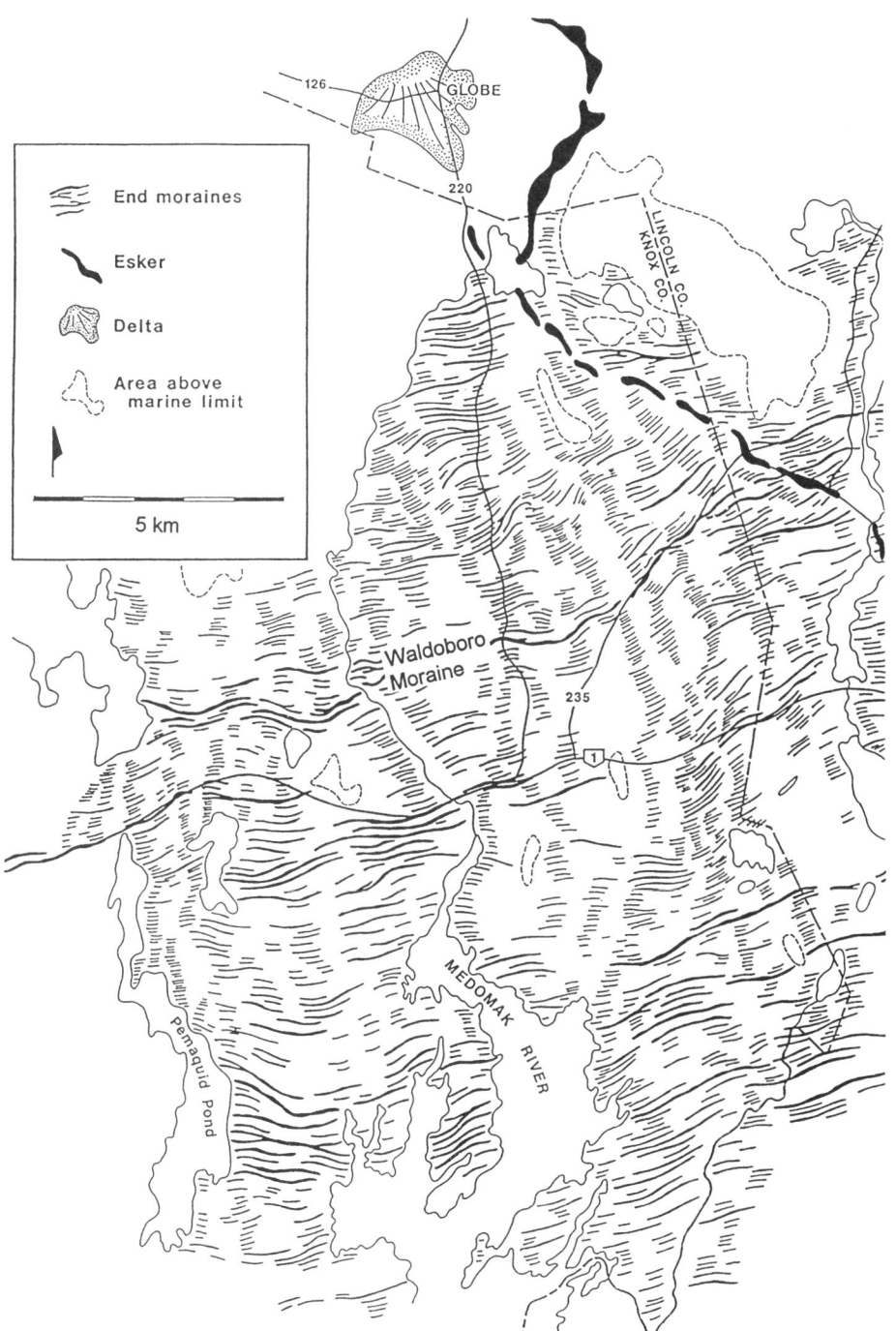

Figure 8. Distribution of morainal banks (end moraines) in central coastal Maine.

variable. Large flat-topped morainal banks are prominently developed in eastern and in southwestern coastal Maine. Ice-frontal deltas are common in positions of major glacial meltwater drainages along the limit of marine submergence (Smith, 1982; Thompson et al., 1989).

3. The axes of morainal banks are oriented parallel to one another and are aligned perpendicular to the direction of last ice movement as recorded by sediment fabrics and glacial striations (e.g., Smith, 1977b, 1981, 1994a, 1994b). There is little evidence of significant crosscutting relationships among morainal banks, but where they occur (Fig. 9), the relationship appears to reflect local pinning, differential calving causing more rapid and irregular retreat toward areas of deeper water (Smith, 1982).

4. On both large scale and small scales, morainal bank ridges conform to topography, and follow contours into valleys. That is, morainal banks curve upstream toward valley centers and toward the axes of eskers that mark the location of conduits

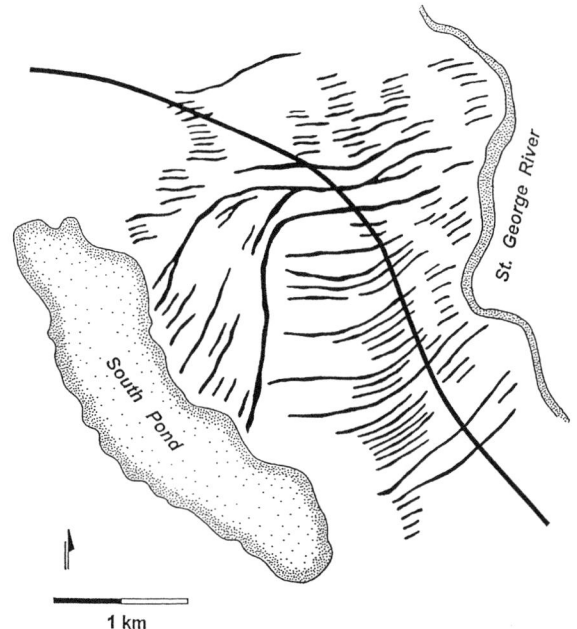

Figure 9. Detail of crosscutting relationships of morainal banks (black curvilinear lines) in vicinity of Warren (Knox County), Maine. Crosscutting relations produced by minor readvance of ice tongue in valley of South Pond (base map: Waldoboro East, Maine U.S. Geological Survey 7.5-minute topographic quadrangle).

Figure 10. Detail of concave-downvalley pattern of morainal banks in St. George River valley (Knox County). Figure illustrates pattern of morainal banks produced by development of small-scale calving embayment. Solid lines are 100 ft (~30 m) contour (base map: Union, Maine, U.S. Geological Survey 7.5-minute topographic quadrangle).

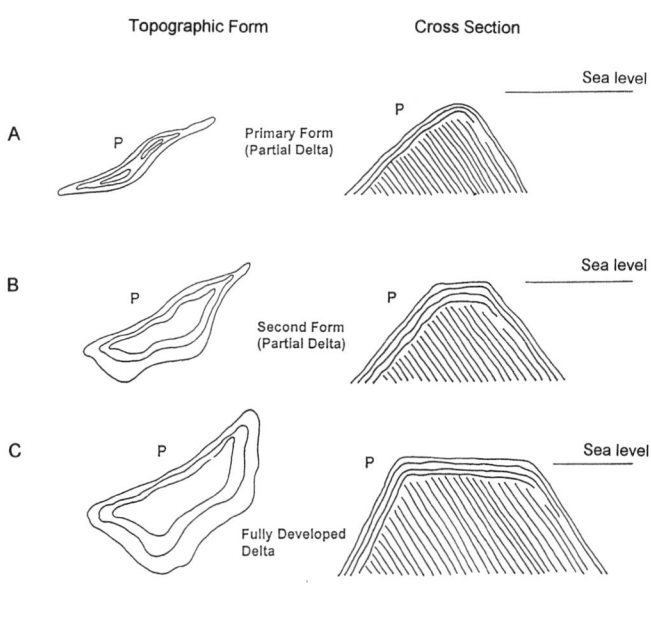

Figure 11. Varieties of ice-frontal delta. A, Primary form: sharp-crested ridge, not constructed to sea level. B, Second form: narrow, flat-topped ridge, constructed to sea level. C, Fully developed delta: broad, flat-topped surface, constructed to sea level (after Gluckert, 1975).

that discharged sediment and water at the grounding line (Fig. 10).

Evaluating the distribution of morainal banks leads us to the following general conclusions regarding how they formed.

The Laurentide ice margin made its final (late Wisconsinan) retreat in coastal Maine by tidewater calving into the Gulf of Maine. Shoaling occurred as the grounding line approached the limit of marine submergence, which resulted in slowed retreat and the buildup of larger morainal bank forms. Morainal banks are found seaward of the marine limit and exhibit interfingering relationships (Smith, 1981) between morainal and marine sediments (the Presumpscot Formation).

There is a lack of evidence for major climatically induced readvances during the retreat of last ice from the coastal zone, suggesting that retreat was consistent and uninterrupted.

Local variations in calving speed and retreat rates reflect changes in ice-flow dynamics (e.g., basal shear stress, internal strain rate) responding to local variations in subglacial topography and corresponding grounding-line water depth (see discussion in Van der Veen, 1997). Local variation around pinning points resulted in the development of calving embayments along the retreating ice margin. This condition determined not only the pattern of morainal banks, but also their spacing and their morphology (sharp crested vs. flat topped).

## Morphology of morainal banks

In simplest terms, the morainal banks of coastal Maine have two distinctly different end-member morphologies. Most of the banks are simple sharp-crested linear ridges, whereas others are flat-topped ridges that grade into true delta forms. The sharp-crested ridges are typical of features that have been described by many authors in a variety of geologic and geographic settings with a variety of different names (e.g., see Elson, 1968).

The sharp-crested morainal banks are generally found in clusters of 2–30 or more. The greatest numbers and largest sizes tend to be found in stream valleys, lowlands, or against the lower slopes of topographic highs (Smith, 1982; Jong, 1980; Stemen, 1979). Morainal banks vary in height from <1 m to >7 m (local relief above covering sediments), but most are commonly 3–4 m in height; width is typically 6–18 m, and most are between 12 and 18 m (Jong, 1980). Individual segments are generally 0.5–1.0 km in length, and seldom exceed 1.5 km in length, although some ridges are as much as 3 km long (Jong, 1980). Cross-sectional profiles of the morainal banks are predominantly asymmetric; distal slopes are consistently steeper ($20°$–$35°$) than are proximal slopes ($15°$–$30°$). As a general rule, asymmetry increases with the size of the morainal bank. This is converse to the occurrence of morainal banks in southeast Alaska, where distal slopes are typically more gentle (Powell, 1981, 1991; Hunter et al., 1996b, 1996c). The occurrence in coastal Maine may be attributed to wave erosion being concentrated on the distal slopes during marine regression.

The morainal banks of coastal Maine display a diversity in ridge form similar to that of the cross-valley moraines described by Andrews and Smithson (1966) on Baffin Island. The common forms include: (1) simple linear, (2) s-shaped, (3) bifurcate (branched), and (4) joined (adjacent ridges joined by transverse ridge). Commonly, morainal bank segments consist of interconnected conical or subconical mounds of glacial sediment rather than smooth-crested ridges. Bedrock is often exposed within morainal bank segments.

Flat-topped morainal banks (the moraine banks of Goldthwait, 1925, 1949, and Leavitt and Perkins, 1935) range in form from linear features that are marginally flat topped (Merriland Ridge; Fig. 5) to well-developed delta forms with well-defined delta plain surfaces and steep distal slopes (Pineo Ridge; Fig. 6). These features are morphologically and sedimentologically similar to the Salpuasselka moraines of Finland (Gluckert, 1975, 1977; Smith, 1982). Gluckert (1975, 1977; Fig. 11) described a sequence of features, which he termed "ice-frontal deltas," that are transitional between sharp-crested morainal banks (primary form) and deltas (second form, fully developed form). In southwestern Maine, Smith (1982) described this transition in considering the occurrence and genesis of Merriland Ridge (Fig. 5), the Bragdon Road morainal bank, and the Bauneg Beg morainal bank (ice-contact delta).

In eastern Maine, the Pineo Ridge complex and associated features (Fig. 6) display a sequence of morainal bank features that developed along an interlobate (calving embayment) ice margin, similar to the features described for southwestern Maine. Sharp-crested morainal banks (primary forms of Gluckert; Fig. 11) are transitional to flat-topped morainal banks (second forms of Gluckert; e.g., Columbia Falls delta), and then to ice-frontal deltas (fully developed forms of Gluckert; e.g., Pineo Ridge). Pineo Ridge is a complex delta feature (Miller, 1986; Ashley et al., 1991) that continued to develop headward (northwestward) as ice withdrew above the marine limit.

Throughout the coastal zone, flat-topped morainal banks occur as ice-frontal deltas, formed at the position of the inland limit of marine submergence (e.g., Crossen, 1991; Thompson et al., 1989). These features will not be treated further here, except to indicate that they generally represent the fully-developed delta form of Gluckert (1975, 1977).

The size and shape of morainal banks result from changing conditions of water depth, the available sediment supply and delivery rate, and the duration of grounding-line stillstand. The sharp-crested morainal banks are most common near the present coastline and throughout the central coastal zone, where they record conditions of relatively deep water and short intervals of grounding-line stillstand or readvance (Smith, 1976; Smith and Andersen, 1975; Thompson and Smith, 1978). Conversely, the larger flat-topped morainal banks reflect conditions of relatively shallow water and prolonged grounding-line stillstands (Smith, 1982). Morphology of morainal banks in this latter group is also a function of the degree of development of a full delta form (Gluckert, 1975, 1977). The morainal bank form is a function of water depth and duration of stillstand of the grounding line.

## Composition of morainal banks

The stratigraphic makeup of morainal banks in coastal Maine reflects the variability of sedimentary processes involved in their formation. Some morainal banks consist entirely of diamicton, typically remobilized and resedimented (Bingham, 1981; Figs. 12, 13, and 14), but these are relatively rare features. Some morainal banks consist entirely of stratified sediments (Fig. 15), which are also relatively rare. The great majority of morainal banks are composed of varying admixtures of diamicton and stratified coarse to fine clastic sediments that are often glacially deformed (Figs. 16 and 17).

Detailed studies of the internal character of morainal banks in coastal Maine were conducted by Hunter (1989), Smith and Hunter (1989), Retelle and Bither (1989), and Hunter et al. (1996c). As a result of these studies, it has been possible to define major lithofacies within morainal banks as: (1) diamicton; (2) gravel; and (3) sand, each with several subfacies. Evaluation and interpretation of these lithofacies lead to the conclusion that many, if not most, of the morainal banks in coastal

Figure 12. Small sharp-crested morainal bank in east-central coastal Maine. View is to west along the axis of a moraine that is ~3 m high. Ice-proximal slope is to right (north). Note pronounced asymmetry of morainal bank.

Figure 13. Diamicton composing morainal bank illustrated in Figure 12. Much of diamicton in this morainal bank has been remobilized by gravity flow (flow till). Knife, 22 cm. for scale.

Figure 14. Small morainal bank in central coastal Maine. This morainal bank is composed entirely of diamicton that has been dumped at ice margin. Large boulder in center foreground is ~2.5 m in greatest diameter.

Figure 15. Morainal bank in vicinity of Waldoboro moraine. View is obliquely toward proximal slope of morainal bank. Stratified glacimarine sediments have been recumbently folded; overturning is toward distal slope of morainal bank. Clipboard for scale.

Figure 16. Large morainal bank in eastern coastal Maine. View is along strike of bank axis. Diamicton and stratified glacifluvial sediments have been folded and sheared by ice push from left (north). Exposed face is ~16 m high.

Figure 17. Morainal bank in southwestern coastal Maine. View is along strike of bank axis. Stratified glaciofluvial sediments have been folded and sheared by ice push from left (north).

Maine are composed of facies associations (assemblages) listed in Figure 18 (see Smith and Hunter, 1989; Retelle and Bither, 1989; Hunter et al., 1996c).

Recent studies conclude that most morainal banks consist of sediments that are derived from stream discharges at the grounding line of a marine-based ice sheet accompanying marine transgression (Smith, 1982; Smith and Hunter, 1989; Hunter et al., 1996c; Retelle and Bither, 1989). The coarse and fine stratified deposits of the morainal banks were originally described by Smith (1982) as subaqueous outwash (Rust and Romanelli, 1975), and subsequently as subaqueous fan deposits (Smith and Hunter, 1989). Recognition of distinct facies associations led to the conclusion that much of the sediment composing morainal banks consists of remobilized (gravity flow) sediment (Smith and Hunter, 1989; Hunter et al., 1996c). Diamicton within the morainal banks generally makes up a carapace over stratified sediments or is folded or sheared into the gravel-sand-silt lithofacies in the cores of the banks (Smith, 1982; Smith and Hunter, 1989).

The deposits of most morainal banks display significant glaciotectonic deformation, in addition to sediment disruption by gravity-flow processes and subsequent compaction (dewatering) processes (e.g., Hunter et al., 1996c). Glaciotectonic deformation typically involves low-angle thrusting and recumbent folding of sediments. The sense of deformation is invariably normal to the axes of morainal banks and always from proximal to distal slopes of morainal banks.

Models that describe the sedimentary processes involved in the formation of morainal banks have evolved significantly over the past two decades, as the study of the internal character of the morainal banks became more detailed and comprehensive. Current models (Smith and Hunter, 1989; Retelle and Bither, 1989; Hunter et al., 1996c) depict morainal bank formation in terms of the development of submarine fans at points of sediment and water discharge along the grounding line. Sediment gravity flows redistribute sediment within developing fans, while minor fluctuations of the grounding line deform the entire sedimentary package and modify the initial morainal bank morphology. Where water is shallow, fans may develop to produce partial deltas or full deltas once a morainal bank has aggraded to sea level, as suggested by Gluckert (1975, 1977; Fig. 11) for portions of the Salpausselka, and by Smith (1982) for morainal banks in southwestern coastal Maine.

These models have been further refined through the study of modern tidewater glaciers with tidewater margins in southeastern Alaska (Powell, 1981, 1991; Hunter et al., 1996a, 1996b; Seramur et al., 1997). The analogy between the formation of morainal banks in coastal Maine and similar features in southeastern Alaska was discussed by Hunter et al. (1996c). Depositional models based on analogy with modern tidewater glaciers in the tidewater environment focus on the role of meltwater jet dynamics at the tidewater terminus (Fig. 19). The meltwater jet trajectory near the seafloor at the terminus and the rate of jet decay away from the subglacial conduit are the primary controls on the development of the facies architecture near the fan apex in a developing bank. Sedimentary facies within this setting, as applied to the morainal banks of coastal Maine, include those that are produced directly by ice (diamicton facies), those related to the meltwater jet (gravel facies, sand facies, fine facies), and those developed as a result of sediment remobilization (high-density gravity flow). Within this depositional framework, the morainal bank process-based facies associations shown in Figure 20 have been recognized and can be applied to most of the morainal banks that have been studied in the Maine coastal zone (Hunter et al., 1996c).

## DISCUSSION

### Pattern of last ice retreat

The pattern defined by traces of morainal banks throughout the coastal zone not only reflects the influence of bathymetry and proximity to stream effluences, but faithfully outlines the successive positions of the retreating ice margin (grounding line) of the last ice sheet as it withdrew from coastal Maine. This fact was recognized by Stone (1899) more than a century ago; since then, the delineation of morainal banks as discrete features across the coastal zone has resulted in a somewhat different interpretation of ice-marginal positions than those proposed by Stone or by Katz and Keith (1917). The distribution of morainal banks, coupled with their limited length, makes it difficult to trace any single ice-marginal position for any great distance through the coastal zone. However, patterns of morainal banks within clusters (Thompson and Borns, 1985) clearly indicate the general orientation of ice-marginal positions across the coastal zone (Fig. 8).

The pattern of last ice retreat from coastal Maine is best considered in sections, as follows: (1) eastern coastal Maine, (2) the central coastal zone, and (3) southwestern coastal Maine. A composite model for ice retreat can then be constructed from evaluation of features in these sections.

### Eastern coastal Maine

The morainal banks of eastern coastal Maine consist of clusters of numerous small sharp-crested ridges (DeGeer moraines) and a number of large flat-topped morainal banks (stratified moraines, fans, and deltas). Both large and small sharp-crested ridges are dominant forms from the present coastline to

---

Glacial diamicton association
Jet contact association
Buoyant jet association
High density sediment gravity flow association

Figure 18. List of process-based facies associations.

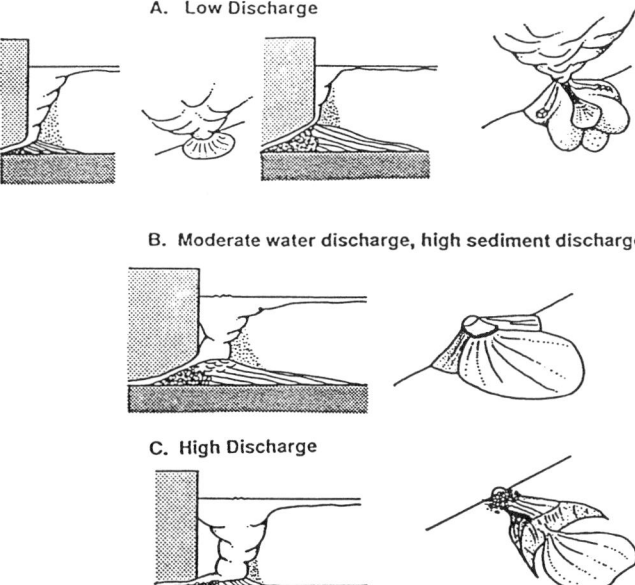

Figure 19. Model of meltwater plumes and fan types that develop under different discharge conditions (from Powell, 1990). A, At low discharges, fan grows through coarse traction deposition near efflux with sediment gravity flows and suspension settling introducing sediment to distal fan environments. B, Under higher sediment discharge conditions, jet remains in traction, producing rapid sedimentation deposits in sheets or crude scour-and-fill structures. C, At high discharge, migrating barchanoid bars can be produced at detachment zone

---

Diamicton/proximal fan facies association
Subaqueous (submarine) fan and plain facies association
Subaqueous (submarine) end moraine facies association
Marine (glaciomarine) mud association, the shallow marine facies assemblage of Retelle and Bither (1989)
Marine delta association

---

Figure 20. List of facies associations.

a position near the marine limit. East of Pineo Ridge (Fig. 6), small sharp-crested morainal banks occur from the present coastline to the marine limit. In the vicinity of Pineo Ridge and the Columbia Falls delta, sharp-crested moraines are replaced along the marine limit by large flat-topped morainal banks that record extended periods of glacial stillstand and sediment accumulation. East of the Columbia Falls delta (Fig. 6), morainal banks trend uniformly east-northeast to west-southwest. The head of the Columbia Falls delta and the north (ice proximal) side of Pineo Ridge have the same trend. The morainal banks to the south and west of Pineo Ridge trend north-northeast to south-southwest, and are clearly crosscut by Pineo Ridge. This situation led some (Borns and Hughes, 1977) to consider Pineo Ridge as the record of a significant glacial readvance. Subsequent study of the Pineo Ridge system (Miller, 1986; Ashley et al., 1991) indicates that the crosscutting relationship of the moraines was caused by the establishment of two ice lobes that were produced during the development of a small calving embayment, where more rapid calving led to the generation of a lobate form in the vicinity of Pineo Ridge (Ashley et al., 1991).

The large flat-topped morainal banks of the eastern coastal zone head in a network of extensive esker systems (Fig. 21) and related ice-contact deposits that can be traced northward for a considerable distance above the marine limit (Ashley et al, 1991). The eskers are, without question, the source of sediment that makes up not only the large morainal banks, but also many of the sharp-crested morainal banks that occur between Pineo Ridge and the present coastline. The pattern of esker occurrence is also important in the reconstruction of patterns of the last ice retreat from the coastal zone.

To the west and northwest of Pineo Ridge, morainal banks have not been mapped (Thompson and Borns, 1985). Independent study indicates that morainal banks occur in significant numbers west of Pineo Ridge into the Penobscot River embayment. These morainal banks display trends that record a gradual shift from the north-northeast to south-southwest orientation observed south and west of Pineo Ridge to a northwest to southeast orientation of morainal banks in the vicinity of the eastern Penobscot River valley. There is no suggestion of a persistent north-northeast to south-southwest trend of morainal banks, as might be expected if a calving embayment continued to develop west of Pineo Ridge, thereby isolating an ice mass in eastern Maine from the remainder of ice retreating across the coastal zone west of the Penobscot River.

The pattern of esker systems (Fig. 21) in eastern coastal Maine likewise suggests that last ice retreat was not characterized by persistent lobation and subsequent development of a separate ice lobe in the area east of the Penobscot River. Whether we regard the eskers as being short-lived time-transgressive features (Hebrand and Amark, 1989) or the deposits of continuous well-integrated tunnel systems (Shreve, 1972, 1985), it is reasonable to expect that lobation of the retreating ice-sheet margin would be recorded in deflection of trends of the esker systems to conform with the shape of the progressively developing ice margin (and thus points of stream discharge). Such is not the case in coastal Maine. Although there are several well-defined branched esker systems that terminate in the coastal zone, each system (Thompson and Borns, 1985) appears to maintain a persistent direction of tunnel development throughout its mapped occurrence within the coastal zone. This is clearly the case for the eskers of the eastern coastal zone (see also Ashley et al., 1991).

Superimposition of mapped patterns of clusters of morainal banks and esker systems (Fig. 21) places constraints on the reconstruction of ice-marginal positions during the period of ice withdrawal from the eastern coastal zone. In our opinion, it is highly unlikely, given available geomorphological and sedimentological field evidence, that a well-developed separate ice lobe occurred in any significant and sustained fashion much beyond the western edge of Pineo Ridge and the Deblois Plain (Miller, 1986).

### *Central coastal zone*

Morainal banks are most abundant in the central coastal zone, where they are predominantly sharp crested (Fig. 8). Here, the flat-topped forms occur only as deltas at the marine limit. Morainal bank size varies from the small DeGeer-type forms to large features like the Waldoboro moraine.

The trends of morainal bank axes are uniformly east-northeast to west-southwest, oblique to the present coastline. Slight departures from this trend are local, and reflect the development of small-scale calving embayments within stream valleys. The persistent trend of morainal banks throughout the central coastal zone indicates that while there was development of ice-marginal embayments in the Penobscot and Kennebec River valleys, there was no strongly embayed ice margin during the last retreat through the central coastal zone.

Esker systems in this part of the coastal zone are more irregular in orientation that they are elsewhere (Fig. 21), particularly in the area between the Penobscot and Kennebec River valleys (Thompson and Borns, 1985). In several instances, esker ridges parallel the trends of morainal banks. For the most part, however, esker segments define integrated drainage systems that are part of either the Kennebec-Androscoggin system or the Penobscot system.

Superimposition of esker trends with trends of morainal banks in the central coastal zone (Fig. 21) indicates that ice retreat through this part of Maine was regular, with the progressive development of calving embayments in the valleys of the Penobscot and Kennebec Rivers.

### *Southwestern coastal Maine*

The occurrence of morainal banks in southwestern coastal Maine is very similar to that for morainal banks in eastern coastal Maine. Sharp-crested forms predominate from the present coast to a position near the marine limit. At the marine limit, flat-topped morainal banks predominate (Figs. 5 and 6). The transition from sharp-crested forms to flat-topped forms reflects decreasing water depth and retreat rates as the ice margin approached the marine limit. Submarine fans evolved into partial delta (Fig. 11) or full delta forms (Merriland Ridge, Bragdon Road morainal bank). Large (pitted) deltas (Bauneg Beg Pond morainal bank) were constructed at the marine limit, as ice became land based.

The Bauneg Beg Pond morainal bank complex heads in several esker systems and glaciofluvial valley fills, much like the network that heads Pineo Ridge in eastern Maine (Figs. 6 and 21). The extensive area of ice-contact deposits north and west of Bauneg Beg Pond indicates that ice retreat above the marine limit was accomplished, at least in large part, by downwasting. This condition may reflect the influence of the White Mountains in partitioning ice in western Maine.

The trends of morainal bank axes in southwestern coastal Maine are uniformly east-northeast to west-southwest, to east to west. That trend does not change from the present coast to the marine limit. Throughout this part of the coastal zone, morainal banks can be traced to the marine limit, where they are buried by deltas constructed into the late glacial sea at the marine limit.

Germane to a consideration of the pattern of last ice retreat, as reconstructed from the pattern of morainal banks, is the existence of two general models have been published depicting the pattern of last ice retreat from coastal Maine. One model is based primarily on limiting radiocarbon dates (Thompson and Borns, 1985, after Davis and Jacobson, 1985; Denton and Hughes, 1981). The other is based primarily on morphological evidence of features (e.g., morainal banks, eskers, striations) mapped in the field (Smith, 1982, 1985). The two models differ significantly in how they depict the magnitude of lobation of the retreating ice margin, and the implied rates of ice retreat (see following discussion). At the crux of this disparity is the fact that old dates (>13 ka) have been documented for shells in basal (?) marine sediments in the upper reaches of the Kennebec (Embden Pond; Stuiver and Borns, 1975) and the Penobscot River drainages. In order to fit these dates into the general collection of dates from the coastal zone, Thompson and Borns (1985) depicted an ice margin at 13 ka that is prominently lobate in both the eastern and southwestern parts of the state, with a broad reentrant coincident with the limit of marine submergence in the central coastal zone. The problem of trying to delineate patterns of ice retreat in conformance with radiocarbon dates continues today. What appears to be lost in that approach is a consideration of available geomorphological and sedimentological evidence that would suggest a more regular (less lobate) pattern of ice retreat.

### *Rates of last ice retreat*

On the assumption that the nature of glacial sedimentation and morainal bank construction in the Maine coastal zone is broadly similar to that in modern tidewater glacial conditions (Hunter et al., 1996b, 1996c), we propose that the morainal banks of coastal Maine were formed during annual summer-retreat–winter-advance cycles. If it is accepted that the morainal banks, at least in general terms, represent annual constructional events, it is possible to employ the spacing of morainal banks in any area to approximate rates of ice retreat.

Jong (1980) determined that the average spacing between morainal banks in central coastal Maine (Lincoln and Knox Counties) is ~70 m, implying a rate of ice-margin retreat of 0.07 km/yr. Subsequent work (Smith, 1982) suggests that mo-

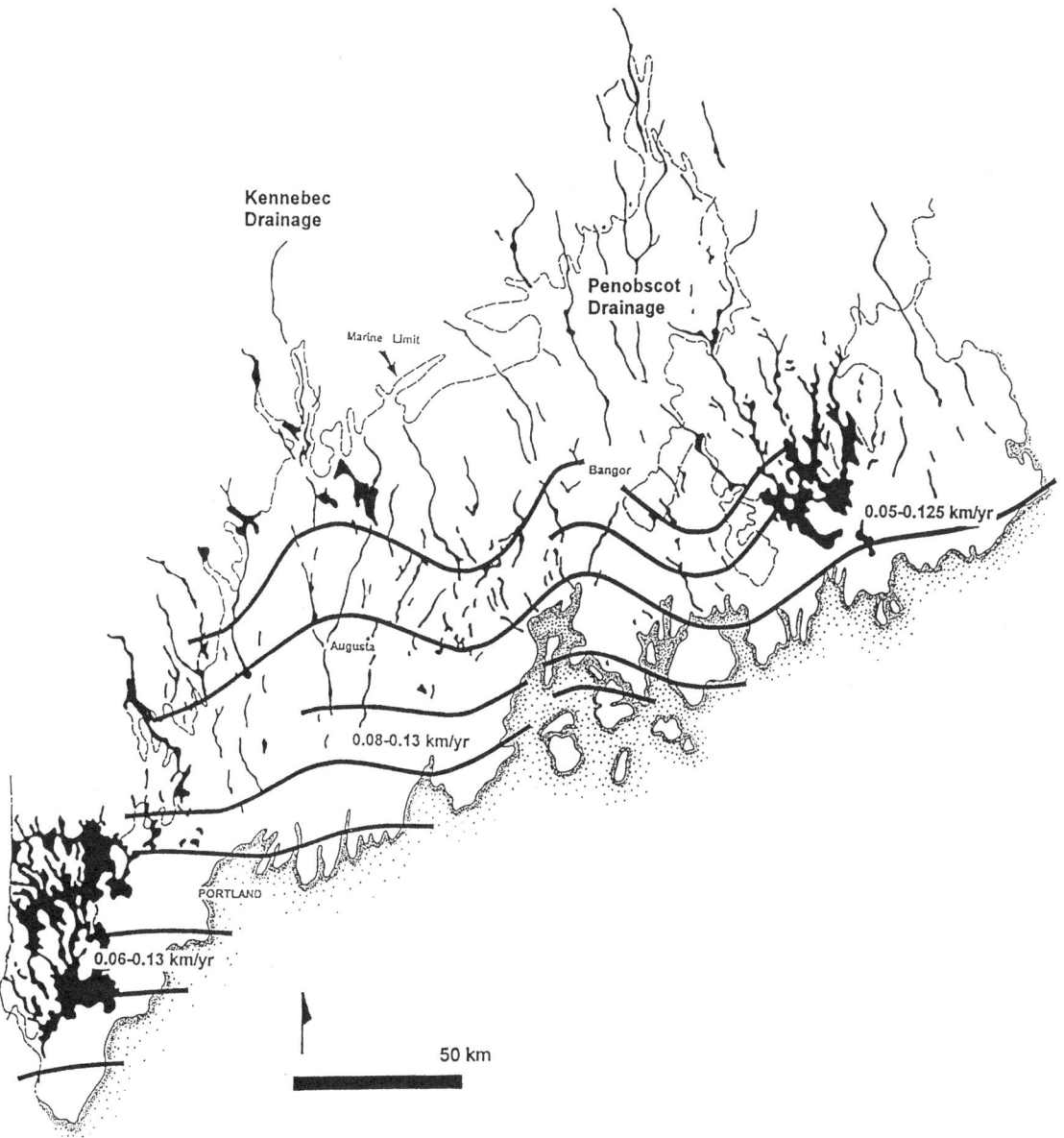

Figure 21. Simple model for last ice retreat from coastal Maine. Figure shows distribution of eskers and related glaciofluvial deposits in Maine coastal zone (after Thompson and Borns, 1985). Proposed ice-frontal positions (heavy east-west–trending lines) are based on patterns of morainal banks and trends of eskers in Maine coastal zone. Ice-frontal positions, as drawn, have no time significance. Numbers represent rates of ice retreat throughout Maine coastal zone. Rates are based on average moraine spacing within each section of coastal zone. Inland marine limit is indicated by thin dashed line.

rainal bank spacing is less regular when a larger portion of the central coastal zone is considered, and that 70 m is a minimal spacing for the morainal banks. As a result, we propose that retreat rates vary between 0.07 km/yr and 0.13 km/yr in the central coastal zone. Wider morainal bank spacing, hence a greater retreat rate, is found in major valleys, and within the Penobscot-Kennebec embayment. Spacing of morainal banks decreases regularly toward the marine limit. In general, it is then reasonable to suggest that ice retreat throughout much of the Penobscot-Kennebec embayment was on the order of 0.10–0.13 km/yr, and that rates of marginal retreat decreased to ~0.07 km/yr toward the marine limit, both east and west of the embayment and at the head of the embayment.

With regard to retreat rates, larger (e.g., the Waldoboro moraine) morainal banks are interspersed between groups of smaller morainal banks throughout the central coastal zone (Figs. 2 and 8). Spacing of these morainal banks is less regular than that of the smaller morainal banks, ranging between 1 km

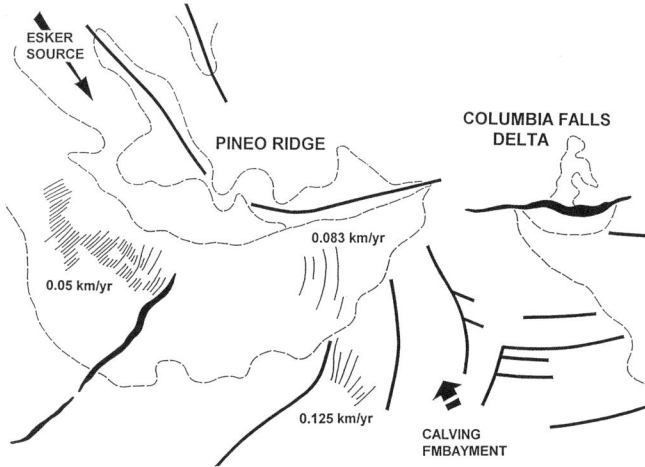

Figure 22. Suggested rates of ice retreat in vicinity of Pineo Ridge. In general, retreat rates decrease toward marine limit.

and 3 km. These larger morainal banks very likely record intervals of prolonged stillstand or readvance of the generally retreating grounding line. It is therefore likely that these rates of retreat should be regarded as minimal rates.

Rates of ice retreat in eastern coastal Maine (Fig. 22) vary between 0.05 km/yr and 0.125 km/yr. There is a clear trend toward closer spacing of morainal banks and a reduced rate of ice retreat in the vicinity of the marine limit.

Hunter (1989; Hunter et al., 1996c) documented retreat rates in southwestern Maine of between 0.06 km/yr and 0.13 km/yr, the slowest rates (0.06–0.07 km/yr) being close to the marine limit. This same trend occurs throughout Cumberland and York Counties (Smith, 1980), where retreat rates are estimated to range from 0.11 km/yr near the present coast to 0.09 km/yr near the marine limit.

The rates of retreat determined for marine-grounded ice (0.05–0.07 km/yr) are very similar to rates (0.07 km/yr) proposed by Thompson (1998) for terrestrial ice retreat in western Maine and rates (0.05 km/yr) proposed by Dorion (1998) for terrestrial ice retreat in northern and central Maine. Apparently, adjustment within the margin zone of the retreating ice sheet was accomplished relatively quickly in the transition from marine-based to terrestrially based conditions.

## CONCLUSIONS

As a general model for last retreat of Laurentide from coastal Maine, we propose the following. (1) Ice withdrew roughly parallel to the present coastline, with annual summer-retreat–winter-advance cycles that produced the morainal banks of the coastal zone. (2) Small and large calving embayments developed along the ice margin in major stream valleys, the most prominent occurring in the central Penobscot-Kennebec lowland. (3) Rates of ice retreat were greatest (0.13 km/yr) where water was deepest and calving rates were high (the central coastal zone). (4) Rates of ice retreat diminished (0.05–0.07 km/yr) in the vicinity of the marine limit, and were maintained at approximately the same level above the marine limit. (5) As retreat rates decreased and ice withdrew above the marine limit in eastern and southwestern Maine, ice continued to retreat at a relatively rapid rate in the central coastal zone until it also withdrew above the marine limit.

The idea of a simple model of ice retreat (Fig. 21) from coastal Maine may accommodate what appears to be a serious problem in reconciling existing radiocarbon dates with available geomorphological and sedimentological field evidence. Given the proposed retreat rates and the general sequence of proposed events, it is reasonable to suggest that the time involved between the attainment of terrestrially based ice conditions in eastern and southwestern Maine and the attainment of similar conditions in central Maine could be between 1000 and 1500 yr. Until such time as issues of the reliability of uncorrected marine radiocarbon dates and marine reservoir effects (Ridge et al., 1998; this volume) can be resolved, we suggest that patterns of ice retreat based on radiocarbon chronologies be considered very carefully.

The work of Stone (1899) produced a model of the pattern of last ice retreat (Fig. 3) that is remarkably similar to that proposed in this chapter. This pattern of ice retreat is characterized by its simplicity, ice retreat generally paralleling the present coastline, and progressing without significant ice lobation or partitioning of separate ice masses. Note that Stone's model was based entirely on geomorphological and sedimentological field evidence, without benefit of modern techniques and procedures.

## ACKNOWLEDGMENTS

We thank Walter Anderson and Harold Borns, Jr., for the insight and dedicated effort that led to the recognition of the importance of surficial geology in Maine, and the implementation of the surficial mapping program for Maine. Field work was facilitated by logistical support from the Maine Geological Survey and by assistance from the survey staff, notably Woodrow Thompson and Thomas Weddle. The development of ideas regarding the morainal banks has benefited from discussions with Harold Borns, Jr., Woodrow Thompson, Thomas Weddle, Michael Retelle, and Ross Powell, among others. Assistance in the field was provided by Lynne Hunter. Partial funding for field work was provided by the Houk Fund of Ohio University. We thank Jeffrey Strasser and an anonymous reader for their critical reviews, and Mark Hardenberg for editorial suggestions.

## REFERENCES CITED

Andrews, J.T., and Smithson, E.B., 1966, Till fabrics of the cross-valley moraines, north-central Baffin Island, Northwest Territories, Canada: Geological Society of America Bulletin, v. 77, p. 271–292.

Ashley, G.M., Boothroyd, J.C., and Borns, H.W., Jr., 1991, Sedimentology of late Pleistocene (Laurentide) deglacial-phase deposits, eastern Maine; an example of a temperate marine grounded ice margin, *in* Anderson, J.B., and Ashley, G.M., eds., Glacial marine sedimentation; paleoclimatic significance: Geological Society of America Special Paper 261, p. 107–125.

Bingham, M.P., 1981, The structure and origin of washboard moraines and related marine sediment in southeastern coastal Maine [M.S. thesis]: Orono, University of Maine, 79 p.

Bloom, A.L., 1960, Late Pleistocene changes of sea level in southwestern Maine: Augusta, Maine Geological Survey, 134 p.

Borns, H.W., Jr., 1973, Late Wisconsin fluctuations of the Laurentide ice sheet in southern and eastern New England, *in* Black, R.F., et al., eds., Wisconsinan stage: Geological Society of America Memoir 136, p. 37–45.

Borns, H.W., Jr., 1975, Reconnaissance surficial geology of the Columbia Falls quadrangle, Maine: Maine Geological Survey Open-File Report 75-1, scale 1:62 500.

Borns, H.W., Jr., 1980, Glaciomarine geology of the eastern coastal zone: Orono, Maine, American Quaternary Association, 6th Biennial Conference Field Trip Guide, Trips A and D, 18 p.

Borns, H.W., Jr., and Andersen, B., 1982, Reconnaissance surficial geology of the Cherryfield quadrangle, Maine: Maine Geological Survey, Open-File Report 82-2, scale 1:62 500.

Borns, H.W., Jr., and Hughes, T., 1977, The implications of the Pineo Ridge readvance in Maine: Géographie Physique et Quaternaire, v. 31, p. 203–206.

Crossen, K.J., 1991, Structural control of deposition by Pleistocene tidewater glaciers, Gulf of Maine, *in* Anderson, J.B., and Ashley, G.M., eds., Glacial marine sedimentation; paleoclimatic significance: Geological Society of America Special Paper 261, p. 127–135.

Davis, R.B., and Jacobson, G.L., 1985, Late-glacial and early Holocene landscapes in northern New England and adjacent areas of Canada: Quaternary Research, v. 23, p. 341–368.

Denton, G.H., and Hughes, T.J., eds., 1981, The last great ice sheets: New York, Wiley-Interscience, 484 p.

Dorion, C.C., 1998, Style and chronology of deglaciation in central and northern Maine: Geological Society of America Abstracts with Programs, v. 30, no. 1, p. 15.

Elson, J.A., 1968, Washboard moraines and other minor moraine types, *in* Fairbridge, R.W., ed., The encyclopedia of geomorphology: New York, Reinhold Book Corporation, p. 1213–1219.

Gluckert, G., 1975, The second Salpausselka at Karkkila, southern Finland: Geological Society of Finland Bulletin 47, p. 45–53.

Gluckert, G., 1977, On the Salpausselka ice-marginal formations in southern Finland: Zietschrift für Geomorphologie, v. 27, p. 79–88.

Goldthwait, L., 1925, The geology of New Hampshire: New Hampshire Academy of Sciences Handbook No. 1, 86 p.

Goldthwait, L., 1949, Clay survey, 1948: Maine Development Commission Report of the State Geologist 1947–1948, p. 63–69.

Hebrand, M., and Amark, M., 1989, Esker formation and glacier dynamics in eastern Skane and adjacent areas, southern Sweden: Boreas, v. 18, p. 67–81.

Hildreth, C.T., 1990, Surficial geology of the Biddeford quadrangle, Maine: Maine Geological Survey Open-File Report 90-36, scale 1:24 000.

Hughes, T., 1987, Ice dynamics and deglaciation models when ice sheets collapsed, *in* Ruddiman, W.F., and Wright, H.E., Jr., eds., North America and adjacent oceans during the last deglaciation: Boulder, Colorado, Geological Society of America, Geology of North America, v. K-3, p. 183–220.

Hunter, L.E., 1989, Late Wisconsinan deglaciation of the Bar Mills 7.5-minute quadrangle, York County, Maine [M.S. thesis]: Athens, Ohio University, 161 p.

Hunter, L.E., Powell, R.D., and Lawson, D.E., 1996a, Flux of debris transported by ice at three Alaskan tidewater glaciers: Journal of Glaciology, v. 42, p. 123–135.

Hunter, L.E., Powell, R.D., and Lawson, D.E., 1996b, Morainal bank sediment budgets and their influence on the stability of tidewater termini of valley glaciers entering Glacier Bay, Alaska, U.S.A.: Annals of Glaciology, v. 22, p. 211–216.

Hunter, L.E., Powell, R.D., and Smith, G.W., 1996c, Facies architecture and grounding-line fan processes of morainal banks during the deglaciation of coastal Maine: Geological Society of America Bulletin, v. 108, p. 1022–1038.

Jong, R.S., 1980, Small push moraines in central coastal Maine [M.S. thesis]: Athens, Ohio University, 75 p.

Katz, F.J., and Keith, A., 1917, The Newington moraine, Maine, New Hampshire, and Massachusetts: U.S. Geologicial Survey Professional Paper 108-B, p. 11–29.

Kelley, J.T., Kelley, A.R., Belknap, D.F., and Shipp, R.C., 1986, Variability in the evolution of two adjacent bedrock-framed estuaries in Maine, *in* Wolfe, D.A., ed., Estuarine variability: New York, Academic Press, p. 21–42.

Leavitt, H.W., and Perkins, E.H., 1935, A survey of road materials and glacial geology of Maine: Volume 11, Glacial geology of Maine: Maine Technology Experiment Station Bulletin 30, 232 p.

Miller, S.B., 1986, History of the glacial landforms in the Deblois region, Maine [M.S. thesis]: Orono, University of Maine, 74 p.

Powell, R.D., 1981, A model for sedimentation by tidewater glaciers: Annals of Glaciology, v. 2, p. 129–134.

Powell, R.D., 1984, Guide to the glacial geology of Glacier Bay, Southeast Alaska: Anchorage, Alaska Geological Survey, 85 p.

Powell, R.D., 1990, Glacimarine processes at grounding-line fans and their growth to ice-contact deltas, *in* Dowdeswell, J.A., and Scourse, J.D., eds., Glacimarine environments, processes and sediments: Geological Society [London] Special Publication 53, p. 53–73.

Powell, R.D., 1991, Grounding-line systems as second-order controls on fluctuations of tidewater termini of temperate glaciers, *in* Anderson, J.B., and Ashley, G.M., eds., Glacial marine sedimentation: Paleoclimatic significance: Geological Society of America Special Paper 261, p. 75–93.

Retelle, M.J., and Bither, K.M., 1989, Late Wisconsinan glacial and glaciomarine sedimentary facies in the lower Androscoggin Valley, Topsham, Maine, *in* Tucker, R.D., and Marvinney, R.G., eds., Studies in Maine geology, Volume 6: Quaternary geology: Augusta, Maine Geological Survey, p. 33–52.

Ridge, J.C., Canwell, B.A., Callahan, J.W., Cook, G.J., Nicholson, R.S., Toll, N.J., and Kelly, M., 1998, Varve, paleomagnetic, and radiocarbon chronologies for the deglaciation of northern New England: Geological Society of America Abstracts with Programs, v. 30, no. 1, p. 70.

Rust, B.R., and Romanelli, R., 1975, Late Quaternary subaqueous outwash deposits near Ottawa, Canada, *in* Jopling, A.V., and McDonald, B.C., eds., Glaciofluvial and glaciolacustrine sedimentation: Society of Economic Paleontologists and Mineralogists Special Publication 23, p. 177–192.

Seramur, K.C., Powell, R.D., and Carlson, P.R., 1997, Evaluation of conditions along the grounding line of temperate marine glaciers: An example from Muir Inlet, Glacier Bay, Alaska: Marine Geology, v. 140, p. 307–327.

Shreve, R.L., 1972, Movement of water in glaciers: Journal of Glaciology, v. 11, p. 205–214.

Shreve, R.L., 1985, Late Wisconsin ice-surface profile calculated from esker paths and types, Katahdin esker system, Maine: Quaternary Research, v. 23, p. 27–37.

Smith, G.W., and Andersen, B., 1975, Reconnaissance surficial geology of the Waldoboro East quadrangle, Maine: Maine Geological Survey Open-File Report 75-25, scale 1:24 000.

Smith, G.W., 1976, Reconnaissance surficial geology of the Waldoboro West quadrangle, Maine: Maine Geological Survey Open-File Report 76-39, scale 1:24 000.

Smith, G.W., 1977a, Reconnaissance surficial geology of the Berwick quadrangle, Maine: Maine Geological Survey Open-File Report 77-9, scale 1:62 500.

Smith, G.W., 1977b, Reconnaissance surficial geology of the Kennebunk quadrangle, Maine: Maine Geological Survey Open-File Report 77-13, scale 1:62 500.

Smith, G.W., and Thompson, W.B., 1977, Reconnaissance surficial geology of the Buxton quadrangle, Maine: Maine Geological Survey Open-File Report 77-19, scale 1:62 500.

Smith, G.W., 1980, End moraines and glaciofluvial deposits, Cumberland and York Counties, Maine: Maine Geological Survey Open-File Map, scale 1:126 720.

Smith, G.W., 1981, The Kennebunk glacial advance: A reappraisal: Geology, v. 9, p. 250–253.

Smith, G.W., 1982, End moraines and the pattern of last ice retreat from central and south coastal Maine, in Larson, G.J., and Stone, B.D., eds., Late Wisconsinan glaciation of New England: Dubuque, Iowa, Kendall/Hunt Publishing, p. 195–209.

Smith, G.W., 1985, Chronology of late Wisconsinan deglaciation of coastal Maine, in Borns, H.W., et al., eds., Late Pleistocene history of northeastern New England and adjacent Quebec: Geological Society of America Special Paper 197, p. 29–44.

Smith, G.W., 1990a, Surficial geology of the North Berwick quadrangle, Maine: Maine Geological Survey Open-File Report 90-40, scale 1:24 000.

Smith, G.W., 1990b, Surficial geology of the Wells quadrangle, Maine: Maine Geological Survey Open-File Report 90-41, scale 1:24 000.

Smith, G.W., 1994a, Surficial geology of the Gorham quadrangle, Maine: Maine Geological Survey Open-File Report 94-6, scale 1:24 000.

Smith, G.W., 1994b, Surficial geology of the Somersworth quadrangle, Maine: Maine Geological Survey Open-File Report 94-8, scale 1:24 000.

Smith, G.W., and Hunter, L.E., 1989, Late Wisconsinan deglaciation of coastal Maine, in Tucker, R.D., and Marvinney, R.G., eds., Studies in Maine geology, Volume 6: Quaternary geology: Augusta, Maine Geological Survey, p. 13–32.

Smith, G.W., Stemen, K.S., and Jong, R., 1982, The Waldoboro moraine and related glaciomarine deposits, Lincoln and Knox Counties, Maine: Maine Geology, v. 2, p. 33–44.

Stemen, K.S., 1979, Glacial stratigraphy of portions of Lincoln and Knox Counties, Maine [M.S. thesis]: Athens, Ohio University, 67 p.

Stone, G.H., 1899, The glacial gravels of Maine and their associated deposits: U.S. Geological Survey Monograph 34, 499 p.

Stuiver, M., and Borns, H.W., Jr., 1975, Late Quaternary marine invasion in Maine: Its chronology and associated crustal movement: Geological Society of America Bulletin, v. 86, p. 99–104.

Thompson, W.B., 1998, Deglaciation of western Maine and the northern White Mountains: Geological Society of America Abstracts with Programs, v. 30, no. 1, p. 79.

Thompson, W.B., and Borns, H.W., Jr., 1985, Surficial geologic map of Maine: Augusta, Maine Geological Survey, scale 1:500 000.

Thompson, W.B., and Smith, G.W., 1978, Reconnaissance surficial geology of the Union quadrangle, Maine: Maine Geological Survey Open-File Report 78-22, scale 1:24 000.

Thompson, W.B., and Smith, G.W., 1988, Pleistocene stratigraphy of the Augusta and Waldoboro areas, Maine: Maine Geological Survey Bulletin 27, 36 p.

Thompson, W.B., Crossen, K.J., Borns, H.W., Jr., and Anderson, B.G., 1989, The glaciomarine deltas of Maine and their relation to late Pleistocene–Holocene crustal movements, in Anderson, W.A., and Borns, H.W., Jr., eds., Neotectonics of Maine; studies of seismicity, crustal warping, and sea-level change: Maine Geological Survey Bulletin 40, p. 43–67.

Van der Veen, C.J., ed., 1997, Calving glaciers, Report of a Workshop, February 28–March 2, 1997: Columbus, Ohio, Byrd Polar Research Center Report 15, 194 p.

Manuscript Accepted by the Society June 9, 2000

Geological Society of America
Special Paper 351
2001

# Atmospheric $^{14}C$ chronology for late Wisconsinan deglaciation and sea-level change in eastern New England using varve and paleomagnetic records

**John C. Ridge**
*Department of Geology, Tufts University, Medford, Massachusetts 02155, USA; e-mail: jack.ridge@tufts.edu*
**Brandy A. Canwell**
*Department of Education, Tufts University, Medford, Massachusetts 02155, USA*
**Meredith A. Kelly**
*Geologische Institut, Universitat Bern, Baltzerstrasse 1, 3012-CH, Bern, Switzerland*
**Sharon Z. Kelley**
*19 North Main Street, Westford, Massachusetts 01886, USA*

## ABSTRACT

An atmospheric $^{14}C$ chronology has been formulated for the deglaciation of southern New Hampshire and Maine that allows the first direct comparison of atmospheric $^{14}C$ ages with the uncorrected marine $^{14}C$ ages for deglaciation and isostatic adjustment in this region. The new chronology is based on $^{14}C$ ages from terrestrial plant macrofossils in the Connecticut Valley that are used to calibrate the New England varve chronology and its associated paleomagnetic declination record. A correlation of the declination record from New England with a similar record from New York shows identical $^{14}C$ ages where independent $^{14}C$ analyses are available from both areas. The correlation of varves in the Connecticut and Merrimack Valleys for more than 550 yr, as originally proposed by Ernst Antevs, is valid, but is also confirmed by similar paleomagnetic declinations in the two sequences at precisely the time of a unique western maximum in declination (35°–40° W). Varves that represent the onlapping basal part of the Merrimack Valley sequence establish the age of deglaciation in northern Lake Merrimack as 12.5–12.4 ka ($^{14}C$). Varves in Maine have been correlated to the New England varve chronology and have paleomagnetic declinations that are consistent with the construction of ice-recessional positions from the Merrimack Valley to Maine and the eastward extension of the atmospheric $^{14}C$ chronology to an area of marine ice recession. The new $^{14}C$ chronology is ~1.0–1.5 k.y. younger on average than uncorrected $^{14}C$ ages from marine fossils and lacustrine bulk sediment samples that have previously been used to date deglaciation and isostatic uplift. The marine $^{14}C$ ages are inferred to be anomalous due to a marine reservoir effect and the influences of glacial meltwater in the late Wisconsinan glaciomarine environment of the Gulf of Maine.

Ridge, J.C., Canwell, B.A., Kelly, M.A., and Kelley, S.Z., 2001, Atmospheric $^{14}C$ chronology for late Wisconsinan deglaciation and sea-level change in eastern New England using varve and paleomagnetic records, *in* Weddle, T.K., and Retelle, M.J., eds., Deglacial History and Relative Sea-Level Changes, Northern New England and Adjacent Canada: Boulder, Colorado, Geological Society of America Special Paper 351, p. 171–189.

# INTRODUCTION

Over the past decade both marine and ice-core records have provided evidence of a complex pattern of climate change at the close of the last glaciation (Heinrich, 1988; Fairbanks, 1989, 1990; Bard et al., 1990, 1993; Broecker, 1990; Keigwin et al., 1991; Bond et al., 1992; Dansgaard et al., 1993; Edwards et al., 1993; Keigwin and Lehman, 1994; Bond and Lotti, 1995; Cuffey et al., 1995; Stuiver et al., 1995; Clark et al., 1996; Adkins et al., 1998). These records provide evidence of changes in deep ocean circulation, sudden pulses of freshwater to the ocean surface, ocean surface cooling events, drastic flips in climate over time spans as short as a decade, periods of rapid and reduced sea-level rise, and iceberg discharge (Heinrich) events. However, the causes of climatic events at the century and millennial scales, and the extent to which they are linked to ice-sheet dynamics and ocean and atmospheric circulation changes, remain uncertain. Direct evidence of the role of ice sheets, as either responding to changing climate, or by their instability being the dynamic systems that stimulated abrupt climate changes, has remained elusive. Although needed, a comparison of terrestrial glacial events in North America (readvances and recessions) and marine and ice-core records has not been possible with an appropriate resolution to resolve cause and effect relationships for terrestrial, marine, and climatic events. At the heart of this problem are uncertainties in comparing the ages of climatic and oceanic events dated with layer counting in ice cores, $^{14}C$ ages from marine fossils in ocean cores, and ages for terrestrial and nearshore glacial events derived from $^{14}C$ analyses of terrestrial and glaciomarine fossils.

Uncertainties with the ages of glacial events are especially prevalent in New England (Fig. 1), where the timing of glacial events has been inferred from uncorrected $^{14}C$ ages of marine fossils and lake-bottom bulk sediment samples. Marine $^{14}C$ ages have been used verbatim without the subtraction of a marine reservoir correction (Stuiver and Borns, 1975; Hughes et al., 1985; Smith, 1985; Thompson and Borns, 1985; Koteff et al., 1993). It has not been possible, without a direct comparison of uncorrected marine $^{14}C$ ages and an atmospheric $^{14}C$ time scale, to estimate the magnitude of a reservoir correction. However, marine reservoir corrections must be considered when dealing with any marine $^{14}C$ analysis, and they are especially important when dealing with fossils from an arctic marine environment. Errors of 400–750 yr for modern shells from Norway, Spitsbergen, Greenland, and northern Canada (Mangerud, 1972; Hjort, 1973; Mangerud and Gullikson, 1975) and 700–800 yr for Younger Dryas–age planktonic forams in the North Atlantic (Bard et al., 1994; Birks et al., 1996) have been documented. Complicating the application of a standard marine reservoir correction to $^{14}C$ ages for glaciomarine fossils is the variation of reservoir effects in both time and space (Bard, 1988). Reservoir errors are amplified by the effects of sea ice cover and ocean circulation changes that reduce exchanges between the atmosphere and ocean and produce greater errors in ice age fossils than in modern arctic fossils. In addition, Rodrigues (1988, 1992), Hillaire-Marcel (1988), and Anderson (1988) identified a $^{14}C$ error in shells collected from glaciomarine deposits of the Champlain Sea that appears to be related to an anomalous age for glacial meltwater in a restricted, poorly mixed marine environment. Non-reservoir corrected $^{14}C$ ages for shells from a meltwater-rich environment near the highest shoreline of the Champlain Sea at Ottawa are as much as 1.4 k.y. older than uncorrected $^{14}C$ ages for shells from the oldest full marine deep-water sediment in the center of the St. Lawrence basin (Rodrigues, 1992). The fossil-bearing glaciomarine deposits of eastern New England were also deposited in a somewhat restricted embayment in the Gulf of Maine behind shallower water areas or possibly behind an emergent area of the outer continental shelf while eustatic sea level was low. However, late Pleistocene shells from the Gulf of Maine have never been sys-

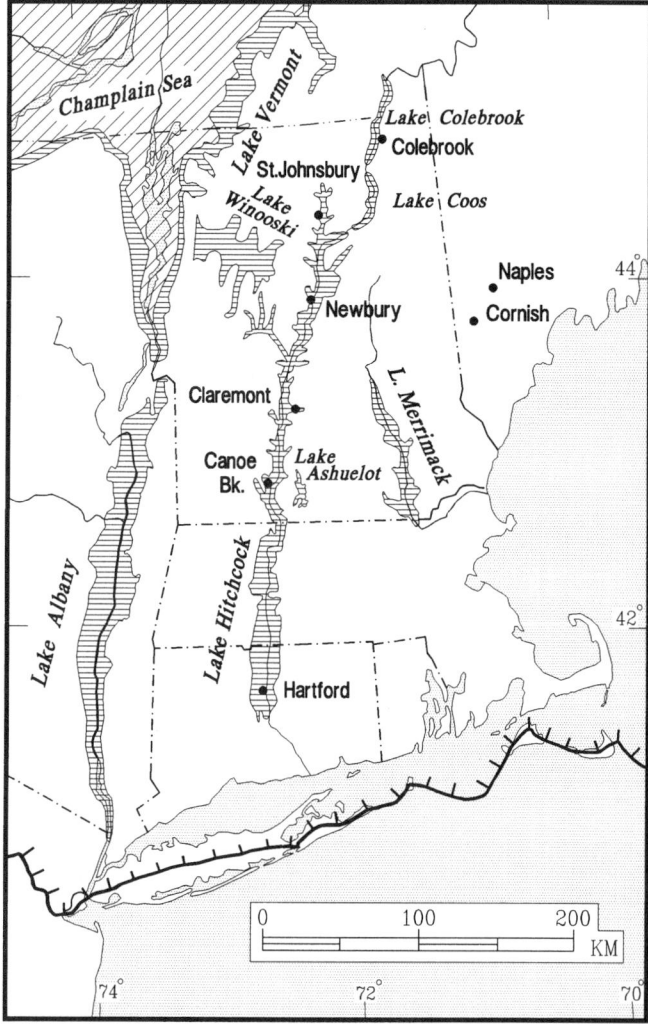

Figure 1. Location of late Wisconsinan glacial lakes in New England and sites discussed in text. Ice margin extending across southern New England is maximum extent of late Wisconsinan ice.

tematically tested for errors associated with either marine reservoir or meltwater effects.

Scattered $^{14}C$ ages from bulk sediment lake bottom samples, the only available terrestrial $^{14}C$ ages prior to the development of accelerator mass spectrometer (AMS) $^{14}C$ analysis, have been used to infer ages of deglaciation across much of New England (Davis et al., 1980; Davis and Ford, 1982; Davis and Jacobson, 1985; Stone and Borns, 1986). Bulk sediment samples may have anomalous old ages associated with the incorporation of aquatic plants as well as other organic materials that do not obtain their carbon directly from the atmosphere (Shotton, 1972; Oeschger et al., 1985; Andrée et al., 1986; Wohlfarth et al., 1993; Wohlfarth, 1996). In modern arctic lakes there also appear to be large $^{14}C$ errors due to multiple nonatmospheric sources of carbon (Abbott and Stafford, 1996). Evidence of similar errors in Pleistocene samples has been found at some places in New England. A sample with aquatic plant remains from Colebrook, New Hampshire (Fig. 1), has yielded a $^{14}C$ age that is 9 k.y. older than nonaquatic plants from the same beds (Table 1; Miller and Thompson, 1979). Bulk sediment samples from a few lakes in New England and adjacent Canada have yielded ages 2–8 k.y. older than from basal materials in surrounding lakes (Mott, 1975, 1981; Karrow and Anderson, 1975; Davis and Davis, 1980; Davis et al., 1995; Lini et al., 1995; see Ridge et al., 1999, for review). In these situations errors are easily recognized because of their magnitude, but there is no simple test for errors <2 k.y. caused by the same processes. $^{14}C$ ages from plant macrofossils recovered from basal lake sediment have yielded significantly younger ages than most bulk sediment analyses (Thompson et al., 1996, 1999). The $^{14}C$ ages for nonaquatic plant macrofossils combined with new $^{14}C$ ages from varves in the Connecticut Valley suggest that there may be systematic errors of as much as 1.5 k.y. associated with bulk sediment samples in New Hampshire and Vermont (Ridge et al., 1999). The AMS $^{14}C$ dating of nonaquatic plant macrofossils provides an opportunity to eliminate errors associated with bulk sediment samples, but these dates have not yet been widely applied to the age of deglaciation in New England. Before an accurate chronology for deglaciation can be proposed for comparison to marine and ice-core records, a firmly established atmospheric $^{14}C$ chronology for ice recession must be formulated that eliminates errors associated with marine and lacustrine bulk sediment samples.

This chapter presents a comparison of the deglacial chronology of northeastern New England based on uncorrected $^{14}C$ ages for marine fossils with $^{14}C$ ages from nonaquatic plant macrofossils. The approach taken is to use multiple techniques in establishing a complete atmospheric $^{14}C$ chronology for deglaciation across southern New Hampshire and Maine. In particular, the exact correlation of varve and paleomagnetic chro-

TABLE 1. $^{14}C$ AGES FROM VARVES IN THE CONNECTICUT VALLEY

| Laboratory number | Age ($^{14}C$ yr B.P.) | $\delta^{13}C$ (‰)* | New England varve number | Material dated | Reference |
|---|---|---|---|---|---|
| **Canoe Brook, Vermont (Ridge and Larsen, 1990)** | | | | | |
| GX-25735 | 12 660 ± 50 | −28.9 | 5858 | Woody twigs and *Dryas* leaves (AMS) | New |
| GX-14231 | 12 355 ± 75 | −27.2 | 6150 | Bulk sample of silt and clay with nonaquatic leaves and twigs | Ridge and Larsen, 1990 |
| GX-14780 | 12 455 ± 360 | −27.6 | 6150 | Handpicked nonaquatic leaves and twigs, mostly *Dryas* and *Salix* | Ridge and Larsen, 1990 |
| CAMS-2667 | 12 350 ± 90 | — | 6150 | *Salix* twig (AMS) | Norton Miller, 1993, personal commun. |
| GX-14781 | 12 915 ± 175 | −27.1 | 6156 | Bulk sample of silt and clay with fragments of peat and gyttja | Ridge and Larsen, 1990 |
| **Amherst, Massachusetts** | | | | | |
| Beta-124780 | 12 370 ± 120 | −27.1 | 5761–5768 | Plant fragments from contorted zone (AMS) | Rittenour, 1999 |
| **Newbury, Vermont (site 73 of Antevs, 1922)** | | | | | |
| GX-23765 | 11 530 ± 95 | −27.0 | 7435–7452 | Woody twig (AMS) | Ridge et al., 1999 |
| GX-23766 | 11 045 ± 70 | −27.5 | 8206 | Woody twig (AMS) | Ridge et al., 1999 |
| GX-23640 | 10 940 ± 70 | −26.8 | 8357 | Woody twig (AMS) | Ridge et al., 1999 |
| GX-23641 | 10 080 ± 580 | −26.7 | 8498–8500 | Woody twig | Ridge et al., 1999 |
| GX-23767 | 10 685 ± 70 | −26.3 | 8504 | Woody twig (AMS) | Ridge et al., 1999 |
| GX-23642 | 10 040 ± 230 | −26.5 | 8542–8544 | Chunk of wood | Ridge et al., 1999 |
| GX-23643 | 10 440 ± 520 | −26.8 | 8652–8662 | 2 woody twigs | Ridge et al., 1999 |
| **East Barnet, Passumpsic River Valley, Vermont** | | | | | |
| GX-26456 | 11 220 ± 50 | −27.1 | 7754 | Woody twig (AMS) | New |
| **Columbia Bridge, Vermont (Miller and Thompson, 1979)** | | | | | |
| WIS-961 | 11 540 ± 110 | −29.0 | Unknown (>7400) | Wood fragments | Miller and Thompson, 1979 |
| WIS-919 | 11 390 ± 115 | −27.5 | Unknown (>7400) | Wood fragments | Miller and Thompson, 1979 |
| WIS-925 | 20 500 ± 250 | — | Unknown (>7400) | *Potamogeton* leaves and other plant remains | Miller and Thompson, 1979 |

*Hyphens indicate that $\delta^{13}C$ values were not measured.

nologies in the Connecticut and Merrimack Valleys has allowed the transfer of the atmospheric $^{14}$C chronology of the Connecticut Valley across the region (Ridge et al., 1999). Previously, the age of deglaciation could only be inferred from a few $^{14}$C ages of basal lacustrine bulk sediment samples and inferences based on uncorrected marine $^{14}$C ages associated with ice recession on the coast of Maine.

We first review and update the pertinent parts of the New England varve chronology (Antevs, 1922, 1928) and its relationship to deglaciation. A much more detailed review of the New England varve chronology in northern New England was given in Ridge et al. (1999). Following the varve summary is a discussion of the $^{14}$C calibration of the varve chronology and the application of paleomagnetic stratigraphy developed from lacustrine sediments across New York and New England. A correlation of marine and terrestrial ice-recession positions and new varve and paleomagnetic results from southern Maine are then discussed, leading to a comparison of marine and terrestrial chronologies.

## VARVE CHRONOLOGY IN NEW ENGLAND

In the 1920s, Ernst Antevs (1922, 1928) established the New England (NE) varve chronology by meticulously matching varve sections in the Connecticut Valley drainage system from as far south as Hartford, Connecticut, northward to St. Johnsbury, Vermont (Passumpsic Valley), as well as other drainage basins (Figs. 1 and 2A; see Ridge et al., 1999, for more detailed review). Except for the tail ends of the sequence, and two intervals in Massachusetts of <20 yr each, Antevs formulated his varve stratigraphy from multiple overlapping records (Fig. 2B). The two intervals in Massachusetts have since been duplicated in cores taken on the campus of the University of Massachusetts at Amherst (Rittenour, 1999). Antevs (1922, 1928) assembled two large master or normal varve sequences, one from sections south of Claremont, New Hampshire, called the lower Connecticut Valley varves (NE varves 2701–6352, including matched sequences in the Hudson, Ashuelot, and Merrimack Valleys), and one from sections north of Claremont called the upper Connecticut Valley varves (NE varves 6601–7750, including matched sequences from Lake Winooski). At Newbury, Vermont, Antevs (1922, site 73) counted an additional 750 thin couplets that represent the potential for extending the upper Connecticut Valley varves to at least NE varve 8500 (Ridge and Toll, 1999).

Antevs (1922) could not find a clear match between the lower and upper Connecticut Valley varves and inferred that the sequences were separate in time as a result of a 600 yr stillstand of the receding ice sheet at Claremont; however, field evidence for this long halt in ice recession is lacking in the Claremont region. Recent visual and computer analysis reveals a crude similarity of the two sequences (lower Connecticut varve 6012 = upper Connecticut varve 6601; Ridge et al., 1996, 1999), despite megavarves (to 4 m) in the upper Connecticut Valley varves and nearly an order of magnitude difference in couplet thickness between the two sequences. The local influences of ice proximal and deltaic deposition at the mouths of large rivers at Claremont and farther north apparently obscured the record of regional weather patterns in the lowest 200–300 varves of the upper Connecticut Valley sequence, rendering them largely useless for regional correlation. The correlation and overlap of the two Connecticut Valley sequences are now supported by independent $^{14}$C ages and paleomagnetic chronologies from both sequences (Ridge et al., 1999).

An important part of Antevs (1922, 1928) chronology was the regional correlation of varve records in the Connecticut Valley and all available records from lakes in other drainage systems (Figs. 1 and 2). Antevs found exact year to year correlations between varves in the Connecticut Valley and varves in the lower and upper Hudson Valley (NE varves 2701–3170 and 5501–5800), the Ashuelot Valley in southwestern New Hampshire (NE varves 5804–5879), the Merrimack Valley (5709–5749 and 5771–6352), and Lake Winooski from Montpelier to Waterbury, Vermont (NE varves 7059–7288). The most important correlation for the purposes of this paper is the correlation between the Connecticut and Merrimack Valleys (Fig. 3) that provides an exact annual correlation of deglacial events in the two drainage basins. We have been able to correlate three new varve sections (MER1-3) in the Merrimack Valley (Fig. 4) with both the Merrimack and lower Connecticut Valley sequences of Antevs (1922).

## DEGLACIATION OF THE MERRIMACK VALLEY

Antevs (1922) was able to relate the deposition of varves in the Merrimack Valley to the timing of deglaciation at sections where varves overlie till or become very thick and sandy downward in the base of a section, indicating close proximity to till or bedrock and deposition very shortly after deglaciation. Antevs (1922) did not report the exact thickness of all basal varves, but at two sections he reported varve thicknesses that reached 75 cm. There are five valley side sections (Antevs, 1922) showing an onlapping basal varve stratigraphy that allow the construction of a south to north plot of the approximate age of deglaciation in terms of NE varve years (Fig. 5). Although this plot does not eliminate the possibility of older basal varves in deeper basins that occur in the center of the Merrimack Valley, the rapid sedimentation rates for basal varves combined with the generally low relief of the valley suggest that the onlapping ages of basal varves are a valid approximation of the age of deglaciation (Ridge et al., 1999). In addition, exposures of till and bedrock and generally shallow depths (<30 m) to bedrock beneath terraces in the northern Lake Merrimack basin (Penacook area of Pendleton, 1995) indicate that older basal varves cannot be hidden in deep basins along the axis of the Merrimack Valley. Deglaciation occurred at a rate of 100 m/varve yr during

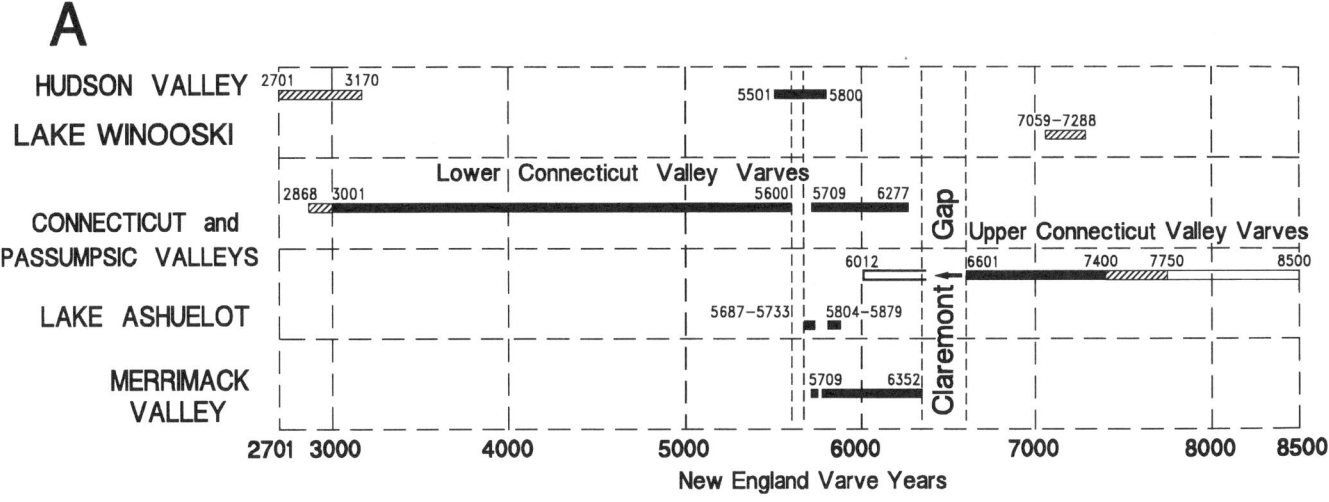

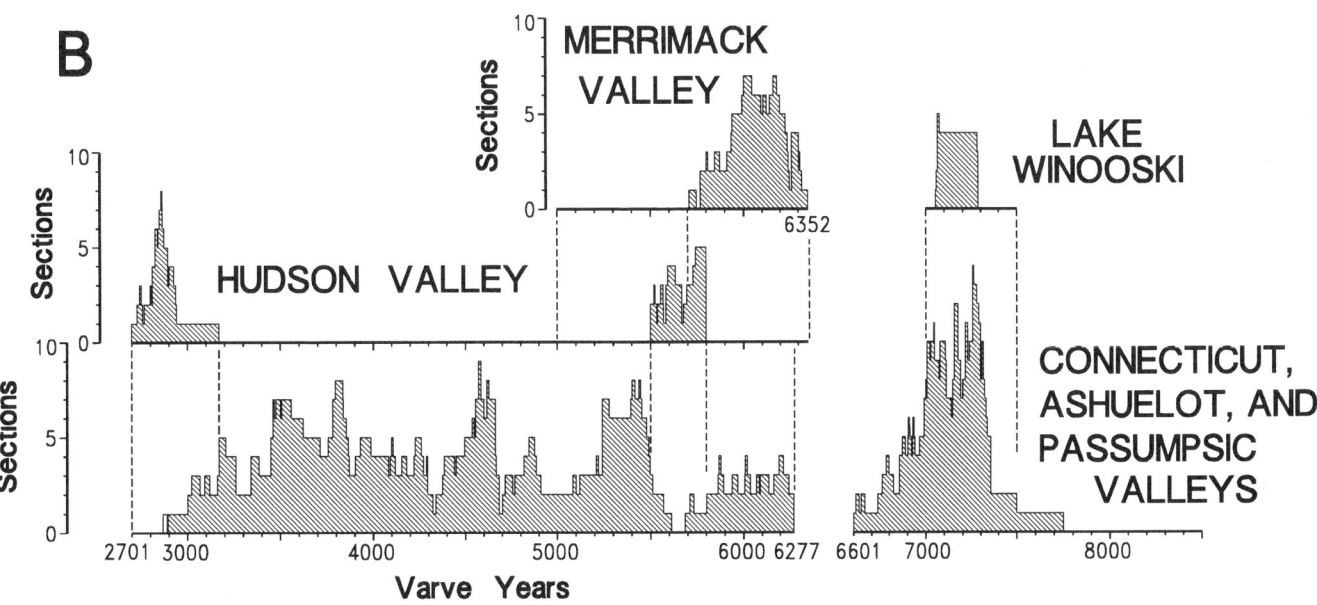

Figure 2. A, Time spans in New England varve years (younger to right) for varve sequences that were compiled by Antevs (1922, solid bars; 1928, scribed bars) to form New England varve chronology. Open bar at end of upper Connecticut Valley varves are varves that were counted but not measured at Newbury, Vermont (Antevs, 1922). Position of upper Connecticut Valley varves, where they are inferred to be correlative to lower Connecticut Valley varves, is indicated with arrow and open box. B, Number of sections (outcrops) matched by Antevs (1922, 1928) to construct New England varve chronology.

NE varve yr 5700–6000, and deglaciation can be no younger than NE varve yr 6150 at Franklin, New Hampshire (Fig. 5). Unfortunately, no varve sections from southern Lake Merrimack have been matched to the varve chronology that can reveal the age and pattern of deglaciation relative to the varve stratigraphy in southernmost New Hampshire and northern Massachusetts.

## $^{14}$C CALIBRATION OF THE NEW ENGLAND VARVE CHRONOLOGY

### Calibration of the lower Connecticut Valley varve sequence

The $^{14}$C calibration of the lower Connecticut Valley varve sequence was accomplished at Canoe Brook, Vermont (Ridge

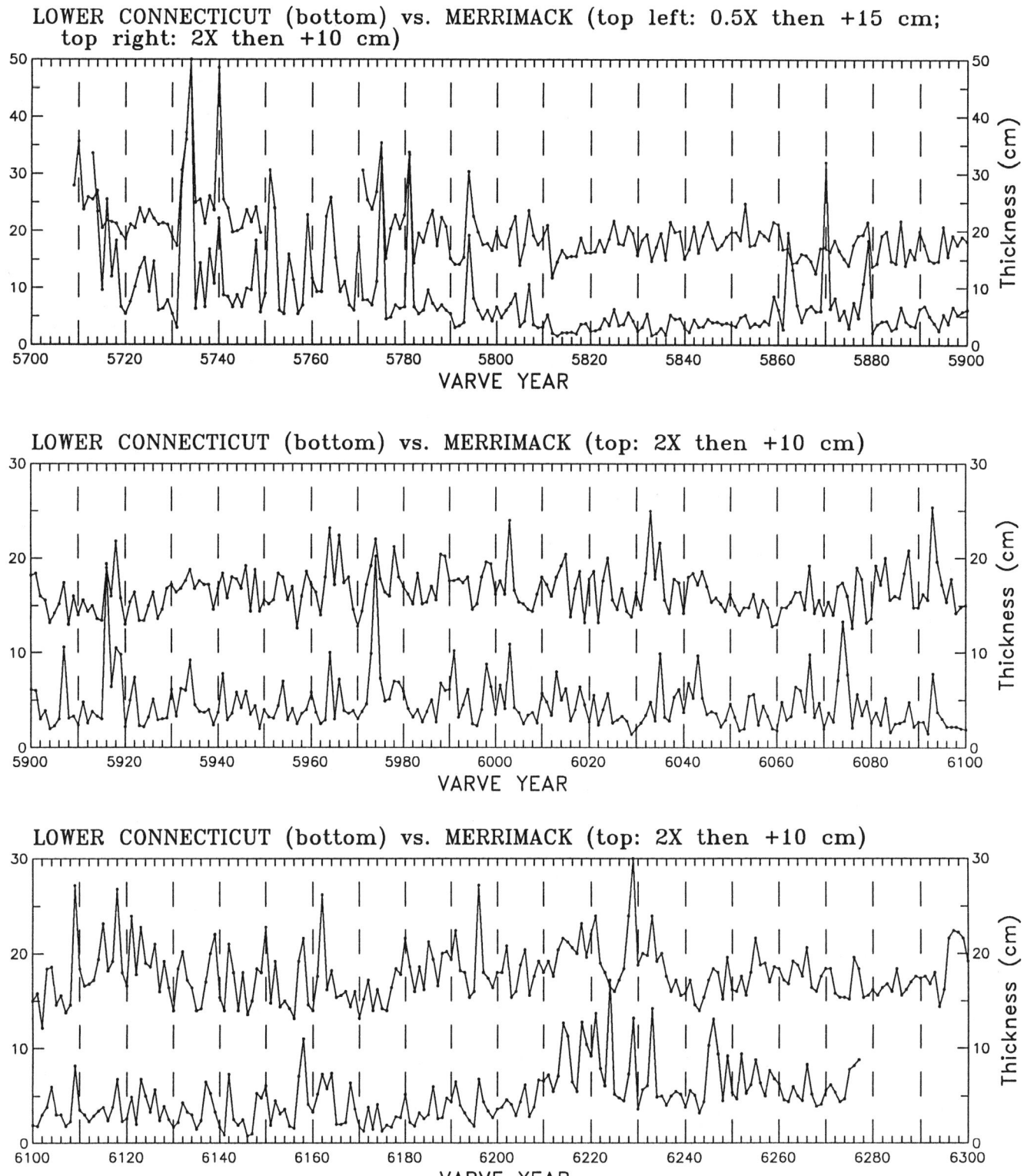

Figure 3. Correlation of lower Connecticut and Merrimack Valley normal varve records of Antevs (1922). Varve records from Merrimack Valley are scaled and then offset at their new scale as indicated at top of each plot. Varves plotted on upper axis of each graph are thicker than shown and are off scale.

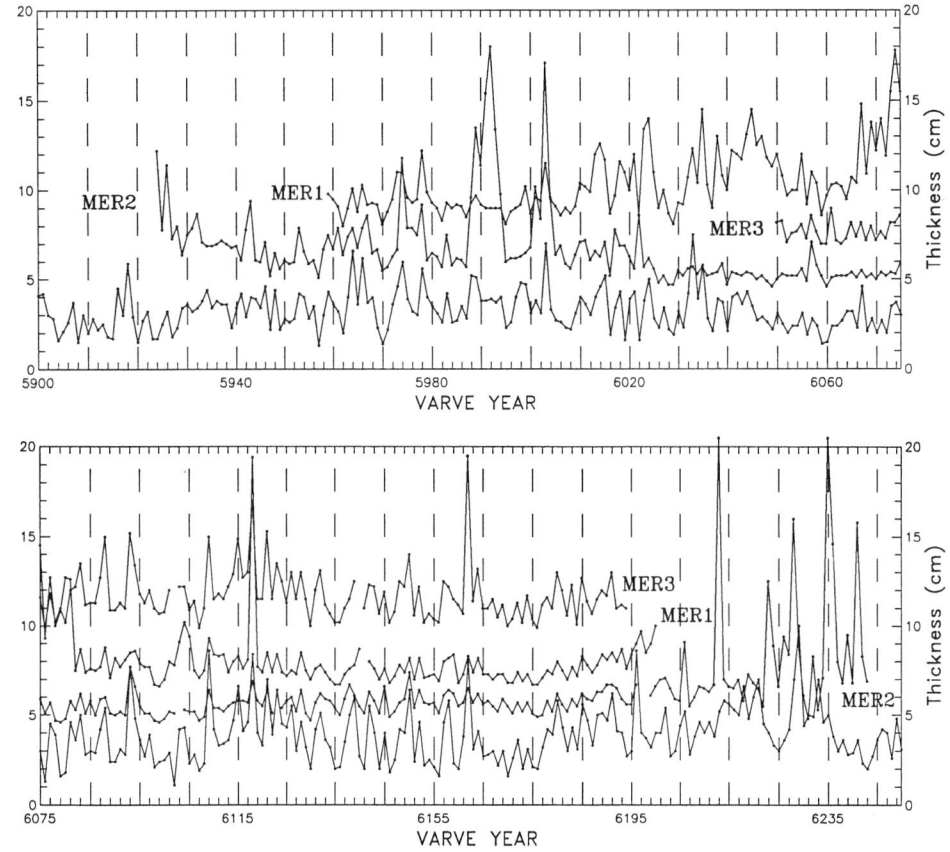

Figure 4. New varve records from Merrimack Valley (MER1-3) matched to Merrimack Valley normal record (Antevs, 1922). All varve records are plotted at same scale, but new sections are offset. Offsets on upper plot: MER1, +7 cm; MER2, +4 cm; MER3, +6 cm. Offsets on lower plot: MER1, +6 cm; MER2, +4 cm; MER3, +9 cm. Upper part of MER2 section (New England varves 6210–6243) does not match Merrimack Valley normal curve because of local deltaic sedimentation that introduced thick coarse to medium sand beds to otherwise thin varve sequence. Varves that are off scale on MER2 plot are much thicker (>30 cm) than shown.

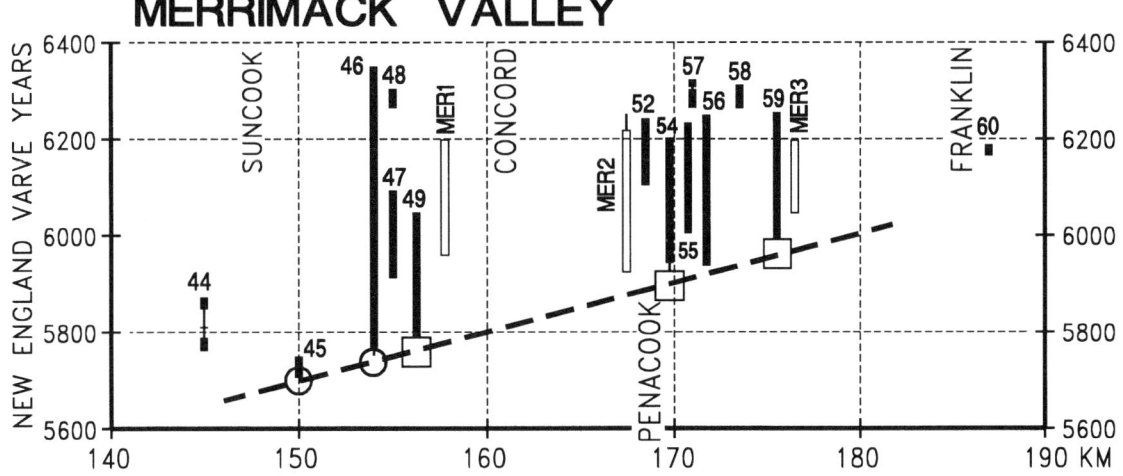

Figure 5. Time spans of varve sections in Merrimack Valley from south (left) to north along axis of valley. Distance is from arbitrary point to south in Massachusetts. Numbered sections (solid) are from Antevs (1922); MER sections (open) are new (Fig. 4). Circles at bottoms of sections indicate sites where varves were found overlying till or bedrock; boxes indicate basal varve sequences that are interpreted to be close to base of varve section because of very thick downward-thickening couplets and their downward-coarsening sandy character. Heavy dashed line is interpreted time-transgressive age of deglaciation inferred from ages of onlapping basal varves.

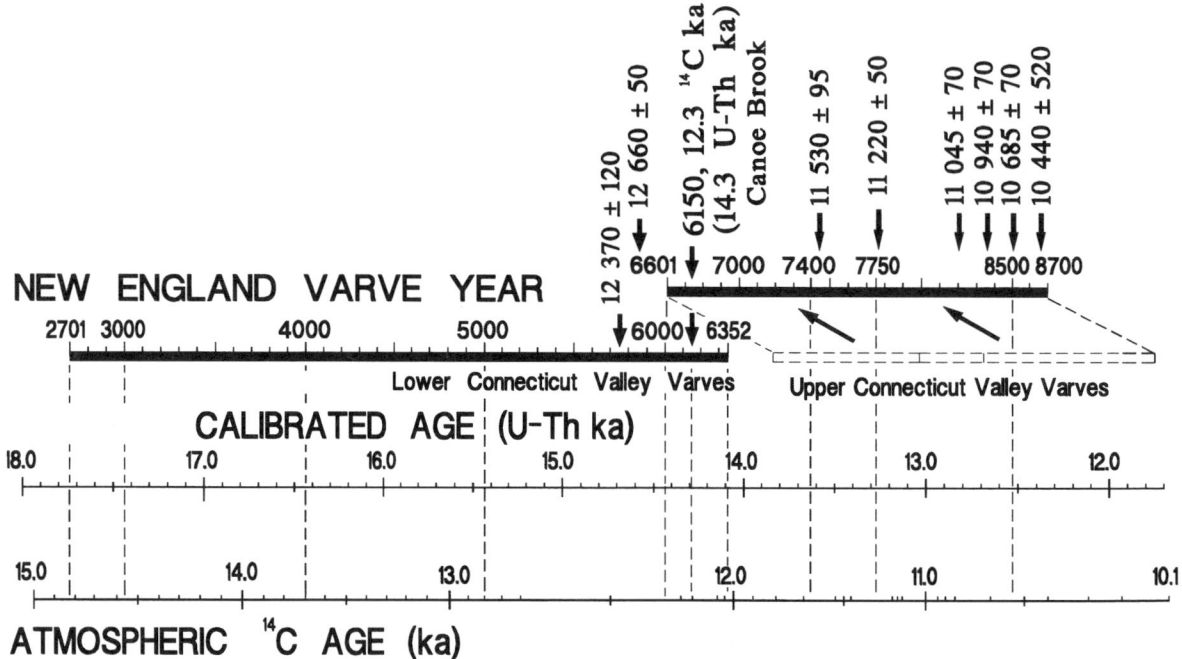

Figure 6. Atmospheric $^{14}$C and calibrated (U-Th or calendar) time scales applied to lower and upper Connecticut Valley varves of Antevs (1922, 1928) based on $^{14}$C ages at Canoe Brook (12.3 ka) and results of CALIB 4.0 computer program (Stuiver and Reimer, 1993; Stuiver et al., 1998). Division lines on all time scales are every 100 yr. Note nonlinear behavior (compressions and plateau) of $^{14}$C time scale from 13 to 10.5 $^{14}$C ka. Inferred ages for New England varves 7510–8700 are based on recounting of Newbury section originally studied by Antevs (1922). Some $^{14}$C ages from Newbury are shown for comparison and have been used to test Canoe Brook calibration point.

and Larsen, 1990; N. Miller, 1993, personal commun.), where three $^{14}$C ages on nonaquatic plant macrofossils (Table 1) allowed an estimate of 12.3 ka for the age of NE varve 6150 (Fig. 6). An analysis of macrofossils from varve 6150 revealed no vascular aquatic plant remains and only very minor amounts of wet-soil sedges, the only macrofossils found in varve 6150 that could cause anomalously old $^{14}$C ages (N. Miller, 1998, personal commun.). A $^{14}$C age of 12.3 ka was chosen (Ridge et al., 1999), rather than an average age of 12.4 ka (Ridge and Larsen, 1990; Ridge et al., 1996; Ridge, 1997), to prevent any of the individual $^{14}$C ages at Canoe Brook from being older than its enclosing varve. This choice also allows for a lag of 50 yr between the time the plant macrofossils were alive and the time of their lacustrine deposition. It is important to stress that the assignment of NE varve 6150 to 12.3 ka (ages herein are $^{14}$C, unless noted) is a maximum limiting age estimate, and the varves may be younger because it is not possible with the existing data to evaluate the full magnitude of lags in deposition. Lags in deposition of macrofossils have been an obstacle to accurately calibrating varve sequences in Sweden (Wohlfarth et al., 1995). A $^{14}$C age of 12.6 ka from NE varve 5858 at Canoe Brook (Table 1) is in agreement with the calibration based on ages from NE varve 6150.

An additional $^{14}$C age of 12.9 ka was obtained at Canoe Brook from fragments of ripped-up peat and gyttja in NE varve 6156 (Table 1). These materials, especially the gyttja, have been found to be inaccurate for calibrating varve and lacustrine sequences to atmospheric $^{14}$C yr (Oeschger et al., 1985; Andrée et al., 1986; Wohlfarth et al., 1993; Wohlfarth, 1996) because they may contain the remains of aquatic plants, especially algae, that receive their carbon from sources other than the atmosphere. At Canoe Brook they also represent pond sediment that was eroded and redeposited in the Connecticut Valley varves, and there is a lag in time between when the gyttja was originally formed and when it was redeposited in the varves. Additional support for the 12.3 ka age of varve 6150 is based on the correlation of paleomagnetic records across New England and New York and is discussed later.

The CALIB 4.0 computer program (Stuiver and Reimer, 1993; Stuiver et al., 1998) was used to infer $^{14}$C ages for varves above and below NE varve 6150, assuming that varve and calibrated (U-Th or calendar) years are of equal duration (Fig. 6). (The CALIB program is available from the website of the Quaternary Isotope Laboratory at the University of Washington: http://depts.washington.edu/qil/.) This program corrects for the nonlinear behavior of the $^{14}$C time scale and the offset between $^{14}$C and calendar years as a result of the secular variation of atmospheric $^{14}$C. During the time period discussed herein (13–10.5 ka; Fig. 6), there are significant compressions of the $^{14}$C time scale relative to varve and calibrated (U-Th) years at 12.4–

12.0 ka (NE varve yr 6100–6350), 11.6–11.4 ka (NE varve yr 7500–7600), and 11.3–10.7 ka (NE varve yr 7850–8200). There is also a prominent $^{14}$C plateau at 12.6–12.4 ka (NE varves 5200–6100) when $^{14}$C years changed very slowly as compared to varve and calibrated years. The end of this plateau (12.4 ka) closely corresponds to the beginning of a period of rapid sea-level rise (meltwater pulse 1a) that was underway by 14.2 ka (12.25 [$^{14}$C] ka; Adkins et al., 1998). This time also appears to correspond to a sudden decrease in the ventilation of deep water in the North Atlantic (Adkins et al. 1998), suggesting a link between climatic and oceanographic events and nonlinear features of the $^{14}$C time scale.

*Calibration of the upper Connecticut Valley varve sequence*

The $^{14}$C calibration of the upper Connecticut Valley varves has been accomplished at Newbury, Vermont (Antevs, 1922, site 73) and at a section in the Passumpsic Valley. A careful recounting of the varves at Newbury in duplicate sets of overlapping cores (Ridge and Toll, 1999; Ridge et al., 1999) has allowed a detailed count of varves above where the section matches the upper Connecticut sequence of the New England varve chronology (to NE varve 7510). Antevs (1922) was not able to produce an accurate count or measurement of varves in the upper part of this exposure during his field study. Some couplets in the section are simply too thin (<2 mm) or indistinct to count accurately without sample collection, careful drying to improve the contrast between silt and clay layers, and magnification. The new count was done on two sets of overlapping PVC cores (60 m long, 7.6 cm diameter) with the aid of magnified video images of partially dried cores and computerized measurement and counting. The new count shows that the section has varves numbering to NE varve 8679 +35/−20 after accounting for uncertainties in the interpretation of annual couplets. Varves at the top of the measured sequence that become thicker and sandier upward are overlain by nonvarved laminated lacustrine sand and appear to represent the gradual shoaling of lake water.

Five AMS and three conventional $^{14}$C ages were obtained on woody twigs and wood from Newbury and the Passumpsic Valley (Table 1) and allow a test of the Canoe Brook calibration combined with the correlation of the upper and lower Connecticut Valley varve sequences (Table 1; Fig. 6). All of the AMS $^{14}$C ages are in general agreement with the predicted atmospheric $^{14}$C time scale and provide support for the correlation of the lower and upper Connecticut Valley varves and the 12.3 ka age for NE varve 6150 at Canoe Brook. Only one of the conventional $^{14}$C ages is in close agreement with the AMS $^{14}$C ages, but the conventional ages have large precision parameters that make them difficult to use for a specific assignment of atmospheric $^{14}$C ages to the varve sequence. All of the $^{14}$C samples from both Canoe Brook and Newbury have $\delta^{13}$C values of −27.5‰ to −26.3‰, as expected for terrestrial nonaquatic plants. As with the $^{14}$C ages from Canoe Brook it is important to stress that the $^{14}$C ages at Newbury provide only a maximum limiting age estimate for the upper Connecticut Valley varves. Some Newbury $^{14}$C ages that are slightly younger than the predicted age of the varve sequence suggest that the varves may be slightly younger than the atmospheric $^{14}$C time scale applied here (Fig. 6).

## PALEOMAGNETIC STRATIGRAPHY

An additional stratigraphic tool applied to varves in New England and New York is the measurement of paleomagnetic records. Remanent declination records were recorded for the New England varve chronology that provided an unparalleled time scale for the first studies on the secular variation of remanent magnetization in sediments (McNish and Johnson, 1938; Johnson et al., 1948). By matching varve measurements to the New England varve chronology, recounting the varves at Newbury, and collecting paleomagnetic samples of known varve age, Johnson et al. (1948) assembled a remanent declination record for the entire varve sequence (Fig. 7). However, their laboratory experiments convinced them to disregard inclination measurements and their work was done prior to the development of alternating field demagnetization (Zijderveld, 1967), a standard technique now employed to remove unstable remanence from detrital sediment.

With modern techniques, Verosub (1979a, 1979b) reformulated the declination record of NE varves 3150–5500 and added an inclination record (Fig. 7). In the past 10 years one of us (Ridge) and students at Tufts University have completely reformulated the declination and inclination records for the remaining upper part of the lower Connecticut Valley sequence (NE varves 5500–6250, Table 2) and most of the upper Connecticut Valley sequence (NE varves 6960–8500; Ridge et al., 1999). In addition, three new varve sections in the Merrimack Valley (Fig. 4; Table 2) have yielded paleomagnetic records that match those in the Connecticut Valley at correlative positions in the lower Connecticut Valley sequence (Fig. 7). The updates of the original work by Johnson et al. (1948) reveal that their results are essentially correct with only minor differences from the more recent data. Declination records for the top of the lower Connecticut varve sequence and the base of the upper Connecticut sequence indicate an overlap in the paleomagnetic records without a break in time that provides support for the proposed correlation of the two varve records (Figs. 6 and 7).

A declination record similar to the one in New England has been constructed for late Wisconsinan deglaciation in central and northern New York and western Vermont (Fig. 7), which has been useful for testing interregional correlations (Brennan et al., 1984; Ridge et al., 1990, 1991, 1999; Pair et al., 1994; Ridge, 1997). This record lacks the precise placement of paleomagnetic data versus a varve chronology and instead relies on the arrangement of samples according to relative age by stratigraphic position, onlapping relationships of basal varves, and the relative ages of glacial lakes and events across

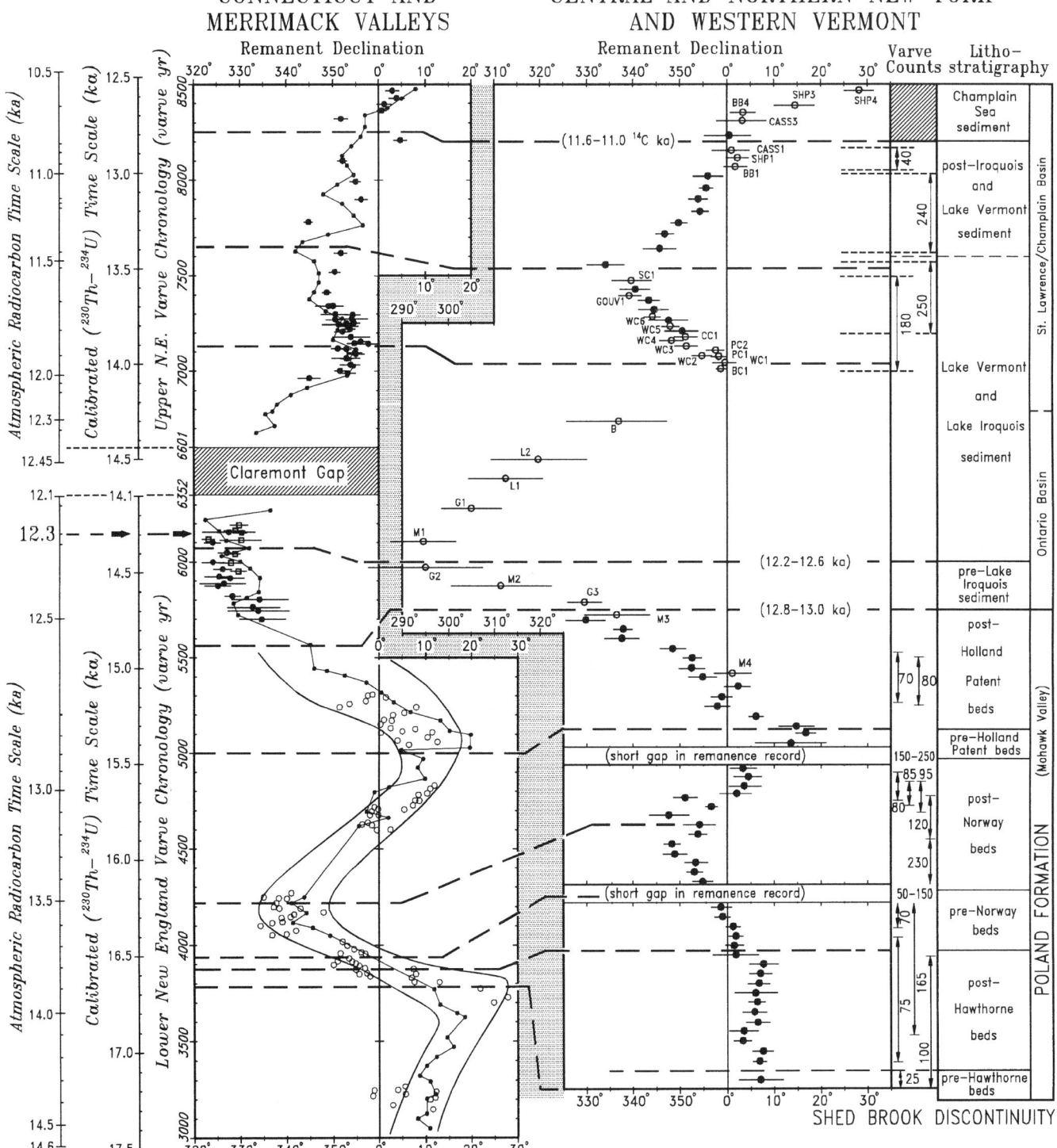

Figure 7. Correlation of late Wisconsinan declination records from varves in New England and New York. Data from Connecticut and Merrimack valleys are plotted versus New England varve chronology, and data from central and northern New York and western Vermont are plotted on time scale relative to stratigraphic position and glacial lakes and events. New England data sources: small dots and tie line, Johnson et al. (1948); open circles, Verosub (1979a); solid circles with error bars in lower New England varve chronology, new data from Connecticut Valley (see Table 2); open boxes with error bars, new data from Merrimack Valley (see Table 2); solid circles with error bars in upper New England varve chronology, Ridge et al. (1999). New York data sources: Poland Formation in Mohawk Valley from Ridge et al. (1990); open circles with labels in eastern Ontario basin from Brennan et al. (1984); open circles with labels in St. Lawrence basin from Pair et al. (1994); solid circles from Champlain Valley from Ridge et al. (1999). Age of Lake Iroquois is based on Muller and Prest (1985), Muller et al. (1986), and Muller and Calkin (1993). Age of Champlain Sea invasion is based on Rodrigues (1988, 1992) and Anderson (1988).

*Atmospheric $^{14}C$ chronology for late Wisconsinan deglaciation and sea-level change in eastern New England*

TABLE 2. PALEOMAGNETIC MEASUREMENTS FROM SOUTHERN NEW HAMPSHIRE AND MAINE

| Sample name | Number of horizons/ total specimens | Range of NE varve yr | Declination (°) | Inclination (°) | Intensity (mA/m) | $\alpha_{95}$ (°) |
|---|---|---|---|---|---|---|
| **Canoe Brook, Dummerston, Vermont-Connecticut Valley (Ridge and Larsen, 1990)** | | | | | | |
| CAN1 | 5/14 | 5687–5702 | 334.8 | 27.7 | 33.8 | 4.9 |
| CAN2 | 4/10 | 5732–5741 | 334.0 | 37.0 | 39.1 | 6.6 |
| CAN3 | 4/18 | 5793–5803 | 334.4 | 44.3 | 54.3 | 6.3 |
| CAN4 | 3/11 | 5867–5879 | 325.3 | 41.8 | 38.2 | 2.9 |
| CAN5 | 4/11 | 5931–5941 | 325.7 | 52.9 | 47.2 | 3.0 |
| CAN6 | 4/10 | 5990–6001 | 324.6 | 52.5 | 55.3 | 2.4 |
| CAN7 | 5/12 | 6045–6056 | 327.2 | 56.2 | 65.3 | 2.1 |
| CAN8 | 5/13 | 6086–6101 | 324.3 | 57.0 | 44.0 | 1.9 |
| CAN9 | 5/9 | 6153–6165 | 327.8 | 48.8 | 69.9 | 5.8 |
| CAN10 | 4/11 | 6169–6178 | 330.7 | 52.4 | 87.4 | 1.5 |
| **Mill Brook, Putney, Vermont-Connecticut Valley** | | | | | | |
| MIL1 | 6/12 | 5763–5770 | 333.1 | 37.3 | 82.5 | 5.8 |
| MIL2 | 6/12 | 5815–5826 | 328.5 | 45.2 | 54.2 | 1.7 |
| MIL3 | 5/9 | 5873–5883 | 326.6 | 43.7 | 68.3 | 5.0 |
| MIL4 | 6/12 | 5916–5926 | 327.9 | 51.8 | 68.3 | 3.2 |
| MIL5 | 6/12 | 5961–5973 | 326.6 | 56.1 | 88.7 | 2.1 |
| **Soucook, New Hampshire-East side of Merrimack Valley north of Soucook River** | | | | | | |
| MER1 | 6/12 | 6105–6118 | 330.5 | 59.0 | 81.6 | 4.5 |
| **Hayward Brook, Penacook, New Hampshire-East side of Merrimack Valley** | | | | | | |
| MER2-A | 6/12 | 5939–5964 | 330.4 | 57.1 | 128.1 | 1.7 |
| MER2-B | 6/12 | 5986–6005 | 328.1 | 54.0 | 114.4 | 2.6 |
| MER2-C | 6/12 | 6032–6053 | 327.9 | 58.1 | 155.3 | 1.1 |
| MER2-D | 6/12 | 6105–6122 | 323.9 | 56.7 | 149.1 | 1.4 |
| MER2-E | 6/12 | 6148–6160 | 328.0 | 57.5 | 188.1 | 0.9 |
| MER2-F | 7/12 | 6186–6200 | 329.8 | 54.6 | 162.7 | 1.9 |
| **Bryant Brook, New Hampshire-7 km north of Penacook, east side of Merrimack Valley** | | | | | | |
| MER3 | 4/8 | 6159–6169 | 328.7 | 52.0 | 155.7 | 4.6 |
| **Cornish, Maine-Ossipee River Valley** | | | | | | |
| CORN | 13/26 | 6084–6102 | 319.3 | 35.6 | 78.3 | 4.2 |
| **Naples, Maine-P and K Sand and Gravel Co., Crooked Brook Valley** | | | | | | |
| NAP1 | 6/18 | lower varves (31–50)* | 342.8 | 54.9 | 507.9 | 1.7 |
| NAP2 | 3/12 | upper varves (8–13)* | 343.2 | 53.4 | 482.0 | 1.7 |

*Note:* All results are after alternating field demagnetization at 30 mT. Average sample volume is 7.0 cm³. Samples are listed from bottom (first) to top (last) for each sample site.
$\alpha_{95}$ is the 95% cone of confidence about the mean direction. NE, New England.
*Recorded as varves above base of unit.

the region. As in New England, the declination record is nearly complete with only minor gaps that do not interrupt the pattern of secular change in declination.

The declination records for New England and New York cover the same period of deglaciation and have been correlated based on a comparison of relative amplitudes of declination maxima on the two records (Fig. 7). A unique feature in both records is a strong westward maximum in declination to 35°–60°W (325°–300°) that occurs in the Canoe Brook section in the Connecticut Valley (NE varves 5900–6150) and at the onset of Lake Iroquois in New York. Independent estimates of the $^{14}C$ age of this western declination maximum are similar in both records. The onset of Lake Iroquois in central New York occurred ca. 12.6–12.2 ka (Muller and Prest, 1985; Muller et al., 1986; Muller and Calkin, 1993) and NE varves 5900–6150 have an age of 12.5–12.3 ka at Canoe Brook. The coincidence of independent $^{14}C$ ages in correlative declination records provides strong support for the Canoe Brook calibration point at 12.3 ka in the lower Connecticut Valley varves.

An additional $^{14}C$ test of the paleomagnetic correlation comes from a comparison of the lacustrine to marine transition associated with the invasion of the Champlain Sea in New York and Ontario and correlative paleomagnetic stratigraphy at Newbury, Vermont. The lacustrine to marine transition occurs at about the time of a change from western to eastern declination (0° in Fig. 7; Pair et al., 1994) and has been given an age of 11.0–11.6 ka based on uncorrected $^{14}C$ ages from marine shells of the Champlain Sea (Rodrigues, 1988, 1992) and $^{14}C$ ages for the pollen stratigraphy of southern Ontario (Anderson, 1988). A marine reservoir correction, when applied to the Champlain

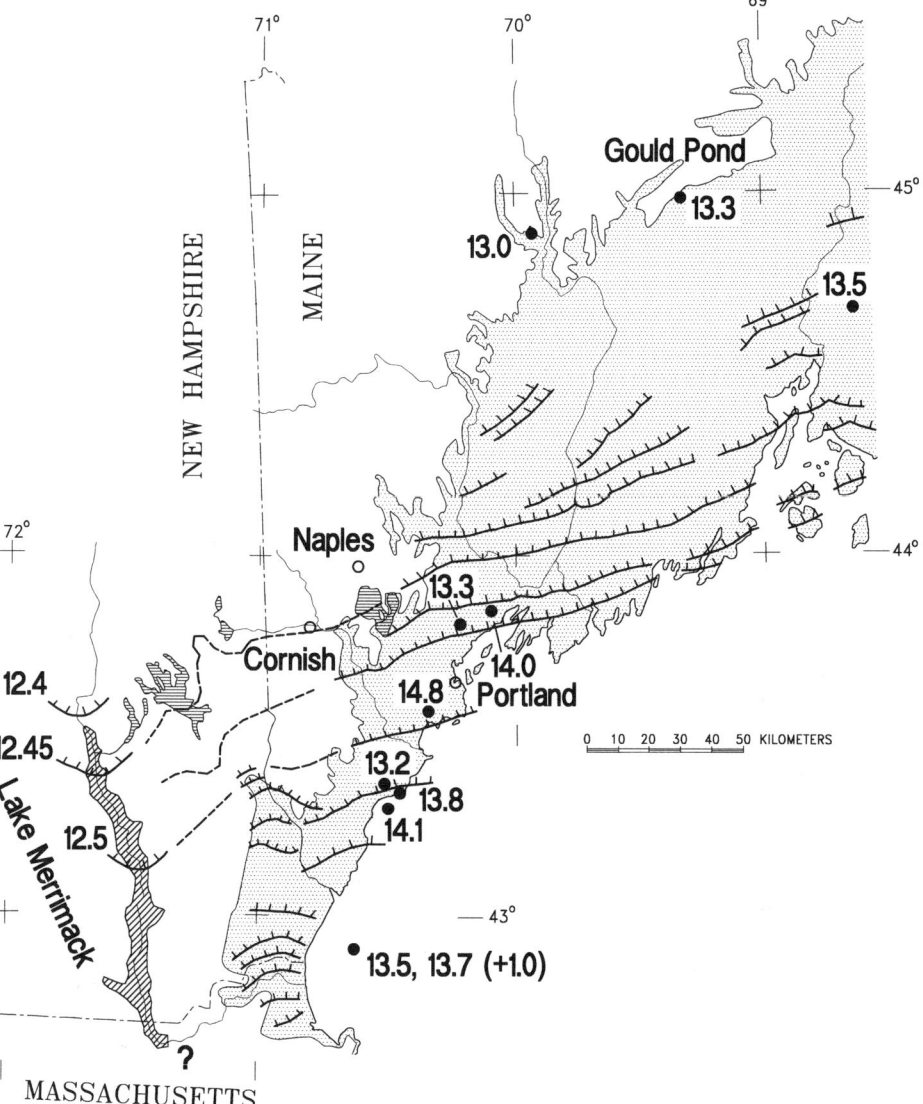

Figure 8. Inferred pattern of deglaciation across southeastern New Hampshire and southern Maine based on ice marginal features in Maine (Smith, 1980, 1982, 1985; Thompson and Borns, 1985; Smith and Hunter, 1989) and coastal New Hampshire and Massachusetts (Koteff et al., 1993). Ice margins and their atmospheric $^{14}$C ages (in ka) for northern Lake Merrimack basin are inferred from analysis of varves (Fig. 5) and atmospheric $^{14}$C time scale for New England varve chronology (Fig. 6). Solid circles and numbers in Maine and at core site in ocean are uncorrected $^{14}$C ages (in ka) from marine fossils and lacustrine bulk sediment samples. Dashed lines indicate correlation between ice margins in Merrimack Valley and areas to east and have been constructed to fit intervening topography. Shaded area shows region inundated by marine water during ice recession, as inferred from Smith (1982), Thompson and Borns (1985), and Koteff et al. (1993).

Sea $^{14}$C ages, can only make the age of the Champlain Sea invasion younger. The 0° declination occurs in NE varves 8000–8400 at Newbury that, according to the $^{14}$C time scale applied to the Connecticut Valley varves, have an age of 11.1–10.6 ka. Unfortunately, the quality of the Newbury paleomagnetic data is insufficient to further resolve the position of the 0° declination in the Connecticut Valley varve stratigraphy. There is also a compression of the $^{14}$C time scale at this time that makes a precise determination of the $^{14}$C age of the 0° declination at Newbury difficult. However, independent $^{14}$C ages for the paleomagnetic records overlap and provide support, although in this case not very precise, for the $^{14}$C calibration of the upper Connecticut Valley varves (Fig. 6).

Remanent inclination records for the upper Connecticut Valley varves and post-12.2 ka events in New York and western Vermont are similar, support the conclusions here, and are discussed elsewhere (Ridge et al., 1999). Severe compaction-related flattening of inclination in samples from the western Mohawk Valley (Ridge et al., 1990) and eastern Ontario basin of New York (Brennan et al., 1984) has rendered the New York inclination record useless for correlation prior to about 12.2 ka. Therefore, it has not been possible to formulate a correlation of the lower Connecticut Valley varve sequence with older varves in New York based on inclination records.

## DEGLACIATION OF SOUTHERN NEW HAMPSHIRE AND SOUTHERN MAINE

An atmospheric $^{14}$C chronology of deglaciation of southern New Hampshire and Maine has been formulated (Fig. 8) based on results compiled from varve and paleomagnetic correlations between the Merrimack and Connecticut Valleys, recent U.S. Geological Survey (USGS) mapping in southern Maine and New Hampshire (Koteff et al., 1993), and maps of ice-marginal features in southern Maine (Smith, 1980, 1982, 1985; Thompson and Borns, 1985). A connection was inferred between existing USGS data in southeastern New Hampshire (Koteff et al., 1993) and ice-front positions in the Merrimack Valley to reasonably fit the intervening topography (dashed lines in Fig. 8). The atmospheric $^{14}$C chronology for deglaciation of the Merrimack Valley (12.5–12.4 ka) is based on the analysis of the varve year ages of basal varves (NE varve 5700–6100) in the Merrimack Valley (Fig. 5) and the atmospheric $^{14}$C calibration of the New England varve chronology based on the $^{14}$C ages at Canoe Brook (Fig. 6; Table 1). No materials for $^{14}$C dating have been found in varves in the Merrimack Valley and no varve sections in the southern Lake Merrimack basin have been found that would allow an analysis of deglaciation in southernmost New Hampshire and Massachusetts. It is also important to realize that the deglaciation of the northern Lake Merrimack basin was during a prominent $^{14}$C plateau when 100 yr on the $^{14}$C time scale (12.5–12.4 ka) corresponds to ~400 varve or calibrated years (Fig. 6).

The reconstructed pattern of ice recession across southern Maine and New Hampshire does not fit previous models of deglaciation based on pollen stratigraphy and lake bottom bulk sediment $^{14}$C ages (Davis and Jacobson, 1985) or numerical ice-sheet modeling (Fastook and Hughes, 1982; Hughes et al., 1985). The proposed east-west–trending deglacial ice-front positions across the region, however, conform to the pattern of ice recession depicted by "ground truth" features such as ice-marginal deposits in southern Maine (Smith, 1980, 1982, 1985; Thompson and Borns, 1985; Smith and Hunter, 1989), southeastern New Hampshire (Koteff et al., 1993), and eastern Massachusetts (Stone and Peper, 1982). There is an obvious disparity between existing marine and bulk sediment $^{14}$C ages in Maine and the atmospheric $^{14}$C chronology for deglaciation in Lake Merrimack (Fig. 8), which is discussed later.

## VARVE AND PALEOMAGNETIC RESULTS FROM SOUTHERN MAINE

New varve sections have been analyzed in southern Maine as a test of the deglaciation chronology. Both varve measurements and paleomagnetic data from Cornish and Naples, Maine (Fig. 8), support the proposed correlation of ice-front positions across the region. The most conclusive new data come from Cornish in the Ossipee River Valley, where a section of ~40 varves (Fig. 9A) directly overlies ice-proximal sand and gravel and grades upward into lacustrine sand (Holland, 1986). The varves at this section closely approximate the time of deglaciation at Cornish. An undisturbed sequence of 40 varves was measured and very closely matches the Merrimack and lower Connecticut Valley varve sequences of Antevs (1922, NE varves 6077–6117, Fig. 10). In addition, 26 paleomagnetic samples from 13 different horizons within the sequence yielded an average declination of 41°W (319°, Table 2), indicating a unique correlation with the strong western declination maximum that appears in declination records from New England and New York (Fig. 7). The proposed construction of ice-front positions and both the varve and paleomagnetic correlations of the Cornish sequence with the Merrimack Valley varves are consistent with an atmospheric $^{14}$C age of 12.45 ka for deglaciation at Cornish (Fig. 8).

At Naples in the Crooked Brook valley, sandy varves are exposed over till in a kettle filling (Fig. 9B). The middle of the section is interrupted by a massive sand bed that is the result of mass movement, and the varves are separated into lower (59 varves) and upper (28 varves) sequences. Because of frequent bedding disturbances and varve thickness changes across the face of the outcrop, it was not expected that the Naples sequence would produce a reliable varve record for matching to established chronologies, and no unequivocal match with any of Antevs' (1922, 1928) sequences was found. The varves at Naples are younger than the varves at Cornish because of their more northward occurrence and a possible delay in deposition at the base of a kettle hole. Along with this time constraint it was hoped that paleomagnetic samples could at least provide a test of the age of the sequence. Groups of paleomagnetic samples from horizontal beds in both the upper and lower varves yielded a declination of 16°–18°W (344°–342°) and an upsection secular change of about 1.5° to the east (Table 2). An eastward shifting declination of 18°–16°W in New England (Fig. 7) that postdates the deposits at Cornish corresponds to a NE varve age no older than 6200–6300 (Merrimack) or 6800–6900 (upper Connecticut). The paleomagnetic record at this section does not provide an unequivocal correlation to the New England varve chronology. However, it is consistent with an age of deglaciation at Naples of ca. 12.2 ka (250 $^{14}$C yr and 200–300 varve yr after Cornish) and is consistent with our ice-recession reconstruction (Fig. 8).

## IMPLICATIONS OF THE ATMOSPHERIC $^{14}$C CHRONOLOGY

There are several implications that the new atmospheric $^{14}$C chronology has for the history of deglaciation, sea-level change, and postglacial isostatic adjustment in eastern New England. The new chronology is based on only $^{14}$C ages of nonaquatic plant macrofossils and yields a chronology that is younger than previous chronologies that were formulated using uncorrected $^{14}$C ages derived from lake-bottom bulk sediment and marine fossils. Because a marine reservoir effect or the uptake of aquatic carbon did not influence the macrofossil samples from the Connecticut Valley varve sequences, the resulting ages are considered more representative of atmospheric $^{14}$C ages.

Figure 9. Varves from southern Maine. Knife for scale in both images is 27 cm. A, Varve couplets with thick clay (dark, winter) beds and laminated silt and sand (summer) beds (New England varves 6098–6102) at Cornish, Maine. B, Varve section in P and K Sand and Gravel Co. pit at Naples, Maine, showing part of lower varve sequence with thin dark clay (winter) beds and laminated silt and sand (summer) beds.

### Atmospheric chronology versus uncorrected marine and lacustrine bulk sediment ages

The new atmospheric $^{14}C$ chronology for deglaciation in southern New Hampshire shows an obvious disparity with uncorrected $^{14}C$ ages from lacustrine bulk sediment and marine fossils related to ice recession in the coastal area of southern Maine (Table 3; Fig. 8). All of the marine $^{14}C$ ages and relevant lake-bottom bulk sediment $^{14}C$ ages are 700–2200 yr older than ages inferred by the atmospheric $^{14}C$ chronology of this chapter. Further, uncorrected marine $^{14}C$ ages in southern Maine are not consistent with each other, which may be partly due to some marine fossils being from sediments that probably postdate deglaciation by several hundred years. Some ages were also obtained prior to the development of AMS $^{14}C$ measurement and improved laboratory techniques. However, another potential source of differences among the marine $^{14}C$ ages is the variation of both marine reservoir and meltwater errors in space and time and in fossils from different paleowater depths. We interpret our atmospheric $^{14}C$ chronology to indicate that the uncorrected marine $^{14}C$ ages along the southern coast of Maine have, at least in part, an error due to the combination of a reservoir effect and potential influences of glacial meltwater in the marine environment of the Gulf of Maine during deglaciation. Using the oldest uncorrected marine $^{14}C$ ages from ice-proximal marine sediments, a combined reservoir and meltwater correction on the order of 1–2 k.y. is necessary to make the atmospheric and uncorrected marine $^{14}C$ ages compatible.

### Sea-level change and postglacial isostatic adjustment

If the atmospheric $^{14}C$ chronology developed here is correct, a reexamination of comparisons between global events and the deglaciation of the coastal area of New England is required. Deglaciation of the coastal area of Maine and New Hampshire would have occurred entirely after Heinrich event 1 (15–14 ka). At 14 ka the edge of the ice sheet was still south of the present Maine coast and would have come on land in Massachusetts. The coastal area of Maine was deglaciated and underwent its first isostatic uplift during a period of rapid sea-level rise associated with the beginning of meltwater pulse 1a (between 12.6 and 11.7 ka; Fairbanks, 1989; Bard et al., 1990, 1993, 1995; Clark et al., 1996; Adkins et al., 1998).

Koteff et al. (1993) recognized evidence for rapid sea-level rise prior to the onset of postglacial isostatic uplift along the coast of southern Maine and New Hampshire. They estimated

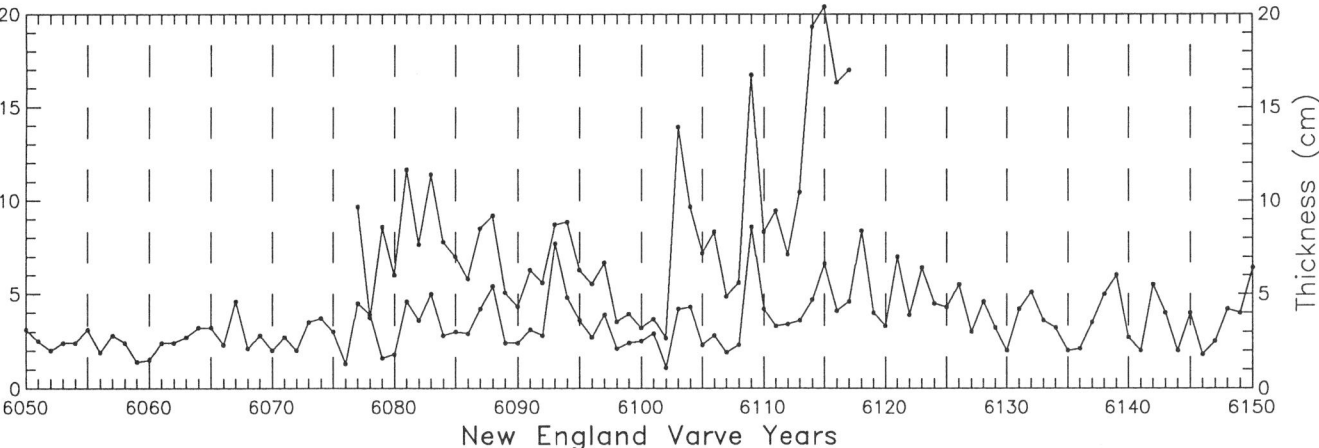

Figure 10. Correlation of varves at Cornish, Maine (top), with Merrimack Valley normal record (bottom) of Antevs (1922). Varves at Cornish are plotted at 0.67X.

TABLE 3. $^{14}$C AGES RELEVANT TO THE DEGLACIATION OF SOUTHERN COASTAL MAINE AND NEW HAMPSHIRE

| Age ($^{14}$C yr BP) | Laboratory number | Location | Material dated | Reference |
|---|---|---|---|---|
| 13 020 ± 240 | Y-1477 | Embden Pond, Embden, Maine | Shells from marine sediment | Stuiver and Borns, 1975 |
| 13 200 ± 120 | Y-2208 | Kennebunk, Maine | Shells from deformed marine sediment | Stuiver and Borns, 1975 |
| 13 830 ± 100 | QL-192 | Great Hill, Kennebunk, Maine | Marine shells from between tills | J.T. Andrews, unpub. in Smith, 1985 |
| *11 790 ± 395 | SI-5370 | Gould Pond, Dexter, Maine | Bulk sample of basal lacustrine sediment with plant remains | Anderson et al., 1992 |
| 13 280 ± 410 | SI-5371 | Gould Pond, Dexter, Maine | Carbonate fraction of marine silt and clay containing shells. Just below glaciomarine limit | Anderson et al., 1992 |
| *25 280 ± 1010 | SI-5372 | Gould Pond, Dexter, Maine | Carbonate fraction of basal marine silt and clay containing shells. Just below glaciomarine limit | Anderson et al., 1992 |
| 13 300 ± 50 | OS-4419 | Pownal, Maine | *Mytilus edulis* in shallow water marine sediment at 60 m above modern sea level | Weddle and Retelle, 1995 |
| 13 500 ± 300 | I-5639 | Moulton Pond, Bucksport, Maine | Bulk sample of lake-bottom sediment | Davis et al., 1975 |
| 13 540 ± 120 | AA-4855 | Continental shelf, New Hampshire | *Portlandia arctica* shells inferred to post-date deglaciation by 1000 yr. | Birch, 1990 |
| 13 670 ± 145 | AA-2572 | Continental shelf, New Hampshire | *Portlandia arctica* shells inferred to post-date deglaciation by 1000 yr. | Birch, 1990 |
| 14 045 ± 95 | AA-10164 | Freeport, Maine | *Portlandia arctica* shell | Weddle, 1994 |
| 14 090 ± 450 | PITT-0743 | Offshore of Prouts Neck, Maine | *Macoma baltica* shell, 21 m below modern sea level | Kelley et al., 1992 |
| 14 820 ± 105 | AA-10166 | Scarborough, Maine | Marine shell | Weddle et al., 1994 |

*Not shown in Figure 8.

that a rise of 7–10 m may have occurred over a period of 50–100 yr ca. 14.0 ka based on uncorrected marine $^{14}$C ages and strandline data from coastal Maine and New Hampshire. The rapid introduction of freshwater to the ocean was suggested by Koteff et al. (1993) to explain rapid changes in oxygen isotope ratios in planktonic forams ca. 14.0 ka in a core from off the coast of Portugal (Bard et al., 1987). However, recent investigation of a later oxygen isotope event of even greater magnitude in the same Portugal core (12.6–11.7 ka) does not support the claim that oxygen isotope anomalies in cores from the North Atlantic necessarily represent large volumes of meltwater. The anomaly at 12.6–11.7 ka, while it appears to be coincident with meltwater pulse 1a, indicates a discharge of freshwater from North America that would have been <10% of the 24 m of water introduced to the ocean from 12.6 to 11.7 ka (Clark et al., 1996). In addition, studies of sea-level change from corals have yet to resolve a major sea-level rise event ca. 14.0 ka (Fairbanks, 1989; Bard et al., 1990, 1993). There also appears to be no reason to completely ignore a reservoir or meltwater correction of marine $^{14}$C ages along the coast of New England, especially when comparing the New England ages to marine records from the northeastern Atlantic or tropical corals where a standard reservoir correction of 400–500 yr has been applied. Our proposed atmospheric $^{14}$C chronology places the period of rapid sea-level rise along the coast of New England ca. 12.5 ka, an age that is considerably younger than that proposed by Koteff et al. (1993). Our age closely corresponds to the beginning of meltwater pulse 1a that was well underway by 14.2 ka (calendar; 12.3 $^{14}$C ka) with a rate of sea-level rise of 27 m/k.y. at that time (Adkins et al., 1998). Meltwater pulse 1a is also an event that is well documented by studies of tropical corals as well as other marine data (Fairbanks, 1989; Bard et al., 1989, 1990, 1993; Clark et al., 1996), and now appears to coincide in time with a sudden decrease in the ventilation of deep water in the North Atlantic (Adkins et al., 1998).

Evidence was presented by Koteff et al. (1993) for delayed isostatic uplift until after the period of rapid sea-level rise on the coast of New England. Our atmospheric $^{14}$C model indicates that isostatic uplift was delayed until at least 12.5 ka. This delay is in agreement with an apparent delay in uplift in the Champlain Valley. The highest shoreline for the last glacial lake in the Champlain Valley (Fort Ann Phase of lake Vermont) has a steep water plane (1 m/km) that dips only slightly more than the highest water plane of the Champlain Sea (0.9 m/km; Parent and Occhietti, 1988). The invasion of the Champlain Sea occurred ca. 11.6–11.0 ka (Rodrigues, 1988, 1992; Anderson, 1988). Given the minor amount of tilting (0.1 m/km) that occurred prior to this event, uplift appears to have been initiated only shortly before the marine invasion and was probably delayed until at least some time after 12.0 ka.

Additional evidence for delayed isostatic uplift along the southern coast of Maine until 12.0 ka comes from Gould Pond, where terrestrial materials have been recovered immediately above marine deposits. Basal lacustrine sediment with unspecified plant fossils at Gould Pond (location in Fig. 8) has a bulk $^{14}$C age of 11.8 ka and directly overlies marine silt and clay with shells that have an uncorrected $^{14}$C age of 13.3 ka (Table 3). Gould Pond is very close to the marine limit, where a lacustrine environment replaced an initial marine environment as soon as regression was initiated by isostatic uplift. Despite potential errors related to a bulk sediment sample that could produce an anomalous old age, the lacustrine sediment, which very nearly records the beginning of isostatic uplift, has an age younger than 12.0 ka. In southern Maine, near Portland (Fig. 8), is another locality where terrestrial, nonaquatic plant fossils have been recovered from glaciomarine sediment. However, a 1700 yr range of $^{14}$C ages for spruce needles, cones, and logs, and the occurrence of the fossils in the base of glaciomarine clay ~65 m below the glaciomarine limit (Hyland et al., 1978; Anderson et al., 1990) are perhaps best explained by emplacement of the fossils by mass movement that followed marine regression (T. Weddle, 1998, personal commun.).

## CONCLUSIONS

The approach we have taken in developing an atmospheric $^{14}$C chronology for the deglaciation of eastern New England has been one that eliminates ambiguities associated with uncorrected $^{14}$C ages from lacustrine bulk sediment and marine fossils. Work done prior to the development of AMS $^{14}$C analyses relied heavily on uncorrected marine and lacustrine bulk sediment $^{14}$C ages, because large samples of nonaquatic plants in sediment closely associated with deglaciation were not available. With the advent of AMS $^{14}$C analysis it has been possible to obtain atmospheric $^{14}$C analyses from very small samples of nonaquatic plant macrofossils. At present, there appears to be no reason to use uncorrected marine $^{14}$C ages verbatim given the well-documented corrections necessary to account for a marine reservoir effect (Mangerud, 1972; Hjort, 1973; Mangerud and Gullikson, 1975; Bard, 1988; Bard et al., 1994; Birks et al., 1996) and potential meltwater influences in a glaciomarine environment (Sutherland, 1986; Hillaire-Marcel, 1988; Rodrigues, 1988, 1992). The most important result of our findings is that in terms of atmospheric $^{14}$C years, eastern New England was deglaciated ~1–1.5 k.y. later than depicted by previous models based on uncorrected $^{14}$C ages for lacustrine bulk sediment and marine fossils (Smith, 1985; Davis and Jacobson, 1985; Thompson and Borns, 1985). The new atmospheric $^{14}$C chronology represents a more accurate way of comparing deglacial events in New England with global climatic events because it eliminates uncertainties associated with previous $^{14}$C ages and it allows a more accurate calibration of $^{14}$C ages to calendar years.

In addition to using only atmospheric $^{14}$C ages, we have incorporated the use of other correlation techniques, varve chronology and paleomagnetism, that allow a wider application of the $^{14}$C chronology. Varve chronology and paleomagnetism are techniques that can be applied over most of the New England landscape because glacial lakes that formed at the close of the last glaciation were common and varves can be studied in outcrop and from cores in many drainage basins. The combined use of atmospheric $^{14}$C ages, varve chronology, and paleomagnetism has also created a system of independent tests of correlations with much greater precision than was previously possible. The rebirth of the New England varve chronology, ~70 yr after its original formulation by Antevs (1922, 1928), appears to provide a means of correlation with unprecedented resolution.

## ACKNOWLEDGMENTS

This study was partly supported by Gene Boudette, the New Hampshire State Geologist, and funds for student research from the Geology Department at Tufts University. A Tufts University Faculty Research Grant provided additional funding for $^{14}$C ages. We thank Woody Thompson, Carol Hildreth, Tom Weddle, David Franzi, Richard Pendleton, and Lee Gaudette for helping us locate many varve sections and for their assistance in the field. Carl Koteff, Gail Ashley, Julie Brigham-Grette, and Mark Besonen provided many provocative discussions in the field. We also thank Don Pair, Tom Weddle, and Al Werner for their very helpful reviews of the manuscript.

## REFERENCES CITED

Abbott, M.B., and Stafford, T.W., 1996, Radiocarbon geochemistry of modern and ancient arctic lake systems, Baffin Island, Canada: Quaternary Research, v. 45, p. 300–311.

Adkins, J.F., Cheng, H., Boyle, E.A., Druffel, E.R.M., and Edwards, R.L., 1998, Deep-sea coral evidence for rapid change in ventilation of the deep North Atlantic 15,400 years ago: Science, v. 280, p. 725–728.

Anderson, R.S., Miller, N.G., Davis, R.B., and Nelson, R.E., 1990, Terrestrial fossils in the marine Presumpscot Formation: Implications for late Wisconsinan paleoenvironments and isostatic rebound along the coast of Maine: Canadian Journal of Earth Sciences, v. 27, p. 1241–1246.

Anderson, R.S., Jacobson, G.L., Jr., Davis, R.B., and Stuckenrath, R., 1992, Gould Pond, Maine: Late-glacial transitions from marine to upland environments: Boreas, v. 21, p. 359–371.

Anderson, T.W., 1988, Late Quaternary pollen stratigraphy of the Ottawa Valley–Lake Ontario region and its application in dating the Champlain Sea, in Gadd, N.R., ed., The late Quaternary development of the Champlain Sea basin: Geological Association of Canada Special Paper 35, p. 205–224.

Andrée, M., Oeschger, H., Siegenthaler, U., Riesen, T., Möll, M., Ammann, B., and Tobolski, K., 1986, $^{14}$C dating of plant macrofossils in lake sediment: Radiocarbon, v. 28, p. 411–416.

Antevs, E., 1922, The recession of the last ice sheet in New England: American Geographical Society Research Series no. 11, 120 p.

Antevs, E., 1928, The last glaciation with special reference to the last ice sheet in North America: American Geographical Society Research Series no. 17, 292 p.

Bard, E., 1988, Correction of accelerator mass spectrometry $^{14}$C ages measured in planktonic foraminifera: Paleoceanographic implications: Paleoceanography, v. 3, p. 635–645.

Bard, E., Arnold, M., Maurice, P., Duprat, J., Moyes, J., and Duplessy, J.-C., 1987, Retreat velocity of the North Atlantic polar front during the last deglaciation determined by $^{14}$C accelerator mass spectrometry: Nature, v. 328, p. 791–794.

Bard, E., Hamelin, B., Fairbanks, R.G., and Zindler, A., 1990, Calibration of the $^{14}$C time scale over the past 30,000 years using mass spectrometric U-Th ages from Barbados corals: Nature, v. 345, p. 405–410.

Bard, E., Arnold, M., Fairbanks, R.G., and Hamelin, B., 1993, $^{230}$Th-$^{234}$U and $^{14}$C ages obtained by mass spectrometry on corals: Radiocarbon, v. 35, p. 191–199.

Bard, E.O., Arnold, M., Mangerud, J., Paterne, M., Labeyrie, L., Duprat, J., Mélières, M.-A., Sønstegaard, E., and Duplessy, J.-C., 1994, The North Atlantic atmosphere-sea surface $^{14}$C gradient during the Younger Dryas climatic event: Earth and Planetary Science Letters, v. 126, p. 275–287.

Birch, F.S., 1990, Radiocarbon dates of Quaternary sedimentary deposits on the inner continental shelf of New Hampshire: Northeastern Geology, v. 12, p. 218–230.

Birks, H.H., Gulliksen, S., Haflidason, H., Mangerud, J., and Possnert, G., 1996, New radiocarbon dates from the Vedde Ash and the Saksunarvatn Ash from western Norway: Quaternary Research, v. 45, p. 119–127.

Bond, G., and Lotti, R., 1995, Iceberg discharges into the North Atlantic on millennial time scales during the last glaciation: Science, v. 267, p. 1005–1010.

Bond, G., Heinrich, H., Broecker, W., Labeyrie, L., McManus, J., Andrews, J., Huon, S., Jantschik, R., Clasen, S., Simet, C., Tedesco, K., Klas, M., Bonani, G., and Ivy, S., 1992, Evidence for massive discharges of icebergs into the North Atlantic Ocean during the last glacial period: Nature, v. 360, p. 245–249.

Brennan, W.J., Hamilton, M., Kilbury, R., Reeves, R.L., and Covert, L., 1984, Late Quaternary secular variation of geomagnetic declination in western New York: Earth and Planetary Science Letters, v. 70, p. 363–372.

Broecker, W.S., 1990, Salinity history of the North Atlantic during the last deglaciation: Paleoceanography, v. 5, p. 459–467.

Clark, P.U., Alley, R.B., Keigwin, L.D., Licciardi, J.M., Johnsen, S.J., and Wang, H., 1996, Origin of the first global meltwater pulse following the last glacial maximum: Paleoceanography, v. 11, p. 563–577.

Cuffey, K.M., Clow, G.D., Alley, R.B., Stuiver, M., Waddington, E.D., and Saltus, R.W., 1995, Large arctic temperature change at the Wisconsin-Holocene glacial transition: Nature, v. 270, p. 455–458.

Dansgaard, W., Johnsen, S.J., Clausen, H.B., Dahl-Jensen, D., Gundestrup, N.S., Hammer, C.U., Hvidberg, C.S., Steffensen, J.P., Sveinbjörnsdottir, A.E., Jouzel, J., and Bond, G., 1993, Evidence for general instability of past climate from a 250-kyr ice core record: Nature, v. 364, p. 218–220.

Davis, M.B., and Ford, M.S., 1982, Sediment focusing in Mirror Lake, New Hampshire: Limnology and Oceanography, v. 27, p. 137–150.

Davis, M.B., Spear, R.W., and Shane, L.C.K., 1980, Holocene climate of New England: Quaternary Research, v. 14, p. 240–250.

Davis, P.T., and Davis, R.B., 1980, Interpretation of minimum-limiting radiocarbon dates for deglaciation of Mount Katahdin area, Maine: Geology, v. 8, p. 396–400.

Davis, P.T., Dethier, D.P., and Nickmann, R., 1995, Deglaciation chronology and late Quaternary pollen records from Woodford Bog, Bennington County, Vermont: Geological Society of America Abstracts with Programs, v. 27, no. 1, p. 38.

Davis, R.B., and Jacobson, G.L., Jr., 1985, Late glacial and early Holocene landscapes in northern New England and adjacent areas of Canada: Quaternary Research, v. 23, p. 341–368.

Davis, R.B., Bradstreet, T.E., Stuckenrath, R., Jr., and Borns, H.W., Jr., 1975, Vegetation and associated environments during the past 14,000 years near Moulton Pond, Maine: Quaternary Research, v. 5, p. 435–465.

Edwards, R.L., Beck, J.W., Burr, G.S., Donahue, D.L., Chappell, J.M.A., Bloom, A.L., Druffel, E.R.M., and Taylor, F.W., 1993, A large drop in atmospheric $^{14}$C/$^{12}$C and reduced melting in the Younger Dryas, documented with $^{230}$Th ages of corals: Science, v. 260, p. 962–967.

Fairbanks, R.G., 1989, A 17,000-year glacio-eustatic sea level record: Influence of glacial melting rates on the Younger Dryas event and deep-ocean circulation: Nature, v. 342, p. 637–642.

Fairbanks, R.G., 1990, The age and origin of the "Younger Dryas Climate Event" in Greenland ice cores: Paleoceanography, v. 5, p. 937–948.

Fastook, J.L., and Hughes, T., 1982, A numerical model for reconstruction and disintegration of the late Wisconsin glaciation in the Gulf of Maine, in Larson, G.J., and Stone, B.D., eds., Late Wisconsinan glaciation of New England: Dubuque, Iowa, Kendall/Hunt Publishing, p. 229–242.

Heinrich, H., 1988, Origin and consequences of cyclic ice rafting in the northeast Atlantic Ocean during the past 130,000 yr: Quaternary Research, v. 29, p. 142–152.

Hillaire-Marcel, C., 1988, Isotopic composition ($^{18}$O, $^{13}$C, $^{14}$C) of biogenic carbonates in Champlain sea sediments, in Gadd, N.R., ed., The late Quaternary development of the Champlain Sea basin: Geological Association of Canada Special Paper 35, p. 177–194.

Hjort, C., 1973, A sea correction for East Greenland: Geologiska Föreningens i Stockholm Förhandlingar, v. 95, p. 132–134.

Holland, W., 1986, Features associated with the deglaciation of the upper Saco and Ossipee River basins, northern York and southern Oxford Counties, Maine, in Newberg, D.W., ed., Guidebook for field trips in southwestern Maine: 78th Annual Meeting New England Intercollegiate Geological Conference: Lewiston, Maine, Bates College, p. 98–123.

Hughes, T., Borns, H.W., Jr., Fastook, J.L., Hyland, M.R., Kite, J.S., and Lowell, T.V., 1985, Models of glacial reconstruction and deglaciation applied to Maritime Canada and New England, in Borns, H.W., Jr., et al., eds., Late Pleistocene history of northeastern New England and adjacent Quebec: Geological Society of America Special Paper 197, p. 139–150.

Hyland, F., Thompson, W.B., and Stuckenrath, R., 1978, Late Wisconsinan wood and other tree remains in the Presumpscot Formation, Portland, Maine: Maritime Sediments, v. 14, p. 103–120.

Johnson, E.A., Murphy, T., and Torreson, O.W., 1948, Pre-history of the Earth's magnetic field: Terrestrial Magnetism and Atmospheric Electricity (now Journal of Geophysical Research), v. 53, p. 349–372.

Karrow, P.F., and Anderson, T.W., 1975, Palynological studies of lake sediment profiles from SW New Brunswick: Discussion: Canadian Journal of Earth Sciences, v. 12, p. 1808–1812.

Keigwin, L.D., and Lehman, S.J., 1994, Deep circulation change linked to Heinrich Event 1 and Younger Dryas in a mid-depth North Atlantic core: Paleoceanography, v. 9, p. 185–194.

Keigwin, L.D., Jones, G.A., Lehman, S.J., and Boyle, E.A., 1991, Deglacial meltwater discharge, north Atlantic deep circulation, and abrupt climate change: Journal of Geophysical Research, v. 96, p. 16811–16826.

Kelley, J.T., Dickson, S.M., Belknap, D.F., and Stuckenrath, R., Jr., 1992, Sea-level change and the introduction of late Quaternary sediment to the southern Maine inner continental shelf, in Fletcher, C.H., and Wehmiller, J.F., eds., Quaternary coasts of the United States: Marine and lacustrine systems: SEPM (Society for Sedimentary Geology) Special Publication 48, p. 23–34.

Koteff, C., Robinson, G.R., Goldsmith, R., and Thompson, W.B., 1993, Delayed postglacial uplift and synglacial sea levels in coastal central New England: Quaternary Research, v. 40, p. 46–54.

Lini, A., Bierman, P.R., and Lin, L., 1995, Stable carbon isotopes in post-glacial lake sediments: A technique for timing the onset of primary productivity and verifying AMS 14-C dates: Geological Society of America Abstracts with Programs, v. 27, no. 6, p. 58.

Mangerud, J., 1972, Radiocarbon dating of marine shells including a discussion of apparent age of recent shells from Norway: Boreas, v. 1, p. 143–172.

Mangerud, J., and Gulliksen, S., 1975, Apparent radiocarbon ages of recent marine shells from Norway, Spitsbergen, and Arctic Canada: Quaternary Research, v. 5, p. 263–274.

McNish, A.G., and Johnson, E.A., 1938, Magnetization of unmetamorphosed varves and marine sediments: Terrestrial Magnetism and Atmospheric Electricity (now Journal of Geophysical Research), v. 43, p. 401–407.

Miller, N.G., and Thompson, G.G., 1979, Boreal and western North American plants in the Pleistocene of Vermont: Journal of the Arnold Arboretum, v. 60, p. 167–218.

Mott, R.J., 1975, Palynological studies of lake sediment profiles from southwestern New Brunswick: Canadian Journal of Earth Sciences, v. 12, p. 273–288.

Mott, R.J., 1981, Palynology of southeastern Quebec, in Shilts, W.W., ed., Surficial geology of the Lac Mégantic Area, Québec: Geological Survey of Canada Memoir 397, p. 99–102.

Muller, E.H., and Calkin, P.E., 1993, Timing of Pleistocene events in New York State: Canadian Journal of Earth Sciences, v. 30, p. 1829–1845.

Muller, E.H., and Prest, V.K., 1985, Glacial lakes in the Ontario Basin, in Karrow, P.F., and Calkin, P.E., eds., Quaternary evolution of the Great Lakes: Geological Association of Canada Special Paper 30, p. 212–229.

Muller, E.H., Franzi, D.A., and Ridge, J.C., 1986, Pleistocene geology of the western Mohawk Valley, New York, in Cadwell, D.H., ed., The Wisconsinan Stage of the first geological district, eastern New York: New York State Museum Bulletin 455, p. 143–157.

Oeschger, H., Andrée, M., Möll, M., Riesen, T., Siegenthaler, U., Ammann, B., Tobolski, K., Bonani, B., Hofmann, H.J., Morenzoni, E., Nessi, M., Suter, M., and Wölfli, W., 1985, Radiocarbon chronology of Lobsigensee. Comparison of materials and methods, in Lang, G., ed., Swiss lake and mire environments during the last 15,000 years: Dissertationes Botanicae 87, p. 135–139.

Pair, D.L., Muller, E.H., and Plumley, P.W., 1994, Correlation of late Pleistocene glaciolacustrine and marine deposits by means of geomagnetic secular variation, with examples from northern New York and southern Ontario: Quaternary Research, v. 42, p. 277–287.

Parent, M., and Occhietti, S., 1988, Late Wisconsinan deglaciation and Champlain Sea invasion in the St. Lawrence Valley, Quebec: Géographie Physique et Quaternaire, v. 42, p. 215–246.

Pendleton, R.M., 1995, Deglacial history of the Penacook quadrangle, and Holocene evolution of the upper Merrimack River, New Hampshire [Master's thesis]: Durham, University of New Hampshire, 107 p.

Ridge, J.C., 1997, Shed Brook discontinuity and Little Falls Gravel: Evidence for the Erie Interstade in central New York: Geological Society of America Bulletin, v. 109, p. 652–665.

Ridge, J.C., and Toll, N.J., 1999, Are late-glacial climate oscillations recorded in varves of the upper Connecticut Valley, northeastern United States?: Geologiska Foreningens i Stockholm Forhandlingar, v. 121 (Pt. 3), p. 187–193.

Ridge, J.C., and Larsen, F.D., 1990, Re-evaluation of Antevs' New England varve chronology and new radiocarbon dates of sediments from glacial Lake Hitchcock: Geological Society of America Bulletin, v. 102, p. 889–899.

Ridge, J.C., Brennan, W.J., and Muller, E.H., 1990, The use of paleomagnetic declination to test correlations of late Wisconsinan glaciolacustrine sediments in central New York: Geological Society of America Bulletin, v. 102, p. 26–44.

Ridge, J.C., Franzi, D.A., and Muller, E.H., 1991, Late Wisconsinan, pre-Valley Heads glaciation in the western Mohawk Valley, central New York, and its regional implications: Geological Society of America Bulletin, v. 103, p. 1032–1048.

Ridge, J.C., Thompson, W.B., Brochu, M., Brown, S., and Fowler, B., 1996, Glacial geology of the upper Connecticut Valley in the vicinity of the lower Ammonoosuc and Passumpsic Valleys of New Hampshire and Vermont, in Van Baalen, M.R., ed., Guidebook to field trips in northern New Hampshire and adjacent regions of Maine and Vermont: 88th Annual Meeting of the New England Intercollegiate Geological Conference: Cambridge, Massachusetts, Harvard University, p. 309–340.

Ridge, J.C., Besonen, M.R., Brochu, M., Brown, S., Callahan, J.W., Cook, G.J., Nicholson, R.S., and Toll, N.J., 1999, Varve, paleomagnetic, and $^{14}$C chronologies for late Pleistocene events in New Hampshire and Vermont, U.S.A.: Géographie Physique et Quaternaire, v. 53, p. 79–107.

Rittenour, T.M., 1999, Drainage history of glacial Lake Hitchcock, northeastern USA [Ph.D. thesis]: Amherst, University of Massachusetts, 179 p.

Rodrigues, C.G., 1988, Late Quaternary invertebrate faunal associations and chronology of the western Champlain Sea basin, in Gadd, N.R., ed., The late Quaternary development of the Champlain Sea basin: Geological Association of Canada Special Paper 35, p. 155–176.

Rodrigues, C.G., 1992, Successions of invertebrate microfossils and the late Quaternary deglaciation of the central St. Lawrence Lowland, Canada and United States: Quaternary Science Reviews, v. 11, p. 503–534.

Shotton, F.W., 1972, An example of hard water error in radiocarbon dating of vegetable matter: Nature, v. 240, p. 460–461.

Smith, G.W., 1980, End moraines and glaciofluvial deposits, Cumberland and York Counties, Maine: Augusta, Maine Geological Survey, scale 1:125 000.

Smith, G.W., 1982, End moraines and the pattern of last ice retreat from central and south coastal Maine, in Larson, G.J., and Stone, B.D., eds., Late Wisconsinan glaciation of New England: Dubuque, Iowa, Kendall/Hunt Publishing, p. 195–209.

Smith, G.W., 1985, Chronology of late Wisconsinan deglaciation of coastal Maine, *in* Borns, H.W., Jr., et al., eds., Late Pleistocene history of northeastern New England and adjacent Quebec: Geological Society of America Special Paper 197, p. 29–44.

Smith, G.W., and Hunter, L.E., 1989, Late Wisconsinan deglaciation of coastal Maine, *in* Tucker, R.D., and Marvinney, R.G., eds., Studies in Maine geology, Volume 6: Quaternary geology: Augusta, Maine Geological Survey, p. 13–32.

Stone, B.D., and Borns, H., Jr., 1986, Pleistocene glacial and interglacial stratigraphy of New England, Long Island, and adjacent Georges Bank and Gulf of Maine, *in* Sibrava, V., et al., eds., Quaternary glaciations in the Northern Hemisphere: New York, Pergamon Press, p. 39–52.

Stone, B.D., and Peper, J.D., 1982, Topographic control of the deglaciation of eastern Massachusetts: Ice lobation and the marine incursion, *in* Larson, G.J., and Stone, B.D., eds., Late Wisconsinan glaciation of New England: Dubuque, Iowa, Kendall/Hunt Publishing, p. 145–166.

Stuiver, M., and Borns, H.W., Jr., 1975, Late Quaternary marine invasion in Maine: Its chronology and associated crustal movement: Geological Society of America Bulletin, v. 86, p. 99–104.

Stuiver, M., and Reimer, P.J., 1993, Extended $^{14}$C data base and revised CALIB 3.0 $^{14}$C age calibration program: Radiocarbon, v. 35, p. 215–230.

Stuiver, M., Grootes, P.M., and Braziunas, T.F., 1995, The GISP2 $\delta^{18}$O climate record of the past 16,500 years and the role of the sun, ocean, and volcanoes: Quaternary Research, v. 44, p. 341–354.

Stuiver, M., Reimer, P.J., Bard, E., Beck, J.W., Burr, G.S., Hughen, K.A., Kromer, B., McCormac, F.G., van der Plicht, J., and Spark, M., 1998, INTCAL98 radiocarbon age calibration 24,000–0 cal BP: Radiocarbon, v. 40, p. 1041–1083.

Sutherland, D.G., 1986, A review of Scottish marine shell radiocarbon dates, their standardization and interpretation: Scottish Journal of Geology, v. 22, p. 145–164.

Thompson, W.B., and Borns, H.W., Jr., eds., 1985, Surficial geologic map of Maine: Augusta, Maine Geological Survey, scale 1:500 000.

Thompson, W.B., Fowler, B.K., Flanagan, S.M., and Dorian, C.C., 1996, Recession of the Late Wisconsinan ice sheet from the northwestern White Mountains, N.H., *in* Van Baalen, M.R., ed. Guidebook to field trips in northern New Hampshire and adjacent regions of Maine and Vermont: 88th Annual Meeting of the New England Intercollegiate Geological Conference: Cambridge, Massachusetts, Harvard University, p. 203–234.

Thompson, W.B., Fowler, B.K., and Dorian, C.C., 1999, Deglaciation of the northwestern White Mountains, New Hampshire: Géographie Physique et Quaternaire, v. 53, p. 59–77.

Verosub, K.L., 1979a, Paleomagnetism of varved sediments from western New England: Secular variation: Geophysical Research Letters, v. 6, p. 245–248.

Verosub, K.L., 1979b, Paleomagnetism of varved sediments from western New England: Variability of the recorder: Geophysical Research Letters, v. 6, p. 241–244.

Weddle, T.K., 1994, Surficial geology of the Freeport 7.5-minute quadrangle, Cumberland County, Maine: Maine Geological Survey Open-File 94-5, 11 p., scale 1:24 000.

Weddle, T.K., and Retelle, M.J., 1995, Glaciomarine deposits of the late Wisconsinan Casco Bay sublobe of the Laurentide ice sheet, *in* Hussey, A.M., and Johnston, R.A., eds., Guidebook to field trips in southern Maine and adjacent New Hampshire: 87th Annual Meeting of the New England Intercollegiate Geological Conference: Brunswick, Maine, Bowdoin College, p. 173–193.

Weddle, T.K., Lowell, T.V., and Dorian, C.C., 1994, Glacial geology of the Penobscot River basin between Millinocket and Medway, *in* Hanson, L.S., and Caldwell, D.W., eds., Guidebook to field trips in north-central Maine: 86th Annual Meeting of the New England Intercollegiate Geological Conference: Augusta, Maine Geological Survey, p. 193–212.

Wohlfarth, B., 1996, The chronology of the last termination: A review of radiocarbon-dated, high-resolution terrestrial stratigraphies: Quaternary Science Reviews, v. 15, p. 267–284.

Wohlfarth, B., Björck, S., Possnert, G., Lemdahl, G., Brunnberg, L., Ising, J., Olsson, S., and Svensson, N.-O., 1993, AMS dating Swedish varved clays of the last glacial/interglacial transition and the potential/difficulties of calibrating Late Weichselian 'absolute' chronologies: Boreas, v. 22, p. 113–128.

Wohlfarth, B., Björck, S., and Possnert, G., 1995, The Swedish time scale: A potential calibration tool for the radiocarbon time scale during the late Weichselian: Radiocarbon, v. 37, p. 347–359.

Zijderveld, J.D.A., 1967, Demagnetization of rocks: Analysis of results, *in* Collinson, D.W., et al., eds., Methods in paleomagnetism: Amsterdam, Elsevier Publishing, p. 254–286.

MANUSCRIPT ACCEPTED BY THE SOCIETY JUNE 9, 2000

# Deglaciation and relative sea-level chronology, Casco Bay Lowland and lower Androscoggin River valley, Maine

Michael J. Retelle
*Department of Geology, Bates College, Lewiston, Maine 04240, USA*
Thomas K. Weddle
*Maine Geological Survey, State House Station 22, Augusta, Maine 04333, USA*

## ABSTRACT

Retreat of the late Wisconsinan Casco Bay sublobe of the Laurentide ice sheet and relative sea-level history in the Casco Bay Lowland and lower Androscoggin River valley, Maine, is represented by deglacial-phase deposits, including glaciomarine deltas, fans, stratified end moraines, and fossiliferous glaciomarine mud, as well as nearshore deposits associated with synglacial sea level and postglacial uplift. Deglaciation was systematic, but interrupted by minor readvances in the region along an east-northeast–trending lobate ice margin, oblique to the present coastline, and is documented by a consistent alignment of ice-marginal deposits.

Radiocarbon ages on marine shells (uncorrected for the marine-reservoir effect) and on terrestrial organics imply that southwestern coastal Maine was deglaciated by 15 ka, the Casco Bay region by 14 ka, and the White Mountain foothills by 13 ka. At a site along the present Maine coast in Phippsburg, isostatic adjustment had begun by 13.6 ka, and by this time relative sea level had already fallen from a maximum local marine limit of 64 m above sea level (asl) to 49 m asl. A relative sea-level curve for the Casco Bay Lowland and lower Androscoggin River valley is similar in form to the regional sea-level curve for Maine, but differs because it indicates that deglaciation and emergence of southwestern Maine occurred earlier than previously reported.

Radiocarbon ages on marine shells are the most common data used to infer age of deglaciation in coastal Maine. The deglaciation of the Casco Bay Lowland occurred between 14 and 13 ka, during the time of Heinrich event 1 in the North Atlantic Ocean. By 12.8 ka, ice was north of the study area and isostatic emergence had caused relative sea level in the lower Androscoggin Valley to drop to 45 m asl. The Brunswick sand plain, a coastal braid-plain delta, had begun forming by the time the marine regression dropped to 35 m asl. On the basis of the local sea-level curve, the sand plain formed 12.5 to 12 ka, within the time span of meltwater pulse 1-A, when falling relative sea level may have stabilized as a response to the meltwater-induced eustatic sea-level rise.

Retelle, M.J., and Weddle, T.K., 2001, Deglaciation and relative sea-level chronology, Casco Bay Lowland and lower Androscoggin River valley, Maine, *in* Weddle, T.K., and Retelle, M.J., eds., Deglacial History and Relative Sea-Level Changes, Northern New England and Adjacent Canada: Boulder, Colorado, Geological Society of America Special Paper 351, p. 191–214.

# INTRODUCTION

The recession of the late Wisconsinan Casco Bay sublobe of the Laurentide ice sheet in southwestern Maine is represented by deposits of a marine-based ice sheet, including glaciomarine sediments that accumulated in an ice-marginal sea, as well as nearshore deposits associated with sea level and postglacial isostatic rebound. The timing of late glacial events and ice retreat in the study area is defined by detailed mapping and stratigraphically controlled radiocarbon age estimates on materials found in association with ice-marginal positions and emergence deposits (Weddle et al., 1993; Weddle and Retelle, 1995, 1998). The altitudes of ice-marginal deltas and marine-limit indicators, such as raised beaches, deposited during ice retreat provide a record of sea-level fluctuations during this time, and characterize the nature of postglacial uplift and emergence of the region (Thompson et al., 1989; Koteff et al., 1993; Barnhardt et al., 1995).

In this chapter we provide revised late Wisconsinan ice retreat and relative sea-level chronologies for a marine-based sector of the southeastern margin of Laurentide ice sheet on the southwestern Maine coast in the Casco Bay area. The revised chronologies, based on recent detailed surficial mapping and stratigraphic studies, allow for comparison with offshore records in Maine (Barnhardt et al., 1997; Schnitker et al., this volume) and onshore and offshore records of marine-based ice retreat in eastern Maine and adjacent Nova Scotia (Dorion et al., this volume; Stea et al., this volume). In addition, the Casco Bay chronology provides a link for comparison with terrestrial-based ice retreat through central New England and the White Mountains (Ridge et al., 1999, and this volume; Thompson et al., 1999).

# LATE WISCONSINAN GLACIATION, DEGLACIATION, AND RELATIVE SEA LEVEL

The southwestern margin of the Laurentide ice sheet in New England had a lobate geometry controlled by subglacial topography (Stone and Borns, 1986). The western Gulf of Maine was inundated by ice from the Charles-Merrimack lobe in southwestern Maine (which in its southern extent formed the Cape Cod Bay lobe) and the South Channel lobe (Stone and Borns, 1986, p. 48).

Laurentide ice reached its late Wisconsin maximum extent on Georges Bank on the continental shelf shortly after 22 ka (Stone and Borns, 1986; King, 1996). Several recently proposed models describe the ice configuration at the maximum late Wisconsinan extent and during early stages of retreat in the Gulf of Maine. A regional deglaciation model proposes the existence of the Gulf of Maine marine ice stream fed by inland ice that occupied the lowland valleys of the region (Hughes et al., 1985). At its maximum extent, the ice stream fed a floating ice tongue over several of the deeper basins in the gulf. Based on examination of seismic facies and microfossil evidence in sediment cores, Belknap and Shipp (1991) proposed that during initial deglaciation of the outer continental shelf in the Gulf of Maine, the ice shelf was warm based and pinned in several places on bedrock topographic highs in the gulf.

Schnitker (1986) estimated that initial deglaciation of the central Gulf of Maine, in the Wilkinson basin, occurred prior to 17.6 ka. To the northeast, Gipp and Piper (1989) and King (1996) also noted the deglaciation of the central Emerald basin south of Nova Scotia between 17.7 and 18 ka, although sediments interpreted as ice proximal are found on the margin of the basin as late as 15 ka.

Hughes et al. (1985) and Hughes (1987) proposed that ice streams and floating ice shelves were replaced by a calving bay in the gulf by 14 ka. Belknap and Shipp (1991) and Bacchus (1993) described a glaciomarine unit on the inner continental shelf that drapes bedrock and is overlain by a ponded glaciomarine mud. They interpreted this stratigraphy to represent the shift from a floating ice margin to a calving embayment during deglaciation. As ice retreat progressed, calving along the floating margin eventually gave way to a grounded ice margin with calving along a tidewater cliff, similar to the conditions reported from glaciomarine tidewater settings in southeastern Alaska (Molnia, 1983; Phillips et al., 1991; Powell, 1991).

Alternatively, Oldale (1989) suggested that ice was grounded at its maximum position in the western gulf and that the sediments were produced by meltwater flux from point sources at the grounded tidewater terminus of the ice sheet. Rapid deglaciation of the South Channel and Cape Cod Bay lobes may have occurred from development of a concave ice margin by a calving front over the Wilkinson basin in the western Gulf of Maine (Stone and Peper, 1982).

Regardless of the mode of ice retreat on the continental shelf, numerous studies of glacial deposits exposed on land in the state of Maine and adjacent regions have demonstrated ice-contact deposits such as moraines and grounding line fans (cf. Powell, 1991) interbedded with glaciomarine sediments; from this we infer that deglaciation of the coastal lowland by a grounded tidewater glacier was contemporaneous with marine submergence of the region (Borns and Hagar, 1965; Stuiver and Borns, 1975; Thompson, 1982; Smith, 1985; Retelle and Bither, 1989; Ashley et al., 1991). Lowell and Borns (1994) suggested that the effectiveness of a calving embayment in the major valleys in Maine was likely impeded by topography and shallow water depths of the transgressing sea, and that deglaciation may have been more influenced by climate at that time. By 14–12 ka, due to waning of the ice, smaller sublobes of the retreating ice sheet occupied the topographic lowlands; we propose the name Casco Bay sublobe for the lobate ice mass that occupied the Casco Bay Lowland region. (Fig. 1A).

Shoreline features, such as deltas and beaches, that define the local marine limit (the upper level of submergence) increase in elevation from the coast to the interior of the state in response to the former increased isostatic load of the late Wisconsinan ice (Thompson et al., 1989). In the field area, marine limit in-

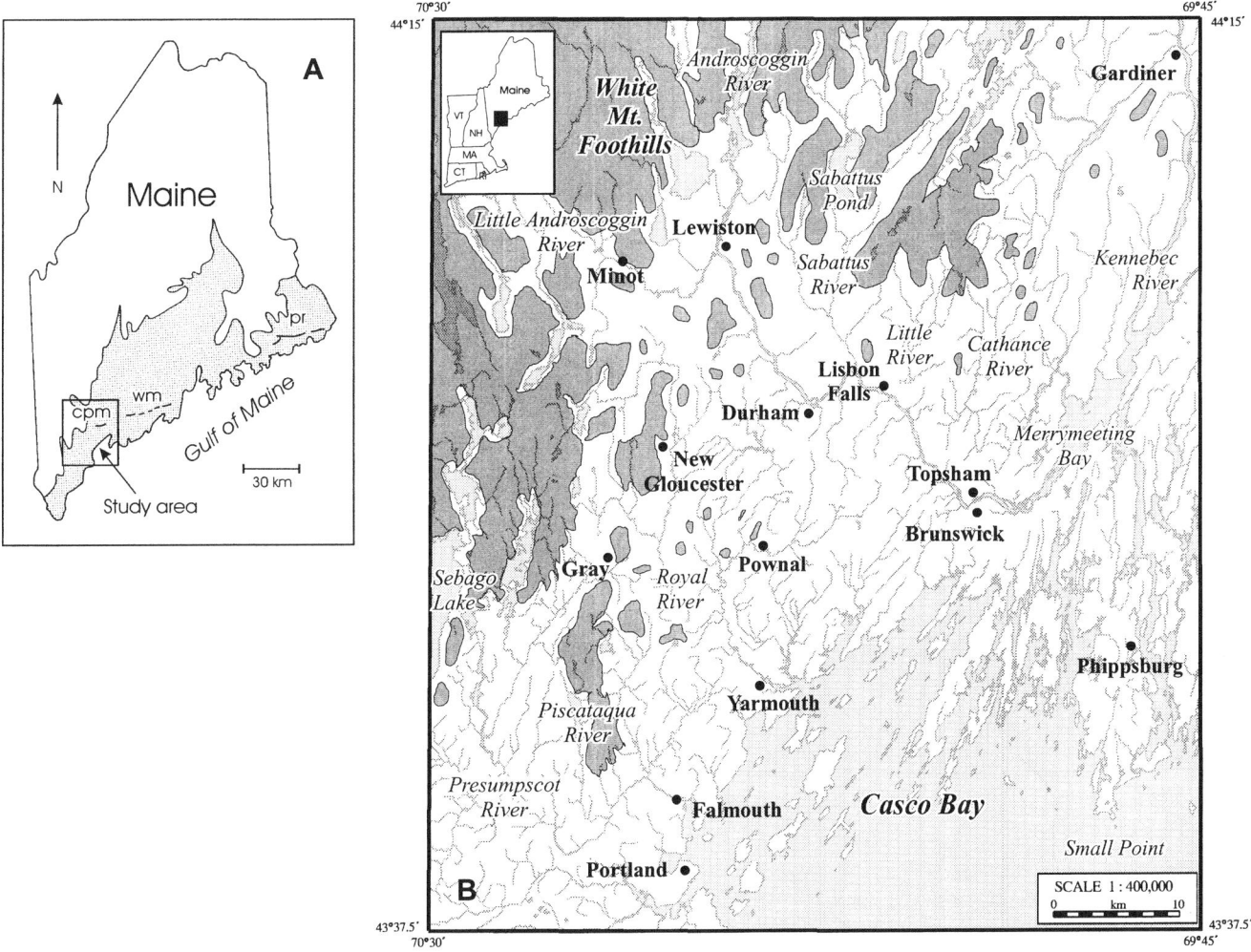

Figure 1. A, location map of study area; gray shaded region is area below marine limit. Dark lines represent trace of ice-marginal position at locations of specific glacial deposits (cpm—Cox Pinnacle moraine; wm—Waldoboro moraine; pr—Pineo Ridge delta and end-moraine complex). B, Detailed location map of study area and place names, lower Androscoggin River valley–upper Casco Bay region, southwestern Maine (river basins italicized). Gray shaded region is area above 100 m elevation above sea level.

creases from ~75 m along the outer coast south of Phippsburg to more than 100 m in the upper Androscoggin River valley north of Lewiston (Thompson et al., 1989). The marine limit was established in synchrony with ice retreat; thus the oldest marine sediments date across the lowland from 14 to 13 ka.

The marine transgression was quickly followed by regression of the synglacial sea as relative sea level dropped due to rapid glacioisostatic uplift (Bloom, 1960, 1963; Borns and Hagar, 1965; Schnitker, 1974; Stuiver and Borns, 1975; Belknap et al., 1987; Crossen, 1984, 1991; Kelley et al., 1992; Barnhardt et al., 1995). Koteff et al. (1993) estimated that emergence in coastal central New England (including southwestern Maine) began sometime between 14 and 13.3 ka. Smith (1985) and Smith and Hunter (1989) estimated maximum marine inundation of the coastal lowland by 12.6 ka and emergence of the present landmass of southwestern Maine by 11.5 ka. The postglacial lowstand of the sea extended onto the inner continental shelf to a depth of −55 m by 10.5 ka (Barnhardt et al., 1997).

## FIELD AREA

The Casco Bay Lowland, as described herein, extends from the present shore of Casco Bay to the foothills of the White Mountains in west-central Maine (Fig. 1B). The lowland is framed east and west by bedrock ridges and hills. To the west it is bordered by an elongate ridge that extends from Gray south to Falmouth. The eastern margin of the lowland is bounded by a broad hilly complex that forms the watershed between the Cathance and Kennebec Rivers.

In this chapter the terms coastal lowland and central uplands are used to describe the main physiographic elements of the regions, following the work of Denny (1982) and Hanson and Caldwell (1989), who characterized the regional physiography in relation to lithologic and structural controls. The coastal lowland includes a broad area from Yarmouth to Topsham, where the average relief is <150 m. The lowest relief of this area includes a marine sediment-filled trough that extends inland from the coast to the northwest up the axis of the present Androscoggin River valley. The coastal lowland also extends inland in several prominent recesses into the highlands, following major glacial and postglacial drainage systems of the Piscataqua and Royal Rivers in the west, the Little Androscoggin River and its tributaries in the center of the lowland, and the Androscoggin, Sabattus, Little, and Cathance Rivers in the east. In general, the rivers flow south to southeasterly across the lowland, oblique to the northeast-southwest regional structural trend of the igneous and metamorphic rocks of the Casco Bay region (Osberg et al., 1985).

Inland, the hilly terrain of the central uplands increases to elevations exceeding 300 m and total relief increases to >375 m along the transition to the foothills. Numerous lakes, repetitively scoured by successive Pleistocene ice advances, occupy this zone along the transition to the upland, which extends into the eastern portion of the Carboniferous Sebago pluton, where high-grade metamorphic rocks are in contact with the pluton (Osberg et al., 1985).

## METHODS

Detailed surficial mapping was conducted at a scale of 1:24 000 for a number of 7.5-minute quadrangles in the study area as part of the U.S. Geological Survey–Maine Geological Survey STATEMAP project (Smith, 1994; Bernotavicz, 1995; Bernotavicz and Dubois, 1995; Bolduc et al., 1994; Retelle, 1994, 1997a, 1997b; Weddle, 1994, 1997a, 1997b; Marvinney, 1996; Thompson, 1997; Weddle et al., 1999). Detailed stratigraphic logs were made at selected sites, especially where samples were collected for radiocarbon dating. Sites noted in previous studies (Stuiver and Borns, 1975; Attig, 1975; Smith, 1985) were resurveyed and logged to determine the stratigraphic context of shell collections. Sampling was particularly targeted at recovery of in situ valves of mollusc species such as *Portlandia arctica* in stratigraphic positions close to ice-contact sediments in order to best delimit ice retreat from the site. A second focus was made on targeting samples in regressive deposits that provided estimates on postglacial sea level. At several sites on the Phippsburg Peninsula sediment cores containing glaciomarine-lacustrine sediment transitions (containing shell, organic matter, and other dateable materials) were recovered from modern lakes and inlets by percussion and modified Livingstone coring techniques (Coequyt, 1997; Hubeny, 1997). All dated samples reported in this chapter are in radiocarbon years before present, uncorrected for a marine-reservoir effect (Mangerud and Gullickson, 1975; Bard et al., 1990a, 1994; Arnold, 1995; Bondevik et al., 1999).

## SURFICIAL DEPOSITS

The progressive retreat of the marine-based ice sheet through the coastal lowland region in Maine and adjacent regions is marked by numerous ice-marginal deposits (Fig. 2), interbedded with marine sediments deposited during transgression across the isostatically depressed proglacial foreland (Bloom, 1960; Thompson and Borns, 1985). Numerous stillstands or minor readvances of the ice are represented by end moraines, grounding-line fans, and deltas in the lowland. Sedimentation at point sources along the inland margin of the synglacial sea is marked by eskers and large ice-contact and outwash-fed glaciomarine deltas (Thompson et al., 1989; Crossen, 1991). The deltaic and fan sediments are interbedded in their distal margins with fossiliferous marine mud (Bloom, 1960; Borns and Hagar, 1965; Smith, 1982; Thompson, 1982), which also blankets low-elevation ice-contact deposits (moraines and fans) in the lowlands. Shoreline and nearshore features and deposits mark the position of relative sea level during postglacial isostatic adjustment of the coastal lowland. The deposits commonly contain marine macrofossils and microfossils, which are further utilized for paleoenvironmental reconstruction through faunal and geochemical analysis of fossil assemblages (cf. Kreutz, 1994; Dorion, 1997; Kaplan, 1999) and geochronology through radiocarbon analysis of the carbonate shell material (Stuiver and Borns, 1975; Smith, 1985; Dorion et al., this volume).

### Ice-marginal deposits

**End moraines.** There are thousands of moraine ridges in the coastal lowland and they occur in clusters of parallel landforms, many concealed beneath younger ice-marginal deposits or blanketed by fine-grained glaciomarine mud (Smith, 1982; Thompson, 1982). The principal characteristic shared by most moraines is their ridge morphology, representing a linear alignment interpreted as a former ice-marginal position (Fig. 2). In general, most moraines in southwestern Maine are relatively small, and are referred to as ribbed, washboard, or DeGeer moraines (Smith, 1982, 1985; Retelle and Bither, 1989; Hunter, 1990; Hunter et al., 1996). Many of the moraines are composed of diamicton and glaciotectonized stratified drift (Fig. 3A). The larger landforms, commonly referred to as stratified end moraines (Borns, 1973; Ashley et al., 1991), consist of stratified sand and gravel interbedded with minor amounts of diamicton. The stratified moraines usually contain interstratified materials folded and thrust faulted by overriding ice or ice shove. The many moraines and the intimate stratigraphic relationship with glaciomarine sediment indicate that (1) the grounded ice margin was active during its northward retreat, and (2) ice retreat was synchronous with marine incursion of the coastal lowland (Smith, 1982; Hughes et al., 1985).

The Cox Pinnacle moraine (CPM in Figs. 1A and 2; Weddle, 1997a) is larger and more continuous over a greater dis-

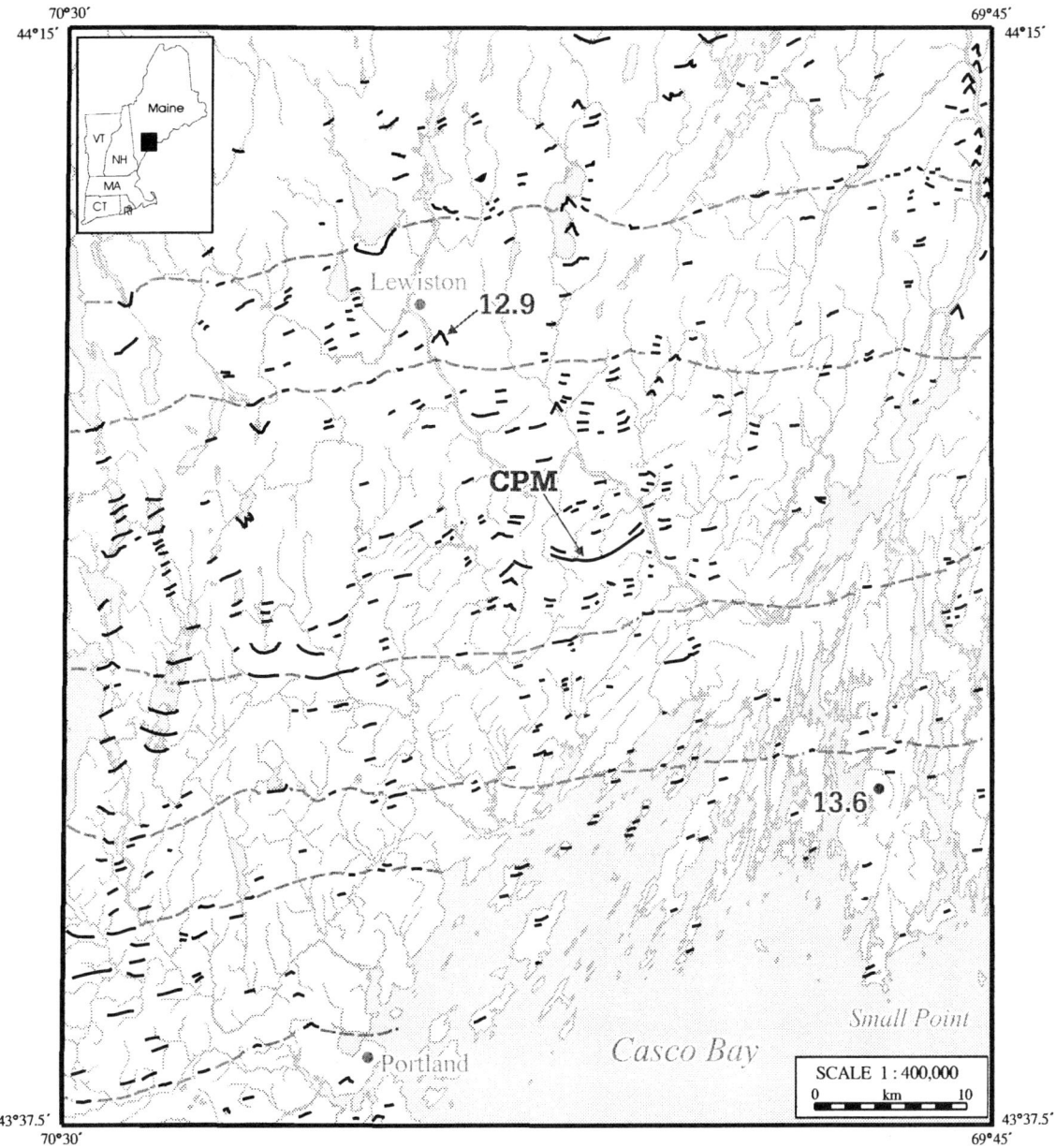

Figure 2. Schematic map with inferred trend of ice margins (gray dashed lines). Black segments are selected ice-marginal positions based on moraines, ice-contact slope on deltas and glaciomarine fans, and head of outwash. Numbers are sites with radiocarbon age estimate (13.6 and 12.9 regressive ages). CPM is Cox Pinnacle moraine.

tance than most other end moraines in southwestern Maine, with the exception of the Waldoboro moraine (Thompson and Borns, 1985). The gently lobate form can be traced for ~6 km across the landscape and delineates the ice-marginal position where the retreating ice margin was pinned on the bedrock upland between the Androscoggin River and Royal River valleys. On the upland, the moraine is composed of sandy diamicton and large boulders are present along its crest (Fig. 3B). This moraine is exposed in gravel pits to the east in the lowland of the Androscoggin River valley, where it is composed of till and deformed subaqueous outwash deposits (Weddle, 1997a).

Collectively, the moraines in the lower Androscoggin valley and adjacent Sebago and Kennebec lowlands indicate that the ice margin retreated in a north-northwest direction into the interior. In general, the average moraine-crest trend is east-northeast, although some lobate and interlobate forms show local variation in the crest trend (Fig. 2). The compilation of moraine trends in Figure 2 also indicates that ice retreat was not greatly influenced by the Androscoggin valley. The bedrock surface in the central axis of the valley extends as deep as $-30$ m (Bither, 1989), which with a marine limit of ~90 m indicates a paleowater depth of 120 m. The retreating ice margin was

Figure 3. A, Exposure in stratified end moraine (cf. Ashley et al., 1991), Knight pit, Pownal; view is to northwest. Fine- to medium-grained sand (f-ms) is overlain by deformed sand (ds) with apparent shear sense from left to right. Deformed sand is overlain by diamicton (d), emplaced by minor readvance of ice margin. Diamicton is overlain by glaciomarine mud (gm) and regressive nearshore deposit (nd). Shovel handle (circled) in lower right corner is 60 cm long. B, Cox Pinnacle moraine; boulder-strewn low-relief ridge beneath power lines, Durham. Utility pole on left is approximately 14 m high.

apparently pinned on topographic highs and did not form a deeply recessed calving embayment in the narrow valley.

***Eskers.*** The term esker is used to describe the coarse-grained sediments from a predominantly high energy glaciofluvial source that delivers sediments to the ice-proximal zones of fans and deltas (Gustavson and Boothroyd, 1975; Sollid and Carlson, 1984; Warren and Ashley, 1994, 1997; Levasseur, 1995; Ashley et al., 1991). These deposits occur as distinct, sinuous ridges in valleys (Thompson and Borns, 1985), as feeder tails on ice-proximal sides of deltas (Thompson et al., 1989), and as coarse-grained cores of fans and deltas, where the supporting ice has retreated and a delta or fan has prograded basinward over its former conduit (Sollid and Carlson, 1984; Warren and Ashley, 1994, 1997).

***Glaciomarine fans.*** Glaciomarine fans originate at the mouth of the esker or meltwater conduit and grade distally to the seafloor (cf. Sharpe, 1988). The fans, also referred to as submarine outwash (Rust and Romanelli, 1975), subwash fans (Burbidge and Rust, 1988), and grounding-line fans (Powell, 1990), are irregular mound, cone, or fan-shaped features that commonly drape or flank the moraine ridges. Over time, particularly if the retreat of the grounded ice margin is halted at a pinning point on a subglacial topographic high or a valley constriction, a submarine fan may aggrade vertically and prograde distally and along the ice margin (Fyfe, 1990). Fans may evolve into more massive and extensive deposits that approach and may eventually even reach contemporaneous sea level. In this latter and most developed case, the fan may eventually become a delta with a subaerially exposed topset plain (Powell, 1990; Slayton, 1993). Isotopic analyses by Kreutz (1994) on marine fossils from eastern Maine suggest that paleoceanographic conditions support an ice-marginal sedimentation model similar to that described by Powell (1990) for southeastern Alaska, where meltwater emanating from the base of the ice sheet rises directly to the surface as a plume over cold, saline basin waters. In the ice-proximal zone, coarse-grained sediments are deposited adjacent to conduit sources as grounding-line fans. In distal regions, overflow plumes transport fine-grained sediment, which settles from suspension (Cowan and Powell, 1990).

The fan sediments represent the transition between glaciofluvially dominated processes associated with the ice-tunnel environment, and processes of the proglacial marine basin. Consequently, materials in the fans show complex stratigraphic relationships both parallel to the ice margin and distally from the former tunnel mouth. Sediments in the proximal zone include coarse-grained gravelly stratified materials (Fig. 4, A and B) supplied by subglacial meltwater currents, and diamicton that may originate from slope failure and downslope movement from the adjacent moraine, by ice thrust (Retelle and Bither, 1989; Ashley et al., 1991; Warren and Ashley, 1994, 1997), or squeezed from under the ice (Barnett and Holdsworth, 1974). Medial and distal portions of the fans are commonly fining-upward sequences (Retelle and Bither, 1989) that grade transitionally upward from coarse proximal fan sediments at the base to fine-grained distal sediments. In medial and distal zones, the major lithofacies include rhythmically bedded sand and mud that grade to the muddy seafloor (Fig. 4, C–F). These sediments, termed cyclopsams and cyclopels (Mackiewicz et al., 1984), were likely deposited by suspension from a highly turbid but buoyant suspended sediment plume (Cowan and Powell, 1990).

Resedimentation of the deposits, particularly in both the proximal and medial portions of the fan, produces grain flows and debris flows of varying textural composition. The fining-upward fan sequence may represent ice retreat with removal of the ice-tunnel sediment source, lateral switching of a distribution channel on the fan lobe (Burbridge and Rust, 1988), or pulsing of fine-grained sediment to blanket previously deposited coarse fan sediments (D.R. Sharpe, 1999, personal commun.).

***Deltas.*** The origin and stratigraphy of glaciomarine deltas in Maine were described and categorized by Thompson et al. (1989). Although they are widely distributed across the coastal lowland, the largest deltas are located at or near the inland marine limit in southwestern and eastern Maine. Many are localized where the ice margin was temporarily pinned against bedrock ridges and other topographic highs. Exposures in deltas show that many are Gilbert type deltas, with horizontal fluvial topset beds (delta-plain deposits) overlying inclined foreset beds deposited on the prograding delta front (Fig. 4, C and G). These deltas are typical of environments where coarse-grained sediments are rapidly deposited in basins of sufficient depth to produce a steeply sloping delta front (Nemec, 1990; Postma, 1990). As a rule, in glaciomarine deltas, the elevation of the stratigraphic contact between the topset and foreset beds is commonly utilized to provide an estimate of the paleowater plane to which the deltas were graded. Thompson et al. (1989) and Koteff et al. (1993) surveyed the topset-foreset contact elevations of a number of ice-contact glaciomarine deltas in Massachusetts, New Hampshire, and Maine to define the plane of the upper limit of marine submergence, or marine limit for the region that includes the field area.

### Fine-grained glaciomarine sediments

A glaciomarine mud (Fig. 5) was named the Presumpscot Formation in Maine by Bloom (1960, 1963); the unit was mapped and extended into New Hampshire by Koteff (1991). Subsurface data and surface exposures show that this unit directly overlies bedrock, till, fans, and end moraines, and is interbedded with medial and distal components of fans and deltas. It can be massive or layered, containing outsized clasts and ice-rafted debris. The mud is also fossiliferous, with a rich assemblage ranging from foraminifers and diatoms through mollusc shells to marine vertebrates. Based on associated fossil assemblages, the mud is considered a late Pleistocene cold-water marine unit (Bloom, 1963). The marine mud has a blue-gray color when fresh, and an olive-gray color when weathered. Fractures in the weathered Presumpscot Formation may also be stained with iron-manganese. The weathered fine-grained sediment was previously considered a different unit from the blue-gray mud, deposited by different glacial advances (Trefethen et al., 1947). However, Leavitt and Perkins (1935), Goldthwait (1949, 1951), Caldwell (1959), and Bloom (1960, 1963) attributed the color difference to postglacial oxidation. Kelley (1989) and Mayer (1990) discussed the mineralogy of the Presumpscot Formation from both offshore and onshore samples.

The Presumpscot Formation was deposited by glaciofluvial discharges into the sea, the winnowed fine-grained fraction settling out as glaciomarine mud (processes discussed in Retelle and Bither, 1989). A sandy facies of the glaciomarine Presumpscot Formation overlies the fine-grained facies (Thompson, 1982). The contact between the facies may be sharp, although it is more commonly gradational. The informal term "sandy member of the Presumpscot Formation" has been used to describe a mappable unit (Weddle, 1987; Smith, 1988; Hildreth, 1990; Hunter, 1990; Retelle, 1991) associated with marine regressive deposits, found stratigraphically above the massive mud of the Presumpscot Formation (sensu stricto). Alternatively, where the unit is thick enough to be mapped separately, the term "nearshore deposit" is better suited for these shallow-

Figure 4 (on this and facing page). A, Ice-tunnel deposit, north end of Marr Point delta (Fig. 7); view is to north. Channel deposits are composed of crudely stratified alternating clast-supported, silt-coated boulder-gravel and pebble-gravel layers, overlain by poorly sorted weakly stratified gravel. B, Well-stratified, ice-proximal inclined bedding of interbedded pebble-sand and sand layers. Note boulder-gravel layer (coarse-grained debris-flow deposit) in upper part of photo; view is to east (Williams pit, Pittston; cf. Thompson and Smith, 1988; photo by W.B. Thompson). C, Medial fan deposits in coarse to medium sand with low-angle, inclined bedding. Beds include normally graded units (Dube pit, Lewiston). D, Low-angle, inclined beds of planar-bedded medium-fine sand, silt, and gravelly sand within glaciomarine fan. Antidune-like structures above knife (20 cm length) are draped by medium-fine sand beds, which are cut by channel infilled with medium-fine sand. Some massive (diffusely bedded) sand occurs near top of section. Paleoflow is left to right (foresets in dune and fluid-drag structure below knife; Whorff pit, Topsham). E, Undeformed, rhythmically bedded sand (light) and mud (dark) overlying deformed beds (beneath 14-cm long pen); medial to distal fan deposits. Deformation is from syndepositional dewatering (Webber pit, Topsham; photo by W.B. Thompson). F, Thinly bedded silty sand (light) and mud (dark); distal fan deposits (crude rhythmites). Small dropstone is above knife (20 cm length) handle (Bisson pit, Topsham; cf. Retelle and Bither, 1989). G, East Gray delta topset-foreset contact; paleoflow of foresets is to southeast (view is to north); McKin Superfund site (photo by Rebecca Hewett).

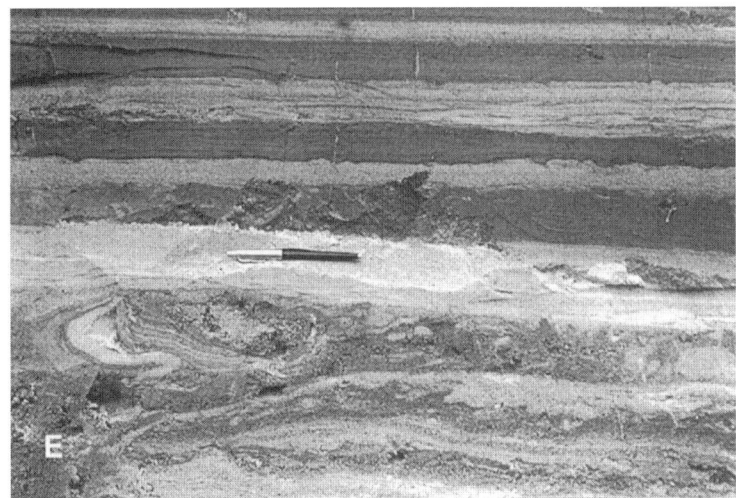

water or wave-reworked deposits associated with marine transgression and regression.

### Shallow-marine deposits

After the ice margin and sediment source retreated, shallow or nearshore marine sediments were deposited (1) as sea level overstepped the landforms when eustatic sea-level rise was greater than isostatic rebound, or (2) as sea level fell around the landforms due to isostatic rebound exceeding eustatic sea-level rise. The shallow-marine facies (Smith, 1985; Retelle and Bither, 1989) contains a range of lithofacies and biofacies ranging from well-sorted tidal to subtidal sand to coarse bouldery lag deposits or lagoonal mud. The deposits include a variety of morphostratigraphic units such as beaches, spits, tombolos, subtidal sand bodies, distal outwash deltas, and an extensive and nearly ubiquitous veneer of wave-reworked sediments. The deposits also display a wide range of textural maturity, reflecting the source landform being subjected to wave reworking and the available energy (Fig. 6, A and B).

Distal outwash deltas formed where meltwater streams entered the sea some distance from the ice margin (Thompson et al., 1989; cf. Tary et al., this volume). Isostatic uplift was in progress across much of the region as the ice withdrew inland from the marine limit. Broad sand plains are present in Maine, many found at the inland extent of the marine limit and sometimes associated with ice-marginal deposits (Thompson et al., 1989). They range in elevation from 136 m above sea level (asl) to 36 m asl (Weddle and Retelle, 1995). Some distal deltas are found well seaward of the inland marine limit, and of these the Brunswick sand plain is the farthest from the inland marine limit, and is found at the lowest elevation (Weddle, 1994; Weddle and Retelle, 1995). The plain surface grades from about 36 m to 18 m asl (surface gradient 1.8 m/km). Shallow excavations in the surface of the plain reveal fluvial trough cross-beds of fine- to medium-grained sand, in places containing mud rip-up

Figure 5. Well-bedded, glaciomarine mud (Presumpscot Formation), Bunganuc Bluff, Brunswick (photo by W.B. Thompson). White bar is 1 m long.

clasts and mud drapes, typical of a braided-stream environment. The plain morphology and sedimentology classify it as a coastal braid-plain delta (McPherson et al., 1987; Nemec, 1990; Postma, 1990; Blair and McPherson, 1994) that formed as braided streams in the late glacial Androscoggin River valley entered the regressive sea.

These braided-stream deposits unconformably overlie very low angle foreset beds. The foresets overlie a coarsening-upward sequence of massive and laminated mud grading upward to interbedded sand and glaciomarine mud, described previously as the sandy Presumpscot Formation. This muddy unit beneath the braid-plain delta represents a transition from a distal glaciomarine and submarine-plain environment to shallowing conditions during marine regression. Subsequent to the deposition of the massive glaciomarine mud, existing units were reworked by the marine regression and nearshore sediments were deposited. In the Brunswick area, these deposits are best exposed at eroded coastal bluffs (Weddle et al., 1993; Weddle, 1994; Crider, 1998), and are reported from detailed geotechnical logs from the Superfund site at the Naval Air Station in Brunswick (Draft Final Feasibility Study, ABB Environmental Services Inc., November 1991, Portland, Maine). Crider (1998) studied the Brunswick sand plain using ground-penetrating radar imaging, and determined an overall south- to east-dipping clinoform reflector orientation, interpreted as the foreset bedding of the braid-plain delta.

*Morphostratigraphic assemblages*

Examples of fans and deltas and minor moraines in the study area are represented in Figure 7. The collective assemblage forms a beaded morphology (Sollid and Carlson, 1984; Caldwell et al., 1985) between two larger Gilbert-type deltas along the east side of Sabattus Pond. The two deltas and the series of smaller, elongate lobe-shaped fans between the two larger landforms represent distinct ice-marginal positions where submarine outwash emanated from the ice-tunnel system during the retreat of the ice. As many as four or five small moraines flank each of the beads along the eastern side of the valley. The moraine ridges have a general east-west alignment, are convex down-valley, and often are clustered in the valley pinned to higher hillocks on the adjacent valley upland to the east. There are as many as 40 minor moraine ridges, mapped in the 2-km-long valley east of Sabattus Pond, with an average spacing of 50–100 m between the discrete ridge crests (Bernotavicz, 1994). These moraines are small to moderately large, and usually occur in clusters; they are not more than 3–10 m high, 30–50 m wide, and to several hundred meters in length, but their internal composition is unknown.

The two deltas in Figure 7 have different gross morphology but share some noteworthy characteristics. The relatively flat surface elevation of each delta is ~100 m asl. Both have an elongate northern end, most prominent on the Marr Point delta. A ridge traceable along Pleasant Hill Road enters the northern end of the Pleasant Hill delta just east of Round Pond (a kettle pond). These features represent the position of the ice tunnel at or near the ice margin, where the sediment emanated from the tunnel to build the deposits. The Pleasant Hill delta is a kidney-shaped form, more lobate than the Marr Point delta, which is more wedge shaped. The Pleasant Hill delta likely represents laterally migrating and coalescing delta or fan lobes along the ice margin. The Marr Point delta may represent a delta depos-

Figure 6. A, Tangential cross-bedding in coarse-medium sand nearshore deposit that overlies poorly exposed glaciomarine mud (located at meter-scale bar in photo). Below mud and scale bar are distal deposits from glaciomarine fan, represented by layered sand and mud at base of exposure (MDOT pit, Topsham; cf. Weddle and Retelle, 1995; photo by J.T. Kelley). B, Distal beds of Sabbathday Pond delta, New Gloucester. Interbedded fine sand and silt layers are bioturbated. Dark and light layers represent cyclical deposition, possibly daily tidal cycles. Shovel handle is 60 cm long.

ited with little or no lateral migration of the tunnel feeder source. Bernotavicz (1994) noted a change in orientation of end moraines near the Marr Point delta front and suggested that the shift from a roughly east-west trend (parallel to the ice margin), to a northwest-southeast trend reflects opening and enlargement of the margin near the ice tunnel, thus allowing lateral expansion of the Marr Point delta. During deglaciation, as the ice margin retreated in the Sabattus Pond lowland, it was pinned to the east at Sabattus Mountain. The shift in moraine trends also represents the influence of the highland as a pinning point.

## DEGLACIATION AND RELATIVE SEA-LEVEL CHRONOLOGIES

### Ice-marginal recession

New evidence from this and other studies indicates that the Laurentide ice margin retreated from the Gulf of Maine to the position of the present coast south of Portland prior to 14.8 ka (Dorion, 1997), into the Casco Bay lowland by ca. 14 ka, and into the interior near Lewiston at least by 13 ka.

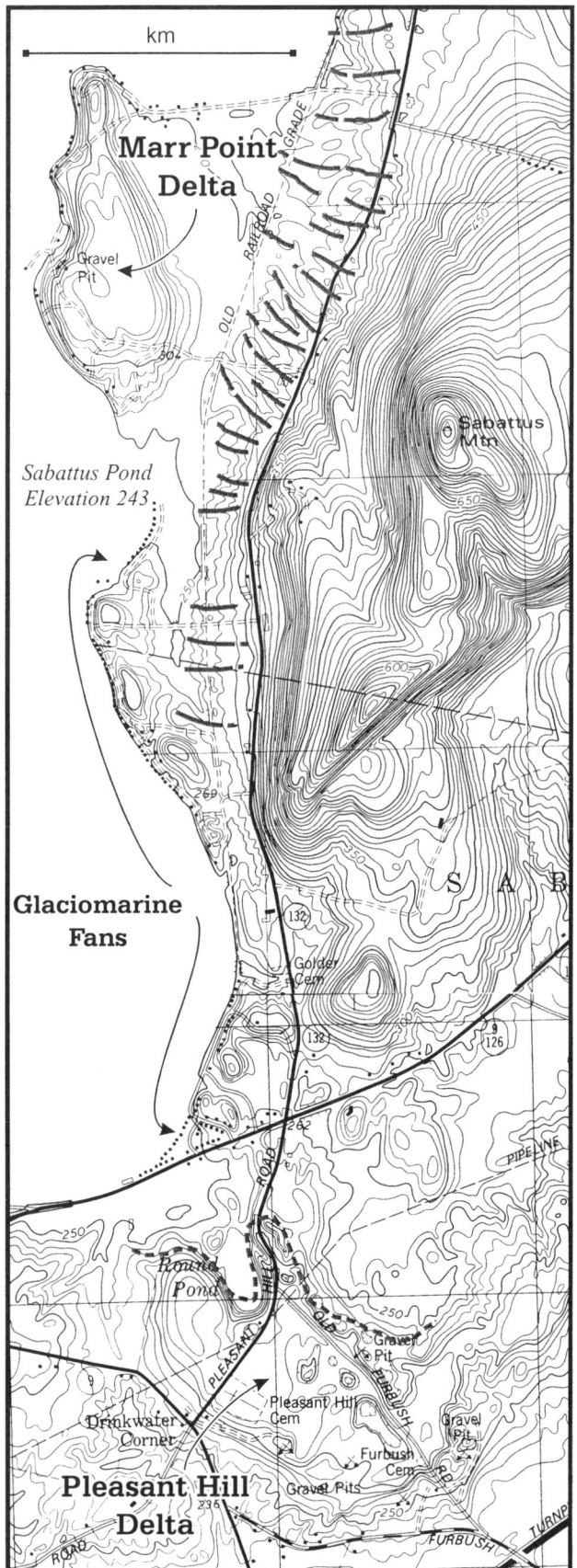

Figure 7. Pleasant Hill and Marr Point glaciomarine deltas and glaciomarine fans between deltas (east shore of Sabattus Pond). Broad-dashed lines trace selected moraine crests; short-dashed line traces ice-contact slope of Pleasant Hill delta (contour interval is 10 ft). North is toward top in figure. Note ice-block depressions (kettles) in Pleasant Hill delta top. Topset-foreset contact is measured in Pleasant Hill delta at 98.5 m asl (Thompson et al., 1989). Photo in Figure 4A is from narrow, northern end of Marr Point delta (Lisbon Falls North and Monmouth 7.5′ quadrangles, Maine).

The oldest reported published radiocarbon dated marine shells that provide minimum estimates for ice recession in southwestern Maine include an age estimate of 14 820 ± 105 $^{14}$C yr B.P., determined from *Portlandia arctica* in marine mud from Scarborough (Dorion, 1997), and an age obtained 21 km offshore by Kelley et al. (1992) of 14 090 ± 450 $^{14}$C yr B.P. on *Macoma balthica* from a core of glaciomarine sediment. A radiocarbon age of 13 830 ± 100 $^{14}$C yr B.P. (Thompson and Borns, 1985) from Great Hill in Kennebunk on broken and whole mollusc shells (*Mytilus edulis* valves; J.T. Andrews, 1998, personal commun.) has been used as an important ice-marginal age in several recent reconstructions. However, the stratigraphic context of the dated sample is not well understood (cf. Smith, 1984, 1985).

Figure 8 shows lithostratigraphic columns of sites in the Casco Bay Lowland where $^{14}$C dated shells and organic matter from sediment cores are used to detail ice retreat and relative sea-level history (Table 1). Marine mollusc shells submitted for age estimates are in several facies assemblages: (1) ice proximal, where in situ paired valves in coarse-grained fan and delta deposits allow for the earliest estimate of ice retreat; (2) distal glaciomarine, where mollusc shells were recovered from massive fine-grained deep-water glaciomarine mud that is not closely interfingered with ice-marginal or shoreline deposits; and (3) offlap facies, where shells are recovered from nearshore deposits and generally provide an estimate for falling relative sea level. In addition, dateable materials from transition zones in several lake and marine sediment cores from the Phippsburg Peninsula (Hubeney, 1997; Coequyt, 1997) are used to provide timing of basin emergence (cf. Hafsten, 1983) from the postglacial sea.

New evidence from $^{14}$C dated ice-marginal and emergent nearshore deposits suggests that the retreating ice margin was shoreward of the present coast in Casco Bay by ca. 14 ka (Figs. 2 and 9). Weddle (1997a) and Weddle et al. (1993) reported analyses from the Seymour pit in Freeport, where valves of *Portlandia arctica*, dated as 14 045 ± 95 $^{14}$C yr B.P. (AA-10164, Weddle et al., 1993) were found in glaciomarine mud overlying a sharp contact with glaciotectonized ice-proximal deposits. Coarse-grained ice-contact deposits and overlying glaciomarine mud that contain the molluscs are deformed by high-angle normal faults. The sample, however, yielded a $\delta^{13}$C value of −9.1‰ necessitating resubmission of *Portlandia arctica* shells from the site. At the time, the first age estimate

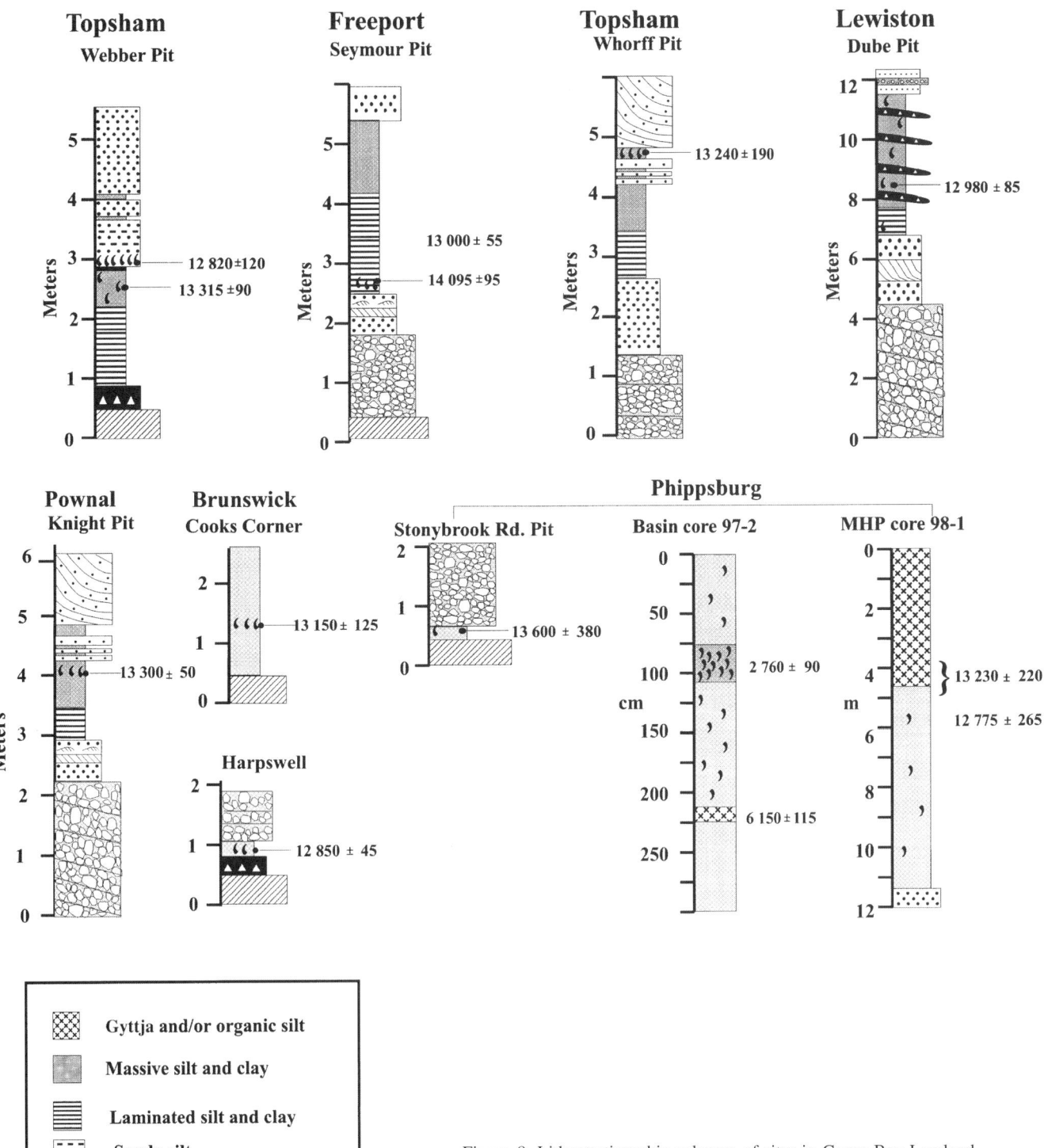

Figure 8. Lithostratigraphic columns of sites in Casco Bay Lowland where $^{14}$C age analyses on shells and organic matter are used to detail ice retreat and relative sea-level history (Retelle and Bither, 1989; Slayton, 1993; Weddle et al., 1993; Weddle and Retelle, 1995). Locations of sections are shown in Figure 9.

TABLE 1. RADIOCARBON AGE ESTIMATES, LOWER AND ANDROSCOGGIN VALLEY-UPPER CASCO BAY

| Sample* | Age ($^{14}$C yr B.P.) | Site location | Elevation (masl) | Material | $\delta^{13}$C |
|---|---|---|---|---|---|
| AA10164 | 14 045 ± 95 | Seymour Pit | 56 | Portlandia arctica | −9.1 |
| OS18899 | 13 000 ± 55 | Seymour Pit | | | −1.15 |
| GX21939† | 13 600 ± 380 | Stonybrook Road Pit | 49 | Mytilus edulis | −0.3 |
| AA10162 | 13 315 ± 90 | Webber Pit | 45 | Portlandia arctica | −2.3 |
| OS4419† | 13 300 ± 50 | Knight Pit | 61 | Mytilus edulis | −0.96 |
| GX21632† | 13 240 ± 190 | Whorff Pit | 58 | Mesodesma arctatum | 1.4 |
| GX20774 | 13 100 ± 125 | Cooks Corner | 15 | Hiatella arctica | 1.1 |
| AA10165 | 12 890 ± 85 | Dube Pit | 100 | Portlandia arctica | −1.2 |
| OS2348† | 12 850 ± 45 | Harpswell | 25 | Zirphaea crispata | 0.35 |
| GX22922 | 13 230 ± 220 | MHP core 98-1 | 20 | Bulk organics | −23.5 |
| AA25090† | 12 775 ± 265 | MHP core 98-1 | 19 | Sculpin mandibular | −12.7 |
| GX22763† | 6 160 ± 115 | Basin core 97-2 | −7 | Bulk organics | −26.2 |
| GX22764† | 2 760 ± 90 | Basin core 97-2 | −2 | Shell material | 2.1 |

*Note:* masl, meters above sea level.
*Samples listed above include those previously unreported. AA10164 and OS18899 are from the same site and elevation; the younger age supersedes the older age because of the discrepancy of $\delta^{13}$C value. GX22763 and GX22764, and AA25090 and GX22922 are from a single core but from different elevations in the cores.
†Samples used in Casco Bay Lowland relative sea-level curve (see Fig. 10).

obtained from the pit (14 045 ± 95 $^{14}$C yr B.P.) was the oldest $^{14}$C date in the lowland, providing the best estimate for time of deglaciation of the coastal zone. However, the resubmitted sample from the same site yielded an age of 13 000 ± 55 $^{14}$C yr B.P. (OS-18899; $\delta^{13}$C = −1.15‰). Although the second sample is significantly younger, it was recovered from the massive glaciomarine mud that was likely deposited after the ice margin retreated from the site, and thus corresponds with earlier deglaciation.

Similarly, stratigraphy and age estimates from four other sites, which are well below their projected marine limit elevations (Thompson et al., 1989), indicate that ice had retreated into the interior before 13 ka (Webber and Whorff pits, Topsham; Knight pit, Pownal; and Stonybrook Road, Phippsburg, Figs. 8 and 9). At the Webber pit, paired *Portlandia arctica* valves dated as 13 315 ± 90 $^{14}$C yr B.P. were recovered from the massive, distal glaciomarine mud facies that overlies ice-contact fan deposits (Retelle and Bither, 1989). The fossiliferous distal mud is overlain by regressive sand and mud in which upright in situ barnacles attached to paired *Mytilus edulis* shells at the base of this regressive unit provided an age of 12 820 ± 120 $^{14}$C yr B.P. (Retelle and Bither, 1989), providing a minimum regressive sea-level estimate for the site.

Two previously reported sites in the lower Androscoggin Valley (Stuiver and Borns, 1975; Smith, 1985) were revisited for this study. These sites are located in a complex of gravel pits, including the Whorff pit. A new accelerator mass spectrometer (AMS) $^{14}$C analysis from the Whorff pit provides an older age for shells found clearly in a nearshore stratigraphic association, and supersedes the younger previously reported ages for this site as an estimate for the onset of emergence. The sample (*Mesodesma arctatum*) was found below a thick regressive spit deposit in muddy sand that also contained large clasts with attached barnacles, and provided an age estimate of 13 240 ± 190 $^{14}$C yr B.P.

In the Knight pit in Pownal, *Mytilus edulis* valves in fine-grained sediments underlying coarse, sandy regressive sediments yielded an age of 13 300 ± 50 $^{14}$C yr B.P. The mussel shells were slightly abraded and worn, and numerous valves were disarticulated and may have been transported or slumped downslope to the depositional site from shallower water.

The earliest emergence age is from the Stonybrook Road site on west-central Phippsburg Peninsula (Figs. 2 and 9). At this site, *Mytilus edulis* valves were sampled from massive mud underlying a poorly sorted nearshore (regressive) gravel and yielded an age of 13 600 ± 380 $^{14}$C yr B.P. This site is located at 45 m elevation, and the marine limit for this part of the peninsula is ~64 m (Thompson et al., 1989; Weddle and Retelle, 1995, 1998); thus the stratigraphy at the site clearly postdates ice retreat, establishment of marine limit, and initial emergence.

By 13 ka or earlier, the ice margin had retreated ~20 km northwest to the vicinity of Lewiston and the lowland to the south was emerging. At an ice-contact glaciomarine fan-delta complex located in Lewiston, *Portlandia arctica* shells in a section of marine mud overlying ice-contact gravel provided an age estimate of 12 980 ± 85 $^{14}$C yr B.P. (Figs. 8 and 9). The shells were recovered from glaciomarine mud separated from the underlying ice-contact gravel by a distinct sharp contact, which likely represents the shift from ice marginal to distal sedimentation. The *Portlandia* valves are part of a more diverse invertebrate fossil assemblage in the upper sediments that includes *Hiatella arctica* and *Mytilus edulis* shells and barnacle plates, found in both undeformed mud and included as fragments in debris-flow deposits interbedded with the distal marine mud. The debris flows most likely originated from slumping of coarse-grained gravel and sand that mixed with prodelta mud on the eastern flank of the delta after ice retreated from the ice-marginal position to the north of the pit. The Portlandia shells are found 28 m below the local marine limit, defined by the

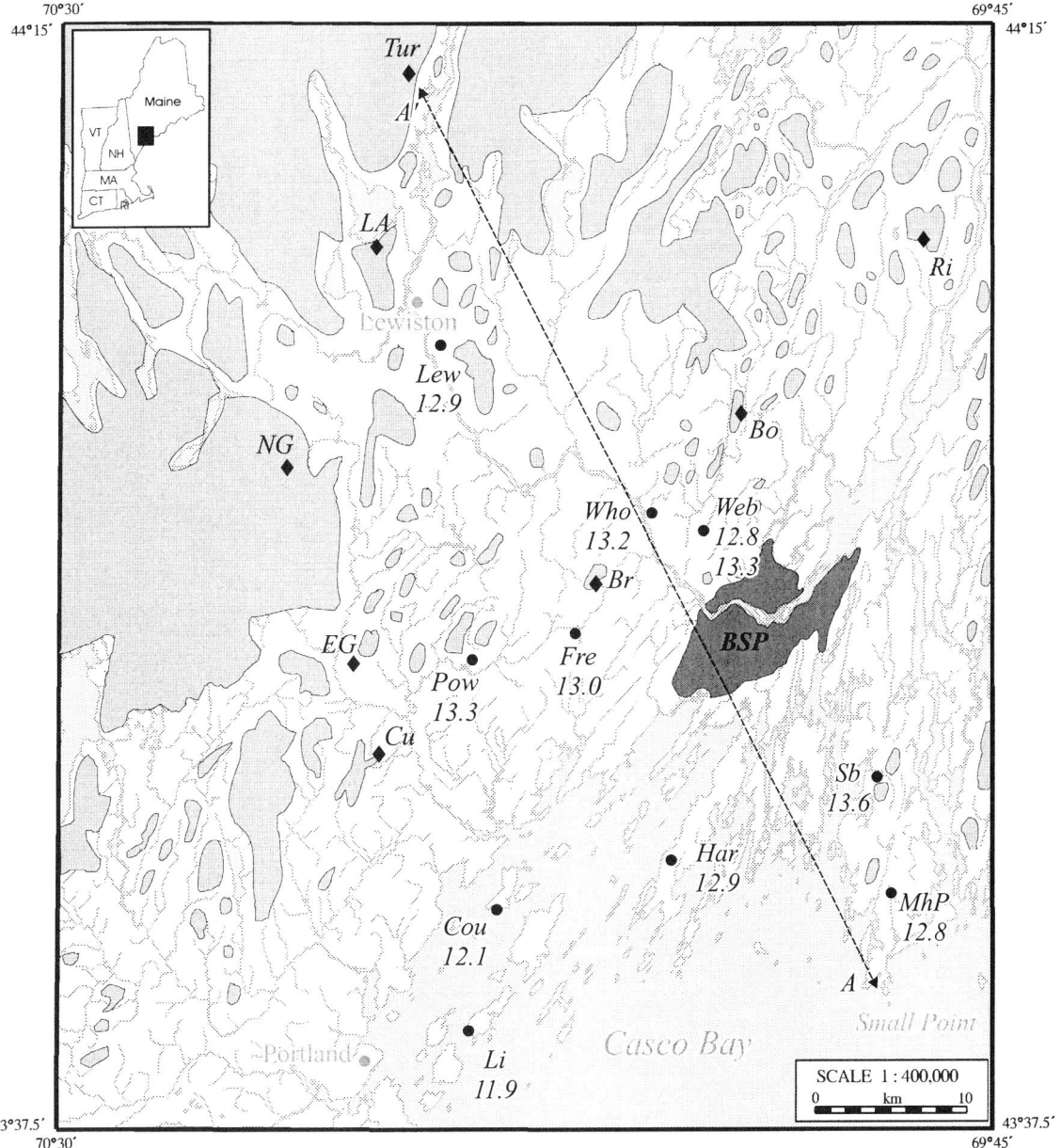

Figure 9. Location map of sites with radiocarbon ages (solid black circles) and sites of glaciomarine deltas and marine shoreline deposits at or near marine limit elevation (diamonds); refer to Table 1 for ages. Light gray shading represents areas above marine limit (approximate elevation range, 60 m in southeast corner to 110 m in northwest corner). Dark gray shading represents Brunswick sand plain (BSP; elevation range 36–18 m above sea level). Other abbreviations: $^{14}$C dated sites—Cousins Island (Cou), Freeport (Fre), Harpswell (Har), Lewiston (Lew), Long Island (Li), Meetinghouse Pond (MhP), Pownal (Pow), Stonybrook Road (Sb), Webber pit (Web), Whorff pit (Who); undated highstand marine shoreline deposits and glaciomarine deltas—Bowdoinham (Bo), Brunswick (Br), Cumberland (Cu), East Gray delta (EG), Lake Auburn delta (LA), New Gloucester delta (NG), Richmond (Ri), Turner delta (Tur).

delta surface at ~100 m asl. The fossils were buried by glaciomarine mud and debris flows when relative sea level was above their depositional elevation. The shell age must postdate deglaciation only by a few hundred years.

## Relative sea level during deglaciation

Figure 10 shows the deglaciation phase of a local relative sea-level curve for the southern part of the field area based on

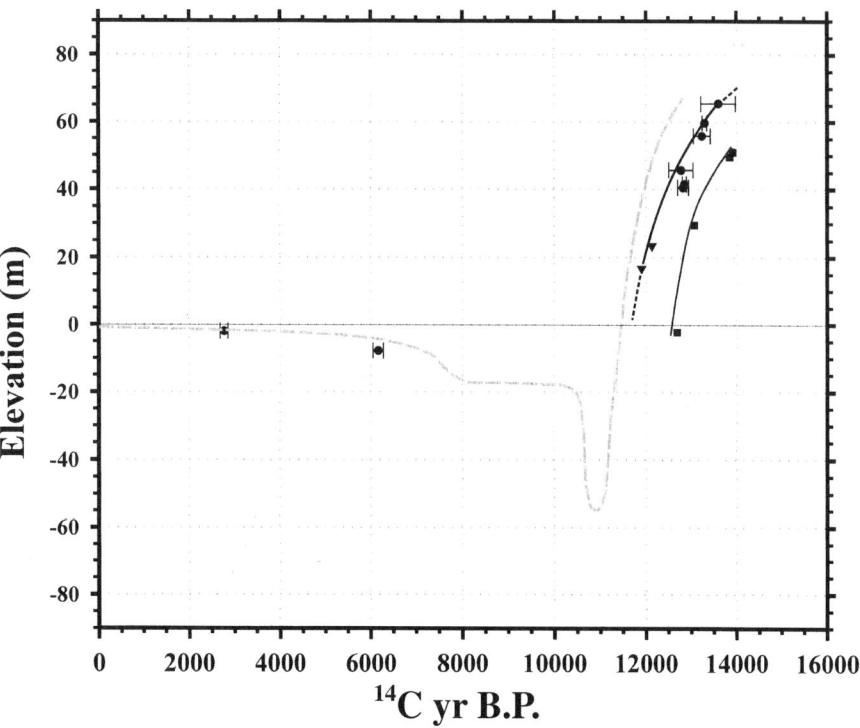

Figure 10. Relative sea-level curves from Maine. Circles with error bars are Casco Bay Lowland data (refer to Table 1); inverted triangles are data from islands in Casco Bay (Stuiver and Borns, 1975; Smith, 1985). Squares are eastern Maine data (Dorion, 1997). Gray dashed line is regional curve for Maine (Barnhardt et al., 1995). Age-estimate elevations are projected to local common values of rebound (80 m for Casco Bay Lowland; 54 m for eastern Maine).

new radiocarbon-dated samples from the Casco Bay Lowland. Figure 10 also shows a regional curve utilizing data from the Maine coastal lowland and offshore sediments from the Gulf of Maine (Barnhardt et al., 1995), and new deglacial-phase data from eastern Maine (Kreutz, 1994; Dorion, 1997; Kaplan, 1999) fit to a curve based on the projected marine limit in that region (Thompson et al., 1989). Two previously reported ages (Stuiver and Borns, 1975; Smith, 1985) from regressive deposits on the islands of Casco Bay also are included in Figure 10.

The new control points that constrain the local Casco Bay Lowland curve are from age estimates of shells recovered from marine mud underlying or interbedded with nearshore deposits, or are from AMS-dated levels in cores that closely estimate the emergence of an isolation lake (Table 1; Fig. 8). In addition to the early emergence age estimates described previously, and an age analysis from Harpswell (12 850 ± 45 $^{14}$C yr B.P.) from marine mud under gravel interpreted as coarse-grained beach deposits, each of the analyses on shells are from intertidal or nearshore species that underlie from 1 to 1.5 m of sediments, indicating a limited range of paleowater depth and the usefulness of the sample as relative sea-level indicators. Other new samples recovered in field work in this study include the Cooks Corner sample (13 150 ± 125 $^{14}$C yr B.P.), and the lower date from the Webber pit in Topsham (13 315 ± 90 $^{14}$C yr B.P.). However, because these samples were recovered from the middle of the glaciomarine mud unit, and are not good indicators of intertidal or nearshore conditions, they were not utilized in the Casco Bay Lowland local curve. As expected, the postglacial emergence phase of the Casco Bay Lowland curve and the sea-level curve from eastern Maine roughly parallel the land-based portion of the regional sea-level curve of Barnhardt et al. (1995), although the local curves are less steep. Two ages from marine sediment cores from a marine inlet (Hubeny, 1998), taken from terrestrial organic matter (peat) and marine mollusc shell fragments, plot close to the late Holocene transgressive portion of the Barnhardt et al. (1995) curve. The most similar feature in all three curves is the pattern of a rapid postglacial emergence phase (cf. Andrews, 1970). The rate of emergence determined from the Casco Bay Lowland curve averages 30 m per 1 k.y., roughly equivalent to the regional curve. The curves all show a slight curvilinear trend in their uppermost segments representative of slow or restrained initial rebound (cf. Andrews, 1970). The main distinctions of the Casco Bay Lowland curve are that it displays earlier initial emergence than the regional curve, and slightly greater total emergence than the eastern Maine and regional curve. Both local curves are constructed from data from only the southern coastal portion of the lowlands, thus an even greater total emergence would be expected in curves produced from inland marine sectors.

A preliminary equidistance plot (Fig. 11) shows the proximal-distal relationship between dated and undated ice-marginal features and marine shoreline and nearshore deposits distal to the retreating ice margin (Retelle, 1994; Weddle and Retelle, 1995; Weddle, 1997a, 1997b). Such plots are commonly used to correlate strandlines, the geometries of which reflect the pattern of regional postglacial crustal warping. In fiord regions of the Canadian arctic, Fennoscandia, or Great Britain (Andrews, 1970; Devoy, 1983; Gray, 1983; Sissons,

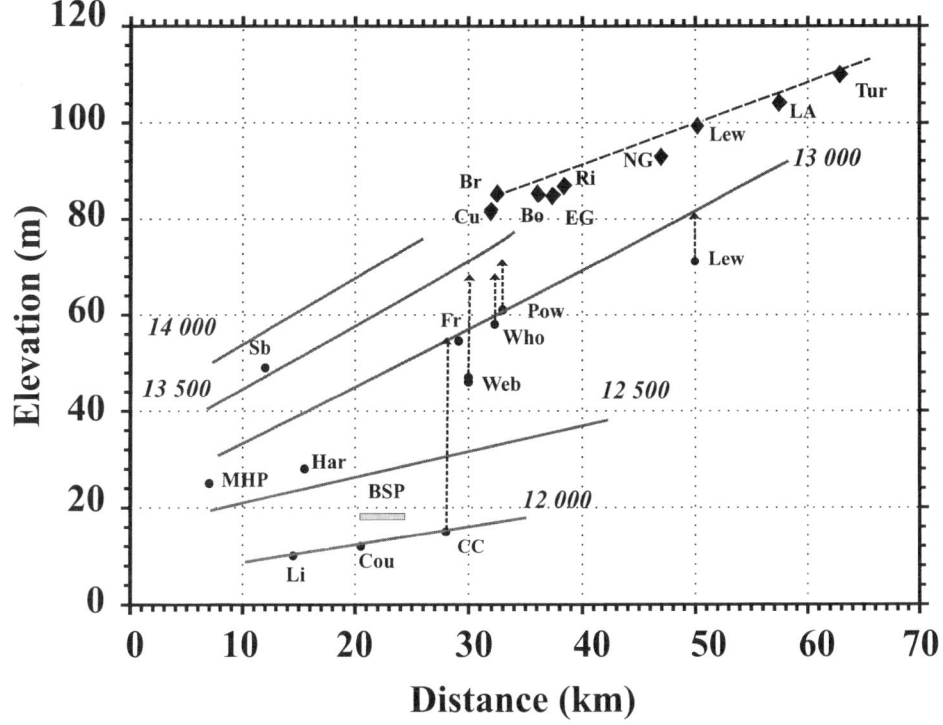

Figure 11. Elevation plotted against distance from Small Point (axis) to Turner, Maine (A-A' in Figure 9). Dashed line is approximate projected elevation of marine limit; heavy lines are isochrones. Black diamonds are sites of ice-contact glaciomarine deltas and marine shoreline deposits at or near marine-limit elevation (abbreviations as in Fig. 9). Black solid circles are sites of dated marine fossils (refer to Table 1). Dashed arrows above selected sites project to elevation of contemporaneous relative sea level, indicating former water depth over sample site.

1983; Ingolfsson et al., 1995; Snyder et al., 1997), segments of related shoreline features of the same age are more easily correlated than the deposits in the Casco Bay Lowland. The shoreline deposits in this study area are generally scattered and discontinuous, mostly capping fans deposited during ice retreat in the central lowland.

In Figure 11, ice-marginal features, including marine-limit beaches and ice-contact deltas, plot on a plane rising inland at ~0.7 m/km. As expected, the provisional strandlines for 14, 13.5, and 13 ka show a slightly higher gradient than the 12.5 and 12 ka strandlines (Fig. 11). The two lower strandlines also provide an indication of the timing and extent of marine deposits during offlap or regression, showing that by ca. 12 ka the sea had regressed to the Brunswick area. Furthermore, the deposition of the Brunswick sand plain, although not yet dated directly by radiocarbon samples, is constrained during regression between 12.5 and 12 ka.

The relative sea-level record in the Casco Bay Lowland indicates that sea level fell, or conversely, isostatic rebound began, concomitant with ice retreat. Koteff et al. (1993) determined that rapid uplift in southern New England and in southwestern Maine was delayed until after 14 ka. If this was the case, the postglacial rebound pattern (tilt) in the Casco Bay Lowland would be similar to that of coastal New Hampshire (Koteff et al., 1993) However, if rapid uplift had begun by the time the ice margin was in the lower Androscoggin River valley, the postglacial gradient of the water plane (as represented by topset-foreset contacts in the study area) should be less steep than that in coastal New Hampshire and southwestern Maine

(0.85 m/km). An estimate of postglacial tilt in the lower Androscoggin River valley using elevations of delta tops from topographic maps and topset-foreset contact measurements from Thompson et al. (1989) is between 0.63 and 0.71 m/km. Farther east in the Kennebec River valley region, isostatic uplift has tilted the plane to the southeast, and it now has an average slope of 0.53 m/km (Thompson et al., 1989). The west to east progressively lower gradient of the postglacial tilt in Maine supports the concept that rapid uplift was underway when the ice margin had reached the central coastal region of Maine.

## DISCUSSION AND SUMMARY

### Regional deglaciation

A number of scenarios have been proposed to depict ice retreat in the Maine coastal region utilizing available geochronological evidence (Bloom, 1960; Stuiver and Borns, 1975; Denton and Hughes, 1981; Davis and Jacobson, 1985; Hughes et al., 1985; Smith, 1985; Dyke and Prest, 1987). Bloom (1960) provided the first $^{14}$C-dated chronology and estimated that initial marine submergence of the coastal lowland began at 12 100 $^{14}$C yr B.P., and by 11 800 $^{14}$C yr B.P. the sea had transgressed to its maximum inland extent. Stuiver and Borns (1975) cited several dates in the 13 ka range in the south coastal zone, including a date of 13 200 $^{14}$C yr B.P. at Kennebunkport, which indicated that ice had to be onshore prior to 13 200 $^{14}$C yr B.P. Stuiver and Borns (1975) further noted that the maximum extent of the synglacial sea was attained by ca. 12 700 $^{14}$C yr B.P.

Denton and Hughes (1981), Davis and Jacobson (1985), and Thompson and Borns (1985) presented a model that featured a major calving bay in the Penobscot Valley by 13 ka where the synglacial sea had reached its inland maximum extent. The model also shows that by 13 ka the southern and central Kennebec and Androscoggin valley regions were ice free, while a long lobate pendant of ice remained in the western interior of the state above the marine limit. The residual ice in the western highlands was separated from ice in the east (at or near Pineo Ridge) by a prominent calving embayment in the Penobscot valley. This widely used model was constrained by geochronology at several key sites in the state. In the west, basal ages were used from several radiocarbon-dated lacustrine sediment core sites at Sinkhole Pond and Poland Spring Pond in the foothills west of the Casco Bay Lowland (Davis and Jacobson, 1985) that provided basal ages of 12 710 ± 125 $^{14}$C yr B.P. and 12 860 ± 325 $^{14}$C yr B.P., respectively. In the upper Kennebec Valley at Embden Pond, a 13 020 ± 240 $^{14}$C yr B.P. age estimate was obtained on marine shells from raised marine sediments (Stuiver and Borns, 1975). At Gould Pond, situated below marine limit (90 m asl) between the Kennebec and Penobscot Rivers in the west-central interior, an age of 13 280 ± 410 $^{14}$C yr B.P. was obtained on the carbonate fraction of basal marine sediments recovered in a piston core (Anderson et al., 1992).

Smith (1985) provided a thorough review of available geochronological evidence defining ice-retreat scenarios in the Maine coastal lowland. In a composite model reflecting previously published (mostly from Stuiver and Borns, 1975) and more recently obtained $^{14}$C analyses, his revised ice margin reached the present coastline by 13.8 ka, the Phippsburg Peninsula by 13 ka, and the inland marine limit by 12.7 ka with a less prominent calving bay developed in the Penobscot Valley. Smith (1985) underscored the problem of ice present at or near Kennebunkport from ca. 13.8 to 13.2 ka while having Embden Pond in the upper Kennebec valley ice free by 13 ka.

Dorion (1997) reported age estimates from *Portlandia arctica* valves from glaciomarine mud in Scarborough (14 820 ± 105 $^{14}$C yr B.P.), ~30 km south of the study area, suggesting that the southernmost coastal lowland in Maine was ice free by 14.8 ka. Subsequent retreat northward into the Casco Bay Lowland from the southern part of the state is documented by consistent alignment of ice-marginal positions (moraines, deltas, and fans) that trend almost east-west, at a shallow angle to the present coast inland of Casco Bay (Smith, 1980; Thompson and Borns, 1985; Fig. 2). Data presented here suggest that Laurentide ice had retreated into the Casco Bay Lowland by 14 ka. At present, the exact position of the 14 ka margin is unknown; however, several lines of evidence suggest that ice had receded into the lowland by that time. Nearshore deposits on the Phippsburg Peninsula at 49 m asl dated as 13.6 ka and ~15 m below local marine limit suggest deglaciation and inital emergence prior to 13.6 ka. Similarly, fossiliferous glacialmarine and nearshore deposits in the central Androscoggin Valley dated as 13–13.3 ka from Freeport, Topsham, and Pownal are also well below local marine limit and as far as 20 km from the present coast.

A prominent stillstand during retreat through the Androscoggin Valley is marked by the Cox Pinnacle moraine in the Lisbon-Durham Falls area. The 6-km-long moraine is laterally correlative to other smaller moraine ridges and ice-contact features to the west and to a large fan and moraine complex to the east in Topsham (Retelle and Bither, 1989). Although the timing of emplacement of the moraine is not known at present, ice retreat from the grounding line at Topsham is older than 13.3 ka.

By at least 13 ka, the ice margin had retreated to the Lewiston area. The persistent east-west moraine orientation in the axis of the Androscoggin Valley and lack of a calving-bay geometry indicate that rapid calving though the center of the lowland was probably impeded by bedrock topography that trends oblique to the moraine crests. Northeast-trending bedrock strike ridges in the lower valley and along the present coast may have served to ground the ice margin at valley constrictions and on numerous sills. Ice-marginal features, such as grounding-line fans and deltas, occur on a number of these pinning points (Ireland, 1990; Lowell and Borns, 1994; Slayton, 1993; Weddle, 1997a, 1997b; Weddle et al., 1999).

The timing of the late Wisconsinan ice retreat from the Gulf of Maine into the Casco Bay Lowland is synchronous with ice retreat detailed in the eastern Maine coastal region and adjacent Maritime Canada. Dorion (1997, this volume) and Kaplan (1999) report AMS $^{14}$C ages on marine shells that indicate that the present coast south of Cutler was ice free by ca. 14 000 $^{14}$C yr B.P. Two major moraines, the Pond Ridge and the Pineo Ridge moraine system, occur within 20 km of the coast (cf. Lepage, 1982; Borns and Hughes, 1977; Kaplan, 1999) and mark the stabilization and readvance of the retreating grounded marine-based ice margin. Kaplan (1999) cited marine shell ages constraining these moraine positions between 13 800 and 13 350 $^{14}$C yr B.P. and further argued that paleoecological proxy information indicates that the moraines formed during cold climatic and oceanographic conditions.

Stea et al. (1998) documented evidence for three late Wisconsinan ice advances in Nova Scotia. The oldest advance, the Escuminac phase, represents the maximum stand on the Scotia shelf break ca. 22–19 ka. Two subsequent phases, the Scotia and Chignecto, represent reorganization of ice flow during retreat from the maximum. Stea et al. (1998) noted that retreat from the Scotia phase probably occurred between 15 and 13 ka, the grounded marine ice margin being close to the present coastline by ca. 14 ka. Ice flow during the Chignecto phase was most likely from local residual ice caps of the Scotia phase ice. Stea et al. (1998) and Stea and Mott (1998) tentatively correlated moraines associated with this phase (ca. 13–12.5 ka) to the Port Huron advance of the midwestern United States. In similar fashion, Borns and Hughes (1977) previously correlated the Pineo Ridge readvance, then dated as 13–12 ka, to the Port Huron readvance. This well-documented readvance in the upper Great Lakes area is dated as ca. 12.9 ka (Eschman and Mickelson, 1986; Mickleson et al., 1983; Johnson et al., 1997). The revised age estimates of the Pond Ridge and Pineo Ridge moraine belts in eastern Maine (between 13.8 and 13.35 ka) are older than the Port Huron readvance. However, these ages, as cited by Kaplan (1999), are uncorrected for marine reservoir effect. The application of the commonly used marine reservoir correction of 450 yr to the age estimate of 13 350 $^{14}$C yr B.P. on marine shells north of the Pineo Ridge moraine system would indicate that the age of emplacement of the moraine should be older than 12.9 ka.

*Correlation with other paleoenvironmental records.* During the past decade, syntheses of detailed studies of ice-sheet history, glacioisostasy, and eustatic sea level, and analysis of high-resolution deep-sea cores from the North Atlantic and polar ice cores have provided a clearer understanding of environmental changes that occurred during the last deglaciation. These records detail the history of rapid disintegration of the last ice sheets and complex interactions in the ocean-ice-atmosphere systems (Broecker and Denton, 1989; Andrews and Peltier, 1989; Fairbanks, 1989; Bard et al., 1990b; Bond et al., 1992; Bond and Lotti, 1995; Dowdeswell et al., 1995; Lehman and Kegwin, 1992; Mayewski et al., 1994; Dansgaard et al., 1993; Lowell et al., 1995). The revision of the ice retreat and sea-level histories in the Casco Bay Lowland allows speculation on their correlation with the high-latitude marine sediment and ice cores and, in particular, the cyclic millenial-scale Bond cycles evident in high-latitude marine sediment cores that are punctuated by detrital carbonate or Heinrich layers (Bond et al., 1992). The bracketing radiocarbon ages that frame Heinrich event 1 (H1) in the sediment cores in the North Atlantic region are 13 200 to 14 500 $^{14}$C yr B.P (Bond et al., 1992, 1993; Bond and Lotti, 1995; Andrews, 1998; note that these ages are quoted from the literature where an ocean reservoir correction of −400 to −450 yr has been applied). The uncorrected $^{14}$C ages place H1 between ca. 13 600 and 15 000 $^{14}$C yr B.P., and the uncorrected $^{14}$C ages from glaciomarine deposits in southern Maine and ice-marginal ages in eastern Maine (Dorion et al., this volume; Kaplan, 1999) indicate that retreat from the present Gulf of Maine into the coastal lowland was likely initiated during H1.

The interplay of several mechanisms may be responsible for initiating ice retreat through southern coastal Maine. These include the response of the Laurentide ice to the rapid warming that follows the main ice buildup in the cooling phase of H1 (cf. MacAyeal, 1993; Broecker, 1994), or uncoupling and rapid calving of ice in the Gulf of Maine due to relative sea-level rise associated with isostatic depression, or eustatic sea-level rise due to the flux of icebergs and meltwater at the culmination of the warming.

Andrews (1998) postulated that local relative sea-level rise due to isostatic loading along the tidewater margin of the Hudson Strait ice stream may have initiated uncoupling of the ice from its bed, which may have led to massive discharge of ice through Hudson Strait into the North Atlantic; this in turn may have subsequently caused a 1–5 m rise in global sea level (cf. Anderson and Thomas, 1991). A sea-level rise of this magnitude from the Hudson Strait discharge may have subsequently triggered ice outflow from other North Atlantic sources (cf. Broecker, 1994; Andrews, 1998).

Deglaciation of the Laurentide ice sheet occurred during an overall regime of steadily rising postglacial sea level interrupted by two major meltwater pulses at 12.5 and 9 ka (Fairbanks, 1989; Bard et al., 1990b). However, Locker et al. (1996) demonstrated prominent eustatic sea-level rises associated with drowned paleoshoreline deposits in Florida, where sea-level rise occurred in rapid steps between ca. 13.5 and 14.5 ka (ages were reported with a −400 yr correction); this rise, if uncorrected, correlates with the initiation of ice retreat in southwestern and eastern Maine.

Subsequent to ice retreat, the synglacial sea that occupied the coastal lowland regressed in response to rapid postglacial rebound. The marine regression in Maine was not monotonic, but was likely influenced by eustatic sea-level rise, punctuated by periods of rapid increase (Bard et al., 1996; Blanchon and Shaw, 1995; Fairbanks, 1989; Locker et al., 1996). The interplay between postglacial rebound and abrupt sea-level rise is seen geomorphically where prominent nearshore landforms, including regressive outwash deltas, beaches, and terraces, mark apparent stillstands of sea level during the marine regression (Weddle and Retelle, 1995).

The most prominent geomorphic evidence for such a stillstand in the study area is the Brunswick sand plain, an expansive, regressive coastal braid-plain delta, which likely formed between 12.5 and 12 ka. The deposition of this regressive delta sand plain was likely influenced by a combination of several factors: (1) increased discharge by the late to postglacial Androscoggin River, (2) loss of capacity where the late to postglacial Androscoggin River exited from a confined valley to an unconfined valley, (3) reworking of coarse-grained ice-marginal sediments deposited during ice retreat through the lowland, and (4) the plain being deposited during a period when falling relative sea level may have stabilized as a response to a rising eustatic sea level, balancing glacioisotatic uplift long enough for the coastal braid delta to form. In any model, the age of the plain is constrained by ages on *Hiatella arctica* shells (13 100 ± 125 $^{14}$C yr B.P) found beneath the outer edge of the plain at Cooks Corner in Brunswick (A.M. Hussey II, 1995, personal commun.), and by ages ca. 12 800 $^{14}$C yr B.P. at sites beneath regressive deposits in nearby Harpswell and Topsham, but at elevations above that of the sand plain. In context with these maximum ages, the local relative sea-level curve, and the emergence equidistance data, the plain formed sometime between 12.5 and 12 ka.

Between 12.6 and 11.7 ka eustatic sea level rose by as

much as 20 m, an event termed meltwater-pulse 1A (MWP-1a; Fairbanks, 1989; Bard et al., 1996; Adkins et al., 1998). The timing of the rise was reported by these authors, who applied a −400 yr low-latitude marine reservoir correction factor to radiocarbon age estimates. Even with this factor applied to the ages of the Brunswick sand plain, the time of formation of the plain is during MWP-1a.

The proposed terrestrial evidence of a stillstand of relative falling sea level in Maine is ~1 k.y. younger than the rapid sea-level rise reported by Koteff et al. (1993) for southwestern Maine and eastern New Hampshire. Further field data and more radiocarbon ages are required to test this proposal for the time and mode of formation of the Brunswick sand plain.

The revised deglacial and relative sea-level history presented in this study is based primarily on new radiocarbon age estimates on marine shells (uncorrected for marine reservoir effect), and it conflicts with recent evidence from central New England presented by Ridge (1997, and this volume). An inherent problem with marine shell dates is that relative to the atmospheric radiocarbon chronology, they are generally considered too old due to meltwater and marine reservoir effect (Mangerud and Gullickson, 1975; Arnold, 1995; Wohlfarth, 1996). On the basis of a comparison of a deglacial chronology for central New York and southern New England, Ridge et al. (1998) proposed that a correction as great as 1.5 k.y. may be necessary for marine shell age estimates in Maine. Ridge (1997) established a deglacial chronology determined by a correlation of paleomagnetic declination records and varve sequences of southern New England and central New York, linked by radiocarbon ages from varve sections in the Connecticut River valley in Vermont (Ridge and Larsen, 1990; Ridge et al., 1999), and led him to propose correlations between inferred ice-marginal positions in the Merrimack River valley of southeastern New Hampshire with inferred ice-marginal positions in southwestern Maine (Ridge et al., 1998). Specifically, the record in Maine is based on varve record sections in southwestern Maine that appear to correlate with the varve chronology of the Merrimack and Connecticut valleys (Canwell, 1997). Paleomagnetic results from the Maine site correspond to the signature of a unique paleomagnetic event recorded in New Hampshire, Vermont, and New York, and radiocarbon age estimated as ca. 12.5 ka. The time of deglaciation at the sites in Maine is estimated to be just prior to the deposition of the varves, hence slightly before 12.5 ka.

It is most probable that some marine reservoir correction is needed to correlate between shell-based chronologies and those based on terrestrial macrofossils. However, at present a reasonable reservoir correction suitable for deglacial through postglacial time reflecting changing postglacial composition of the Gulf of Maine waters has not been determined. The revised deglacial and sea-level chronologies compiled for coastal Maine in this study are, however, internally consistent and demonstrate earlier ice recession and emergence through a major portion of the coastal lowland than previously presented.

## ACKNOWLEDGMENTS

We are grateful for the support of the Maine Geological Survey (Robert Marvinney, State Geologist, and Walter Anderson, State Geologist Emeritus). The manuscript was improved by reviews by Jack Ridge and Dave Sharpe. Funding from the National Science Foundation EPSCOR program (NSFCAOCE 801015) supported field work and radiocarbon age analyses. Support for student research and radiocarbon dates that contributed to various aspects of the project was obtained from various sources at Bates College. Field discussions with numerous colleagues have contributed to the evolution of this project over the years. In particular the contributions of Gail Ashley, Dan Belknap, Kathy Bither, Jon Boothroyd, Hal Borns, Chris Dorion, Bob Johnston, Joe Kelley, Carl Koteff, Bob Oldale, Jack Ridge, Geoff Smith, Janet Stone, and Woodrow Thompson are greatly appreciated.

## REFERENCES CITED

Adkins, J.F., Cheng, H., Boyle, E.A., Druffel, E.R.M., and Edwards, R.L., 1998, Deep-sea coral evidence for rapid change in ventilation of the deep North Atlantic 15,400 years ago: Science, v. 280, p. 725–728.

Anderson, J.B., and Thomas, M.A., 1991, Marine ice-sheet decoupling as a mechanism for rapid, episodic sea-level change: The record of such events and their influence on sedimentation: Sedimentary Geology, v. 70, p. 87–104.

Anderson, R.S., Jacobson, G.L., Davis, R.B., and Stuckenrath, R., 1992, Gould Pond, Maine: Late-glacial transitions from marine to upland environments: Boreas, v. 21, p. 359–371.

Andrews, J.T., 1970, A geomorphological study of post-glacial uplift with particular reference to Arctic Canada: Institute of British Geographers Special Publication 2, 156 p.

Andrews, J.T., 1998, Abrupt changes (Heinrich events) in late Quaternary North Atlantic marine environments: A history and review of data and concepts: Journal of Quaternary Science, v. 13, p. 3–16.

Andrews, J.T., and Peltier, W.R., 1989, Quaternary geodynamics in Canada, in Fulton, R.J., ed., Quaternary geology of Canada and Greenland: Ottawa, Ontario, Geological Survey of Canada, Geology of Canada, no. 1, p. 541–572.

Arnold, L.D., 1995, Conventional radiocarbon dating, in Rutter, N.W., and Catto, N.R., eds., Dating methods for Quaternary deposits: Geological Association of Canada, GEOtext 2, p. 107–123.

Ashley, G.M., Boothroyd, J.C., and Borns, H.W., Jr., 1991, Sedimentology of late Pleistocene (Laurentide) deglacial-phase deposits, eastern Maine; an example of a temperate marine grounded ice-sheet margin, in Anderson, J.B., and Ashley, G.M., eds., Glacial marine sedimentation; paleoclimatic significance: Geological Society of America Special Paper 261, p. 107–125.

Attig, J.W., Jr., 1975, Quaternary stratigraphy and history of the Androscoggin River valley, Maine [M.S. thesis]: Orono, University of Maine, 56 p.

Bacchus, T.S., 1993, Late Quaternary stratigraphy and evolution of the eastern Gulf of Maine [Ph.D. thesis]: Orono, University of Maine, 347 p.

Bard, E., Hamelin, B., Fairbanks, R.G., and Zindler, A., 1990a, Calibration of the $^{14}$C timescale over the past 30,000 years using mass spectrometric U-Th ages from Barbados corals: Nature, v. 345, p. 405–410.

Bard, E., Hamelin, B., and Fairbanks, R.G., 1990b, U-Th ages obtained by mass spectrometry in corals from Barbados: Sea level during the past 130,000 years: Nature, v. 346, p. 456–458.

Bard, E., Arnold, M., Mangerud, J., Paterne, M., Labeyrie, L., Duprat, J.,

Mélières, M.-A., Sønstegaard, E., and Duplessy, J.C., 1994, The North Atlantic atmosphere-sea surface $^{14}$C gradient during the Younger Dryas climatic Event: Earth and Planetary Science Letters, v. 126, p. 275–287.

Bard, E., Hamelin, B., Arnold, M., Montaggioni, L., Cabioch, G., Faure, G., and Rougerie, F., 1996, Deglacial sea-level record from Tahiti corals and the timing of global meltwater discharge: Nature, v. 382, p. 241–244.

Barnett, D.M., and Holdsworth, G., 1974, Origin, morphology, and chronology of sublacustrine moraines, Generator Lake, Baffin Island, Northwest Territories, Canada: Canadian Journal of Earth Sciences, v. 11, p. 380–408.

Barnhardt, W.A., Gehrels, W.R., Belknap, D.F., and Kelley, J.T., 1995, Late Quaternary relative sea-level change in the western Gulf of Maine: Evidence for a migrating glacial forebulge: Geology, v. 23, p. 317–320.

Barnhardt, W.A., Belknap, D.F., and Kelley, J.T., 1997, Stratigraphic evolution of the inner continental shelf in response to late Quaternary relative sea-level change, northwestern Gulf of Maine: Geological Society of America Bulletin, v. 109, p. 612–630.

Belknap, D.F., and Shipp, R.C., 1991, Seismic stratigraphy of the glacial marine units, Maine inner shelf, in Anderson, J.B., and Ashley, G.M., eds., Glacial marine sedimentation; Paleoclimatic significance: Geological Society of America Special Paper 261, p. 137–157.

Belknap, D.F., Andersen, B.G., Anderson, R.S., Anderson, W.A., Borns, H.W., Jr., Jacobson, G.L., Kelley, J.T., Shipp, R.C., Smith, D.C., Stuckenrath, R., Jr., Thompson, W.B., and Tyler, D.A., 1987, Late Quaternary sea-level changes in Maine, in Nummedal, D., et al., eds., Sea-level fluctuation and coastal evolution: Society of Economic Paleontologists and Mineralogists Special Publication 41, p. 71–85.

Bernotavicz, A.A., 1994, Glacial and postglacial history of the Sabattus valley, Sabattus, Maine [B.A. thesis]: Lewiston, Maine, Bates College, 141 p.

Bernotavicz, A.A., 1995, Surficial geology of the Portland East quadrangle, Maine: Maine Geological Survey Open-File 95-72, scale 1:24 000.

Bernotavicz, A.A., and Dubois, M., 1995, Surficial geology of the South Harpswell quadrangle, Maine: Maine Geological Survey Open-File 95-73, scale 1:24 000.

Bither, K.M., 1989, Quaternary stratigraphy and hydrologic significance of paleodrainage patterns within the lower Androscoggin River valley, Durham, Maine [M.S. thesis]: Durham, University of New Hampshire, 129 p.

Blair, T.C., and McPherson, J.G., 1994, Alluvial fans and their natural distinction from rivers based on morphology, hydraulic processes, sedimentary processes, and facies assemblages: Journal of Sedimentary Research, v. A64, p. 450–499.

Blanchon, P., and Shaw, J., 1995, Reef drowning during the last deglaciation: Evidence for catastrophic sea-level rise and ice-sheet collapse: Geology, v. 23, p. 4–8.

Bloom, A.L., 1960, Late Pleistocene changes of sea level in southwestern Maine: Maine Geological Survey Department of Economic Development, 143 p.

Bloom, A.L., 1963, Late Pleistocene fluctuations of sea level and postglacial crustal rebound in coastal Maine: American Journal of Science, v. 261, p. 862–879.

Bolduc, A.M., Thompson, W.B., and Meglioli, A., 1994, Surficial geology of the North Windham quadrangle, Maine: Maine Geological Survey Open-File 94-2, scale 1:24 000.

Bond, G., and Lotti, R., 1995, Iceberg discharges into the North Atlantic on millennial time scales during the last glaciation: Science, v. 267, p. 1005–1009.

Bond, G., Heinrich, H., Broecker, W., Labeyrie, L., McManus, J., Andrews, J., Huon, S., Jantschik, R., Clasen, S., Simet, C., Tedesco, K., Klas, M., Bonani, G., and Ivy, S., 1992, Evidence for massive discharges of icebergs into the North Atlantic Ocean during the last glacial period: Nature, v. 360, p. 245–249.

Bond, G., Broecker, W., Johnsen, S., McManus, J., Labeyrie, L., Jouzel, J., and Bonani, G., 1993, Correlations between climate records from north Atlantic sediment and Greenland ice: Nature, v. 365, p. 143–147.

Bondevik, S., Birks, B.H., Gulliksen, S., and Mangerud, J., 1999, Late Weichselian marine $^{14}$C reservoir ages at the western coast of Norway: Quaternary Research, v. 52, p. 104–114.

Borns, H.W., Jr., 1973, Late Wisconsinan fluctuations of the Laurentide Ice Sheet in southern and eastern New England, in Black, R.F., et al., eds., The Wisconsinan stage: Geological Society of America Memoir 136, p. 37–45.

Borns, H.W., Jr., and Hagar, D.J., 1965, Late-glacial stratigraphy of a northern part of the Kennebec River valley, western Maine: Geological Society of America Bulletin, v. 76, p. 1233–1250.

Borns, H.W., and Hughes, T.J., 1977, The implications of the Pineo Ridge readvance in Maine: Géographie Physique et Quaternaire, v. 31, p. 203–206.

Broecker, W.S., 1994, Massive iceberg discharges as triggers for global climate change: Nature, v. 272, p. 421–424.

Broecker, W.S., and Denton, G.H., 1989, The role of ocean-atmosphere reorganizations in glacial cycles: Geochimica et Cosmochimica Acta, v. 53, p. 2465–2501.

Burbridge, G.H., and Rust, B.R., 1988, A Champlain Sea subwash fan at St. Lazare, Quebec, in Gadd, N.R., ed., The late Quaternary development of the Champlain Sea basin: Geological Association of Canada Special Paper 35, p. 47–61.

Caldwell, D.W., 1959, Glacial lake and glacial marine clays of the Farmington area, Maine: Maine Geological Survey Bulletin 10, 48 p.

Caldwell, D.W., Hanson, L.S., and Thompson, W.B., 1985, Styles of deglaciation in central Maine, in Borns, H.W., Jr., et al., eds., Late Pleistocene history of Northeastern New England and adjacent Quebec: Geological Society of America Special Paper 197, p. 45–57.

Canwell, B.A., 1997, The construction of varve and paleomagnetic chronologies for the deglaciation of southern Maine: estimating the radiocarbon marine reservoir effect [B.S. thesis]: Medford, Massachusetts, Tufts University, 21 p.

Coqueyt, A., 1997, Late Holocene marine transgression and sea-level change at Meetinghouse Pond, Phippsburg Peninsula, Maine [B.S. thesis]: Lewiston, Maine, Bates College, 77 p.

Cowan, E.A., and Powell, R.D., 1990, Suspended sediment transport and deposition of cyclically interlaminated sediment in a temperate glacial fjord, Alaska, U.S.A., in Dowdeswell, J.A., and Scourse, J.D., eds., Glacimarine environments: Processes and sediments: Geological Society [London] Special Publication 53, p. 75–89.

Crider, H.B., 1998, Late Pleistocene development of the Brunswick, Maine sand plain and adjacent paleochannel [M.S. thesis]: Boston, Massachusetts, Boston University, 219 p.

Crossen, K.J., 1984, Glaciomarine deltas in southwestern Maine: Formation and vertical movements [M.S. thesis]: Orono, University of Maine, 121 p.

Crossen, K.J., 1991, Structural control of deposition by Pleistocene tidewater glaciers, Gulf of Maine, in Anderson, J.B., and Ashley, G.M., eds., Glacial marine sedimentation; paleoclimatic significance: Geological Society of America Special Paper 261, p. 127–135.

Dansgaard, W., Johnsen, S.J., Clausen, H.B., Dahl-Jensen, D., Gundestrup, N.S., Hammer, C.U., Hiidberg, C.S., Steffensen, J.P., Sveinbjornsdottir, A.E., Jouzel, J., and Bond, G., 1993, Evidence for general instability of past climate from a 250-kyr ice-core record: Nature, v. 364, p. 218–220.

Davis, R.B., and Jacobson, G.L., 1985, Late glacial and early Holocene landscapes in northern New England and adjacent areas of Canada: Quaternary Research, v. 23, p. 341–368.

Denny, C.S., 1982, Geomorphology of New England: U.S. Geological Survey Professional Paper 1208, 18 p.

Denton, G.H., and Hughes, T., 1981, The last great ice sheets: New York, Wiley, 484 p.

Devoy, R.J., 1983, Late Quaternary shorelines in Ireland: An assessment of their implications for isostatic land movement and relative sea-level changes, in Smith, D.E., and Dawson, A.G., eds., Shorelines and isostasy: Institute of British Geographers Special Publication 16, p. 227–254.

Dorion, C.C., 1997, An updated high resolution chronology of deglaciation and accompanying marine transgression in Maine [M.S. thesis]: Orono, University of Maine, 147 p.

Dowdeswell, J.A., Maslin, M.A., Andrews, J.T., and McCave, I.N., 1995, Iceberg production, debris rafting, and the extent and thickness of Henrich layers (H-1, H-2) in North Atlantic sediments: Geology, v. 23, p. 301–304.

Dyke, A.S., and Prest, V.K., 1987, Late Wisconsinan and Holocene history of the Laurentide Ice Sheet: Géographie Physique et Quaternaire, v. 41, p. 237–264.

Eschman, D.F., and Mickelson, D.M., 1986, Correlation of glacial deposits of the Huron, Lake Michigan, and Green Bay lobes in Michigan and Wisconsin, in Sibrava, V., et al., eds., Quaternary glaciations in the Northern Hemisphere: Quaternary Science Reviews, v. 5, p. 53–58.

Fairbanks, R.G., 1989, A 17,000-year glacio-eustatic sea level record: Influence of glacial melting rates on the Younger Dryas event and deep-ocean circulation: Nature, v. 342, p. 637–642.

Fyfe, G.J., 1990, The effect of water depth on ice-proximal sedimentation: Salpausselka I, southern Finland: Boreas, v. 19, p. 147–164.

Gipp, M.R., and Piper, D.J.W., 1989, Chronology of late Wisconsinan glaciation, Emerald Basin, Scotian Shelf: Canadian Journal of Earth Sciences, v. 26, p. 333–335.

Goldthwait, L., 1949, Clay survey—1948, in Report of the State Geologist, 1947–1948: Augusta, Maine Development Commission, p. 63–69.

Goldthwait, L., 1951, The glacial-marine clays of the Portland-Sebago Lake region, in Report of the State Geologist, 1949–1950: Augusta, Maine Development Commission, p. 24–34.

Gray, J.M., 1983, The measurement of shoreline altitudes in areas affected by glacioisostasy, with particular reference to Scotland, in Smith, D.E., and Dawson, A.G., eds., Shorelines and isostasy: Institute of British Geographers Special Publication 16, p. 97–128.

Gustavson, T.C., and Boothroyd, J.B., 1987, A depositional model for outwash, sediment sources, and hydrologic characteristics, Malaspina Glacier, Alaska: A modern analog for the southeastern margin of the Laurentide Ice Sheet: Geological Society of America Bulletin, v. 99, p. 187–200.

Hafsten, U., 1983, Biostatigraphical evidence for Late Weichselian and Holocene sea-level changes in southern Norway, in Smith, D.E., and Dawson, A.G., eds., Shorelines and isostasy: British Institute of Geographers Special Publication 16, p. 161–182.

Hanson, L.S., and Caldwell, D.W., 1989, The lithologic and structural controls on the geomorphology of the mountainous areas in north-central Maine, in Tucker, R.D., and Marvinney, R.G., eds., Studies in Maine geology, Volume 5, Quaternary studies: Augusta, Maine Geological Survey, p. 147–167.

Hildreth, C.T., 1990, Surficial geology of the Biddeford quadrangle, Maine: Maine Geological Survey Open-File 90-36, scale 1:24 000.

Hubeny, J.B., 1997, Late Quaternary stratigraphy and geologic history of The Basin, Phippsburg, Maine [B.S. thesis]: Lewiston, Maine, Bates College, 87 p.

Hughes, T., 1987, Ice dynamics and deglaciation models when ice sheets collapsed, in Ruddiman, W.F., and Wright, H.E., Jr., eds., North America and adjacent oceans during the last deglaciation: Boulder, Colorado, Geological Society of America, Geology of North America, v. K-3, p. 183–220.

Hughes, T., Borns, H.W., Jr., Fastook, J.L., Hyland, M.R., Kite, J.S., and Lowell, T.V., 1985, Models of glacial reconstruction and deglaciation applied to Maritime Canada and New England, in Borns, H.W., Jr., et al., eds., Late Pleistocene history of northeastern New England and adjacent Quebec: Geological Society of America Special Paper 197, p. 139–150.

Hunter, L.E., 1990, Surficial geology of the Bar Mills quadrangle, Maine: Maine Geological Survey, Open-File 90-34, scale 1:24 000.

Hunter, L.E., Powell, R.D., and Smith, G.W., 1996, Facies architecture and grounding-line fan processes of morainal banks during the deglaciation of coastal Maine: Geological Society of America Bulletin, v. 108, p. 1022–1038.

Ingolfsson, O., Norddahl, R., and Haflidason, H., 1995, Rapid isostatic rebound in southwestern Iceland at the end of the last glaciation: Boreas, v. 24, p. 245–259.

Ireland, M.K.D., 1990, Application of geophysical techniques and facies logging in a complex glacial-marine setting in Lisbon, Maine [B.S. thesis]: Lewiston, Maine, Bates College, 76 p.

Johnson, W.H., Hansel, A.K., Bettis, E.A., III, Karrow, P.F., Larson, G.J., Lowell, T.V., and Schneider, A.F., 1997, Late Quaternary temporal and event classifications, Great Lakes region, North America: Quaternary Research, v. 47, p. 1–12.

Kaplan, M.R., 1999, Retreat of a tidewater margin of the Laurentide ice sheet in eastern coastal Maine between ca. 14 000 and 13 000 $^{14}$C yr B.P.: Geological Society of America Bulletin, v. 111, p. 620–632.

Kelley, J.T., 1989, A preliminary analysis of the mineralogy of glaciomarine mud from the western margin of the Gulf of Maine: Northeastern Geology, v. 11, p. 141–151.

Kelley, J.T., Dickson, S.M., Belknap, D.F., and Stuckenrath, R., Jr., 1992, Sea-level change and the introduction of late Quaternary sediment to the southern Maine inner continental shelf, in Fletcher, C.H., and Wehmiller, J.F., eds., Quaternary coasts of the United States: Marine and lacustrine systems: SEPM (Society for Sedimentary Geology), 48, p. 23–34.

King, L.H., 1996, Late Wisconsinan ice retreat from the Scotian Shelf: Geological Society of America Bulletin, v. 108, p. 1056–1067.

Koteff, C., 1991, Surficial geologic map of parts of the Rochester and Somersworth quadrangles, Strafford County, New Hampshire: U.S. Geological Survey Map I-2265, scale 1:24 000.

Koteff, C., Robinson, G.R., Goldsmith, R., and Thompson, W.B., 1993, Delayed postglacial uplift and synglacial sea levels in coastal central New England: Quaternary Research, v. 40, p. 46–54.

Kreutz, K.J., 1994, Paleoceanographic conditions of the late Wisconsinan marine transgression and regression in eastern coastal Maine; stable isotopic evidence [M.S. thesis]: Orono, University of Maine, 113 p.

Leavitt, H.W., and Perkins, E.H., 1935, A survey of road materials and glacial geology of Maine, Volume II, Glacial geology of Maine: Maine Technological Experiment Station Bulletin 30, 232 p.

Lehman, S.J., and Keigwin, L.D., 1992, Sudden changes in North Atlantic circulation during the last deglaciation: Nature, v. 356, p. 757–762.

Lepage, C., 1982, The composition and origin of the Pond Ridge moraine, Washington County, Maine [M.S. thesis]: Orono, University of Maine, 74 p.

Levasseur, D., 1995, Les eskers: essai de synthesis bibliographique: Géographie Physique et Quaternaire, v. 49, p. 459–479.

Locker, S.D., Hine, A.C., Tedesco, L.P., and Shinn, E.A., 1996, Magnitude and timing of episodic sea-level rise during the last deglaciation: Geology, v. 24, p. 827–830.

Lowell, T.V., and Borns, H.W., Jr., 1994, Maine's calving bay?: Geological Society of America Abstracts with Programs, v. 26, no. 3, p. 57.

Lowell, T.V., Heusser, C.J., Anderson, B.G., Moreno, P.I., Hauser, A., Heusser, L.E., Schluchter, C., Marchant, D.R., and Denton, G.H., 1995, Interhemispheric correlation of late Pleistocene glacial events: Science, v. 269, p. 1541–1549.

MacAyeal, D.R., 1993, Binge/purge oscillations of the Laurentide Ice Sheet as a cause of the North Atlantic's Henrich Events: Paleoceanography, v. 8, p. 775–784.

Mackiewizc, N.E., Powell, R.D., Carlsson, P.R., and Molnia, B.F., 1984, Interlaminated ice-proximal glacimarine sediments in Muir Inlet, Alaska: Marine Geology, v. 57, p. 133–147.

Mangerud, J., and Gullickson, S., 1975, Apparent radiocarbon ages of recent marine shells from Norway, Spitsbergen and Arctic Canada: Quaternary Research, v. 5, p. 263–273.

Marvinney, C.L., 1996, Surficial geology of the North Pownal quadrangle, Maine: Maine Geological Survey Open-File 96-15, scale 1:24 000.

Mayer, L.M., 1990, Mineralogy and pore water chemistry of Presumpscot clays: Maine Geological Survey Open-File 90-23, 3 p.

Mayewski, P.A., Meeker, L.D., Whitlow, S., Twickler, M.S., Morrison, M.C., Bloomfield, P., Bond, G.C., Alley, R.B., Gow, A.J., Grootes, P.M., Meese, D.A., Ram, M., Taylor, K.C., and Wumkes, W., 1994, Changes in atmospheric circulation and ocean ice cover over the North Atlantic during the last 41,000 years: Science, v. 263, p. 1747–1751.

McPherson, J.G., Shanmugam, G., and Moiola, R.J., 1987, Fan-deltas and braid-deltas: Varieties of coarse-grained deltas: Geological Society of America Bulletin, v. 99, p. 331–340.

Mickleson, D.M., Clayton, L., Fullerton, D.S., and Borns, H.W., Jr., 1983, The late Wisconsin glacial record of the Laurentide Ice Sheet in the United States, in Porter, S.C., ed., Late-Quaternary environments of the United States, Volume 1, The late Pleistocene: Minneapolis, University of Minnesota Press, p. 3–37.

Molnia, B.F., 1983, Subarctic glacial-marine sedimentation; a model, in Molnia, B.F., ed., Glacial-marine sedimentation: New York, Plenum Press, p. 95–144.

Nemec, W., 1990, Deltas—Remarks on terminology and classification, in Collela, A., and Prior, D.B., eds., Coarse-grained deltas: International Association of Sedimentologists Special Publication 10, p. 3–12.

Oldale, R.M., 1989, Timing and mechanisms for the deposition of the glaciomarine mud in and around the Gulf of Maine: A discussion of alternative models, in Tucker, R.D., and Marvinney, R.G., eds., Studies in Maine geology, Volume 5; Quaternary studies: Augusta, Maine Geological Survey, p. 1–10.

Osberg, P.H., Hussey, A.M., II, and Boone, G.M., 1985, Bedrock geologic map of Maine: Augusta, Maine Geological Survey, scale 1:500 000.

Phillips, A.C., Smith, N.D., and Powell, R.D., 1991, Laminated sediments in prodeltaic sediments, Glacier Bay Alaska, in Anderson, J.B., and Ashley, G.M., eds., Glacial marine sedimentation; paleoclimatic significance: Geological Society of America Special Paper 261, p. 51–60.

Postma, G., 1990, Depositional architecture and facies of river and fan deltas: A synthesis, in Collela, A., and Prior, D.B., eds., Coarse-grained deltas: International Association of Sedimentologists Special Publication 10, p. 3–12.

Powell, R.D., 1990, Glacimarine processes at grounding-line fans and their growth to ice-contact deltas, in Dowdeswell, J.A., and Scourse, J.D., eds., 1990, Glacimarine environments: Processes and sediments: Geological Society [London] Special Publication 53, p. 53–73.

Powell, R.D., 1991, Grounding-line systems as to second-order controls on fluctuations of tidewater termini of temperate glaciers, in Anderson, J.B., and Ashley, G.M., eds., Glacial marine sedimentation: Paleoclimatic significance: Geological Society of America Special Paper 261, p. 75–93.

Retelle, M.J., 1991, Surficial geology of the Old Orchard Beach quadrangle, Maine: Maine Geological Survey Open-File 91-4 scale 1:24 000.

Retelle, M.J., 1994, Surficial geology of the Cumberland Center quadrangle, Maine: Maine Geological Survey Open-File 94-9, scale 1:24 000.

Retelle, M.J., 1997a, Surficial geology of the Yarmouth quadrangle, Maine: Maine Geological Survey Open-File 97-1, scale 1:24 000.

Retelle, M.J., 1997b, Surficial geology of the Raymond quadrangle, Maine: Maine Geological Survey Open-File 97-57, scale 1:24 000.

Retelle, M.J., and Bither, K.M., 1989, Late Wisconsinan glacial and glaciomarine sedimentary facies in the lower Androscoggin Valley, Topsham, Maine, in Tucker, R.D., and Marvinney, R.G., eds., Studies in Maine geology, Volume 6, Quaternary studies: Augusta, Maine Geological Survey, p. 33–51.

Ridge, J.C., 1997, Shed Brook discontinuity and Little Falls gravel: Evidence for the Erie interstade in central New York: Geological Society of America Bulletin, v. 109, p. 652–665.

Ridge, J.C., and Larsen, F.D., 1990, Reevaluation of Antev's New England varve chronology and new radiocarbon dates of sediments from glacial Lake Hitchcock: Geological Society of America Bulletin, v. 102, p. 889–899.

Ridge, J.C., Canwell, B.A., Callahan, J.W., Cook, G.J., Nicholson, R.S., Toll, N.J., and Kelley, M., 1998, Varve, paleomagnetic, and radiocarbon chronologies for the deglaciation of northern New England: Geological Society of America Abstracts with Programs, v. 30, no. 1, p. 70.

Ridge, J.C., Besonen, M.R., Brochu, M., Brown, S.L., Callahan, J.W., Cook, G.J., Nicholson, R.S., and Toll, N.J., 1999, Varve, paleomagnetic, and $^{14}C$ chronologies for late Pleistocene events in northern New Hampshire and Vermont (U.S.A.), in Thompson, W.B., et al., eds., Late Quaternary history of the White Mountains, New Hampshire and adjacent southeastern Quebec: Géographie Physique et Quaternaire, v. 53, p. 79–107.

Rust, B.R., and Romanelli, R., 1975, Late Quaternary subaqueous deposits near Ottawa, Canada, in Jopling, A.V., and McDonald, B.C., eds., Glaciofluvial and glaciolacustrine sedimentation: Society of Economic Paleontologists and Mineralogists Special Publication 23, p. 177–192.

Schnitker, D., 1974, Postglacial emergence of the Gulf of Maine: Geological Society of America Bulletin, v. 85, p. 491–494.

Schnitker, D., 1986, Ocean basin with a past: Explorations: University of Maine Journal of Research, v. 3, p. 29–37.

Sharpe, D.R., 1988, Glaciomarine fan deposition in the Champlain Sea, in Gadd, N.R., ed., The late Quaternary development of the Champlain Sea basin: Geological Association of Canada Special Paper 35, p. 63–82.

Sissons, J.B., 1983, Shorelines and isostasy in Scotland, in Smith, D.E., and Dawson, A.G., eds., Shorelines and isostasy: Institute of British Geographers Special Publication 16, p. 209–226.

Slayton, J.T., 1993, Sediments and stratigraphy of a glacialmarine ice-contact landform in the Androscoggin Valley, Lewiston, Maine [B.S. thesis]: Lewiston, Maine, Bates College, 84 p.

Smith, G.W., 1980, End moraines and glaciofluvial deposits, Cumberland and York Counties, Maine: Maine Geological Survey Open-File EMDG, scale 1:126 720.

Smith, G.W., 1982, End moraines and the pattern of last ice retreat from central and south coastal Maine, in Larson, G.J., and Stone, B.D., eds., Late Wisconsinan glaciation of New England: Dubuque, Iowa, Kendall/Hunt Publishing Co., p. 195–210.

Smith, G.W., 1984, Glaciomarine sediments and facies associations, southern York County, Maine, in Hanson, L.S., ed., Guidebook for field trips in the coastal lowlands, Boston Massachusetts to kennebunk, Maine: New England Intercollegiate Geological Conference, 76th Annual Meeting, Danvers, Massachusetts: Salem, Massachusetts, Salem State College, p. 352–369.

Smith, G.W., 1985, Chronology of late Wisconsinan deglaciation of coastal Maine, in Borns, H.W., Jr., et al., eds., Late Pleistocene history of northeastern New England and adjacent Quebec: Geological Society of America Special Paper 197, p. 29–44.

Smith, G.W., 1988, Surficial geology of the Portsmouth quadrangle, Maine: Maine Geological Survey Open-File 88-5, scale 1:24 000.

Smith, G.W., 1994, Surficial geology of the Gorham quadrangle, Maine: Maine Geological Survey Open-File 94-6, scale 1:24 000.

Smith, G.W., and Hunter, L.E., 1989, Late Wisconsinan deglaciation of coastal Maine, in Tucker, R.D., and Marvinney, R.G., eds., Studies in Maine geology, Volume 6, Quaternary studies: Augusta, Maine Geological Survey, p. 13–32.

Snyder, J.A., Forman, S.L., Mode, W.N., and Tarasov, G.A., 1997, Postglacial relative sea-level history: Sediment and diatom records of emerged coastal lakes, north central Kola Peninsula, Russia: Boreas, v. 26, p. 329–346.

Sollid, J.L., and Carlson, A.B., 1984, DeGeer moraines and eskers in Parvik, north Norway: Striae, v. 20, p. 55–61.

Stea, R.R., and Mott, R.J., 1998, Deglaciation of Nova Scotia: Stratigraphy and chronology of lake sediment cores and buried organic sections: Géographie Physique et Quaternaire, v. 52, p. 1–21.

Stea, R.R., Piper, D.J.W., Fader, G.B.J., and Boyd, R., 1998, Wisconsinan glacial and sea-level history of Maritime Canada and the adjacent continental shelf: A correlation of land and sea events: Geological Society of America Bulletin, v. 110, p. 821–845.

Stone, B.D., and Borns, H.W., Jr., 1986, Pleistocene glacial and interglacial stratigraphy of New England, Long Island, and adjacent Georges Bank and the Gulf of Maine, *in* Sibrava, V., et al., eds., Quaternary glaciations in the Northern Hemisphere: Quaternary Science Reviews, v. 5, p. 39–52.

Stone, B.D., and Peper, J.D., 1982, Topographic control of the deglaciation of eastern Massachusetts: Ice lobation and the marine incursion, *in* Larson, G.J., and Stone, B.D., eds., Late Wisconsinan glaciation of New England: Dubuque, Iowa, Kendall/Hunt, p. 145–166.

Stuiver, M., and Borns, H.W., Jr., 1975, Late Quaternary marine invasion in Maine; its chronology and associated crustal movement: Geological Society of America Bulletin, v. 86, p. 99–104.

Thompson, W.B., 1982, Recession of the late Wisconsinan ice sheet in coastal Maine, *in* Larson, G.J., and Stone, B.D., eds., Late Wisconsinan glaciation of New England: Dubuque, Iowa, Kendall/Hunt, p. 211–228.

Thompson, W.B., 1997, Surficial geology of the Portland West quadrangle, Maine: Maine Geological Survey Open-File 97-51, scale 1:24 000.

Thompson, W.B., and Borns, H.W., Jr., 1985, Surficial geologic map of Maine: Augusta, Maine Geological Survey, scale 1:500 000.

Thompson, W.B., and Smith, G.W., 1988, Pleistocene stratigraphy of the Augusta and Waldoboro areas, Maine: Maine Geological Survey Bulletin 27, 36 p.

Thompson, W.B., Crossen, K.J., Borns, H.W., Jr., and Anderson, B.G., 1989, Glaciomarine deltas of Maine and their relation to late Pleistocene–Holocene crustal movements, *in* Anderson, W.A., and Borns, H.W., Jr., eds., Neotectonics of Maine—Studies in seismicity, crustal warping, and sea-level change: Augusta, Maine Geological Survey, p. 43–67.

Thompson, W.B., Fowler, B.K., and Dorion, C.C., 1999, Deglaciation of the northwestern White Mountains, New Hampshire, *in* Thompson, W.B., et al., eds., Late Quaternary history of the White Mountains, New Hampshire and adjacent southeastern Quebec: Géographie Physique et Quaternaire v. 53, p. 59–77.

Trefethen, J.M., Allen, H., Leavitt, L., Miller, R.N., and Savage, C., 1947, Preliminary report on marine clays, *in* Report of the State Geologist, 1945–1946: Augusta, Maine Development Commission, p. 11–47.

Warren, W.P., and Ashley, G.M., 1994, Origins of ice-contact stratified ridges (eskers) of Ireland: Journal of Sedimentary Research, v. A64, p. 433–449.

Warren, W.P., and Ashley, G.M., eds., 1997, Ice-contact sedimentation: Processes and deposits: Quaternary Science Reviews, v. 16, 819 p.

Weddle, T.K., 1987, Reconnaissance surficial geology of the Norridgewock quadrangle: Maine Geological Survey, Open-File 87-23, scale 1:62 500.

Weddle, T.K., 1994, Surficial geology of the Freeport quadrangle, Maine: Maine Geological Survey Open-File 94-5, scale 1:24 000.

Weddle, T.K., 1997a, Surficial geology of the Lisbon Falls South quadrangle, Maine: Maine Geological Survey Open-File 97-49, scale 1:24 000.

Weddle, T.K., 1997b, Surficial geology of the Gray quadrangle, Maine: Maine Geological Survey Open-File 97-58, scale 1:24 000.

Weddle, T.K., and Retelle, M.J., 1995, Glaciomarine deposits of the late Wisconsinan Casco Bay Sublobe of the Laurentide Ice Sheet, *in* Hussey, A.M., II, and Johnston, R.J., eds., Guidebook to field trips in southern Maine and adjacent New Hampshire: New England Intercollegiate Geological Conference, 87th Annual Meeting, Brunswick, Maine: Dubuque, Iowa, Times Mirror Higher Education Group, p. 173–194.

Weddle, T.K., and Retelle, M.J., 1998, Deglacial style and relative sea-level chronology, Casco Bay lowlands to White Mountain Foothills, southwestern Maine: Geological Society of America Abstracts with Programs, v. 30, no. 1, p. 83.

Weddle, T.K., Koteff, C., Thompson, W.B., Retelle, M.J., and Marvinney, C.L., 1993, The late glacial marine invasion of coastal central New England (northeastern Massachusetts–southwestern Maine): Its ups and downs, *in* Cheney, J.T., and Hepburn, J.C., eds., Field trip guidebook for the northeastern United States: Amherst, University of Massachusetts, Department of Geology and Geography Contribution 67, p. I1–I32.

Weddle, T.K., Normand, A.E., and Bernotavicz, A.A., 1999, Surficial geology of the Lisbon Falls North quadrangle, Maine: Maine Geological Survey Open-File 99-9, scale 1:24 000.

Wohlfarth, B., 1996, The chronology of the last termination: A review of radiocarbon-dated, high-resolution terrestrial stratigraphies: Quaternary Science Reviews, v. 15, p. 267–284.

MANUSCRIPT ACCEPTED BY THE SOCIETY JUNE 9, 2000

# Stratigraphy, paleoceanography, chronology, and environment during deglaciation of eastern Maine

**Christopher C. Dorion***
*Department of Geological Sciences and Institute for Quaternary Studies, University of Maine, Orono, Maine 04469, USA*

**Gregory A. Balco**
*Department of Geological Sciences and Quaternary Research Center, University of Washington, Seattle, Washington 98195-1310, USA*

**Michael R. Kaplan**
*Department of Geology and Geophysics, University of Wisconsin-Madison, 1215 West Dayton Street, Madison, Wisconsin 53706, USA*

**Karl J. Kreutz**
*Department of Geological Sciences and Institute for Quaternary Studies, University of Maine, Orono, Maine 04469, USA*

**James D. Wright**
*Department of Geological Sciences Wright Laboratory, Rutgers, The State University of New Jersey, Piscataway, New Jersey 08854, USA*

**Harold W. Borns, Jr.**
*Department of Geological Sciences and Institute for Quaternary Studies, University of Maine, Orono, Maine 04469, USA*

## ABSTRACT

The eastern Maine coastal lowland contains glaciomarine sediments, deposited between ca. 14 000 and 12 500 $^{14}$C yr B.P. and now exposed above sea level. These deposits provide the opportunity to compare the dynamics of the retreating Laurentide ice sheet with the paleoceanography of the Gulf of Maine. We studied the geomorphology, sedimentology, faunal assemblages, isotope geochemistry, and chronology of these sediments. Faunal assemblages as well as molluscan and foraminiferal $\delta^{18}$O results indicate that deglaciation took place in arctic to subarctic conditions, characterized by water temperatures of 0–2°C and strong density stratification in which water of normal marine salinity was present at the grounding line and meltwater was confined to the surface layer. Radiocarbon age estimates from ice-proximal sediments indicate that ice retreated through the coastal portion of the study area at 10–40 m/yr between ca. 14 000 and ca. 13 000 $^{14}$C yr B.P. The retreat rate then increased to 100–150 m/yr, and the remainder of the study area was fully deglaciated by ca. 12 500 $^{14}$C yr B.P. This abrupt increase in retreat rate may be correlative with widespread North Atlantic warming ca. 12 800–12 700 $^{14}$C yr B.P. However, it was not accompanied by any change in the paleoceanographic conditions in the Gulf of Maine. Thus, accelerated ice recession must have been a response to atmospheric warming or internal ice dynamics rather than oceanographic changes.

---

*Current address: C.C. Dorion Geological Services, 79 Bennoch Road, Orono, Maine 04473, USA.

Dorion, C.C., Balco, G.A., Kaplan, M.R., Kreutz, K.J., Wright, J.D., and Borns, H.W., Jr., 2001, Stratigraphy, paleoceanography, chronology, and environment during deglaciation of eastern Maine, *in* Weddle, T.K., and Retelle, M.J., eds., Deglacial History and Relative Sea-Level Changes, Northern New England and Adjacent Canada: Boulder, Colorado, Geological Society of America Special Paper 351, p. 215–242.

# INTRODUCTION

A complicated series of changes in climate, ocean circulation, and ice dynamics marked the termination of the last glaciation. Correlation of events taking place in these various spheres is of basic importance to the study of global climate. The Maine coastal lowland, the subject of this study (Fig. 1), exposes a sequence of glaciomarine sediments that span the time period between deglaciation of the area and the marine regression that took place hundreds to thousands of years later. These deposits record both the history of ice recession through this region and the paleoceanography of the deglacial Gulf of Maine. Thus they can potentially be used to directly correlate ice dynamics with contemporaneous oceanographic conditions, and so link terrestrial records of ice retreat and marine paleoclimate records. In addition, they are unusual among glaciomarine sediments in that they are well exposed above sea level, and can be examined in great detail without the expense and difficulty of offshore core sampling.

## Physiography of eastern Maine

The eastern Maine coastal lowlands, the subject of this study, are low-relief surfaces with average elevations of 20–50 m above sea level (asl) (Fig. 1). The Passamaquoddy lowland rises gently from the present coastline to the east-west–trending Tomah Ridge, which rises to 300 m asl, and contains scattered hills rising to 100–150 m asl. The Penobscot lowland is framed between the Passadumkeag-Mattawamkeag uplands to the east and the Appalachian mountains to the west. Both lowlands are underlain by a variety of Paleozoic metasedimentary rocks of low to medium metamorphic grade, intruded by late Paleozoic granitic and gabbroic plutons (Osberg et al., 1985).

## Deglaciation and relative sea-level history of coastal Maine

Radiocarbon ages of buried soils and marine sediments indicate that the Laurentide ice sheet advanced into Maine from Quebec between ca. 30 000 and 24 000 $^{14}$C yr B.P. (McDonald and Shilts, 1971; Anderson et al., 1992; Dorion, 1997, Table 4). The ice sheet continued its southward advance, and reached its terminal position at the Browns and Georges Banks, Martha's Vineyard, Long Island, and northern New Jersey between 20 000 and 22 000 $^{14}$C yr B.P. (King, 1996; Bothner and Spiker, 1980; Sirkin and Stuckenrath, 1980; Tucholke and Hollister, 1973). The ice margin began to retreat from the Scotian Shelf prior to 18 000 $^{14}$C yr B.P. (King, 1996); from Stellwagen basin in the southwestern Gulf of Maine before 18 900 $^{14}$C yr B.P. (Tucholke and Hollister, 1973); and from northern New Jersey by 18 390 ± 200 to 18 570 ± 250 $^{14}$C yr B.P. (Cotter et al., 1986).

The sedimentary record of the deglaciation of the Gulf of Maine consists of a thick sequence of rhythmically laminated glaciomarine mud (Belknap and Shipp, 1991). Oldale (1989) and Oldale et al. (1990) explained these deposits by suspension settling from turbid, sediment-laden meltwater plumes debouching at the glacial grounding line as the ice margin steadily retreated from the Georges Bank to the present coastline between 18 000 and 11 000 $^{14}$C yr B.P. However, Hughes et al. (1985), Belknap et al. (1989), and Schnitker and Jorgensen (1990), in an effort to explain the lateral continuity of certain prominent seismic reflectors over tens to hundreds of kilometers, proposed that this mud was deposited as basal debris melted out from a temperate, floating ice shelf that entirely covered the Gulf of Maine between ca. 18 000 and 14 000 $^{14}$C yr B.P., and did not begin to melt seasonally until ca. 14 000 $^{14}$C yr B.P. These authors did not propose an explicit chronology of grounding line retreat across the region, but their model appears to require an early, rapid grounding line retreat from the Georges Bank to near the present shoreline ca. 18 000 $^{14}$C yr B.P., followed by a period of relative stability between 18 000 and 14 000 $^{14}$C yr B.P. A combination of floating ice shelves and grounded ice was present on the Scotian Shelf to the east (King, 1996). The western portion of the shelf contained "... ice shelves that were pinned on the banks surrounding the basins. Further retreat led to open-water conditions across the basins, surrounded by grounded ice with floating front margins that led to till-tongue and moraine development on the basin flanks" (King, 1966, p. 1065). Channelized meltwater deposits on the eastern portion of the shelf suggest predominantly grounded ice (Boyd et al., 1988; King, 1993, 1994).

Stuiver and Borns (1975) and Smith (1985) attempted to determine the chronology of ice retreat north of the present shoreline through the eastern Maine coastal lowlands. Both studies concluded that deglaciation of the coastal lowlands took place between ca. 13 200 and 12 600 $^{14}$C yr B.P. These studies, however, were handicapped by a paucity of radiocarbon ages and some uncertainty regarding the context of the dated material. Smith (1985, p. 41) qualified his results by noting that "The number of dates directly related to positions of the retreating ice margin is severely limited . . . no detailed work has been done on either the paleontology/paleoecology or the sedimentology of the glacial-marine succession . . . [no effort was made] to relate ice retreat in coastal Maine to the behavior of ice elsewhere." We attempt to address some of these deficiencies in this study.

At the time of deglaciation in eastern Maine, global eustatic sea level was ~110 m below present (Fairbanks, 1989). However, the landscape was sufficiently depressed by the weight of the ice sheet that relative sea level at the present coastline stood ~70 m above present (Stone, 1899; Barnhardt et al., 1995; Thompson et al., 1989). Marine transgression was concomitant with deglaciation, the ice retreated in continuous contact with the ocean, and the Penobscot and Passamaquoddy lowlands were embayments of the De Geer sea (which included the present Gulf of Maine and the area inland from the present coastline

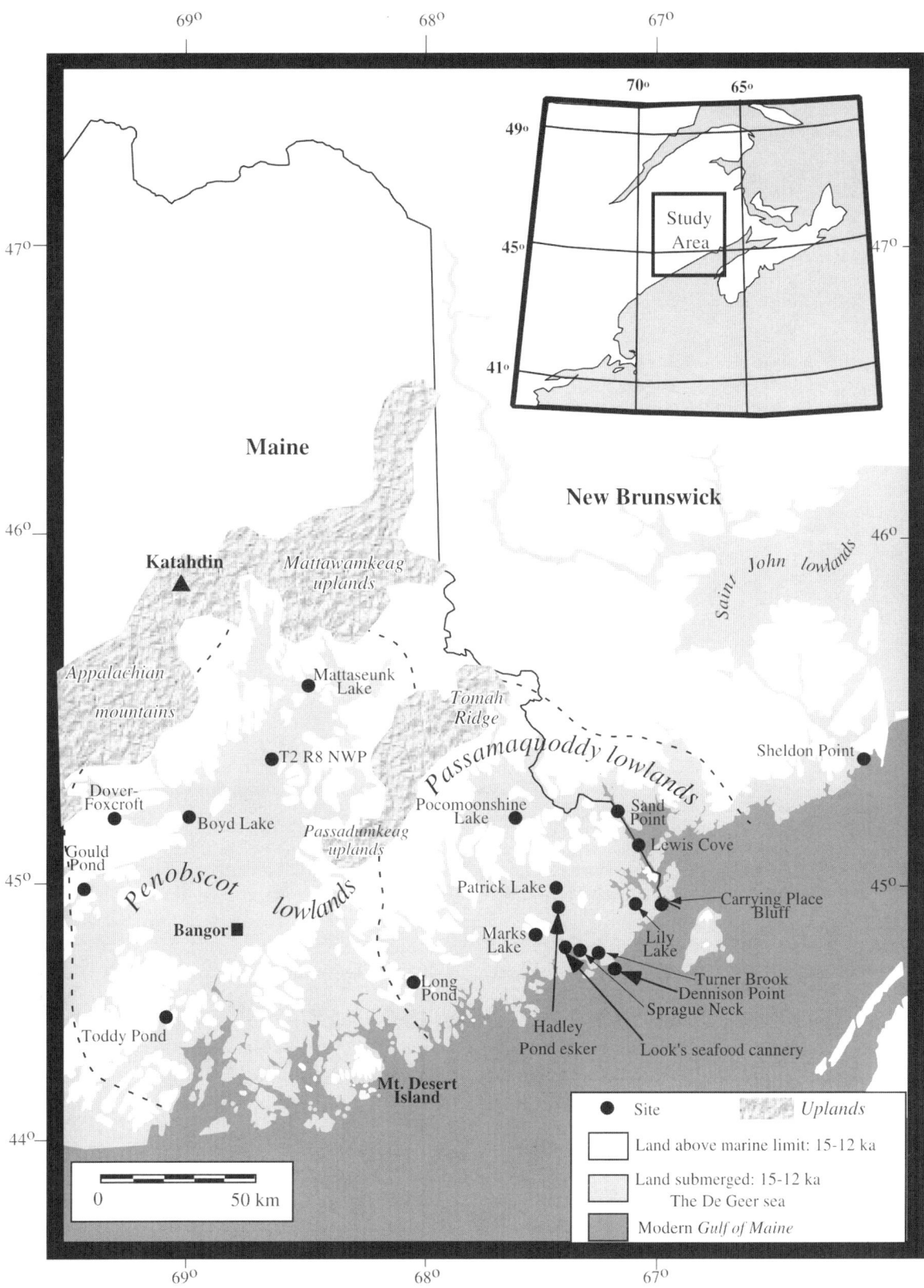

Figure 1. Location map showing study area, field sites, and hypothesized maximum extent of De Geer sea ca. 13 000 $^{14}$C yr B.P. Marine limit is calculated from delta topset-foreset contacts in Thompson et al. (1989).

218                                                                 C.C. Dorion et al.

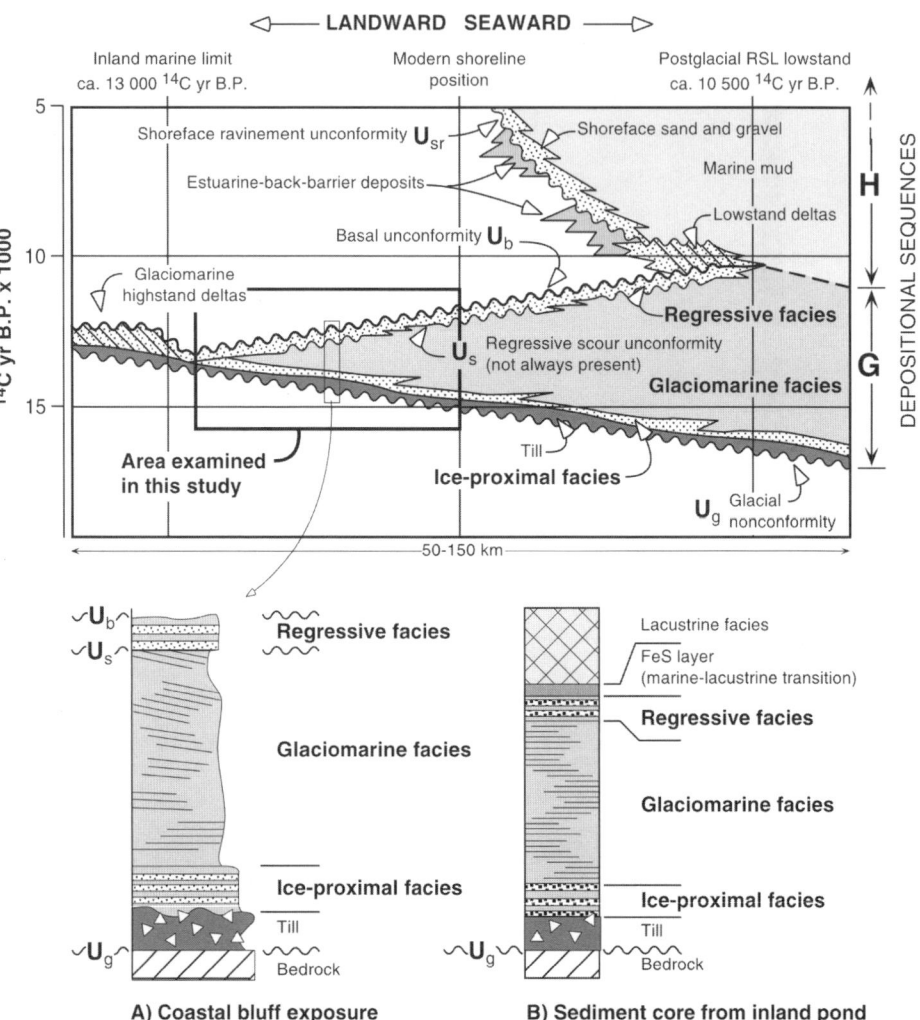

Figure 2. Chronostratigraphic diagram adapted from Barnhardt et al. (1995) and Belknap and Shipp (1991), showing sequence-stratigraphic context of glaciomarine facies on Maine coastal lowland and inner shelf during glacial (G) and Holocene (H) time. Stratigraphic nomenclature follows Barnhardt et al. (1995), with two exceptions. First, they do not differentiate regressive glaciomarine facies and regressive scour unconformity Us. Second, they refer to our ice-proximal facies as "stratified drift." We retain their term "basal unconformity" despite fact that this unconformity is at top of all of our stratigraphic sections. A, Representative stratigraphic section from coastal bluff exposure. B, Stratigraphy of inland ponds differs in that Us and Ub are absent, and lacustrine sedimentation continued after relative sea-level fall.

that was submerged during late glacial time; Lougee and Lougee, 1976). At this time the study area resembled the present deeply indented Maine coastline. The low hills in the Passamaquoddy lowland formed an island archipelago; the Penobscot lowland contained fewer islands and was connected to the open ocean only by a narrow strait south of Bangor. By the time the ice margin retreated above the marine limit, isostatic rebound had outpaced eustatic sea-level rise. After 12 500 $^{14}$C yr B.P., relative sea level fell rapidly to a lowstand of −60 m at 10 500 $^{14}$C yr B.P. before beginning a gradual rise which continues today (Barnhardt, 1994; Barnhardt et al., 1995).

The glaciomarine sediments in eastern Maine thus compose a depositional sequence deposited between deglaciation and accompanying marine transgression, and later marine regression. This package of sediment (sequence G, for glacial, of Belknap and Shipp, 1991, and Barnhardt et al., 1995) is bounded on the bottom by a basal nonconformity produced by glacial erosion and on top by a regressive unconformity formed by subaerial erosion after relative sea level fall (Fig. 2). Because both bounding surfaces are time transgressive, the time span

recorded in these sediments increases in a seaward direction, and glaciomarine sediments at low elevation near the present coastline preserve the longest record of paleoceanographic conditions that is available on land. Near the coast in eastern Maine, this sequence was deposited between ca. 14 000 and 12 500 $^{14}$C yr B.P. (Dorion, 1997). In southwestern coastal Maine, this interval is longer, covering ca. 15 000–11 500 $^{14}$C yr B.P. (Anderson et al., 1990; Weddle and Retelle, 1998; Barnhardt et al., 1995). These sediments are coeval with major oceanographic changes in the North Atlantic between Heinrich event 1 ca. 14 000 $^{14}$C yr B.P. (Bond and Lotti, 1995) and the abrupt warming and presumed onset of North Atlantic thermohaline circulation at 12 700 $^{14}$C yr B.P. (Lehman and Keigwin, 1992), and we chose to study them because of the potential to directly correlate the behavior of the margin of the Laurentide ice sheet with these events.

The stratigraphy and geomorphology of the glaciomarine deposits in eastern Maine are well studied. Ashley et al. (1991, p. 108) summarized the deposits in the Passamaquoddy lowlands:

A variety of subparallel ice-front deposits, collectively called stratified moraines, mark the lobate margin of the ice sheet as it retreated in contact with the rising sea. The largest moraines . . . have been correlated laterally over tens of kilometers . . . (Leavitt and Perkins, 1935; Borns, 1980; Smith, 1981, 1982). These moraines are geomorphologically and sedimentologically complex, consisting of ice-tunnel deposits, submarine fans, and flat-topped deltas. These deposits commonly interfinger laterally into, or are overlain by, a fossiliferous glacial-marine deposit (Presumpscot Formation; Bloom, 1960; Smith et al., 1979; Jong, 1980). Many of these moraines display evidence of deformation by active ice . . . Smaller and more discontinuous DeGeer-type moraines (Hoppe, 1959), with spacing of about 100 m, occur between and subparallel to the larger moraines and consist mainly of stratified marine fan deposits with minor debris-flow diamict.

More than 100 single and amalgamated glacial-marine deltas have been mapped in Maine . . . (Thompson et al., 1989). Some deltas are draped over, or have been modified to become part of, the end moraine complexes . . . but others occur as isolated landforms. All were built into the sea and thus mark the position of the sea at the time of formation . . . Nearly all the deltas are associated with segmented eskers that record the former presence of ice-tunnel systems that fed sediment to deltas at the ice margin (Thompson and Borns, 1985; Shreve, 1972; 1985a; 1985b).

Deposits in the lower Penobscot valley are similar (Fig. 3), although the large stratified moraines and minor DeGeer moraines are absent in the upper Penobscot valley (Thompson and Borns, 1985). Throughout the study area, linear ice-marginal deposits trend east-northeast–west-southwest while flow indicators and large esker systems trend approximately northwest to southeast, in most cases nearly normal to end moraines. Smith (1985) and Shreve (1985a, 1985b) concluded from this pattern of landforms that the ice margin maintained a consistent shape and surface slope as it gradually retreated to the north-northwest. Smith (1985) further concluded that the ice margin remained relatively straight because neither large- nor small-scale erosional features showed evidence of flow convergence into marine reentrants. Kaplan (1999), however, mapped a portion of the study area in detail and found evidence for small-scale (1–5 km wide, perhaps 1 km long) reentrants in areas of deeper water.

In general, the stratigraphy in eastern Maine below the marine limit is as follows: Polished and striated Paleozoic bedrock is overlain by dense lodgment till. This may be overlain by a variety of ice-proximal deposits, including subaqueous outwash (Rust and Romanelli, 1975), mass movement, and debris-flow gravity-driven deposits (Middleton and Hampton, 1976), including flowtill (Hartshorn, 1958), and glaciotectonized material. These ice-proximal deposits are in turn mantled by the thick ice-proximal to ice-distal glaciomarine mud of the Presumpscot Formation. The Presumpscot Formation is often covered by a coarse unit made up of material reworked and deposited by nearshore processes during marine regression. We recognize three primary glaciomarine facies in this study: (1) an ice-proximal facies consisting of subaqueous outwash, glaciotectonized material, flowtill, and/or gravity-driven sediment flows, (2) the ice-distal glaciomarine mud facies of the Presumpscot Formation, and (3) a coarse-grained regressive facies.

Several authors have attempted to reconstruct the depositional environment of the Presumpscot Formation and its offshore equivalent (Emerald Silt of King and Fader, 1986; facies GM of Belknap and Shipp, 1991). Bloom (1960, 1963) studied its mollusc fauna and inferred that the oceanographic conditions were subarctic, similar to those found in the Labrador Sea today (0–12 °C). Other workers, however, pointed to the extremely high sedimentation rates required by the large glaciomarine deltas and argued that conditions at the ice margin were temperate, analogous to the modern Gulf of Alaska (5–14 °C; Ashley et al., 1991; Oldale, 1989). Schnitker and Jorgenson (1990) studied the offshore record in cores from the Wilkinson basin in the southwestern Gulf of Maine, and found that the $\delta^{18}O$ of benthic foraminifera was between 3‰ and 4‰ from ca. 17 000 to 12 500 $^{14}C$ yr B.P. In addition, they reported a near total absence of diatoms. Schnitker and Jorgenson (1990) and Bacchus (1993) used these data to argue in favor of the floating ice-shelf model, theorizing that the ice shelf excluded diatoms but an influx of warm and saline Atlantic slope water (presumably above 5 °C) made conditions favorable for foraminifera and explained the high $\delta^{18}O$ values.

## This study

In this chapter we take advantage of well-exposed glaciomarine sections in eastern Maine to examine the sedimentological, paleontological, and geochemical character of the glaciomarine sediments in eastern Maine, and reconstruct the paleoenvironmental conditions at the ice margin. We use this information to better understand the context and significance of radiocarbon ages from this area, and present a revised chronology of ice recession based on new accelerator mass spectrometry (AMS) radiocarbon ages.

We studied 16 sites that expose a part or the entire succession of ice-marginal glaciomarine, ice-distal glaciomarine, and regressive facies (Table 1; Fig. 1). At cutbanks of rivers, eroding coastal bluffs, and a few gravel pits, much of the sequence can be directly sampled in natural or anthropogenic exposures. However, the regressive unconformity at such sites typically cuts out a significant amount of the glaciomarine facies. We addressed this problem by obtaining sediment cores from lakes and ponds throughout the study area. Lakes and ponds, being enclosed basins, were protected from shoreface erosion during marine regression. Thus, the regressive unconformity is minimal or absent, and these basins preserve a nearly complete record of deglaciation, glaciomarine sedimentation, shallowing, and transition to freshwater conditions.

## METHODS

We excavated natural and artificial exposures by hand, and obtained lake sediment cores using a modified Wright-Livingston piston corer (Wright, 1967). The description of lithostrat-

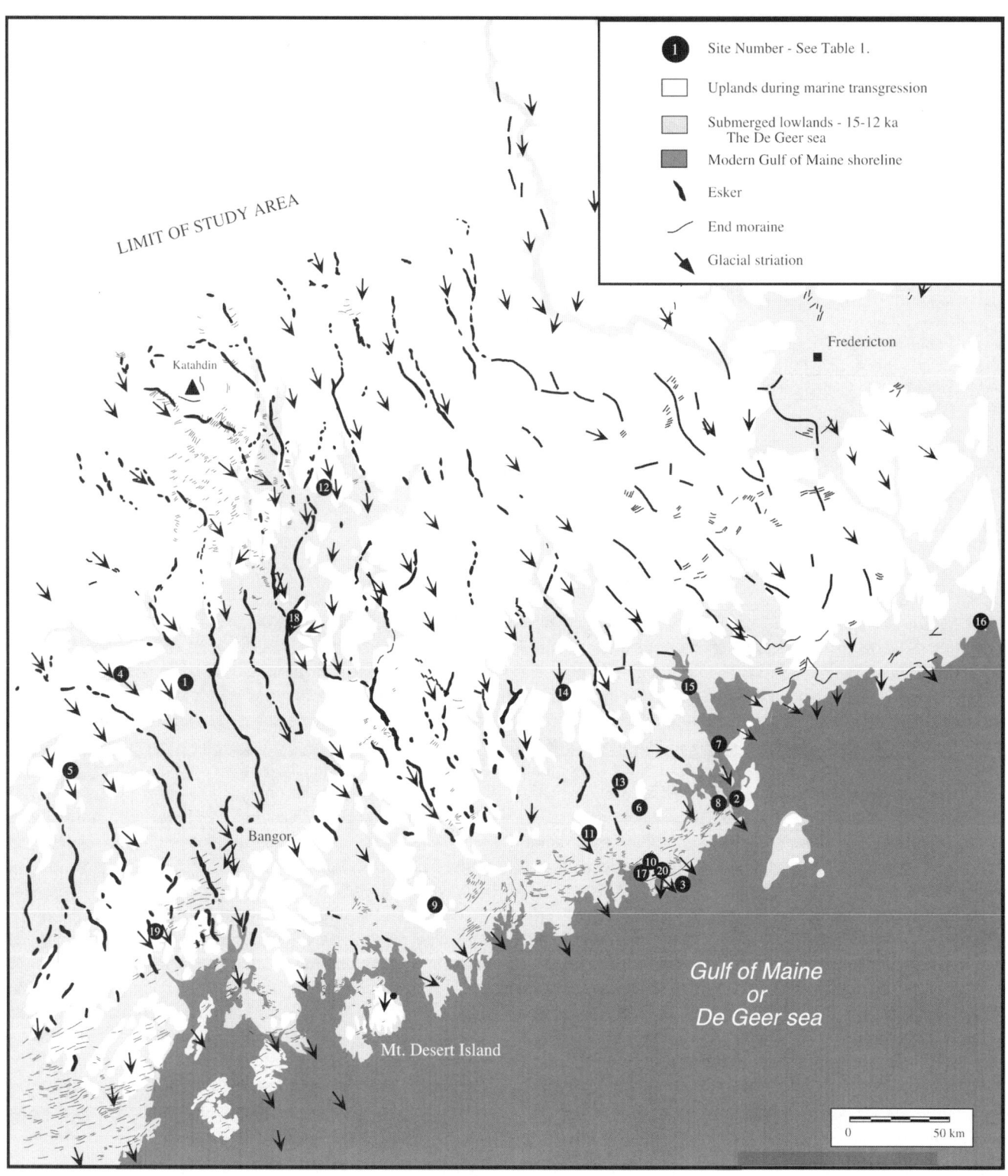

Figure 3. Site locations (see Table 1), surficial geologic features, and extent of inland marine limit. Eskers, end moraines, and glacial striae data are modified after Thompson and Borns (1985) and Borns et al. (1987). Marine limit is calculated from delta topset-foreset contact elevations in Thompson et al. (1989).

## TABLE 1. SITE NUMBERS AND ATTRIBUTES; RADIOCARBON SAMPLE SUBMISSION DATA

| Site no. | Site name | Latitude (N) | Longitude (W) | Sample elevation (masl) | Stratigraphic setting | Date class | Accession number | Reported age ($^{14}$C yr B.P.) | *Reservoir corrected age | **Calendar age | Material | $\delta^{13}C$ (‰) | $\delta^{18}O$ (‰) | Reference | Comments |
|---|---|---|---|---|---|---|---|---|---|---|---|---|---|---|---|
| 1 | Boyd Lake | 45°10.11' | 68°55.25' | 82.95–83.05 | IP (ice-proximal) | I | AA-9293 | 13 075 ± 90 | 12 675 ± 90 | 14 910 ± 210 | *Portlandia arctica* (J. E. Gray) | 0.4 | 3.01 | | Ice-proximal facies in lake core. |
| 2 | Carrying Place Bluff | 44°48.47' | 66°58.46' | 0.20 | IP | I | AA-8213 | 13 150 ± 140 | 12 750 ± 140 | 15 030 ± 265 | *Hiatella arctica* (dwarfed form) | 0.15 | 3.52 | | Lake surface 92 masl. Ice-proximal subaqueous outwash. Coastal bluff. |
| | Carrying Place Bluff | 44°48.47' | 66°58.46' | 0.20 | IP | I | OS-7135 | 13 350 ± 50 | 12 950 ± 50 | 15 360 ± 140 | *Hiatella arctica* (dwarfed form) | 0.58 | 3.48 | | Ice-proximal subaqueous outwash. Coastal bluff. |
| | Carrying Place Bluff | 44°48.47' | 66°58.46' | 2.00–2.30 | GM (Glaciomarine) | N.D.ª | OS-2075 | 13 800 ± 80 | 13 400 ± 80 | 16 030 ± 150 | *Nucula tenuis* (Montagu) | 1.36 | 3.09 | | Upper glaciomarine facies. Coastal bluff. |
| 3 | Dennison Point | 44°38.30' | 67°14.30' | 1.00–2.00 | IP | I | OS-2154 | 14 000 ± 85 | 13 600 ± 85 | 16 290 ± 140 | *Macoma calcarea* (Gmelin) | −0.58 | 3.87 | Kaplan (1994) | Ice-proximal marine diamicton. Coastal bluff. |
| 4 | Dover—Foxcroft | 45°12.15' | 69°10.51' | 105.00 | GM + 65 cm | II | OS-11022 | 13 550 ± 60 | 13 150 ± 60 | 15 670 ± 140 | *Macoma balthica* (Linnaeus) | −1.09 | −2.94 | | Lower glaciomarine facies. |
| 5 | Gould Pond | 44°59.33' | 69°19.09' | 68.74–68.84 | GM | II | AA-7463 | 13 290 ± 85 | 12 890 ± 85 | 15 260 ± 190 | *P. arctica* | −0.51 | 0.83 | | Lower glaciomarine facies in lake core. |
| 5 | Gould Pond | 44°59.33' | 69°19.09' | 68.90–69.05 | GM | II | SI-5371 | 13 280 ± 410 | 12 880 ± 410 | 15 240 ± 640 | *P. arctica* | N.D.ª | N.D.ª | Anderson et al. (1992) | Lake surface 90 masl. |
| 6 | Hadley Pond Esker | 44°48.25' | 67°24.40' | 46.00 | GM | III | OS-2155 | 12 800 ± 50 | 12 400 ± 50 | 14 510 ± 170 | *H. arctica* | 0.56 | N.D.ª | Kaplan (1994) | Upper glaciomarine facies. |
| 7 | Lewis Cove | 45°02.02' | 67°06.27' | 2.90 | GM | III | OS-2659 | 12 900 ± 50 | 12 500 ± 50 | 14 650 ± 170 | *N. tenuis* | 2.34 | 4.5 | | Upper glaciomarine facies. |
| 8 | Lily Lake | 44°49.42' | 66°06.06' | 23.70–23.80 | R (Regressive) | N.D.ª | OS-2660 | 13 000 ± 65 | 12 600 ± 65 | 14 800 ± 1860 | *Mytilus edulis* | 0.59 | −0.34 | | Regressive facies in lake core. Lake surface 32 masl. |
| | Lily Lake | 44°49.42' | 66°06.06' | 23.70–23.80 | R (Regressive) | N.D.ª | OS-2681 | 13 200 ± 80 | 12 800 ± 80 | 15 110 ± 190 | *Desmarestia ligulata* | −19.72 | § | | Regressive facies in lake core. Lake surface 32 masl. |
| | Lily Lake | 44°49.42' | 66°06.06' | 15.98–16.18 | IP | I | OS-2151 | 13 350 ± 50 | 12 950 ± 50 | 15 360 ± 140 | *H. arctica* (dwarfed form) | 0.15 | 3.2 | Kaplan (1994) | Ice-proximal sediment from lake core. Lake surface 32 masl. |
| 9 | Long Pond | 44°35.43' | 68°01.23' | 34.40–35.70 | R | III | OS-3466 | 12 950 ± 120 | 12 550 ± 120 | 14 720 ± 240 | *N. tenuis*, *P. arctica* | −17.96 | Periostracum | | Ice-proximal sediment from lake core. Pond surface 54 masl. |
| 10 | Look's Seafood Cannery | 44°42.28' | 67°18.48' | −0.70 | GM | III | OS-2152 | 12 900 ± 50 | 12 500 ± 50 | 14 650 ± 170 | *N. tenuis* | −7.66 | N.D.ª | Kaplan (1994) | Upper glaciomarine facies. |
| 11 | Mark's Lake | 44°45.27' | 67°30.18' | 21.15 | GM basal | III | OS-3161 | 13 300 ± 65 | 12 900 ± 65 | 15 280 ± 160 | *N. tenuis* | 1.29 | 3.94 | | Basal glaciomarine facies in lake core. Lake surface 41 masl. |
| 12 | Mattaseunk Lake | 45°35.24' | 68°22.41' | 89.30–89.35 | GM upper | III | OS-1322 | 13 450 ± 75 | 13 050 ± 75 | 15 520 ± 160 | *P. arctica* | −16.8 | Periostracum | | Upper glaciomarine facies in lake core. Lake surface 99 masl. |

*(continued)*

## TABLE 1. SITE NUMBERS AND ATTRIBUTES; RADIOCARBON SAMPLE SUBMISSION DATA (continued)

| Site no. | Site name | Latitude (N) | Longitude (W) | Sample elevation (masl) | Stratigraphic setting | Date class | Accession number | Reported age ($^{14}$C yr B.P.) | *Reservoir corrected age | **Calendar age | Material | $\delta^{13}$C (‰) | $\delta^{18}$O (‰) | Reference | Comments |
|---|---|---|---|---|---|---|---|---|---|---|---|---|---|---|---|
| | Mattaseunk Lake | 45°35.24' | 68°22.41' | 89.30–89.35 | GM upper | III | OS-4382 | 10 550 ± 70 | † | 12 480 ± 150 | Terr. and aquatic veg. | −19.36 | § | | Upper glaciomarine facies in lake core. |
| | Mattaseunk Lake | 45°35.24' | 68°22.41' | 89.30–89.35 | GM upper | III | OS-14149 | 10 350 ± 120 | † | 12 230 ± 200 | Terr. and aquatic veg. | −20.14 | § | | Upper glaciomarine facies in lake core. |
| 13 | Patrick Lake | 44°52.39' | 67°23.07' | 20.70–21.00 | GM | II | OS-3465 | 13 400 ± 95 | 13 000 ± 95 | 15 440 ± 180 | Seaweed | −19.33 | § | | Lower glaciomarine facies in lake core. Lake surface 39 masl. |
| 14 | Pocomoonshine Lake | 45°07.31' | 67°32.20' | 11.60–12.70 | GM | II | OS-2661 | 13 200 ± 60 | 12 800 ± 60 | 15 110 ± 170 | P. arctica | −0.07 | 3.27 | | Lower glaciomarine facies in lake core. Lake surface 40 masl. |
| 15 | Sand Point | 45°08.21' | 67°07.45' | 23.80 | GM | II | OS-2663 | 13 700 ± 70 | 13 300 ± 70 | 15 890 ± 140 | N. tenuis | 1.92 | 2.67 | | Lower glaciomarine facies in submarine fan. Gravel pit. |
| 16 | Sheldon Point | 45°13.23' | 66°06.44' | 0.50–2.00 | GM | II | AA-8216 | 13 255 ± 125 | 12 855 ± 125 | 15 200 ± 240 | Seaweed | −22.6 | § | | Lower glaciomarine facies in moraine. Coastal bluff. |
| | Sheldon Point | 45°13.23' | 66°06.44' | 0.50–2.00 | GM | II | AA-8217 | 13 221 ± 94 | 12 821 ± 94 | 15 150 ± 200 | P. arctica | N.D.# | Periostracum | | Lower glaciomarine facies in moraine. Coastal bluff. |
| 17 | Sprague Neck | 44°39.51' | 67°19.07' | 0.50 | GM | III | AA-7462 | 13 370 ± 90 | 12 970 ± 90 | 15 390 ± 180 | P. arctica | −1.47 | 3.21 | | Glaciomarine facies on flank of moraine. Coastal bluff. |
| | Sprague Neck | 44°39.51' | 67°19.07' | 1.50–1.55 | R | N.D.# | AA-8219 | 12 425 ± 110 | 12 025 ± 110 | 14 020 ± 190 | M. edulis | 0.27 | 1.27 | | Regressive facies. Coastal bluff. |
| | Sprague Neck | 44°39.51' | 67°19.07' | 1.50–1.55 | R | N.D.# | AA-8218 | 12 930 ± 85 | 12 530 ± 85 | 14 690 ± 200 | D. ligulata | N.D. | § | | Regressive facies. Coastal bluff. |
| 18 | T2 R8 NWP | 45°22.04' | 68°32.55' | 76.00 | GM | II | OS-3160 | 13 300 ± 65 | 12 900 ± 65 | 15 280 ± 160 | N. tenuis | 0.31 | § | | Lower glaciomarine facies in submarine fan. Gravel pit. |
| 19 | Toddy Pond | 44°32.33' | 69°03.25' | 71.00–71.35 | GM basal | I | OS-2662 | 13 000 ± 60 | 12 600 ± 60 | 14 800 ± 180 | Elphidium excavatum (Terquem) forma clavata | −1.24 | 1.95 | | Basal glaciomarine facies in pond core. Pond surface 87 masl. |
| 20 | Turner Brook | 44°40.07' | 67°15.00' | 4.90 | IP (lower GM) | I | AA-7461 | 13 810 ± 90 | 13 410 ± 90 | 16 040 ± 160 | N. tenuis | −1.77 | 2.9 | | Ice-proximal facies in moraine. Coastal bluff. |
| | Turner Brook | 44°40.07' | 67°15.00' | 1.10 | IP (upper GM) | I | OS-1314 | 13 650 ± 55 | 13 250 ± 55 | 15 820 ± 130 | M. calcarea | −2.01 | 3.37 | | Ice-proximal facies in moraine. Coastal bluff. |

*Note:* masl—meters above sea level.
*The reservoir corrected age is a 400 yr subtraction from the reported value.
†Sample was terrestrial and limnic organic material therefore no reservoir factor.
§Oxygen isotope analysis not performed due to material type.
#N.D.—Not Determined
**Calibration to calendar years using CALIB 3.0.3c after Stuiver and Reimer (1993) using the marine data set which has a built-in 400 yr reservoir correction.

| Symbol | Facies Code | Interpretation |
|---|---|---|
| | **Diamicton** | |
| ▲▲ | Dmm | Diamicton, matrix supported | Glacially emplaced lodgment or ablation till |
| ▲▲~ | Dmm (r) | Diamicton (resedimented) | Flowtill |
| ▲▲z | Dmm (s) | Diamicton (sheared) | Glacially tectonized till |
| | **Gravel** | |
| ○○○ | Gms | Gravel, matrix supported | High energy, fluctuating velocities |
| ○○○ | Gm | Gravel, clast supported | Grain flows or surf zone deposits |
| | **Sand** | |
| ≡ | Sh | Sand, planar laminated in horizontal beds | Upper and lower flow regime |
| ≈ | Sm | Sand, planar cross-bedded, planar bounding surface | Scour and fill channels Point bar deposits Megaripples |
| ~ | Sr | Sand ripples (all types) | Lower flow regime |
| | **Fines** | |
| ≡ | Fl | Rhythmically laminated mud and/or sand | Distal glaciomarine or nearshore deposits |
| ≡○ | Fl (d) | Rhythmically laminated mud and/or sand (with dropstones) | Distal glaciomarine or nearshore deposits |
| | Fm | Fines, massive, fossiliferous | Bedding obliterated by secondary depositional processes or bioturbation |
| | Fm (d) | Fines, massive, (with dropstones) | Bedding obliterated by secondary depositional processes. Original texture maintained |
| | **Contacts** | |
| ▬ | Erosional | |
| ○○ | Conformable | |
| ∼∼ | Loaded | |

Figure 4. Lithofacies coding used herein for glaciomarine sediments. Modified after Folk (1974), Reineck and Singh (1980), Eyles et al. (1983), and Boggs (1987).

igraphic units used in this chapter follows a scheme modified after that of Eyles et al. (1983) (Fig. 4). We separated macrofossils and microfossils from sediment by disaggregating samples of unweathered material with water, KOH, and/or commercial dispersant, depending on whether the sample was destined for $^{14}$C or $\delta^{18}$O analysis; picked macrofossils and benthic foraminifera from the >125 μm size fraction; and identified them by comparison with published descriptions of Gulf of Maine benthic fauna (Schnitker, 1975; Ellis, 1984; Cotter, 1985).

Faunal abundances vary widely between sites and between units at a given site. For each unit at a given site, we report a list of identifiable macrofauna. If sufficient microfauna were present (n > 25) we report the quantitative species composition; if not, we report only a list of species. A few sites contained sufficient microfauna to allow reconstruction of stratigraphic changes in species assemblage; however, these locations were rare and for consistency we report only the aggregate species composition at each site.

We performed stable isotope analyses on both macrofauna (bivalves) and microfauna (benthic foraminifera) from cores and outcrops. Bivalve samples consisted of whole valves which were identified and crushed and ground into a powder to homogenize each sample. Then, ~50–100 mg of each powdered sample was loaded into a stainless steel boat for analysis. Each sample of benthic foraminifera consisted of ~10 specimens selected from the >125 μm size fraction, which were then loaded into a stainless steel boat and lightly crushed. We reacted both macrofaunal and microfaunal samples in phosphoric acid at 90 °C in an AutoCarb peripheral attached to a Prism II mass spectrometer. All values in this study are reported to the Peedee belemnite (PDB) standard by normalizing the NBS-19 and NBS-20 values analyzed during each sample run to −4.14 and 1.06 for $\delta^{18}O$ and $\delta^{13}C$, respectively (Coplen et al., 1983). The precision (1σ) of the standards analyzed during the sample analyses was 0.06 and 0.05 for $\delta^{18}O$ and $\delta^{13}C$, respectively.

We obtained AMS radiocarbon age estimates on mollusc shell and periostracum, seaweed, terrestrial organic material, and in one case on foraminifera (Table 1). At most sites, we sampled molluscs that were articulated and in life position, and dated a single valve. We washed all adhering sediment from shell samples with tap water and did not use deflocculants. At some sites where the carbonate in mollusc shells had been dissolved, we used mollusc periostracum. We examined samples of seaweed and other organic material under a binocular microscope for evidence of rootlets or other modern contamination, and discarded suspicious material.

We took all possible steps to minimize the likelihood that chronological inconsistencies in this study are the result of sample contamination. Previous studies of the Presumpscot Formation found that some mollusc samples used for radiocarbon dating were contaminated by postdepositional replacement of original aragonitic shell material by calcite in isotopic equilibrium with groundwater (Stuiver and Borns, 1975; Dorion, 1997). We made every possible effort to avoid this problem. We sampled only fauna that were fully enclosed by unoxidized, unweathered sediment. We discarded samples that showed visual evidence of mineralogical alteration or displayed anomalous $\delta^{13}C$ or $\delta^{18}O$ values, which could indicate reequilibration with younger groundwater. If we found anomalous isotopic values, we used x-ray diffraction (XRD) analysis to check for alteration of aragonite to calcite, and discarded altered samples. We tested for the presence of bedrock-derived carbonate in lake sediments by reaction with HCl and loss-on-ignition analysis (Bengtsson and Enell, 1986), and failed to find discernible evidence of reworked old carbon.

Due to the many uncertainties regarding oceanographic conditions during the last termination at middle latitude ice-marginal areas, the appropriate marine reservoir correction for our radiocarbon ages is unknown. Thus, $^{14}C$ ages presented in the text do not include a reservoir age correction unless specifically noted. In Table 1 we have applied the standard North Atlantic marine reservoir correction of 400 yr for ease of comparison to published data sets from other North Atlantic localities.

## RESULTS

### Sedimentology of glaciomarine facies

**Carrying Place Bluffs.** The portion of the study area that is below the marine limit is almost entirely covered by 3 m to as much as 40 m of glaciomarine sediment. Ice-contact landforms such as end moraines, deltas, fans, eskers, and other coarse-grained deposits cover only a small fraction of the total area. The typical glaciomarine sequence is well exposed at Carrying Place Bluff, along the present shoreline in extreme eastern Maine (Figs. 1 and 5). Polished and striated volcaniclastic siltstone and breccia of the Silurian Quoddy Formation (Gates, 1975) is overlain by a compact light gray diamicton containing faceted and striated clasts of local bedrock. We interpret this as lodgment till of latest Wisconsinan age.

A dark gray (N4), fine to medium sandy, matrix-supported gravel 30 cm thick overlies the lodgment till. We interpret this unit as ice-proximal subaqueous outwash deposited immediately adjacent to the grounding line (Rust and Romanelli, 1975; Smith, 1985).

The ice-proximal unit is conformably overlain by 195 cm of laminated mud and fine sand. This unit is pale red (2.5YR 6/2) to dark gray (N4) and contains sparse cobbles and pebbles, presumably dropstones, that often deform underlying bedding. The muddy beds are 1–2 cm thick and are separated by sand laminae <1 cm thick. This unit is the Presumpscot Formation of Bloom (1960). We interpret it here as glaciomarine mud deposited by settling from turbid meltwater overflows and minor melt-out from icebergs. Although the Presumpscot Formation is gray (5Y 5/1) to bluish gray (5B 5/1) in most of coastal Maine, it is reddish in this area, presumably because of the presence of hematite and jasper derived from maroon basalts, shales, and siltstones of the Leighton and Eastport Formations (Gates, 1975), located 5 km north of Carrying Place Bluff.

Other studies (Belknap et al., 1989; Bacchus, 1993) divide the glaciomarine unit into a variety of ice-proximal and ice-distal facies; we believe that this distinction is gradational at best and consider the glaciomarine mud unit as a single facies. Our facies divisions are the same as those described by Elverhøi et al. (1989) for modern tidewater glacier environments in the Barents Sea, and are similar to those proposed by Dethier et al. (1995) for glaciomarine sediments in the Puget Lowland.

Figure 5. Natural coastal exposure at Carrying Place Bluff shows glaciomarine mud and fine sand (Presumpscot Formation) draped over striated bedrock knob. Dark gray lodgment till directly overlies bedrock. Bottom 15 cm of laminated Presumpscot Formation has high sand content and unique dwarfed molluscan fauna—ice-proximal facies. Occasional dropstones disrupt bedding in ice-distal facies. This unit is horizontally truncated by shoreface sand and gravel deposited during marine regression. Shovel is 100 cm long.

The ice-distal marine mud unit is unconformably overlain by 50 cm of gently dipping beds of pebbly sand and gravel containing 2–10-cm-thick beds of openwork pebble gravel, matrix-supported pebbly sand, and coarse sand. The lower surface of this unit truncates bedding in the underlying ice-distal glaciomarine facies (Fig. 5), and we interpret it as the regressive scour unconformity (Fig. 2). This overlying unit is part of the regressive facies, and we interpret it as a nearshore or beach deposit formed at least in part by reworking of the underlying glaciomarine facies during relative sea-level fall. It is not clear whether this unit can be considered glaciomarine: it may have been derived solely from reworking of existing material, or may contain sediment transported to the shoreline by meltwater streams draining a terrestrial ice margin. Dorion (1997) estimated that the regressive scour unconformity at this site represents a gap of ~2 k.y., based on the absence of intertidal deposits and the timing of marine regression in the area.

*Pond Ridge moraine.* The stratigraphy of the glaciomarine section within the major stratified moraines is more complicated. LePage (1982) studied the stratigraphy of Pond Ridge moraine in detail and found that the glaciomarine mud was interbedded with a variety of ice-proximal facies and glaciotectonically emplaced material, which she interpreted as evidence of deposition by an active ice margin. We reinvestigated several sites at Pond Ridge moraine in order to obtain radiocarbon age estimates, samples for isotope analysis, and microfauna, and observed the same stratigraphy. Striated and polished bedrock is exposed at eroding bluffs on the present shoreface at Turner Brook (Fig. 6), and nearby test borings also reached bedrock. Stiff gray diamicton, presumably lodgment till, overlies bedrock in the bluff exposure and two test borings (King, 1979). The third boring penetrated a muddy unit of uncertain significance between bedrock and till.

The remainder of the Turner Brook bluff exposure contains a number of units of texturally variable sediment. Lodgment till directly overlying bedrock is covered by 15 cm of sandy gravel, which is then conformably overlain by several meters of fine-grained marine mud of the Presumpscot Formation. This sequence of ice-proximal to ice-distal facies is identical to that at Carrying Place Bluff. At this site, however, a variety of glacially deposited units overlie and are interbedded with the glaciomarine mud facies. These include graded and reverse-graded beds of gravel, laminites of sand and mud, strongly bimodal stony mud, massive matrix-supported sandy gravel, plane-bedded sand and gravel, and muddy diamicton containing striated and faceted clasts. This stratigraphy indicates that the ice margin fluctuated one or more times during the construction of the moraine, resulting in interbedded ice-proximal (coarse-grained) and ice-distal (fine-grained) units.

This package of sediments was deposited at varying distances from the ice margin and underwent varying degrees of sorting. We interpret the sandy gravel directly above the till as subaqueous outwash deposited directly at the ice margin, and the fine-grained mud as ice-distal sediment deposited by settling from suspension. The other units range from glaciotectonic material emplaced directly by ice to subaqueous debris-flow deposits deposited by mass movement at some distance from the grounding line. The entire coarse-grained section here can be described as resedimented diamicton or flowtill (e.g., Hartshorn, 1958). We believe that this flowtill formed at a similar distance from the ice margin as the subaqueous outwash observed here and at Carrying Place Bluff, and refer to these units collectively as the coarse ice-marginal facies.

In section D at Turner Brook (Fig. 6), an upper glaciomarine mud unit is sandwiched between coarse-grained units. It contains deformed bedding and discontinuous, sheared sand lenses. This unit is either a glaciotectonically emplaced unit reflecting reworking of previously deposited glaciomarine mud

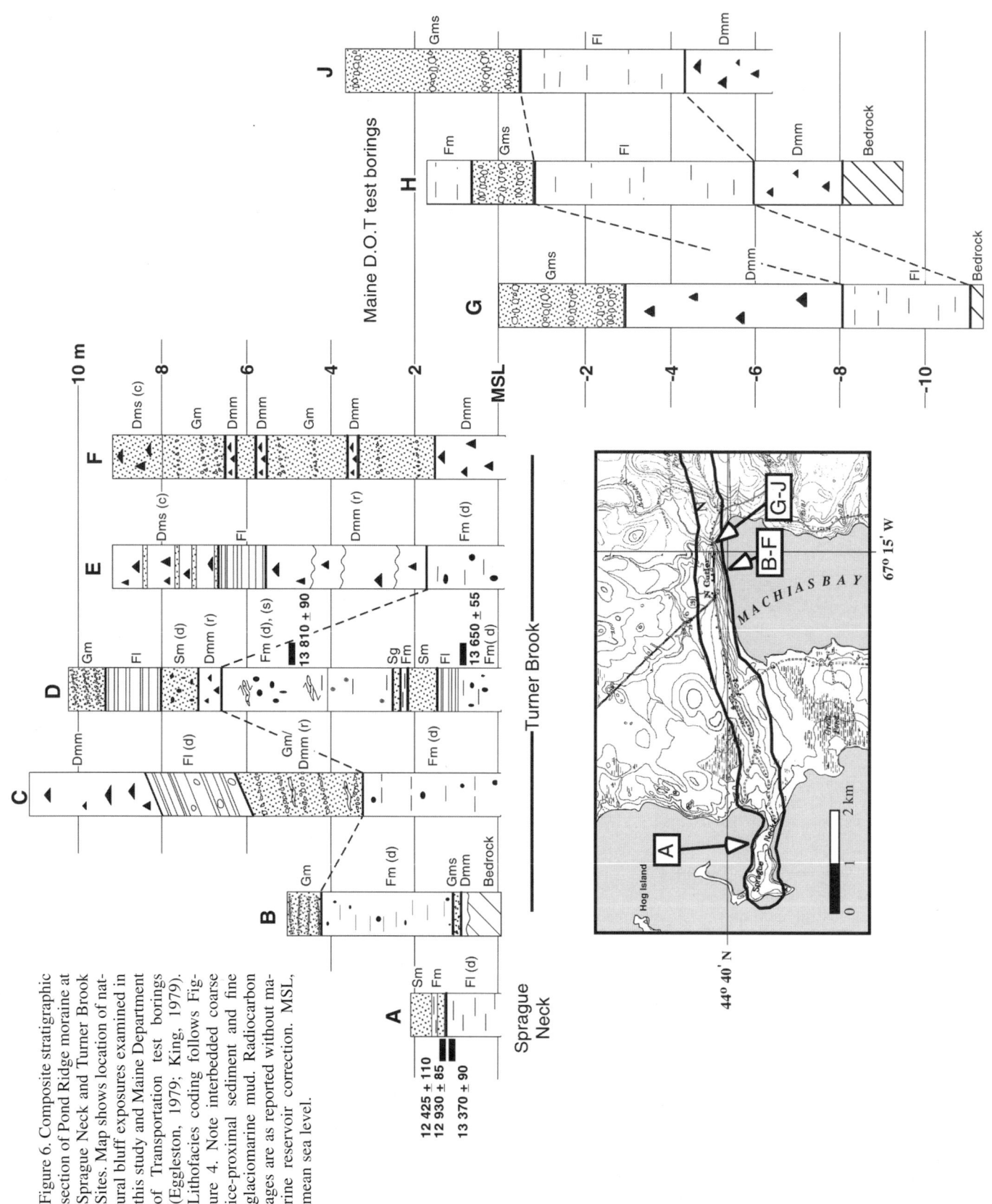

Figure 6. Composite stratigraphic section of Pond Ridge moraine at Sprague Neck and Turner Brook Sites. Map shows location of natural bluff exposures examined in this study and Maine Department of Transportation test borings (Eggleston, 1979; King, 1979). Lithofacies coding follows Figure 4. Note interbedded coarse ice-proximal sediment and fine glaciomarine mud. Radiocarbon ages are as reported without marine reservoir correction. MSL, mean sea level.

by an advancing ice margin or a mass movement down the distal slope of the end moraine.

At Sprague Neck, on the proximal side of Pond Ridge moraine (Fig. 6) the glaciomarine mud facies is conformably overlain by, first, mud with interbedded sand beds and laminae that thicken progressively upsection; and, second, massive to crossbedded sand and gravely sand. These two units contain an intertidal faunal assemblage (discussed in the following), and we interpret them as a regressive facies recording gradual shallowing and increasing influence of tidal processes, followed by reworking and deposition of coarse material at the shoreface. In contrast to Carrying Place Bluff, this site was sheltered from the open ocean by the crest of Pond Ridge moraine, and we do not identify a significant regressive unconformity.

At some moraine exposures, the coarse-grained regressive facies is texturally similar to parts of the coarse-grained ice-proximal facies. We separated the two facies on the basis of their distinctly different faunal assemblages (discussed in the following) and by their stratigraphic relationship. At the large moraines, such as Pond Ridge, that were directly exposed to the open ocean during deglaciation, the moraine crest was eroded and flattened by wave action. Reworked material was washed over the moraine as relative sea-level fell and the moraine crest was cut down. Thus the regressive deposits are usually thickest on the landward side of these moraines, and exposures here display a unique stratigraphy of large, convex-upward beds. More landward beds truncate more seaward beds, indicating that the entire landform was cut down by several centimeters at each overwash event, and the resulting sediment was washed into the lee of the moraine ridge. This represents an extreme example of shoreface erosion during relative sea-level fall.

*Lily Lake.* All of the lakes and ponds that we cored in this study exposed a full record of ice-proximal to ice-distal facies. The core from Lily Lake (Fig. 7) contains the best example of this sequence. The core refused in dense, stony sediment which we interpret as lodgment till. This is overlain by 30 cm of graded fine to medium sand beds which we interpret as ice-marginal subaqueous outwash. This unit grades upward into 9 m of laminated mud and very fine sand of the ice-distal glaciomarine facies. The glaciomarine unit contains occasional distorted or steeply dipping beds in contact with 10–20-cm-thick beds of matrix-supported diamicton. We interpret these as syndepositional slumps.

At the top of the glaciomarine mud facies in the Lily Lake core, the sand beds thicken and coarsen gradually over 1 m, becoming coarse sand at the top. This coarsening unit contains an intertidal faunal assemblage (discussed in the following) and presumably records gradual shallowing and increasing tidal influence. It is overlain by 30 cm of finely laminated gray mud, and then by a black FeS unit that records the onset of meromixis and conversion of the basin to freshwater. This unit terminates the marine section.

Although the coarse-grained regressive unit in this core shows some reworking by tidal processes, we do not identify a significant regressive unconformity in this core. The gradual transition from laminated muds to progressively sandier laminae and finally coarse sand is analogous to the stratigraphy at Sprague Neck and most likely reflects increased inwash of reworked sediment from areas surrounding the basin that were within the intertidal zone. In most of the other lakes and ponds we examined in this study, no coarse-grained regressive facies is present, indicating that these basins were entirely protected from tidal and shoreface erosion during regression.

### Fauna of glaciomarine facies

We sampled lodgment till at several sites and found it to be barren of fossils. At one site, Boyd Lake, we hammered a 10-cm-diameter Russian corer 30 cm into a dense lodgment till beneath the ice-proximal marine facies. Careful sieving and binocular examination of the till produced no vegetative or faunal remains.

The coarse ice-proximal facies is barren of fauna at four sites (Tables 2 and 3). At Lily Lake and Carrying Place Bluff, if contains abundant dwarf *Hiatella arctica*, and at Boyd Lake it contains dwarf *Portlandia arctica*. These molluscs were articulated, with intact periostracum, and showed no evidence of transport or reworking. We found microfauna in this facies at three sites. *Elphidium excavatum* forma *clavatum* dominates the foraminiferal assemblage. *E. excavatum* f. *excavatum* is also common, and individuals of several other species were present. Ostracodes, predominantly *Cytheromorpha macchesneyi*, are present at several sites.

This faunal assemblage is consistent with an ice-proximal glaciomarine environment. It suggests water temperatures near 0 °C and high turbidity with fluctuating salinities. Hald et al. (1994) found that *E. excavatum* f. *clavatum* was most common at water temperatures between $-2$ and 1 °C, and found an analogous fauna dominated by this species near the grounding lines of high arctic tidewater glaciers in the Kara and Barents Seas. The arctic ostracode *Cytheromorpha macchesneyi* (synonymous with *Roundstonia macchesneyi*) dominates modern nearshore environments from the Beaufort and Bering seas (Brouwers et al., 2000). They reported that this species is found in greatest abundance in water depths <10 m and variable sea water temperatures and salinities. South of this frigid climactic zone, in Norton Sound, the species rarely occurs. Based on the dominance of this species in assemblages from the glaciomarine sediments of Maine (Table 3), we infer that arctic sea water temperatures prevailed together with fluctuating salinities during deglaciation. Both *H. arctica* and *P. arctica* have wider temperature ranges, but are common in arctic glaciomarine environments: *P. arctica* in particular is abundant in ice-marginal environments with a high sedimentation rate (Ockelmann, 1958; Dyke et al., 1996).

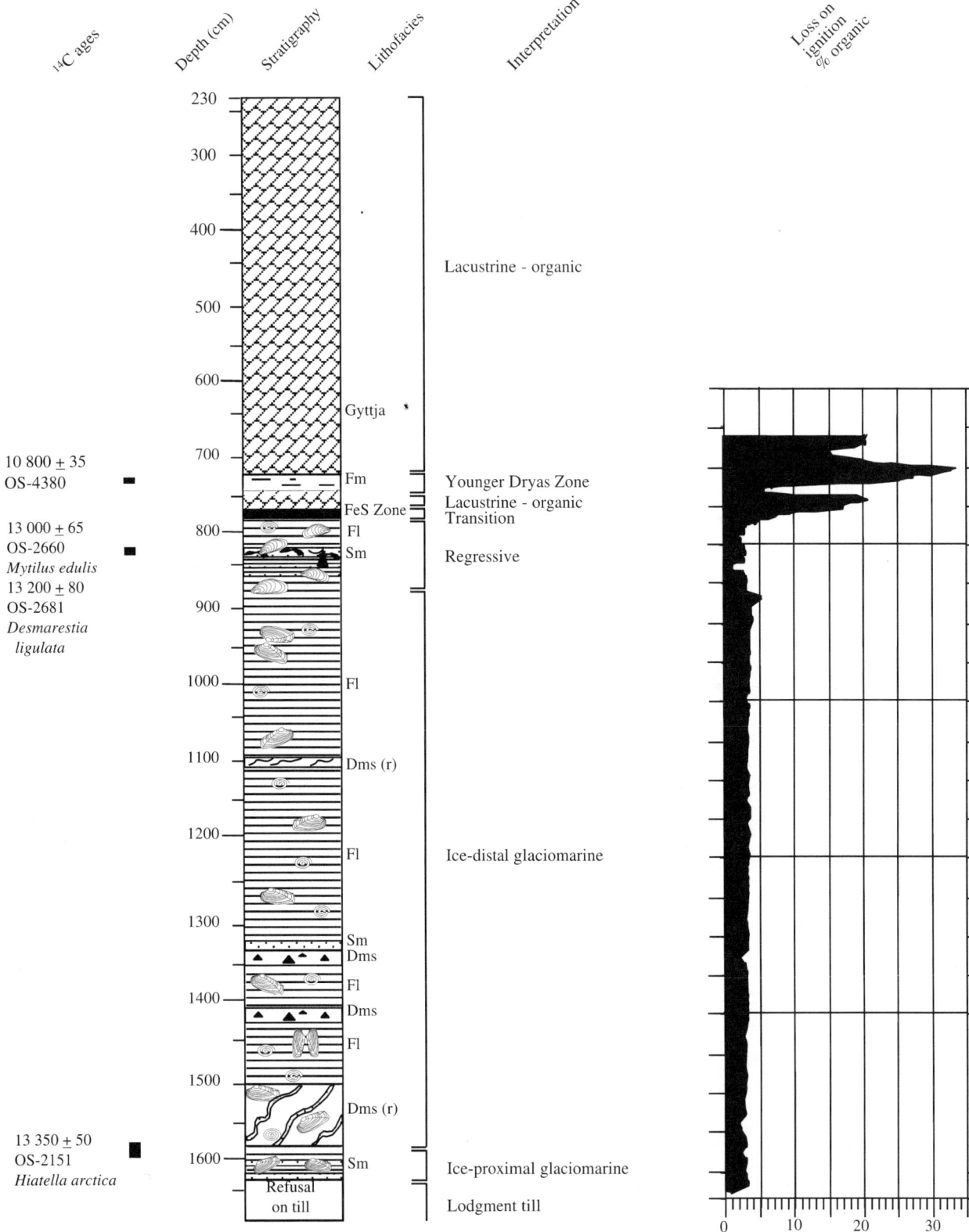

Figure 7. Lily Lake core log, loss on ignition, and $^{14}C$ ages. Core penetrated 8 m of ice-proximal, ice-distal, and regressive marine deposits. Younger Dryas zone at 730 cm is prominent minerogenic layer correlative with numerous other cores from Maine and maritime Canada (e.g., Deevey, 1951; Stea and Mott, 1989; Mayle et al., 1993; Dorion, 1997). Radiocarbon ages are as reported without marine reservoir correction.

TABLE 2. MACROFAUNA OF GLACIOMARINE DEPOSITS

| Site number | Site name | Ice-proximal facies | Ice-distal glaciomarine mud facies | Coarse regressive facies |
|---|---|---|---|---|
| **Macrofauna** | | | | |
| 1 | Boyd Lake | *Portlandia arctica* (dwarf) | *P. arctica* (dwarf), *Mytilus edulis* (fragments) | Facies not present |
| 2 | Carrying Place Bluff | *Hiatella arctica* (dwarf) | *P. arctica*, *Nucula tenuis* | Barren |
| 3 | Dennison Point | Not examined | *Macoma calcarea* | Facies not present |
| 4 | Dover-Foxcroft | Facies not present | *P. arctica*, *Macoma balthica*, *M. edulis* (fragments) | Not examined |
| 5 | Gould Pond | Facies not present | *P. arctica* | Facies not present |
| 6 | Hadley Pond Esker | Facies not present | *H. arctica* | Facies not present |
| 7 | Lewis Cove | Not examined | *N. tenuis M. calcarea, M. edulis* (disarticulated), *Yoldia myalis* | *M. edulis, N. tenuis* |
| 8 | Lily Lake | *H. arctica* (dwarf) | *H. arctica, M. edulis* (disarticulated) | *M. edulis, Desmarestia ligulata* |
| 9 | Long Pond | Barren | *P. arctica* (periostracum), *N. tenuis* (periostracum), *Dryas integrifolia*, other terrestrial plant species | Facies not present |
| 10 | Look's Seafood Cannery | Facies not present | *N. tenuis* | Facies not present |
| 11 | Mark's Lake | Facies not present | *N. tenuis* | Facies not present |
| 12 | Mattaseunk Lake | Barren | *P. arctica* (periostracum), insect parts, Characeae oospores, woody debris, *D. integrifolia*, *Potamogeton* sp. | Facies not present |
| 13 | Patrick Lake | Barren | *P. arctica* (periostracum), unidentified seaweed, *D. integrifolia* | Facies not present |
| 14 | Pocomoonshine Lake | Facies not present | *P. arctica, N. tenuis, M. edulis* (fragments) | Facies not present |
| 15 | Sand Point* | Not examined | *N. tenuis* | Not examined |
| 16 | Sheldon Point† | Facies not present | *P. arctica*, unidentified seaweed | Facies not present |
| 17 | Sprague Neck | Facies not present | *P. arctica, Astarte undata* | *M. edulis, D. ligulata* |
| 18 | T2 R8 NWP | Facies not present | *M. calcarea, H. arctica, N. tenuis, Balanus* sp. | Facies not present |
| 19 | Toddy Pond | Barren | *P. arctica* | Facies not present |
| 20 | Turner Brook | Not examined | *M. calcarea, N. tenuis, P. arctica, Y. myalis* | Barren |

*Gadd (1973) gave a comprehensive list of fauna collected at this site.
†Nicks (1991) gave a comprehensive list of data collected at this site.

The glaciomarine mud facies contains a diverse marine fauna (Tables 2 and 3). We found a variety of molluscs, usually articulated and in life position, at nearly all sites, most commonly *H. arctica, P. arctica, Nucula tenuis, Macoma calcarea,* and *M. balthica*. Most of these molluscs are arctic to subarctic species characteristic of glaciomarine environments found well to the north of Maine at present (Dyke et al., 1996). This assemblage is consistent with previous studies of Presumpscot Formation fauna such as those of Stuiver and Borns (1975), Cotter (1985), and Bloom (1960, 1963). However, Dyke et al. (1996, p. 137), in their compilation of $^{14}$C dated mollusc assemblages younger than 20 000 $^{14}$C yr B.P. from North American waters, cautioned that the characteristic late glacial mollusc assemblages in Maine and Atlantic Canada "are not strictly analogous to modern or Holocene assemblages near the subarctic-arctic boundary in that they include more temperate elements." Thus the molluscan assemblage in the glaciomarine facies is best explained by an arctic to subarctic glaciomarine environment, but does not entirely exclude the possibility that water temperatures were more temperate.

Occasionally, acidic groundwater flowing through the glaciomarine mud facies has completely dissolved mollusc shell material, leaving only periostracum behind. At several sites we observed intact, articulated periostraca of *P. arctica* and *N. tenuis* preserved in life position.

The foraminiferal assemblage in the glaciomarine facies (Tables 2 and 3) is consistently dominated by *E. excavatum* f. *clavatum*. The most common ostracode is *C. macchesneyi*, although other species are present in varying abundance. A microfauna dominated by *C. macchesneyi* and *E. excavatum* f. *clavatum* strongly indicates arctic to subarctic glaciomarine conditions, with cold (1 to $-2$ °C), turbid, and normal salinity water (Hald et al., 1994). This suggests that this facies was deposited in an arctic to subarctic glaciomarine environment.

At three sites, we found marine fossils interbedded with terrestrial organic debris within the glaciomarine mud facies, an unusual association that has also been reported in glaciomarine sediments in central Maine (Anderson et al., 1992; Jager, 1996; Donner, 1995; Dorion, 1997). At Patrick Lake and Long Pond (Fig. 1), this unit contained *P. arctica* and *N. tenuis* with leaves of *Dryas integrifolia*, an arctic tundra shrub common in recently deglaciated upland landscapes (Anderson et al., 1992). This indicates that the nearby uplands were vegetated shortly after deglaciation, and upland plant material was incorporated in glaciomarine sediments. At Mattaseunk Lake

## TABLE 3. MICROFAUNA OF GLACIOMARINE DEPOSITS

| Site number | Site name | Ice-proximal facies | Ice-distal glaciomarine mud facies | Coarse regressive facies |
|---|---|---|---|---|
| **Microfauna** | | | | |
| 1 | Boyd Lake | Not examined | Foraminifera (n = 78):<br>66% *Elphidium excavatum* f. *clavatum*<br>32% *E. excavatum* f. *lidoensis*<br>2% *Polymorphina* sp. | Facies not present |
| 2 | Carrying Place Bluff | Barren | Ostracodes (n = 110):<br>50% *Cytheromorpha macchesneyi*<br>29% *Cytheropteron elaeni*<br>9% *Jonesia simplex*<br>6% *Cytheropteron suzdalyskyi*<br>5% *Acanthocythereis dunelmensis* | Not examined |
| 3 | Dennison Point* | Not examined | Foraminifera (n = 122):<br>85% *E. excavatum* f. *clavatum*<br>8% *E. excavatum* f. *excavatum*<br>4% *E. excavatum* f. *lidoensis*<br>3% *Cassidulina reniformae*<br>Ostracodes:<br>*C. macchesneyi*<br>*C. suzdalskyi*<br>*C. elaeni*<br>*Semicytherura complanata*<br>*J. simplex*<br>*A. dunelmensis* | Facies not present |
| 4 | Dover-Foxcroft | Facies not present | Barren | |
| 5 | Gould Pond† | Facies not present | *E. excavatum* f. *clavatum* | Facies not present |
| 6 | Hadley Pond Esker | Facies not present | Not examined | Facies not present |
| 7 | Lewis Cove | Not examined | Not examined | Not examined |
| 8 | Lily Lake§ | Barren | Foraminfera (n = 184):<br>91% *E. excavatum* f. *clavatum*<br>7% *E. excavatum* f. *excavatum*<br>6% other; includes *E. excavatum* f. *lidoensis*, *Triloculina trincarinata*, *Cyclogyra* sp. | Foraminifera:<br>*E. excavatum* f. *clavatum* |
| 9 | Long Pond | Barren | Foraminifera:<br>*E. excavatum* f. *clavatum*<br>*E. excavatum* f. *lidoensis* | Facies not present |
| 10 | Look's Seafood Cannery | Facies not present | Not examined | Facies not present |
| 11 | Mark's Lake | Facies not present | Foraminifera:<br>*E. excavatum* f. *clavatum* (dominant)<br>*E. excavatum* f. *excavatum*<br>*Buccella frigida*<br>*C. reniformae*<br>*Polymorphina* sp.<br>*Cyclogyra* sp.<br>*Triloculina tricarinata*<br>Ostracods:<br>*C. macchesneyi*<br>*C. suzdalskyi*<br>Other: juvenile *Nuculana* sp. | Facies not present |

(Fig. 1), we found *P. arctica* in the upper glaciomarine facies with *D. integrifolia*, woody debris, and seeds of the freshwater lacustrine taxa *Characeae* and *Potamogeton*. Mixing of freshwater and marine taxa in a marine environment occurs today along the present Maine coast. Stream-fed coastal embayments are often dammed for several years by stony beach ridges constructed during winter storms. During this time, lacustrine and fen vegetation colonizes the ponded wetland landward of the beach ridge. Upon later breaching of the barrier, terrestrial and lacustrine material are washed into the nearshore and incorporated into marine sediments.

The coarse-grained regressive facies contains an entirely different fauna from the units below it. We found abundant valves of the intertidal mollusc *Mytilus edulis* in this unit. The valves are usually disarticulated, nested, and concentrated in coarse-grained lag deposits, although we found some in life position at Sprague Neck. This suggests reworking by tidal processes. The arctic seaweed *Desmarestia ligulata*, which is also found in the intertidal zone, is common in this facies. This fauna is consistent with our interpretation of this facies as an intertidal or shoreface deposit. At several sites, we found fragments of *M. edulis* and unidentified seaweed in the glaciomarine mud facies. Presumably these were transported from shallower water depths prior to deposition. We only searched for microfauna in the regressive facies at one site, Lily Lake, where we found the foraminifer *E. excavatum*.

## TABLE 3. MICROFAUNA OF GLACIOMARINE DEPOSITS (continued)

| Site number | Site name | Ice-proximal facies | Ice-distal glaciomarine mud facies | Coarse regressive facies |
|---|---|---|---|---|
| 12 | Mattaseunk Lake | Barren | Foraminifera (n = 448):<br>  31% *E. excavatum* f. *clavatum*<br>  34% *E. excavatum* f. *excavatum*<br>  33% *E. excavatum* f. *lidoensis*<br>  2% *Polymorphina* sp.<br>Terrestrial plant debris | Facies not present |
| 13 | Patrick Lake | Barren | Foraminifera:<br>  *E. excavatum* f. *clavatum* (dominant)<br>  *B. frigida*<br>  *C. reniformae*<br>  *Quinqueloculina seminulum* | Facies not present |
| 14 | Pocomoonshine Lake | Facies not present | Foraminifera:<br>  *E. excavatum* f. *clavatum*<br>  *C. reniformae*<br>  *T. tricarinata*<br>Ostracodes:<br>  *C. macchesneyi*<br>  *C. suzdalskyi*<br>  *J. simplex* | Facies not present |
| 15 | Sand Point | Not examined | Foraminifera:<br>  *E. excavatum* f. *clavatum*<br>  *B. frigida*<br>Ostracodes:<br>  *C. suzdalskyi* | Not examined |
| 16 | Sheldon Point | Facies not present | Not examined | Facies not present |
| 17 | Sprague Neck | Facies not present | Not examined | Not examined |
| 18 | T2 R8 NWP | Facies not present | Foraminifera:<br>  *E. excavatum* f. *clavatum* (dominant)<br>  *E. excavatum* f. *excavatum*<br>  *Q. seminulum*<br>Ostracodes:<br>  *C. suzdalskyi* | Facies not present |
| 19 | Toddy Pond | Barren | Foraminifera (n = 516):<br>  81% *E. excavatum* f. *clavatum*<br>  15% *B. frigida*<br>  1% *E. excavatum* f. *excavatum*<br>  1% *E. excavatum* f. *lidoensis*<br>  1% *Polymorphina* sp.<br>  1% *Lagena* sp. | Facies not present |
| 20 | Turner Brook | Not examined | Foraminfera (n = 188):<br>  93% *E. excavatum* f. *clavatum*<br>  2% *C. reniformae*<br>  2% *E. excavatum* f. *excavatum*<br>  2% *Haynesina orbiculare*<br>  1% *Cyclogyra* sp.<br>Ostracods:<br>  *C. macchesneyi* | Not examined |

*Kaplan (1999).
†Cotter (1985) and Anderson et al. (1992) presented a comprehensive list of fauna at this site.
§Kaplan (1994).

## *Isotopic character of glaciomarine facies*

Fauna from the ice-proximal and ice-distal glaciomarine mud facies have similar $\delta^{18}O$ and $\delta^{13}C$ values (Table 4). Both subtidal molluscs and benthic foraminifera have $\delta^{18}O$ values between 2‰ and 4‰ at all sites. This indicates water of normal marine salinity, with no dilution by meltwater. Hillaire-Marcel (1988) estimated that Laurentide ice sheet meltwater in the Champlain Sea had a $\delta^{18}O$ of $-16$‰. If this were the case in Maine as well, even a 5‰ decrease in salinity from the normal marine value of 35‰ would result in negative $\delta^{18}O$ values.

Other estimates of the average isotopic composition of the Laurentide ice sheet based on the global benthic foraminiferal $\delta^{18}O$ record (Fischer, 1992; Mix and Ruddiman, 1984) suggest that the $\delta^{18}O$ of Maine meltwater was probably closer to $-35$‰. In this case, the effect of reduced salinity on our measured $\delta^{18}O$ results would be even more pronounced, and we would expect strongly negative $\delta^{18}O$ values. Because we do not observe negative $\delta^{18}O$ values, we conclude that fauna in the ice-proximal and ice-distal glaciomarine facies were living in water of normal marine salinity.

Kreutz (1994) also analyzed serial samples from subtidal

## TABLE 4. COMPREHENSIVE ISOTOPE RESULTS BY SEDIMENTOLOGIC FACIES

| Site/Facies/Species | Number of specimens | Number of analyses | Average δ¹⁸O (o/oo) | δ¹⁸O standard deviation (o/oo) | Average δ¹³C (o/oo) | δ¹³C standard deviation (o/oo) |
|---|---|---|---|---|---|---|
| Boyd Lake | | | | | | |
|   Ice-proximal facies | | | | | | |
|     *Portlandia arctica* | 3 | 3 | 3.01 | 0.06 | −0.2 | 0.02 |
| Turner Brook | | | | | | |
|   Glaciomarine mud facies (glaciotectonized) | | | | | | |
|     *P. arctica* | 11 | 16 | 2.92 | 1.25 | −1.34 | 1.19 |
|     *Nucula tenuis* | 34 | 38 | 2.9 | 0.54 | −1.77 | 0.83 |
|     *Yoldia myalis* | 1 | 9 | 2.92 | 0.36 | −1.46 | 0.72 |
| Denison Point | | | | | | |
|   Glaciomarine mud facies | | | | | | |
|     *Macoma calcarea* | 25 | 10 | 3.88 | 0.25 | 0.49 | 1.69 |
| Carrying Place Bluff | | | | | | |
|   Ice-proximal facies (0–20 cm) | | | | | | |
|     *H. arctica* | 16 | 9 | 3.52 | 0.19 | 0.37 | 0.39 |
|     *P. arctica* | 12 | 5 | 3.39 | 0.03 | 0.08 | 0.3 |
|   Glaciomarine mud facies (>30 cm) | | | | | | |
|     *H. arctica* | 32 | 27 | 3.23 | 0.19 | 0.21 | 0.41 |
|     *P. arctica* | 42 | 19 | 3.17 | * | −1.54 | * |
|     *N. tenuis* | 9 | 3 | 3.57 | * | −1.07 | * |
| Sprague Neck | | | | | | |
|   Glaciomarine mud facies (0.45–1.10 m) | | | | | | |
|     *P. arctica* | 16 | 11 | 3.35 | 0.17 | −0.65 | 0.63 |
|     *Astarte undata* | 1 | 1 | 3.06 | * | 1.91 | * |
|   Regressive facies (1.25 m) | | | | | | |
|     *Mytilus edulis* | 11 | 11 | 1.27 | 0.51 | 0.27 | 0.79 |
| Lewis Cove | | | | | | |
|   Glaciomarine mud facies | | | | | | |
|     *M. calcarea* | 3 | 1 | 3.88 | * | 0.26 | * |
|     *N. tenuis* | 15 | 8 | 4.5 | 0.21 | 2.34 | 0.47 |
|     *M. edulis* | 4 | 9 | −0.67 | 2.65 | −0.25 | 1.02 |
|   Regressive facies | | | | | | |
|     *M. edulis* | 5 | 8 | 1.45 | 1.11 | −0.16 | 0.17 |
| Lily Lake | | | | | | |
|   Ice-proximal facies (below 1598 cm) | | | | | | |
|     *H. arctica* | ca. 10 | 6 | 1.99 | 1.7 | −0.17 | 0.53 |
|     *E. excavatum* | ca. 30 | 3 | 1.94 | 0.53 | −1.9 | 0.11 |
|   Glaciomarine mud facies (1598–940 cm) | | | | | | |
|     *H. arctica* | ca. 10 | 55 | 3.08 | 1.03 | 0.15 | 1.1 |
|     *M. edulis* | 4 | 4 | −0.37 | 0.87 | −0.49 | 0.29 |
|     *E. excavatum* | ca. 30 | 55 | 1.92 | 0.35 | −2 | 0.47 |
|   Regressive facies (above 940 cm) | | | | | | |
|     *M. edulis* | * | 15 | 0.5 | 1.85 | −0.39 | 0.59 |
|     *E. excavatum* | * | 1 | 1.15 | * | −0.81 | * |
| Sand Point, NB | | | | | | |
|   Glaciomarine mud facies | | | | | | |
|     *N. tenuis* | 1 | 3 | 2.67 | 0.18 | 1.92 | 0.29 |
| Marks Lake | | | | | | |
|   Glaciomarine mud facies | | | | | | |
|     *N. tenuis* | 1 | 3 | 3.94 | 0.08 | 0.93 | 0.03 |
| Pocomoonshine Lake | | | | | | |
|   Glaciomarine mud facies | | | | | | |
|     *P. arctica/N. tenuis* (aggregated) | * | * | 3.27 | * | * | * |
| Toddy Pond | | | | | | |
|   Glaciomarine mud facies | | | | | | |
|     *E. excavatum* | ca. 150 | 3 | 1.95 | 0.13 | −1.24 | 0.46 |
| Gould Pond | | | | | | |
|   Glaciomarine mud facies | | | | | | |
|     *P. arctica* | 9 | 9 | 0.83 | 0.16 | −0.51 | 0.31 |
|     *E. excavatum* | ca. 15 | 1 | 1.23 | 0.09 | −1.25 | 0.11 |
| T2 R8 NWP | | | | | | |
|   Glaciomarine mud facies | | | | | | |
|     *M. calcarea* | 1 | 1 | 1.96 | * | −0.9 | * |
|     *Balanus* sp. | 1 | 1 | 2.49 | * | −0.02 | * |
|     *H. arctica* | 1 | 1 | 0.57 | * | −1.48 | * |
|     *N. tenuis* | 1 | 1 | * | * | 0.31 | * |
| Dover—Foxcroft | | | | | | |
|   Glaciomarine mud facies | | | | | | |
|     *Macoma balthica* | 3 | 3 | −2.94 | 0.13 | −1.09 | 0.08 |

*Not reported.

molluscs found in the ice-proximal and ice-distal glaciomarine mud. He found that $\delta^{18}O$ values did not vary by more than 1.5‰ throughout the life of the organism. Thus, seasonal variations in salinity were small, indicating that meltwater influx was minimal at the grounding line where the molluscs were living. It is most likely that isotopically depleted glacial meltwater was confined to the sea surface, as, for example, Pfirman and Solheim (1989) observed along the tidewater grounding line of the Nordaustlandet ice cap.

At Lily Lake (Fig. 7), $\delta^{18}O$ values in the lowest part of the ice-proximal facies were ~1‰ lower than values in the overlying ice-distal glaciomarine mud facies. This suggests some meltwater influence very close to the grounding line. At Carrying Place Bluff, however, $\delta^{18}O$ values in the ice-proximal and ice-distal glaciomarine mud facies are indistinguishable, and at both sites, $\delta^{18}O$ values are constant throughout the glaciomarine section. Thus, full-salinity ocean water was present very close to the grounding line, and bottom water temperature and salinity were stable as the ice retreated.

The $\delta^{13}C$ values for the ice-proximal and glaciomarine facies are generally between −2‰ and 2‰, not significantly different from normal marine values. Hillaire-Marcel (1988) interpreted depletion of $\delta^{13}C$ in mollusc shells from the late glacial Champlain Sea as an indication of meltwater influx. We do not observe any uniform $\delta^{13}C$ depletion, which is consistent with the normal marine salinity indicated by $\delta^{18}O$ results. The $\delta^{13}C$ values do have high variability at some sites. This variability may reflect fractionation effects related to salinity and pH changes at the sites; however, the relatively constant salinity indicated by our $\delta^{18}O$ results makes this unlikely. Alternatively, the observed $\delta^{13}C$ variability may be the result of varying influx of terrestrial organic matter, possibly associated with meltwater or runoff. Although carbon in glacial meltwater is virtually nonexistent compared to dissolved carbon in the ocean, we observed terrestrial plant macrofossils in the glaciomarine facies at several sites (see following discussion), indicating that the exposed land areas near some of the sites were vegetated shortly after deglaciation and some terrestrial carbon was available. Because $\delta^{13}C$ values are influenced by metabolic effects and sedimentary organic matter as well as the carbon composition of the water, however, we do not interpret these variations as strong evidence of freshwater influx.

The $\delta^{18}O$ values indicate that the paleotemperature of bottom water during deposition of the ice-marginal and glaciomarine mud facies was between 0 and 2 °C (assuming salinity >30‰; Epstein et al., 1953). This estimate is consistent with the paleotemperature we estimate from the faunal assemblage in these units.

We also analyzed valves of *M. edulis*, presumably transported from the intertidal zone, which were deposited within the ice-distal glaciomarine facies at Lewis Cove and Lily Lake. Both sites are flanked by steep hillsides extending upward into what would have been the intertidal zone during deglaciation. Presumably *M. edulis*, an intertidal species, was washed downslope and deposited with the deeper water fauna. Valves of *M. edulis* at Lewis Cove had average $\delta^{18}O$ values of −0.67‰, significantly lower than the average of 4.5‰ from the subtidal species *N. tenuis* at the same location. If salinity were constant, a temperature difference of 19 °C between surface and bottom waters would be required to explain this difference (Epstein et al., 1953). Because such a temperature gradient would be extremely unlikely, surface-water salinity must have been reduced. We expect that surface meltwater would have been between 0 and 2 °C as well (e.g., the modern East Greenland Current is ~2 °C). Thus, if the $\delta^{18}O$ of Laurentide ice sheet meltwater was ~−35‰, surface-water salinity of 28‰–30‰ is necessary to account for the observed *M. edulis* $\delta^{18}O$.

The isotopic character of the regressive facies is somewhat different. Mean $\delta^{18}O$ values for *M. edulis* from this unit are 1‰–1.5‰ (Table 4), and variability is much greater than observed in the older units. If salinity were constant between deposition of the glaciomarine mud and the regressive unit, the 2‰–3‰ difference would require surface water warming of 15 °C. This is extremely unlikely, which implies some dilution of surface water by isotopically lighter runoff. Because *M. edulis* occupies a wide variety of intertidal and nearshore microenvironments, it is difficult to assess the importance of regional meltwater input relative to localized surface runoff in controlling these results. It is interesting that *M. edulis* from within the ice-distal glaciomarine mud facies has lower $\delta^{18}O$ than *M. edulis* from the regressive facies; this may reflect a greater meltwater supply during early glaciomarine sedimentation, and a decrease in meltwater supply by the time of relative sea-level fall.

The core from Lily Lake provides a high-resolution record of the entire period of marine submergence at that location. Radiocarbon ages from the top and bottom of the glaciomarine mud unit (Fig. 7) indicate that the lake was deglaciated ca. 13 350 ± 50 $^{14}C$ yr B.P. (Kaplan, 1999) and relative sea-level fall isolated the lake basin slightly after 13 000 $^{14}C$ yr B.P. The $\delta^{18}O$ values for the subtidal mollusc *H. arctica* and the benthic foraminifer *E. excavatum* are remarkably constant throughout this entire time period, with two exceptions. First, *E. excavatum* $\delta^{18}O$ increases rapidly in the lowest 20 cm of the core; this probably reflects the presence of meltwater close to the grounding line, and the rapid incursion of cold, saline, normal marine water soon after deglaciation. Second, near the top of the core, $\delta^{18}O$ values for *H. arctica* and *M. edulis* begin to decrease before any sedimentological evidence of shallowing occurs. This could indicate warming of bottom water, but is better explained by isotopically depleted surface water extending below the intertidal zone and signaling shallowing before the appearance of coarser sediment. Thus, it appears that oceanographic conditions were remarkably consistent during the period represented at Lily Lake, and the changes in isotopic composition primarily reflect retreat of the grounding line and local hydrographic changes during relative sea-level fall.

At Carrying Place Bluff, $\delta^{18}O$ values are constant throughout the entire period of glaciomarine sedimentation. Age estimates at this site range from 13 800 ± 80 to 13 150 ± 140 $^{14}$C yr B.P. (no reservoir correction), but an age inversion makes it impossible to determine the exact time period represented.

## DISCUSSION

### Paleoceanography and paleoenvironment at the ice margin

Both the ice-proximal facies and the ice-distal glaciomarine mud facies contain cold-water, ice-proximal faunal assemblages and $\delta^{18}O$ values indicating cold water (0–2 °C) of normal marine salinity. The ice-proximal facies is coarse grained, contains sedimentary structures indicative of deposition by grain flow, underflow, and mass-wasting processes, and has a sparser, less diverse faunal assemblage. The ice-distal glaciomarine facies has a more diverse fauna and contains fine sand and mud laminae diagnostic of deposition by settling from turbid overflows.

We interpret the sedimentology, stratigraphy, morphology, fauna, and isotope geochemistry of these units as indicating deposition near a grounded, tidewater ice margin. Meltwater issuing from the base of the ice margin deposited coarse sediment at the grounding line, then rose immediately to the surface and formed turbid, reduced-salinity, overflow plumes. Ice-distal sedimentation occurred by settling of sediment entrained in these turbid overflows. In addition, the presence of dropstones shows that ablation took place by iceberg calving as well as melting. The high $\delta^{18}O$ values of benthic fauna and the observed $\delta^{18}O$ depletion in intertidal relative to benthic fauna within this unit indicate that strong density stratification prevailed near the ice margin, with bottom water of normal marine salinity and somewhat dilute surface water. This model of circulation is similar to that observed by Powell (1990) and Powell and Molnia (1989) at tidewater glaciers in Alaska, and by Pfirman and Solheim (1989) and Elverhøi et al. (1989) in the high arctic.

Wassenaar et al. (1988) interpreted isotope analyses of glaciomarine fauna from Champlain Sea deposits to indicate that meltwater and seawater were well mixed at the ice margin, with a horizontal salinity gradient from 0‰ at the grounding line to 35‰ well out to sea. Our results, in contrast, indicate that little mixing of meltwater and seawater occurred, and meltwater was confined to the surface layer.

Molluscan $\delta^{18}O$ values from the glaciomarine facies in the study area are similar to approximately contemporaneous benthic $\delta^{18}O$ values of 2.6‰–3.8‰ in cores from the Wilkinson basin in the offshore Gulf of Maine (Schnitker and Jorgenson, 1990). This suggests that the same bottom-water mass was present both in the deep basins and inland of the present coast during deglaciation. Schnitker and Jorgenson (1990), however, proposed that these high $\delta^{18}O$ values could be explained by warm, saline Atlantic slope water entering the Gulf of Maine via the Northeast Channel. We reject this hypothesis based on the good agreement of paleotemperature estimates from both $\delta^{18}O$ values and faunal assemblages, both of which indicate temperatures close to 0 °C.

Studies of the glaciomarine sediments in eastern Maine and neighboring New Brunswick (Ashley et al., 1991; Nicks, 1991) characterized conditions at the ice margin as temperate, analogous to present-day conditions in the Gulf of Alaska, based on the high sedimentation rates required to produce large glaciomarine deltas. Our results from the glaciomarine section in eastern Maine, which is approximately contemporaneous with many of the glaciomarine deltas (e.g., Thompson et al., 1989), indicate that this was not the case. Water temperatures were near 0 °C, more analogous to modern conditions in the Beaufort or Barents Seas and Svalbard area than in the Gulf of Alaska. This agrees with floral, palynological, and geological evidence from southern New England that indicate arctic to subarctic climatic conditions at the same time. For example, Stone and Ashley (1992) and Ashley and Peteet (1999) found ice-wedge casts, pingo scars, eolian deposits, other periglacial features, and pollen and macrofossil assemblages diagnostic of an arctic to subarctic climate with subfreezing mean annual temperatures that persisted for several thousand years after regional deglaciation.

### Stratigraphic context of radiocarbon dates

We obtained nine radiocarbon ages from the ice-proximal facies or the base of the glaciomarine section (Table 1). These include *H. arctica* from the subaqueous outwash unit at Carrying Place Bluff and Lily Lake; other molluscs from glaciomarine mud intercalated between coarse-grained ice-proximal units at the Turner Brook site on Pond Ridge moraine; *M. calcarea* from an ice-proximal unit sandwiched between flowtill or lodgment till at Dennison Point; and a variety of fauna from the basal 50 cm of the marine section at Boyd Lake, Marks Lake, and Toddy Pond.

All of these ages are closely associated with coarse-grained ice-proximal deposits. The sedimentology and isotope geochemistry of the ice-marginal facies indicates that coarse sediment was deposited near the grounding line as meltwater rose rapidly to the sea surface. Thus we interpret these samples to closely estimate the time of ice recession from these sites.

We obtained another nine age estimates from seven sites on material from the lowest 0.5–2 m of the glaciomarine mud facies (Dover-Foxcroft, Gould Pond, Patrick Lake, Pocomoonshine Lake, Sand Point, Sheldon Point, T2 R8 NWP; Table 1). These sediments were deposited some time after ice retreat. In an effort to estimate sedimentation rates for the glaciomarine section and determine the accuracy of these ages, we dated multiple samples from the glaciomarine section at Lily Lake and Carrying Place Bluff. At Carrying Place Bluff, an age inversion makes it impossible to estimate the sedimentation rate (Fig. 8). At Lily Lake (Fig. 7), 8 m of sediment was deposited in 100–400 yr, a sedimentation rate of 2–8 cm/yr. In addition,

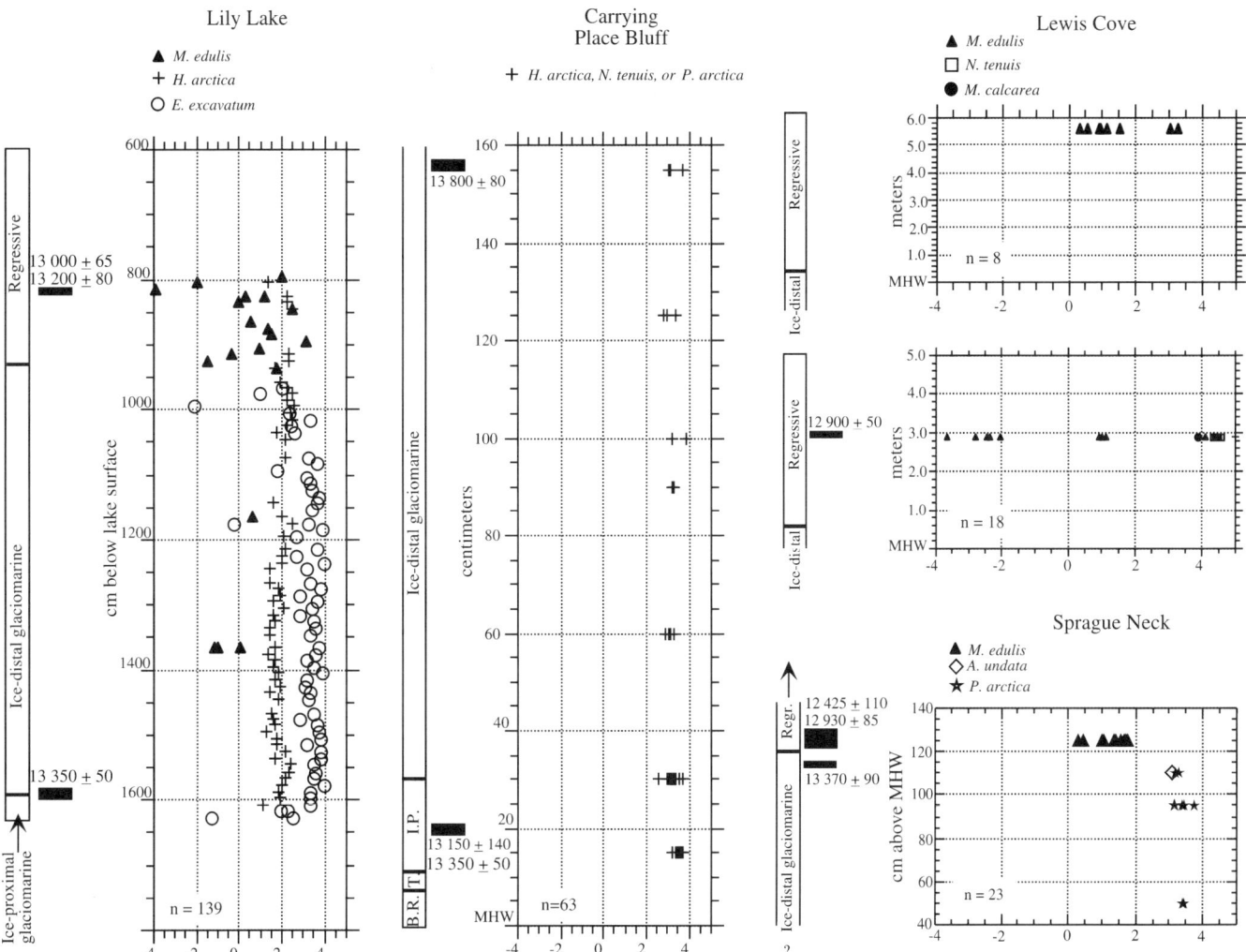

Figure 8. Relationship between glaciomarine facies, $^{14}$C ages, and $\delta^{18}$O from macrofauna and microfauna from 4 sites. Radiocarbon ages are as reported, without marine reservoir correction. B.R.–bedrock; T.–till; I.P.–ice proximal; MHW–mean high water; Regr.–regressive.

we obtained an age of 13 300 ± 65 $^{14}$C yr B.P. from the base of the glaciomarine facies at Mark's Lake. The outlet of Mark's Lake (41 m asl) is 9 m higher than Lily Lake (32 m asl), so glaciomarine sedimentation must have ended at Mark's Lake first. We obtained ages of 13 000 ± 65 and 13 200 ± 80 $^{14}$C yr B.P. on the regressive facies at Lily Lake. Because glaciomarine sedimentation at Mark's Lake also must have ended before this time, the 5 m of glaciomarine mud above the dated material at Mark's Lake must have accumulated in <430 yr, indicating that the sedimentation rate there was at least 1.2 cm/yr.

In contrast, comparison of an age we obtained from a core at Gould Pond with a previously dated core by Anderson et al. (1992) indicates that the sedimentation rate there may have been as low as 0.1 cm/yr. However, this pair of age estimates spans the marine-lacustrine transition. Because the sedimentation rate probably dropped dramatically when the lake basin was isolated from the ocean and thus the source of glaciomarine sediment, we believe the estimates from Lily Lake and Mark's Lake are the best available approximations for the rate of glaciomarine sediment accumulation. On the basis of these estimates, we interpret the five ages from the lowest 2 m of the glaciomarine facies as close minimum ages for deglaciation, probably deposited within 100–200 yr of ice recession from the sites. Because we observe fauna living directly at the glacial grounding line (Pond Ridge moraine, Dennison Point, and Carrying Place Bluffs), we infer that there was no reason for a delay in colonization anywhere else in the De Geer sea, and the lowest fauna found in pond cores were living at the grounding line.

At the remaining six sites, we dated material from higher in the glaciomarine facies. At Long Pond and Mattaseunk Lake, the lowest datable material occurred just below the marine-lacustrine transition. At Sprague Neck, Lewis Cove, Hadley Pond Esker, and Look's Seafood Cannery, the bottom of the glaciomarine unit is not exposed. We interpret these ages as minimum ages for deglaciation.

We also dated shells and seaweed from the regressive facies at Lily Lake and Sprague Neck. Although these ages are not useful for estimating the timing of deglaciation, they estimate the time of shallowing due to marine regression and termination of glaciomarine sedimentation at these sites.

Most attempts to correlate marine and terrestrial records apply a marine reservoir correction of 320–440 $^{14}$C yr to radiocarbon ages on marine fauna (e.g., Bard et al., 1993). We attempted to determine the actual marine reservoir effect at Mattaseunk Lake by dating both *P. arctica* and terrestrial vegetation at the same level in the glaciomarine mud facies, and obtained one age of 13 450 ± 75 $^{14}$C yr B.P. on *P. arctica* and two ages of 10 550 ± 70 and 10 350 ± 120 $^{14}$C yr B.P. on terrestrial vegetation. Because (1) relative sea level was ~60 m below Mattaseunk Lake at 10 500 $^{14}$C yr B.P. (Barnhardt et al., 1995) and thus glaciomarine sedimentation cannot have been underway there at that time; (2) the ages on terrestrial vegetation conflict with the well-dated Younger Dryas stratigraphic marker horizon (Stea and Mott, 1989; Mayle et al., 1993) located 1 m higher in the core; and (3) there is no other evidence to suggest a reservoir age of 3 k.y. in any part of the North Atlantic, we conclude that the terrestrial ages are incorrect. This question requires further investigation.

Terrestrial radiocarbon ages from basal lacustrine sediments in upland ponds within and near the study area (Mott, 1975; Lowell, 1980; Davis and Jacobson, 1985; Lurvey, 1999) are consistently 600–800 $^{14}$C yr younger than nearby basal glaciomarine ages reported in this study. In light of this evidence and studies indicating that the North Atlantic marine reservoir correction changed rapidly during deglaciation and may have been significantly larger than 400 yr (Bard et al., 1993), we believe that a marine reservoir correction of ~700 yr is most appropriate for the $^{14}$C age determinations in this study. However, we use the standard Atlantic reservoir correction of 400 yr in the following discussion for comparison between our chronology and other North Atlantic marine ages.

### Chronology of ice retreat from eastern Maine

The history of ice recession through the Passamaquoddy lowlands is fairly well defined. Evidence from moraines, eskers, and striations throughout this portion of the study area indicate that the ice margin remained relatively straight, with only small-scale calving embayments, and retreated steadily to the north-northwest (Kaplan, 1994, 1999). Two age estimates from the ice-proximal facies in the coastal portion of the study area (Fig. 9; see also Kaplan, 1999) indicate that the ice margin receded from the present coast at Dennison Point (400 yr reservoir-corrected 13 600 ± 85 $^{14}$C yr B.P.) to Pond Ridge moraine (Turner Brook site 3 km up-ice, 400 yr reservoir-corrected 13 250 ± 55 and 13 410 $^{14}$C yr B.P) at ~10 m/yr. These ages are similar to the oldest ages at Carrying Place Bluff, suggesting that the ice margin was approximately linear and parallel to the present coastline. This agrees with the glacial and geological evidence, although the average rate of 10 m/yr does not reflect rapid fluctuations in the glacial grounding line, which almost certainly occurred during overall recession (Kaplan, 1999). Age estimates at Patrick Lake and Marks Lake indicate that the ice retreated from Pond Ridge moraine to these lakes at ~30–40 m/yr. Thus, the rate of ice recession near the present coastline was similar to the overall rate of recession from the terminal position of the Laurentide ice sheet on the Georges Bank across the entire Gulf of Maine to the present shoreline (~40 m/yr; calculated by comparing age estimates in this study with Tucholke and Hollister, 1973).

Ice recession accelerated dramatically north of Patrick Lake. Between Patrick Lake and Pocomoonshine Lake, the retreat rate approached 150 m/yr. This increase in the rate of ice recession ca. 12 900–13 000 $^{14}$C yr B.P. (400 yr reservoir corrected) coincides with a significant geomorphological transition. At approximately the latitude of Patrick Lake, east-southeast–west-northwest–trending end moraines become increasingly segmented. Within 1–2 km, end moraines are replaced by submarine fans. Another 1 km to the north these give way to a zone of esker-fed deltas that extend north to the inland marine limit (Thompson and Borns, 1985).

The history of ice recession in the Penobscot Valley is much less clear. Basal ages from sites in the Penobscot Valley do not become younger to the northwest, as do those in the Passamaquoddy lowlands. In fact, basal ages at Toddy Pond and Boyd Lake are younger than those at Gould Pond, Dover-Foxcroft, and T2 R8 NWP, which is inconsistent with the basic assumption of northward ice retreat common to all previous studies. Eskers and most striations in the Penobscot Valley consistently trend north-northwest–south-southeast (with the exception of southwest striae in Orrington, 20 km south of Bangor; see Fig. 3). Past studies (Shreve, 1985a, 1985b; Smith, 1985) have interpreted this evidence to indicate that the ice margin remained relatively straight across this area as it retreated steadily to the north-northwest. Our ages from the Penobscot Valley are not at all consistent with this model, and suggest instead that deglaciation of the entire valley was essentially instantaneous within the resolution of radiocarbon dating. It is only certain that the entire region deglaciated rapidly between 13 100 and 12 700 $^{14}$C yr B.P. (400 yr reservoir corrected), and our radiocarbon chronology is not sufficiently accurate to resolve the shape of the ice margin during this recession. Although our ages do not preclude the simple "window-shade" style retreat inferred by past glacial and geologic studies (Shreve, 1985a, 1985b; Smith, 1985), they are more consistent with widespread disintegration of the ice sheet within large marine embayments.

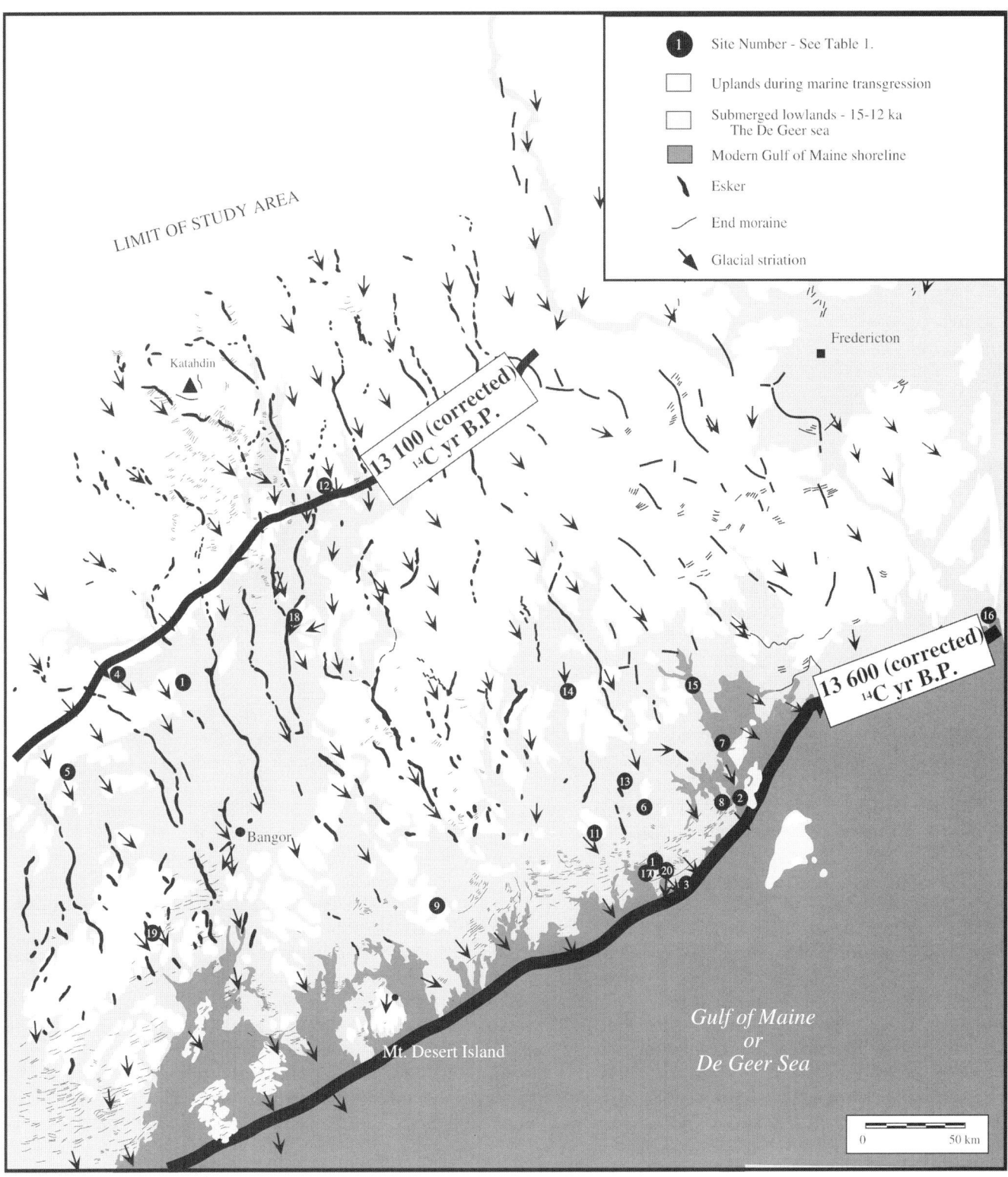

Figure 9. Map showing isolines of southern margin of Laurentide ice sheet. Surficial geology after Thompson and Borns (1985) and Borns et al. (1987). Isoline pattern is based on bifurcating system of eskers, orientation of end moraines and molded landforms, youngest set of striations, and oldest reported $^{14}$C age from each site. Ages on this figure include marine reservoir correction of 400 yr.

## Relationship between climate, paleoceanography, and ice behavior

Our numerical ages indicate that the rate of ice recession increased dramatically shortly after the ice margin retreated north of the present shoreline. This increase in retreat rate coincides with a change in the nature of ice-marginal landforms from moraines to esker-fed deltas. With a 400 year reservoir correction applied to our age estimates, this transition took place ca. 13 000 $^{14}$C yr B.P. With our preferred reservoir correction of 700 yr, it took place at 12 700 $^{14}$C yr B.P., which is correlative with rapid warming at the beginning of the Bolling chronozone in European terrestrial records (Walker, 1995).

Calibration of our marine-based numerical ages to the terrestrial radiocarbon chronology (Stuiver and Reimer, 1993; Fig. 10) indicates that initial deglaciation of the study area and formation of the Pond Ridge moraine took place slightly before 15.9 ka. Ice-core records from Greenland (Stuiver et al., 1995; Fig. 10) indicate full-glacial conditions at that time. Thus, deglaciation of the entire Gulf of Maine, as well as the coastal portion of the study area, took place well before atmospheric warming at the beginning of the Bolling chronozone. The rapid increase in the rate of ice recession north of Patrick Lake took place ca. 15–15.3 ka. Thus, if the correct marine reservoir effect is only 400 yr, accelerated ice recession in coastal Maine led atmospheric warming in Greenland by 300–500 yr. However, if our preferred reservoir correction of 700 yr is correct, the two events would have been synchronous ca. 14.7 ka.

Correlation of our chronology to the North Atlantic marine record is also complicated by the uncertainty in the marine reservoir effect. If the reservoir correction was similar in both coastal Maine and the central North Atlantic, initial deglaciation of the study area and the formation of the Pond Ridge moraine occurred simultaneously with enhanced ice discharge into the North Atlantic during Heinrich event H1 (Bond and Lotti, 1995). The acceleration of ice retreat in the study area ca. 13 000 $^{14}$C yr B.P. (using a 400 yr reservoir correction) apparently took place after the termination of H1 ca. 13 300 $^{14}$C yr B.P., and before North Atlantic warming and the onset of North Atlantic Deep Water circulation ca. 12 700–12 800 $^{14}$C yr B.P. (Lehman and Keigwin, 1992). In light of the uncertainty in the reservoir effect, however, it is impossible to determine which of these events were sychronous.

Regardless of whether accelerated ice recession in Maine was synchronous with large-scale warming of the North Atlantic atmosphere and ocean ca. 12 700–12 800 $^{14}$C yr B.P., our isotope data do not show any warming of the Gulf of Maine during ice retreat across the study area. The continuous $\delta^{18}$O record from Lily Lake, which overlaps the acceleration of ice retreat in the study area (and depending on the reservoir correction used, may overlap the onset of Bolling warming), indicates a consistent water temperature throughout the period of marine submergence at this site. Furthermore, $\delta^{18}$O values of benthic molluscs from throughout the study area do not covary with either sample age or north-south location (Table 4; Fig. 10) and show consistently cold and saline oceanographic conditions. Despite the large uncertainty in the reservoir effect, our chronology indicates that the total glaciomarine sedimentary record in the study area probably overlaps the abrupt Bolling warming: ages from the regressive facies at Sprague Neck indicate that glaciomarine sedimentation lasted until ca. 12 000 $^{14}$C yr B.P. (400 yr reservoir corrected). Thus, our isotope data from the entire period of glaciomarine sedimentation clearly indicate that accelerated ice recession was not associated with any change in oceanographic conditions, and further suggest that regional North Atlantic warming at 12 700–12 800 $^{14}$C yr B.P. did not affect oceanographic conditions in the Gulf of Maine.

## CONCLUSIONS

The sedimentology, faunal assemblage, and isotope geochemistry from glaciomarine sediments in eastern Maine indicate that deglaciation took place in an arctic to subarctic environment. The environment at the grounding line was strongly density stratified: bottom water was ~0–2 °C and near normal marine salinity, and meltwater was confined to the surface layer. These paleoceanographic conditions persisted throughout the period of marine submergence. In contrast, the chronology of ice retreat across the study area shows a significant increase in the rate of ice retreat coincident with a geomorphological transition from moraines to esker-fed deltas. This event, which was followed by rapid deglaciation of the entire inland portion of the study area within only 300–400 yr, may have been synchronous with abrupt atmospheric and oceanographic warming of the North Atlantic region at the onset of the Bolling chronozone ca. 12 700–12 800 $^{14}$C yr B.P., as well as with the onset of North Atlantic deep water circulation at this time. However, this correlation is speculative due to the uncertainty in the marine reservoir effect.

Notwithstanding the uncertainties in correlating events in eastern Maine with those in the North Atlantic, our isotope data indicate that the increase in the rate of ice recession was not associated with an increase in water temperature in the Gulf of Maine. Thus, it appears that North Atlantic warming and ventilation ca. 12 700–12 800 $^{14}$C yr B.P. was not the cause of accelerated recession of the Laurentide ice sheet in Maine. Furthermore, relative sea level was falling during deglaciation of the study area, which means that catastrophic grounding-line instability during relative sea-level rise at a calving ice margin (a proposed cause for rapid collapse of marine-based ice sheets; Weertman, 1974), cannot explain accelerated ice retreat in Maine. Thus, we suggest that accelerated ice retreat must have been a response to atmospheric warming or internal ice dynamics rather than oceanographic changes.

The uncertainties in our chronology make it impossible to differentiate between atmospheric warming and internal ice dy-

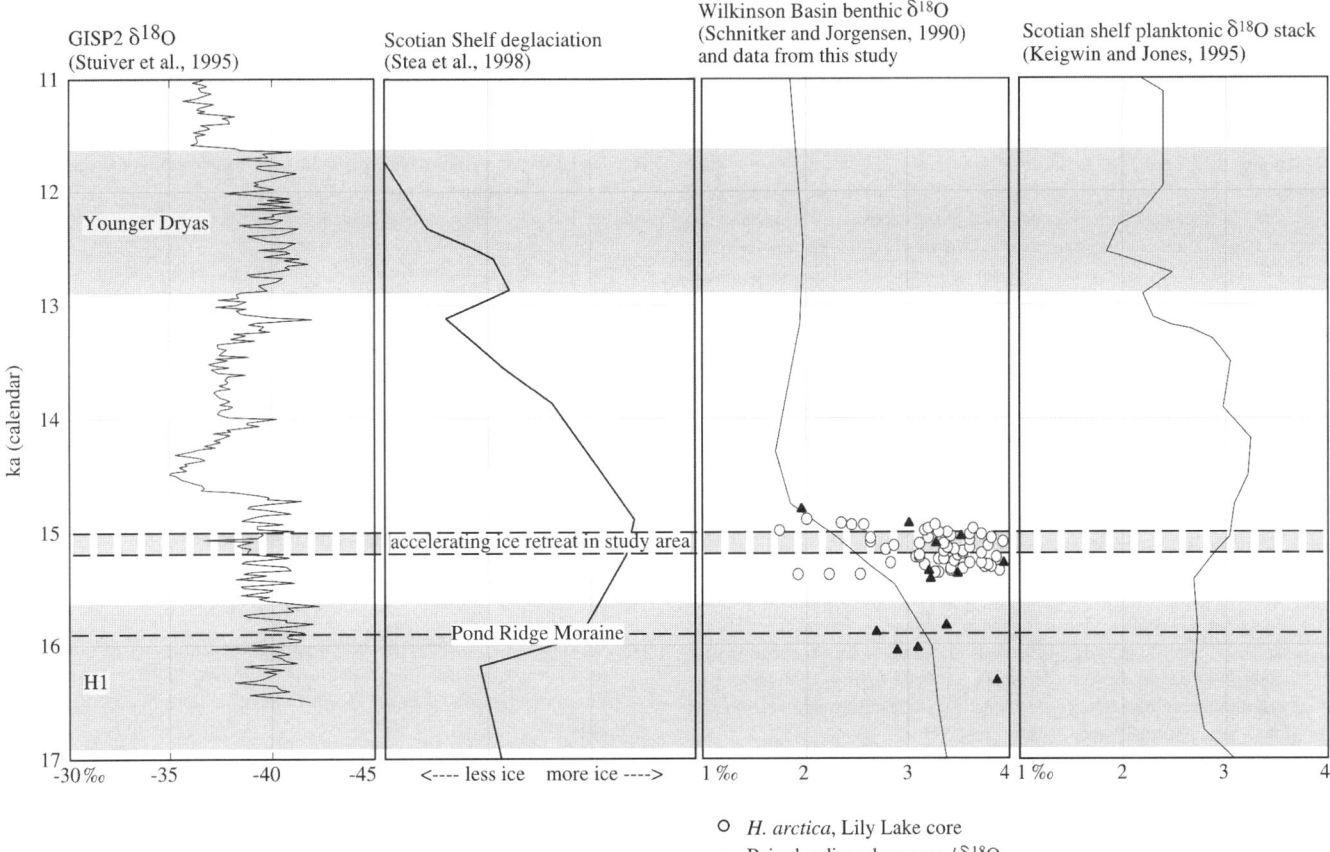

Figure 10. Relationship of events in eastern Maine to records of deglaciation and paleoclimate from Gulf of Maine region and North Atlantic. All records are calibrated to common calendar year time scale using CALIB 3.0.3c marine data set from Stuiver and Reimer (1993), which reflects late glacial marine reservoir correction of ~ 400 $^{14}$C yr. Oxygen isotope data from Lily Lake (Fig. 8) are calibrated using linear least-squares age model fit to three radiocarbon ages from core. Data from Schnitker and Jorgensen (1990) are calibrated by linear interpolation between their reported radiocarbon ages.

namics as the cause of accelerated retreat. These uncertainties can only be resolved by efforts to better establish the appropriate marine reservoir correction for the late glacial Gulf of Maine. In addition, obtaining records similar to the Lily Lake record from low-elevation ponds in southern Maine, where the period of marine submergence extended until 11 500 $^{14}$C yr B.P. (Anderson et al., 1990), would significantly improve our correlation between oceanographic events in Maine and those elsewhere.

## ACKNOWLEDGMENTS

This study was funded by National Science Foundation (NSF) EPSCoR grant RII-8922105 to the University of Maine and by the Maine Geological Survey. Radiocarbon ages in this study were performed by the National Ocean Sciences Accelerator Mass Spectrometry (AMS) laboratory at Woods Hole under NSF Cooperative Agreement OCE-801015 and by the Arizona AMS Facility. Julie Brigham-Grette and Woodrow Thompson provided helpful reviews of the manuscript.

We greatly appreciate the assistance of many faculty and students at the University of Maine, employees of the Maine Geological Survey, and a wide variety of other interested people, who provided equipment and supplies, field and lab assistance, advice, and interpretation. In particular, we are grateful for the generous support given us by the University of Maine Paleoecology Laboratory.

## REFERENCES CITED

Anderson, R.S., Miller, N.G., Davis, R.G., and Nelson, R.E., 1990, Terrestrial fossils in the marine Presumpscot Formation: Implications for late Wisconsinan paleoenvironments and isostatic rebound along the coast of Maine: Canadian Journal of Earth Sciences, v. 27, p. 1241–1246.

Anderson, R.S., Jacobson, G.L., Jr., Davis, R.B., and Stuckenrath, R., 1992, Gould Pond, Maine: Late-glacial transitions from marine to upland environments: Boreas, v. 21, p. 359–371.

Ashley, G.M., and Peteet, D.M., 1999, Chronology and paleoclimate implications from radiocarbon-dated paleobotanical records, Connecticut River Valley, Connecticut: Geological Society of America Abstracts with Programs, v. 31, no. 2, p. A-2.

Ashley, G.M., Boothroyd, J.C., and Borns, H.W., Jr., 1991, Sedimentology of late Pleistocene (Laurentide) deglacial-phase deposits, eastern Maine; an example of a temperate marine grounded ice-sheet margin, in Anderson, J.B., and Ashley, G.M., eds., Glacial marine sedimentation; paleoclimatic significance: Geological Society of America Special Paper 261, p. 107–125.

Bacchus, T.S., 1993, Late Quaternary stratigraphy and evolution of the eastern Gulf of Maine [Ph.D. thesis]: Orono, University of Maine, 347 p.

Bard, E., Arnold, M., Fairbanks, R.G., and Hamelin, B., 1993, $^{230}$Th-$^{234}$U and $^{14}$C ages obtained by mass spectrometry on corals: Radiocarbon, v. 35, p. 191–199.

Barnhardt, W.A., 1994, Late Quaternary relative sea-level change and evolution of the Maine inner continental shelf 12–7 ka B.P. [Ph.D. thesis]: Orono, University of Maine, 196 p.

Barnhardt, W.A., Gehrels, W.R., Belknap, D.F., and Kelley, J.T., 1995, Late Quaternary relative sea-level change in the western Gulf of Maine: Evidence for a migrating glacial forebulge: Geology, v. 23, p. 317–320.

Belknap, D.F., and Shipp, R.C., 1991, Seismic stratigraphy of glacial marine units, Maine inner shelf, in Anderson, J.B., and Ashley, G.M., eds., Glacial marine sedimentation; paleoclimatic significance: Geological Society of America Special Paper 261, p. 137–157.

Belknap, D.F., Shipp, R.C., Kelley, J.T., and Schnitker, D., 1989, Depositional sequence modeling of late Quaternary geolgic history, west-central Maine coast: Studies in Maine Geology, v. 5, p. 29–46.

Bengtsson, D.F., and Enell, M., 1986, Chemical analysis, in Berglund, B.E., ed., Handbook of Holocene paleoecology and paleohydrology: New York, J. Wiley, p. 423–451.

Bloom, A.L., 1960, Late Pleistocene changes of sea level in southwestern Maine: Maine Geological Survey Department of Economic Development, 143 p.

Bloom, A.L., 1963, Late-Pleistocene fluctuations of sealevel and postglacial crustal rebound in coastal Maine: American Journal of Science, v. 261, p. 862–879.

Boggs, S., Jr., 1987, Principles of sedimentology and stratigraphy: Columbus, Ohio, Merrill Publishing, 784 p.

Bond, G.C., and Lotti, R., 1995, Iceberg discharges into the North Atlantic on millennial time scales during the last glaciation: Science, v. 267, p. 1005–1010.

Borns, H.W., Jr., Gadd, N.R., LaSalle, P., Martineau, G., Chauvin, L, Fullerton, D.S., Fulton, R.J., Chapman, W.F., Wagner, W.P., and Grant, D.R., 1987, Quaternary geology, Quebec, 4° × 6° quadrangle: U.S. Geological Survey, Map I-1420 (NL-19), scale 1:1 000 000.

Bothner, M.H., and Spiker, E.C., 1980, Upper Wisconsinan till recovered on the continental shelf southeast of New England: Science, v. 210, p. 423–425.

Boyd, R., Scott, D.B., and Douma, M., 1988, Glacial tunnel valleys and Quaternary history of the outer Scotian shelf: Nature, v. 333, p. 61–64.

Brouwers, E.M., Cronin, T.M., Horne, D.J., and Lord, A.R., 2000, Recent shallow marine ostracods from high latitudes: Implications for late Pliocene and Quaternary paleoclimatology: Boreas, v. 29, p. 127–143.

Coplen, T.B., Kendall, C., and Hopple, J., 1983, Comparison of stable isotope reference samples: Nature, v. 302, p. 236–238.

Cotter, J.F.P., Ridge, J.C., Evenson, E.B., Sevon, W.D., Sirkin, L., and Stuckenrath, R., 1986, The Wisconsinan history of the Great Valley, Pennsylvannia and New Jersey, and the age of the "Terminal Moraine", in Cadwell, D.H., ed., The Wisconsinan Stage of the First Geological District, eastern New York: New York State Museum Bulletin 455, p. 22–49.

Cotter, M.P., 1985, Paleoecology of the foraminifera of the Presumscot Formation, Penobscot Valley, Maine [M.S. thesis]: Orono, University of Maine, 319 p.

Davis, R.B., and Jacobson, G.L, Jr., 1985, Late glacial and early Holocene landscapes in northern New England and adjacent areas of Canada: Quaternary Research, v. 23, p. 341–368.

Deevey, E.S., Jr., 1951, Late-glacial and post-glacial pollen diagrams from Maine: American Journal of Science, v. 249, p. 177–207.

Dethier, D.P., Pessl, F., Jr., Keuler, R.F., Balzarini, M.A., and Pevear, D.R., 1995, Late Wisconsinan glaciomarine deposition and isostatic rebound, northern Puget Lowland, Washington: Geological Society of America Bulletin, v. 107, p. 1288–1303.

Donner, J., 1995, The effects of islands on the recession of the late Wisconsinan ice sheet margin in the De Geer Sea, Maine [M.S. thesis]: Orono, Universtiy of Maine, 90 p.

Dorion, C.C., 1997, An updated high resolution chronology of deglaciation and accompanying marine transgression in Maine [M.S. thesis]: Orono, University of Maine, 147 p.

Dyke, A.S., Dale, J.E., and McNeely, R.N., 1996, Marine molluscs as indicators of environmental change in glaciated North America and Greenland during the last 18,000 years: Géopgraphie Physique et Quaternaire, v. 50, p. 125–184.

Eggleston, A.E., 1979, Subsurface investigation for the proposed reconstruction of a portion of route 191 in the town of Cutler: Maine Department of Transportation Materials and Research Division Soils Section Report 79-25, 8 p.

Ellis, B.F., 1984, The Ellis and Messina catalogues of micropaleontology. Supplement to the 1940 catalogue of Foraminifera: New York, Museum of Natural History, 30 volumes.

Elverhoi, A., Pfirman, S.L., Solheim, A., and Larssen, B.B., 1989, Glaciomarine sedimentation in epicontinental seas exemplified by the northern Barents Sea: Marine Geology, v. 85, p. 225–250.

Epstein, S., Bukchsbaum, R., Lownestam, H.A., and Urey, H.C., 1953, Revised carbonate-water isotopic temperature scale: Geological Society of America Bulletin, v. 64, p. 1315–1326.

Eyles, N., Eyles, C.H., and Miall, A.D., 1983, Lithofacies types and vertical profile models; an alternative approach to the description and environmental interpretation of glacial diamict and diamictite sequences: Sedimentology, v. 30, p. 393–410.

Fairbanks, R.G., 1989, A 17,000-year glacio-eustatic sea level record: Influence of glacial melting rates on the Younger Dryas event and deep-ocean circulation: Nature, v. 342, p. 637–642.

Fisher, D.A., 1992, Possible ice-core evidence for a fresh water melt water cap over the Atlantic Ocean in the early Holocene, in Bard, E., and Broecker, W.S., eds., The last deglaciation: Absolute and radiocarbon chronologies, NATO ASI Series 1, Volume 2: New York, Springer-Verlag, 344 p.

Folk, R.L., 1974, Petrology of sedimentary rocks: Austin, Texas, Hemphill, 184 p.

Gadd, N.R., 1973, Quaternary geology of southwest New Brunswick with particular reference to Fredericton area: Geological Survey of Canada Paper 71-34, 31 p.

Gates, O., 1975, Geologic map and cross sections of the Eastport quadrangle, Maine: Maine Geological Survey Open-File Map and Report GM-3, Scale 1:48 000, 19 p.

Hald, M., Steinsund, P.I., Dokken, T., Korsun, S., Polyak, L., and Aspeli, R., 1994, Recent and late Quaternary distribution of *Elphidium excavatum* f. *clavatum* in arctic seas, in Sejrup, H.P., and Knudsen, K.L., eds., Late Cenozoic benthic foraminifera: Taxonomy, ecology, and stratigraphy: Cushman Foundation Special Publication No. 32, p. 141–153.

Hartshorn, J.H., 1958, Flowtill in southeastern Massachusetts: Geological Society of America Bulletin, v. 69, p. 477–482.

Hillaire-Marcel, C., 1988, Isotopic composition ($^{18}$O, $^{13}$C, $^{14}$C) of biogenic carbonates in Champlain Sea sediments, in Gadd, N.R., ed., The late Quaternary development of the Champlain Sea Basin: Geological Association of Canada Special Paper 35, p. 177–194.

Hughes, T., Borns, H.W., Jr., Fastook, J.L., Hyland, M.R., Kite, J.S., and Lowell, T.V., 1985, Models of glacial reconstruction and deglaciation applied to maritime Canada and New England, in Borns, H.W., Jr., et al., eds., Late Pleistocene history of northeastern New England and adjacent Quebec: Geological Society of America Special Paper 197, p. 139–150.

Jager, M.G., 1996, Deglaciation of the lower Carrabassett River valley drainage, Maine [M.S. thesis]: Orono, University of Maine, 137 p.

Kaplan, M., 1994, The deglaciation of southeastern Washington County, Maine [M.S. thesis]: Orono, University of Maine, 111 p.

Kaplan, M., 1999, Retreat of a tidewater margin of the Laurentide ice sheet in eastern coastal Maine between ca. 14,000 to 13,000 $^{14}$C yr B.P.: Geological Society of America Bulletin, v. 111, p. 620–632.

Keigwin, L.D., and Jones, G.A., 1995, The marine record of deglaciation from the continental margin off Nova Scotia: Paleoceanography, v. 10, p. 973–985.

King, A.R., 1979, Subsurface investigation for the replacement of Turner's Mill bridge over Turner Brook in the town of Cutler: Maine Department of Transportation Materials and Research Division Soils Section Report 79-15, 5 p.

King, L.H., 1993, Till in the marine environment: Journal of Quaternary Science, v. 8, p. 347–358.

King, L.H., 1994, Younger Dryas glaciation of the eastern Scotian Shelf: Canadian Journal of Earth Sciences, v. 31, p. 401–417.

King, L.H., 1996, Late Wisconsinan ice retreat from the Scotian Shelf: Geological Society of America Bulletin, v. 108, p. 1056–1067.

King, L.H., and Fader, G.B.J., 1986, Wisconsinan glaciation of the Atlantic continental shelf of southeast Canada: Geological Survey of Canada Bulletin 363, 72 p.

Kreutz, K.J., 1994, Paleoceanographic conditions of the late Wisconsinan marine transgression and regression in eastern coastal Maine: Stable isotopic evidence [M.S. thesis]: Orono, University of Maine, 113 p.

Lehman, S.J., and Keigwin, L.D., 1992, Sudden changes in North Atlantic circulation during the last deglaciation: Nature, v. 356, p. 757–762.

Lepage, C.A., 1982, The composition and origin of the Pond Ridge moraine, Washington County, Maine [M.S. thesis]: Orono, University of Maine, 124 p.

Lougee, R.J., and Lougee, C.R., 1976, Late-glacial chronology: New York, Vantage Press, 553 p.

Lowell, T.V., 1980, Late Wisconsinan ice extent in Maine, evidence from Mt. Desert Island and the St. John River area [M.S. thesis]: Orono, University of Maine, 91 p.

Lurvey, L.K., 1999, An investigation of a glacial landform transition in southeastern Maine [M.S. thesis]: Orono, University of Maine, 79 p.

Mayle, F.E., Levesque, A.J., and Cwynar, L.C., 1993, Accelerator-mass-spectrometer ages for the Younger Dryas event in Atlantic Canada: Quaternary Research, v. 39, p. 355–360.

McDonald, B.C., and Shilts, W.W., 1971, Quaternary stratigraphy and events in southeastern Quebec: Geological Society of America Bulletin, v. 82, p. 683–698.

Middleton, G.V., and Hampton, M.A., 1976, Subaqueous sediment transport and deposition by sediment gravity flows, in Stanley, D.J., and Swift, D.J.P., eds., Marine sediment transport and environmental management: New York, John Wiley & Sons, p. 197–218.

Mix, A.C., and Ruddiman, W.F., 1984, Oxygen-isotope analyses and Pleistocene ice volume: Quaternary Research, v. 20, p. 1–20.

Mott, R.J., 1975, Palynological studies of lake sediment profiles from southwestern New Brunswick: Canadian Journal of Earth Sciences, v. 12, p. 273–288.

Nicks, L.P., 1991, The study of the glacial stratigraphy and sedimentation of the Sheldon Point moraine, Saint John, New Brunswick: New Brunswick Department of Natural Resources and Energy, Mineral Resources, Open-File Report 91-12, 171 p.

Ockelmann, W.K., 1958, The zoology of East Greenland: Marine Lamellibranchiata: Copenhagen, C.A. Reitzels Forlag, 256 p.

Oldale, R.N., 1989, Timing and mechanisms for the deposition of the glaciomarine mud in and around the Gulf of Maine: A discussion of alternative models: Studies in Maine Geology, v. 5, p. 1–10.

Oldale, R.N., Williams, R.S., Jr., and Colman, S.M., 1990, Evidence against a late Wisconsinan ice shelf in the Gulf of Maine: Quaternary Science Reviews, v. 9, p. 1–13.

Osberg, P.H., Hussey, A.A., and Boone, G.M., 1985, Bedrock geologic map of Maine: Augusta, Maine Geological Survey, scale 1:500 000.

Pfirman, S.L., and Solheim, A., 1989, Subglacial meltwater discharge in the open-marine tidewater glacier environment: Observations from Nordaustlandet, Svalbard Archipelago: Marine Geology, v. 86, p. 265–281.

Powell, R.D., 1990, Glacimarine processes at grounding-line fans and their growth to ice-contact deltas, in Dowdeswell, J.A., and Scourse, J.D., eds., Glacimarine environments: Processes and sediments: Geological Society [London] Special Publication 53, p. 53–73.

Powell, R.D., and Molnia, B.F., 1989, Glacimarine sedimentary processes, facies and morphology of the south-southeast Alaska shelf and fjords: Marine Geology, v. 85, p. 359–390.

Reineck, H.E., and Singh, I.B., 1980, Depositional sedimentary environments: Heidelberg, Germany, Springer-Verlag, 551 p.

Rust, B.D., and Romanelli, R., 1975, Late Quaternary subaqueous outwash deposits near Ottawa, Canada, in Jopling, A.V., and McDonald, B.C., eds., Glaciofluvial and glaciolacustrine sedimentation: Society of Economic Paleontologists and Mineralogists Special Publication 23, p. 177–192.

Schnitker, D., 1975, Late glacial to recent paleoecology of the Gulf of Maine, 1st International Symposium on Continental Margin Benthic Foraminifera Part B: Paleoecology and biostratigraphy: Maritime Sediments Special Publication 1, p. 385–392.

Schnitker, D., and Jorgensen, J.B., 1990, Late glacial and Holocene diatom successions in the Gulf of Maine: Response to climatologic and oceanographic change, in Garbary, D.J., and South, G.R., eds., Evolutionary biogeography of the marine algae of the North Atlantic, NATO ASI Series, Volume G 22: Berlin, Springer-Verlag, p. 35–53.

Shreve, R.L., 1985a, Esker characteristics in terms of glacier physics, Katahdin esker system, Maine: Geological Society of America Bulletin, v. 96, p. 639–646.

Shreve, R.L., 1985b, Late Wisconsinan ice-surface profile calcuated from esker paths and types, Katahdin esker system, Maine: Quaternary Research, v. 23, p. 27–37.

Sirkin, L.A., and Stuckenrath, R., 1980, The Port Washingtonian warm interval in the northern Atlantic coastal plain: Geological Society of America Bulletin, v. 91, p. 332–336.

Smith, G.W., 1985, Chronology of late Wisconsinan deglaciation of coastal Maine, in Borns, H.W., Jr., et al., eds., Late Pleistocene history of northeastern New England and adjacent Quebec: Geological Society of America Special Paper 197, p. 29–44.

Stea, R.R., and Mott, R.J., 1989, Deglaciation environments and evidence for glaciers of Younger Dryas age in Nova Scotia, Canada: Boreas, v. 18, p. 167–187.

Stea, R.R., Piper, D.J.W., Fader, G.B.J., and Boyd, R., 1998, Wisconsinan glacial and sea-level history of maritime Canada and the adjacent continental shelf: A correlation of land and sea events: Geological Society of America Bulletin, v. 110, p. 821–845.

Stone, G.H., 1899, The glacial gravels of Maine and their associated deposits: U.S. Geological Survey Monograph 34, 499 p.

Stone, J.R., and Ashley, G.M., 1992, Ice-wedge casts, pingo scars, and the drainage of glacial Lake Hitchcock, in Robinson, P., and Brady, J.B., eds., New England Intercollegiate Geological Conference guidebook for field trips in the Connecticut valley region of Massachusetts and adjacent states, v. 2, p. 305–331.

Stuiver, M., and Borns, H.W., Jr., 1975, Late Quaternary marine invasion in Maine: Its chronology and associated crustal movement: Geological Society of America Bulletin, v. 86, p. 99–104.

Stuiver, M., and Reimer, P.J., 1993, CALIB 3.0.3A: University of Washington Quaternary Research Center.

Stuiver, M., Grootes, P.M., and Braziunas, T.F., 1995, The GISP2 $\delta^{18}$O climate record of the past 16,500 years and the role of the sun, ocean, and volcanoes: Quaternary Research, v. 44, p. 341–354.

Thompson, W.B., and Borns, H.W., Jr., 1985, Surficial geologic map of Maine: Augusta, Maine Geological Survey, scale 1:500 000.

Thompson, W.B., Crossen, K.J., Borns, H.W., Jr., and Andersen, B.G., 1989, Glaciomarine deltas of Maine and their relation to late Pleistocene–Holocene crustal movements, *in* Anderson, W.A., and Borns, H.W., Jr., eds., Neotectonics of Maine: Studies in seismicity, crustal warping, and sea-level change: Maine Geological Survey Bulletin 40, p. 43–68.

Tucholke, B.E., and Hollister, C.D., 1973, Late Wisconsinan glaciation of the southwestern Gulf of Maine: New evidence from the marine environment: Geological Society of America Bulletin, v. 84, p. 3279–3296.

Walker, M.J.C., 1995, Climatic changes in Europe during the last glacial/interglacial transition: Quaternary International, v. 28, p. 63–76.

Wassenaar, L., Brand, U., and Terasmae, J., 1988, Geochemical and paleoecological investigations using invertebrate macrofossils of the late Quaternary Champlain Sea, Ontario and Quebec, *in* Gadd, N.R., ed., The late Quaternary development of the Champlain Sea Basin: Geological Association of Canada Special Paper 35, p. 195–205.

Weddle, T.K., and Retelle, M.J., 1998, Deglacial style and relative sea-level chronology, Casco Bay lowlands to White Mountain foothills, southwestern Maine: Geological Society of America Abstracts with Programs, v. 30, no. 1, p. 83.

Weertman, J., 1974, Stability of the junction of an ice sheet and an ice shelf: Journal of Glaciology, v. 13, p. 3–11.

Wright, H.E., Jr., 1967, A square-rod piston sampler for lake sediments: Journal of Sedimentary Petrology, v. 37, p. 975–976.

MANUSCRIPT ACCEPTED BY THE SOCIETY JUNE 9, 2000

Geological Society of America
Special Paper 351
2001

# Late Wisconsinan glacial dynamics, deglaciation, and marine invasion in southern Québec

**Serge Occhietti**
*Geotop, Université du Québec à Montréal, CP 8888 Centre-ville, Montréal, Québec H3C 3P8, Canada*
**Michel Parent**
*Geological Survey of Canada, 2535 Boulevard Laurier, C.P. 7500, Sainte-Foy, Québec G1V 4C7, Canada*
**William W. Shilts**
*Illinois State Geological Survey, 615 East Peabody Drive, Champaign, Illinois 61820-6964, USA*
**Jean-Claude Dionne**
*Département de géographie, Université Laval, Saint-Foy, Québec G1K 7P4, Canada*
**Étienne Govare**
*Département de géographie, Université du Québec à Montréal, CP 8888 Centre-ville, Montréal, Québec H3C 3P8, Canada*
**Dominique Harmand**
*Département de géographie, Université de Nancy II, Nancy, France*

## ABSTRACT

Deglaciation patterns of the Laurentide ice sheet in southern Québec were related to climatic and nonclimatic factors. Thinning of the ice sheet and thermolatitudinal ice retreat are directly linked to the global warming at the end of late Wisconsinan time, between 17 ka and 11 ka. However, correlations between regional deglacial events and global climatic oscillations during that period have yet to be established, except for the St. Narcisse Moraine event, which has been assigned to the Dryas III, and perhaps for the reactivation of Laurentide ice in the middle Chaudiere Valley area during an older cold event. Nonclimatic factors also played a major role on the deglaciation of the region. After the last glacial maximum, the Laurentide ice sheet began to decrease, and the St. Lawrence corridor channelized a major ice stream, the St. Lawrence ice stream, which became a major feature of the southeast sector of the ice sheet. The St. Lawrence ice stream is a flow convergence zone caused by a combination of ice dynamics and topographic factors and rapid ablation at its terminus. The head of the flow convergence migrated deeply into the Laurentide ice sheet and caused thinning of adjacent ice masses. As a consequence of this accelerated ablation, an Appalachian sector became differentiated from the main ice sheet. Regionally, the terminus of the ice stream was a calving bay that retreated along the Laurentian channel to the mouth of the Saguenay fjord. The ice stream and the deglaciated estuary generated the well-known flow reversal along much of the northern margin of the Appalachian sector. In addition to these generalized deglaciation processes, local and regional topographic features influenced the ice dynamics and the final deglaciation patterns.

---

E-mails: Occhietti, ochietti.serge@uqam.ca; Parent, parent@gsc.nrcan.gc.ca; Shilts, shilts@geoserv.isgs.uiuc.edu; Dionne, ggr@ggr.ulaval.ca; Govare, e.govare@sympatico.ca; Harmand, harmand@clsh.univ-nancy2.fr

Occhietti, S., Parent, M., Shilts, W.W., Dionne, J.-C., Govare, É., and Harmand, D., 2001, Late Wisconsinan glacial dynamics, deglaciation, and marine invasion in southern Québec, *in* Weddle, T.K., and Retelle, M.J., eds., Deglacial History and Relative Sea-Level Changes, Northern New England and Adjacent Canada: Boulder, Colorado, Geological Society of America Special Paper 351, p. 243–270.

## INTRODUCTION

The southeastern margin of the Laurentide ice sheet covered a physiographically diverse region (Figs. 1, A, and B, and 2A): the northern Appalachian highlands and uplands (New England, New Brunswick, Nova Scotia, Newfoundland and southern Québec), the St. Lawrence corridor (which includes the gulf and estuary of the St. Lawrence River as well as the St. Lawrence Lowland), and the deeply indented southern margin of the Laurentian highlands (Fig. 1). This landscape promoted deglacial styles that differed substantially from region to region: while concentric thermolatitudinal retreat patterns (retreat rate depending on the distance from the center of the main ice dome, in this case the New Quebec dome [Fig. 1A], and on insolation) predominated in New England until ca. 13 000 yr B.P. (Larson and Stone, 1982; Borns et al., 1985; Dyke and Prest, 1987) (Fig. 2, A–E), the ice sheet contracted into a series of partly interconnected ice divides, domes, and caps in regions surrounding the Gulf of St. Lawrence (Grant, 1989; Stea et al., 1998) (Fig. 2, C–F).

These contrasting styles are further compounded by the difficulty of connecting the terrestrial-based $^{14}$C chronology of New England's deglacial events (Ridge et al., 1999, and Ridge, this volume) with the marine-based $^{14}$C chronology of events in southern Québec (Thompson et al., 1996, 1999), which raises the question of local $^{14}$C reservoir effects in the Goldthwait and Champlain seas. Thus, in spite of its more northerly position and its proximity to the New Quebec dome, $^{14}$C chronologies have suggested that southern Québec was deglaciated well before adjacent Maine (e.g., Borns, 1985). However, this problem has been largely resolved as a result of a recent reassessment of the deglacial chronology in southeastern Québec (Parent and Occhietti, 1999). Although the complexity of the regional physiography has several disadvantages, it has favored the deposition of lacustrine, glaciolacustrine, and marine sediments during ice-sheet growth and decay (Occhietti, 1990), and this has in turn provided several opportunities for dating deglacial events directly. Given that regional deglaciation took place over a very short time interval, between about 13 500 and 10 500 yr B.P. correlations between brief, poorly dated or even undated local events must be made cautiously. Previous syntheses on the deglaciation of southern Québec include those of McDonald (1968), Gadd et al. (1972), Parent et al. (1985), Chauvin et al. (1985), Parent and Occhietti (1988), Occhietti (1989), and LaSalle and Chapdelaine (1990), in addition to a broader perspective provided by Dyke and Prest (1987). Since then, new work has expanded our regional knowledge of ice-flow data into poorly known or unknown regions. South of the St. Lawrence Valley, new information comes mainly from Blais (1989), Plouffe (1997), Shilts (1997), Rappol (1993), and Hétu (1998); north of the St. Lawrence River, new data and interpretations have come from Dionne (1994), Dionne and Occhietti (1996), Occhietti et al. (1996, 1997), Govare (1995), Bonenfant (1993), Gagnon (1994), Lanoie (1995), Bernatchez (1997), and Fournier (1998). Unpublished research by students at the Université du Québec à Montréal in the Beauce and Charlevoix regions under the supervision of Occhietti have also made significant contributions. The need to revisit several key aspects of the deglaciation of southern Québec and to propose new relationships with adjacent regions of Atlantic Canada and northern New England arose from this body of new work.

Through integration of currently available data, the objectives of this chapter are (1) to present the overall setting and main controlling factors of the last deglaciation and to reassess its main paradigms, (2) to review the currently available $^{14}$C dates and to establish the current limitations of the chronological framework of deglaciation, and (3) to propose a sequence and chronology of deglacial events in southern Québec, in relationship with deglacial events recorded in adjacent regions. We integrate knowledge and data drawn from our work in the middle and upper estuary of the St. Lawrence River and in adjacent regions.

## METHODOLOGY

Former ice-flow patterns and stages of glacial recession in the varied regional settings are reconstructed from directional data and lithofacies observations.

### Directional data from former ice flows

Ice-flow direction is primarily established from glacial striations. In the Appalachians, erosional glacial microforms (see Laverdière and Guimont, 1980; Lortie and Martineau, 1987) are abundant on fine-grained rocks, and crag-and-tail features were used to assign ice-flow directions (Lamarche, 1971; Lortie and Martineau, 1987; Parent, 1987; Shilts, 1997). In the Laurentians, where crag and tails are scarce, other types of erosional marks must be used, such as nailhead striae, lunate fractures, and stoss and lee features. Crosscutting relationships between sets of glacial striations and clast fabric variations between till sheets give the relative chronology of ice flows. Till clast fabrics have been measured by recording the orientation and plunge of the A axis of prolate and bladed clasts, and the B axis of bladed clasts. They give significant results in the Appalachians and Laurentians and confirm ice-flow directions identified by erosional features (McDonald, 1969; Parent, 1987; Blais, 1989; Fournier, 1998), whereas they give unclear information in the St. Lawrence Valley (Bernier and Occhietti, 1991). Till clast lithology and matrix geochemistry were extensively applied by Shilts (1973, 1981) and Parent (1987). Till clast lithology has allowed Blais (1989), Plouffe (1997), and Fournier (1998) to confirm the northward ice-flow reversal in areas that are 45–70 km apart.

Using these methods of investigation, Fournier (1998) showed that the latest local ice flows or ice movements, for example, the latest local southward ice flow in the valleys of the western Charlevoix region, are recorded by ice-front fea-

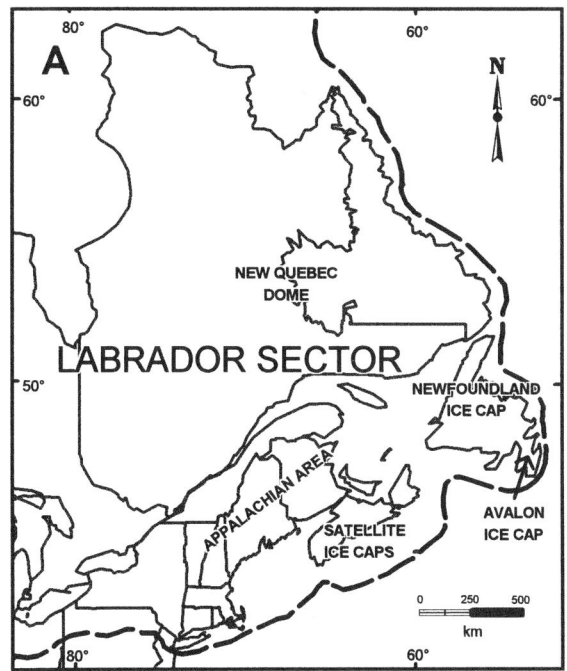

Figure 1. A, Location of New Quebec dome and adjacent Appalachian area. B, Southern Québec sites referred to in text and location of Appalachian northern piedmont.

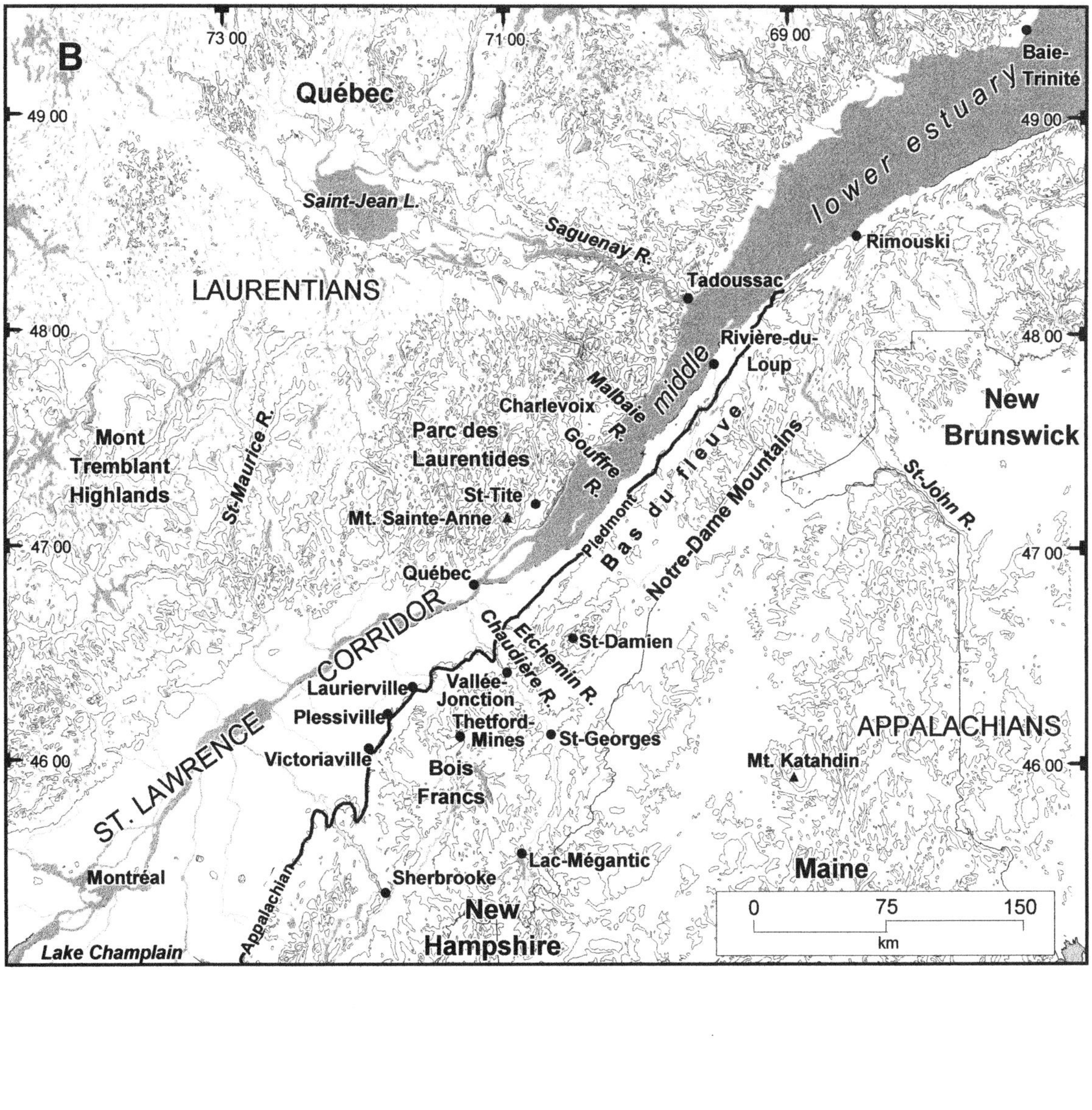

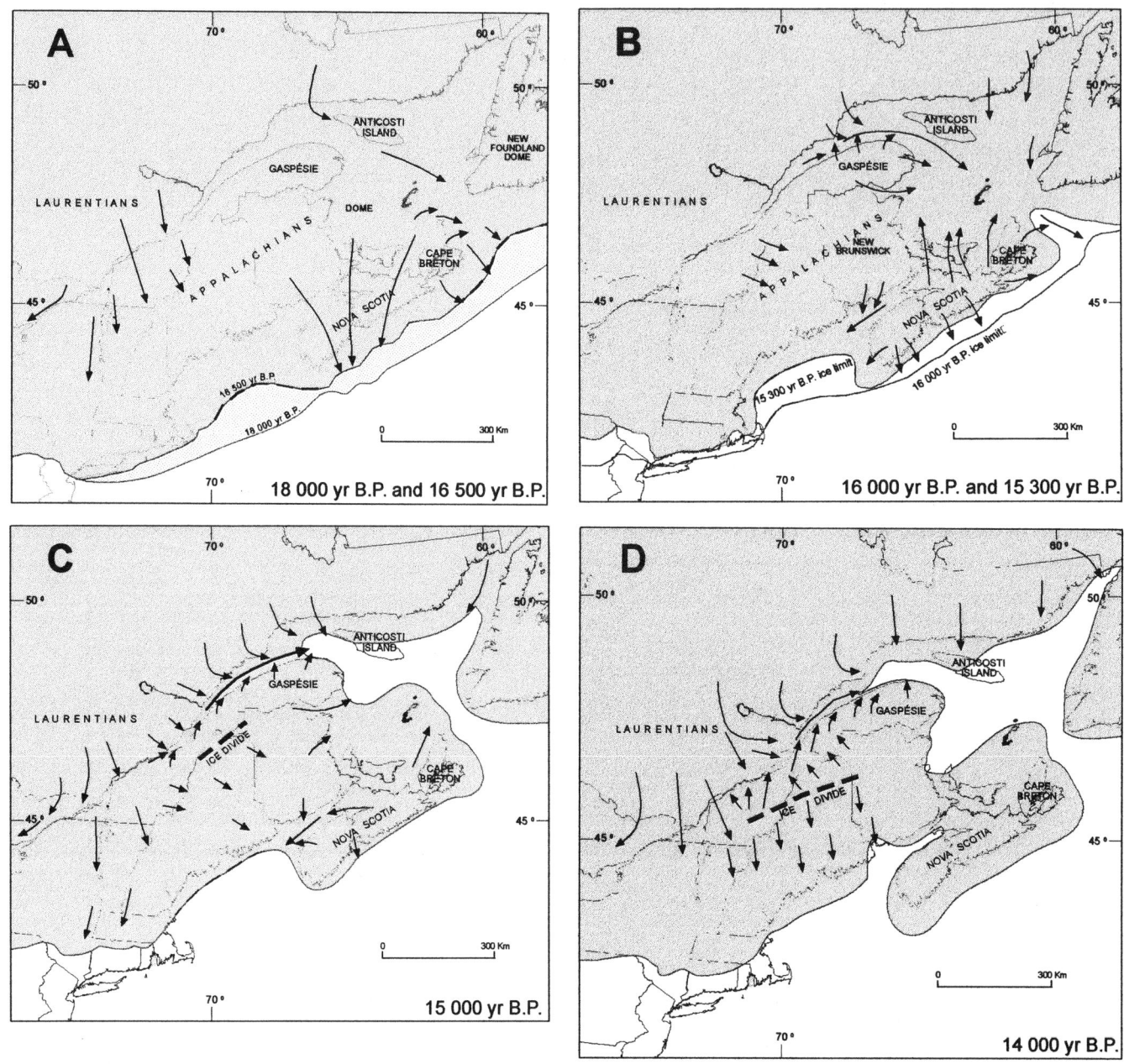

Figure 2 (on this and facing page). Tentative reconstruction of late Wisconsinan ice dynamics and glacial extent in southeast margin of Laurentide ice sheet. Glacial episodes in Gulf of St. Lawrence area are based on Stea et al. (1998). Sequence of events (A–F) is explained in text.

tures with no related glacial striae or till fabrics. This seems to be frequently the rule when stagnant ice masses remained in local or regional depressions.

### Directional data from ice-marginal stratified drift

With the exception of the Saint-Narcisse Moraine, most of the moraine belts in southern Québec (Fig. 3) consist of discontinuous bodies of ice-contact and outwash deposits. The morphologies of eskers, outwash fans and deltas, and ice-front ridges, together with intralandform facies relationships and paleocurrent indicators, give a good idea of the position of former wasting ice masses (Parent, 1987; Blais, 1989). Reworked beds at the top and on the sides of the ice-front accumulations may lead to misleading conclusions on the direction of paleocurrents, as explained by Chauvin et al. (1985).

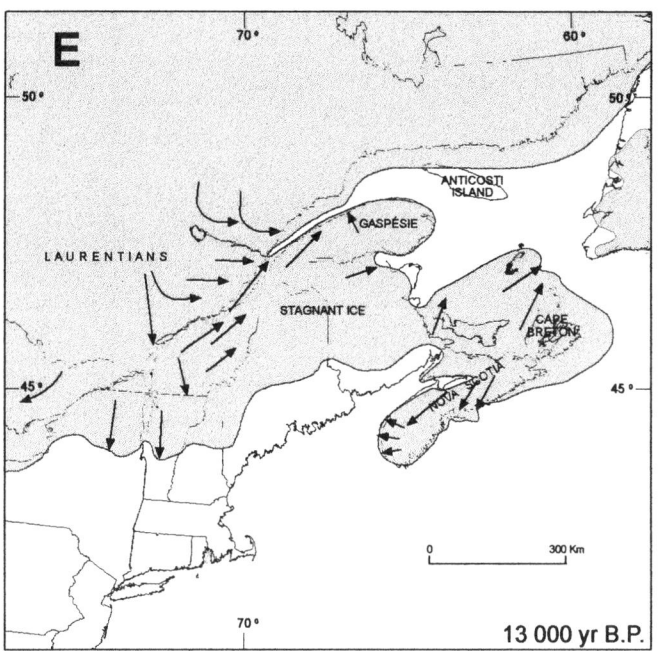
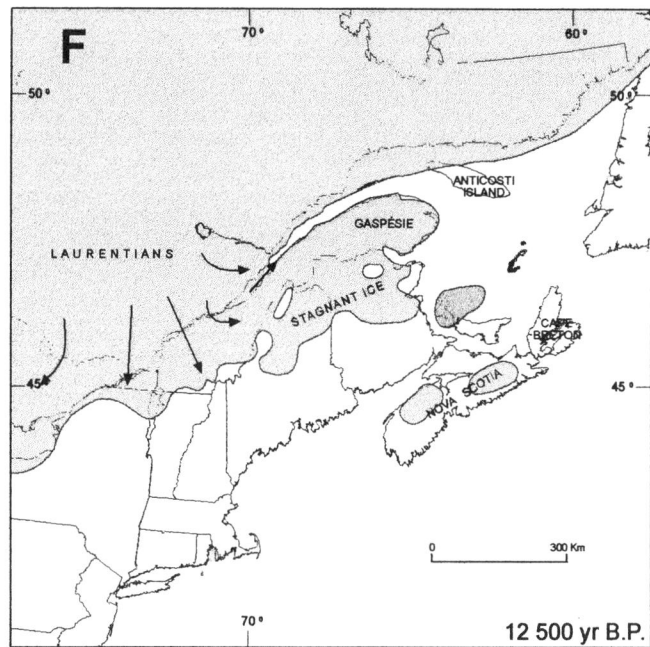

Figure 2. (*continued*)

## *General approach*

With few exceptions, the published data have been generally considered reliable, even if the methods of investigation have varied from one study to another. The major difficulty encountered in this synthesis was to separate factual data from the attached interpretations, mainly because regional models have evolved during the past 35 yr (Table 1) in parallel with global paradigms. We used a multianalytical approach (Occhietti, 1990) to reassess and integrate available data. The basic rule has been that an interpretation may be considered valid when most of the data (facies, stratigraphic setting, morphology, and chronology) converge toward a common conclusion, which is coherent with the current paradigms. While preparing this synthesis, it became obvious that many of the $^{14}$C ages were unreliable. Therefore, after a statement of the processes and the present paradigms, a critical analysis of the available $^{14}$C ages is included (Table 2) before the section on the reconstruction of the sequence of deglacial events in southern Québec.

## CONTROLS OF THE ICE DYNAMICS AND RECESSION OF THE SOUTHEASTERN MARGIN OF THE LAURENTIDE ICE SHEET

The mode of deglaciation of part of the southeastern margin of the Laurentide ice sheet was controlled by a series of semiindependent climatic and nonclimatic factors.

## *Climatic factors*

***Global warming after the last glacial maximum.*** The global warming trend that began ca. 17 ka (Fig. 4) and continued into the Bölling interval (see GRIP curve, Johnsen et al., 1992; Dansgaard et al., 1993; Grootes et al., 1993; Stuiver et al., 1995) signaled the onset of generally negative mass-balance conditions for the Laurentide ice sheet. We think that this resulted in the gradual transition from a dome-shaped ice sheet, such as the Antarctic ice sheet, to a much flatter, plateau-like ice sheet, much like modern-day Greenland ice sheet (Llibourtry, 1965). This thinning of the ice sheet had several consequences in its marginal areas, including (1) increasing topographic control on ice-flow patterns, (2) accelerated development of ice streams, as in the Greenland ice sheet, (3) migration of ice divides and glacioisostatic loading centers, and (4) several other glaciodynamic readjustments. Although generalized thermolatitudinal ice retreat was the main deglaciation process along the southern margin of the Laurentide ice sheet, several nonclimatic factors were also involved.

***Climatic fluctuations.*** Global climatic fluctuations during late Wisconsinan time have now been well documented through palynostratigraphic records as well as through marine and ice-core records (Dansgaard et al., 1993; Bond et al., 1993; Grootes et al., 1993; Stuiver et al., 1995; Stea et al., 1998; Stea and Mott, 1998; Yu and Eicher, 1998). Although these climatic events were not rigorously correlative from region to region, their overall effects on the mass balance on the eastern sector

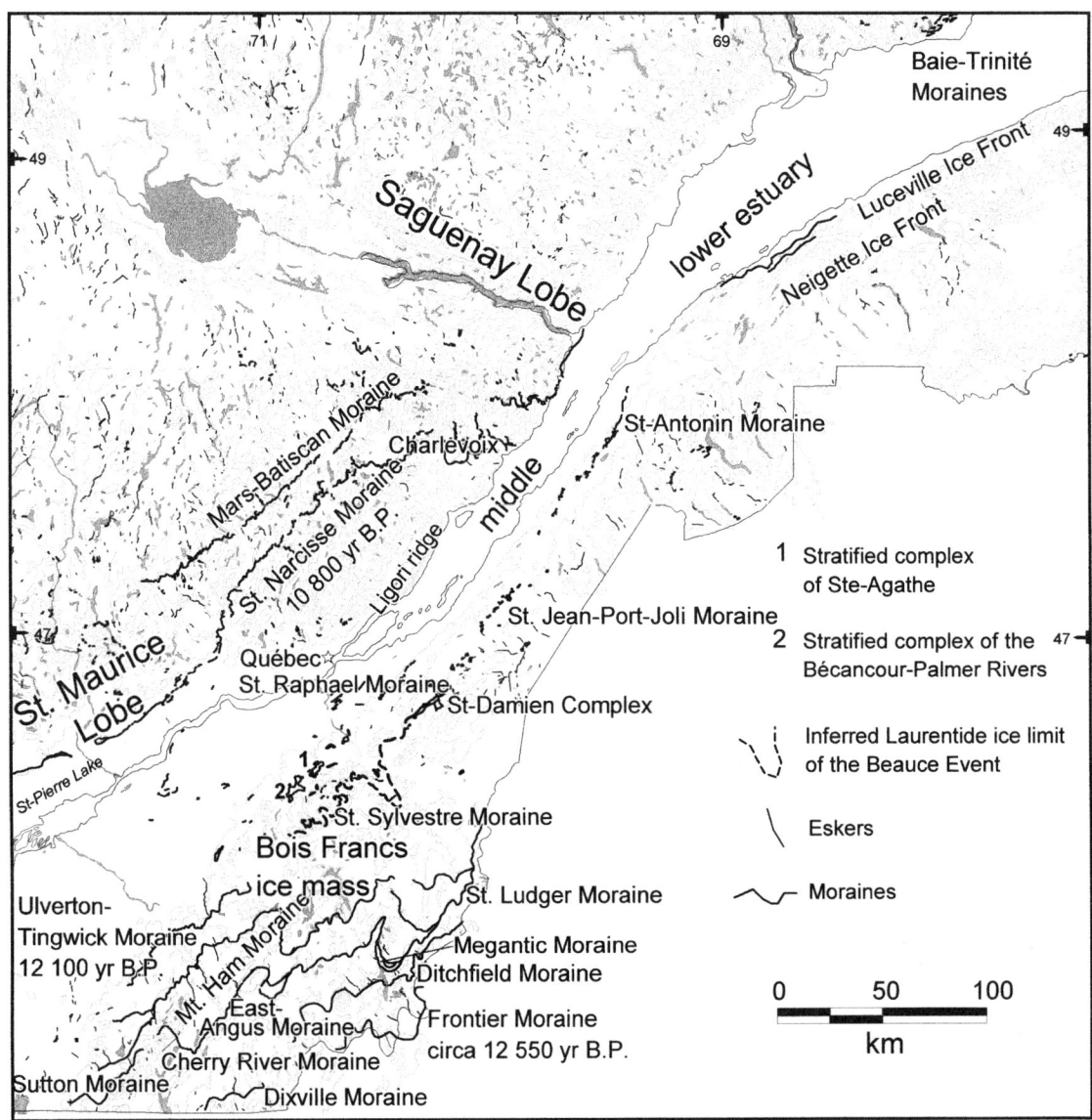

Figure 3. Recessional ice-front features and moraine belts in southern Québec.

of the Laurentide ice sheet, on rates of ice-marginal retreat, or on the snow budget of peripheral domes such as the Appalachians are roughly coincident. Changes in storm-track patterns induced by these climatic fluctuations may have caused changes in snow accumulation patterns and hence on glacial outflow centers. Increased snow accumulation in peripheral regions may also explain why satellite ice caps persisted in the Appalachians after the Bölling-Alleröd interval (this hypothesis remains to be confirmed). Correlations between stages of deglaciation of the eastern Laurentide ice sheet and global climatic fluctuations have generally been difficult to establish, except perhaps the correlation between the Saint-Narcisse moraine and the Dryas III (LaSalle, 1966; LaSalle and Elson, 1975). The persistence of tundra conditions after deglaciation in southeastern Québec (Richard, 1994a) has greatly hindered the recognition of such climatic fluctuations in palynostratigraphic records. For example, in spite of efforts by several researchers (Richard, 1977, 1994a, 1994b; Mott, 1977), the Alleröd-Dryas cold episode dated as 12 570 ± 220 yr B.P. (Geological Survey of Canada [GSC] GSC-419) and recognized by LaSalle (1966) and Terasmae and LaSalle (1968) at Mont Saint-Hilaire could not be identified in other pollen records. That this episode occurred almost in the middle of the Champlain sea episode did not help. In this chapter we postulate that, once nonclimatic signals (e.g., the St. Lawrence ice stream or calving bay ablation) are filtered out and the long response time of the Laurentide ice sheet is taken into account (e.g., global decreasing budget of the ice sheet), most of the ice-marginal fluctuations and most of the glaciodynamic changes are actually responses to the main global climatic fluctuations. These correlations have been made

TABLE 1. EVOLVING CONCEPTS AND MODELS OF DEGLACIATION IN SOUTHERN QUÉBEC

| Processes and late glacial features | Authors |
|---|---|
| Appalachian northward ice striations | Chalmers, 1898; Clark, 1937; Cooke, 1937; Lamarche, 1971, 1974 |
| Ice divide in the Appalachians | Shilts, 1981 |
| • Quebec Ice Divide | Shilts, 1981 |
| • Northwest Maine Ice Divide | Lowell, 1985 |
| South to north ice retreat in Eastern Townships of Québec | McDonald, 1967, 1968; Shilts, 1970, 1981; Parent et al., 1985 |
| Stagnant ice over the Bois Francs area | Parent and Occhietti, 1988 |
| Late Laurentide ice reactivation in the Beauce area | Gadd, 1964b |
| • After deglaciation of the Chaudière Valley (Highland Front Moraine System) | Blais, 1989; Blais and Shilts, 1992a, 1992b |
| • With stagnant ice in the Beauce area (Beauce event) | This chapter |
| Evolving concept of the glacial features along the Appalachian piedmont | |
| • Highland front moraine system (informal) | Gadd, 1964a |
| • Highland Front Moraine System | Gadd et al., 1972 |
| • Subdivision in diachronous features | |
| —eskers and Sutton Moraines | Prichonnet, 1984 |
| —eskers, Ulverton-Tingwick Moraine | Parent, 1987; Parent and Occhietti, 1988 |
| —recessional diachronous features in the lower Chaudière Valley, described previously by Gadd, 1964a, 1978 and Blais, 1989: St-Sylvestre Moraine, parts of the St-Raphael Moraine, Bécancour-Palmer drift complex | This chapter |
| —recessional diachronous features in the Bas du Fleuve area: | Chauvin et al., 1985 |
| St-Damien Moraine complex | Chauvin et al., 1985 |
| St-Antonin Moraine and St-Jean-Port-Joli Moraine | Lee, 1962; Chauvin et al., 1985 |
| St-Raphael Moraine (features partly described by Gadd, 1964a; Chauvin et al., 1985) | This chapter |
| Evolving model of deglaciation in the middle estuary of St. Lawrence between the Saguenay River and Québec City (western arm of Goldthwait Sea) | |
| • Calving bay model applied to the Gaspé area | Antevs, 1925; Lebuis and David, 1977 |
| • Calving bay hypothesis developed | Thomas, 1977 |
| • and applied to the St. Lawrence Valley, based on old $^{14}$C ages from marine shells in the Ottawa area (these $^{14}$C ages are now rejected) | Gadd, 1980 |
| • Upstream limit of the calving bay in the St-Antonin Moraine area and model of deglaciation along the Appalachian piedmont in the upper part of the middle estuary | Chauvin et al., 1985, and see Rappol, 1989 |
| • Converging ice toward the St. Lawrence ice stream and slow ice retreat in the middle estuary (no calving bay model) | Occhietti et al., 1996, 1997; this chapter |
| —eastward and northeastward striations interpreted as the result of converging ice flow along the north shore of the middle and upper estuary | Lanoie, 1995; Occhietti et al., 1996, 1997; Fournier, 1998 |
| —upstream migration of the convergence head of the St. Lawrence | Occhietti et al., 1996, 1997; Parent and Occhietti, 1999 |
| —northward ice flow in the north shore of the St. Lawrence ice stream | Lanoie, 1995; Fournier, 1998; Occhietti et al., 1996, 1997; Paradis and Bolduc, 1999 |
| —late ice in the northern part of the middle estuary | Occhietti et al., 1996, 1997; Fournier, 1998; Dionne and Occhietti, 1996 |

possible by new detailed studies as well as by several new $^{14}$C dates. We are aware of the long response time of large glaciers to external forcings such as insolation or atmospheric-oceanic circulation, yet we estimate that these delayed responses are generally less than or of the same duration as the error term of $^{14}$C conventional ages.

We postulate that in eastern North America, the marked warming of the Bölling followed by the Alleröd had several consequences: increasingly negative mass balance, ice-marginal thinning, increased rates of ice retreat, and increased meltwater flux and sedimentation in ice marginal areas. Climatic coolings, such as the Dryas I and II and intra-Alleröd cold episode (Stuiver et al., 1995; Yu and Eicher, 1998), may be recorded by a slower ice-marginal retreat, a halt, or a readvance with corresponding changes of ice-marginal lithofacies assemblages. Correlations between regional features and global changes are based on several internally consistent and converging criteria.

### Nonclimatic factors

Deglaciation of the southeastern margin of the Laurentide ice sheet was also strongly influenced by several nonclimatic, mainly topographic, factors.

**Glacial context from the glacial maximum to topographically controlled deglaciation.** The southeastern margin of the Laurentide ice sheet, including the Appalachian sector, the St. Lawrence corridor, and the margin of the Laurentian highlands were sufficiently far from the New Quebec dome to form an almost flat plateau during the earliest warming stages ca. 17 ka. The overall deglacial context is that of a receding ice margin with flow lines initially connected with the Labrador sector and

## TABLE 2. RADIOCARBON AGES

| Laboratory number | Corrected age (yr B.P. published or calculated $\delta^{13}C = 0‰$) | Conventional age (yr B.P. published or calculated $\delta^{13}C = -25‰$) | Calibrated age B.P. (2 σ)* | Locality | Dated material | Reference |
|---|---|---|---|---|---|---|
| **North coast of Gaspé Peninsula** | | | | | | |
| QU-1117 | N.P.# | 13 420 ± 220 | 15 550 +670, −1230 | Anse-au-Griffon | Unidentified shells | Allard and Tremblay, 1981 |
| GSC-2511 | 12 500 ± 140 | 12 910 ± 140 | 14 340 +1170, −630 | Anse-au-Griffon | Unidentified shells | Allard and Tremblay, 1981 |
| Beta-60895 | N.P.# | 13 020 ± 90 | 14 700 +860, −980 | Ruisseau-à-Rebours | Unidentified shells | Hétu, 1998 |
| UQ-1084 | N.P.# | 13 850 ± 200 | Non valid | Ruisseau-à-Rebours | Unidentified shells | cf. Hétu, 1998 |
| GSC-2376 | 12 700 ± 170 | 13 110 ± 170 | 14 590 +1170, −870 | Rivière-au-Renard | Unidentified shells | Allard and Tremblay, 1981 |
| GSC-4545 | 12 200 ± 110 | 12 610 ± 110 | 14 100 +1140, −430 | Nouvelle | | Veillette and Cloutier, 1993 |
| QU-275 | N.P.# | 13 890 ± 160 | Non valid | Nouvelle | | Lebuis and David, 1977 |
| QU-83 | N.P.# | 13 580 ± 350 | 15 730 +930, −1500 | Sainte-Félicité | *Hiatella arctica* | Lebuis and David, 1977 |
| QU-84 | N.P.# | 13 450 ± 470 | 15 580 +1180, −1850 | Sainte-Félicité | *Hiatella arctica* | Lebuis and David, 1977 |
| QU-85 | N.P.# | 13 540 ± 300 | 15 680 +830, −1370 | Baie-des-Capucins | Unidentified shells | Lebuis and David, 1977 |
| **Rimouski area** | | | | | | |
| QU-264 | N.P.# | 13 360 ± 320 | 15 500 +840, −1360 | Neigette valley | *Hiatella arctica* | Locat, 1977 |
| GSC-4726 | 12 700 ± 130 | 13 110 ± 130? | 14 590 +1110, −870 | | Unidentified shells | Rappol, 1993 |
| GSC-1186 | 12 600 ± 160 | 13 010 ± 160 | 14 715 +910, −1000 | | *Hiatella arctica* | Dionne, 1972 |
| Beta-42949 | N.P.# | 11 990 ± 80 | 13 450 +360, −405 | | *Hiatella arctica* | cf. Hétu, 1998 |
| GSC-4698 | 13 900 ± 170 | 14 310 ± 170? | 16 570 +620, −580? | Rimouski | Unidentified shells | Rappol, 1993 |
| Beta-47286 | N.P.# | 12 640 ± 90 | 14 110 +1140, −430 | Bic | Unidentified shells | Dionne and Coll, 1995 |
| GSC-4707 | 12 400 ± 100 | 12 810 ± 100 | 14 300 +1110, −590 | | Unidentified shells | Rappol, 1993 |
| Beta-58564 | N.P.# | 11 230 ± 150 | 12 865 +875, −720 | | Unidentified shells | cf. Hétu, 1998 |
| UL-1 193 | N.P.# | 14 170 ± 150 | Non valid | Saint-Fabien | Unidentified shells | Dionne and Coll, 1995 |
| QU-271 | N.P.# | 13 390 ± 690 | 15 520 +1650, −2090 | | *Hiatella arctica* | Locat, 1977 |
| TO-4637 | N.P.# | 13 240 ± 90 | 15 380 +430, −1160 | | *Mya truncata* | Dionne and Coll, 1995 |
| Beta-48532 | N.P.# | 12 640 ± 90 | 14 110 +1140, −430 | | *Mya t. + Hiatella a.* | Dionne and Coll, 1995 |
| Beta-28296 | N.P.# | 12 570 ± 210 | 14 090 +1220, −650 | | *Mya t. + M. c. + H. a.* | Dionne and Coll, 1995 |
| QU-270 | N.P.# | 12 300 ± 260 | 13 820 +1280, −760 | Mitis valley | Unidentified shells | Locat, 1977 |
| UQ-1081 | | 13 400 ± 200 | 15 530 +630, −1200 | | *Hiatella arctica* | Prichonnet, 1995 |
| **Trois-Pistoles Rivière-du-Loup area** | | | | | | |
| GSC-102 | 12 720 ± 170 | 13 130 ± 170 | 14 560 +1230, −830 | Trois-pistoles | | Lee, 1962 |
| TO-948 | N.P.# | 12 450 ± 160 | 13 930 +1200, −495 | Rivière-du-Loup | | Rappol, 1993 |
| TO-947 | N.P.# | 11 720 ± 160 | 13 160 +630, −290 | Rivière-du-Loup | | Rappol, 1993 |
| **Bas du Fleuve, upstream from Rivière-du-Loup** | | | | | | |
| UL-1116 | N.P.# | 11 320 ± 130 | 12 900 +850, −520 | Kamouraska | *Macoma calcarea* | Dionne and Occhietti, 1996 |
| QU-565 | N.P.# | 10 920 ± 90 | 12 520 +370, −950 | Mont-Carmel | Unidentified shells | Lortie and Guilbault, 1984 |
| QU-492 | N.P.# | 10 900 ± 150 | 12 540 +355, −1050 | Morigeau | *Balanus hameri* | cf. Dionne and Occhietti, 1996 |
| QU-403 | N.P.# | 10 790 ± 120 | 12 000 +870, −610 | Trois-Saumons | *Balanus* sp. | Lortie and Guilbault, 1984 |
| **Québec City area** | | | | | | |
| GSC-1533 | 12 400 ± 160 | 12 810 ± 160 | 14 300 +1160, −600 | Charlesbourg | *Portlandia arctica* | cf. LaSalle and Shilts, 1993 |
| QU-93 | N.P.# | 12 230 ± 250 | Non valid | St-Henri-de-Lévis | *Hiatella arctica* | cf. LaSalle and Shilts, 1993 |
| Beta-11587 | 10 670 ± 70 | 11 080 ± 80 | 12 740 +180, −970 | St-Henri-de-Lévis | id. | This chapter |
| GSC-1235 | 11 600 ± 160 | 12 010 ± 160 | 13 460 +390, −420 | N-D-des-Laurentides | *Mya truncata* | cf. LaSalle and Shilts, 1993 |
| GSC-1295 | 11 200 ± 160 | 11 610 ± 160 | 13 140 +640, −470 | Ste-Anne-de-Beaupré | *Balanus* sp. | cf. LaSalle and Shilts, 1993 |
| GSC-1232 | 11 100 ± 160 | 11 510 ± 160 | 13 000 +760, −350 | Beauport | *Balanus* sp. | cf. LaSalle and Shilts, 1993 |
| **Southeastern area of Champlain Sea** | | | | | | |
| I-13342 | N.P.# | 11 700 ± 170 | 13 160 +630, −470 | Warwick, Québec | *Hiatella arctica* | cf. Parent and Occhietti, 1988 |
| GSC-505 | 11 880 ± 180 | 12 290 ± 180 | 13 820 +485, −500 | L'avenir, Québec | *Macoma balthica* | Lowdon and Blake, 1970 |
| GSC-936 | 12 000 ± 230 | 12 410 ± 230 | 13 950 +1230, −550 | L'avenir, Québec | id. (50% leached) | Lowdon and Blake, 1970 |
| GSC-2338 | 11 900 ± 120 | 12 310 ± 120 | 13 820 +470, −410 | Peru, New York | *Macoma balthica* | Lowdon and Blake, 1979 |
| GSC-2366 | 11 800 ± 150 | 12 210 ± 150 | 13 660 +610, −450 | Plattsburg, New York | *Macoma balthica* | Lowdon and Blake, 1979 |
| **Terrestrial plant material** | | | | | | |
| I-8841 | | 11 400 ± 340 | 13 410 +1570, −750 | Shefford, Québec | | Richard, 1977 |
| GSC-1294 | | 11 200 ± 160 | 13 155 +620, −470 | Lac Dufresne, Québec | | Gadd et al., 1972 |
| GSC-1248 | | 11 200 ± 200 | 13 155 +640, −485 | Bondary Pond, Maine | | Gadd et al., 1972 |
| GSC-2282 | | 11 100 ± 180 | 13 130 +630, −480 | Lac Colin, Québec | | Mott, 1977 |
| GSC-420 | | 11 020 ± 330 | 13 010 +800, −850 | Barnston Lake, Québec | | Mott, 1977 |
| GSC-1289 | | 11 000 ± 240 | 13 000 +760, −610 | Mont Sainte-Cécile, Québec | | Gadd et al., 1972 |
| I-8141 | | 10 880 ± 160 | 12 920 +250, −490 | Albion, Québec | | Richard, 1977 |

*Calibration using DOS/Windows/NT 4 CALIB (Stuiver and Reimer, 1993; Stuiver and Brazunias, 1998; Stuiver et al., 1998).
#N.P. = not published.

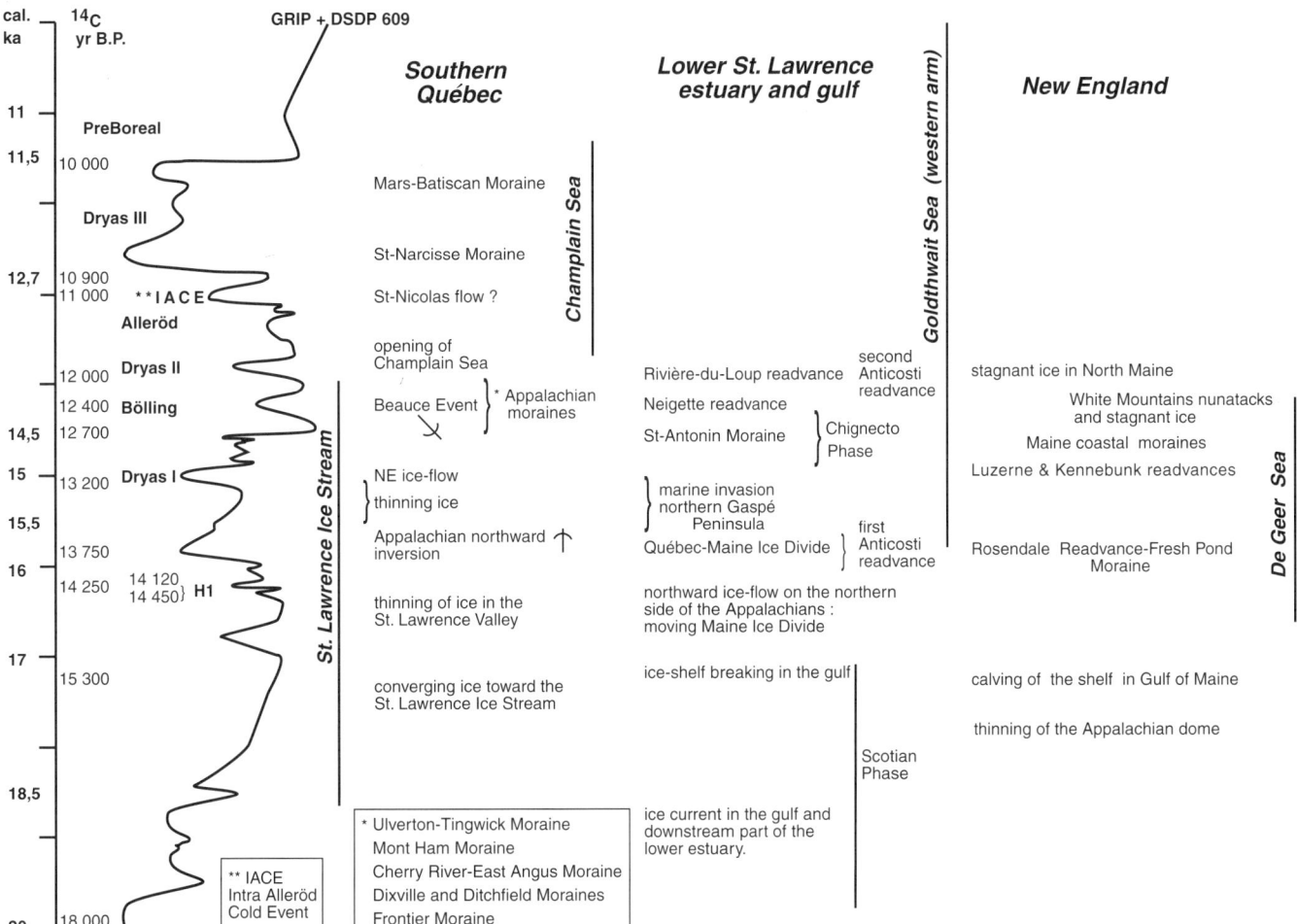

Figure 4. Sequence of events in southern Québec related to events in New England and Gulf of St. Lawrence and lower estuary regions. Arrows indicate ice-flow directions. Calibrated time scale and climatic curve are established from Bond et al. (1993), Johnsen et al. (1992), and Dansgaard et al. (1993). Time scale ($^{14}$C yr B.P.) is mainly based on corrected ages from marine shells ($\delta^{13}$C = 0‰). Possible time discrepancies may be due to local reservoir effects in postglacial Champlain and Goldthwait seas and to anomalous ages from basal lacustrine sediments.

subsequently with the New Quebec dome after the differentiation of the Appalachian ice masses (Stea et al., 1998), and, after the deglaciation of southern Québec, of the Hudson Bay ice mass (e.g., Dyke and Prest, 1987).

Subsequent deglaciation was characterized by: (1) generalized glacial thinning, with coeval restrained glacioisostatic rebound; (2) the increasing role of subglacial topography on glacial dynamics; and (3) the local emergence of nunataks (Mount Katahdin in Maine; Caldwell et al., 1985) followed by emergence of the highlands of the Appalachian ridge (Boundary mountains; Borns and Calkin, 1977) and Monteregian hills in southern Québec (LaSalle, 1966).

*Major topographic control in the southeastern margin of the Laurentide ice sheet: The St. Lawrence corridor.* Below a certain ice thickness and in conjunction with accelerated calving in the Gulf of St. Lawrence, a major northeast-trending ice stream formed within the ice sheet (Occhietti et al., 1996, 1997; Parent and Occhietti, 1999).

Diachronously, between ca. 17 and 14 ka (from 15 000 to 12 000 $^{14}$C yr B.P., Fig. 2, B–F), the head of the St. Lawrence ice stream migrated 1000 km along the axis of the St. Lawrence corridor deep into the Laurentide ice sheet. This major feature of the Laurentide ice sheet was characterized by flow rates that were at least one order of magnitude higher than in adjacent ice masses, similar to present flow rates in Greenland (Lliboutry, 1965). This accelerated ablation regime (ice stream and iceberg calving) caused substantial thinning within the catchment area of the ice stream, particularly on the northwest flank of the Appalachian uplands (Genes et al., 1981; Lowell, 1985) and along the southern margin of the Laurentian highlands (Fournier, 1998). Instead of the expected concentric marginal ablation pattern in the southeastern margin of the Laurentide ice sheet (concentric to the New Quebec dome), the formation of the St. Lawrence ice stream favored the progressive isolation of Appalachian ice masses in New Brunswick and northern Maine, in the Gaspé Peninsula, and in the Notre Dame Mountains of southern Québec (Fig. 2, B–F).

The ice-mass disequilibrium generated by the St. Lawrence ice stream strengthened the role of regional topographic control on regional glacial dynamics and increased the sensitivity of some ice margins to climatic fluctuations.

***Role of regional topography in southern Québec.*** Mostly allochthonous ice masses over the Appalachians were progressively isolated from the New Quebec dome, at first as a result of an ice-flow reversal (Figs. 2C and 5D) and then by a marine embayment that cut them off from the main ice sheet (Fig. 2E). This two-fold sequence of events occurred in rapid succession between 14 000 and 12 000 yr B.P., from the Gaspé Peninsula to the Bois-Francs region (Allard and Tremblay, 1981; Lebuis and David, 1977; David and Lebuis, 1985; Locat, 1977; Genes et al., 1981; Lowell, 1985; Lowell et al., 1986, 1990; Newman et al., 1985; Rappol, 1993; Martineau and Corbeil, 1983; Chauvin et al., 1985; Dionne, 1977) (Fig. 2, C–F). This separation, initially caused by the dynamics of the St. Lawrence ice stream, led to the formation of independent ice domes in the Gaspé Peninsula, in New Brunswick, in northern New England, and adjacent Québec (Lowell et al., 1986; Stea et al., 1998) (Fig. 2C). The latter dome became increasingly autonomous as a result of recession through coastal Maine and ice flow toward the St. Lawrence at its northern margin. The eastern part of this dome remained connected with the New Brunswick dome (Rappol, 1989), while its western part was characterized by rapid ablation and by flow convergence toward the Hudson-Champlain Valley (Connally and Sirkin, 1973; Connally, 1982; Hughes et al., 1985) (Fig. 2D). This late differentiation of the ice mass over the Appalachians was the result of topographic, glaciodynamic, and climatic factors. It is possible that higher rates of snow accumulation on the New England dome caused initially reduced rates of ablation of this disconnected Laurentide ice. However, at the height of the Bölling warm interval (ca. 12 700–12 000 yr B.P.) the overall budget must have been strongly negative. The proposed mode of deglaciation in the Appalachians is that of a low mountain ice cap, the sectors of which were differentiated on the basis of distance to Laurentide ice masses, particularly in the western sector, and according to regional topography. Hence the high ridges of the White and Notre Dame Mountains progressively and discontinuously disconnected ice masses on their northwestern flank in Québec from the inactive masses on their southeastern flank in Maine (Shilts, 1981; Genes et al., 1981; Lowell, 1985). Three sectors can be recognized on the northwestern flank of the mountains (Fig. 3): (1) southwestern ice masses that were still connected with the main Laurentide ice sheet, (2) the Bois Francs residual ice cap, which was characterized by widespread downwasting and stagnation (Parent and Occhietti, 1988; LaSalle and Chapdelaine, 1990), and (3) the northeastern sector, which was affected by late glacial flow toward the St. Lawrence River.

During the last deglaciation, thinning of the ice along the Appalachian piedmont favored early deglaciation and incursion of the western arm of the Goldthwait Sea between the main ice sheet and the piedmont. This mode of glacial retreat does not require the development of a calving bay across the axis of the corridor; on the contrary, it implies calving along the axis of the corridor. This mode of deglaciation came to an end when Lake Candona finally drained along the Appalachian piedmont near Plessisville-Laurierville (Figs. 1 and 6D), allowing marine waters to enter the central St. Lawrence lowlands (Parent and Occhietti, 1988, 1999). Further evidence for this sequence of events is provided by the occurrence of channelled gravels on the piedmont, by the absence of marine clays in that part of the piedmont, and by the presence of a northeast-trending recessional moraine ridge below the piedmont (Dubé, 1971). The major water gap of the Chaudière Valley adds further local complications in the overall deglaciation pattern.

The southern margin of the Laurentian highlands includes the Mont Tremblant highlands, the Saint-Maurice Valley, the Parc des Laurentides highlands, and the Saguenay fjord (Fig. 1). This physiographic context favored faster ice-marginal retreat in areas downglacier (south) of the highlands while arcuate lobes were maintained in the two main valleys (Occhietti, 1980; Govare, 1995). The St. Lawrence ice stream interfered with this topographic context because flow convergence toward the estuary from the eastern flank of the Parc des Laurentides highlands (Charlevoix) resulted in accelerated thinning while the ice sheet remained thick on the western flank of these highlands and in the Saint-Maurice Valley. On the north shore of the middle and upper estuary, major valleys (Fig. 1) of the Saguenay (Dionne and Occhietti, 1996), Malbaie (Poulin, 1977), du Gouffre (Govare, 1995), and Saint-Maurice (Occhietti, 1980) favored the persistence of late glacial lobes that readvanced locally, thus modifiying significantly the mode of deglaciation of the St. Lawrence Valley. These lobes and outlet glaciers provide a much more sensitive record of the mass balance of the Laurentide ice sheet than the ice masses in the St. Lawrence corridor.

Deglaciation patterns within the St. Lawrence corridor are poorly known, partly because many deglacial features were subsequently buried or partly buried by marine sediments or were extensively reworked during the episode. Three nonclimatic factors controlled deglacial dynamics: ablation by the ice stream, the topography along the margin of the Laurentian highlands, and glacial isostatic rebound. The northward pattern of thermolatitudinal glacial retreat implies that glacial isostatic rebound began earlier along the Appalachian piedmont than along the southern edge of the Laurentians. Glacial thinning generated by the ice stream also caused early crustal unloading along the axis of the St. Lawrence corridor and a delayed response from east to west (Dionne, 1977; Locat, 1977; Lebuis and David, 1977; Parent, 1987; Dionne, 1988; Dionne and Coll, 1995). During cold climatic intervals, reduced rates of glacial eustatic

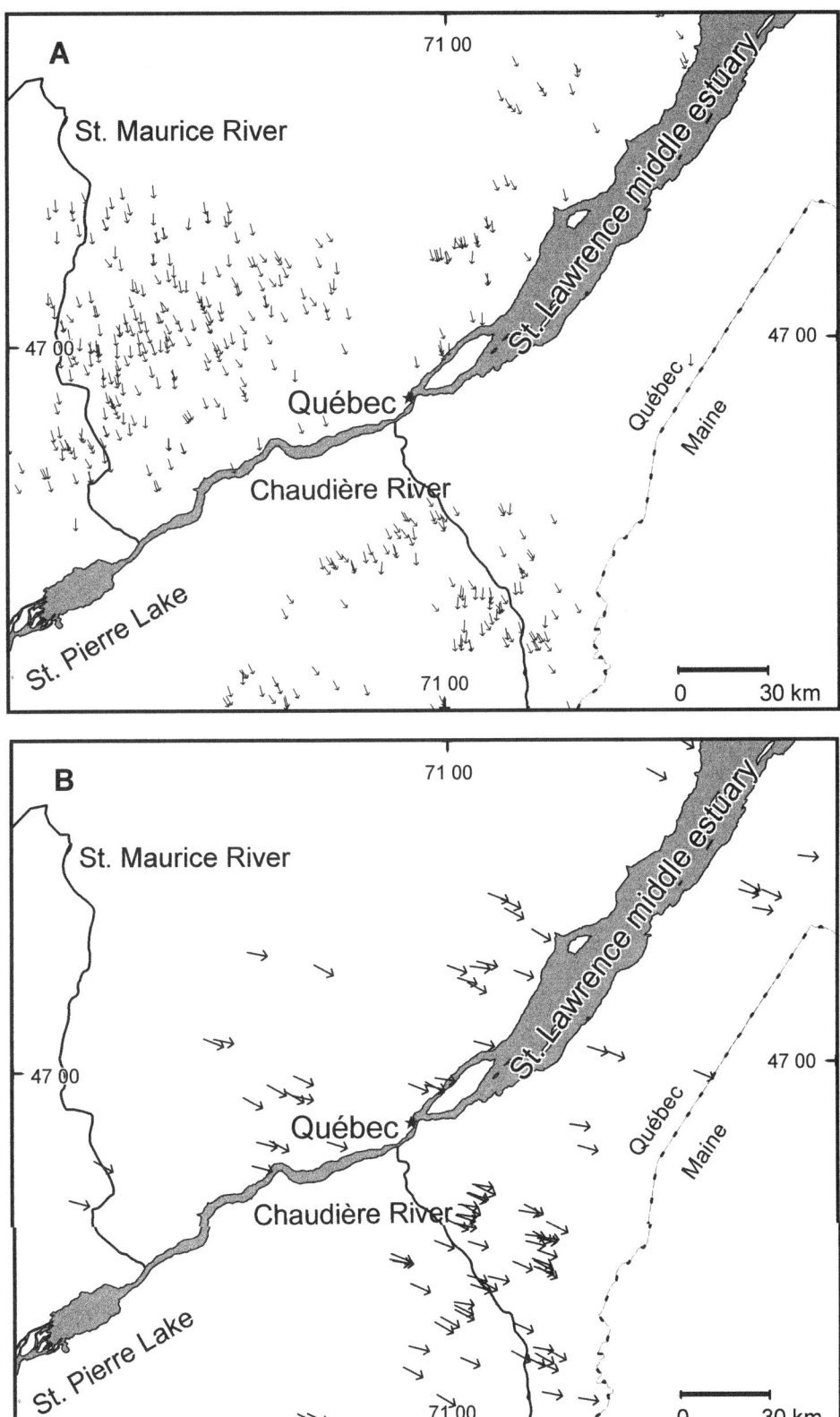

Figure 5 (this and next two pages). Diachronous groups of oriented glacial striations in southern Québec, from unpublished data banks of Atlas du Québec (Occhietti), Geological Survey of Canada in Québec (Parent and Bolduc), and Fournier (1998). Glacial striae included in data banks are collected from maps, theses, memoirs, and published and unpublished reports dating back a century.

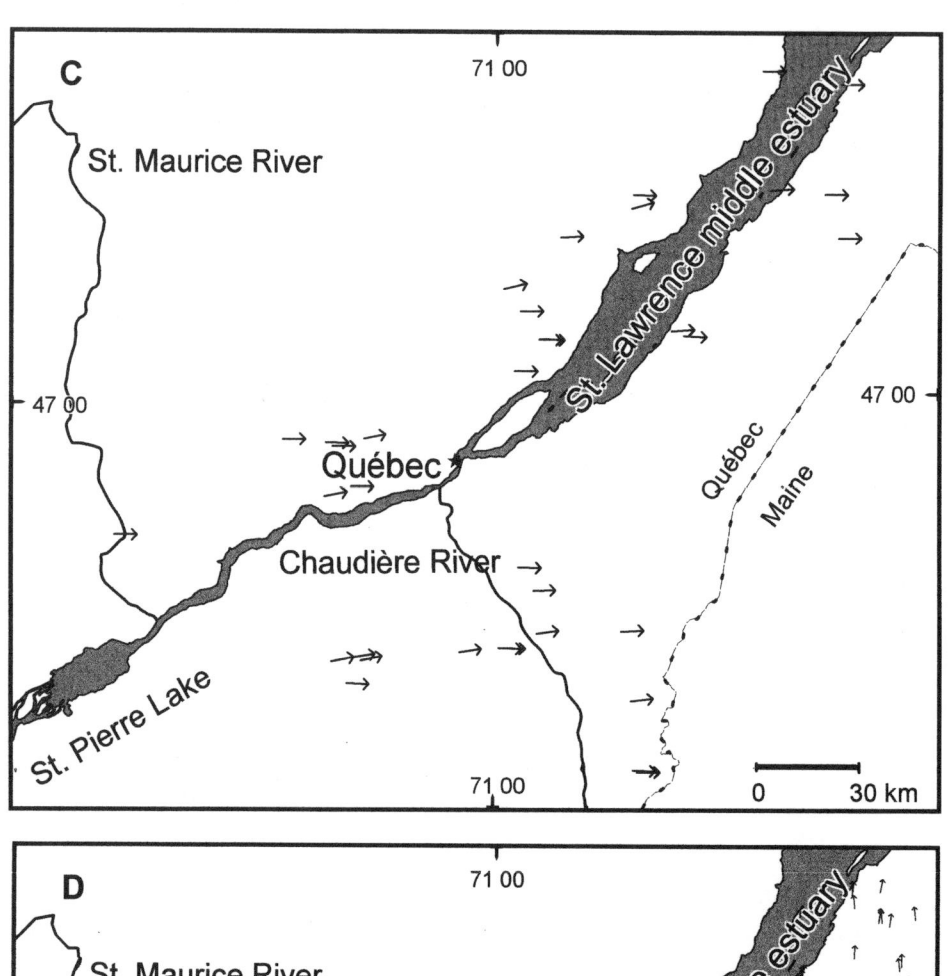
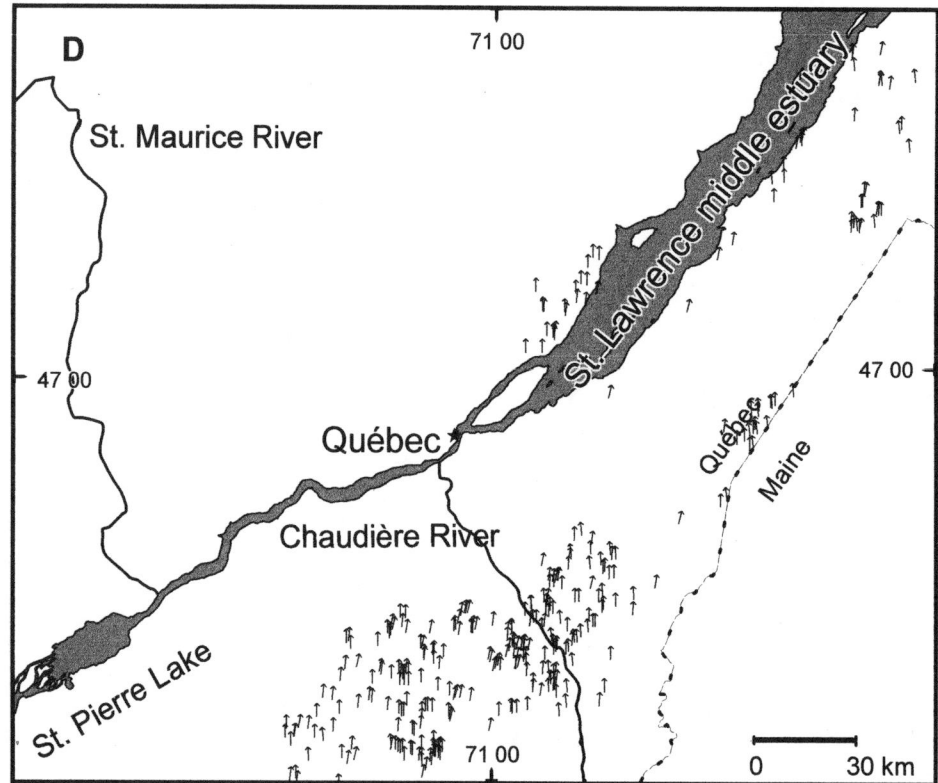

Figure 5. (*continued*)

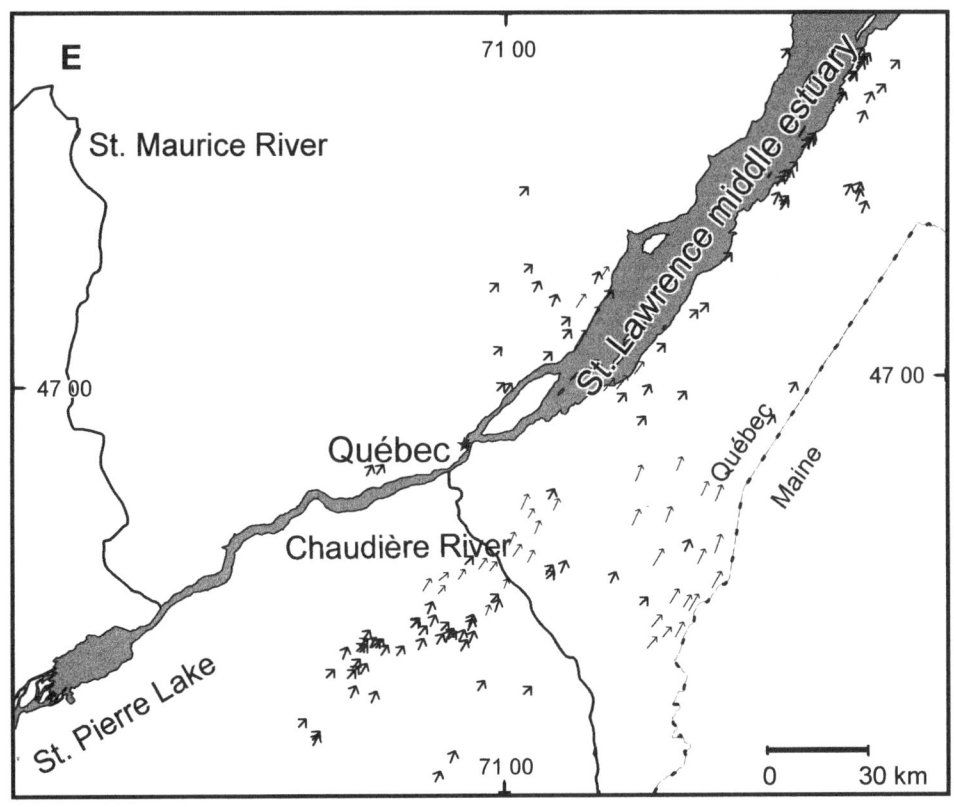
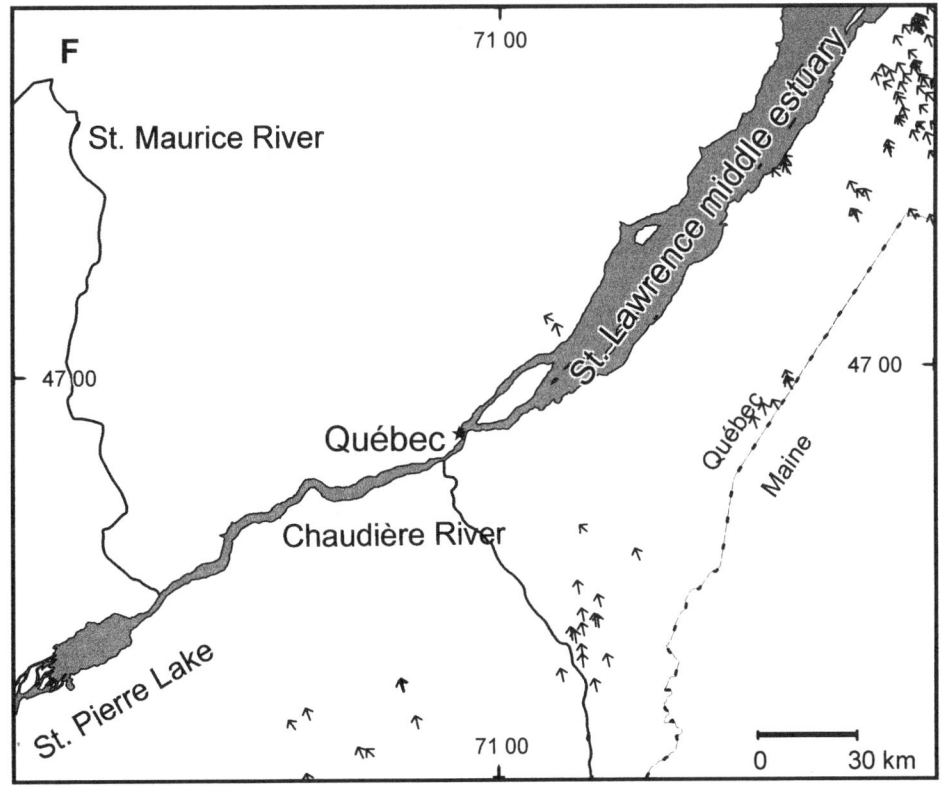

Figure 5. (*continued*)

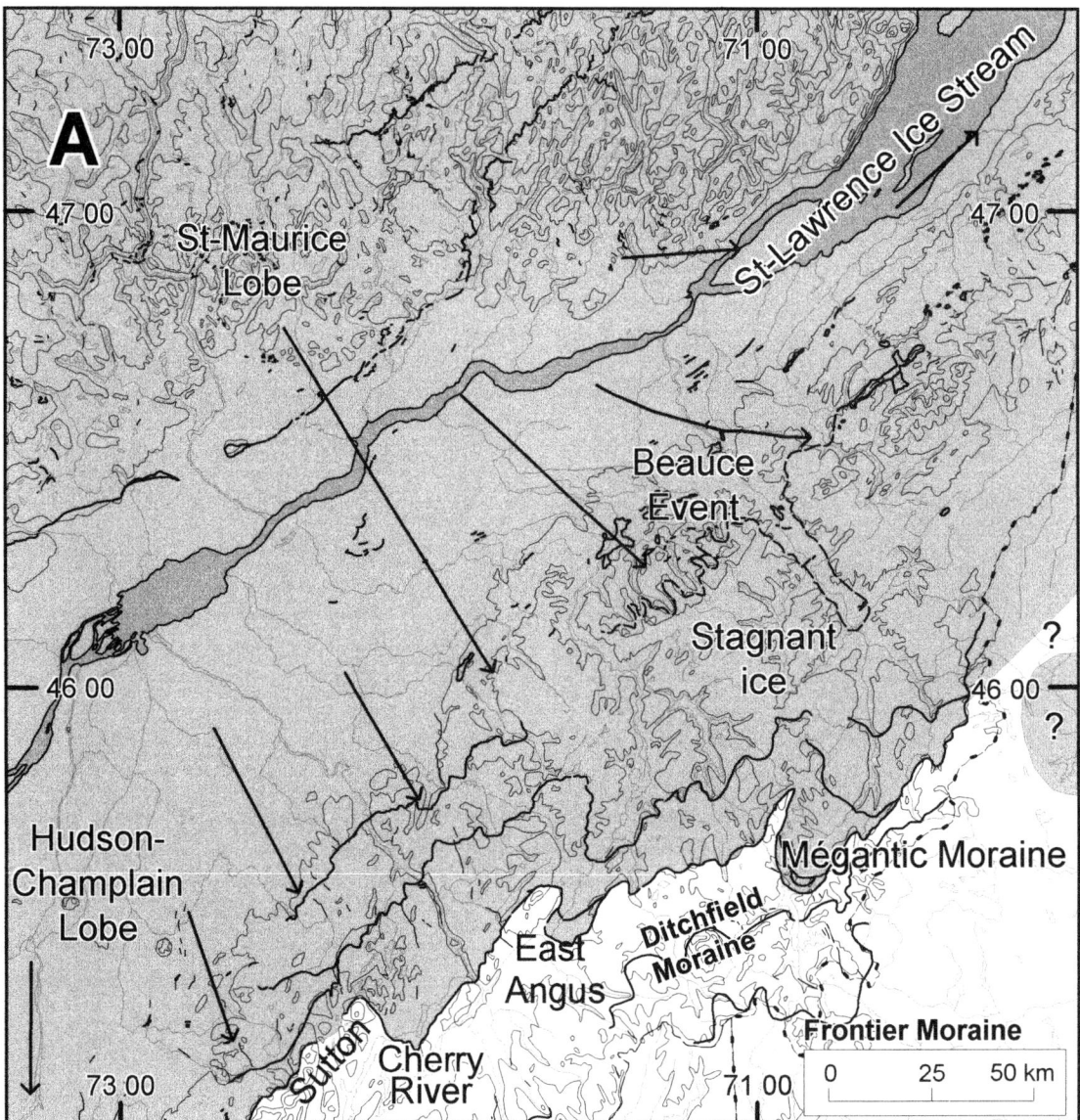

Figure 6 (this and next three pages). Reconstruction of deglaciation episodes in southern Québec. A, Episode of Sutton, Cherry River, East Angus, and Megantic ice front, ca. 12 325 yr B.P. Reactivation of Laurentide ice over part of Appalachians (Beauce event) is tentatively related to this episode, which was characterized by minor oscillations or pauses of local lobes in East Angus and Megantic depressions, and probably by thin inactive ice over Bois Francs area. B, Episode of Mont Ham Moraine, ca. 12 200 yr B.P. Ice was actively connected to Laurentide ice sheet in southwestern part and passively connected in Bois-Francs and Beauce areas. Emergence of nunataks in Notre-Dame Mountains is inferred (deglaciated areas might have been larger). High-level local lakes (areas in dark gray) inundated several deglaciated Appalachian valleys.

rise favored the grounding of floating ice margins. This factor, which certainly played a role during the Saint-Narcisse episode (Occhietti, 1980), has yet to be quantitatively documented.

*Role of local late glacial dynamics.* In nearly all regions, the latest phases of deglaciation were marked by the increased influence of local topographic effects and by changes in the physical properties of the ice masses. Local glacial dynamics became very sensitive to the migration of outflow centers and ice divides within the New Quebec dome as well as within Appalachian ice masses. Thinning at glacial margins led to the emergence of nunataks, to the progressive isolation of stagnant ice masses on parts of the highlands, to the sliding of ice masses toward valleys, and to the persistence of outlet glaciers or lobes in several valleys. The latest glacial event recorded in these valleys was the deposition of ice-contact or proglacial sediment bodies (e.g., the moraine belts in the Appalachians of southern

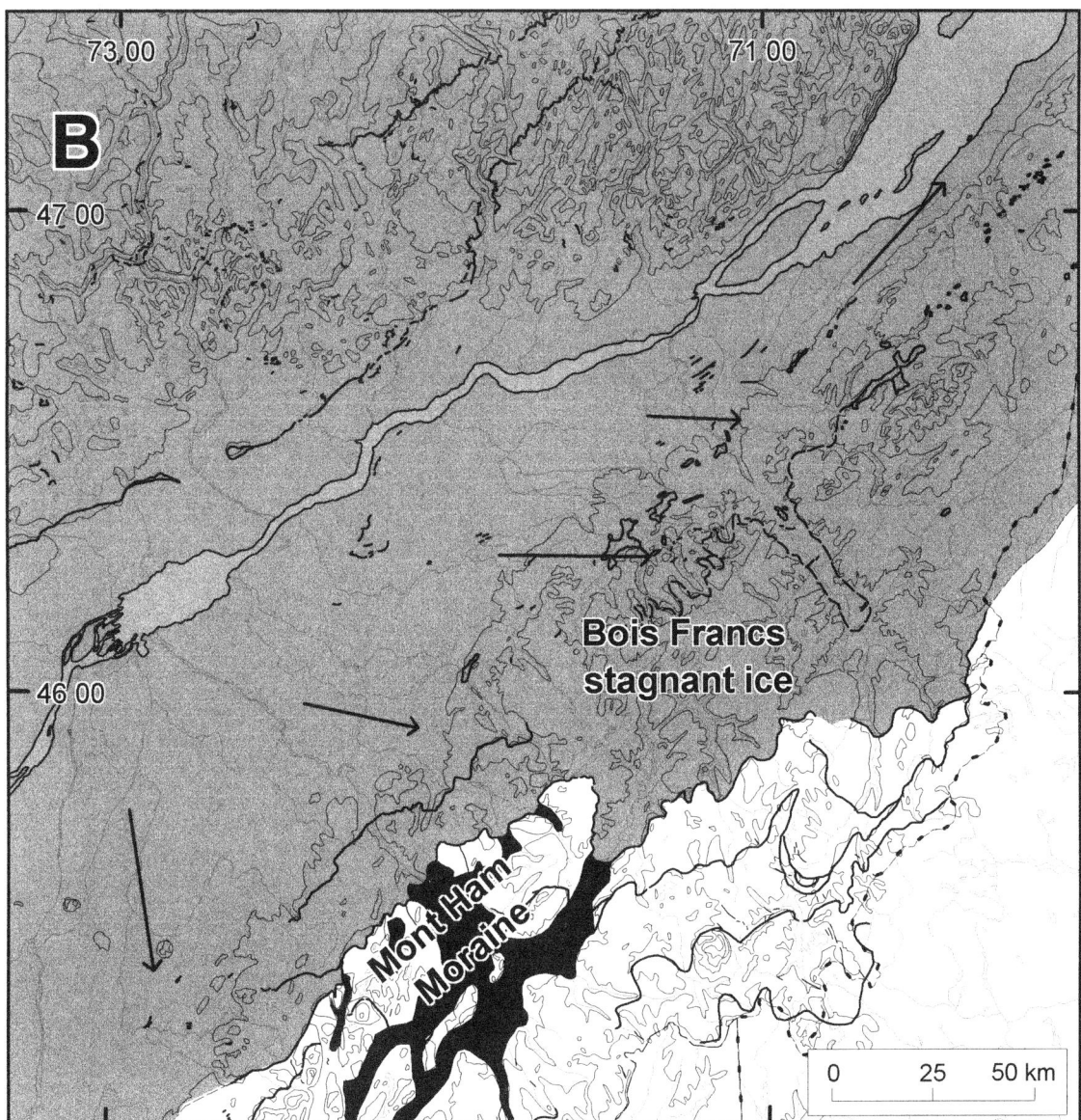

Figure 6. (*continued*)

Québec; McDonald, 1967; Shilts, 1981; Parent, 1987). The passage of cold-based to warm-based ice conditions led to increased rates of drift deposition. In such a context, the ultimate glacial fluctuations (active or passive ice) recorded by ice-frontal sediment bodies did not necessarily lead to the formation of corresponding glacial striae (e.g., in western Charlevoix; Fournier, 1998). There are numerous examples of valleys or basins that were not filled by glaciofluvial, glaciolacustrine, or marine sediments, thus implying the persistence of local stagnant ice bodies (e.g., lakes in the River Saint-Maurice middle reaches [Occhietti, 1980], and in the Bois Francs area).

## ST. LAWRENCE ICE STREAM AND REASSESSMENT OF THE CALVING BAY PROCESS

Stea et al. (1998) inferred that a local ice divide had already formed over Nova Scotia during the Scotian phase (18–15 ka) and that deglaciation of the Gulf of St. Lawrence began ca. 15 500 yr B.P. Regional deglaciation of the gulf was dominated by ice-shelf breaking and calving-bay dynamics, and this continued until ice margins were reactivated during the Chignecto phase, between 13 000 and 12 500 yr B.P. in Nova Scotia (Stea et al., 1998) and in western Anticosti Island (Gratton et al.,

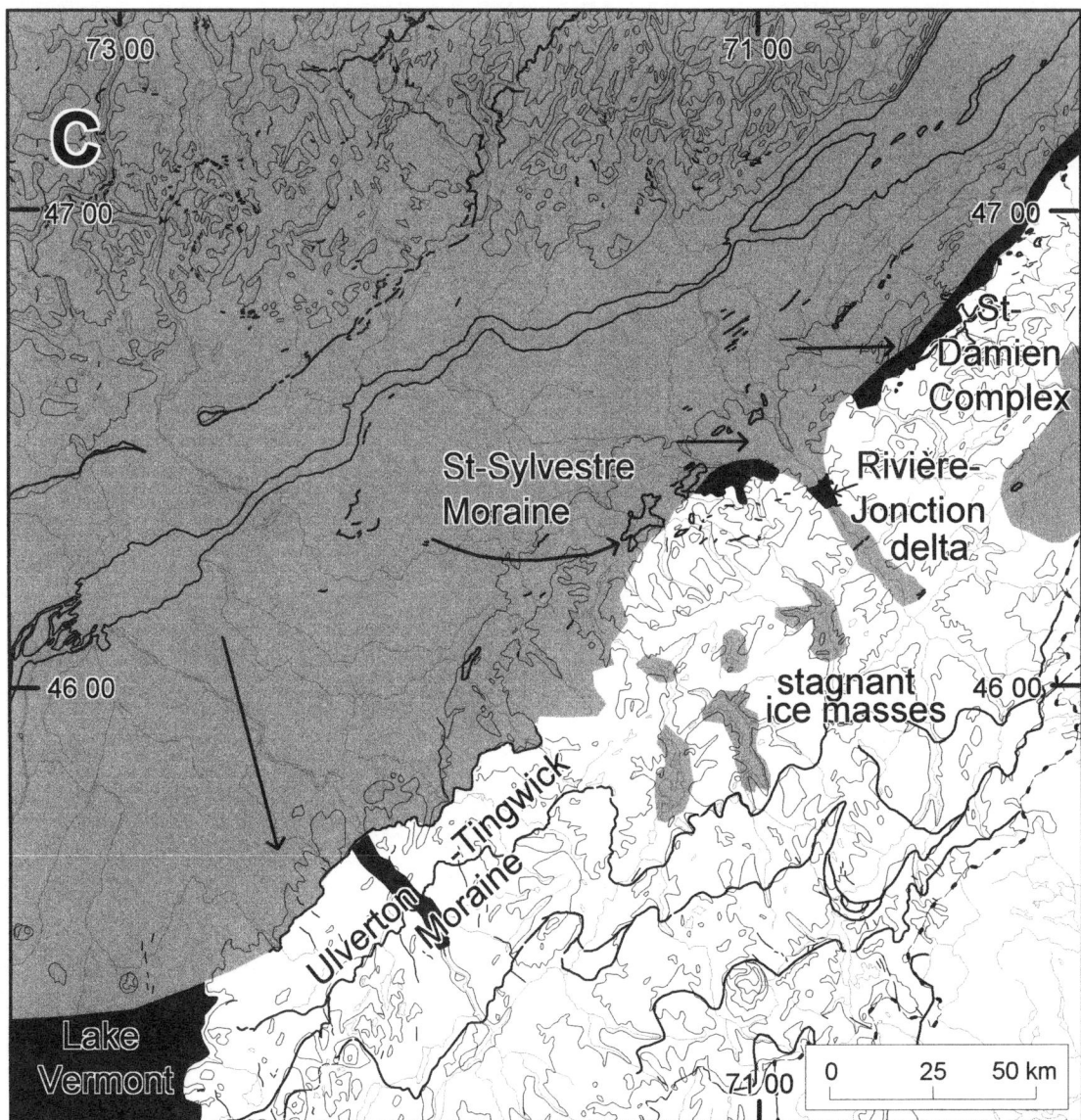

Figure 6. (*continued*). C, Episode of Ulverton-Tingwick Moraine, ca. 12 125 yr B.P. Notre Dame Mountains were probably deglaciated and residual stagnant ice masses remained in Bois Francs area. Saint-Sylvestre Moraine is tentatively related to this recessional episode. Glacial lake between Appalachian highlands and Laurentide ice is inferred from outwash deposits. D, Paleogeographical setting prior to Champlain Sea invasion, ca. 12 000 yr B.P. (age from marine shells). Laurentide ice abutted Appalachian piedmont in Bois Francs area. St-Raphael Moraine was last Laurentide ice-front feature before opening to marine waters of lower Chaudière area. Vast Bécancour-Palmer Rivers ice-contact sediments may have been deposited in Chaudière-Etchemin glacial lake, at margin of meltwater flowing along piedmont. Some remnant ice patches were probably remaining in local depressions of Appalachians.

1984; Painchaud et al., 1984). Several $^{14}$C dates suggest that the northern coastline of the Gaspé Peninsula was deglaciated between 13.9 and 13 ka (Table 2). On the north shore of the gulf, at Baie-Trinité, two series of end moraines were emplaced on the Pointe-des-Monts peninsula (Dredge, 1983; Dubois and Dionne, 1985; Vincent, 1989). The inner moraine has an estimated age of about 9 500 yr B.P. and is at the western end of the 800-km-long Québec North Shore Moraine system (Dubois and Dionne, 1985). Although the outer moraines at Baie-Trinité (Fig. 3) have yet to be dated directly, it seems reasonable to assume that they were emplaced during the preceding cold event, the Younger Dryas (10 800–10 300 yr B.P.), and may thus correlate with the Saint-Narcisse moraine, related to the Younger Dryas by LaSalle (1966). The Baie-Trinité moraines

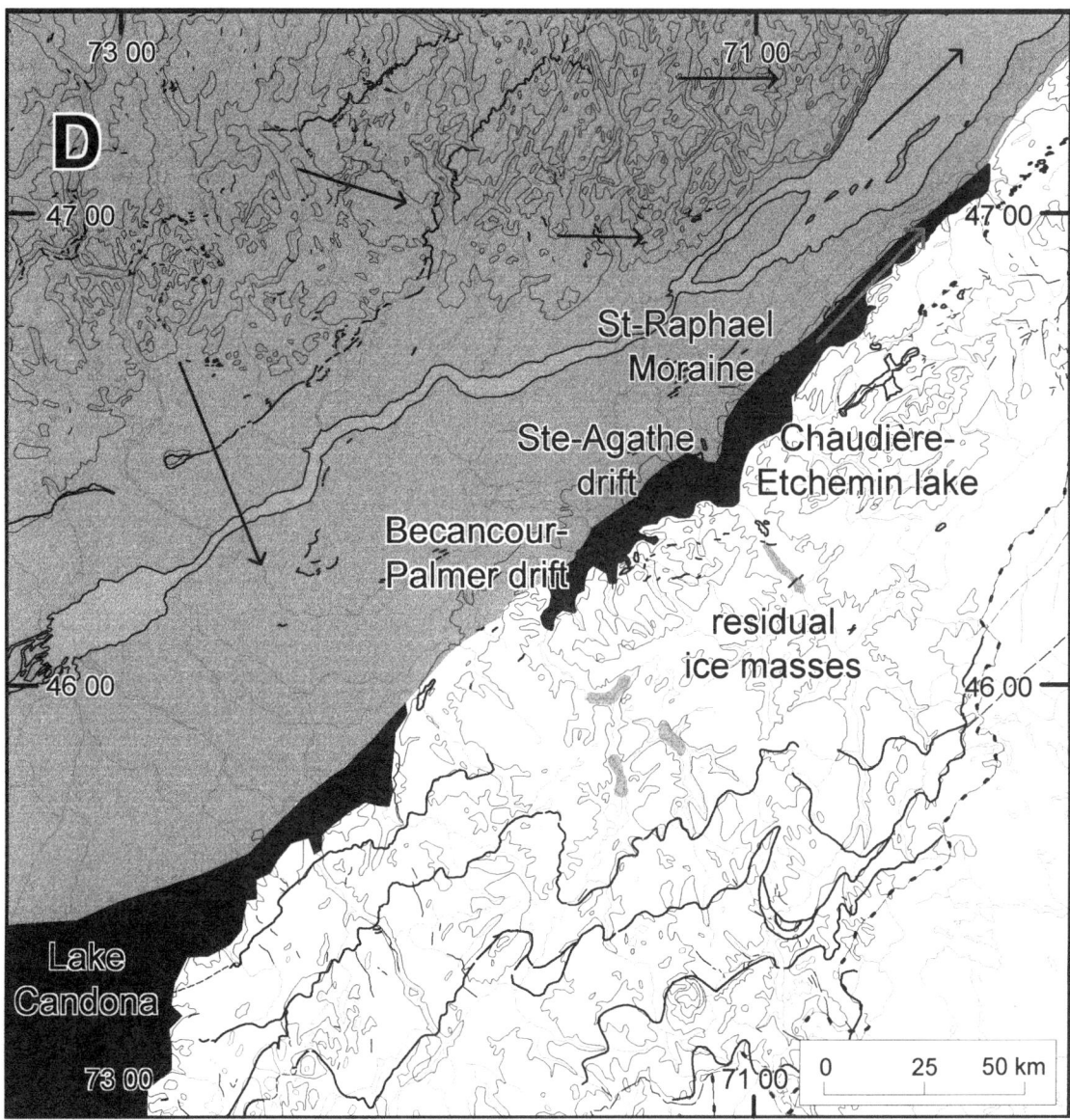

Figure 6. (*continued*)

indicate that the margin of the Laurentide ice sheet remained grounded until about 10 500 yr B.P. on the north shore of the gulf and lower estuary (Bernatchez, 1997).

Although a precise chronology has yet to emerge, the coupled St. Lawrence ice stream–calving bay framework includes several phases. These have been established on the basis of many earlier contributions (e.g., Hughes et al., 1985; Grant, 1989; Occhietti, 1989; Stea et al., 1998; Stea and Mott, 1998), as follows.

1. Initially (18 000–15 500 yr B.P.), the ice stream was channeled within the Laurentian trough to the edge of the continental shelf (Fig. 2, A and B).

2. An important ice-shelf breakup event ca. 15 500 yr B.P. is recorded by a large increase of hematite-coated grains in an Atlantic core (Bond and Lotti, 1995); this may also signal the onset of rapid calving-bay recession within the Laurentian trough (Stea et al., 1998). This early convergence of ice flow from Laurentide and Appalachian sources toward the maritime estuary may explain the relatively late age of its deglaciation (Fig. 2, B and C).

3. Between 14 000 and 13 000 yr B.P., the margin of the Laurentide ice sheet became grounded in western Anticosti Island (Gratton et al., 1984; Painchaud et al., 1984), but must have formed an ice shelf in the deeper parts of the lower estuary as the grounding line migrated upvalley quickly (Fig. 2, D and E). Deglaciation of the northern coast of the Gaspé Peninsula was favored by the warming effect of the piedmont. The ice stream must have had an arcuate outer margin because rates of

recession were higher on the south shore than on the north shore (Fig. 2E). This mode of ice retreat, which continued up to the head of the Laurentian trough, just downstream from the mouth of the Saguenay fjord, explains the difference in the rate of deglaciation between the north and south shores of the estuary, as well as significant thinning of the ice cover in highlands upglacier, such as the south margin of the Laurentians (Charlevoix) and the northern margin of the Appalachians.

4. In the middle estuary, where the Laurentian trough broke into narrow submarine valleys, available data (Dionne, 1972; Chauvin et al., 1985; Rappol, 1993) suggest that (1) a late marine incursion took place ca. 12 000 yr B.P. in a sea arm along the south shore, between the Appalachian piedmont and the ice front, and (2) the ice front retreated northward while maintaining glacial streaming in deeper channels of the middle estuary, followed by stagnation (Fournier, 1998). This model of deglaciation contrasts sharply with the previously held model of a migrating calving bay proposed by Thomas (1977) and Gadd (1980). Ice-contact sediment bodies flanking the north side of the ice stream provide key evidence favoring the northward retreating ice-front model, including: (1) paleocurrents in a glaciofluvial fan at Saint-Tite, at the west end of Charlevoix, trend toward northwest, and (2) ice-contact sediments overlying Precambrian terrain at the foot of nearby Mont Sainte-Anne contain sedimentary clasts with a southern provenance (Fournier, 1998).

## GLACIAL FEATURES ALONG THE APPALACHIAN PIEDMONT AND NEW PARADIGMS

In his early reconnaissance work (Table 1), Gadd (1964a, p. 1251) prudently introduced the concept of a " . . . highland front morainic system . . . along the north-facing flank of the Appalachian Highland of southern Quebec. . . . but this is not proposed as a formal name because stratigraphic work may show that another name, or indeed several names, may be more appropriate. . . ." In the Beauce area, he distinguished "Older Ice-Front Features . . . between St. Philémon (Dorchester County) . . . and Leeds Village between altitudes of about 900 and 1200 feet . . ." (Gadd, 1964a, p. 1253). He extended the highland front morainic system from the vicinity of Mount Shefford, near Granby, to Rivière-du-Loup and included the St. Antonin Moraine of Lee (1962). Later, according to the prevailing paradigm of that time, this system was thought as one of the most prominent positions of a northward-retreating front of the ice sheet (Gadd et al., 1972). After the rediscovery of the northward-trending glacial striae (Chalmers, 1898; Clark, 1937) in the Thetford Mines area by Lamarche (1971, 1974), Gadd (1978) designated the moraine and ice-front features deposited on the western side of the Chaudière Valley as the type area of the Highland Front moraine system. He included in this morphotype the deposits previously referred to as the "features older than the Highland Front morainic system." Later, in the lower St. Lawrence region, Chauvin et al. (1985) proposed to distinguish the Saint-Damien Moraine complex and the Saint-Jean-Port-Joli Moraine, both deposited by meltwater flowing toward north and northeast from detached Appalachian ice masses, from the classical Saint-Antonin Moraine. They also restricted the name Highland Front Moraine to a small part of the ice-front features described by Gadd (1964a). In areas between the Chaudière Valley and the southwestern edge of the Appalachians toward the Lake Champlain Valley, detailed work has shown that the sediment bodies that had been grouped under the name of Highland Front moraine system were in fact a diachronous series of eskers and ice-contact sediment bodies (Prichonnet, 1984; Parent, 1987). For this reason, Parent and Occhietti (1988) proposed that a new name, the Saint-Sylvestre Moraine, be applied to the moraine belt constructed across the Chaudière Valley that was previously identified by Gadd as the older part of the Highland Front moraine system.

### Field data and reassessment of the published data

Previous studies cited herein provide evidence indicating that a synchronous ice front did not exist along the Appalachian piedmont and that the Highland Front moraine system concept should be abandoned, as Parent and Occhietti (1988) had concluded. The Saint-Antonin Moraine (Fig. 3) seems to be the oldest moraine feature along the piedmont and an interlobate construction (Rappol, 1993) or at least an Appalachian moraine (Dionne, 1972; Chauvin et al., 1985). The Saint-Jean-Port-Joli, Saint-Damien (Chauvin et al., 1985), and Ulverton-Tingwick (Parent, 1987) moraines refer to separate, unconnected field features (Fig. 3). In the Beauce area, a careful analysis of the published data (Gadd, 1964b, 1978; Gadd et al., 1972; LaSalle et al., 1976, 1977a, 1977b; Gauthier, 1975; Chauvin et al., 1985; Lortie and Martineau, 1987; Blais and Shilts, 1989, 1992a, 1992b) and recent field work confirm the following.

1. In response to readjustments within the Laurentide ice sheet and St. Lawrence ice stream, southeastward and eastward ice flow (here called the Beauce event) was reestablished in part of the area covered by the northward-trending striae (Figs. 3, 4, and 6A). On the eastern side of the Chaudière Valley, the limit of the Beauce event is recorded by ice-front sediments (Blais, 1989) and the St-Damien Moraine complex (Blais and Shilts, 1992a).

2. Several ice-front features and fluvioglacial sediment bodies are related to several recessional stages of the ice, after the Beauce event.

3. Several features indicate northward-, southward-, eastward-, and westward-flowing meltwaters.

4. Two generations of east-southeast– and southeast-trending glacial striae can be observed (Fig. 5, A and B). The older generation predates the northward flow event and can be observed mainly on the east side of the Chaudière Valley (Blais and Shilts, 1992b); the younger one postdates the northward flow event and is observed on the piedmont on each side of the valley (in Fig. 5A). This younger ice flow is evidenced by two

sets of striations; a set of striations to the southeast observed on the piedmont, on the west side of the Chaudière Valley (Fig. 5A), and a set to the east-southeast and true east (Fig. 5, A, B, and C), observed mainly on the east side of the valley. The southeast phase indicates a strong Laurentide ice sheet ice movement; the true east phase records active ice flowing toward the estuary and is related to the last phase of the St. Lawrence ice stream.

5. In the Chaudière Valley, several features indicate a northward-retreating ice front from St-Georges to Vallée-Jonction (Blais, 1989; Blais and Shilts, 1992a, 1992b).

This complex setting indicates that several events occurred in the same restricted area. We propose to name this Laurentide ice sheet reactivation in the Chaudière-Etchemin area the Beauce event (Fig. 6A). We also propose to restrain the Saint-Sylvestre Moraine to the short moraine ridges located south of Saint-Patrice-de-Beaurivage, 4 km north of Saint-Sylvestre village (Fig. 3). These till and ice-contact ridges were deposited at the margin of a receding ice front (Fig. 6A). In agreement with Gadd (1978), the Laurentide ice sheet reactivation reached a limit farther south than the Saint-Sylvestre Moraine (Figs. 3 and 6A). This means that (1) the maximum extent of the Laurentide ice during the Beauce event is not recorded by a distinct ice-front feature on the west side of the Chaudière Valley; and (2) this vague southern limit may be due to the coalescence of reactivated Laurentide ice with inactive remnant ice masses on the Beauce area. This setting does not exclude nunataks during or shortly after the Beauce event (Fig. 6B). On the eastern side of the Chaudière Valley, the St-Damien Moraine complex, which was previously related to an Appalachian ice-front feature (Chauvin et al., 1985), would correspond to a complex accumulated at the Laurentide ice sheet ice front, away from the stagnant ice mass over the Notre-Dame Mountains (Fig. 6C) or with no more ice over these mountains, or to an interlobate complex between the Laurentide ice sheet reactivated ice and the Appalachian stagnant ice.

The former local Highland Front moraine system in the first original definition of Gadd (1964a) (distinct from the Older Ice-Front Features), which marks the last recessional positions of the ice front, is subdivided into two parts:

1. On the eastern side of the Chaudière Valley, discontinuous ice-front features built by southward water flow are recorded by LaSalle et al. (1977a). We propose to name this discontinuous aligned forms between the Etchemin River and Saint-Raphael village the Saint-Raphael Moraine (Fig. 3). Three local ice-front features located to the southwest of this alignment, on both sides of the Chaudière Valley (Scott-Jonction and Saint-Bernard), and 6 km to the west, apparently belong to this moraine.

2. Farther southwest (Fig. 3), north and northwest of the Saint-Sylvestre Moraine, two large areas (about 10 × 2.5 km) of stratified drift, respectively near the confluence of the Bécancour and Palmer Rivers (referred to as the stratified complex of the Bécancour-Palmer Rivers in Fig. 3), and on both sides of the Pilkars River (east of Sainte-Agathe, referred to as the stratified complex of Ste-Agathe in Fig. 3) are not currently associated with the Saint-Raphael Moraine. In the Bécancour-Palmer drift complex, sedimentary structures indicate a water flow to the east. This complex could be a vast fan built by meltwaters flowing from the west and funneled between the Appalachian piedmont and the margin of the Laurentide ice sheet. This type of deposition is compatible with the late west to east ice flow indicated by ice striations and with a water body ponded between the ice-sheet front and the Appalachian piedmont. This water body was inferred by Blais (1989) from stratified silts and rhythmites (observed also locally by Gadd, 1978) and subaqueous outwash deposits. The extent and water level of this Chaudière-Etchemin glacial lake have changed from the beginning to the late phase of the Beauce event.

The Saint-Raphael Moraine is the latest stage of damming the Chaudière and Etchemin Valleys, before the opening to the marine waters (Fig. 6D). The absence of clayey till and/or shelly till in the area precludes a glacial readvance after a marine invasion in the Québec City area (Blais, 1989). The relative level of the Chaudière-Etchemin lake was at about 200 m above sea level when the Vallée-Jonction delta and the Bécancour-Palmer drift complex were built.

## CRITICAL ANALYSIS OF $^{14}$C AGES

In southern Québec, because the ice front was commonly in contact with marine water, deglacial events can be dated directly. Datable fossil material (Table 2) has been obtained from diamictons associated with moraine complexes, marine sediments, peat, and gyttja in early deglaciated areas. Nevertheless, the rising eustatic level associated with the rapid glacioisostatic rebound left few traces (e.g., paleoshorelines) on the shores of the middle estuary downstream from the Chaudière Valley and along the Bois Francs piedmont.

### Comparison of nonnormalized, normalized, and calibrated dates

The $^{14}$C dates published by different laboratories are directly comparable when these dates are measured from terrestrial vegetal matter (wood, peat, gyttja). For fossil marine carbonate, the Geological Survey of Canada (GSC) laboratory has published $^{14}$C ages normalized to $\delta^{13}$C = 0‰. From a chronologic point of view in $^{14}$C yr B.P., these dates are directly comparable to conventional radiocarbon dates derived from vegetal matter. However, almost all other laboratories publish shell dates normalized to $\delta^{13}$C = $-25$‰ (according to conventions stipulated in Stuiver and Polach [1977] for reporting $^{14}$C dates), and consequently, the reservoir effect correction gives $^{14}$C ages that are 410 yr older with respect to ages from terrestrial plant material. More recently, in order to be consistent with the reporting procedures of most of the other laboratories, the GSC Laboratory has added a normalized age ($\delta^{13}$C =

−25‰) to its published shell dates, and has undertaken to update its existing shell date database, using $\delta^{13}C$ measurements when available. Usually, when $\delta^{13}C$ ratios are not available, a $\delta^{13}C$ value = 0‰ is ascribed to the dated shells by the laboratories, because the $^{13}C/^{12}C$ ratio for marine shells of Goldthwait and Champlain seas varies within a limited range of about $\delta^{13}C$ = 0‰ ± 2‰ (Hillaire-Marcel, 1981).

For these reasons, in Table 2, we have indicated the $^{14}C$ corrected dates with $\delta^{13}C$ = 0‰ from measured or estimated values, the published or calculated ages normalized to $\delta^{13}C$ = −25‰ from measured or estimated values and the calibrated ages using the CALIB 4.0 program (Stuiver and Reimer, 1993; Stuiver et al., 1998).

## Specific setting of the postglacial Champlain and Goldthwait seas

Retreat of the glacial front in the marine basin, stratification of the water column, variation in meltwater flux, gradual glacioisostatic regression of the Champlain and Goldthwait seas, and large carbonate outcrops (a source of dead carbon) caused the isotopic composition of marine shells to vary spatially and temporally (Hillaire-Marcel, 1981; Rodrigues and Vilks, 1994). These uncertainties in estimating the exact reservoir effect in postglacial seas explain some old ages of marine shells that are at odds with other geological data. Furthermore, global marine reservoir ages get apparently older for Younger Dryas (700 yr instead of 400 yr; Austin et al., 1995). Nevertheless, all these uncertainties do not explain the important discrepancies between the ages of different laboratories, as shown by redating some units in Table 2. It seems that published ages by the GSC, TO (Toronto), and Beta laboratories are more consistent with other past results.

## Timing of deglacial events in southern Québec

Because appropriate local marine $^{14}C$ reservoir corrections have yet to be established for the postglacial seas, the estimated ages for deglacial events in southern Québec cannot be firmly established. In the absence of local marine $^{14}C$ reservoir corrections for the Champlain and Goldthwait seas, the ages used in this chapter are $^{14}C$ ages of vegetal matter ($\delta^{13}C$ = −25‰) and corrected ages of marine shells ($\delta^{13}C$ = 0‰), keeping in mind that the ages of postglacial marine shells might be 200–700 yr too old. In order to give an idea of the real time framework (Lowell and Teller, 1994), the sequence of events of the studied area is also referred to calibrated years B.P. in Figure 4 and Table 2.

On the northern shore of Gaspésie, available $^{14}C$ ages for early marine faunas form a group of 12 500 to 13 200 yr B.P. corrected ages (i.e., 12 900 to 13 600 conventional ages) which seem reliable (cf. references in Table 2).

In the Rimouski area, the oldest age from the GSC laboratory, 13 900 yr B.P. (GSC-4698; Rappol, 1993), seems at odds with the other ages. It seems that there were two episodes of marine incursion: (1) an older one ca. 13 000 yr B.P., related to the marine invasion of the northern shore of Gaspésie (Luceville ice front; Hétu, 1998) and (2) a younger one, ca. 12 700–12 400 yr B.P. (Neigette readvance of the Appalachian ice; Hétu, 1998).

In the Trois-Pistoles–Rivière-du-Loup area, early marine incursion, dated as 12 700 yr B.P. (Lee, 1963), may have been followed by a glacial readvance ca. 11 900 yr B.P. (Rappol, 1993).

In the Bas du Fleuve area, between Rivière-du-Loup and the Chaudière Valley, there are surprisingly no dates older than 11 000 yr B.P. (Table 2; Dionne and Occhietti, 1996), whereas the cluster of 12 000–11 500 yr B.P. ages upstream from this area (Parent and Occhietti, 1999), along the Appalachian piedmont, is distinctly older.

In the Québec City area, one age, from Charlesbourg, is older than the 12 000 yr B.P. (12 400 ± 160 yr B.P., GSC-1533; LaSalle et al., 1977b) from a marine unit on a carbonate bedrock. The age of the Saint-Henri marine shells is revised to 10 670 ± 70 yr B.P. (Beta-115870; Table 2). On the northern shore of the estuary, there are a few ages slightly older than 11 000 yr B.P. in the Saguenay area (Dionne and Occhietti, 1996), and in the Québec City area (LaSalle and Shilts, 1993).

The crucial age in southern Québec is the age of the beginning of the Champlain Sea invasion in the St. Lawrence Valley, ca. 12 000 yr B.P. (Parent and Occhietti, 1988). Given this estimated age and the 103 yr varve record at the Rivière Landry stratotype (Danville Varves; Parent, 1987; Parent and Occhietti, 1999), and because postglacial Lake Candona (Parent and Occhietti, 1988; St. Lawrence Lake of Rodrigues, 1992; see discussion in Parent and Occhietti, 1999) drained when the ice front had retreated about 20 km northwest of this site, the estimated minimum rate of ice retreat is about 200 m/yr. The age of the Ulverton-Tingwick Moraine, which is 5 km southeast of the Rivière Landry site, may be estimated at 12 100 yr B.P. (Parent and Occhietti, 1999). The age of the earlier moraine belts may also be inferred on the basis of a retreat rate of 200 m/yr. The maximum age of the Dixville Moraine, which is 80 km southeast of the Ulverton-Tingwick Moraine, may thus be estimated as ca. 12 500 yr B.P. Applying the same procedure yields an approximate age of 12 200 yr B.P. for the Mont Ham Moraine, 12 325 yr B.P. for the Cherry River–East-Angus–Megantic Moraine, and between 12 600 and 12 550 yr B.P for the Frontier Moraine (Parent and Occhietti, 1999).

Local marine reservoir effects may explain the discrepancies between the southern Québec chronology and the chronology of deglacial features in adjacent northern New England. The marine-based age of 12 500 yr B.P. for the Dixville Moraine is close to the ages assigned to the Bethlehem Moraine (12 400 yr B.P.; Thompson et al., 1996, 1999) and the Success Moraine (Gerath et al., 1985) on the basis of terrestrial dates. If a rate of ice retreat of about 200 m/yr was applied to the 80 km distance separating the Dixville and Bethleem moraines, the offset between the two chronologies would be on the order of 400–500 yr, which gives us an indication of what the local marine reservoir might be. Nevertheless, the downwasting of the Appalachian ice masses over the White Mountains occurred in a very short time at the beginning of the Bölling warm phase and the time span between the two moraines is not easy to evaluate. Moreover, the $^{14}$C ages from basal organic beds of lakes are not reliable (Richard et al., 1997; Ridge et al., 1999) and are minimum ages. Therefore, estimating the local marine reservoir effect of the early phase of Champlain and Goldthwait seas requires further investigation, and the estimated ages of the moraine belts in southern Québec are probably too old by ca. 400–500 yr.

At present, the application of a standard marine reservoir correction reduces the time gap between the inferred age of deglaciation and establishes the time of the onset of organic deposition in lakes and ponds of southeastern Québec and near Maine uplands, ca. 11 140 ± 120 yr B.P. (TO-6330).

## SEQUENCE OF DEGLACIATION EVENTS IN SOUTHERN QUÉBEC

### Events recorded downstream from the mouth of the Saguenay River

***Late Wisconsinan glacial maximum ca. 20–18 ka.*** From glacial striations and other directional data, the ice flowed toward the south sector during the last glacial maximum, with a southeast flow in the eastern areas (Charlevoix, Bas du Fleuve, Beauce areas) and a southwest flow in the western areas (upper St. Lawrence Valley) (Figs. 2A, 4, and 5A).

***Early distal stage of the ice current in the St. Lawrence corridor (16 500–15 300 yr B.P.)*** During this period, the Gulf of St. Lawrence remained completely covered (Stea et al., 1998) by either grounded ice or by an ice shelf. There is no direct regional evidence of the early phase of the St. Lawrence ice stream. According to Stea et al. (1998), a major ice shelf breaking is recorded in the Gulf of St. Lawrence ca. 15 300 yr B.P. From the ice-front position recorded on Anticosti Island (Gratton et al., 1984), we think that the St. Lawrence ice stream was one of the major ice-drainage features of the Laurentide ice sheet before the shelf breaking, at least in the lower part of the St. Lawrence estuary (Fig. 2B). The ice in the gulf was allochthonous, fed by ice from the adjacent and upstream areas.

***Accelerated ablation phase related to the St. Lawrence ice stream and to the global warming phase between 15 500 and 14 000 yr B.P.*** The following period is characterized by a global warming trend that was interrupted by cold phases (Fig. 4). Between 15 300 and 14 000 yr B.P. connected ice domes over Nova Scotia and probably over New Brunswick (Rampton et al., 1984) became increasingly independent from the Laurentide ice sheet, and glacial ablation was accelerated by ice funneling into major valleys (Stea et al., 1998) (Fig. 2, C and D). This trend toward ice-cap isolation and ice funneling also occurred in southern Québec, as a result of the overall thinning of the southeast margin of the ice sheet. We infer that the ice stream was most efficient in the lower and middle estuary after the ice shelf breaking (15 300 yr B.P.) and before the major ice readvance over the western part of Anticosti Island, ca. 13 500 yr B.P. (Painchaud et al., 1984; Gratton et al., 1984). This means that the accelerated thinning process over the northern side of the Appalachians and the southern side of the Laurentians (Occhietti et al., 1996, 1997) occurred before 14 000 yr B.P. (Fig. 2D).

***Cold event between 13 800 and 13 500 yr B.P.*** The GRIP core from Greenland records two cold phases, between 13 800 and 12 500 yr B.P. (Bond et al., 1993), separated by a warming phase. In the lower estuary, fluctuations of the Laurentide ice sheet margin are recorded on Anticosti Island. Gratton et al. (1984, p. 238) wrote (translated from French): " . . . after (28 000 yr B.P.), ice is thinning on the intervales and a glaciomarine material is sedimented. . . . In the main valleys, abundant melt waters bring fluvioglacial deposits in the regressive sea, as low as the present sea level. Then, a minor local ice-readvance occurs, followed by a retreat with resulting deposition of interstratified till and fluvioglacial deposits. This oscillation is followed by a major readvance of the ice over all the land, which lays down (a) typical pale brown till. . . . The readvance is synchronous to a second marine invasion, the ice is grounded everywhere but below 20 m. An intense calving at the ice front limited the extent of the ice sheet. Below 20 m, in the main valleys, a glaciomarine sediment is deposited, . . . . circa 13 500 yr B.P . . . The base of (the Goldthwait Sea) sequence of deep water (sediments) is dated at 13 000 yr B.P." From these data, the Anticosti readvance (first Anticosti readvance in Fig. 4) seems to be related to the pre-13 500 yr B.P. cold phase.

Painchaud et al. (1984) and Gratton et al. (1984) documented the Anticostian readvance (second Anticosti readvance in Fig. 4) as 12 000 yr B.P. and related it to a "glacio-dynamic readjustment of the ice sheet." No fluctuations of the Laurentide ice sheet margin associated with the 13 000–12 500 yr B.P. cold phase (the Chignecto phase in Nova Scotia) are recorded on Anticosti Island, either the ages of the two events (Anticosti major first readvance and Chignecto phase) are questionable, or the response of the ice masses in the southeast margin of the ice sheet to the global cold fluctuations are not widely extended.

***The warm event between 13 500 and 13 100 yr B.P.*** The upstream extent of the postglacial sea along the Appalachian coast of the lower estuary (from the Gaspé Peninsula to the opposite side of the Saguenay mouth) occurred before or by 13 000 yr B.P. as documented by several $^{14}$C ages (Table 2, references). At that time, southern Québec was still glaciated (Fig. 2E).

***Cold event between 13 000 and 12 500 yr B.P. related to the Chignecto phase in Nova Scotia and dynamics of the Appalachian ice masses.*** In the Atlantic provinces, the ice flow phase 4 or Chignecto phase records a glaciodynamic event that occurred between 13 200 and 12 500 yr B.P. (Fig. 2E) and is correlated to the Port Huron event in the Great Lakes area (Stea et al., 1998). At that stage, the dome over Nova Scotia and the Magdalen Shelf started to dismantle into several interconnected but radially flowing ice masses. During or at the end of the Chignecto phase, the Baie des Chaleurs reentrant was deglaciated, and the ice masses over New Brunswick and the Gaspé Peninsula, still partly connected, had distinct ice dynamics. Deglaciation of southern Québec began.

### Sequence of late glacial and deglaciation events in southern Québec

***Convergence of the ice toward the St. Lawrence ice stream and accelerated ablation.*** Because of the invasion of the western arm of Goldthwait Sea into the head of the lower estuary as early as 13 000 yr B.P. it must be assumed that most of the effects of the ice stream had already been felt upstream on the eastern side of the Parc des Laurentides highlands (Charlevoix) and the northern side of the Appalachians. The flow convergence of the ice stream and the Appalachian ice mass migrated from the gulf to the upper estuary in several thousand years, probably from 16 500 to 13 000 yr B.P. This progressive convergence is documented by glacial striae toward the southeast (Fig. 5B) and later to the east in the Charlevoix and Québec City area (Fig. 5C) and by northward striae along the Appalachian side from Rimouski to the Bois Francs area (northeast sector of Fig. 5D). As early as the Scotian phase, between ca. 19 and 17 ka, ice began to funnel along major bedrock valleys in the Maritime provinces (Stea et al., 1998).

***Phase of thin ice in the St. Lawrence Valley and middle estuary.*** This phase is inferred from the following episode of northward ice flow extending from the Beauce area as far north as an area located on the Canadian shield in the Saint-Tite basin (western Charlevoix; Occhietti et al., 1996, 1997; Fournier, 1998) and the Québec City (Paradis and Bolduc, 1999) areas. The phase of thin ice is the result of a major regional ablation by funneling of the ice, and of a dominant warming trend during and after the Scotian phase. We tentatively give an age of 14 000–13 900 yr B.P. to the end of this ice dynamics episode, when the ice reached its lowest thickness. Striations to the east in Charlevoix and in the Québec City area (north area of Fig. 5C), northward striations along the Appalachian piedmont, and a few northeast striations on the southern shore of the middle estuary (Dionne, 1972, and this chapter) are related to this episode.

***Major northward Appalachian ice-flow reversal.*** This late glacial reversal is well documented by striations and rat-tails in the Beauce area (Fig. 5D) and in the eastern part of the southern Appalachians of Québec (Lamarche, 1971, 1974; Lortie and Martineau, 1987; Rappol, 1993). This reversal was a powerful event because it caused a northward ice flow (Fig. 5D) in the small Saint-Tite basin, in western Charlevoix (Fig. 1). There, northward striations on the bedrock surface and till fabrics and sedimentary clasts in the upper till deposited over the Precambrian bedrock indicate that the ice flowing downstream into the estuary was pushed in the small basin and was thrust over the Ligori Ridge, which is the outer edge of the Laurentians (Fournier, 1998; Fig. 3). The local Ligori Ridge event is documented by crosscutting northeastward striations over eastward striations (Lanoie, 1995; Fournier, 1998). In the Beauce area, the northward glacial striations are observed almost as far south as the Québec-Maine boundary (Shilts, 1981; Fig. 5D). The Quebec Ice Divide (Fig. 2D) was defined by Shilts (1981) and is the equivalent to the late phase of the moving North Maine Ice Divide (Fig. 2C) defined later by Lowell and Kite (1986) and Lowell et al. (1986). Westward striations mapped by Lortie and Martineau (1987) mark the western limit of the reversal zone. The strong Appalachian reversal is considered as a major event that followed the slower northward convergence of the previous phase. It seems to have been a reequilibration event of the Appalachian ice masses. The ice over southern Québec and northern Maine flowed toward thinner ice in the middle and upper estuary. This event is pre-Bölling in age and happened after the warmer phase that ended by 13 900 yr B.P. (Fig. 4). A tentative 13 800–12 900 yr B.P. age bracket is proposed, and the event could be more or less coeval of the Anticosti Island phase or the Chignecto phase. Heavy snow falls on the Maine–New Hampshire–southern Québec ice mass might have amplified the strength of the reversal, reinforcing the St. Lawrence ice stream in the middle estuary.

***Reequilibration of the Maine–New Hampshire and southern Québec ice mass and strong ice stream in the middle estuary.*** Blais (1989) and Shilts (1997) identified northeastward striations in the upper part of the middle Chaudière Valley (Fig. 5E) that are younger than the major northward striae. These are related to ice flowing from a high over the upper Chaudière–Thetford Mines area or from a source farther southwest. This event indicates a strong ice current in the St. Lawrence middle estuary documented by north-northeast to northeast striations (Fig. 5E) in the Bois Francs, the Appalachian piedmont east of the Chaudière River and upper Etchemin areas, over the Precambrian bedrock of Ligori Ridge, and throughout the islands and the southern coast of the middle estuary (Dionne, 1972, and this chapter). In the inland of western and central Charlevoix, eastward glacial striae (Charlevoix area in Fig. 5C) record the convergence toward the estuary and the ice stream (Lanoie, 1995; Fournier, 1998). This ice-dynamics episode indicates an unstable ice mass in the southwest, over the Sherbrooke area or farther southwest over the Montreal–lower Lake Champlain lowlands. This event precludes early marine invasion in the St. Lawrence Valley and middle estuary and discredits the validity of the old $^{14}C$ ages on marine shells from the Ottawa Valley (12 700 ± 100 yr B.P., GSC-2151, revised to 12 180 ± 90 yr B.P., TO-245; see Table 4.12 and discussion in Occhietti, 1989, p. 376). Despite the lack of direct dating, this event is likely correlative to the Port Huron Stade in southern Ontario and the Chignecto phase in the Maritime provinces. The most probable age bracket is 13 200–12 700 yr B.P. and we propose to correlate this event to the Oldest Dryas (Dryas I) of Europe. This event caused a rapid thinning of the ice over the Bois Francs and upper Beauce area.

*Local ice flow in the middle Chaudière area.* Late striations toward the north-northwest (Fig. 5E) are observed in the Chaudière Valley and laterally as far as 20 km on each side of this valley (Blais and Shilts, 1992a, 1992b; Shilts, 1997). Funneled ice from the upper Beauce indicates a reequilibration and accelerated thinning of the related Appalachians ice masses. The ice was still active enough to leave erosional marks on the bedrock. This short transitional episode is probably coeval with a general thinning of the ice in the St. Lawrence Valley, after the inland progression of the head of convergence of the ice stream as far as the border of the Saint-Maurice basin (Parent and Occhietti, 1999). It marks the beginning of the dismantlement of the ice masses over the White Mountains Range.

*Rapid disintegration of the southern Québec Appalachian ice masses during the Bölling phase.* At the start of the final deglaciation, the southern Appalachian ice masses in Québec were in three different settings. These settings (Fig. 6A) are inferred from the episodes of deglaciation that followed. The western side of the Appalachian ice mass was connected to the Hudson–Lake Champlain lobe of the Laurentide ice sheet. The central part was the termination of the Saint-Maurice lobe, which remained connected to and was fed by the ice sheet. The eastern part, from the Bois Francs to the Bas du Fleuve area (downstream to the vicinity of Rivière du Loup) was passively connected to but not fed by the ice sheet. The thinned ice over this area became stagnant and melted rapidly. The different recessional morainic belts in the southern Appalachians of Québec (Fig. 3) were mainly described by McDonald (1966, 1967, 1968, 1969), Shilts (1970, 1981), Prichonnet (1984), and Parent (1987) (see syntheses in Parent and Occhietti, 1988, 1999).

The Frontier Moraine (Shilts, 1981) (Figs. 3 and 6A), phase a, provides evidence for the early shrinkage of the ice masses over the White Mountains. To the north, it marks the terminus of the Laurentide ice sheet. To the south, stagnant ice masses must have persisted in the local depressions, much as described by Dorion (1998) in the adjacent uplands of northwestern Maine.

Phase b is the recession of the ice front from the south to the north, evidenced by the Dixville-Ditchfield moraine belt. These ice-front features (Figs. 3 and 6A) are transverse to the southwest-northeast structural trend of the Appalachians. Their outline is strongly lobate and influenced by the topography, lobes often existing in the valleys and reentrants on adjacent uplands. A large lobe occupied the Lake Champlain basin. The moraine belt disappears eastward in the White Mountains of northwestern Maine, where the ice was probably stagnant.

The ice front of the Sutton–Cherry River–East Angus–Megantic moraine belt and the Beauce event, phase c, was strongly lobate due to strong topographic influence (Fig. 6A), which indicates that ice-marginal slopes were fairly gentle. The Beauce event is tentatively considered a temporal equivalent of the Sutton–Cherry River–East-Angus–Megantic moraine belt. The beginning of the event could be related to ice readvances identified on the Appalachian ice front in the Bas du Fleuve (Rappol, 1993) and in the Rimouski area (Rappol, 1993; Hétu, 1998), where the Neigette readvance is dated as 12 400 yr B.P. (marine shells). The Beauce event is related to a drastic change in the Laurentide ice sheet budget and/or to a significant glaciodynamic reequilibration of the New Quebec ice dome. We tentatively suggest a relation with the intra-Bölling cold event, ca. 12 400 yr B.P., or with the Dryas II cold event.

Strong lobation of the ice front during phase d, including the Mont Ham and Saint-Ludger moraines (Fig. 6B) and the earlier phase supports the hypothesis that a nearly stagnant ice mass had already been isolated over the Bois Francs, upper and middle Chaudière Valley and the upper ridge (Notre-Dame Mountains) of the Bas du Fleuve region. Nunataks had probably formed in the Notre-Dame Mountains.

During the short phase of complex ice retreat phase e, 100–200 yr in duration, the remaining ice masses over the southern Appalachians of Québec were dismantled from the south to the north, with probably some local remnant stagnant ice masses in the Bois Francs area. In the southwest part, the ice front remained connected to the Montreal–Lake Champlain and Saint-Maurice lobes and remained active while receding between the time of deposition of the Mont Ham and Ulverton-Tingwick moraines (Fig. 6C). In the central part, the Bois Francs residual ice mass (Parent and Occhietti, 1988) remained passively connected to the Laurentide ice sheet until a few centuries before the opening of Champlain Sea. In the Chaudière Valley and the Bas du Fleuve areas, the ice masses became progressively disconnected from the ice sheet and from other residual ice masses over the northwestern plateau of Maine. A buttonhole (an elongated deglaciated zone between the stagnant ice over the Appalachians and the active ice in the St. Lawrence corridor) on the Appalachian piedmont opened in the lower middle Chaudière and Etchemin Valleys and was filled in by lacustrine water, as evidenced by glaciofluvial outwash deposits with a northward depositional direction at an elevation of 370 m above sea level (Blais, 1989).

According to Parent and Occhietti (1999), phases a–e occurred during a short time span, roughly between 12 550 and 12 150 yr B.P., i.e., during the Bölling warm phase. During this period of time, the Laurentide ice sheet over the southern and eastern sides of the Parc des Laurentides highlands (the Quebec City and western and central Charlevoix areas) became thinner. As evidenced by the rapid deglaciation that followed, the global warming climatic trend of the Allerod did not allow the New Quebec dome to rebuild the ice masses over areas formerly subjected to the rapid ablation and thinning due to the convergence toward the ice stream.

Phase f comprises the ultimate Ulverton-Tingwick ice front and recession stage of the Beauce event. The age of the Ulverton-Tingwick ice-front episode (Fig. 6C) can be established.

From the number of glacial varves at the Rivière Landry section (Parent, 1987; Parent and Occhietti, 1999), the ice front receded from the Ulverton-Tingwick position more than 120 yr prior to the opening, ca. 12 000 yr B.P., of the St. Lawrence Valley to marine waters. In the Beauce area, we tentatively relate the Saint-Sylvestre Moraine to this episode. The age of 12 150 yr B.P. dates the end of the presence of Laurentide ice over the southwestern Appalachians. Glacial Lake Candona inundated the deglaciated lowlands of the St. Lawrence Valley and valleys in the Appalachians (Parent and Occhietti, 1988).

Phase g is the rapid deglaciation along the Appalachian piedmont, south of the middle estuary. Along the northeastern Appalachian piedmont, the ultimate episode of Laurentide ice retreat is the Saint-Raphael Moraine (Fig. 6D). After the retreat from this ice-front position, the western arm of Goldthwait Sea extended rapidly along the southern shore of the middle estuary of St. Lawrence up to the lacustrine buttonhole of the Chaudière-Etchemin Rivers. The Bécancour-Palmer stratified drift accumulation could be related to this phase. The deglaciated reentrant between the Appalachian piedmont and the retreating Laurentide ice sheet margin extended rapidly upstream up to the Bois Francs piedmont, where the ice-sheet front still abutted against the piedmont (Fig. 6D).

Phase h is the Champlain Sea invasion. The Saint-Maurice lobe remained active while some residual ice may have persisted on the Bois Francs plateau, in the Chaudière Valley, and in the northern side of the middle estuary. Then, from field evidence in the Plessisville-Laurierville area and the Rivière Landry section (Parent, 1987), the opening of the St. Lawrence Valley, upstream from Quebec City, to the marine waters occurred in several short steps:

(1) thinning of the Laurentide ice sheet ice front along the piedmont between Victoriaville and Laurierville, by the inferred piedmont thermal effect; (2) overflow of the lacustrine waters of lake Candona through a spillway inset into the ice and parallel to the piedmont; (3) phase of mixing fresh and marine waters, as recorded by the transitional unit in the Rivière Landry section (Parent and Occhietti, 1999); and (4) full marine invasion in the St. Lawrence Valley with an active and grounded Laurentide ice front retreating northward, as recorded by a moraine parallel to the piedmont in the Laurierville area (Dubé, 1971). The age of the marine incursion is close to 12 000 yr B.P.

The deglaciation events between the beginning of the Champlain Sea invasion and the Saint-Narcisse Moraine event (ca. 10 800–10 400 yr B.P.) are poorly known. Some events are locally identified in the Saint-Nicolas area (LaSalle and Shilts, 1993), in the lower Chaudière Valley (LaSalle et al., 1977b), in the lower Saint-Maurice Valley (Occhietti, 1980; Parent and Occhietti, 1988), in the Quebec City area (LaSalle et al., 1972), and in the Charlevoix area (Hardy, 1970; Rondot, 1974; Govare, 1995; Dionne and Occhietti, 1996; Fournier, 1998). A tentative overview would be premature due to the lack of data.

The age and significance of the Saint-Narcisse episode are being reassessed by one of us (Occhietti). Moraine ice-front features that are younger than the Saint-Narcisse Moraine have been identified in the Charlevoix area (Govare, 1995), in the Parc des Laurentides highlands (Bolduc, 1995), and in the Saint-Maurice basin (Occhietti, 1980), informally named the Mars-Batiscan moraine.

## CONCLUSION

Within the general context of the increasingly negative mass balance of the Laurentide ice sheet during late Wisconsinan warming, the deglaciation of southern Québec was preceded by extensive thinning generated by the St. Lawrence ice stream. Before the shelf breakup phase, ca. 15 500 yr B.P. in the Gulf of St. Lawrence, ice from a large area within the Laurentide ice sheet was redirected toward the Laurentian trough. The head of the catchment area of the ice stream migrated 1000 km toward the inner ice sheet, while its terminus became a calving bay. This calving bay in turn began migrating rapidly within the southeastern margin of the Laurentide ice sheet. The retreating terminus remained grounded along the north shore of the Gulf of St. Lawrence, notably on Anticosti Island, while the margin of the Appalachians along the Gaspé Peninsula was deglaciated. This glacial streaming and calving process explains the gradual isolation of Appalachian ice masses from the Gaspé Peninsula to the Chaudière Valley. These ice masses remained initially coalescent with the ice stream, but became gradually detached along the northern margin of the Appalachians.

The deglaciated reentrant was invaded by an early arm of the Goldthwait Sea ca. 13 200–12 500 yr B.P. (corrected ages with $\delta^{13}C$ = 0‰, i.e., 13 600–12 900 yr B.P. conventional ages) along the north shore of the Gaspé Peninsula and reached the vicinity of Rivière-du-Loup ca. 12 700 yr B.P. The ice stream also caused the ice-flow reversal in the northern Appalachians, from the Gaspé Peninsula to the Chaudière Valley. The southernmost limit of the striae generated by the flow reversal is known as the Quebec Ice Divide. The flow reversal caused by the disequilibrium within Appalachian ice masses was coeval with the northward ice-flow reorientation that was recently recognized in the Charlevoix and Quebec City regions. It is possible that this disequilibrium was strengthened by a climatic fluctuation that favored higher rates of snow accumulation on Appalachian ice masses. At about the same time, the Hudson-Champlain lobe and the western part of the Appalachian ice masses remained actively connected with the Laurentide ice sheet.

The northward flow event in the Appalachians was apparently followed by an eastward readjustment in part of the St. Lawrence Valley and adjacent Beauce. This succession of late glacial reorientations is not compatible with an early marine incursion in the middle estuary and Quebec City region.

It was only after these events that the deglaciation of southern Québec began, at the beginning of the Alleröd, ca. 12 700 yr B.P. in the Rivière-du-Loup area and ca. 12 550 yr B.P. in the New Hampshire–Maine–Québec border area. Appalachian ice masses receded within a period of less than 700 yr, with regionally distinct styles. Deglaciation of the Hudson-Champlain lobe and southwestern Québec Appalachians, which had remained connected with the main ice sheet, was characterized by the deposition of a series of mainly recessional morainic belts from the southeast to the northwest. In southern Québec, these belts are the Frontier, Dixville-Ditchfield, Cherry River–East Angus–Megantic, Mont Ham, and Ulverton-Tingwick moraines, which were deposited within a time period of ca. 400 yr. In the central Bois-Francs uplands, in the Beauce region and in the Notre-Dame Mountains, residual ice masses became stagnant and downwasted gradually. On the Appalachian piedmont in the region of Chaudière and Etchemin Valleys, where Appalachian ice masses had thinned out considerably, a phase of glaciodynamic reactivation within the coalescent Laurentide ice took place. This event, which we call the Beauce event, caused the reorientation of ice-flow patterns toward the east-southeast in parts of the area that had been previously affected by the northward ice flow. Because the Appalachian and Laurentide ice masses had remained coalescent prior to and during this readjustment, the southern margin cannot be identified on the west side of the Chaudière Valley; on the east side, the St-Damien Moraine complex seems to have been built at the Laurentide ice sheet ice front and close to the Appalachian stagnant ice. The glacial retreat that followed in the Chaudière Valley and adjacent regions was characterized mostly by a southeast to northwest retreating ice front, the deposition of the Saint-Sylvestre Moraine, the subaquatic outwash at Vallée-Jonction, and then by the Saint-Raphael Moraine. A short-lived glacial lake, contemporaneous with Lake Candona to the west, was dammed by the retreating ice margin in the Chaudière and Etchemin Valleys. The incursion of marine waters along the Appalachian piedmont between Rivière-du-Loup and the Bois-Francs region seems to have been extremely rapid. It took place not earlier than 12 000 yr B.P. (from $^{14}$C ages of marine shells), probably ca. 11 500 yr B.P. (in $^{14}$C ages of nonmarine material), and was preceded by the drainage of Lake Candona along the Appalachian piedmont. The margin of the Laurentide ice sheet subsequently retreated from south to north, with an ice margin extending nearly parallel to the axis of the St. Lawrence corridor.

## ACKNOWLEDGMENTS

We thank Benoit Poulin and Francine Robert, from the Département de Géographie (Atlas du Québec project), Université du Québec à Montréal, who patiently helped to prepare and revise the figures, and David Franzi, Grahame Larson, Woodrow Thompson, Thomas Weddle, and Mike Retelle, who kindly reviewed the manuscript and offered helpful comments and suggestions that improved the text. This chapter is contribution 2000192 of the Geological Survey of Canada.

## REFERENCES CITED

Allard, M., and Tremblay, G., 1981, Observations sur le Quaternaire de l'extrémité orientale de la péninsule de Gaspé, Québec: Géographie Physique et Quaternaire, v. 35, p. 105–125.

Antevs, E., 1925, Retreat of the last ice sheet in eastern Canada: Geological Survey of Canada Memoir 146, 142 p.

Austin, W.E.N., Bard, E., Hunt, J.B., Kroon, D., and Peacock, J.D., 1995, The $^{14}$C age of the Icelandic Vedde Ash: Implications for Younger Dryas marine reservoir age corrections: Radiocarbon, v. 37, p. 53–62.

Bernatchez, P., 1997, Géomorphologie et environnements quaternaires du bassin de la rivière aux Anglais, région de Baie-Comeau: Étude de la formation de dépôts coquilliers [M.S. thesis]: Québec, Canada, Université de Sherbrooke, 197 p.

Bernier, F., and Occhietti, S., 1991, Nouvelle séquence glaciaire antérieure aux Sédiments de Saint-Pierre, Sainte-Anne-de-la-Pérade, Québec: Géographie Physique et Quaternaire, v. 45, p. 101–110.

Blais, A., 1989, Lennoxville glaciation of the middle Chaudiere and Etchemin valleys, Beauce region, Quebec [M.S. thesis]: Ontario, Canada, Carleton University, 124 p.

Blais, A., and Shilts, W.W., 1989, Surficial geology of Saint-Joseph-de-Beauce map area, Chaudière River Valley, Québec, in Current Research, 1989: Geological Survey of Canada Paper 89-1B, p. 137–142.

Blais, A., and Shilts, W.W., 1992a, Surficial geology of the St-Zacharie and Ste-Justine areas, Québec: Geological Survey of Canada Open-File 2536, 2 sheets, scale 1:50 000.

Blais, A., and Shilts, W.W., 1992b, Surficial geology of the Beauceville and St-Joseph areas, Québec: Geological Survey of Canada Open-File 2537, 2 sheets, scale 1:50 000.

Bolduc, A.M., 1995, Landforms in the Laurentians of southern Quebec: Implications for the deglaciation history ot the Laurentide Ice Sheet: Canadian Quaternary Association–Canadian Geomorphology Research Group, CANQUA-CGRG joint meeting, St. John's, Newfoundland, Program, Abstracts and Field Guides, p. CA5.

Bond, G.C., and Lotti, R., 1995, Iceberg discharges into the North Atlantic on millennial time scales during the last glaciation: Science, v. 267, p. 1005–1010.

Bond, G., Broecker, W., Johnsen, S., McManus, J., Labeyrie, L., Jouzel, J., and Bonani, G., 1993, Correlations between climate records from North Atlantic sediments and Greenland ice: Nature, v. 365, p. 143–147.

Bonenfant, R., 1993, Chronologie des événements post-glaciaires à l'Holocène dans la basse vallée du Gouffre (Charlevoix) [M.S. thesis]: Québec, Canada, Université Laval, 148 p.

Borns, H.W., Jr, 1985, Changing models of deglaciation in northern New England and adjacent Canada, in Borns, H.W., Jr, et al., eds., Late Pleistocene history of northeastern New England and adjacent Quebec: Geological Society of America Special Paper 197, p. 135–138.

Borns, H.W., Jr., and Calkin, P.E., 1977, Quaternary glaciation, west-central Maine: Geological Society of America Bulletin, v. 88, p. 1773–1784.

Borns, H.W., Jr., LaSalle, P., and Thompson, W.B., 1985, Late Pleistocene history of northeastern New England and adjacent Québec: Introduction: Geological Society of America Special Paper 197, p. vii–ix.

Caldwell, D.W., Hanson, L.S., and Thompson, W.B., 1985, Styles of deglaciation in central Maine, in Borns, H.W., Jr., et al., eds., Late Pleistocene history of northeastern New England and adjacent Quebec: Geological Society of America Special Paper 197, p. 45–58.

Chalmers, R., 1898, Report on the surface geology and auriferous deposits of southeastern Québec: Geological Survey of Canada Annual Report, v. 10, part J, 160 p.

Chauvin, L., Martineau, G., and LaSalle, P., 1985, Deglaciation of the Lower St. Lawrence region, Québec, in Borns, H.W., Jr., et al., eds., Late Pleistocene history of northeastern New England and adjacent Quebec: Geological Society of America Special Paper 197, p. 111–123.

Clark, T.H., 1937, Northward moving ice in southern Quebec: American Journal of Science, v. 34, p. 215–220.
Connally, G.C., 1982, Deglacial history of western Vermont, in Larson, J.G., and Stone, B.D., eds., Late Wisconsinan glaciation of New England, A proceeding volume of the Symposium: Dubuque, Iowa, Kendall/Hunt Publishing Company, 242 p.
Connally, G.G., and Sirkin, L.A., 1973, Wisconsinan history of the Hudson-Champlain Lobe, in Black, R.F., The Wisconsin Stage: Boulder, Colorado, Geological Society of America Memoir 136, p. 47–69.
Cooke, H.C., 1937, Further note on northward moving ice: American Journal of Science, v. 34, p. 221.
Dansgaard, W., Johnsen, S., Clausen, H., Dahl-Jensen, D., Gundestrup, N., Hammer, C., Hvidberg, C., Steffensen, J., Sveinbjornsdottir, A., Jouzel, J., and Bond, G., 1993, Evidence for general instability of past climate from a 250-kyr ice core record: Nature, v. 364, p. 218–220.
David, P.P., and Lebuis, J., 1985, Glacial maximum and deglaciation of western Gaspé, Québec, Canada, in Borns, H.W., Jr., et al., eds., Late Pleistocene history of northeastern New England and adjacent Quebec: Geological Society of America Special Paper 197, p. 85–109.
Dionne, J.-C., 1972, Le Quaternaire de la région de Rivière-du-Loup/Trois-Pistoles, côte sud du Saint-Laurent: Québec, Environnement Canada, Centre de recherches forestières des Laurentides, Rapport d'information Q-F-X-27, 95 p.
Dionne, J.-C., 1977, La mer de Goldthwait au Québec: Géographie Physique et Quaternaire, v. 31, p. 61–80.
Dionne, J.-C., 1988, Holocene relative sea-level fluctuations in the St. Lawrence estuary, Québec, Canada: Quaternary Research, v. 29, p. 233–244.
Dionne, J.-C., 1994, Les erratiques lointains de l'embouchure du Saguenay, Québec: Géographie Physique et Quaternaire, v. 48, p. 179–194.
Dionne, J.-C., and Coll, D., 1995, Le niveau marin relatif dans la région de Matane (Québec), de la déglaciation à nos jours: Géographie Physique et Quaternaire, v. 49, p. 363–380.
Dionne, J.-C., and Occhietti, S., 1996, Aperçu du Quaternaire à l'embouchure du Saguenay, Québec: Géographie Physique et Quaternaire, v. 50, p. 5–34.
Dorion, C.C., 1998, Style and chronology of deglaciation in central and northern Maine: Geological Society of America Northeastern Section, 33rd Annual Meeting, Portland, Maine, Abstracts with Programs, p. 15.
Dredge, L.A., 1983, Surficial geology of the Sept-Îles area, Quebec North Shore: Geological Survey of Canada Memoir 408, 40 p.
Dubé, J.C., 1971, Géologie des dépôts meubles, région de Lyster, Comtés de Mégantic, Lotbinière, Nicolet et Arthabaska: Gouvernement du Québec, ministère des Richesses naturelles, 12 p.
Dubois, J.-M., and Dionne, J.-C., 1985, The Québec North Shore Moraine system: A major feature of Late Wisconsin deglaciation, in Borns, H.W., Jr., et al., eds., Late Pleistocene history of northeastern New England and adjacent Quebec: Geological Society of America Special Paper 197, p. 125–133.
Dyke, A.S., and Prest, V.K., 1987, Late Wisconsinan and Holocene history of the Laurentide Ice Sheet: Géographie Physique et Quaternaire, v. 41, p. 237–263.
Fournier, M., 1998, Stratigraphie des dépôts quaternaires et modalités de déglaciation au Wisconsinien supérieur dans le Charlevoix occidental, Québec [M.S. thesis]: Montréal, Université du Québec, 147 p.
Gadd, N.R., 1964a, Moraines in the Appalachian region of Quebec: Geological Society of America Bulletin, v. 75, p. 1249–1254.
Gadd, N.R., 1964b, Surficial geology, Beauceville map area: Geological Survey of Canada Paper 64, 3 p.
Gadd, N.R., 1978, Surficial geology of Saint-Sylvestre map-area, Quebec: Geological Survey of Canada Paper 77, 9 p.
Gadd, N.R., 1980, Late-glacial ice-flow patterns in eastern Ontario: Canadian Journal of Earth Sciences, v. 17, p. 1439–1453.
Gadd, N.R., McDonald, B.C., and Shilts, W.W., 1972, Deglaciation of southern Québec: Geological Survey of Canada Paper 71-47, 19 p.
Gagnon, M., 1994, Cartographie, lithostratigraphie et paléogéographie des dépôts quaternaires de la région de Saint-Raymond-de-Portneuf [M.S. thesis]: Montréal, Université du Québec, 74 p.
Gauthier, C.R., 1975, Déglaciation d'un secteur des rivières Chaudière et Etchemin [M.S. thesis]: Montréal, McGill University, 180 p.
Genes, A.N., Newman, W.A., and Brewer, T.B., 1981, Late Wisconsinan glaciation models of northern Maine and adjacent Canada: Quaternary Research, v. 16, p. 48–65.
Gerath, R.F., Fowler, B.K., and Haselton, G.M., 1985, The deglaciation of the northern White Mountains of New Hampshire, in Borns, H.W., Jr., et al., eds., Late Pleistocene history of northeastern New England and adjacent Quebec: Geological Society of America Special Paper 197, p. 21–28.
Govare, É., 1995, Géomorphologie et paléoenvironnements de la région de Charlevoix, Québec, Canada [Ph.D. thesis]: Montréal, Université de Montréal, 425 p.
Grant, D.R., 1989, Quaternary geology of the Atlantic Appalachian region of Canada, in Fulton, R.J., ed., Quaternary geology of Canada and adjacent Greenland: Boulder, Colorado, Geological Society of America, Geology of North America, v. K-1, p. 393–440.
Gratton, D., Gwyn, Q.H.J., and Dubois, J.M., 1984, Les paléoenvironnements sédimentaires au Wisconsinien moyen et supérieur, île d'Anticosti, Golfe du Saint-Laurent, Québec: Géographie Physique et Quaternaire, v. 38, p. 229–242.
Grootes, P.M., Stuiver, M., White, J.W.C., Johnsen, S., and Jouzel, J., 1993, Comparison of oxygen isotope records from the GISP2 and GRIP Greenland ice cores: Nature, v. 366, p. 552–554.
Hardy, L., 1970, Géomorphologie glaciaire et post-glaciaire de St-Siméon à St-François d'Assises (Comtés de Charlevoix Est et de Chicoutimi) [M.S. thesis]: Québec, Université Laval, 112 p.
Hétu, B., 1998, La déglaciation de la région de Rimouski, Bas-Saint-Laurent (Québec): Indices d'une récurrence glaciaire dans la Mer de Goldthwait entre 12 400 et 12 000 BP: Géographie Physique et Quaternaire, v. 52, p. 325–347.
Hillaire-Marcel, C., 1981, Paléo-océanographie isotopique des mers post-glaciaires du Québec: Palaeogeography, Palaeoclimatology, Palaeoecology, v. 35, p. 63–119.
Hughes, T., Borns, H.W., Fastook, J.L., Hyland, J.S., Kite, J.S., and Lowell, T.V., 1985, Models of glacial reconstruction and deglaciation applied to Maritime Canada and New England, in Borns, H.W., Jr., et al., eds., Late Pleistocene history of northeastern New England and adjacent Quebec: Geological Society of America Special Paper 197, p. 139–150.
Johnsen, S.J., Clausen, H.B., Dansgaard, W., Fuhrer, K., Gundestrup, N., Hammer, C.U., Iversen, P., Jouzel, J., Stauffer, B., and Steffensen, J.P., 1992, Irregular glacial interstadials recorded in a new Greenland ice core: Nature, v. 359, p. 312–313.
Lamarche, R.Y., 1971, Northward moving ice in the Thetford Mines area of southern Québec: American Journal of Science, v. 271, p. 383–388.
Lamarche, R.Y., 1974, Southestward, northward and westward ice movement in the Asbestos area of southern Québec: Geological Society of America Bulletin, v. 85, p. 465–470.
Lanoie, J., 1995, Les écoulements glaciaires du Wisconsinien supérieur en Charlevoix occidental [M.S. thesis]: Montréal, Université du Québec, 83 p.
Larson, J.G., and Stone, B.D., 1982, Late Wisconsinian glaciation of New England, a Proceeding Volume of the Symposium: Dubuque, Iowa, Kendall/Hunt Publishing Company, 242 p.
LaSalle, P., 1966, Late Quaternary vegetation and glacial history in the St. Lawrence lowlands, Canada: Leidse Geologische Mededelingen, v. 38, p. 91–128.
LaSalle, P., and Chapdelaine, C., 1990, Review of late-glacial and Holocene events in the Champlain and Goldthwait Seas areas and arrival of man in eastern Canada, in Lasca, N.P., and Donahue, J., eds., Archeological geology of North America: Geological Society of America Centennial Special Volume 4, p. 1–19.

LaSalle, P., and Elson, J.A., 1975, Emplacement of the St. Narcisse moraine; a climatic event in eastern Canada: Quaternary Research, v. 5, p. 621–625.

LaSalle, P., and Shilts, W.W., 1993, Younger Dryas-age readvance of Laurentide ice into the Champlain Sea: Boreas, v. 22, p. 25–37.

LaSalle, P., Hardy, L., and Poulin, P., 1972, Une position du front glaciaire au nord et au nord-est de la ville de Québec: Québec, Ministère des Richesses naturelles, Rapport S-135, 8 p.

LaSalle, P., Martineau, G., and Chauvin, L., 1976, Géologie des sédiments meubles d'une partie de la Beauce et du bas Saint-Laurent: Québec, Ministère des Richesses naturelles, Rapport DPV-438, 13 p.

LaSalle, P., Martineau, G., and Chauvin, L., 1977a, Dépôts morainiques et stries glaciaires dans la région de Beauce-Monts Notre-Dame-Parc de Laurentides: Québec, Ministère des Richesses naturelles, Rapport DPV-515, 22 p.

LaSalle, P., Martineau, G., and Chauvin, L., 1977b, Morphologie, stratigraphie et déglaciation dans la région de Beauce-Monts Notre-Dame-Parc de Laurentides: Québec, Ministère des Richesses naturelles, Rapport DPV-516, 74 p.

Laverdière, C., and Guimont, P., 1980, Le vocabulaire de la géomorphologie glaciaire, IX. Terminologie illustrée des formes mineures d'érosion glaciaire: Géographie Physique et Quaternaire, v. 34, p. 363–377.

Lebuis, D., and David, P.P., 1977, La stratigraphie et les événements du Quaternaire de la partie occidentale de la Gaspésie, Québec: Géographie Physique et Quaternaire, v. 31, p. 275–296.

Lee, H.A., 1962, Surficial geology of Rivière-du-Loup/Trois-Pistoles Area, Québec: Geological Survey of Canada Paper 61-32, 2 p.

Llibourty, L., 1965, Traité de glaciologie, Tome II, Glaciers Variations du climat Sols gelés: Paris, Masson, p. 429–1040.

Locat, J., 1977, L'émersion des terres dans la région de Baie-des-Sables/Trois-Pistoles, Québec: Géographie Physique et Quaternaire, v. 31, p. 297–306.

Lortie, G., and Guilbault, J.-P., 1984, Les diatomées et les foraminifères de sédiments marins post-glaciaires du Bas Saint-Laurent (Québec). Une analyse comparée des assemblages: Le Naturaliste canadien, v. 111, p. 297–310.

Lortie, G., and Martineau, G., 1987, Les systèmes de stries glaciaires dans les Appalaches du Québec: Québec, Ministère de l'Énergie et des Ressources, Rapport DV 85-10, 45 p.

Lowdon, J.A., and Blake, W., Jr., 1970, Geological Survey of Canada radiocarbon dates IX: Radiocarbon, v. 12, p. 46–86.

Lowdon, J.A., and Blake, W., Jr., 1979, Geological Survey of Canada radiocarbon dates XIX: Geological Survey of Canada Paper 79-7, 58 p.

Lowell, T.V., 1985, Late Wisconsin ice-flow reversal and deglaciation, northwestern Maine, in Borns, H.W., Jr., et al., eds., Late Pleistocene history of northeastern New England and adjacent Quebec: Geological Society of America Special Paper 197, p. 71–83.

Lowell, T.V., and Kite, J.S., 1986, Glaciation style of northwestern Maine, in Kite, J.S., et al., eds., Contributions to the Quaternary of northern Maine and adjacent Canada: Maine Geological Survey Bulletin, v. 37, p. 53–68.

Lowell, V.T., and Teller, T.J., 1994, Radiocarbon vs. calendar ages of major late glacial hydrological events in North America: Quaternary Science Reviews, v. 13, p. 801–803.

Lowell, T.V., Becker, D.A., and Calkin, P.E., 1986, Quaternary stratigraphy in northwestern Maine: A progress report: Géographie Physique et Quaternaire, v. 40, p. 71–84.

Lowell, V.T., Kite, J.S., Calkin, P.E., and Halter, E.F., 1990, Analysis of small-scale erosional data and a sequence of late Pleistocene flow reversal, northern New England: Geological Society of America Bulletin, v. 102, p. 74–85.

Martineau, G., and Corbeil, P., 1983, Réinterprétation d'un segment de la moraine de Saint-Antonin, Québec: Géographie Physique et Quaternaire, v. 37, p. 217–222.

McDonald, B.C., 1966, Surficial geology, Richmond-Dudswell, Quebec: Geological Survey of Canada Map 4-1966, scale 1: 63 360.

McDonald, B.C., 1967, Pleistocene events and chronology in the Appalachian region of southeastern Quebec, Canada [Ph.D. thesis]: New Haven, Connecticut, Yale University, 161 p.

McDonald, B.C., 1968, Deglaciation and differential postglacial rebound in the Appalachian region of southeastern Quebec: Journal of Geology, v. 76, p. 664–677.

McDonald, B.C., 1969, Surficial geology of La Patrie–Sherbrooke area, Quebec, including Eaton River watershed: Geological Survey of Canada Paper 67-52, 21 p.

Mott, R.J., 1977, Late-Pleistocene and Holocene palynology in southeastern Quebec: Géographie Physique et Quaternaire, v. 31, p. 139–149.

Newman, W.A., Genes, A.N., and Brewer, T., 1985, Pleistocene geology of northeastern Maine, in Borns, H.W., Jr., et al., eds., Late Pleistocene history of northeastern New England and adjacent Quebec: Geological Society of America Special Paper 197, p. 59–71.

Occhietti, S., 1980, Le Quaternaire de la région de Trois-Rivières-Shawinigan, Québec. Contribution à la paléogéographie de la vallée moyenne du Saint-Laurent et corrélations stratigraphiques: Paléo-Québec, v. 10, 227 p.

Occhietti, S., 1989, Quaternary geology of St. Lawrence Valley and adjacent Appalachian subregion, in Fulton, R.J., ed., Quaternary geology of Canada and Greenland: Boulder, Colorado, Geological Society of America, Geology of North America, v. K-1, p. 350–388.

Occhietti, S., 1990, La lithostratigraphie du Quaternaire de la vallée du Saint-Laurent: méthode, cadre conceptuel et séquences sédimentaires: Géographie Physique et Quaternaire, v. 44, p. 137–145.

Occhietti, S., Dionne, J.C., Govare, É., and Rondot, J., 1996, Écoulements glaciaires en Charlevoix et dans l'estuaire moyen du Saint-Laurent: 8th congress, Association québécoise pour l'étude du Quaternaire, Programme et Résumés: Sainte-Foy, Québec, Université Laval, p. 57.

Occhietti, S., Govare, É., Richard, P.J.H., Dionne, J.C., Bolduc, A.M., Rondot, J., and Fournier, M., 1997, Late-glacial ice dynamics and events in the St. Lawrence valley, middle estuary and adjacent Laurentians: 8th Biennial Congress, Canadian Quaternary Association (CANQUA), Montréal, Québec, Canada, Programme-Abstracts, p. 47–49.

Painchaud, A., Dubois, J.-M., and Gwyn, Q.H.J., 1984, Déglaciation et émersion des terres de l'ouest de l'île d'Anticosti, Golfe du Saint-Laurent, Québec: Géographie Physique et Quaternaire, v. 38, p. 93–111.

Paradis, S.J., and Bolduc, A.M., 1999, Mouvement glaciaire vers le nord sur le piémont laurentien dans la région de Québec, in Current Research 1999: Paper 1999-D, Geological Survey of Canada p. 1–7.

Parent, M., 1987, Late Pleistocene stratigraphy and events in the Asbestos-Valcourt region, southeastern Québec [Ph.D. thesis]: London, University of Western Ontario, 320 p.

Parent, M., and Occhietti, S., 1988, Late Wisconsinan deglaciation and Champlain Sea invasion in the St. Lawrence valley, Québec: Géographie Physique et Quaternaire, v. 42, p. 215–246.

Parent, M., and Occhietti, S., 1999, Late Wisconsinan deglaciation and glacial lake development in the Appalachians of southeastern Quebec, in Fowler, B.K., et al., eds., Late Quaternary History of the White Mountains, New Hampshire and adjacent southeastern Québec: Géographie Physique et Quaternaire, v. 53, p. 117–135.

Parent, M., Dubois, J.-M., Bail, P., Larocque, A., and Larocque, G., 1985, Paléogéographie du Québec méridional entre 12 500 et 8 000 ans BP.: Recherches Amérindiennes au Québec, v. 15, p. 17–37.

Plouffe, D., 1997, Glaciation et déglaciation au Wisconsinien supérieur dans les Monts Notre-Dame, Québec; Chronologie des écoulements glaciaires [M.S. thesis]: Montréal, Université du Québec, 118 p.

Poulin, P., 1977, Le complexe morainique de Saint-Narcisse dans le secteur sud de la rivière Malbaie. Interprétation paléoclimatique par l'analyse pollinique [M.A. thesis]: Québec, Université Laval, 83 p.

Prichonnet, G., 1984, Réévaluation des systèmes morainiques du sud du Québec (Wisconsinien supérieur): Commission Géologique du Canada Étude 83-29, 20 p.

Prichonnet, G., 1995, Géologie et géochronologie postglaciaire dans la région limitrophe de la Gaspésie et du Bas-Saint-Laurent, Québec: Commission Géologique du Canada Bulletin 488, 69 p.

Rampton, V.N., Gauthier, R.C., Thibault, J., and Seaman, A.A., 1984, Quaternary geology of New Brunswick: Geological Survey of Canada Memoir 416, 77 p.

Rappol, M., 1989, Glacial history and stratigraphy of the northwestern New Brunswick: Géographie Physique et Quaternaire, v. 43, p. 191–206.

Rappol, M., 1993, Ice flow and glacial transport in Lower St. Lawrence, Québec: Geological Survey of Canada Paper 90-19, 28 p.

Richard, P.J.H., 1977, Végétation tardiglaciaire au Québec méridional et implications paléoclimatiques: Géographie Physique et Quaternaire, v. 31, p. 161–176.

Richard, P.J.H., 1994a, Postglacial paleophytogeography of the eastern St. Lawrence River watershed and the climatic signal of the pollen record: Palaeogeography, Palaeoclimatology, Palaeoecology, v. 109, p. 137–163.

Richard, P.J.H., l994b, Wisconsinan late-glacial environmental change in Quebec: A regional synthesis: Journal of Quaternary Science, v. 9, p. 65–170.

Richard, P.J.H., Veillette, J.J., Larouche, A.C., Hétu, B., Gray, J.T., and Gangloff, P., 1997, Chronologie de la déglaciation en Gaspésie: nouvelles données et implications: Géographie Physique et Quaternaire, v. 51, p. 163–184.

Ridge, J.C., Besonen, M.R., Brochu, M., Brown, S.L., Callaghan, J.W., Cook, G.J., Nicholson, R.S., and Toll, N.J., 1999, Varve, paleomagnetic, and $^{14}$C chronologies for late Pleistocene events in New Hampshire and Vermont (USA), in Fowler, B.K., et al., eds., Late Quaternary History of the White Mountains, New Hampshire and adjacent southeastern Québec: Géographie Physique et Quaternaire, v. 53, p. 79–106.

Rodrigues, C.G., 1992, Successions of invertebrate microfossils and the late Quaternary deglaciation of the Central St Lawrence Lowland, Canada and United States: Quaternary Science Reviews, v. 11, p. 503–534.

Rodrigues, C.G., and Vilks, G., 1994, The impact of glacial lake runoff on the Goldthwait and Champlain Seas: The relationship between glacial lake Agassiz runoff and the Younger Dryas: Quaternary Science Reviews, v. 13, p. 923–944.

Rondot, J., 1974, L'épisode glaciaire de Saint-Narcisse dans Charlevoix, Québec: Revue de Géographie de Montréal, v. 28, p. 375–388.

Shilts, W.W., 1970, Pleistocene geology of the Lac Mégantic region, southeastern Québec, Canada [Ph.D. thesis]: Syracuse, New York, Syracuse University, 154 p.

Shilts, W.W., 1973, Glacial dispersal of rocks, minerals and trace elements in Wisconsinan till, southeastern Quebec, Canada, in Black, R.F., et al., eds., The Wisconsinan Stage: Geological Society of America Memoir 136, p. 189–219.

Shilts, W.W., 1981, Surficial geology of the Lac-Mégantic area, Québec: Geological Survey of Canada Memoir 397, 102 p.

Shilts, W.W., 1997, Erosional and depositional stratigraphy of the Appalachians of southeastern Quebec: 8th Biennial Congress, Canadian Quaternary Association (CANQUA), Montréal, Québec, Canada, Programme-Abstracts, p. 72.

Stea, R.R., and Mott, R.J., 1998, Deglaciation of Nova Scotia: Stratigraphy and chronology of lake sediment cores and buried organic sections: Géographie Physique et Quaternaire, v. 52, p. 3–21.

Stea, R.R., Piper, D.J.W., Fader, G.B.J., and Boyd, R., 1998, Wisconsinan glacial and sea-level history of Maritime Canada and the adjacent continental shelf: A correlation of land and sea events: Geological Society of America Bulletin, v. 110, p. 821–845.

Stuiver, M., and Brazunias, T.F., 1998, High-precision radiocarbon age calibration for terrestrial and marine samples: Radiocarbon, v. 40, p. 1127–1151.

Stuiver, M., and Polach, H.A., 1977, Discussion: Reporting of $^{14}$C data: Radiocarbon, v. 19, p. 355–363.

Stuiver, M., and Reimer, P.J., 1993, Extended $^{14}$C data base and revised CALIB 3.0 $^{14}$C age calibration program: Radiocarbon, v. 35, p. 215–230.

Stuiver, M., Grootes, P., and Braziunas, T.F., 1995, The GISP2 $\delta^{18}$O climate record of the past 16,500 years and the role of the sun, ocean and volcanoes: Quaternary Research, v. 44, p. 341–354.

Stuiver, M., Reimer, P.J., Bard, E., Beck, J.W., Burr, G.S., Hughen, K.A., Kromer, B., McCormac, F.G., v. d. Plicht, J., and Spurk, M., 1998, INTCAL98 radiocarbon age calibration 24,000–0 cal BP: Radiocarbon, v. 40, p. 1041–1083.

Terasmae, J., and Lasalle, P., 1968, Notes on late-glacial palynology and geochronology at St. Hilaire, Quebec: Canadian Journal of Earth Sciences, v. 5, p. 249–257.

Thomas, R.H., 1977, Calving bay dynamics and ice sheet retreat up the St. Lawrence Valley system: Géographie Physique et Quaternaire, v. 31, p. 347–356.

Thompson, W.B., Fowler, B.K., Flanagan, S.M., and Dorion, C.C., 1996, Recession of the late Wisconsinan ice sheet from the northwestern White Mountains, New Hampshire, in Van Baalen, M.R., ed., Guidebook to fieldtrips in northern New Hampshire and adjacent regions of Maine and Vermont: Cambridge, Massachusetts, Harvard University, New England Intercollegiate Geological Conference, 88th annual meeting, p. B4-1–B4-32.

Thompson, W.B., Fowler, B.K., and Dorion, C.C., 1999, Deglaciation of the northwestern White Mountains, New Hampshire, in Fowler, B.K., et al., eds., Late Quaternary History of the White Mountains, New Hampshire and adjacent southeastern Québec: Géographie Physique et Quaternaire, v. 53, p. 59–77.

Veillette, J., and Cloutier, M., 1993, Géologie des formations en surface, Gaspésie, Québec: Commission Géologique du Canada Map 1804A, scale 1: 250 000.

Vincent, J.-S., 1989, Quaternary geology of the southeastern Canadian Shield, in Fulton, R.J., ed., Quaternary geology of Canada and Greenland: Boulder, Colorado, Geological Society of America, Geology of North America, v. K-1, p. 249–275.

Yu, Z., and Eicher, U., 1998, Abrupt climate oscillations during the last deglaciation in central North America: Science, v. 282, p. 2235–2238.

MANUSCRIPT ACCEPTED BY THE SOCIETY JUNE 9, 2000

# Relative sea-level changes in the St. Lawrence estuary from deglaciation to present day

**Jean-Claude Dionne**
*Department of Geography and Centre d'Études Nordiques, Université Laval, Québec, Québec G1K 7P4, Canada*

## ABSTRACT

The St. Lawrence estuary was deglaciated between 13 and 12 ka. The south shore was ice free by 12 ka, whereas some areas of the north shore (Baie-St-Paul, La Malbaie, Tadoussac, and Baie-Comeau) were partly ice covered until 11.5–11 ka. On the south shore, however, the northern margin of the remaining Appalachian ice was generally situated close to (2–3 km) the present-day shoreline. Depressed coastal areas were progressively submerged by the Goldthwait Sea. On the south shore, its maximum level ranges between 110 m at Matane and 180 m at Lévis, although the altitude does not increase upstream in a regular fashion as previously thought. A similar pattern characterizes the north shore. The maximum postglacial sea level is 170 m at Baie-Comeau, 140–150 m at Tadoussac, 175 m at Baie-St-Paul, and 190 m at Québec City. Because the relative sea level and the isostatic recovery were not synchronous on both shores, geologic events should be examined separately at least for the first half of the Holocene. On both shores, however, the isostatic recovery was relatively rapid; more than 85% occurred during the first 3 k.y. after deglaciation. On the south shore, the relative sea level was close to present day ca. 8–7 ka; a lowstand (±10 m) occurred between 7 and 6 ka and was followed by the 10–12 m Laurentian transgression between 5.8 and 3.5 ka. Evidence for these two events has been collected from 11 localities between Québec City and Matane. Field work in progress on the north shore between Québec City and Baie-Comeau also provides evidence for these two events. The last important relative sea level fluctuation is related to the Micmac shoreline (±6 m level), an event dated as 2.5–2 ka on both shores except at Tadoussac where it is slightly younger (1.4 ka). The lowstand and the Laurentian transgression, which occurred during the Holocene, are possibly related to the northward migration of the forebulge and to the disintegration of the Laurentide ice sheet remnant in central Québec during the hypsithermal period in southeastern Canada, but may also be related to tectonic activity in an area of high seismicity.

## INTRODUCTION

Although knowledge of the postglacial relative sea level and the isostatic coastal land recovery of the St. Lawrence estuary and gulf has increased considerably during the past two decades (Dionne, 1977, 1997a), there is still a lack of data for extensive sectors on both shores. Work in progress and additional field investigations should help to clarify various questions and to solve some problems of correlation related to deglaciation, postglacial marine submergence, and subsequent land emergence related to isostatic crustal readjustment.

In the St. Lawrence estuary and gulf, research on former sea levels and land emergence was initiated at the end of the nineteenth and beginning of the twentieth centuries by geologists (Chalmers, 1886, 1896, 1904, 1905; Goldthwait, 1911a, 1911b, 1912, 1913, 1914; Fairchild, 1918; Twenhofel and Conine, 1921; Coleman, 1922). However, very little field work was carried out between 1925 and 1960 (Faessler, 1948; Lougee, 1953a, 1953b). Until 1960, the main objective concerned mapping the maximum level reached by the postglacial sea, now called the Goldthwait Sea (Dionne, 1972). There were no data available to determine the chronology of the marine submergence and the rate of isostatic rebound. The first radiocarbon date for a raised shoreline in the St. Lawrence estuary was obtained in 1958 (Lougee and Lougee, 1976). A piece of wood collected in a terrace at about 13 m above sea level (asl), at Baie-Ste-Catherine, near Tadoussac, provided an age of 3150 ± 30 $^{14}$C yr B.P. (Dionne, 1996a).

Most data for sea levels and coastal isostatic recovery were gathered after 1960 (Dionne, 1977), and especially after 1980 (Dionne, 1997a). The first relative sea-level curve for the south shore of the St. Lawrence estuary was based on data by Lee (1962, 1963) and was published by Elson (1969). Since then many curves have been proposed for various regions of the estuary and gulf (Table 1). A major problem for construction of sea-level curves remains insufficient data. Because few areas have been studied in detail, there are not enough reliable $^{14}$C dates, particularly for raised beaches. Other problems are that the maximum level of the Goldthwait Sea is poorly documented along large sectors on both shores of the St. Lawrence estuary, and postglacial submergence of the coastal areas was neither synchronous on both shores, nor necessarily progressive everywhere from northeast to southwest.

**TABLE 1. REFERENCES FOR RELATIVE SEA LEVEL VERSUS ISOSTATIC CURVES FOR THE ST. LAWRENCE ESTUARY AND GULF**

| Location | Reference |
|---|---|
| North shore | Sept-Iles (Dredge, 1971; Dubois, 1979; Hillaire-Marcel, 1979) |
| | Blanc-Sablon (Dubois, 1979; Bigras and Dubois, 1987) |
| | Anticosti Island (Bigras and Dubois, 1987) |
| | Lower, middle and upper North Shore (Dubois, 1979; Bigras and Dubois, 1987) |
| | Grandes-Bergeronnes (Daigneault, 1985; Archambault, 1987) |
| | Baie-St-Paul (Bonenfant, 1993) |
| | La Malbaie (Govare, 1996) |
| | Tadoussac (Dionne and Occhietti, 1996) |
| South shore | Rivière-du-Loup (Elson, 1969) |
| | Baie-des-Sables-Trois-Pistoles (Locat, 1977; Hillaire-Marcel, 1979) |
| | St-Eugène-Trois-Pistoles (Lortie and Guilbault, 1984) |
| | St-Fabien-sur-Mer (Dionne, 1988b) |
| | Montmagny (Dionne, 1988d) |
| | Montmagny-Trois-Pistoles (Bélanger, 1993) |
| | Rimouski (Hétu, 1994) |
| | Matane (Dionne and Coll, 1995) |
| Gaspé Peninsula | North shore (Hillaire-Marcel, 1979; Gray and Hétu, 1981; Gray, 1987) |
| | South shore (Bail, 1983, 1987) |

The purpose of this chapter is to highlight the events related to sea-level changes and coastal land recovery (Dionne, 1988a, 1991), emphasizing the events that occurred during the middle part of the Holocene, particularly on the south shore of the estuary between Québec City and Matane (Fig. 1).

*Sequence of main events*

There are three major periods to consider (Table 2). The first period extends from ca. 13 to 9 ka and is related to the deglaciation of the St. Lawrence Valley and the adjacent coastal land of the Appalachian region to the south and the Laurentides to the north (Occhietti et al., this volume). This period is characterized by a major transgression and subsequent isostatic land recovery. The events were not necessarily synchronous on both shores and from downstream to upstream; for example, Matane was deglaciated and submerged before Québec City; and the coastal area of Charlevoix between Québec City and Baie-St-Paul was deglaciated before the areas of La Malbaie, Tadoussac, Baie-Comeau and Sept-Iles. Also, the isostatic recovery did not increase progressively upstream as suggested in the past; for example it increased from Ste-Anne-des-Monts to Trois-Pistoles (Lebuis and David, 1977; Locat, 1977), but from there it decreased upstream to only 145 m at Rivière-du-Loup (Dionne, 1990) and to 160–165 m at Montmagny (Dionne, 1988d).

This first period can be subdivided into three phases. The first, from 13 to 12 ka, is related to the deglaciation of the St. Lawrence Valley and adjacent coastal areas. Marine coastal submergence occurred when the ice sheet separated into two major ice masses, which remained on the margins of the Appalachian highlands to the south and the Laurentides to the north (Chauvin et al., 1985; David and Lebuis, 1985; Hétu, 1998). Glaciomarine sedimentation and diamictons characterized this period, during which a few deltas were constructed. However, beaches are rare because the shoreline was commonly the ice front. The interval 12–11 ka corresponds with the main Goldthwait Sea transgression. It is characterized by deep-water, fine-grained deposits, shallow-water sandy deposits, sand and gravel beaches, and a few large deltas, and also by a high rate of isostatic recovery (5–10 cm/yr) (Dionne, 1977; Locat, 1977). Macrofauna is relatively abundant in these deposits. Regression prevailed during the third phase, which took place between 11 and 9 ka. This interval was characterized by a slowing rate of isostatic uplift (<1 cm/yr) and an increasing rate of eustatic sea-level rise, and increasing fossiliferous shallow-water deposits and sand and gravel beaches.

The second major period covers the interval between 9 and 6 ka. It is characterized by emergence of land and a lowstand of the relative sea level. This period can be subdivided into two phases. The interval between 9 and 8 ka is not well documented. There are only a few $^{14}$C dates available. We do not know exactly what happened during this interval. However, it is likely that the relative sea level was similar to that of present day or

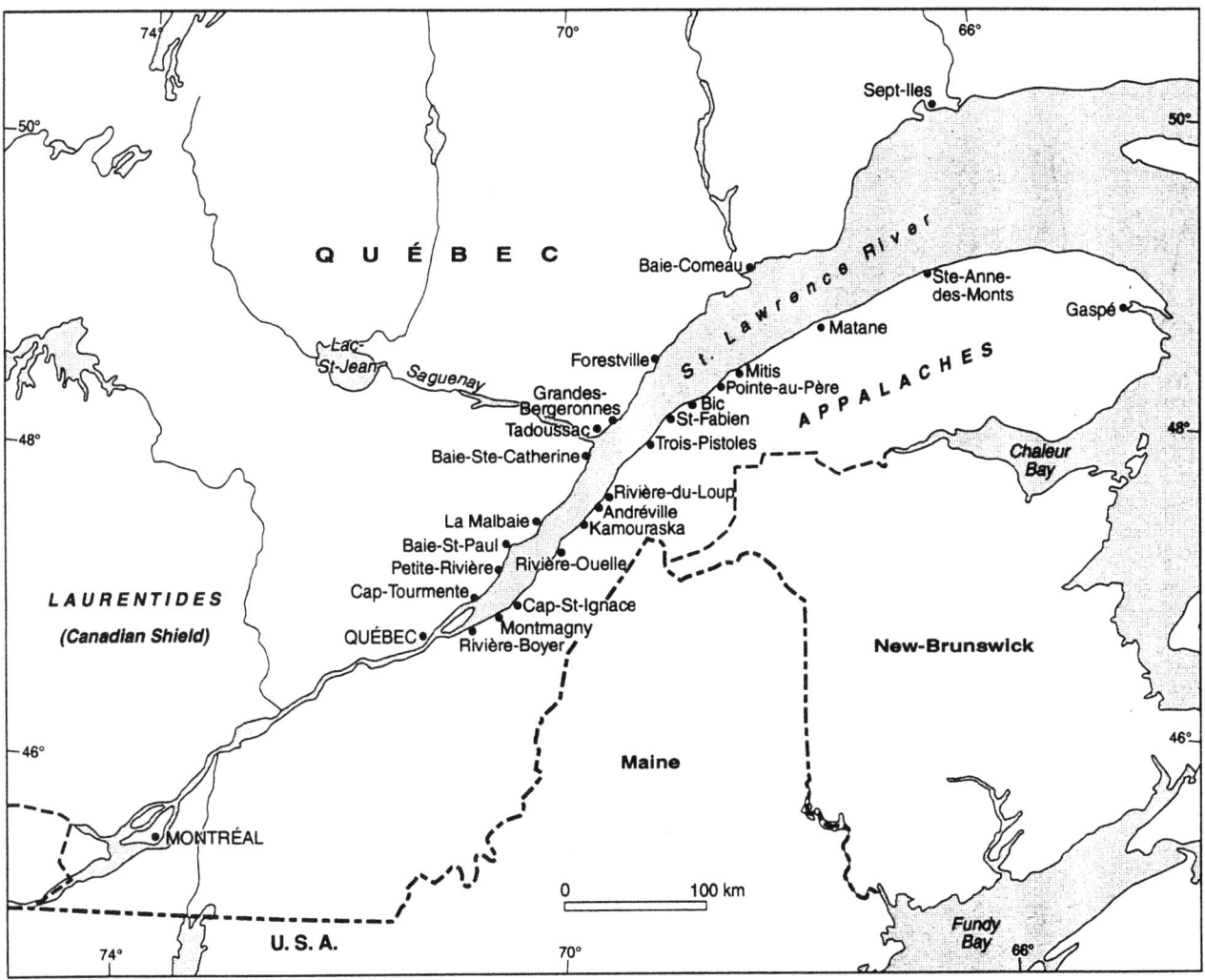

Figure 1. Location map and place names.

was even lower. The second phase, which took place between 8 and 6 ka, is characterized by a low stand of the relative sea level. It can be subdivided into two intervals. Between 8 and 7 ka the sea level was similar to that of today, or was even lower, as suggested by intertidal and infratidal deposits. Between 7 and 6 ka the sea level was lower than at present by at least 10 m and possibly more (Dionne, 1988d).

The third major period, which extends from 6 ka to the present day, was characterized by the Laurentian transgression (± 10 m). This event took place between 5.6 and ~4 ka, and was related to eustatic sea-level rise and possibly also to regional tectonic events associated with the northward migration of the forebulge. It is characterized by transgressive intertidal deposits (Dionne, 1985, 1988c, 1988d, 1997b; Dionne and Coll, 1995). The Laurentian transgression was followed by a phase of regression between 4 and 3 ka related to land recovery. By 3 ka, sea level most probably reached the level of today. Between 2.5 and 1.5 ka a minor transgression (± 2 m) occurred.

This event is called the Mitis stage and is well dated on both shores of the St. Lawrence estuary (Dionne, 1992, 1993). After that event, the land was uplifted at the rate of a few millimeters per century until recently.

The trend of the relative sea level is not well known. Studies made from tide-gauge data are sometimes contradictory (Table 3), indicating emergence, submergence, or stability depending upon localities (Dohler and Ku, 1970; Pirazzoli, 1986; Emery and Aubry, 1991; Anctil and Troude, 1992). However, there is geomorphic evidence at many localities on both shores of the St. Lawrence for a slowly rising sea level (mm/yr) (Dionne, 1986a).

## EVIDENCE FOR MID-HOLOCENE SEA-LEVEL FLUCTUATIONS

There is evidence in many localities on the south shore and even on the north shore of the St. Lawrence estuary for a low

### TABLE 2. DESCRIPTION OF SEQUENCE OF EVENTS, RELATIVE SEA LEVEL, AND ISOSTATIC RECOVERY, SOUTH SHORE OF THE ST. LAWRENCE ESTUARY

1. First phase of late vs. postglacial transgression
   Period: 13–12 ka
   Event related to crust depressed by the ice sheet
   Maximum level: variable but generally increasing upstream
   Sedimentation: mainly glaciomarine silt clay or diamictons; few beaches and few deltas
   Early phase of isostatic recovery, low eustatic sea level
2. Main Goldthwait Sea transgression
   Period: 12–11 ka
   High rate of isostatic recovery
   Eustatic sea level rising
   Sedimentation: deep-water fine-grained deposits; shallow-water sand deposits; a few sand and gravel beaches; a few deltas
3. Main phase of regression
   Period: 11–9 ka
   Rate of isostatic recovery slowing
   Eustatic sea level rising
   Sedimentation: shallow-water deposits; sand and gravel beaches
4. Period of transition: not well documented
   Period 9–8 ka
   Sea level possibly similar to present day
   Possible forebulge migration to the north
5. Lowstand
   Period 8–6 ka (two phases)
   a) 8–7 ka: Sea level similar or lower than today
   Sedimentation: intertidal and/or subtidal deposits
   b) 7–6 ka: Sea level lower than at present;
   Period of erosion; stone pavement formation
6. Laurentian transgression
   Period: 5.6–4 ka
   Rising sea level (transgressive sequence)
   Possible tectonic activity
   Sedimentation: intertidal flat deposits
7. Second phase of regression
   Residual land recovery
   Period: two possibilities
   a) 4 ka to present day with a halt (pause) between 2.5 and 1.5 ka (Mitis stage)
   b) 4 ka–2.5 ka = regression
      2.5–1.5 ka = minor sea-level fluctuation (1–2 m): Mitis stage
      1.5 ka to present day = regression
8. Present day: Sea-level tendency poorly known
   Three hypotheses:
   a) Isostatic land upheaval (few mm/yr)
   b) Period of relative stability
   c) Rising relative sea level (± 1 mm/yr) at least locally
   Geomorphic evidence on south and north shores

### TABLE 3. TREND OF RELATIVE SEA-LEVEL CHANGE FROM TIDE GAUGE DATA, ST. LAWRENCE ESTUARY AND GULF

| Locality | Period | Regression slope (mm/yr) | Type of fluctuation | t-confidence |
|---|---|---|---|---|
| **A. EMERY and AUBREY (1991)** | | | | |
| Ste-Anne-des-Monts | 1970–1980 | 2.2 | Emergence | 0.96 |
| Pointe-au-Père | 1925–1980 | 0.5 | Submergence | 1 |
| Rivière-du-Loup | 1970–1980 | 8 | Emergence | 0.85 |
| St-François (Ile d'Orléans) | 1965–1980 | 2.5 | Submergence | 0.8 |
| Québec | 1965–1980 | 3.2 | Submergence | 0.79 |
| Tadoussac | 1968–1980 | 5.7 | Emergence | 0.72 |
| Harrington Harbour | 1940–1980 | 0.3 | Emergence | 1 |
| **B. PIRAZZOLI (1986)** | | | | |
| Harrington Harbour | 1940–1979 | 0 | Stability | — |
| Pointe-au-Père | 1897–1977 | 0.2 | Emergence | — |
| Québec | 1894–1949 | 0.9 | Emergence | — |
| **C. DOHLER & KU (1970)** | | | | |
| Harrington Harbour | 1939–1970 | 1.2 | Submergence | >90% |
| Pointe-au-Père | 1909–1970 | 0.2 | Emergence | <90% |
| Québec | 1937–1970 | 0.5 | Emergence | >90% |

### TABLE 4. EVIDENCE OF THE LAURENTIAN TRANSGRESSION AND/OR THE HOLOCENE LOWSTAND ON THE SOUTH SHORE OF THE ST. LAWRENCE ESTUARY

| Main Localities | References |
|---|---|
| St-Vallier (riv. Boyer) | Dionne (1985) |
| Anse de Bellechasse | Dionne (1997b, 2000) |
| **Montmagny** (type section) | Dionne (1988d, 1998) |
| | Lortie and Dionne (1990) |
| Cap-St-Ignace | Dionne (1998) |
| Rivière-Ouelle | Dionne (1988c) |
| | Pfalgraf et al. (1996) |
| Rivière-du-Loup | Dionne (1990) |
| St-Fabien-sur-Mer | Dionne (1988b) |
| Cap-à-l'Original (baie des Hahas) | Dionne (1988b) |
| Bic (anse aux Bouleaux) | Dionne (unpublished) |
| Matane | Dionne and Coll (1995) |
| **Secondary localities** (partial evidence) | |
| Québec City (riv. St-Charles) | Samson et al. (1977) |
| Cap-Tourmente | Troude (1985) |
| Kamouraska | McNeely (1989) |
| Andréville (riv. Fouquette) | Dionne (unpublished) |
| Cacouna-Est | Dyck and Fyles (1963) |
| | Dionne (1977) |
| Baie de Mitis | Dionne et Coll (1995) |

sea level prior to 6 ka followed by a transgression of ~10 m amplitude (Table 4).

A complete record of the geologic events that took place from 8 ka to the present is in the Montmagny area. The events were observed in a few vertical sections in the 8–10 m terrace and in the intertidal zone at Montmagny, Cap-St-Ignace, Pointe St-Thomas, and Bellechasse Cove since 1980 (Dionne, 1997b, 1998). The sections studied were all exposed by erosional processes.

### Montmagny Airport section

The most complete sequence of events is preserved in the Montmagny Airport geologic section (Fig. 2A), which is therefore considered to be the type section. Figure 3 summarizes the lithostratigraphy and the geologic events related to sea-level fluctuations during the Holocene.

At the base of the section is a Goldthwait Sea fossiliferous gray silt-clay deposit. Unit 1 is not visible at the base of the cliff but was observed in trenches dug by Dionne in 1985 and 2000, and Guy Lortie in 1986. However, this marine unit is close to the surface (10–15 cm) and sometimes crops out in the intertidal zone. A stone lag (Dionne, 1987) characterizes the surface of the underlying clay unit in some places and is interpreted to represent an erosion surface cut during a lowstand of

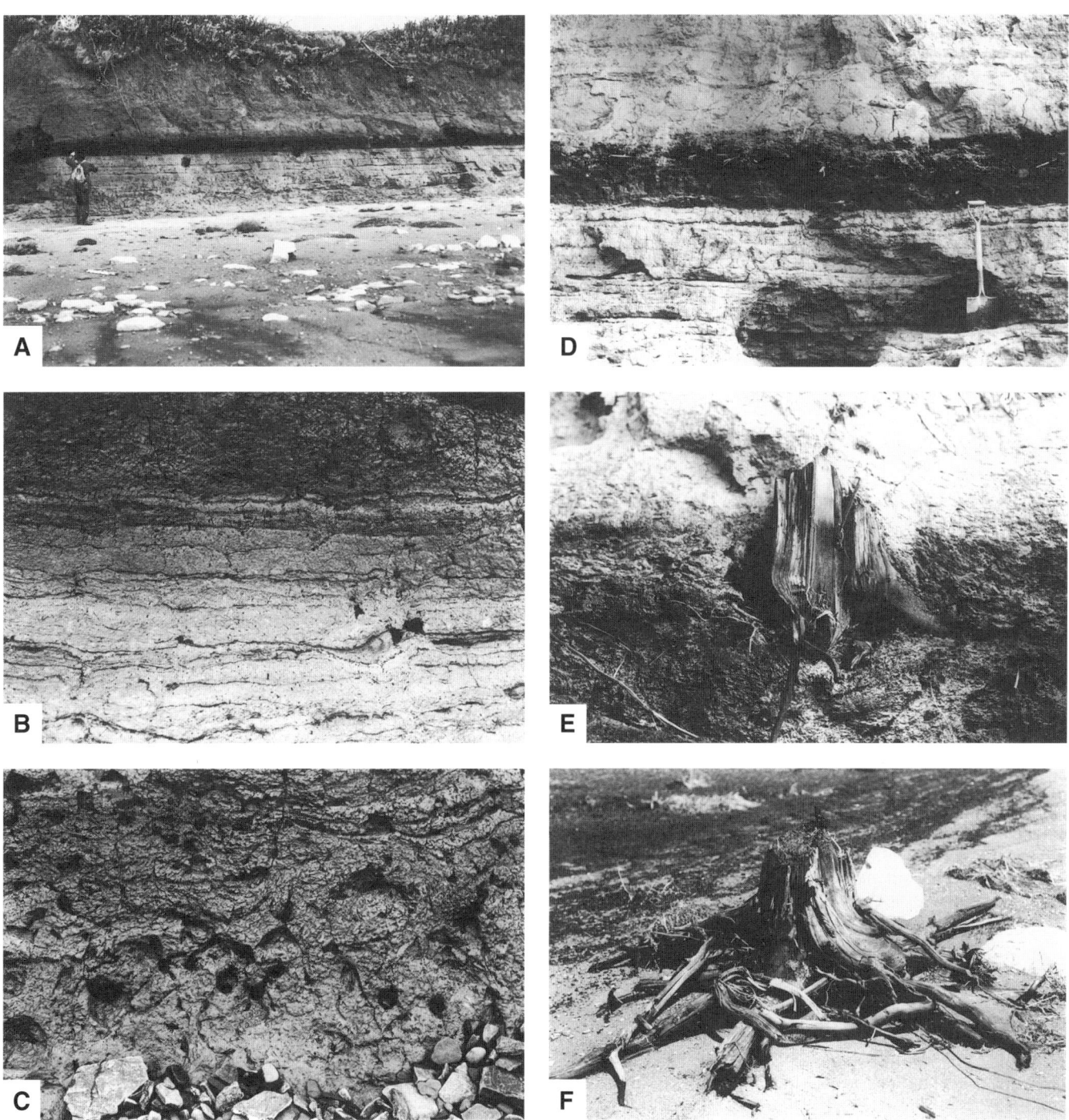

Figure 2. A, General view of Montmagny Airport section showing organic layer (black) between two intertidal units in 8 m terrace (85.6.6: Date photograph was taken, June 6, 1985). B, General aspect of unit 2 at base of Montmagny Airport section showing thinly stratified silt and clay (low tidal-flat facies) dated as 7–8 ka (95.11.8). C, Details of unit 2 showing remains (bulbs) of aquatic plants (95.11.8). D, Details of organic layer embedded into intertidal deposits, Montmagny Airport section (90.5.28). E, Stump in situ in organic layer (unit 3), Montmagny Airport section (96.5.22). F, Stump recently exposed by erosion of organic layer (unit 3) in 8 m terrace, Montmagny Airport section (94.5.15).

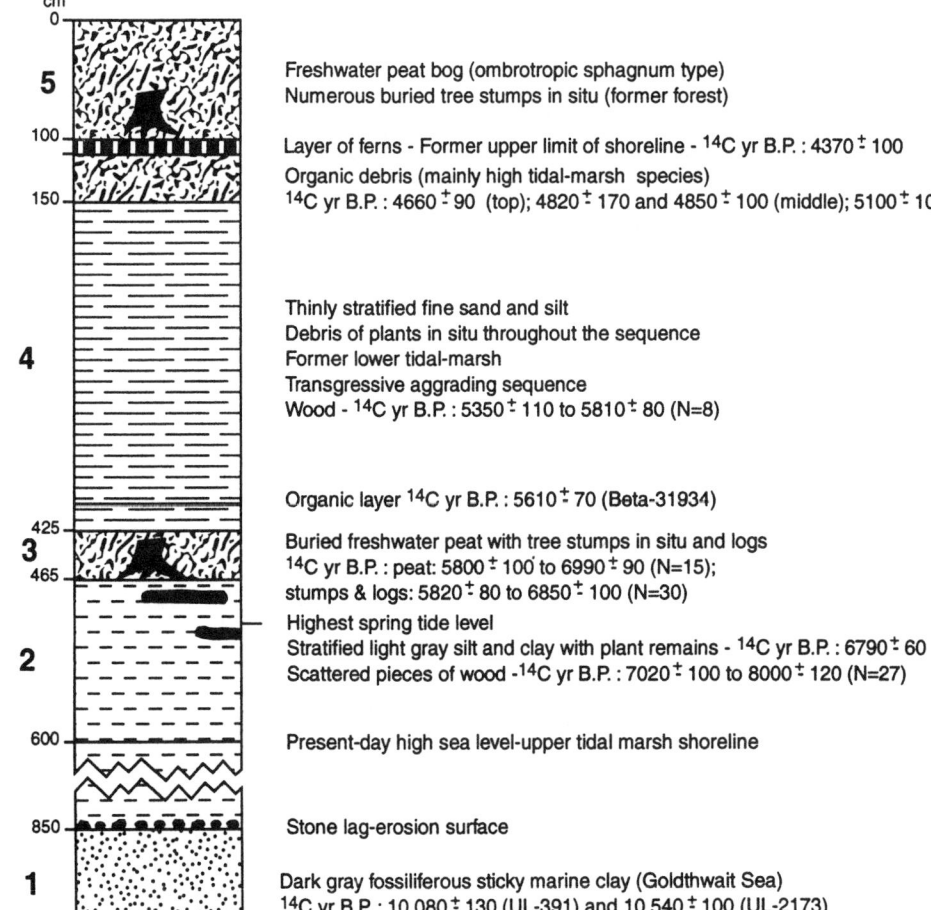

Figure 3. Typical section in 8–10 m terrace at Montmagny, south shore of St. Lawrence estuary.

relative sea level, which occurred during the first half of the Holocene, likely between 9 and 8 ka. The clay contains a variety of shells characteristic of relatively deep water; they provided $^{14}$C dates ranging between 10 and 11 ka (Dionne, 1998).

Unit 2, which unconformably overlies the Goldthwait Sea clay, is a stratified silt and clay deposit (rhythmites), 4 m thick, including thin (1–2 mm) laminae of sand (Fig. 2B). The lower 2 m contain scattered organic debris whereas the upper 2 m contain relatively abundant intertidal plant remains (Fig. 2C), small wood fragments, and even small tree trunks near the top of the unit. Unit 2 does not contain any marine shells but has a few forams and ostracoda, the most abundant species being *Elphidium excavatum forma clavata* and *Cyprideis* sp., respectively. The forams are most probably reworked from older Goldthwait Sea clay deposits (Dionne, 1998). The diatoms were studied by Guy Lortie (Lortie and Dionne, 1990). *Bacterosira fragilis* is a characteristic taxon. Unit 2 has been interpreted as a low intertidal or subtidal deposit unconformably overlying the erosion surface of the underlying marine clay. Radiocarbon dates on wood fragments (N = 27) range from 7020 ± 100 to 8000 ± 120 $^{14}$C yr B.P., whereas plant remains in situ at about the middle of unit 2 released a date of 7570 ± 120 $^{14}$C yr B.P. (Table 5). The altitude of the relative sea level during this episode is not known: however, it was probably similar to or even slightly lower than present-day sea level. Diatom assemblages in the upper part of unit 2 indicate shallowing water or emerging conditions (Lortie and Dionne, 1990).

Unit 3 is mostly an organic layer 50–65 cm thick containing logs and tree stumps in situ (Fig. 2, D–F) The nature of the organic debris and the macrofossils were studied by Alayn Larouche and Michelle Garneau (Dufresne, 1994; Bhiry et al., 2000). The plant species composing the organic layer (Dionne, 1998) indicate clearly that the surface was emerged at the time. Tree stumps and logs in unit 3 (Table 6) indicate a mixed forest of resinous and decidous species, spruce (*Picea* sp.) and tamarack (*Larix laricina*) being the two more common resinous species and ash (*Fraxinus* sp.) dominating in the decidous species. Radiocarbon dates (N = 30) on stumps and logs range from 5820 ± 80 to 6850 ± 100 $^{14}$C yr B.P. (Table 7), whereas $^{14}$C dates (N = 15) on the peat range from 5800 ± 100 to 6990 ± 90 $^{14}$C yr B.P. (Table 8). During this episode the relative sea level was at least 5 m lower than today (Dionne, 1988d).

Unconformably overlying the peat layer is a 3–4-m-thick deposit (unit 4) of fine sand and silt stratified in relatively thin layers (Fig. 4A). The upper 2 m contain abundant roots and

## TABLE 5. RADIOCARBON DATES ON WOOD AND ORGANIC DEBRIS IN UNIT 2 MONTMAGNY AIRPORT SECTION

| Laboratory number | Age ($^{14}$C yr B.P.) | Species |
|---|---|---|
| UL-390 | 7020 ± 100 | Tsuga canadensis |
| Beta-31935 | 7040 ± 80 | |
| Beta-18329 | 7040 ± 100 | Abies balsamea |
| GSC-4862 | 7070 ± 70 | Ulmus sp. |
| Beta-58203 | 7090 ± 70 | |
| Beta-26492 | 7100 ± 50 | |
| Beta-31393 | 7160 ± 90 | |
| Beta-26488 | 7170 ± 80 | |
| Beta-26489 | 7170 ± 80 | |
| Beta-28374 | 7190 ± 70 | Thuya occidentalis |
| Beta-95590 | 7210 ± 80 | |
| Beta-95589 | 7250 ± 70 | |
| UL-442 | 7250 ± 110 | Betula papyrifera |
| UQ-745 | 7260 ± 80 | |
| Beta-31398 | 7290 ± 70 | |
| UQ-977 | 7300 ± 100 | |
| UL-597-2 | 7310 ± 100 | Picea sp. |
| UQ-974 | 7400 ± 200 | |
| Beta-44579 | 7470 ± 60 | |
| UL-298 | 7500 ± 90 | Betula papyrifera |
| UL-736 | 7520 ± 90 | |
| UL-2137 | 7570 ± 120 | Plant remains |
| UQ-746 | 7640 ± 90 | |
| UL-736-2 | 7740 ± 80 | |
| UL-739 | 7800 ± 100 | |
| Beta-32323 | 7820 ± 60 | |
| UL-1431 | 7830 ± 100 | |
| UL-1430 | 8000 ± 120 | Picea sp. |

## TABLE 6. STUMPS AND LOGS IN UNIT 3, MONTMAGNY AIRPORT SECTION

| Resinous | Number | Deciduous | Number |
|---|---|---|---|
| Abies balsamea | 1 | Acer rubrum | 1 |
| Larix laricina | 3 | Acer saccharum | 1 |
| Picea sp. | 7 | Fraxinus sp. | 5 |
| Picea/Larix | 3 | Fraxinus nigra | 3 |
| Pinus strobus | 1 | Ulmus | 2 |
| | | Juglans nigra | 1 |
| | | Tilia americana | 1 |

## TABLE 7. RADIOCARBON DATES ON STUMPS AND LOGS, UNIT 3 OF THE MONTMAGNY AIRPORT SECTION

| Laboratory number | Age ($^{14}$C yr B.P.) | Species |
|---|---|---|
| Beta-18327 | 5820 ± 80 | Picea sp. |
| Beta-18326 | 5830 ± 70 | Abies balsamea |
| Beta-24656 | 5830 ± 80 | Picea sp. Larix laricina |
| UL-747 | 5840 ± 90 | Pinus strobus |
| Beta-45270 | 5850 ± 60 | |
| Beta-32321 | 5940 ± 90 | Tilia americana |
| Beta-18328 | 5950 ± 80 | Larix laricina |
| UL-170 | 5950 ± 90 | Picea sp. |
| Beta-31395 | 5960 ± 70 | Juglans nigra |
| Beta-45267 | 5980 ± 60 | |
| UL-137 | 6000 ± 80 | |
| UL-351 | 6040 ± 120 | Larix laricina |
| I-14290 | 6050 ± 110 | Picea sp. |
| Beta-32319 | 6130 ± 80 | Fraxinus nigra |
| Beta-52290 | 6150 ± 70 | Fraxinus sp. |
| UL-171 | 6150 ± 80 | Picea sp. |
| Beta-52639 | 6220 ± 50 | |
| UQ-969 | 6300 ± 110* | |
| UQ-1179 | 6300 ± 100* | Picea sp. |
| UL-740-2 | 6330 ± 90* | |
| UL-1448 | 6410 ± 80* | |
| UL-137-2 | 6470 ± 80 | |
| Beta-14448 | 6480 ± 120 | Picea sp. |
| UL-441 | 6550 ± 100* | |
| Beta-45269 | 6600 ± 60* | |
| UL-740 | 6660 ± 110* | |
| Beta-24658 | 6680 ± 90* | |
| UL-735 | 6790 ± 90 | Fraxinus sp. |
| UL-734 | 6830 ± 100 | Fraxinus sp. |
| UL-1178 | 6850 ± 100* | Picea sp. |

*Logs.

## TABLE 8. RADIOCARBON DATES ON THE PEAT LAYER, UNIT 3 OF THE MONTMAGNY AIRPORT SECTION

| Laboratory number | Age ($^{14}$C yr B.P.) |
|---|---|
| QU-1328 | 5800 ± 100 |
| UQ-972 | 5800 ± 100 |
| GSC-4919 | 5860 ± 70 |
| UQ-650 | 6010 ± 90 |
| UL-116 | 6040 ± 80 |
| QU-1327 | 6050 ± 100 |
| GSC-4881 | 6100 ± 70 |
| UL-2176 | 6130 ± 70 |
| UL-2175 | 6340 ± 80 |
| UL-115 | 6420 ± 80 |
| UQ-653 | 6590 ± 110 |
| GSC-4870 | 6760 ± 70 |
| QU-1329 | 6760 ± 100 |
| UL-114 | 6800 ± 90 |
| UQ-752 | 6990 ± 100 |

twigs (Fig. 4B) of intertidal plant species in situ including bullrush (Scirpus sp.). It is interpreted as a tidal-flat deposit passing from a low intertidal facies to a low marsh facies. The lower 2 m of unit 4 is mostly devoid of transported organic macrofossils. Only a few pieces of organic material, including tree logs and peat clasts, were observed at the Montmagny Airport section. These organic debris were probably reworked material from unit 3; dated samples (N = 8) provided radiocarbon ages from 5.3 to 5.8 ka, whereas shells in growing position (Macoma balthica) in a section at Pointe St-Thomas were dated as 5.4 and 5.5 ka. The plant remains in growth position in the upper 2 m of unit 4 were dated 5.5 and 5.6 ka. In addition, two $^{14}$C dates on fossil plant rhizomes and roots (Scirpus sp.) at the top of unit 4 provided an age of 4.4 and 5.1 ka. The younger date corresponds to the last phase of the Laurentian transgression. Another characteristic of unit 4 is the absence of ice-rafted coarse debris, suggesting warmer climatic conditions during this episode or the absence of coarse debris in the intertidal zones at the time. This situation contrasts greatly with present-day conditions (Fig. 4C), ice processes being very active (Dionne, 1987). Because unit 4 was deposited over the peat layer, it implies a rise of the relative sea level or a transgression.

At the surface of the 8–10 m terrace, in the eastern area of the Montmagny embayment, and ~200–300 m inland from the cliff margin, there is a peat bog 50–150 cm thick (unit 5), which

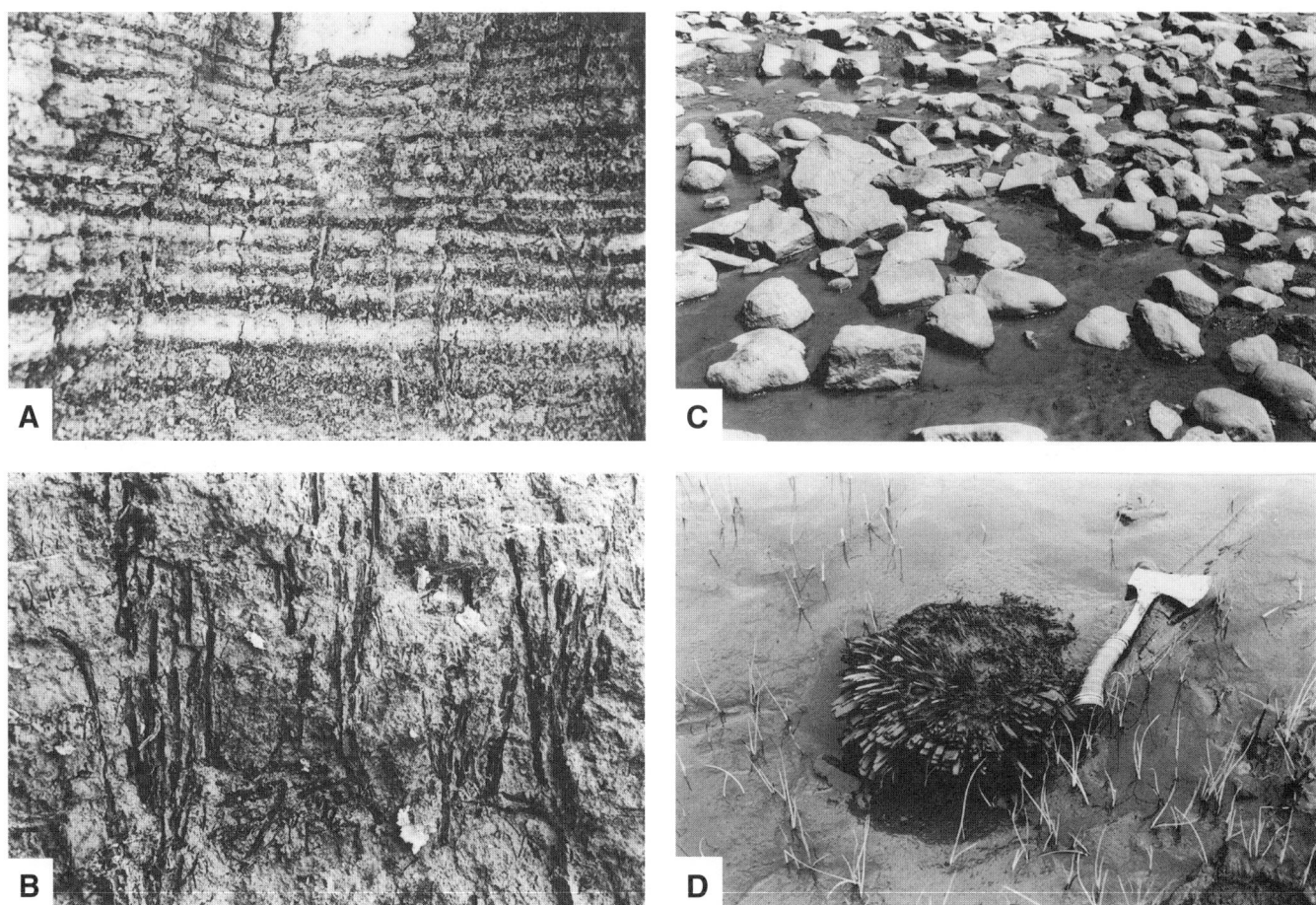

Figure 4. A, Detailed view of unit 4 composed of stratified silt and fine sand (low marsh facies), Montmagny Airport section (96.4.13: Date photograph was taken, April 13, 1996). B, Detailed view of unit 4 showing plant remains (*Scirpus americanus*) in growing position (95.6.13). C, Typical aspects of intertidal low marsh, in May, in front of Montmagny Airport section showing lag of small boulders at surface of underlying Goldthwait sea clay (91.5.13). D, Stump partially buried by fresh mud occurring at seaward edge of intertidal low marsh in area facing Montmagny Airport section (85.6.25).

is dominated by sphagnum species with stumps of spruce (*Picea* sp.) and tamarack (*Larix laricina*) in growth position. The transition between the ombrotrophic bog and the lower 50 cm of unit 5 is characterized by a layer of ferns (*Osmunda* sp.) dated as 4.3 ka, indicating the end of the Laurentian transgression, which was followed by the emergence of the 8–10 m terrace.

### Cap-St-Ignace and Pointe St-Thomas sections

A similar vertical sequence of deposits characterizes a 5–6-m-high cliff at Cap-St-Ignace, about 4 km northeast of the Montmagny exposure (Fig. 5). A large excavation made in the 8 m terrace at that locality in 1993 allowed observation of the organic layer (unit 3), wherein tree stumps and logs occur, at ~1000 m inland from the edge of the cliff. At the Airport section, the organic layer with logs and stumps in situ was observed in a gully as much as 150 m inland.

This organic layer (unit 3) has not yet been observed in the area west of Rivière du Sud. However, at the southwest extremity of the Montmagny embayment (Pointe Saint-Thomas), an exposure in the 4-m-high cliff (Fig. 6) showed shells (*Macoma baltica*) in living position in the intertidal unit emplaced during the Laurentian transgression. Two samples provided $^{14}$C dates of 5.4 and 5.5 ka. The age of unit 4 is thus similar in the three sections examined in the Montmagny area.

### Bellechasse Cove section

At Bellechasse Cove, ~15 km southwest of Montmagny, another section provides evidence for the low sea level between 6 and 7 ka followed by the Laurentian transgression.

The first description of the 8-m-high section (Dionne, 1997b) did not report any organic layer between the clay substrate and the unconformably overlying intertidal deposit because it was not exposed at the time. Three small landslides,

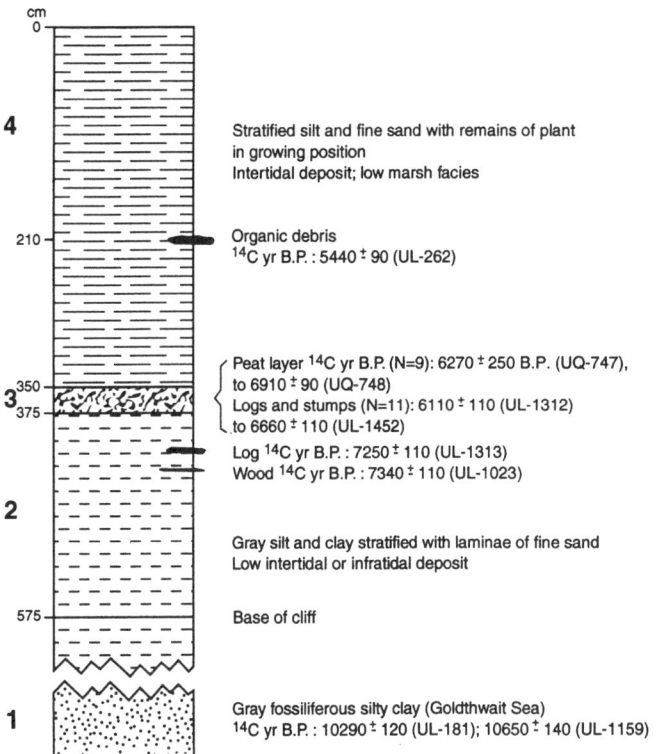

Figure 5. Section in 8 m terrace at Cap-St-Ignace ~4.5 km northeast of Montmagny Airport section.

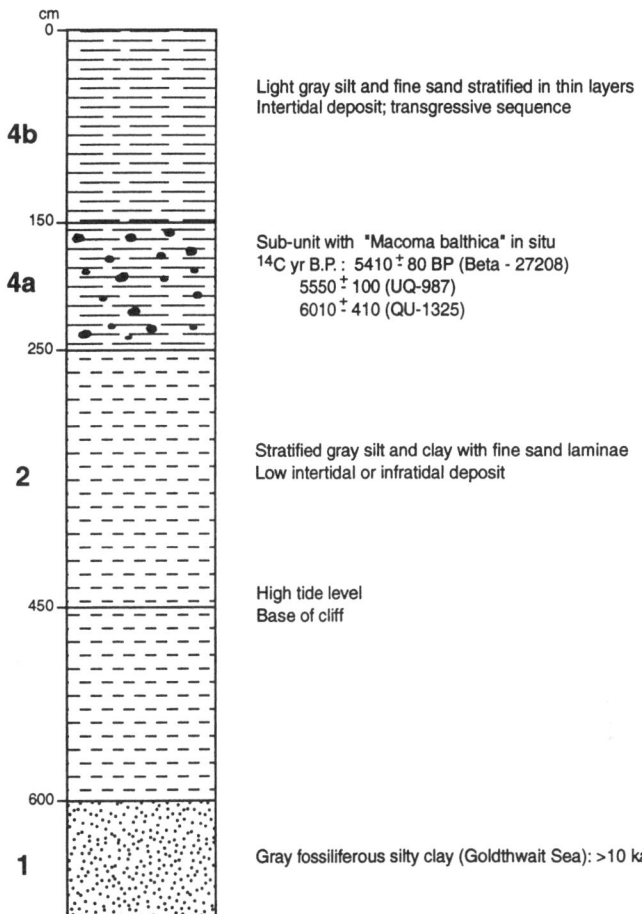

Figure 6. Section in 6 m terrace at Pointe St-Thomas ~6 km west-southwest of Montmagny Airport section.

which occurred in 1998, exposed an organic layer (peat) 25–30 cm thick and many tree trunks. The peat and logs provided radiocarbon dates ranging from 6.4 to 6.8 ka (Fig. 7). These data correlate well with the Montmagny and Cap-St-Ignace sections (Dionne, 2000).

The overlying intertidal deposit (units 3a and 3b) which is 4 m thick, contains abundant plant remains (roots in situ), and a few fragments of wood reworked from the underlying organic layer. The surface of the terrace is 12 m asl, i.e., 2 m higher than at Montmagny.

### Relict forest at mean sea level

Additional evidence for a sea level lower than today is provided by the occurrence of logs and stumps in situ at the seaward margin of the low intertidal marsh (Figs. 4D and 8) (Dionne, 1986b). More than 100 stumps and logs were observed in 5 sites in the area east of Rivière du Sud, at Montmagny. The species are exclusively resinous, mainly spruce (*Picea* sp.) and tamarack (*Larix laricina*) for stumps and white pine (*Pinus strobus*) and hemlock (*Tsuga canadensis*) for logs. Radiocarbon ages (N = 24) range from 5820 ± 80 to 6760 ± 80 $^{14}$C yr B.P. (Table 9). This chronology is thus similar to the age of the peat layer and the associated stumps and logs in the vertical section of the 8–10 m terrace described here. Consequently, sea level at the time was lower than the level of the seaward margin of the modern low intertidal marsh, which is about the geodetic zero of topographic maps, or 2.6 m above the Canadian hydrographic zero (low tide level).

The lowest level reached during the lowstand of the sea is not known. However, at Montmagny a wide erosion surface in the clay deposit of the Goldthwait sea ends seaward in a small scarp at about the 10 m isobath. At La Pocatière, D'Anglejan (1981) reported a similar surface with an entrenched river valley, which was subsequently filled by mid-Holocene sediment. In the middle St. Lawrence estuary, an erosion surface on the bottom (Praeg et al., 1992) suggests a lowstand 25–30 m below present sea level.

## RELATIVE SEA-LEVEL CURVE FOR THE SOUTH SHORE OF THE ST. LAWRENCE ESTUARY

The relative sea-level curve for the south shore of the St. Lawrence estuary proposed about a decade ago (Dionne, 1988a, 1988b, 1988d) showed a double transgression and regression event that contrasted greatly with the previous published curves

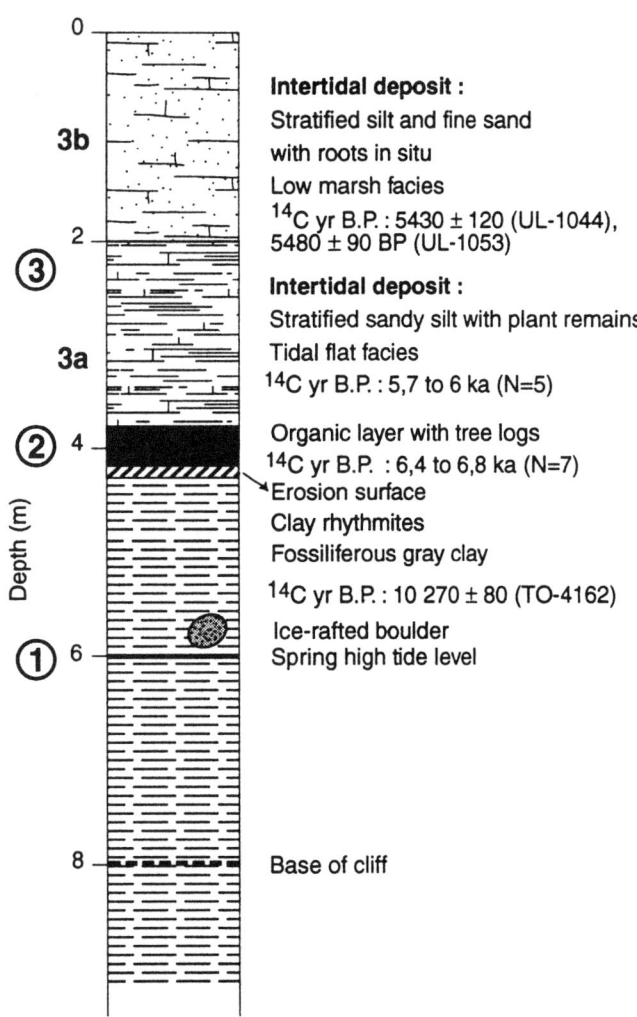

Figure 7. Section in 10 m terrace at Bellechasse cove ~15 km southwest of Montmagny Airport section.

(Elson, 1969; Locat, 1977, Lortie and Guilbault, 1984) and with the geophysical model of Quinlan and Beaumont (1981, 1982). It is worthy of mention that the land uplift curves of Locat (1977) and Lortie and Guilbault (1984) showed a lack of data for the period between 9.5 and 2 ka. This major gap explains some of the deficiencies of the Quinlan and Beaumont model (Dionne, 1991). To summarize, this model suggested a series of four typical curves for southeastern Canada related to glacial and deglacial events and to the forebulge migration. Two proposals were made, one with a maximum and the other with a minimum ice cover.

In the maximum ice model, the St. Lawrence estuary is in zone A and is characterized by a major transgression followed by a continuous isostatic uplift, while in the minimum ice model, the St. Lawrence estuary and gulf fall into zones B and C, which are characterized by a major transgression followed by a regression below present-day sea level. A subsequent progressive rise of the eustatic sea level is postulated for zone B (e.g., Maine coast; Barnhardt et al., 1995), while in zone C the sea level has never been higher than today.

The typical curves for zones A, B, and C do not fit with the new relative sea-level curve for the south shore of the St. Lawrence estuary. The explanation is simply that the events reported ~10 yr ago (Dionne, 1988b, 1988d) were not known at the time the model was conceived.

The new relative sea-level curve for the south shore of the St. Lawrence estuary (Fig. 9) shows a double sequence of transgression and regression during the Holocene. This type of curve, which modifies the current concept, was not known in Canada prior to 1988, although it was suggested by Marthinussen (1962) for Norway; Hjort (1973) has reported similar events for the east coast of Greenland. Curiously, the 8 m Vega transgression of eastern Greenland occurred between 6 and 4.6 ka, which is approximately the time of the Laurentian trans-

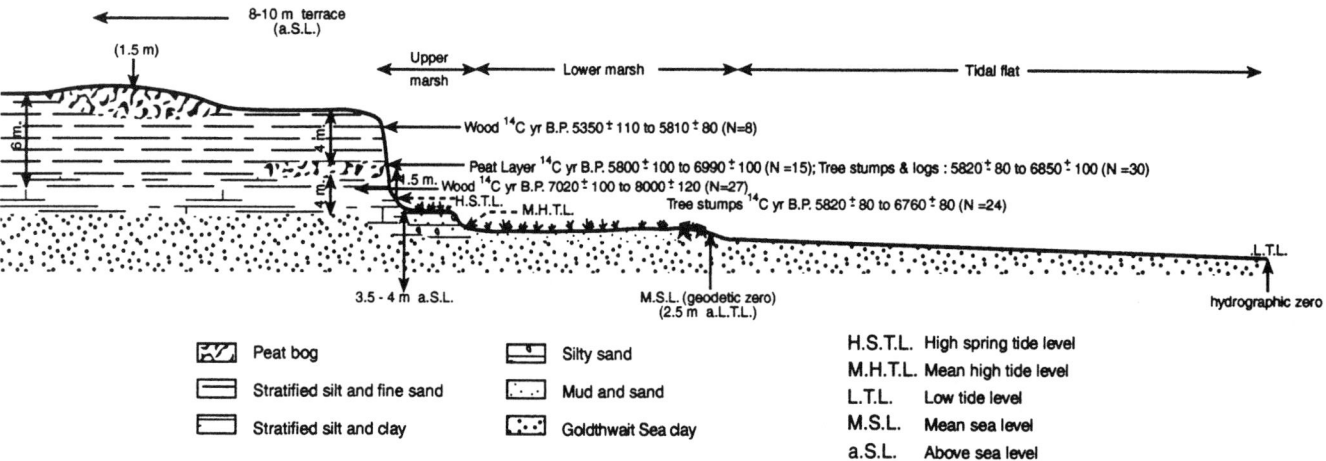

Figure 8. Schematic cross section of shore zone at Montmagny Airport section showing the 8–10 m terrace and emplacement of relict forest at seaward edge of low marsh.

TABLE 9. RADIOCARBON DATES OF STUMPS AND LOGS AT THE EDGE OF THE LOW INTERTIDAL MARSH AT MONTMAGNY

| Laboratory number | Age ($^{14}$C yr B.P.) | Species |
|---|---|---|
| UL-961 | 5820 ± 80* | |
| Beta-31394 | 5950 ± 80* | Picea sp. vs Larix laricina |
| UQ-1143 | 5950 ± 100 | |
| Beta-31397 | 5960 ± 70* | |
| UL-105 | 6040 ± 80 | Tsuga canadensis |
| Beta-29058 | 6110 ± 70 | Picea sp. |
| Beta-45264 | 6120 ± 60 | |
| UQ-1177 | 6150 ± 100 | |
| Beta-44575 | 6170 ± 70 | Picea sp. vs Larix laricina |
| UL-167 | 6170 ± 90 | Larix laricina |
| UL-166 | 6190 ± 90 | Picea sp. |
| UL-169 | 6190 ± 90* | Pinus strobus |
| Beta 44577 | 6200 ± 50 | |
| UL-104 | 6210 ± 80 | |
| UL-162 | 6260 ± 90 | |
| UL-168 | 6270 ± 90 | Larix laricina |
| Beta-45265 | 6280 ± 70* | |
| Beta-44576 | 6340 ± 50* | |
| UL-163-2 | 6380 ± 90 | Picea sp. |
| UL-163 | 6390 ± 80 | Picea sp. |
| UQ-1015 | 6500 ± 100 | |
| UL-164 | 6600 ± 90 | |
| UL-165 | 6610 ± 90 | |
| UL-945 | 6760 ± 80 | Tsuga canadensis |

*Log.

gression in the St. Lawrence estuary. Since 1988, many relative sea-level curves showing a double transgression and regression event during the Holocene have been published for various countries (Pirazzoli, 1991). Table 10 summarizes the data.

It is clear today that the sea-level events that characterize the south shore of the St. Lawrence estuary are not only of local interest, although the geophysical models proposed by Peltier (1987, 1991) for Rivière-du-Loup and Rimouski did not take into account the data for the interval between 9 and 2 ka.

The original relative sea-level curve published about a decade ago (Dionne, 1988d) has been slightly modified to show the minor sea-level variation related to the Mitis terrace event, which occurred ca. 2 ka.

## SEA LEVEL ON THE NORTH SHORE OF THE ESTUARY

There are not enough published data for the north shore of the St. Lawrence estuary to present a satisfactory review of the relative sea-level history, particularly for the period between 9 and 2.5 ka. Some coastal areas were ice free by 12 ka, whereas others were still partly ice covered by 10.5 ka (LaSalle et al., 1977; Chauvin et al., 1985; Dionne and Occhietti, 1996; Govare, 1996).

The isostatic uplift was rapid during the first 2–3 k.y. and decreased significantly after 9–8 ka. Evidence for a mid-Holocene lowstand has been obtained at a few localities (Dionne, 1996a; Dionne and Occhietti, 1996). Work in progress suggests that the Laurentian transgression also occurred on the north shore. There is stratigraphic evidence in the two large deltaic complexes (Betsiamites and Manicouagan-Outardes deltas) in the area southwest of Baie-Comeau.

The Micmac shoreline at several localities (Dionne, 1993, 1996a, 1996b; Bernatchez et al., 2000) correlates well with the same feature on the south shore, indicating a similar water level throughout the St. Lawrence estuary ca. 2 ka. The occurrence locally of tidal deposits overlying beaches suggests a rising sea level followed by shore emergence after 1.5 ka.

## CONCLUSION

In summary, the relative sea level and emergence of the south shore of the St. Lawrence estuary is more complex than previously suggested. Land emergence related to isostatic recovery has been relatively rapid; more than 85% was realized by 8 ka. Because the eustatic sea level was at least 20 m below present-day level, a lowstand occurred in the St. Lawrence estuary. It is not known exactly what level was reached, although the 10 m isobath is most likely. The lowstand was followed by the mid-Holocene Laurentian transgression, which is possibly related to the migration of the forebulge (Barnhardt et al., 1995)

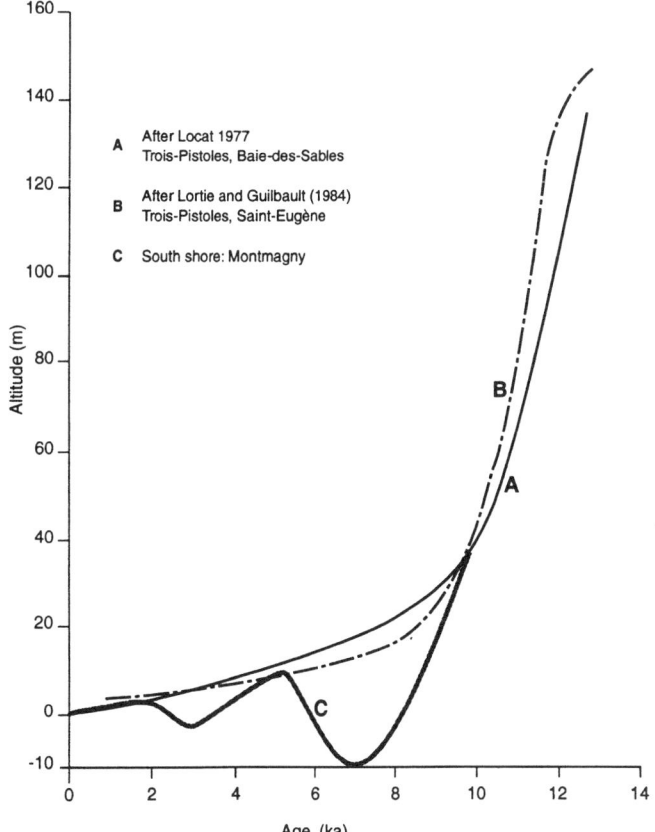

Figure 9. Generalized Holocene relative sea-level curve for south shore of St. Lawrence estuary.

**TABLE 10. PUBLISHED RELATIVE SEA-LEVEL CURVES SHOWING A DOUBLE TRANSGRESSION/REGRESSION EVENT DURING THE HOLOCENE**

| Country | References |
| --- | --- |
| Norway | Marthinussen (1962); Lie et al. (1983); Hafsten (1983); Kaland (1984); Svendsen and Mangerund (1987); Pirazzoli (1991); Hald and Vorren (1993); Corner and Haugane (1993) |
| Russia (Kola peninsula) | Snyder et al. (1996) |
| Spitsbergen | Héquette (1988); Forman (1990); Forman et al. (1987) |
| Iceland | Hansom and Briggs (1991) |
| Ireland | Lambeck (1991a)* |
| Scotland | Lambeck (1991b) |
| Canada | |
| Baffin Is. | Andrews (1989) |
| Vancouver | Clague et al. (1982); Clague and Bobrowsky (1990) Egginton and Andrews (1989) |
| Vancouver Island | Matthews et al. (1970) |
| Québec | Dionne (1988a, 1988b, 1988d, 1998) Dionne and Coll (1995) |

*Field evidence apparently does not support Lambeck theoretical model for relative sea level in northern Ireland (McCabe, 1997).

and coincides with the disintegration of the Laurentide ice sheet in central Québec and with the rising eustatic sea level. It is possible also that the second transgression is partly or fully related to tectonic events in an area of high seismicity (Lamontagne, 1987; Dyke and Peltier, 2000).

The relative sea-level curve for the south shore of the St. Lawrence estuary is thus characterized by a double transgression and regression event also observed in other glaciated countries. In addition, minor relative sea-level fluctuations (1–2 m amplitude) have occurred during the past 3 k.y. (Dionne, 1997b, 1999, 2000). Currently, both shores are apparently in a period of relative stability, although geomorphic evidence suggests subsidence or a slightly rising sea level.

## ACKNOWLEDGMENTS

This contribution was made possible through an operating research grant from the Natural Sciences and Engineering Research Council of Canada (Ottawa). Reviews by Peter B. Clibbon (Université Laval, Québec), J.T. Kelley (University of Maine, Orono), and R.N. Oldale (U.S. Geological Survey, Woods Hole) are fully acknowledged. Figures were drawn at the Department of Geography, Université Laval. Many of the radiocarbon dates were made at the Centre d'études nordiques Laboratory, Université Laval, Québec.

## REFERENCES CITED

Anctil, F., and Troude, J.P., 1992, Étude de la remontée relative des niveaux d'eau de l'estuaire du Saint-Laurent: Canadian Journal of Civil Engineering, v. 19, p. 252–259.

Andrews, J.T., 1989, Quaternary geology of northeastern Canadian shield, in Fulton, R.J., ed., Quaternary geology of Canada and Greenland: Ottawa, Geological Survey of Canada, p. 267–302.

Archambault, M.P., 1987, L'archaïque sur la haute Côte-Nord du Saint-Laurent: Recherches Amérindiennes au Québec, v. 17, p. 101–114.

Bail, P., 1983, Problèmes géomorphologiques de l'emplacement et de la transgression marine pleistocènes en Gaspésie sud-orientale [Ph.D. thesis]: Montréal, McGill University, 148 p.

Bail, P., 1987, Goldthwait Sea sediments, Bonaventure, in Gray, J.T., ed., Quaternary processes and paleoenvironments in the Gaspé Peninsula and Lower St. Lawrence Valley: Ottawa, XIIth INQUA Congress Field Trip C-4, p. 63.

Barnhardt, W.A., Gehrels, W.R., Belknap, D.F., and Kelley, J.T., 1995, Late Quaternary relative sea-level change in the western Gulf of Maine: Evidence for a migrating glacial forebulge: Geology, v. 23, p. 317–320.

Bélanger, C., 1993, Étude géomorphologique des basses terrasses sur la côte sud de l'estuaire laurentien [Ph.D. thesis]: Québec, Université Laval, 237 p.

Bernatchez, P., Dionne, J.C., and Dubois, J.M., 2000, Observations sur la terrasse Mitis à Ragueneau, côte nord de l'estuaire du Saint-Laurent: AQQUA-CGRG Joint Meeting, Montréal, August 22–27, Abstracts of Papers, p. 33.

Bhiry, N., Garneau, M., and Filion, L., 2000, Macrofossil record of a Middle Holocene drop in relative sea level at the St. Lawrence Estuary, Québec: Quaternary Research, v. 54, p. 171–184.

Bigras, P., and Dubois, J.M.M., 1987, Répertoire commenté des datations 14C du nord de l'estuaire et du golfe du Saint-Laurent, Québec et Labrador: Sherbrooke University, Department of Geography Bulletin de Recherche 94–96, 166 p.

Bonenfant, R., 1993, Chronologie des événements post-glaciaires à l'Holocène dans la basse vallée du Gouffre (Charlevoix) [Masters thesis]: Québec, Université Laval, 148 p.

Chalmers, R., 1886, Surficial geology of northern New Brunswick and southeastern Quebec: Geological Survey of Canada Annual Report-1886, Part M, p. 5–42.

Chalmers, R., 1896, Pleistocene marine shorelines on south side of the St. Lawrence Valley: American Journal of Science, ser. 4, v. 1, p. 302–308.

Chalmers, R., 1904, The geomorphic origin and development of the raised shorelines of the St. Lawrence Valley and the Great Lakes: American Journal of Science, ser. 4, v. 18, p. 175–179.

Chalmers, R., 1905, Surface geology of eastern Québec: Geological Survey of Canada Annual Report, v. 16, part A, p. 250–263.

Chauvin, L., Martineau, G., and LaSalle, P., 1985, Deglaciation of the Lower St. Lawrence region, Québec in Borns, H.W., Jr., LaSalle, P., and Thompson, W.B., eds., Late Pleistocene history of northeastern New England and adjacent Quebec: Boulder, Colorado. Geological Society of America Special Paper 197, p. 111–123.

Clague, J.J., and Bobrowsky, P.T., 1990, Holocene sea level change and crustal deformation, southwestern British Columbia, in Current Research, 1990: Geological Survey of Canada Paper 90-1E, p. 245–250.

Clague, J., Harper, J.R., Hebda, R.J., and Howes, D.E., 1982, Late Quaternary sea levels and crustal movements, coastal British Columbia: Canadian Journal of Earth Sciences, v. 19, p. 597–618.

Coleman, A.P., 1922, Physiography and glacial geology of Gaspé Peninsula, Québec: Geological Survey of Canada Bulletin 34, 54 p.

Corner, G.D., and Haugane, E., 1993, Marine-lacustrine stratigraphy of raised coastal basins and postglacial sea-level change at Lyngen and Vanna, Troms, northern Norway: Norsk Geologisk Tidsskrift, v. 73, p. 175–197.

Daigneault, R.A., 1985, Cadre géologique des sites archéologiques DbEj-11 et DbEj-13 de Grandes-Bergeronnes (été 1984): Department of Earth Sciences, Université du Québec à Montréal 34 p.

D'Anglejan, B., 1981, Évolution post-glaciaire et sédiments récents de la plateforme infra-littorale, baie de Sainte-Anne, estuaire du Saint-Laurent, Québec: Géographie Physique et Quaternaire, v. 35, p. 253–260.

David, P.P., and Lebuis, J., 1985, Glacial maximum and deglaciation of western Gaspé, Québec, Canada in Borns, H.W., LaSalle, P., and Thompson, W.B., eds., Late Pleistocene history of northeastern New England and adjacent Quebec: Boulder, Colorado, Geological Society of America Special Paper 197, p. 85–109.

Dionne, J.C., 1972, La dénomination des mers du Post-glaciaire au Québec: Cahiers de Géographie du Québec, v. 16, p. 483–488.

Dionne, J.C., 1977, La Mer de Goldthwait au Québec: Géographie Physique et Quaternaire, v. 31, p. 61–80.

Dionne, J.C., 1985, Observations sur le Quaternaire de la rivière Boyer, Côte sud de l'estuaire du Saint-Laurent: Géographie Physique et Quaternaire, v. 39, p. 35–46.

Dionne, J.C., 1986a, Érosion récente des marais intertidaux de l'estuaire du Saint-Laurent: Géographie Physique et Quaternaire, v. 40, p. 307–323.

Dionne, J.C., 1986b, Tree stumps in situ in the intertidal zone at Montmagny, St. Lawrence estuary, Québec: Geological Society of America Abstracts with Programs, v. 18, no. 1, p. 13.

Dionne, J.C., 1987, Lithologie des cailloux de la baie de Montmagny, côte sud du Saint-Laurent: Géographie Physique et Quaternaire, v. 41, p. 161–169.

Dionne, J.C., 1988a, L'émersion de la côte sud du Saint-Laurent dépuis la dernière déglaciation: GEOS, v. 17, no. 1, p. 18–21.

Dionne, J.C., 1988b, Évidence d'un bas niveau marin à l'Holocène, à Saint-Fabien-sur-Mer, estuaire maritime du Saint-Laurent: Norois, v. 35, p. 19–34.

Dionne, J.C., 1988c, Note sur les variations du niveau marin relatif à Rivière-Ouelle, côte sud du Saint-Laurent: Géographie Physique et Quaternaire, v. 42, p. 83–88.

Dionne, J.C., 1988d, Holocene relative sea-level fluctuations in the St. Lawrence estuary, Québec, Canada: Quaternary Research, v. 29, p. 233–244.

Dionne, J.C., 1990, Observations sur le niveau marin relatif à l'Holocène, à Rivière-du-Loup, estuaire du Saint-Laurent: Géographie Physique et Quaternaire, v. 44, p. 43–53.

Dionne, J.C., 1991, A reappraisal of postglacial uplift, south shore of the St. Lawrence estuary, Québec, in Symposium on late glacial and postglacial events in coastal and adjacent areas, Fredericton, N.B.: Canadian Quaternary Association, Program and Abstracts, p. 20.

Dionne, J.C., 1992, État des connaissances sur la terrasse Mitis: Ligne de rivage Micmac de Goldthwait. 7e Congrès quadriennal de Association Québécoise pour l'étude du Quaternaire (Rouyn-Noranda): Bulletin de l'AQQUA, v. 18, p. 32–33.

Dionne, J.C., 1993, The twenty-foot terrace and sea-cliff of the lower St. Lawrence: Geological Society of America Abstracts with Programs, v. 25, no. 6, p. A-124.

Dionne, J.C., 1996a, La terrasse Mitis à la pointe aux Alouettes, côte nord du moyen estuaire du Saint-Laurent: Géographie Physique et Quaternaire, v. 50, p. 57–72.

Dionne, J.C., 1996b, La basse terrasse à Petite-Rivière (Charlevoix): Un exemple d'activité néotectonique à l'Holocène: Géographie Physique et Quaternaire, v. 50, p. 311–330.

Dionne, J.C., 1997a, Bilan vicennal des connaissances sur la Mer de Goldthwait au Québec: Bulletin de Association Québécoise pour l'étude du Quaternaire, v. 23, p. 6–20.

Dionne, J.C., 1997b, Nouvelles données sur la transgression Laurentienne, côte sud du moyen estuaire du Saint-Laurent, Québec: Géographie Physique et Quaternaire, v. 51, p. 199–208.

Dionne, J.C., 1998, Relative sea-level variations during the Holocene, middle St. Lawrence estuary: Québec, Geological Association of Canada, Guide-Book, Field Trip B-1, 49 p.

Dionne, J.C., 1999, Indices de fluctuations mineures du niveau marin relatif à l'Holocène supérieur, à l'Isle-Verte, côte sud de l'estuaire du Saint-Laurent, Québec: Géographie Physique et Quaternaire, v. 53, p. 277–285.

Dionne, J.C., 2000, Données complémentaires sur les variations du niveau marin relatif, à l'Holocène, à l'anse de Bellechasse, côte sud du moyen estuaire du Saint-Laurent: Géographie Physique et Quaternaire, v. 54, p. 119–122.

Dionne, J.C., and Coll, D., 1995, Le niveau marin relatif dans la région de Matane (Québec), de la déglaciation à nos jours: Géographie Physique et Quaternaire, v. 49, p. 363–380.

Dionne, J.C., and Occhietti, S., 1996, Aperçu du Quaternaire à l'embouchure du Saguenay: Géographie Physique et Quaternaire, v. 50, p. 5–34.

Dohler, G.C., and Ku, L.F., 1970, Presentation and assessment of tides and water level records for geophysical investigations: Canadian Journal of Earth Sciences, v. 7, p. 607–625.

Dredge, L.A., 1971, Late Quaternary sedimentary environments, Sept-Iles, Québec [Masters thesis]: Montréal, McGill University, 102 p.

Dubois, J.M.M., 1979, Environnements quaternaires et évolution post-glaciaire d'une zone côtière en émersion en bordure sud du Bouclier canadien: la moyenne Côte-Nord du Saint-Laurent [Ph.D. thesis]: Ottawa, University of Ottawa, 754 p.

Dufresne, J., 1994, Reconstitution paléoécologique préliminaire d'un site côtier de l'estuaire supérieur du Saint-Laurent, région de Montmagny, Québec [B.A. thesis]: Sherbrooke, Université de Sherbrooke, 75 p.

Dyck, W., and Fyles, J.G., 1963, Geological Survey of Canada radiocarbon dates I and II: Geological Survey of Canada Paper 63-21, 31 p.

Dyke, A.S., and Peltier, W.R., 2000, Forms, response times, and variability of relative sea-level curves, glaciated North America: Geomorphology, v. 32, p. 315–333.

Egginton, P.A., and Andrews, J.T., 1989, Sea levels are changing: GEOS, v. 18, p. 15–22.

Elson, J.A., 1969, Late Quaternary marine submergence of Québec: Revue de Géographie de Montréal, v. 23, p. 242–258.

Emery, K.O., and Aubrey, D.G., 1991, Sea levels, land levels, and tide gauges: New York, Springer-Verlag, 237 p.

Faessler, C., 1948, L'extension maximum de la mer Champlain au nord du Saint-Laurent, de Trois-Rivières à Moisie, in Rapport annuel 1947: Société Provencher d'Histoire Naturelle du Canada, p. 16–28.

Fairchild, H.L., 1918, Post-glacial uplift of northeastern America: Geological Society of America Bulletin, v. 29, p. 187–234.

Forman, S.L., 1990, Post-glacial relative sea-level history of northwestern Spitsbergen: Geological Society of America Bulletin, v. 102, p. 1580–1590.

Forman, S.L., Mann, D., and Miller, G.H., 1987, Late Weichselian and Holocene relative sea-level history of Brøggerhalvøya, Spitsbergen, Svalbard Archipelago: Quaternary Research, v. 27, p. 41–50.

Goldthwait, J.W., 1911a, Raised beaches of southern Québec: Geological Survey of Canada Summary Report-1910, p. 220–233.

Goldthwait, J.W., 1911b, The twenty-foot terrace and sea-cliff of the Lower St. Lawrence: American Journal of Science, ser. 4, v. 32, p. 291–317.

Goldthwait, J.W., 1912, Records of post-glacial changes of level in Quebec and New Brunswick: Geological Survey of Canada Summary Report-1911, p. 296–302.

Goldthwait, J.W., 1913, The post-glacial marine submergence, in Excursion in eastern Québec and the Maritime Provinces: Geological Survey of Canada Guide Book 1, Part 1, 207 p.

Goldthwait, J.W., 1914, Marine shorelines in Eastern Quebec: Geological Survey of Canada Report-1912, p. 357–359.

Govare, E., 1996, Géomorphologie et paléo-environnements de la région de Charlevoix, Québec [Ph.D. thesis]: Montréal, Université de Montréal, 429 p.

Gray, J.T., ed., 1987, Quaternary processes and paleo-environments in the Gaspé Peninsula and the Lower St. Lawrence Valley: 12th INQUA Congress, Field Excursion C-4, Guide Book, 84 p.

Gray, J.T., and Hétu, B., 1981, Geomorphology of the north coast of Gaspésie, subsequent to deglaciation, in Gray, J.T., ed., Weathering zone and the problem of glacial limits: Montréal, AQQUA-CANQUA Conference and field trip in Gaspésie, Québec, p. 106–119.

Hafsten, U., 1983, Shore-level changes in south Norway during the last 13 000 years, traced by biostratigraphical methods and radiometric datings: Norsk Geografisk Tidsskrift, v. 37, p. 63–79.

Hald, M., and Vorren, T.O., 1983, A shore displacement curve from the Tromsø district, north Norway: Norsk Geologisk Tidsskrift, v. 63, p. 103–110.

Hansom, J.D., and Briggs, D.J., 1991, Sea-level change in Vistfirdir, north west Iceland, *in* Maizels, J.K., and Caseldine, C., eds., Environmental changes in Iceland: Past and present: Dordrecht, Kluwer Academic Publishers, p. 79–91.

Héquette, A., 1988, Vues récentes sur l'évolution du Svalbard au Quaternaire: Revue de Géomorphologie Dynamique, v. 37, p. 129–141.

Hétu, B., 1994, Déglaciation, émersion des terres et pergélisol tardiglaciaire dans la région de Rimouski-Québec: Paléo-Québec, v. 22, p. 5–48.

Hétu, B., 1998, La déglaciation de la région de Rimouski, Bas-Saint-Laurent (Québec): Indices d'une récurrence glaciaire dans la mer de Goldthwait entre 12 400 et 12 000 BP: Géographie Physique et Quaternaire, v. 52, p. 325–347.

Hillaire-Marcel, C., 1979, Les mers post-glaciaires du Québec: Quelques aspects [Ph.D. thesis]: Paris, Pierre and Marie Curie University, v. 1, 293 p.

Hjort, C., 1973, The Vega transgression: A hypsithermal event in central East Greenland: Geological Society of Denmark Bulletin, v. 22, p. 25–38.

Kaland, P.E., 1984, Holocene shore displacement and shorelines in Hordaland, western Norway: Boreas, v. 13, p. 203–242.

Lambeck, K., 1991a, Glacial rebound and sea-level changes in the British Isles: Terra Research, v. 3, p. 379–389.

Lambeck, K., 1991b, A model for Devensian and Flandrian glacial rebound and sea-level change in Scotland, *in* Sabadini, R., et al., eds., Glacial isostasy, sea level and mantle rheology: Dordrecht, Kluwer Academic Publishers, p. 33–62.

Lamontagne, M., 1987, Seismic activity and structural features in the Charlevoix region, Quebec: Canadian Journal of Earth Sciences, v. 24, p. 2128–2129.

LaSalle, P., Martineau, G., and Chauvin, L., 1977, Morphology, stratigraphy and deglaciation in Beauce–Notre-Dame Mountains–Laurentide Park area: Quebec Department of Natural Resources Report DPV-516, 74 p.

Lebuis, J., and David, P.P., 1977, La stratigraphie et les événements Quaternaires de la partie occidentale de la Gaspésie, Québec: Géographie Physique et Quaternaire, v. 31, p. 275–296.

Lee, H.A., 1962, Surficial geology of Rivière-du-Lauf area, Québec: Geological Survey of Canada Paper 61-32.

Lee, H.A., 1963, Pleistocene glacial-marine relations, Trois-Pistoles, Québec: Geological Society of America Special Paper 73, p. 195.

Lie, S.E., Stabell, B., and Mangerud, J., 1983, Diatom stratigraphy related to late Weichselian sea-level changes in Sunmøre, western Norway: Norges Geologiske Undersøkelse, no. 380, p. 203–219.

Locat, J., 1977, L'émersion des terres dans la région de Baie-des-Sables/Trois-Pistoles: Géographie Physique et Quaternaire, v. 31, p. 297–306.

Lortie, G., and Dionne, J.C., 1990, Analyse préliminaire des diatomées de la coupe de Montmagny, côte sud de l'estuaire du Saint-Laurent: Géographic Physique et Quaternaire, v. 44, p. 89–95.

Lortie, G., and Guilbault, J.P., 1984, Les diatomées et les foraminifères de sédiments marins post-glaciaires du Bas Saint-Laurent (Québec): Analyse comparée des assemblages: Naturaliste Canadien, v. 111, p. 297–310.

Lougee, R.J., 1953a, A chronology of post-glacial time in eastern North America: Scientific Monthly, v. 76, p. 259–276.

Lougee, R.J., 1953b, The role of upwarping in the post-glacial history of Canada: Revue Canadienne de Géographie, v. 7, p. 3–14.

Lougee, R.J., and Lougee, C.R., 1976, Late glacial chronology: New York, Vantage Press, 553 p.

Marthinussen, M., 1962, C14 datings referring to shore lines, transgression, and glacial substages in Northern Norway: Norges Geologiske Undersøkelse Arbok-1961, no. 215, p. 37–67.

Martineau, G., 1977, Géologie des dépôts meubles de la région de Kamouraska-Rivière-du-Loup: Québec, Ministère des Richesses Naturelles Rapport DPV-545, 17 p.

Mathews, W.H., Fyles, J.G., and Nasmith, H.W., 1970, Postglacial crustal movements in southwestern British Columbia and adjacent Washington State: Canadian Journal of Earth Sciences, v. 7, p. 690–702.

McCabe, A.M., 1997, Geological constraints on geophysical models of relative sea-level change during deglaciation of the western Irish Sea basin: Geological Society of London Journal, v. 154, p. 601–604.

McNeely, R., 1989, Geological Survey of Canada radiocarbon dates XXVIII: Geological Survey of Canada Paper 88-7, 93 p.

Peltier, W.R., 1987, Glacial isostasy, mantle viscosity, and Pleistocene climatic change, *in* Ruddiman, W.P., et al., eds., North America and adjacent oceans during the last deglaciation: Boulder, Colorado, Geological Society of America, Geology of North America, v. K-3, p. 155–182.

Peltier, W.R., 1991, The ICE-3G model of the late Pleistocene deglaciation: construction, verification and application, *in* Sabadini, S., et al., eds., Glacial isostasy, sea-level and mantle rheology: Dordrecht, Kluwer Academic Publishers, p. 95–119.

Pfalgraf, F., Pienitz, R., and Dionne, J.C., 1996, Études des dépôts holocènes de Rivière-Ouelle par l'analyse des diatomées: Québec, 8th AQQUA Congress (Québec City), Program and Abstracts, p. 63.

Pirazzoli, P.A., 1986, Secular trend of relative sea-level (RSL) changes indicated by tide-gauge records: Journal of Coastal Research, Special Issue, no. 1, p. 1–26.

Pirazzoli, P.A., 1991, World atlas of Holocene sea level changes: Amsterdam, Elsevier, 300 p.

Praeg, D., D'Anglejan, B., and Syvitski, J.P.M., 1992, Seismostratigraphy of the middle St. Lawrence estuary: A late Quaternary marine to estuarine depositional/erosional record: Géographie Physique et Quaternaire, v. 46, p. 133–150.

Quinlan, G., and Beaumont, C., 1981, A comparison of observed and theoretical postglacial relative sea level in Atlantic Canada: Canadian Journal of Earth Sciences, v. 18, p. 1146–1163.

Quinlan, G., and Beaumont, C., 1982, The deglaciation of Atlantic Canada as reconstructed from the postglacial relative sea-level record: Canadian Journal of Earth Sciences, v. 19, p. 2232–2246.

Samson, C., Barrette, L., LaSalle, P., and Fortier, J., 1977, Quebec radiocarbon measurements: Radiocarbon, v. 19, p. 96–100.

Snyder, J.A., Korsun, S.A., and Forman, S.L., 1996, Postglacial emergence and the Tapes transgression, north central Kola Peninsula, Russia: Boreas, v. 25, p. 47–56.

Svendsen, J.I., and Mangerud, J., 1987, Late Weichselian and Holocene sea-level history for a cross-section of Western Norway: Journal of Quaternary Science, v. 2, p. 113–132.

Troude, J.P., 1985, Étude du cycle sédimentaire annuel d'un estran à forte sédimentation estivale de l'estuaire du Saint-Laurent [Ph.D. thesis]: Québec, Université Laval, 183 p.

Twenhofel, W.H., and Conine, W.H., 1921, The post-glacial terraces of Anticosti Island: American Journal of Science, ser. 5, v. 1, p. 268–278.

Manuscript Accepted by the Society June 9, 2000

# Index

[Italic page numbers indicate major references]

## A

ablation, 263
   ice stream, 252
   Laurentide ice sheet, 251
abrasion notches, 39
accelerator mass spectrometry, 11, 204, 206
   age estimates, 224
   radiocarbon age dates, 118
acoustic profile, Halibut Channel, 59
acoustic velocity, *55*, 71
   Downing Silt, 69
   Halibut channel, 60
   shell hash, 70
Adolphus Sand, *55*, 60, 61
   fossils, 69
advance, glacial, *69, 72*
age dating, 32. *see also* radiocarbon dating
   amino acid racemization, 11, 37
   errors, *172*
   foraminifers, 16
   Jordan Basin, 17
   radiocarbon, 11, 16, *17*, 36, 72
airgun, 55, 59
amino acid racemization
   age dating, 37
   foraminifer, 11
Androscoggin moraine, 3, 111, 113, 114
Androscoggin River, 117, 209
   valley, 111, 114, 118, 119, 191, 193, *194*, 195, 207, 208
Antarctic ice sheet, 247
Anticosti Island, 257, 259, 263, 264
   ice advance, 263
Appalachian
   highlands, 244
   ice, 39, 256
   mountains, 216, 260
   piedmont, 264
   Ridge, 251
aquifers
   Maine, 111
Ashuelot Valley, varves, 174
Atlantic coast
   reversal points, 44
   shoreline, 47
Avalon Peninsula, 52, 53
   ice caps, 70
   ice dome, 53
Aziscohos Lake, 117

## B

Baffin
   Bay, 22
   Island, 161
   Shelf, 26
Baie des Chaleurs reentrant, 264
Baie-Comeau, sea level, 271
Baie-St-Paul, sea level, 271
Baie-Trinité moraine, 258
Barents ice sheet, 68

Bas du Gleuve, age, 262
Bauneg Beg delta, 153. *see also* Bauneg Beg morainal bank
Bauneg Beg morainal bank, 161, 166. *see also* Bauneg Beg delta
Bay of Fundy, 36, 39
   deglaciation, 39
   reversal points, 44
   sea level rise, 44
beach, 192, 199, 209, 272
   marine limit, 207
   ridge, 136
Bear River, 117
Beauce
   deglaciation, 267
   event, 261, *265*, 267
Beaufort Sea, fossils, 26
Bécancour River, 261
Bécancour-Palmer drift complex, 261
Bellechasse Cove, 274
   section, *278*
benthic fauna, oxygen isotopes, 234
Bethlehem Moraine, age, 262
bioturbation, 21, 60
Bode Lake, core, 227
bog, ombrotrophic, 278
Bois Francs
   piedmont, 261, 266
   plateau, 266
   region, 252
   residual ice cap, 252
   residual ice mass, 265
   striations, 264
   uplands, 267
Bölling
   chronozone, 238
   warm phase, 247, 252, 263, *265*
borehole
   87B-1, 59, *60*
   87B-2, 59, *60*
   87B-4, 59, *60*
   Halibut Channel, 51
   North Sea, 53
   Norwegian Shelf, 53
bottom water
   Gulf of Maine, 21
   paleotemperature, 233
Boundary Mountains, 114, 121
Boundary Pond, 118
bounding surface, 142
Boyd Lake, 234, 236
Bragdon Road delta, 127, 153. *see also* Bragdon Road morainal bank
Bragdon Road morainal bank, 161. *see also* Bragdon Road delta
Branch Brook, 126, 128, 129, 130, 143
Browns Bank, 9, 10, 39, 216
   deglaciation, 20
   ice, 28
   shoreline, 27
Brunswick sand plain, 209
Bryant Hill Pit, 116

bulge, 44
   migration, 53
bulk sediment
   radiocarbon ages, 173, 183
Burin Peninsula, 70
buttonhole
   Chaudière Valley, 265
   Chaudière-Etchemin Rivers, 266
   Etchemin Valley, 265

## C

C Pond Moraine, *114*
Caledonian phase, *72*
CALIB 4.0 computer program, 178
calving, 26, 39, 70, 159, 252, 263
   bay, 192, 208, 243, 257, 259, 266
   Crowell Basin, 29
   embayment, 153, 157, 160, 165, 166, 168, 208, 236
   ice, 9, 16, 22, 28, 30, 32, 146, 156
   ice margin, 127, 238
   iceberg, 234
   tidewater, *160*
Canoe Brook
   radiocarbon dates, 178
   varve sequence, 174
Cap-St-Ignace, 274
Cap-St-Thomas section, *278*
Cape Cod Bay lobe, Laurentide ice sheet, 192
carbon isotopes, *231*
   Québec, *244*
Carrying Place Bluff, *224*, 225, 227, 234, 236
   oxygen isotopes, 233, 234
Casco Bay
   ice retreat, 202
   Laurentide ice sheet, 192
   topography, *194*
Casco Bay lowland, *193, 201*
   deglaciation, 191
   facies, *202*
   ice retreat, 208, 209
   radiocarbon dates, *206*
Casco Bay sublobe, 192
   Laurentide ice sheet, 191, *192*
Cathance River, 193, 194
Chadbourne gravel pit, 117
Champlain Sea, 3, 172, 265
   carbon isotopes, 233
   episode, 248
   invasion, 182, 186, 262, 266
   isotopes, 234
   paleomagnetism, 181
   regression, 262
Champlain Valley, uplift, 186
channel
   fill, 60
   Grand Banks, 55
   Scotian Shelf, 55
Charles-Merrimack lobe, Laurentide ice sheet, 192

285

Charlevoix, 244, 266
  deglaciation, *272*
  striations, 264
Chaudière Valley, 243, 252, 260, 261, 265, 266, 267
  buttonhole, 265
  outwash, 265
Chaudière-Etchemin glacial lake, 261
Chedabucto Bay, 39
Cherry River–East-Angus–Megantic moraine, 267
  age, 262
Chignecto phase, 47, 52, 71, 72, 208, 257, 263, 264
circulation, thermohaline, 218
climate
  change, 216
  fluctuation, 247
clinoform, 125, 136
  reflector, 126, 200
coastal zone
  Sanford-Kennebunk sand plain, *128,* 136
cold event, Greenland, 263
Cold Water Brook, 127
Collins Pond phase, 52, 72
Columbia Falls delta, 165
Connecticut Valley, 171, 210
  paleomagnetism, 174, *183*
  radiocarbon dating, 174
  varve chronology, 121, *174, 183*
  varve sequence, 174, *179,* 182, 183
consolidation test, 55
continental shelf, Newfoundland, 53
Cooks Corner, 209
core
  grain size analysis, 11
  Gulf of Maine, *10*
Cornish
  paleomagnetism, 183
  varves, 183
Cox Pinnacle moraine, *194,* 208
crag-and-tail features, 244
Crooked Brook valley, varves, 183
Crooked River valley, 116
cross beds, 199
crosscutting, 244
Crowell Basin, 15
  calving, 29
  deglaciation, 20
  facies, 14
  ice shelf, 29
crustal tilt, 112, 113
Cupsuptic River valley, 117
Cushman Pond, 118
cyclopels, 197
cyclopsams, 197

### D

Day Brook, 127
De Geer sea, 216, 235
Deblois Plain, 166
debris flow, 70, 219
declination, paleomagnetic, 171, 179, 210

Deering Pond, 118
deformation, morainal bank, 164
deglaciation, 35, 113. *see also* ice retreat
  age, 236
  Bay of Fundy, 39
  chronology, 121
  coastal Maine, *207*
  eastern Maine, *238*
  Gaspé Peninsula, 259
  Gulf of Maine, *9,* 17, *20, 27,* 39, *192*
  Laurentide ice sheet, *201, 247*
  Maine, 153, 171
  New England, *184, 186*
  Québec, *244*
  radiocarbon dating, *173, 182*
  rate, 236
  sand plain, 125
  Sanford-Kennebunk sand plain, 127
  sea level, *205*
  southern New Hampshire, 171, *184*
  St. Lawrence Valley, *272*
  western Maine, *109*
delta, 146, 147, 153, 194, *197,* 208, 218, 224, 246
  Androscoggin River valley, 191
  bank-edge, 39
  braid plain, 200
  eastern Maine, 112
  esker fed, 151, 236, 238
  foresets, 147
  Gilbert type, 115
  glacial, 3
  glaciolacustrine, 110, 111
  glaciomarine, 36, 109, *112,* 127, 129, 197
  ice contact, 114, 116, 117, 128, 140, 146, 147, 151, 153, 207
  ice frontal, 159
  marine, 134, 153
  outwash, 39, 115, 128, 145, 194, 199, 209
  progradation, 140, 144
  reworking, 147
  shallow marine, 148
  western Maine, 112
deluge, Biblical, 1, 2, 7 (*see also* flood)
Dennison Point
  ice margin, 236
  lodgement till, 234
density stratification, 215
depocenter, 130
deposition
  delta, 174
  Halibut Channel, *69*
depositional environment, Presumpscot Formation, *219*
depth, Halibut Channel, 69
diamict, 55, 60, 145
diamicton, 114, 115, 131, 161, 164, 195, 197, 224, 225, 261, 272
  Lily Lake, 227
  sediment
    glaciofluvial, 151
  Turner Brook bluff, 225

diatom, 21, 26, 32
  core, 11
  Georges Basin, 29
  glaciomarine mud, 197
  Gulf of Maine, 18
  productivity, 31
displacement, lithosphere, 44
distal glaciomarine facies, 13, 14, *15,* 202
  environment, *21*
Dixville Moraine, age, 262
Dixville-Ditchfield Moraine, 265, 267
Dover-Foxcroft, 236
Downing Silt, *55,* 60, 69
  fossils, 69
downwasting, 111, 121, 252, 263, 267
drift, 117, 257, 261
  Frontier moraine, 114
  iceberg, 1
  valley, 153
dropstone, 234
drumlin, 128, 129, 133, 140
Dryas III, 248
Dryas phase, 243, 249

### E

eastern Maine, ice retreat, *236*
Eastport Formation, 224
Ellis River basin, 118
Embden Pond, deglaciation, 208
Emerald Basin, 67
  deglaciation, 192
Emerald Silt, 10
eolian deposits, 234
erosion, Halibut Channel, 54
erratics, glacial, 68
errors, radiocarbon dating, *261*
esker, 3, 111, 116, 119, 153, 159, 165, 194, 224, 246, 260
  Casco Bay, *196*
  Penobscot Valley, 219, 236
  system, 112, *165*
Etchemin Valley, 267
  buttonhole, 265
  outwash, 265
  striations, 264

### F

facies, 15
  Carrying Place Bluffs, *224*
  core, 13
  Crowell Basin, 14, 15
  delta, 135
  Georges Basin, 14, 15
  glaciomarine, *224*
  Gulf of Maine, *13*
  ice distal, 225
  Jordan Basin, 14, 15
  Lily Lake, *227*
  mud, 231
  regressive, 225, 227
  Sanford-Kennebunk sand plain, *131,* 142
Failing 1500S drill rig, 55

fan, 146, 164, 197, 224. *see also* submarine fan
  Androscoggin River valley, 191
  glaciomarine, *196*
  grounding line, 196, 208
  outwash, 246
fan-delta complex
  glaciomarine, 204
fauna, sediment, 215
Fennoscandian ice sheet, 68
floating ice shelf, 192
flood. *see also* deluge
  Biblical, 1, 2, 3, 7
floodplain deposit, 136
flow
  homopycnal, 142
  hypopycnal, 142, 144
flow reversal, 266
flowtill, 219, 225
foraminifer, *21, 26*, 32, 70, 71, 231. *see also* fossils
  age, 16, 224
  amino acid racemization, 11
  assemblages, *61*
  benthic, 21
  core, 11
  glaciomarine mud, 197
  Gulf of Maine, *17*
  Halibut Channel, *55*
  Jordan Basin, *17*
  oxygen isotope ratio, 185
forebulge, *44*
  Canada, 47
  migration, 35, *44, 46,* 53, 273, 281
foreset bed, 127, 136, 142, 145, 147, 197, 200
  delta, 147
fossils, 26, 233
  Adolphus Sand, 69
  *Atractocon stonei,* 37
  *Bacterosira fragilis,* 276
  Baffin Bay, 22
  Baffin Shelf, 26
  barnacle plates, 204
  Beaufort Sea, 26
  *Cassidulina reniforme,* 18, 21, 22, 26, 51, 65, 66, 67, *68,* 69, 71, 72
  *Chaetoceros,* 20, 26
  *Cibicides,* 61
  *Cibicides io,* 65
  *Cibicides lobatulus,* 17, 18, 21, 65
  *Cibicides reflugens,* 65
  *Cibicidoides corpulentus,* 65
  *Cibicidoides floridanus,* 65
  *Cibicidoides pseudoungerianus,* 65
  cow tooth, 2
  *Cribroelphidium excavatum,* 51, 65, 66, 67, *68,* 69, 71, 72
  *Cryas integrifolia,* 228
  *Cytheromorpha macchesneyi,* 227, 228
  *D. integrifolia,* 230
  *Desmarestia ligulata,* 230
  Downing Silt, 69
  *Elphidium clavatum,* 22
  *Elphidium excavatum,* 18, 21, 26, 227, 230, 233, 276
  *Elphidium subarcticum,* 66
  *Epistominella nipponica,* 68
  *Eponides pussilus,* 68
  Fram Strait, 22
  *Fursenkoina fusiformis,* 67
  *Glabratellina arcuata,* 65
  *Glabratellina lauriei,* 65
  *Glabratellina wrightii,* 65
  *Gyroidinoides nipponicus,* 59
  *Gyroidinoides quinqueloba,* 59
  *Haynesina pauciloculla,* 51, *66,* 67, 69, 71, 72
  hemlock, 279
  *Heterolepa subhaidingerii,* 65
  *Hiatella arctica,* 204, 209, 227, 228, 233, 234
  *Ioanella turidula,* 68
  *Islandia algida,* 18, 65
  *Islandiella algida,* 21, 51, 61, 68
  *Islandiella helenae,* 18, 26, 66, *67,* 69, 70, 71, 72
  Labrador Shelf, 26
  Lily Lake, 236
  *Lobatula lobatula,* 65, 68
  *Macoma balthica,* 202, 228
  *Macoma calcarea,* 228, 234
  Matteseunk Lake, *228*
  *Mytilus edulis,* 204, 230, 233
  *Neogloboquadrina pachyderma,* 26
  *Nonion pauciloculum,* 66
  *Nonionellina labradorica,* 18, 26
  *Nucula tenuis,* 228, 233
  *Oridosalis umbonatus,* 68
  *Planulina vilksae,* 65
  *Portlandia arctica,* 194, 202, 204, 208, 227, 228, 230, 236
  *Pseudoparella zhengae,* 68
  radiocarbon dates, 202
  *Rosalina globularis,* 65
  *Rotaliella chasteri,* 65
  Sprague Neck, 236
  spruce, 279
  *Stainforthia concava,* 66
  *Stainforthia fusiformis,* 66, 67
  *Stainforthia pauciloculata,* 66
  *Stainforthia rotundata,* 66
  *Stainforthia* spp., 51
  Tail of the Bank, 69
  tamarack, 279
  *Thalassiosira gravida,* 18, 20, 21, 26, 31
  *Thalassiothrix longissima,* 18, 21, 22
  *Tosaia hanzawaia,* 68
  *Tricholhyalus kingi,* 65
  *Turborotalita quinqueloba,* 26
  white pine, 279
Fram Strait, fossils, 22
Frontier Moraine, 114, 265, 267

## G

Gaspé Peninsula, 252, 264, 266
  deglaciation, 258, 259
geomorphic zones, Sanford-Kennebunk sand plain, *136*
geomorphology, sediment, 215
Georges Bank, 9, 10, 216, 236
  deglaciation, 20
  ice, 28
  shoreline, 27
Georges Basin, 32
  diatoms, 29
  facies, 14, 15
  iceberg, 28
  phytoplankton, 30
  wind, 30
Glacial Lake Bethel, 119
Glacial Lake Cambridge, 118
Glacial Lake Cornish, 116
Glacial Lake Mousam, 115
Glacial Lake Town Farm, 116
glaciation, 2, 3, 35
  Maine, 1, 7
  Wisconsinan, 10, 52, 127
glacier
  advance, 47
  ice marginal, 68, 166
  ice proximal, 68
  maximum, 52, 263
  outlet, 256
  retreat, 115, 146 (*see also* ice retreat)
  reversal, 264
  tidewater, 164
glaciolacustrine deposits, 117
global warming, *247*
Goldthwait Sea, 252, 264, 266, 271, 272, 274, 276
  regression, 262
  transgression, 272
Gould Pond, 236
  deglaciation, 208
  sedimentation rate, 235
  uplift, 186
Grafton Notch, 118
Grand Banks, 51, 53
  Drift, *55,* 60, 69
gravel massive, 153
gravitational attraction, forebulge, 44
gravity flow, 164
Green Bank, 53, 71, 72
  ice caps, 71
Greenland ice sheet, 21, 247
GRIP core, Greenland, 263
ground-penetrating radar, 125, 126, 128, 131, 136, 140, 142, 143, 145, 147
  reflectors, 134, 135
grounding, Laurentide ice sheet, 259
grounding line, 54, 151, 153, 156, 160, 164, 194, 208, 215, 224, 233, 234, 235, 238
  fan, 192
  ice cliff, 153
  Laurentide ice sheet, 153
  retreat, 216
  stillstand, 161
Gulf of Alaska, 234

Gulf of Maine, 9, 39, 234
  deglaciation, 39, *216*
  ice retreat, 208
  Laurentide ice sheet, 145
Gulf of St. Lawrence, 9, 39, 67, 71, 244
  deglaciation, *257*
  ice center, 47
  ice sheet, 52
  shoreline, 47
gyttja, 178, 261

## H

Hadley Pond Esker, 235
Halibut Channel, 51, 53, 67, 68
  erosion, 54
  foraminifer, *55*
  geology, *53*
  ice surge, 52
  stratigraphy, *59, 71*
Halifax, forebulge migration, 44
Halifax Harbour
  reversal points, 44
  tide gauge data, 44
Heinrich event 1, 184, 191, 209, 218, 238
Heinrich layer, 209
Highland Front moraine system, 260
highstand, 35, 143
  sea level, 147
Hudson Strait
  core, 17
  ice stream, 209
Hudson Valley, varves, 174
Hudson-Champlain lobe, Laurentide ice sheet, 266, 267
Hudson-Champlain Valley, 252
Hudson–Lake Champlain lobe, Laurentide ice sheet, 265
Huntec Deep Tow System, 10, 59

## I

ICE 3G model, 44
ice advance, *71*
  Anticosti Island, 263
ice cap, 35, 47, 72, 244, 263
  Avalon Peninsula, 70
  Green Bank, 71
  Maine, 1, 7
  St. John River valley, 3
  St. Pierre Bank, 71
ice center, Gulf of St. Lawrence, 47
ice cliff, retreat, 153
ice contact deposit, 130, 246
ice contact features, Sanford-Kennebunk sand plain, 129
ice damming, 111
ice divide, 244
  Nova Scotia, *257*
ice dome, 244, 252
  Avalon Peninsula, 53
  Northumberland Strait, 52
  Nova Scotia, 263, 264
  Prince Edward Island, 52
ice flow, 109, 113, 117, 121
  direction, 244

middle Chaudière area, 265
  reversal, 252
ice front, 260
ice funneling, 263, 264
ice loading, 69
ice margin, 118, 119, 166
  active, 111
  floating, 256
  glacial regime, 68
  Gulf of Maine, 9
  Lake Town Farm, 116
  late Wisconsinan, 39
  recession, 117
  retreat, 110
  sedimentation, 16
ice mass, disequilibrium, 252
ice model
  maximum, 44, 46
  minimum, 44, 46
ice proximal, glacial regime, 68
ice rafting, 26
ice recession, 183. *see also* ice retreat
  eastern Maine, 215
  Passamaquoddy lowlands, *236*
  Patrick Lake, 236
  Pocomoonshine Lake, 236
  rate, 238
ice retreat, 39, 109, 110, 116, 119, 129, 130, 151, 153, 156, 166, 192, 207, 260. *see also* deglaciation; ice recession
  acceleration, 238
  active, 109
  Androscoggin River valley, 193
  Casco Bay Lowland, 209
  Casco Bay sublobe, 192
  continental shelf, 192
  eastern Maine, *236*
  fan sequence, 197
  late Wisconsinan, 109
  Lewiston, 208
  Maine, *164*
  Maritime Canada, 208
  Phippsburg Peninsula, 204
  radiocarbon ages, 112
  rate, *166*, 249
ice sheet, 109, 113, 121, 130. *see also* Laurentide ice sheet
  Jordan Basin, 30
  late Wisconsinan, 119
  receding, 111
ice shelf, 192, 257, 259, 263
  advance, 16
  Crowell Basin, 29
  Gulf of Maine, 9
ice shoving, 116
ice stream, 192, 247, 264
  Northeast Channel, 28
ice thrust, 197
ice tongue, 111, 115, 117, 192
ice tunnel, 197, 200
ice-contact features, 127
ice-front features, 244, 245
ice-marginal features, 206, 207

ice-proximal facies, 202
ice-tunnel system, Pine River esker, 112
ice-wedge casts, 234
iceberg, 9, 22
  drift, 1
  Georges Basin, 29
iceberg calving, 234
iceberg drop deposits, 16
incursion
  age, 266
  marine, 125, 126, 266
index property, 55
inland zone, Sanford-Kennebunk sand plain, *128*, 129, 137
intermediate zone, Sanford-Kennebunk sand plain, *128*, 136
intertidal facies, Montmagny Airport section, 277
Island of Newfoundland, 53
isostacy, 35, 145
  New England, 183
isostatic rebound, 147
isostatic recovery, 46, 125, 271, *272*
isostatic uplift, 186, 280
  Kennebec River valley, 207
  St. Lawrence estuary, 281
isotopes. *see also* age dating; radiocarbon dating
  bivalve, 224
  eastern Maine, 197
  foraminifers, 224
  sediment, 215

## J

Jailhouse delta, 127
Jordan Basin
  age dating, 17
  deglaciation, 20
  facies, 14, 15
  ice sheet, 30
  ice shelf, 30

## K

kame, 3, 153
kame terrace, 111
Kangerdlugssuaq Fjord, 21
Kattegat, 67
Kennebago River valley, 117
Kennebec lowland, 195
Kennebec River, 166, 193
Kennebec River paleodelta, 131, 147
Kennebec River valley, 166
  isostatic uplift, 207
Kennebec Valley, deglaciation, 208
Kennebec-Androscoggin drainage system, 166
kettle, 116, 128, 129, 134, 144, 146, 183
Kezar Falls, 115

## L

La Pocatière, 279
Labrador, 2
Labrador Shelf, fossils, 26
lag deposit, 230, 274

LaHave Clay, 10
lake, glacial, 115, 116, *118,* 119, 186
Lake Candona, 252, 262
   overflow, 266
Lake Champlain basin, 265
Lake Champlain Valley, 260
Lake Winooski, varves, 174
Laurentian Channel, 39, 52
Laurentian highlands, 244, 251, 252
Laurentian transgression, 271, 273, 278, 281
   end, 278
Laurentian trough, 260, 266
Laurentians, 260
Laurentide ice center, 36
Laurentide ice sheet, 9, 32, 39, 47, 52, 114, 119, 127, 128, 143, *145,* 147, *151,* 160, 168, 209, 215, 218, 236, 244, 260, 261, 266
   advance, 216
   Casco Bay, 192
   Casco Bay sublobe, 191
   deglaciation, 282
   geometry, 192
   grounding line, 153
   Gulf of Maine, 145
   morphology, 249
   recession, 238
   southern Québec, 243
   thinning, 266
lee features, 244
Leighton Formation, 224
Lewis Cove, 235
   fossils, 233
Lewiston, ice retreat, 208
Ligori Ridge, 264
Lily Lake, 234, 239
   core, 233
   fossils, 233, 236
   oxygen isotopes, 233, 238
   regressive facies, 235
   sediment age, 235
   stratigraphy, *227*
lineation, structural, 130
Lisbon-Druham Falls area, 208
lithofacies, morainal banks, 151, *161*
lithostratigraphy, Halibut Channel, *55, 71*
Little Androscoggin River, 194
Little Androscoggin River valley, 116
Little Ossipee River basin, 115
   delta, 111
Little River, 194
loading, isostatic, 70
Long Island, 216
Long Pond, 235
Look's Seafood Cannery, 235
Lower Black Pond, 118
lowstand, 47, 127, 143, 271, 272, 273, 274, 279
   nadir, 41, 42
   sea level, 53, 55, 72, 218
   shoreline, 39
   St. Lawrence estuary, 279, 281
lunate features, 244

# M

Magalloway River valley, 117
Magdalen Islands, 39
Magdalen Shelf, 264
magnetic susceptibility, core, 10, 16
Mahoosuc Mountain, 121
Mahoosuc Range, 109, 116, 117, 119
Maine Geological Survey, 111, 112, 194
mantle, rheology, 44
mapping
   Casco Bay, *194*
   western Maine, 109
marine beds, 153
Maritime Canada, 41
   ice retreat, 208
   sea level, 37, 44
Maritime Provinces, 9, 264
   ice sheet, 52
Mark's Lake, 234
   age, 235
Marr Point delta, 200, 201
Mars-Batiscan moraine, ice front features, 266
marsh facies, Montmagny Airport section, 277
Martha's Vineyard, 216
mass balance, 249
   negative, 266
mass movement, 219
Massachusetts, 10
Massachusetts Bog, 117
Matane deglaciation, *272*
Mattaseunk Lake, 235
   age, 236
   fossils, *229*
maximum ice model, 44, 46, 280
MAXWELL viscoelastic Earth model, 44
meltwater, 22, 71, 72, 117, 186, 215, 234, 260
   channel, 72, 109, 110, 117, 125, 147, 148
   flux, 192
   jet dynamics, 164
   plume, 67
   pulse, 184, 185, 186, 191, 209, 210
   stream, 127, 147, 199
   tunnel, 145
Merriland Ridge, 127, 133, 161
   delta, 153
Merrimack River paleodelta, 131, 147
Merrimack Valley, 171
   paleomagnetism, *183*
   varve, 174, *183*
Micmac shoreline, 271, 281
Middle Wisconsinan advance, *72*
minimum ice model, 44, 46, 280
Mitis stage, 273
Mitis terrace event, 281
model
   CALIB 4.0, 178
   ICE 3G, 44
   MAXWELL, 44
Mohawk Valley, paleomagnetism, 182

mollusc, 231
   age, 224
   Gulf of Maine, 17
   oxygen isotopes, 234
Mont Ham moraine, 265
Mont Tremblant highlands, 252
Monte Sainte-Anne, ice contact sediments, 260
Monteregian hills, 251
Montmagny, 274
   Airport section, *274*
   embayment, 277, 278
Montreal–lower Lake Champlain lowlands, 264
morainal bank, 151, *153,* 156, 164
   axes, 159, 166
   coastal Maine, *157*
   distribution, *157,* 164
   flat topped, 161
   morphology, 160, *161*
   Pineo Ridge complex, 161
   sharp crested, 161
   spacing, 167
   trend, 165
moraine, 10, 121, 133, 140, 145, 146, 147, 192, 200, 238
   Androscoggin River valley, 191
   Baie-Trinité, 258
   belt, 267
   bifurcate, 161
   Casco Bay, *194*
   cross valley, 161
   De Geer, 153, 156, 194
   delta, 153
   end, 109, 110, 111, 113, 114, 127, 153, 194, 197, 224, 236
   flat-topped, 165, 166
   hummocky, 114, *115,* 116
   ice-disintegration, 115
   joined, 161
   lift-off, 15, 16
   linear, 54
   minor, 153, 156
   Pointe-des-Monts peninsula, 258
   recessional, 153
   ribbed, 114, 115, 194
   s-shaped, 161
   sharp-crested, 166
   simple linear, 161
   stratified, 153, 219
   system, 260
   terminal, 153
   washboard, 153, 156, 194
morphogenesis, sand plain, 126
morphology, morainal banks, 160, *161*
morphosequence
   ice contact sand and gravel, 109
   sand and gravel, 111, *115,* 116
morphostratigraphy, Sanford-Kennebunk sand plain, *145*
Mount Shefford, 260
Mousam Lake, 115
Mousam River, 126, 128, 129, 130, 143
moutonnees, 2

mud
  Androscoggin River valley, 191
  drape, 200
  glaciomarine, 51, 146, 192, *197,* 200, 216, 219, 225, 227, 228, 234
  marine, 194, 225
  rip-up clasts, 199, 200

## N

nadir, lowstand, 41, 42
nailhead striation, 244
Naples
  paleomagnetism, 183
  varves, 183
Neigette readvance, 265
New Brunswick, 10, 252
  ice dome, 252
New England, 2
  deglaciation, *184, 186*
  paleomagnetism, *179*
  varve chronology, *174*
New Jersey Shelf, 67
New Quebec dome, 244, 252, 256, 265
New York, paleomagnetism, *179*
Newbury
  paleomagnetism, 181
  varves, 179
Newfoundland
  continental shelf, 53
  Grand Banks, 53
Newington moraine, 153, 156
nonconformity, basal, 218
Nordaustlandet ice cap, 233
North Maine Ice Divide, 264
North Sea, 68
  boreholes, 53
Northeast Channel, 10, 234
  ice stream, 28
  slope water, 26
Northumberland Strait ice dome, 52
Norwegian Shelf, borehole, 53
Notre-Dame Mountains, 252
  deglaciation, 267
  nunataks, 265
Nova Scotia, 10
  ice advance, 208
  ice divide, *257*
  ice dome, 263, 264
  sea level, 44, 47
nunatak, 251, 256, 265

## O

offlap facies, 202
Ontario basin, paleomagnetism, 182
ORE Geopulse seismic reflection profile, 10
Ossipee River basin, delta, 111
Ossipee River Valley, varve sequence, 183
outwash, 3, 117, 195, 219, 224, 246
  Chaudière Valley, 265
  Etchemin Valley, 265
  head, 109, 111, 116

subaqueous, 164, 225, 227
submarine, 200
overconsolidation, 69
overwash gravel, 153
oxygen isotopes, 38, *231,* 234. *see also* age dating; radiocarbon dating
  foraminifer, 185

## P

pack ice, 22
paleocurrents, 246
paleomagnetism, 119, 186
  Champlain Sea, 181
  chronology, 174
  Cornish, 183
  declination, 171, 179, 210
  inclination, 179
  Mohawk Valley, 182
  Naples, 183
  New England, *179*
  New York, *179*
  Newbury, 181
  Ontario basin, 182
  southern Maine, *183*
  western Vermont, *179*
Palmer River, 261
palynostratigraphy, 248
  climate, 247
paradigm shift, 1, 2, 7
Parc des Laurentides highlands, 252, 264, 265
  ice-front features, 266
Passadumkeag-Mattawamkeag uplands, 216
Passamaquoddy lowland, 216, 218, 236
  ice recession, *236*
Passumpsic Valley varves, 179
Patrick Lake, 238
  ice recession, 236
peat, 178, 261, 276
  estaurine, 39
Penobscot
  calving bay, 208
  drainage system, 166
  ice recession, 236
  lowland, 216, 218
  River, 165, 166
  valley, 157, 166, 219
Penobscot Bay, moutonnees, 2
Penobscot-Kennebec embayment, 167
periglacial features, 234
peripheral bulge, 44
  migration, 53
Perkins Marsh Brook, 127
Phippsburg Peninsula, 202
phytoplankton, Georges Basin, 30
Pilkars River, 261
Pine River esker, ice-tunnel system, 112
Pineo Ridge, 165, 166
  morainal bank, 161
  moraine, 208
  moraine belt, 209
Pineo Ridge readvance, 209
pingo scars, 234

pinning, 159
  points, 157, 160
Piscataqua River, 194
piston core, 10
Placentia Bay, 53, 68, 70, 71
plankton, 32
Pleasant Hill delta, 200
Plumbago Mountain, 117
plume overflow, 197
Pocomoonshine Lake
  ice recession, 236
Pointe St-Thomas, 274
Pointe-des-Monts peninsula moraine, 258
Poland Spring Pond, 118
  core, 208
pollen, stratigraphy, 181, 183
Pond Ridge moraine, 208, 209, 227, 234
  age, 238
  stratigraphy, *225*
Port Huron
  advance, 208
  event, 264
  readvance, 209
postglacial facies, 14
postglacial tilt, Androscoggin River valley, 207
potholes, 37
Presumpscot Formation, 10, 142, 144, *197, 219, 224,* 228
  isotope age, *224*
  marine mud, 225
Prince Edward Island ice dome, 52
productivity
  bioligical, 67
  diatom, 31
  Gulf of Maine, 22
progradation, delta, 140, 144
proximal glaciomarine facies, 13, 14, *15,* 18
  environment, *21*

## Q

Québec
  Ice Divide, 266
Québec City, 266
  age, 262
  deglaciation, 272
  sea level, 271
  striations, 264
Quoddy Formation, *224*

## R

radar, ground penetrating, 125, 128, 131, 136, 140, 142, 143, 145, 147
radiocarbon dating, *17,* 119, 168, *172,* 234. *see also* age dating
  Androscoggin River valley, 191
  bulk sediment, *173,* 183
  Casco Bay, 194, *202*
  Casco Bay Lowland, *206*
  Connecticut Valley, 174
  deglaciation, *186, 207*
  eastern Maine, 215
  errors, *261*

foraminifer, 55
ice retreat, 112, 166
Jordan Basin, 17
marine reservoir correction, 172
mollusc, 55
Montmagny Airport section, 276
New England, *183*
Northeast Channel, 28
Scarborough mammoth site, 119
varves, 179
rat tails, 264
readvance, 263
   glacial, 156
rebound, 70, 127
   isostatic, 127, 147, 192, 199, 207, 218, 252, 272
   Nova Scotia, 41
recession. *see also* retreat
   ice, 238
   ice margin, 117
redox, 21
reflection, seismic, 51, *55*
reflector, 145
   clinoform, 200
   dipping, 142
   foreset, 144
   ground-penetrating radar, 134, 135, 136
   parallel, 136
regression, 125, 127, 143, 193, 199, 200
   St. Lawrence estuary, 279
reservoir correction, 181, 186, 210, 224, 238, 261, 262, 263
   radiocarbon dating, 172
reservoir effect, 183, 184, 186, 210, 238, 262
retreat. *see also* recession
   ice cliff, 153
reversal points, 44
reworking, delta, 147
rheology, mantle, 44
ridge, ice front, 246
Rimouski
   ice readvance, 265
   striations, 264
Rindgemere Formation, 116
ripple, 136
Rivière du Sud, 278
Rivière Landry
   section, 266
   varve stratotype, 262
   varves, 266
Rivière-du-Loup, 260, 266, 267
Robinson's Head, readvance, 52, 71, 72
Rogen moraine, 115
Rogers Basin, deglaciation, 20
Rose Blanche Bank, 53
Ross Ice Shelf, 21
Round Pond, 200
Royal River, 194
   valley, 195

**S**

Sabattus Mountain, 201
Sabattus Pond, 200
Sabattus Pond lowland, 201
Sabattus River, 194
Sable Island, 39
Sable Island Bank channels, 70
Saco River, 116
   delta, 111
Saco Valley, striations, 113
Saguenay fjord, 243, 252, 260
Saint-Damien Moraine complex, 260
Saint-Jean-Port-Joli Moraine, 260
Saint-Ludger moraine, 265
Saint-Maurice
   basin, 265, 266
   lobe, 265, 266
   Valley, 252
Saint-Narcisse
   episode, 256
   Moraine, 246, 248, 258, 266
Saint-Raphael Moraine, 261, 266
Saint-Sylvestre Moraine, 260, 261, 266
Saint-Tite basin, thin ice, 264
Saint-Tite glaciofluvial fan, 260
salinity, 71, 72, 233, 234
   indicator, 67
   surface, 22
Salmon River Sand, *37*, 38
Salpausselka moraines, 161
sand bar, 136
sand shadow, 136
Sandy Lane, 136
Sanford-Kennebunk sand plain, 125, *126*
   morphostratigraphy, *145*
Sangamon interglacial shorelines, Maritime Canada, 47
Sankaty sand, 38
Saywand's Corner, 134
Scandinavian margin, 68
Scarborough mammoth site, radiocarbon age dates, 119
Scotian ice divide, 39
Scotian phase, 208, 264
Scotian Shelf, 9, 10, 17, 39, 67
   drift, 10
   ice retreat, 216
   sedimentary units, 10
   shoreline, 39
scour, glacial, 10
sea level
   eustasy, 216
   fall, 42, 44, 46, 145, 199, 207
   fluctuation, 35, 192
   history, 210
   Holocene, *40*
   interglacial, 38
   lowstand, 53, 55, 72
   Maine, 47
   New England, 186
   Nova Scotia, 47
   post glacial, 271
   reversal points, 44
   rise, 42, 44, 47, 53, 112, 147, 191, 209, 218, 272 (*see also* transgression)
   St. Lawrence estuary, 280
sea stacks, 37

seaweed, age, 224
Sebago lowland, 195
Sebago pluton, 116, 194
sediment
   fluvioglacial, 260
   glacigenic, 51, 53, 68
   glaciofluvial, 10
   glaciomarine, 10, 14, 32, 36, 39, 53, 54, 60, 65, 69, 70, 71, 156, 157, 192, 215, 216, 218, 219
   ice contact, 260
   ice proximal, 54
   morainal bank, 161
   plume, 197
   postglacial marine, 10
   shallow marine, 199
   wave reworked, 199
sediment volume, Sanford-Kennebunk sand plain, 147
sedimentation, 151
   ice margin, 16
   rate, 70
seismic profile, Halibut Channel, 55
Sewell Ridge, 29
shear strength
   Downing Silt, 69
   undrained, 55
shell hash, 60, 70, 72
shoreline
   Atlantic coast, 47
   Browns Bank, 27
   Georges Bank, 27
   Gulf of St. Lawrence, 47
   lowstand, 39
   Scotian Shelf, 39
Sinkhole Pond, 118
   core, 208
Sir Charles Lyell, journey, 2
Skagerrak, 67
slumping
   ice marginal, 70
   syndepositional, 227
snow budget, 248
South Channel lobe, Laurentide ice sheet, 192
species, assemblage, 224
Speckled Mountain, striations, 113
Spencer Pond, 118
spillway channels, 110
spit, 127, 136, 199
Sprague Neck, 230, 235
   fossils, 236
   stratigraphy, *227*
St-Damien Moraine complex, 267
St. Antonin Moraine, 260
St. John River valley, ice cap, 3
St. Lawrence corridor, 244, 267
   deglaciation, 252
   Laurentide ice sheet, *251*
St. Lawrence estuary
   age, 272
   deglaciation, 271
   ice current, 264
   sea level, 282

St. Lawrence ice stream, *252, 259,* 260, 261, *263,* 264, 266
St. Lawrence Lowland, 119
St. Lawrence River, deglaciation, 244
St. Lawrence River valley, 3
St. Lawrence Valley, 244, 262, 265, 266
    deglaciation, *272*
St. Narcisse Moraine event, 243
St. Pierre Bank, 52, 53, 71, 72
    ice caps, 71
    ice surge, 52
stagnation, 111, 121, 252
STATEMAP project, 111, 194
Stellwagen basin, ice retreat, 216
stillstand, 112, 136, 174, 194, 209, 210
    Androscoggin Valley, 208
Stone, George, *2,* 7, 111
Stony Brook fan, 117
stoss features, 244
stranded shorelines, 127
stratigraphy
    eastern Maine, *219*
    Halibut channel, *59*
    Lily Lake, *227*
    pollen, 181
    Sanford-Kennebunk sand plain, *131*
    Sprague Neck, *227*
    varve, *174*
striation, 37, 110, 116, 117, 224, 244, 257, 260, *264,* 265, 266
    direction, 113
    ice, 261
    Penobscot Valley, 236
    Pond Ridge moraine, 225
    Saco Valley, 113
    Speckled Mountain, 113
submarine fan, 109, 145, 148, 236
    eastern Maine, 112
submergence, 207
    glaciomarine, 153, 157
subsidence, 35
    North Sea, 53
Success Moraine, age, 262
Surplus Pond, 118
Sutton–Cherry River–East-Angus–Megantic moraine belt, 265
swamp, 128

### T

Tadoussac, sea level, 271
Tail of the Bank, 61
    fossils, 69
Tail of the Bank Mud, 67
temperature, Lily Lake, 227
Ten Mile Lake readvance, 52, 72
terrace, 127, 209
    wave cut, 39
thickness, Sanford-Kennebunk sand plain, 130
thinning, ice, 9
tide gauge data, Halifax Harbour, 44

till, 3, 14, 15, 32, 72, 130, 133, 147, 197
    ablation, 113, 114
    apron, 54
    basal, 69, *72*
    clast fabric, 244
    deformation, 69, 71
    Frontier moraine, 114
    Halibut Channel, 59
    hill, 128, 134, 140, 147
    lodgement, 51, 53, 69, *72,* 219, 224, 225, 227, 234
    meltout, 51, 53, 69, *72*
    subglacial, 142
    tongue, 15, 53, *54,* 60, 69, 72
Toddy Pond, 234, 236
Tomah Ridge, 216
tombolo, 199
topography, Sanford-Kennebunk sand plain, 128
topset, 146
    beds, 127, 142, 197
    foreset contact, 207
    plain, 197
transgression, 145, 193, 199, *272,* 273, 274. *see also* sea level, rise
    Goldthwait Sea, 272
    marine, 119
    St. Lawrence estuary, 279
transitional glaciomarine facies, 13, 14, *15*
    environment, *21*
Trois-Pistoles-Rivière-du-Loup, marine incursion, 262
Truxton swell, 30
Turner Brook, 234
    bluff, *225*
two till problem, 3

### U

U.S. Geological Survey, 111, 112, 194
Ulverton-Tingwick
    ice front, 265, 266
    Moraine, 260, 262, 265, 267
unconformity, 71
    erosional, 60
    regressive, 219
    scour, 225
unit BR, Sanford-Kennebunk sand plain, *131,* 135
unit Brw, Sanford-Kennebunk sand plain, *136, 143*
unit GM, Sanford-Kennebunk sand plain, *135, 136,* 137, 142
unit Icd, Sanford-Kennebunk sand plain, *133,* 142
unit SDfc, Sanford-Kennebunk sand plain, *135,* 136, *142,* 145, 147
unit SDff, Sanford-Kennebunk sand plain, *135,* 136, 137, 142, 143, 145, 147
unit T, Sanford-Kennebunk sand plain, *131,* 135, 136, 142

unit Tm
    Sanford-Kennebunk sand plain, *131,* 135, 140, 142
unloading
    crustal, 252
    glacial, 127
uplift
    isostatic, 127, 199
    post glacial, 112
upwelling, 9, 67

### V

varve, 174
    Cornish, 183
    maximum age, 178
    Naples, 183
    Rivière Landry stratotype, 262
    stratigraphy, *174*
varve chronology, 121, 171, 174, *183,* 186
    Connecticut River valley, 121
    Connecticut Valley, *174*
    Maine, *183*
    New England, *174*
    southern Maine, *183*
varve sequence, 119
    central New York, 210
    Connecticut Valley, 182, 183
    Merrimack Valley, 183
    Ossipee River Valley, 183
    southern New England, 210
    upper Connecticut Valley, *179*
Vega transgression, eastern Greenland, 280, 281
Vermont, paleomagnetism, *179*

### W

Waldoboro moraine, 3, 153, 156, 166, 195
warm event, 263
water
    Gulf of Maine, *10*
    mixing, 266
White Mountains, 114, 193, 252, 263
    ice retreat, 265
Whorff pit, 204
Wilkinson Basin, 192, 234
    core, 219
    deglaciation, 20, 192
    ice shelf, 30
Wilkinson-Jeffreys Basin, deglaciation, 20
wind
    Georges basin, 30
    katabatic, 30
Wisconsinan glaciation, 71
Wright-Livingston piston corer, 219

### Y

Younger Dryas, 236, 258, 262
    chronozone, 72
    ice advance, 72